Materials Science and Technology

Volume 4
Electronic Structure and
Properties of Semiconductors

Materials Science and Technology

Materials Science and Technology

A Comprehensive Treatment

Edited by
R.W. Cahn, P. Haasen, E. J. Kramer

Volume 4
Electronic Structure and
Properties of Semiconductors

Volume Editor: Wolfgang Schröter

VCH

Weinheim · New York · Basel · Cambridge

Editors-in-Chief:

Professor R.W. Cahn
University of Cambridge
Dept. of Materials Science
and Metallurgy
Pembroke Street
Cambridge CB2 3QZ, UK

Professor P. Haasen
Institut für Metallphysik
der Universität
Hospitalstraße 3/5
D-3400 Göttingen
Germany

Professor E.J. Kramer
Cornell University
Dept. of Materials Science
and Engineering
Bard Hall
Ithaca, NY 14853-1501, USA

Volume Editor:
Professor W. Schröter
IV Physikalisches Institut der
Georg-August-Universität Göttingen
Bunsenstraße 13–15
D-3400 Göttingen
Germany

This book was carefully produced. Nevertheless, authors, editors and publisher do not warrant the information contained therein to be free of errors. Readers are advised to keep in mind that statements, data, illustrations, procedural details or other items may inadvertently be inaccurate.

Published jointly by
VCH Verlagsgesellschaft mbH, Weinheim (Federal Republic of Germany)
VCH Publishers Inc., New York, NY (USA)

Editorial Directors: Dr. Christina Dyllick-Brenzinger, Karin Sora, Stephen Grollman
Production Manager: Dipl.-Wirt.-Ing. (FH) H.-J. Schmitt
Indexing: W. Borkowski, Schauernheim
The cover illustration shows a semiconductor chip surface and is taken from the journal "Advanced Materials", published by VCH, Weinheim.

Library of Congress Card No.: 90-21936

British Library Cataloguing-in-Publication Data
Materials science and technology: a comprehensive treatment:
Vol. 4. Electronic structure and properties of semiconductors:
A comprehensive treatment.
  I. Schroeter, Wolfgang
  620.1
ISBN 3-527-26817-0

Die Deutsche Bibliothek-CIP-Einheitsaufnahme
Materials science and technology : a comprehensive treatment /
ed. by R.W. Cahn ... - Weinheim ; New York ; Basel ;
Cambridge : VCH.
NE: Cahn, Robert W. [Hrsg.]
Vol. 4. Electronic structure and properties of semiconductors. -
  1991
Electronic structure and properties of semiconductors / vol.
ed.: Wolfgang Schröter. - Weinheim ; New York ; Basel ;
Cambridge : VCH, 1991
  (Materials science and technology ; Vol. 4)
  ISBN 3-527-26817-0 (Weinheim ...)
  ISBN 0-89573-692-6 (New York)
NE: Schröter, Wolfgang [Hrsg.]

Composition, Printing and Bookbinding: Konrad Triltsch, Druck- und Verlagsanstalt GmbH,
D-8700 Würzburg
Printed in the Federal Republic of Germany

Preface to the Series

Materials are highly diverse, yet many concepts, phenomena and transformations involved in making and using metals, ceramics, electronic materials, plastics and composites are strikingly similar. Matters such as transformation mechanisms, defect behavior, the thermodynamics of equilibria, diffusion, flow and fracture mechanisms, the fine structure and behavior of interfaces, the structures of crystals and glasses and the relationship between these, the motion or confinement of electrons in diverse types of materials, the statistical mechanics of assemblies of atoms or magnetic spins, have come to illuminate not only the behavior of the individual materials in which they were originally studied, but also the behavior of other materials which at first sight are quite unrelated.

This continual intellectual cross-linkage between materials is what has given birth to *Materials Science,* which has by now become a discipline in its own right as well as being a meeting place of constituent disciplines. The new Series is intended to mark the coming-of-age of that new discipline, define its nature and range and provide a comprehensive overview of its principal constituent themes.

Materials Technology (sometimes called Materials Engineering) is the more practical counterpart of Materials Science, and its central concern is the processing of materials, which has become an immensely complex skill, especially for the newer categories such as semiconductors, polymers and advanced ceramics but indeed also for the older materials: thus, the reader will find that the metallurgy and processing of modern steels has developed a long way beyond old-fashioned empiricism.

There exist, of course, other volumes and other series aimed at surveying these topics. They range from encyclopedias, via annual reviews and progress serials, to individual texts and monographs, quite apart from the flood of individual review articles in scientific periodicals. Many of these are essential reading for specialists (and those who intend to become specialists); our objective is not to belittle other sources in the cooperative enterprise which is modern materials science and technology, but rather to create a self-contained series of books which can be close at hand for frequent reference or systematic study, and to create these books rapidly enough so that the early volumes will not yet be badly out of date when the last ones are published. The individual chapters are more detailed and searching than encyclopedia or concise review articles, but less so than monographs wholly devoted to a single theme.

The Series is directed toward a broad readership, including not only those who define themselves as materials scientists or engineers but also those active in diverse disciplines such as solid-state physics, solid-state chemistry, metallurgy, construction engineering, electrical engineering and electronics, energy technology, polymer science and engineering.

While the Series is primarily classified on the basis of types of materials and their processing modes, some volumes will focus on particular groups of applications (Nuclear Materials, Biomedical Materials), and others on specific categories of properties (Phase Transformations, Characterization, Plastic Deformation and Fracture). Different aspects of the same topic are often treated in two or more volumes, and certain topics are treated in connection with a particular material (e.g., corrosion in one of the chapters on steel, and adhesion in one of the polymer volumes). Special care has been taken by the Editors to ensure extensive cross-references both within and between volumes, insofar as is feasible. A Cumulative Index volume will be published upon completion of the Series to enhance its usefulness as a whole.

We are very much indebted to the editorial and production staff at VCH for their substantial and highly efficient contribution to the heavy task of putting these volumes together and turning them into finished books. Our particular thanks go to Dr. Christina Dyllick on the editorial side and to Wirt. Ing. Hans-Jochen Schmitt on the production side. We are grateful to the management of VCH for their confidence in us and for their steadfast support.

Robert W. Cahn, Cambridge
Peter Haasen, Göttingen
Edward J. Kramer, Ithaca

April 1991

Preface to Volume 4

This volume has been designed as a text which introduces basic concepts of modern semiconductor physics. A second volume, Volume 16 of this Series, will deal with semiconductor technology. The main idea of our volume is to span the field, from the fundamentals of perfect semiconductors via the physics of defects, which determines the properties of real semiconductors, to that of "artificial" semiconductors like heterostructures, quantum wells, lines and dots, in which interfaces play a dominant role, and finally to amorphous semiconductors.

Although the autors of the various chapters of this volume commence with a description of fundamental phenomena, which are also outlined in standard textbooks, they then develop and describe the mechanisms and concepts used in current semiconductor research. Experimental data are often not presented in their full complexity and detail, rather they are used to exemplify the models described and, as far as they can be reduced to characteristic quantities, are presented in tables and diagrams.

With semiconductors, science has frequently developed in a pronounced interdependency to technology, i.e. the investigation of basic phenomena and the improvement of processing and techniques were intimately related. Today, this interaction is gaining increasing importance for the progress of both science and technology. Therefore, the topics treated in this volume have been selected mostly according to their relevance to semiconductor technology. Consequently, the reader will find seven chapters out of eleven to be concerned with defects in semiconductors, which by their interaction with the host lattice generate the properties and features that are basic to microelectronic and electrooptic devices.

Chapters 1 and 2 have introductory character. The first provides the theoretical principles for understanding the formation of energy bands in periodic and non-periodic solids, and describes the influence of confinements and of perturbations by slowly varying potentials. Chapter 2 explains how experimental observations allow one to study the band structure by optical transitions, carrier scattering and electrical transport, in some selected examples also by nonlinear transport and nonlinear optics.

In contrast to metals, semiconductors crystallize in rather open structures. The diamond lattice provides a tetrahedral interstitial site with the same hard sphere radius and the same symmetry as a normal lattice site. As a consequence, the variety of point defects and reactions between them, e.g. in silicon, are much larger than in typical metals like aluminum. Furthermore, the nearest neighbors of a point defect in silicon do not form a close-packed shell as they do in aluminum and therefore may adjust their positions to lower the total energy of the defect by relaxation or a Jahn-Teller distortion, or they may form negative U-centers or metastable configurations. Simple concepts describe how these features affect the atomic and electronic configuration of the defect. They have evolved from studies of intrinsic point defects in silicon and zinc selenide and are presented in Chapter 3. In Chapter 4 this description is extended to impurities, to nitrogen and transition metals in silicon and compound semiconductors, to chalcogens in silicon, and also to the DX-center in $Al_xGa_{1-x}As$ and thermal donors in silicon. The localization-delocalization puzzle, charge state controlled metastability, and hydrogen passivation are special topics of this chapter. Chapters 5 and 11 present the ingredients of present and future models, describing thermodynamics, transport, high-temperature electronic structure, non-equilibrium response, precipitation and gettering of impurities, especially in silicon but some examples also in germanium and gallium arsenide.

Chapters 6 and 7 deal with dislocations and grain boundaries, respectively, i.e. with intrinsic but extended defects. Questions concerning the core structure, the stability of competing core structures, the existence of dangling bonds in the core, and in the case of dislocations also of their mobility have been the main topics of research for a long time. More recently, the interaction of dislocations with grain boundaries, that of dislocations with point defects, and impurity precipitation at dislocations and grain boundaries have also been incorporated into the description of these defects.

The influence of interfaces, delimiting finite systems, on the properties of the system is greater the smaller the system. Modern device technology and epitaxial growth techniques have exploited this to develop devices and "artificial" materials, whose characteristics are determined by the presence of interfaces. In Chapter 8 the reader is introduced to the key concepts used to characterize the structural properties of interfaces and to the electronic properties as far as they are related to these structures. The interfacial structure becomes more complex as differences in lattice parameter and structure of the two joining crystals increase.

In Chapter 9 the reader will be exposed to problems of quantum transport, which has become possible by exploiting the refined technologies of epitaxial growth and microfabrication. Lithographically patterned $GaAs-Al_xGa_{1-x}As$ heterostructures are used to study the Hall effect in a quasi-one-dimensional electron gas and to develop some of basic concepts of quantum transport.

Since the first demonstration of substitutional doping in 1975, hydrogenated amorphous silicon has become a very promising semiconductor for many applications. As outlined in Chapter 10, the basic properties of this material, the influence of defects, especially that of hydrogen and of coordination defects, on its properties, and the coupling between electronic excitations and structural changes can now be traced back to basic concepts.

I am very grateful to all the authors who took the trouble to write a chapter for this volume. I thank Prof. P. Haasen, Dr. M. Seibt and Prof. H. Feichtinger for many useful proposals and for critical comments. I would also like to thank Dr. Christina Dyllick of VCH for her advice and for the agreeable cooperation.

Wolfgang Schröter
Göttingen, June 1991

Editorial Advisory Board

Contributors to Volume 4

Professor Helmut Alexander
Universität zu Köln
II. Physikalisches Institut
Abteilung für Metallphysik
Zülpicherstraße 77
D-5000 Köln 41
Germany
Chapter 6

Professor Alain Bourret
Centre d'Etudes Nucléaires
de Grenoble
DRF/SPh, 85 X
F-38041 Grenoble Cedex
France
Chapter 7

Dr. Albert M. Chang
AT & Bell Laboratories
Holmdel, NJ 07733-3030
USA
Chapter 9

Professor Helmut Feichtinger
Institut für Experimentalphysik
Karl-Franzens-Universität Graz
Universitätsplatz 5
A-8010 Graz
Austria
Chapter 4

Dr. Dieter Gilles
IV. Physikalisches Institut der
Georg-August-Universität Göttingen
Bunsenstraße 13–15
D-3400 Göttingen
Germany
Chapter 11

Professor Ulrich M. Gösele
Department of Mechanical
Engineering and Materials Science
Duke University
School of Engineering
Durham, NC 27706
USA
Chapter 5

Dr. Robert Hull
AT & Bell Laboratories
Murray Hill, NJ 07974-2070
USA
Chapter 8

Professor Michel Lannoo
Laboratoire d'Etude et Interfaces
ISEN
41 Boulevard Vauban
F-59046 Lille Cedex
France
Chapter 1

Dr. Abbas Ourmazd
AT & Bell Laboratories
Holmdel, NJ 07733
USA
Chapter 8

Dr. Jean Luc Rouvière
Centre d'Etudes Nucléaires
de Grenoble
DRF / SPh, 85 X
F-38041 Grenoble Cedex
France
Chapter 7

Professor Wolfgang Schröter
IV. Physikalisches Institut der
Georg-August-Universität Göttingen
Bunsenstraße 13–15
D-3400 Göttingen
Germany
Chapter 11

Dr. Michael Seibt
IV. Physikalisches Institut der
Georg-August-Universität Göttingen
Bunsenstraße 13–15
D-3400 Göttingen
Germany
Chapter 11

Dr. Robert A. Street
Xerox Palo Alto Research Center
3333 Coyote Hill Road
Palo Alto, CA 94304
USA
Chapter 10

Professor Teh Y. Tan
Department of Mechanical
Engineering and Materials Science
Duke University
Durham, NC 27706
USA
Chapter 5

Professor Helmar Teichler
Institut für Metallphysik
der Universität Göttingen
Hospitalstraße 3–5
D-3400 Göttingen
Germany
Chapter 6

Dr. Jany Thibault
Centre d'Etudes Nucléaires
de Grenoble
DRF / SPh, 85 X
F-38041 Grenoble Cedex
France
Chapter 7

Dr. Raymond T. Tung
AT & Bell Laboratories
Murray Hill, NJ 07974-2070
USA
Chapter 8

Professor Rainer G. Ulbrich
IV. Physikalisches Institut der
Georg-August-Universität Göttingen
Bunsenstraße 13–15
D-3400 Göttingen
Germany
Chapter 2

Professor George D. Watkins
Department of Physics
Sherman Fairchild Center
for Solid State Studies 161
Lehigh University
Bethlehem, PA 18015-3185
USA
Chapter 3

Dr. K. L. Winer
Xerox Palo Alto Research Center
3333 Coyote Hill Road
Palo Alto, CA 94304
USA
Chapter 10

Contents

1 Band Theory Applied to Semiconductors

Michel Lannoo

Laboratoire d'Etude des Surfaces et Interfaces,
Institut Supérieur d'Electronique du Nord, Lille, France

List of Symbols and Abbreviations

a	lattice parameter
$\boldsymbol{a}_j$	basis vector of the unit cell
E_g	bandgap energy
$E_c(\boldsymbol{k})$, $E_v(\boldsymbol{k})$	conduction-band and valence-band energy
e	electron charge
$F(\boldsymbol{r})$	envelope function
$\boldsymbol{G}$	reciprocal lattice vector
$G(E)$	resolvent operator
H	hamiltonian
h	Planck constant
h_{ij}	elements of the hamiltonian
$\boldsymbol{J}$	total angular momentum
$J_{vc}(\omega)$	joint density of states
$\boldsymbol{k}$	wave vector
$\boldsymbol{L}$	orbital angular momentum
M_{vc}	optical matrix element
m^*	effective mass
$n(\varepsilon)$	density of states
$\boldsymbol{p}$	momentum vector
p_{el}, p_h	momentum of electron and hole
p_{ij}	matrix element of the momentum operator
$\boldsymbol{S}$	spin vector
T	kinetic energy
$\boldsymbol{T}$	transfer matrix
$u_{\boldsymbol{k}}(\boldsymbol{r})$	Bloch's function
V	potential
W	perturbation
β	coupling constant
$\varDelta$	splitting energy
ε_n	energy of the state n
$\varepsilon_1(\omega)$, $\varepsilon_2(\omega)$	real and imaginary part of the dielectric constant
λ	wave length
ν	frequency
Σ	self-energy operator
σ	conductivity
$|\chi_{i\mu}\rangle, |\varphi_{i\alpha}\rangle$	atomic states
ψ_n	wave function
Ω_a	atomic volume
ω	frequency
ATA	average t matrix approximation
a-Si, c-Si	amorphous and crystalline silicon
CPA	coherent potential approximation

DOS	density of states
EMA	effective mass approximation
EMT	effective mass theory
EPM	empirical pseudopotential method
EPR	valence pair repulsion
ESR	electron spin resonance
EXAFS	extended X-ray fine structure
LCAO	linear combination of atomic orbitals
LDA	local density approximation
MCPA	molecular coherent potential approximation
TBA	tight binding approximation
UPS	ultraviolet photoemission spectroscopy
VCA	virtual crystal approximation
VSEPR	valence shell electron pair repulsion
XPS	X-ray photoemission spectroscopy

1.1 General Principles

The aim of this chapter is to establish the basic principles of the formation of bands in solids. To do this we start with a one-dimensional square well potential which we consider as a simplified description of an atom. We then bring two such square wells into close contact to simulate the behavior of a diatomic molecule. This can be generalized to include a large number of square wells to illustrate the concept of energy bands corresponding to the one-dimensional model of the free electron gas. These results are shown to be independent of the boundary conditions, allowing us to use Born von Karmann cyclic conditions and to classify the band states in terms of their wave vectors.

Such a simple description does not take into account the spatial variations of the true potential in crystalline solids. However, the free electron gas results can be generalized through the use of Bloch's theorem which allows the classification of all energy bands in terms of their wave vectors, the periodic boundary conditions imposing that the allowed energy values be infinitely close together. This does not hold true, however, in nonperiodic systems such as amorphous solids and glasses. However, we show that, in this case, although general considerations impose the existence of bands, these can contain two types of states with localised or delocalised behavior. This represents the fundamental difference between nonperiodic systems and crystalline materials.

A final situation of practical interest is the case where a slowly varying potential is superimposed on a rapidly varying crystal potential. Such cases can be treated in a simple and very efficient manner by the effective mass theory (EMT), also termed the envelope function approximation. The best known applications of the EMT correspond to hydrogenic impurities in semiconductors and, more recently, to various kinds of systems involving semiconductor heterojunctions, which will be discussed in Sec. 1.7.

1.1.1 From Discrete States to Bands

We want to describe here in the simplest way the basic phenomena which occur when atoms are brought close to each other to form a solid. To do this, the attraction potential of the free atom is represented by a one-dimensional square well potential and the main trends concerning its energy levels and wave functions are analyzed.

We first consider the square well potential of Fig. 1-1, with depth V_0 and width a. We assume a to be of atomic dimensions (i.e., a few Å) and take V_0 to be large enough for the energy ε of the lowest states to be such that $\varepsilon \ll V_0$. In such cases these lowest states are close to those of an infinite potential, i.e., their wave functions ψ_n and energies ε_n are given by:

$$\psi_n \sim \sin k_n x$$

$$\varepsilon_n \sim \frac{\hbar^2}{2m} k_n^2 \qquad (1\text{-}1)$$

$$k_n \sim \frac{n\pi}{a} \qquad (n=1, 2, ...)$$

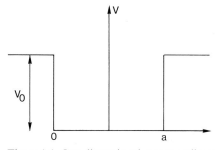

Figure 1-1. One dimensional square well potential of height V_0 and of width a.

the allowed values k_n of the one-dimensional wave vector k practically corresponding to vanishing boundary conditions at $x=0$ and a. For a typical atomic dimension of $a \approx 3\,\text{Å}$, the distance in energy between the lowest two levels is of the order of 10 eV, typical of the values found in atoms. Of course the values of the excited states cannot correspond to what happens to true three-dimensional potentials, but this will not affect the main qualitative conclusions derived below.

Let us now build the one-dimensional equivalent of a diatomic molecule by considering two such potentials at a distance R (Fig. 1-2). We can discuss qualitatively what happens by considering first the infinitely large R limit where the two cells can be treated independently. In this limit, the energy levels of the whole system are equal to those of each isolated well (i.e., $E_1, E_2 \ldots$) but with twofold degeneracy since the wave function can be localised on one or the other subsystem. When R is finite but large, the solutions can be obtained using the first order perturbation theory from those corresponding to the independent wells. This means that the twofold degenerate solutions at $E_1, E_2 \ldots$ will exhibit a shift and a splitting in energy resulting in two sublevels of symmetric and antisymmetric character. This shift and splitting will increase as the distance R decreases. Such a behavior is pictured in Fig. 1-3, from $R = \infty$ to the limiting case where $R = a$. This case is particularly easy

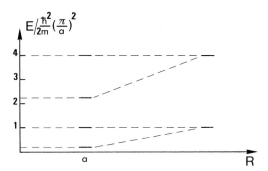

Figure 1-3. Energy levels for the double well of Fig. 1-2 as a function of the distance R between the two wells, a being the width of the individual wells.

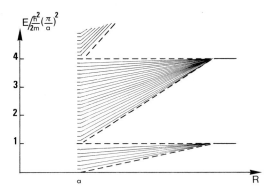

Figure 1-4. Formation of bands for a Krönig-Penney model where a is the well width and R the interwell distance.

to deal with since it corresponds to a single potential well but of width $2a$ instead of a. The consequence is that one again gets the set of solutions Eq. (1-1), but the allowed values of k_n are closer together with an interval $\pi/2a$ instead of π/a. This means that the number of levels is multiplied by two as is apparent in Fig. 1-3.

The generalization to an arbitrary number N of atoms is obvious and is pictured in Fig. 1-4. At large inter-well separation R, the degeneracy of each individual level $E_1, E_2 \ldots$ is N. At closer separation, these levels shift and split into N distinct components. When $R = a$ one recovers one single well of width $N a$, which means that the

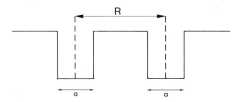

Figure 1-2. The double well potential as a simple example of a molecule.

allowed values of k_n are now separated by $\pi/N a$. For a crystal, N becomes very large (of order 10^7 for a 1 D system) and these allowed values form a "pseudocontinuum", i.e., a set of discrete values extremely close to one another. The same is true of the energy levels whose pseudocontinuums can either extend over the whole range of energies (the case $R = a$ in Fig. 1-4) or more generally $(R > a)$ are built from distinct groups of N levels. These groups of N levels are always contained in the same energy intervals irrespective of the value of N. They are called the allowed energy bands, the forbidden regions being called gaps.

These qualitative arguments can be readily generalized to realistic atomic potentials in three dimensions. One then gets a set of allowed energy bands which contain a number of states equal to the number of atoms N times an integer. These bands are, in general, separated by forbidden energy gaps.

1.1.2 Bloch Theorem for Crystalline Solids

For crystals it is possible to derive fairly general arguments concerning the properties of the allowed energy bands. Before doing this let us come back to our simplified model of N square wells in the situation $R = a$ where one gets a single well of width $N a$. We have seen that the states are completely characterized by the allowed values k_n of the wave number k which are equal to $n\pi/N a$ with $n > 0$. It is to be noted, however, that even this simple model contains complications due to the existence of boundaries or surfaces. It is the existence of these boundaries that lead to solutions having the form of stationary waves $\sin(k x)$ instead of propagating waves $\exp(i k x)$ which are mathematically simpler (the sine function is a combination of two exponentials). Of course the in-

creased mathematical complexity of the vanishing boundary conditions is not very important for this particular one-dimensional problem, but it becomes prohibitive in real cases where one has to deal with the true surfaces of a three-dimensional material.

If one only needs information about volume properties it is possible to avoid this problem by making use of Von Laue's theorem which states that perturbations induced by surfaces only extend a few angströms within the bulk of the material. This means that for large systems, where the ratio of the numbers of surface to volume atoms tends to zero, volume properties can be obtained by any type of boundary conditions even if they do not seem realistic. In that respect, it is best to use Born von Karman's periodic conditions in which one periodically reproduces the crystal under study in all directions, imposing that the wave function has the same periodicity. One then gets an infinite periodic system whose mathematical solutions are propagating waves. For instance, the linear square well problem with $R = a$ becomes a constant potential extending from minus to plus infinity. Its solutions can be taken as:

$$\psi_n = e^{i k_n x}$$

$$\varepsilon_n = \frac{\hbar^2}{2m} k_n^2 \qquad (1\text{-}2)$$

$$k_n = \frac{2 n \pi}{L} \quad (n \geqslant 0 \text{ integer})$$

They are mathematically different from Eq. (1-1) but they lead to the same answer for physical quantities. For instance, the interval between the allowed values of k is larger by a factor of two but this leads to the same density of states (number of levels per unit energy range) because the states are twofold degenerate (states with opposite values of k give the same energy). One can also show that the Fermi energy and electron density are identical in both cases.

The results given by Eq. (1-2) can be generalized to real crystalline solids through the use of the Bloch theorem. Such solids are characterized by a periodically repeated unit cell and, with Born von Karman boundary conditions, they have translational periodicity which greatly simplifies the mathematical formulation of the solutions. Let us then consider a three-dimensional crystal for which one has to solve the one-electron Schrödinger equation:

$$H \psi (r) = E \psi (r), \tag{1-3}$$

where r is the electron position vector. If we call a_j the basis vectors of the unit cell, the fact that the system has translational symmetry imposes that:

$$|\psi (r + a_j)|^2 = |\psi (r)|^2 \tag{1-4}$$

which, for the wave function itself, gives:

$$\psi (r + a_j) = \exp (i \varphi_j) \psi (r) \tag{1-5}$$

If we now consider a translation by a vector R

$$R = \sum_j m_j a_j \tag{1-6}$$

m_j being integers, we automatically get the condition:

$$\psi (r + R) = \exp \left(i \sum_j m_j \varphi_j \right) \psi (r) \tag{1-7}$$

The phase factor in Eq. (1-7) is a linear function of the components of R and can be written quite generally under the simpler form $k \cdot R$, leading to:

$$\psi_k (r + R) = \exp (i k \cdot R) \psi_k (r) \tag{1-8}$$

where k is such that $k \cdot a_j$ is equal to φ_j (in this expression we have indexed the wave function by its wave vector k). To exploit Eq. (1-8), which is a direct consequence of translational invariance, we can write

without the loss of generality:

$$\psi_k (r) = \exp (i k \cdot r) u_k (r), \tag{1-9}$$

which defines a new function $u_k (r)$. The direct application of a translation R to this expression leads to:

$$\psi_k (r + R) = \exp [i k \cdot (r + R)] u_k (r + R), \tag{1-10}$$

and a comparison with Eq. (1-8) automatically leads to:

$$u_k (r + R) = u_k (r), \tag{1-11}$$

i.e., that the function $u_k (r)$ is periodic.

Equations (1-9) and (1-11) constitute Bloch's theorem, which states that the eigenfunctions can be classified with respect to their wave vector k and written as the product of a plane wave $\exp (i k r)$ times a periodic part. They are thus propagating waves in the crystal lattice. This is a generalization of the one-dimensional case discussed above.

If one injects $\psi_k (r)$ into Schrödinger's equation Eq. (1-3), then the eigenvalue will become a continuous function $E (k)$ of the wave vector k. However, the vector k can only take allowed values chosen in such a way that $\psi_k (r)$ satisfies the boundary conditions, which are:

$$\psi_k (r + N_j a_j) = \psi_k (r), \tag{1-12}$$

where N_j is the number of crystal cells along a_j. Using Eq. (1-9) and Eq. (1-11) this imposes the conditions:

$$k \cdot a_j = \frac{2 \pi}{N_j} n_j, \tag{1-13}$$

where n_j is an integer. These allowed values can be expressed more directly with the help of the basis vectors of the reciprocal lattice a_j^* defined by:

$$a_j^* \cdot a_l = 2 \pi \delta_{jl}, \tag{1-14}$$

which leads to:

$$k = \sum_j \frac{n_j}{N_j} a_j^* \qquad (1\text{-}15)$$

This again generalizes the one-dimensional situation described by Eq. (1-2) to any real crystal in 1-, 2-, or 3-dimensions.

As a conclusion Bloch's theorem leads us to write the wave function as a plane wave $\exp(i\mathbf{k} \cdot \mathbf{r})$ modulated by a periodic part. Its energy $E(\mathbf{k})$ is a continuous function of the wave vector $\mathbf{k}$. This one takes discrete values which form a pseudocontinuum. The allowed energies will then be grouped into bands as discussed more qualitatively in the previous section. A final point is that one can show that the energy curves $E(\mathbf{k})$ and Bloch functions $\psi_{\mathbf{k}}(\mathbf{r})$ are periodic in reciprocal space. One then gets the complete information about these quantities from their calculation for $\mathbf{k}$ points lying in one period of the reciprocal lattice. From the symmetry properties $(E(\mathbf{k}) = E(-\mathbf{k}))$ it is better to use the period symmetrical with respect to the origin, called the first Brillouin zone.

1.1.3 The Case of Disordered Systems

Different varieties of solids exist which do not exhibit the long range order characteristic of perfect crystals. They differ qualitatively among themselves by the nature of their disorder. A simple case is the alloy system where one can find either A or B atoms on the sites of a perfect crystalline lattice. This kind of substitutional disorder is typical of the ternary semiconductor alloys $Ga_{1-x}Al_xAs$ where the disorder only occurs on the cationic sites. We shall deal in more detail with this problem in Sec. 1.6. Other cases correspond to amorphous semiconductors and glasses which are characterized by a short range order. For instance, in a-Si, the silicon atoms retain

their normal tetrahedral bonding and bond angles (with some distortions), but there is a loss of long range order. A lot of covalently-bonded systems (with coordination numbers smaller than 4) can be found in the amorphous or glassy state. All have a short range order and moderate fluctuations in bond length, but in some cases they can have large fluctuations in bond angles.

In all these situations, even for the disordered substitutional alloy, it is clear that one cannot make use of Bloch's theorem to classify the band states. We could even ask ourselves if the concept of energy bands still exists. The simplest case demonstrating this point again corresponds to the one-dimensional system with square well potentials. We can simulate a disordered substitutional alloy by considering square wells of width a as before, and of depth V_A for A atoms and V_B for B atoms, with $V_A > V_B$. We also assume that the interwell distance is $R = a$ to get the simplest situation. If the material were purely B then one would get a constant potential which we take as the origin of energies. Thus each time one substitutes an A atom for a B one this results in an extra potential well of width a and depth $V_A - V_B$. Two A neighbors lead to a well of width $2a$, and a cluster of M A neighbors gives a well of width $M a$ (Fig. 1-5). The alloy will consist of a distribution of these potential wells separated by variable distances.

It is interesting to investigate the nature of the states for such a problem. The simplest case corresponds to energies $E > 0$,

Figure 1-5. Simple representation of a disordered binary alloy with square well potentials of width a for one impurity, $2a$ for a pair, etc.

where the states have a propagating behavior and there are solutions at any positive value of E. The existence of boundary conditions due to the fact that the system contains a finite but large number N of atoms will simply transform this continuum of states into a pseudocontinuum. This results in a band of "extended states" for $E > 0$.

The situation for $E < 0$ is drastically different. An isolated impurity A, represented by a single well, gives at least one energy level at E_1, or even more at E_2, E_3, etc. The corresponding wave function is localized and decays exponentially as $\exp(-k|z|)$ where $k = \sqrt{\dfrac{2m}{\hbar^2}|E|}$. Clusters of M atoms lead, as described before, to denser sets of levels in the well since the interval between allowed k values is divided by M. In particular, for a very large cluster, one obtains a pseudocontinuum of levels, the lowest one being infinitely close to the bottom of the potential well. One has thus a statistical distribution of potential wells of varying width giving rise to a corresponding distribution of levels located between $E = 0$ and $E = -(V_A - V_B)$. Again this leads to a pseudocontinuum of levels. Furthermore, there is the possibility of interaction between the potential wells when these are close enough. This also acts in favor of a spread in energy levels. The states in the potential wells can have a more or less localized character depending on the distance between the wells. We shall discuss this problem later.

The conclusion of this simple model is that the occurrence of disorder, at least in somes cases, also leads to the existence of a pseudocontinuum of states, even if Bloch's theorem does not apply. Of course we have only discussed one particular model. We shall later discuss other cases of disorder or randomness which also lead to the existence of well-defined energy bands.

1.1.4 The Effective Mass Approximation (EMA)

Interesting cases that often occur in practice correspond to the application of a potential slowly varying in space to a crystalline solid. Such situations can be handled relatively easily without solving the full Schrödinger equation, by using the so called "effective mass approximation". One major field of application has been the understanding of hydrogenic impurities in semiconductors (for reviews see Bassani et al., 1974; Pantelides, 1978). More recently the same method, often called "the envelope function approximation", has been applied to the treatment of semiconductor heterojunction and superlattices as will be discussed in Sec. 1.7.

1.1.4.1 Derivation of the Effective Mass Approximation for a Single Band

Let us begin with the simplest case of a crystal whose electronic structure can be described in terms of a single energy band, the solution of the perfect crystal Schrödinger equation:

$$H_0 \, \varphi_k(r) = E(k) \, \varphi_k(r) \qquad (1\text{-}16)$$

If a perturbative potential $V(r)$ is applied to the system we can describe a solution of the perturbed system $\psi(r)$ as a linear combination of the perfect crystal eigenstates (k belonging to the first Brillouin zone),

$$\psi(r) = \sum_k a(k) \, \varphi_k(r) \qquad (1\text{-}17)$$

and obtain the unknown coefficients by projecting the new Schrödinger equation,

$$(H_0 + V) \, \psi(r) = E \, \psi(r) \qquad (1\text{-}18)$$

onto the basis states $\varphi_k(r)$. This immediately leads to the set of linear equations:

$$E(k) \, a(k) + \sum_{k'} \langle \varphi_k | V | \varphi_{k'} \rangle \, a(k') = E \, a(k) \qquad (1\text{-}19)$$

At this stage we need to simplify the matrix elements of V, otherwise it is impossible to go further, except numerically. We use the fact that the $\varphi_k(r)$ are Bloch functions,

$$\varphi_k(r) = e^{ik \cdot r} u_k(r) \qquad (1\text{-}20)$$

and express the potential matrix element as:

$$\langle \varphi_k | V | \varphi_{k'} \rangle = \qquad (1\text{-}21)$$
$$= \int V(r) e^{i(k'-k) \cdot r} u_k^*(r) u_{k'}(r) d^3 r$$

In view of the Bloch theorem the product $u_k^* u_{k'}$ is a periodic function of r and we can expand it in a Fourier series,

$$u_k^*(r) u_{k'}(r) = \sum_G C_{k,k'}(G) e^{iG \cdot r} \qquad (1\text{-}22)$$

where G are the reciprocal lattice vectors. The matrix element Eq. (1-21) can thus be expressed exactly as:

$$\langle \varphi_k | V | \varphi_{k'} \rangle = \sum_G C_{k,k'}(G) V(k'+G-k) \qquad (1\text{-}23)$$

where $V(k)$ is the Fourier transform of $V(r)$. At this level we must make some assumptions about $V(r)$. The first one is that it varies slowly in space (i.e., over distances which are large compared to the size of the unit cell). This means that its Fourier transform decreases very rapidly with the modulus of the wave vector, i.e., that one can neglect terms with $G \neq 0$ in Eq. (1-23) and, also, that only terms with $k' \approx k$ will effectively contribute.

We now make the second central assumption of the EMA, that we look for solutions whose energy E is close to a band extremum k_0. If this is so, only states with $k \approx k_0$ will have $a(k)$ sensibly different from zero in Eq. (1-19). This means that one can rewrite Eq. (1-19) using Eq. (1-23) under the approximate form:

$$\qquad (1\text{-}24)$$
$$E(k) a(k) + \sum_{k'} C_{k_0, k_0}(0) V(k'-k) a(k') = 0$$

However, $C_{k_0, k_0}(0)$ has an important property: it is given by the following integral over the crystal volume

$$C_{k_0, k_0}(0) = \int u_{k_0}^*(r) u_{k_0}(r) d^3 r =$$
$$= \int \varphi_{k_0}^*(r) \varphi_{k_0}(r) d^3 r = 1 \qquad (1\text{-}25)$$

in view of the fact that the wave functions $\varphi_k(r)$, are normalized. The final form of the EMA equation is thus

$$\qquad (1\text{-}26)$$
$$E(k) a(k) + \sum_{k'} V(k'-k) a(k') = E a(k)$$

It is interesting to derive a real space equation from this by Fourier transforming Eq. (1-26). To perform this we must take into account the fact that the function $a(k)$ is strongly peaked near k_0. We thus introduce the following Fourier transform:

$$F(r) = \sum_k a(k) \exp(i(k-k_0) \cdot r) \qquad (1\text{-}27)$$

such that $F(r)$ varies slowly in space when $a(k)$ only takes important values in the vicinity of $k = k_0$. To get an equation for $F(r)$ we multiply Eq. (1-26) by $\exp i(k-k_0) \cdot r$ and sum over k, assuming that one makes a negligible error in the potential term by extending the summation over k to the whole space. This leads to the real space equation:

$$\{E(k_0 - i \nabla_r) + V(r)\} F(r) = E F(r) \qquad (1\text{-}28)$$

This is a differential equation in which the operator $k_0 - i \nabla_r$ has been substituted for k in the dispersion relation $E(k)$. As we have seen, the function $F(r)$ is likely to vary slowly with r (or $a(k) \neq 0$ only for $k \approx k_0$) so that one can expand $E(k)$ to the second order in the neighborhood of $k \sim k_0$. Calling α the principal axes of this expansion we have:

$$E(k) \approx E(k_0) + \sum_\alpha \frac{\hbar^2}{2 m_\alpha} (k_\alpha - k_{0\alpha})^2 \qquad (1\text{-}29)$$

which defines the effective masses m_α along direction α. This allows us to rewrite Eq. (1-28) as:

$$\left\{ -\sum_\alpha \frac{\hbar^2}{2\,m_\alpha} \frac{\partial^2}{\partial x_\alpha^2} + V(r) \right\} F(r) =$$

$$= \{E - E(k_0)\} F(r) \qquad (1\text{-}30)$$

which represents the usual form of the EMA equation as derived by many authors (Bassani et al., 1974; Pantelides, 1978).

It is interesting to examine the meaning of the function $F(r)$. To do this we start from the expansion Eq. (1-17) of $\psi(r)$, express the $\varphi_k(r)$ as in Eq. (1-20) and factorize $e^{i k_0 \cdot r}$. This gives

$$\psi(r) = e^{i k_0 \cdot r} \sum_k a(k) e^{i(k-k_0)\cdot r} u_k(r) \quad (1\text{-}31)$$

As $a(k)$ is peaked near k_0, we approximate $u_k(r)$ by its value at k_0, which leads us directly to:

$$\psi(r) \simeq F(r)\, \varphi_{k_0}(r) \qquad (1\text{-}32)$$

This means that $F(r)$ can be rewritten as the product of the Bloch function (which varies over a length typically of the order of the interatomic distances) times a slowly varying "envelope function". The advantage of the EMA is that one directly obtains $F(r)$ from a Schrödinger-like equation involving the effective masses.

1.1.4.2 Applications and Extensions

The first well known use of the EMA was for hydrogenic impurities in semiconductors. If we treat single donor substitutional impurities, like As in Si or Ge for instance, the excess electron will see an attractive potential roughly given by $-e^2/\varepsilon r$ (where ε is the dielectric constant), which one considers as slowly varying. This can stabilize levels in the gap in the proximity of the bottom of the conduction

band E_c. For a single minimum and an isotropic effective mass one gets an hydrogenic-like equation with scaled parameters $e^2 \to e^2/\varepsilon$ and $m \to m^*$. This leads to a set of hydrogenic levels with an effective Rydberg $\frac{m^* e^4}{2\hbar^2 \varepsilon}$ which, for typical values of $m^* \approx 0.1$ and $\varepsilon \approx 10$, becomes of order 14 meV, i.e., fairly small compared to the band gap. This result correctly reproduces the order of magnitude found in experimental data. However, to be truly quantitative, the EMA results must in many cases satisfy the following requirements:

– It must include the effective-mass anisotropy when necessary. This has been done in Faulkner (1968, 1969), one effect being the splitting of p states, for instance.

– It must also properly include the valley-valley interactions when there are several equivalent minima. This can be done, for instance, by first order perturbation theory on degenerate states (since there are as many identical impurity states than there are minima).

With these improvements the EMA theory has achieved considerable success for single donor impurities, especially for excited states (see Bassani et al., 1974; Pantelides, 1978) and Table 1-1 for reviews. Only the ground state is found to depart significantly from the predicted levels at this stage of the theory. This is due to the deviations of the potential from its idealized form $-e^2/\varepsilon r$ in the impurity cell. The corresponding correction is known as the chemical shift. It is also possible to treat more exactly the many-valley interactions by recently derived methods described in Resca and Resta (1979, 1980).

The case of acceptor states derived from the valence band is more complicated. This is due to the threefold degeneracy of the top of the valence band. This means that $\psi(r)$ must be written as a combination of

Table 1-1. Comparison between theoretical and experimental energy spacing (cm^{-1}) for donor impurities in silicon. The theoretical values are taken from Faulkner (1968, 1969). (The spacing between excited states, independent of the ground state position, is more suitable than the observed position of the transition from the ground state to perform a comparison with the theory since this ground state is not hydrogenic.)

Transition	Theory	P	As	Sb
$2p_\pm - 2p_0$	5.11	5.07	5.11	5.12
$3p_0 - 2p_\pm$	0.92	0.93	0.92	0.91
$4p_0 - 2p_\pm$	3.07	3.09	3.10	3.07
$3p_\pm - 2p_\pm$	3.29	3.29	3.28	3.29
$4p_\pm - 2p_\pm$	4.22	4.22	4.21	4.21
$4f_\pm - 2f_\pm$	4.51	4.51	4.49	4.46
$5p_\pm - 2f_\pm$	4.96	4.95	4.94	4.92
$5f_\pm - 2p_\pm$	5.14	5.15	5.14	
$6p_\pm - 2p_\pm$	5.36	5.32	5.32	5.31
$3s - 2p_\pm$	0.65			
$3d_0 - 2p_\pm$	2.65	2.64		
$4s - 2p_\pm$	3.55			
$4f_0 - 2p_\pm$	4.07	4.08		
$5p_0 - 2p_\pm$	4.17	4.17		
$5p_0 - 2p_\pm$	4.77	4.76		4.70
$6h_\pm - 2p_\pm$	5.22	5.52		

the Bloch states $\varphi_{n,k}(r)$ (with $n = 1, 2, 3$) belonging to each of the three energy branches $E_n(k)$. The first part of the derivation proceeds as for a single band and the generalization of Eq. (1-26) becomes:

$$E_n(k) a_n(k) + \sum_{k'} V(k' - k) a_n(k') = E a_n(k)$$
(1-33)

This equation is diagonal in n and is apparently as simple to solve as for the single band extremum. However, the difficulty is to proceed further and transform it to a real space equation as in Eq. (1-28). The reason is that one can no longer define the derivatives of $E_n(k)$ near the valence band maximum at $k = 0$. This will be shown in detail in Section 1.3.2, following Kane's derivation (Kane, 1956, 1957) in which it is shown that the quantities that can be expanded to second order in k are the elements $h_{ij}(k)$ of a 3×3 matrix (6×6 if spin orbit is included (Kane, 1956, 1957; Luttinger and Kohn, 1955; Luttinger, 1956)) whose eigenvalues are the $E_n(k)$. Considering the $a_n(k)$ as the components of a 3-component column vector ($n = 1, 2, 3$) expressed on the basis of the eigenvectors of $h_{ij}(k)$, it is advantageous to rewrite Eq. (1-33) using the natural basis states of $h_{ij}(k)$. This leads to

$$\sum_j h_{ij}(k) a_j(k) + \sum_{k'} V(k' - k) a_i(k) = E a_i(k).$$
(1-34)

Now one can define slowly varying functions $F_j(r)$ by the Fourier transformation Eq. (1-27) from $a_j(k)$ with $k = 0$ and get the generalization of Eq. (1-28):

$$\sum_j h_{ij}(-i\nabla_r) F_j(r) + V(r) F_i(r) = E F_i(r)$$
(1-35)

As the h_{ij} are of second order in k, this represents a set of coupled second order differential equations whose solution will lead to the solution of the envelope functions. Finally, in this approximation, the total wave function $\psi(r)$ becomes, by using the basis set corresponding to h_{ij}:

$$\psi(r) = \sum_j a_j(k) e^{ik \cdot r} u_{jk}(r)$$
(1-36)

which, if only $k \simeq 0$ is involved, becomes

$$\psi(r) \approx \sum_j \varphi_{j0}(r) F_j(r)$$
(1-37)

We do not discuss here the application of the method to hydrogenic acceptor states (details can be found in Bassani et al., 1974; Pantelides, 1978). We shall later see its use in quantum wells and superlattices. In such cases, the application of the envelope function approximation is complicated by the problem of boundary conditions which we discuss in Sec. 1.7.

1.2 The Calculation of Crystalline Band Structures

We have seen that, for crystalline solids, the use of Bloch's theorem allows us to demonstrate quite generally the existence of energy bands. However, in its derivation we have made the implicit assumption that one could write a Schrödinger equation for each electron taken separately. Of course this is in principle not permissible in view of the existence of electron-electron interactions and one should consider the N electron system as a whole. It has been shown that one can generalize the Bloch theorem to the one particle excitations of crystalline many-electron systems. However, when this is done there is no exact method available to calculate these excitations (which correspond to the energy bands) in practice. One is left with approximate methods which are all based on the reduction of the problem to a set of separate one particle equations whose eigenvalues are used to compare the experimental one particle excitations.

First, we thus give a brief account of most one-electron theories that have been used so far: Hartree, Hartree-Fock, local density, etc. We also discuss recent advances which have allowed us to considerably improve the local density results (the so-called "G-W approximation"). Usually, it is not necessary to include core electrons in calculations involving properties of the valence electrons. To achieve this separation one replaces the true atomic potentials by "first principles pseudopotentials" of which we give a short description. All this completely defines the single particle equations and, in Sec. 1.2.2 we present some techniques that can be used for their resolution. These techniques often involve a substantial amount of computation but they can be applied to simple crystals such

as the zinc-blende semiconductors. However, there is a need for simpler empirical methods, either for physical understanding or as simulation tools for more complex systems. We describe two such methods in Sec. 1.2.3: the empirical pseudopotential method (EPM) and the empirical tight binding approximation (TBA).

1.2.1 Ab Initio Theories

We describe here some basic methods that lead to approximate single particle equations. We begin with the Hartree approximation which is the simplest to derive and illustrates the general principles that are applied. We then discuss the Hartree-Fock approximation, local density theory, and its recent improvements via the G-W approximation.

1.2.1.1 The Hartree Approximation

The full N electron Hamiltonian (for fixed nuclei) can be written:

$$H = \sum_{i=1}^{N} h_i + \frac{1}{2} \sum_{i \neq j} \frac{e^2}{r_{ij}} \qquad (1\text{-}38)$$

where the h_i are independent individual Hamiltonians containing the kinetic energy operator of electron i as well as its attraction by the nuclei. The second term in Eq. (1-38) represents the electron-electron interactions, r_{ij} being the distance between electrons i and j. If one could neglect these, the problem would be exactly separable, i.e., one could obtain the solution of the full problem by simply solving the individual Schrödinger equations,

$$h_i \, \varphi_{n_i}(r_i) = \varepsilon_{n_i} \, \varphi_{n_i}(r_i), \qquad (1\text{-}39)$$

the full wave function being a product of individual wave functions (if one forgets for the moment the fact that it must be antisymmetric) and the total energy being the sum of individual energies.

The inclusion of the electron-electron interactions prevents the problem from being separable. However, one can find approximate individual equations by using a trial wave function (in the variational sense) which is of a separated form, i.e., it is a simple product of individual functions. For N electrons this gives

$$\psi\,(r_1 \ldots r_N) = \prod_{n_i} \varphi_{n_i}(r_i) \qquad (1\text{-}40)$$

The unknown wave functions can be obtained by using the variational method, i.e., by minimizing the average value of H with respect to the φ_{n_i}. This leads to a set of individual equations which are the Hartree equations. However, these can be obtained directly by the following simple physical argument. If the problem can be separated, then the Hamiltonian of electron i will consist of the sum of its kinetic energy operator, its potential energy in the field of the nuclei, and its potential energy of repulsion wih other electrons. This leads to the Schrödinger equation:

$$\left\{ h_i + \sum_{j \neq i} \int \frac{e^2}{r_{ij}} |\varphi_{n_j}(r_j)|^2 \, d^3\,r_j \right\} \varphi_{n_i}(r_i) =$$

$$= \varepsilon_{n_i}\,\varphi_{n_i}(r_i) \qquad (1\text{-}41)$$

where the second term in the Hamiltonian represents the average electrostatic repulsion exerted on electron i by all the other electrons. There are as many equations Eq. (1-41) as there are electrons, i.e., N. Each one-electron Hamiltonian contains the wave functions of the other electrons which are unknown. One has thus to proceed by iterations until a self-consistent solution is found, i.e., the wave functions injected into the hamiltonian are the same as the solutions of Eq. (1-41). The Hartree approximation has been used in understanding the basic physics of atoms. It is not refined enough to be used for actual band structure calculations.

1.2.1.2 The Hartree-Fock Approximation

The main drawback of the Hartree approximation is that its wave function is not properly antisymmetrized. If one wants to use a trial function corresponding to independent electrons one cannot make use of a simple product of individual wave functions but instead one must consider a Slater determinant of the form

$$\psi\,(r_1 \ldots r_N) = \qquad (1\text{-}42)$$

$$= \frac{1}{\sqrt{N!}} \begin{vmatrix} \varphi_{n_1}(r_1) & \varphi_{n_2}(r_1) \cdots \varphi_{n_N}(r_1) \\ \varphi_{n_1}(r_2) & \varphi_{n_2}(r_2) \cdots \varphi_{n_N}(r_2) \\ \cdot & \cdot & \cdot \qquad \cdots \ \cdot \\ \varphi_{n_1}(r_N) & \varphi_{n_2}(r_N) \cdots \varphi_{n_N}(r_N) \end{vmatrix}$$

The variational method can be applied in exactly the same way as the Hartree method. However, it is no longer possible to get the one-electron equations using a simple argument. The application of this technique leads directly to:

$$\left\{ h_i + \sum_{j \neq i} \int \frac{e^2}{r_{ij}} |\varphi_{n_j}(r_j)|^2 \, d^3\,r_j \right\} \varphi_{n_i}(r_i) -$$

$$- \sum_{j \neq i} \int \frac{e^2}{r_{ij}} \varphi_{n_j}^*(r_j)\,\varphi_{n_i}(r_j)\,d^3\,r_j\,\varphi_{n_j}(r_i) =$$

$$= E\,\varphi_{n_i}(r_i) \qquad (1\text{-}43)$$

The result is the Hartree contribution plus a correction factor, the exchange term, due to antisymmetry in the electron permutations. It is important to notice that the $\varphi_{n_i}(r_i)$ are "spin orbitals", i.e., products of a spatial part multiplied by a spin function (Slater, 1960). This is fundamental in order to maintain the Pauli principle.

The Hartree-Fock method is not very easy to apply numerically in view of the complexity of the exchange terms. Its application to covalent solids like diamond or silicon (Euwema et al., 1973; Mauger, Lannoo, 1977) leads to the overall correct shape of the energy bands, but a large overestimation of the forbidden gap. For instance, one gets 12 eV and 6 eV for diamond and silicon respectively compared to the experimental values of 5.4 eV and 1.1 eV respectively. Improvements on the Hartree-Fock approximation can be made using what are called correlation effects.

1.2.1.3 The Local Density Approximation

The local density approximation is an extension of the Thomas-Fermi approximation based on the Hohenberg and Kohn theorem (Hohenberg and Kohn, 1964) which shows that the ground state properties of an electron system are entirely determined by the knowledge of its electron density $\varrho(r)$. The total energy of the interacting electron system can be written:

$$E(\varrho) = T(\varrho) + \frac{1}{2} \int \frac{\varrho(r)\varrho(r')}{|r-r'|} \, dr \, dr' + \\ + \int V_{ext}(r)\,\varrho(r)\,dr + E_{xc}(\varrho) \quad (1\text{-}44)$$

where T represents the kinetic energy, the second term gives the electrostatic interelectronic repulsion, V_{ext} is the potential due to the nuclei, and E_{xc} is the exchange correlation energy. A variational solution of the problem (Kohn and Sham, 1965) allows us to derive a set of one particle Schrödinger equations of the form:

$$\left[t + V_{ext}(r) + \int \frac{\varrho(r')\,dr'}{|r-r'|} + V_{xc}(r) \right] \psi_k(r) = \\ = \varepsilon_k \psi_k(r) \quad (1\text{-}45)$$

with t representing the one electron kinetic energy and

$$\varrho(r) = \sum_{occupied\ k} |\psi_k(r)|^2 \quad (1\text{-}46)$$

$$V_{xc}(r) = \frac{\delta E_{xc}[\varrho(r)]}{\delta\varrho(r)}$$

The practical resolution of Eqs. (1-45, 1-46) is, as we will discuss later, usually performed by replacing the atomic potential by pseudopotentials, avoiding the explicit consideration of atomic core states. Originally these pseudopotentials were treated empirically, but now methods exist which allow us to determine them quantitatively from the properties of the free atoms (Hamann et al., 1979). The knowledge of these pseudopotentials plus the local density treatment allows a complete determination of the solutions ψ_k of Eq. (1-45). The wave functions ψ_k can be calculated either using an expansion in plane waves or in orbitals localized on the atoms.

Equations (1-45) and (1-46) have to be solved in a self-consistent way. Up to now this formulation has been exact, the problem is that the quantity E_{xc}, and thus V_{xc}, is not known in general. The local density approximation then is based on the assumption that, locally, the relation between E_{xc} and $\varrho(r)$ is the same as for a free electron gas of identical density, which is known quite accurately. This approximation turns out to give satisfactory results regarding the prediction of the structural properties of molecules and solids. For instance, in solids (either with sp bonds or d bonds, as in transition metals) the cohesive energy, the interatomic distance, and the elastic properties are predicted with a precision better than 5% in general. This remains true for diatomic molecules, except that the binding energy is overestimated by about 0.5 to 1 eV (see Cohen, 1983; Schlüter, 1983) for recent reviews on the subject).

It is tempting to use the differences between the eigenvalues of Eq. (1-45) as particle excitation energies. This is not justified in general as seen by the predicted values for the energy gap $\varepsilon_G = \varepsilon_C - \varepsilon_V$ in semiconductors and insulators, taken as the difference in the energies ε_C of the first empty state and ε_V of the last filled state. The local density value of ε_G is always found to be substantially smaller than the experimental (Hamann, 1979). It is equal to 0.6 eV for silicon instead of 1.2 eV and even vanishes for germanium. The origin of the errors cannot be traced to the use of an overly simplified exchange and correlation potential such as LDA. This has been clearly shown by the almost identical results obtained using an improved exchange-correlation potential. The exchange-correlation term thus cannot be reduced to a simple local potential.

1.2.1.4 Beyond Local Density (the G-W Approximation)

To correct for the deficiencies of local density in defect calculations, one simple approach has been to use a "scissors" operator (Baraff and Schlüter, 1984) which corrects for the band gap error by using a rigid shift of the conduction band states. It was later shown (Sham and Schlüter, 1985, 1986; Perdew and Levy, 1983; Lannoo and Schlüter, 1985) that this procedure was closer to reality than expected, since the one electron exchange correlation potential of the local density formalism must experience a discontinuity across the gap. This is correctly handled by the scissors operator for bulk semiconductors but the applicability of the scissors operator to defect levels is still questionable. In particular, in the case of extended defects such as surfaces, it is doubtful that the defect states are correctly obtained. The advantage of this correction is mainly that it does not add any computational requirements when compared to conventional LDA calculations.

A more sophisticated way of improving the density functional theories is to evaluate the electron self-energy operator $\Sigma(r, r', E)$ (Lannoo and Schlüter, 1985; Hybertsen and Louie, 1985; Godby, 1986).

This Σ contains the effects of exchange and correlation. It is non-local, energy-dependent, and non-hermitian. Its non-hermiticity means that the eigenvalues of the new one particle Schrödinger equation,

$$(t + V_{ext} + V_H)\,\psi_{nk}(r) + \tag{1-47}$$
$$+ \int dr'\, \Sigma(r, r', E_{nk})\, \psi_{nk}(r') = E_{nk}\, \psi_{nk}(r)$$

will generally be complex. The imaginary part gives the lifetime of the quasiparticle, and V_H is the Hartree-potential.

The self-energy operator can be estimated using the G-W approximation (Hedin and Lundquist, 1969). The self-energy is expanded in a perturbation series of the screened Coulomb interaction, W. The first term of the expansion corresponds to the Hartree-Fock approximation. Details can be found in Lannoo and Schlüter (1985); Hybertsen and Louie (1985); Godby et al. (1986, 1987); Hedin and Lundquist (1969).

1.2.1.5 The Pseudopotential Method

The full atomic potentials produce strong divergences at the atomic sites in a solid. These divergences are related to the fact that these potentials must produce the atomic core states as well as the valence states. However, the core states are likely to be quite similar to what they are in the free atom. Thus the use of the full atomic potentials in a band calculation is likely to lead to unnecessary computational complexity since the basis states will have to be chosen in such a way that they describe localized states and extended states at the same time. Therefore, it is of much interest to devise a method which allows us to eliminate the core states, focusing only on the valence states of interest which are easier to describe. This is the basis of the pseudopotential theory.

The pseudopotential concept started with the orthogonalized plane wave theory (Cohen et al., 1970). Writing the crystal Schrödinger equation for the valence states,

$$(T+V)|\psi\rangle = E|\psi\rangle \qquad (1\text{-}48)$$

one has to recognize that the eigenstate $|\psi\rangle$ is automatically orthogonal to the core states $|c\rangle$ produced by the same potential V. This means that $|\psi\rangle$ will be strongly oscillating in the neighbourhood of each atomic core, which prevents its expansion in terms of smoothly varying functions, like plane waves, for instance. It is thus interesting to perform the transformation

$$|\psi\rangle = (1-P)|\varphi\rangle \qquad (1\text{-}49)$$

where P is the projector onto the core states

$$P = \sum_c |c\rangle\langle c| \qquad (1\text{-}50)$$

$|\psi\rangle$ is thus automatically orthogonal to the core states and the new unknown $|\varphi\rangle$ does not have to satisfy the orthogonality requirement. The equation for the "pseudo-state" $|\varphi\rangle$ is:

$$(T+V)(1-P)|\varphi\rangle = E(1-P)|\varphi\rangle \qquad (1\text{-}51)$$

Because the core states $|c\rangle$ are eigenstates of the Hamiltonian $T+V$ with energy E_c, one can rewrite Eq. (1-51) in the form

$$\left\{T+V+\sum_c (E-E_c)|c\rangle\langle c|\right\}|\varphi\rangle = E|\varphi\rangle \qquad (1\text{-}52)$$

The pseudo-wave function is then the solution of a Schrödinger equation with the same energy eigenvalue as $|\psi\rangle$. This new equation is obtained by replacing the potential V by a pseudopotential

$$V_{ps} = V + \sum_c (E-E_c)|c\rangle\langle c| \qquad (1\text{-}53)$$

This is a complex non-local operator. Furthermore, it is not unique since one can add any linear combination of core states to $|\varphi\rangle$ in Eq. (1-52) without changing its eigenvalues. There is a corresponding non-uniqueness in V_{ps} since the modified $|\varphi\rangle$ will obey a new equation with another pseudopotential. This non-uniqueness in V_{ps} is an interesting factor since it can then be optimized to provide the smoothest possible $|\varphi\rangle$, allowing rapid convergence of plane wave expansions for $|\varphi\rangle$. This will be used directly in the empirical pseudopotential method.

Recently, so-called "first principles" pseudopotentials have been derived for use in quantitative calculations (Hamann et al., 1979). First of all, they are ion pseudopotentials and not total pseudopotentials as those discussed above. They are deduced from free atom calculations and have the following desirable properties: (1) real and pseudovalence eigenvalues which agree for a chosen prototype atomic configuration, (2) real and pseudo-atomic wave functions which agree beyond a chosen core radius r_c, (3) total integrated charges at a distance $r > r_c$ which agree (norm conservation), and (4) logarithmic derivatives of the real and pseudo wave functions and their first energy derivatives which agree for $r > r_c$. These properties are crucial for the pseudopotential to have optimum transferability among a variety of chemical environments, allowing self-consistent calculations of a meaningful pseudocharge density.

1.2.2 Computational Techniques

We now will discuss some techniques which allow us to calculate energy bands in practice. They are simply based on an expansion of the eigenfunction of the one-electron Schrödinger equation in some suitable basis functions, plane waves, or localized atomic orbitals. These seem to be, by far, the most commonly adopted methods at this time.

1.2.2.1 Plane Wave Expansion

Plane wave form a particularly interesting basis set for crystalline band structure calculations in conjunction with the use of pseudopotentials. As we shall see below the total pseudopotential can be expressed as a sum of atomic contributions. These consist of a bare ionic part which will be screened by the valence electrons. The resulting atomic pseudopotentials are often assumed to be local, i.e. to be simple functions of the electron position. However, in general, they should be operators having a non-local nature, as discussed later. The best starting point is the expression Eq. (1-9) of the Bloch function in which one makes use of the fact that the function $u_k(r)$ is periodic and can thus be expanded as a Fourier series:

$$u_k(r) = \sum_G u_k(G)\, e^{iG \cdot r} \qquad (1\text{-}54\,\text{a})$$

where G are reciprocal lattice vectors. This leads to the natural plane wave expansion for the wave function:

$$\varphi_k(r) = \sum_G u_k(G)\, e^{i(k+G) \cdot r} \qquad (1\text{-}54\,\text{b})$$

The gain due to the Bloch theorem is that any Bloch state $\varphi_k(r)$ with wave vector k belonging to the first Brillouin zone is obtained by mixing the plane wave $e^{ik \cdot r}$ only with other plane waves whose wave vector is $k+G$, where G is any reciprocal lattice vector, and not with all plane waves with arbitrary wave vectors. Of course, we are only interested in low lying states so that we can truncate the expansion of Eq. (1-54) at some maximum value of $|G|$, which we label G_{max}. The one particle Hamiltonian is thus expressed on this basis as a finite matrix of elements:

$$H_{G,G'}(k) = \left\langle e^{i(k+G) \cdot r} \left| \frac{p^2}{2m} + V \right| e^{i(k+G') \cdot r} \right\rangle \qquad (1\text{-}55)$$

which readily becomes

$$H_{G,G'}(k) = \frac{\hbar^2}{2m} |k+G|^2\, \delta_{G,G'} + V_{G,G'}(k) \qquad (1\text{-}56)$$

where the second term is the potential matrix element. In the case where V is a simple function of r, this matrix element can be reduced to:

$$V_{G,G'}(k) = V(G' - G) \qquad (1\text{-}57)$$

Such plane wave expansions can be used in different contexts. We will later develop an application known as the empirical pseudopotential method (EPM). One can also apply these expansions to first principles calculations. This is formally easy in the local density context where the potential takes a simple form. It becomes more complex, but still can be adapted, in the Hartree-Fock theory or even in the G-W approximation as used in Hybertsen and Louie (1985, 1986) and Godby et al. (1986).

1.2.2.2 Localized Orbital Expansion

In this expansion the wave function is written as a combination of localized orbitals centered on each atom:

$$\psi = \sum_{i,\alpha} C_{i\alpha} \varphi_{i\alpha} \qquad (1\text{-}58)$$

where $\varphi_{i\alpha}$ is the αth free atom orbital of atom i, at position R_i. As each complete set of such orbitals belonging to any given atom forms a basis for Hilbert space, the whole set of $\varphi_{i\alpha}$ is complete, i.e., the $\varphi_{i\alpha}$ are no longer independent and Eq. (1-58) can yield the exact wave function of the whole system. In practice one has to truncate the sum over α in this expansion. In many simplified calculations it has been assumed that the valence states of the system can be described in terms of a "minimal basis set" which only includes free atom states up to the outer shell of the free atom (e.g., 2s and 2p in diamond). It is that description which provides the most appealing physical picture, allowing us to clearly understand the formation of bands from the atomic limit. The "minimal basis set" approximation is also used in most semi-empirical calculations.

When the sum over α in Eq. (1-58) is limited to a finite number, the energy levels E of the whole system are given by the secular equation:

$$\det|H - ES| = 0 \qquad (1\text{-}59)$$

where H is the Hamiltonian matrix in the atomic basis and S the overlap matrix of elements:

$$S_{i\alpha, j\beta} = \langle \varphi_{i\alpha} | \varphi_{j\beta} \rangle \qquad (1\text{-}60)$$

These matrix elements can be readily calculated, especially in the local density theory, and when making use of Gaussian atomic orbitals. The problem, as in the plane wave expansion, is to determine the number of basis states required for good numerical accuracy.

An interesting discussion on the validity of the use of a minimal basis set has been given by Louie (1980). Starting from the minimal basis set $|\varphi_{i\alpha}\rangle$ one can increase the size of the basis set by adding other atomic states $|\chi_{i\mu}\rangle$, called the peripheral states, which must lead to an improvement in the description of the energy levels and wave functions. However, this will rapidly lead to problems related to overcompleteness, i.e., the overlap of different atomic states will become more and more important. To overcome this difficulty, Louie proposes three steps to justify the use of a minimum basis set. These are the following:

1. Symmetrically orthogonalize the states $|\varphi_{i\alpha}\rangle$ belonging to the minimal basis set between themselves. This leads to an orthogonal set $|\bar{\varphi}_{i\alpha}\rangle$.
2. The peripheral states $|\chi_{i\mu}\rangle$ overlap strongly with the $|\bar{\varphi}_{i\alpha}\rangle$. It is thus necessary to orthogonalize them to these $|\bar{\varphi}_{i\alpha}\rangle$, which yield new states $|\bar{\chi}_{i\mu}\rangle$ defined as:

$$|\bar{\chi}_{i\mu}\rangle = |\chi_{i\mu}\rangle - \sum_{j\alpha} |\bar{\varphi}_{j\alpha}\rangle \langle \bar{\varphi}_{j\alpha} | \chi_{i\mu}\rangle \quad (1\text{-}61)$$

3. The new states $|\bar{\chi}_{i\mu}\rangle$ are then orthogonalized between themselves leading to a new set of states $|\bar{\bar{\chi}}_{i\mu}\rangle$.

Louie has shown that, at least for silicon, the average energies of these atomic states behave in such a way that, after step 3, the peripheral states $|\bar{\bar{\chi}}_{i\mu}\rangle$ are much higher in energy and their coupling to the minimal set is reduced. They only have a small (although not negligible) influence, justifying the use of the minimal set as the essential step in the calculation.

The quantitative value of LCAO (linear combination of atomic orbitals) techniques for covalent systems such as diamond and silicon was first demonstrated by Chaney et al. (1971). They have shown that the minimal basis set gives good results for the valence bands and slightly poorer (but still meaningful) results for the lower conduction bands. Such conclusions have been confirmed by Kane (1976), Chadi (1977), and Louie (1980) who worked with pseudopotentials instead of true atomic potentials.

The great asset of the minimal basis set LCAO calculations is that they provide a direct connection between the valence states of the system and the free atom states. This becomes still more apparent with the TBA (tight binding approximation) which we shall later discuss and which allows us to obtain extremely simple, physically sound descriptions of many systems.

1.2.3 Empirical Methods

Up to very recently, first principles theories, sophisticated as they may be, could not accurately predict the band structure of semiconductors. Most of the understanding of these materials was obtained from less accurate descriptions. Among these, empirical theories have played (and still play) a very important role since they allow us to simulate the true energy bands in terms of a restricted number of adjustable parameters. There are essentially two distinct methods of achieving this goal: the tight binding approximation (TBA) and the empirical pseudopotential method (EPM).

1.2.3.1 The Tight Binding Approximation

This can be understood as an approximate version of the LCAO theory. It is generally defined as the use of a minimal atomic basis set neglecting interatomic overlaps, i.e., the overlap matrix defined in Eq. (1-60) is equal to the unit matrix. The secular equation thus becomes

$$\det |H - EI| = 0 \qquad (1\text{-}62)$$

where I is the unit matrix. The resolution of the problem then requires the knowledge of the Hamiltonian matrix elements. In the empirical tight binding approximation these are obtained from a fit to the bulk band structure. For this, one always truncates the Hamiltonian matrix in real space, i.e., one only includes interatomic terms up to first, second, or, at most, third nearest neighbors. Also, in most cases one makes use of a two-center approximation as discussed by Slater and Koster (1954). In such a case, all Hamiltonian matrix elements $\langle \varphi_{i\alpha} | H | \varphi_{j\beta} \rangle$ can be reduced to a limited number of independent terms which we can call $H_{\alpha\beta}(i,j)$ for the pair of atoms (i,j) and the orbitals (α, β). On an

"s, p" basis, valid for group IV, III–V, and II–VI semiconductors, symmetry considerations applied to the two-center approximation only give the following independent terms:

$$H_{\alpha, \beta}(i,j) = H_{ss}(i,j),\ H_{s\sigma}(i,j),\ H_{\sigma s}(i,j)$$
$$H_{\sigma\sigma}(i,j),\ H_{\pi\pi}(i,j) \qquad (1\text{-}63)$$

where $H_{\sigma\pi}$ is strictly zero in a two-center approximation and s stands for the s orbital, σ the p orbital along axis i,j with the positive lobe in the direction of the neighboring atom, and π a p orbital perpendicular to the axis i,j. With these conventions, all matrix elements are generally negative.

Similar considerations apply to transition metals with s, p, and d orbitals. Simple rules obtained for the $H_{\alpha\beta}(i,j)$ in a nearest neighbor's approximation are given in Harrison (1980). They are based on the use of free atom energies for the diagonal elements of the tight binding hamiltonian. On the other hand, the nearest neighbor's interactions are taken to scale like d^{-2} (where d is the interatomic distance) as determined from the free electron picture of these materials which will be discussed later. For s, p systems, this gives

$$H_{ss} = -\frac{10}{d^2}\ \text{eV} \qquad H_{s\sigma} = -\frac{10.8}{d^2}\ \text{eV}$$
$$(1\text{-}64)$$
$$H_{\sigma\sigma} = -\frac{16.9}{d^2}\ \text{eV} \qquad H_{\pi\pi} = -\frac{4.8}{d^2}\ \text{eV}$$

where d is expressed in Å.

Such parameters nicely reproduce the valence bands of zinc-blende semiconductors but poorly describe the band gap and badly describe the conduction bands. Improvements on this description have been attempted by going to the second nearest neighbors (Talwar and Ting, 1982) or by keeping the nearest neighbors treatment as it is but adding one s orbital (labelled s*) to the minimal basis set (Vogl and Hjalmarson, 1983). The role of this latter orbital is to simulate the effect of higher energy d orbitals which have been shown to be essential for a correct simulation of the conduction band. The quality of such a fit can be judged from Fig. 1-6, which shows that the lowest conduction bands are reproduced much more correctly. Fairly recently it has also been shown that the replacement of s* by true d orbitals improves the simulation in a striking manner (Priester).

1.2.3.2 The Empirical Pseudopotential Method

One advantage of the tight binding approximation is that it provides a natural way of relating the electronic properties of a solid to the atomic structure of its constituent atoms. As will be discussed later, it is also the most appropriate way to calculate the properties of disordered systems. However, for crystalline semiconductors the empirical pseudopotential method

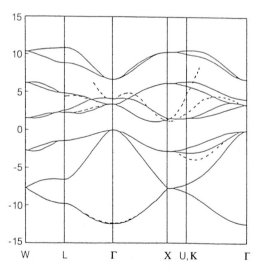

Figure 1-6. Comparison of the sp³ s* description of silicon (- - -) with a more sophisticated calculation (——). The vertical axis represents energies in eV, the horizontal axis the wave vector along symmetry axes in the Brillouin zone. Note that the valence band is practically perfectly reproduced.

seems to be the most efficient way to get a good overall description of both the valence and the conduction bands.

The basis of this method is the plane wave expansion of the wave function given by Eq. (1-54) plus the use of a smooth pseudopotential. The matrix to be diagonalized was derived in Eq. (1-56) but now we pay more attention to the potential matrix elements. In EPM one assumes that the self-consistent crystal pseudopotential can be written as a sum of atomic contributions, i.e.,

$$V(r) = \sum_{j,\alpha} v_\alpha (r - R_j - r_\alpha) \qquad (1\text{-}65)$$

where j runs over the unit cells positioned at R_j and α is the atom index, the atomic position within the unit cell being given by r_α. Let us first assume that the v_α are ordinary functions of r, or, in other words, that we are dealing with local pseudopotentials. In that case the matrix elements [Eq. (1-57)] of V between plane waves become:

$$\langle k + G | V | k + G' \rangle = \qquad (1\text{-}66)$$

$$= \frac{1}{\Omega} \sum_\alpha e^{i(G'-G)\cdot r_\alpha} \int v_\alpha(r) \, e^{i(G'-G)\cdot r} \, d^3 r$$

where Ω is the volume of the unit cell. Suppose that there can be identical atoms in the unit cell. Then the sum over α can be expressed as a sum over groups β of identical atoms with position specified by a second index γ (i.e. $r_\alpha = r_{\gamma\beta}$). Calling n the number of atoms in the unit cell we can write

$$\langle k + G | V | k + G' \rangle =$$

$$= \sum_\beta S_\beta (G' - G) \, v_\beta (G' - G) \qquad (1\text{-}67)$$

where $S_\beta(G)$ and $v_\beta(G)$ are, respectively, the structure and form factors of the corresponding atomic species, defined by

$$S_\beta(G) = \frac{1}{n} \sum_{\gamma\beta} e^{i G \cdot r_{\gamma\beta}}$$

and   (1-68)

$$v_\beta(G) = \frac{n}{\Omega} \int v_\beta(r) e^{iG \cdot r} d^3 r$$

In practice, the empirical pseudopotential method treats the form factors $v_\beta(G)$ as disposable parameters. In the case where the $v_\beta(r)$ are smooth potentials their transforms $v_\beta(G)$ will rapidly decay as a function of $|G|$ so that it may be a good approximation to truncate them at a maximum value of G_c. For instance, the band structure of tetrahedral covalent semiconductors like Si can be fairly well reproduced using only the three lower Fourier components $v(|G|)$ of the atomic pseudopotential. There are thus two cut-off values for $|G|$ to be used in practice: one, G_M, limits the number of plane waves and thus the size of the Hamiltonian matrix; the other one, G_c, limits the number of Fourier components of the form factors. We shall later give some practical examples.

The use of a local pseudopotential is not fully justified since, from Eq. (1-52), it involves, in principle, projection operators. It can be approximately justified for systems with s and p electrons. However, when d states become important, e.g., in the conduction band of semiconductors, it is necessary to use an operator form with a projection operator on the $1 = 2$ angular components.

1.3 Comparison with Experiments for Zinc-Blende Materials

In this section we apply methods which allow us to understand the general features of the band structure of zinc-blende materials. We detail, to some extent, simple models based on tight binding or the empirical pseudopotential. We also discuss briefly the results of the most sophisticated recent calculations. We then concentrate on the general treatment of the band structure near the top of the valence band, using the k-p perturbation theory as in the work by Kane (1956; 1957). In the third part, we examine the optical properties of these materials putting particular emphasis on excitonic states. Finally, we give a comparison of predicted band structures with photo-emission and inverse photo-emission data, in the light of what was done recently on the basis of non-local empirical pseudopotentials.

1.3.1 The General Shape of the Bands

In this section we start from two points of view: (i) the molecular or bond orbital model derived from tight binding and (ii) the nearly free electron picture. We show that both give rise to similar qualitative results at least for the valence bands.

1.3.1.1 The Tight Binding Point of View

Consider an A-B compound in the zinc blende structure where the atoms have tetrahedral coordination. The minimal atomic basis for such systems consists of one s and three p orbitals on each atom. One could solve the Hamiltonian matrix in this basis set and get the desired band structure directly. However, one can get much more insight into the physics by performing a basis change such that, in the new basis set, some matrix elements of the Hamiltonian will be much larger than the others. This will allow us to proceed by steps, treating first the dominant elements and then looking at the corrections due to the others. This is the general basis of molecular (Harrison, 1973; Lannoo and Decarpigny, 1973) or bond orbital models, two names for the same description.

The natural basis change is to build sp³ hybrids of the form

$$\varphi_{ij} = \frac{\varphi_{s,i} + \sqrt{3}\,\varphi_{p,ij}}{2} \qquad (1\text{-}69)$$

where $\varphi_{s\,i}$ is the s orbital of atom i and $\varphi_{p,ij}$ is one of its p orbitals pointing from i to one of its nearest neighbors j. By doing this for each atom, each bond in the system will be characterized by a pair of strongly overlapping hybrids φ_{ij} and φ_{ji} as shown in Fig. 1-7. It is clear that the dominant inter-

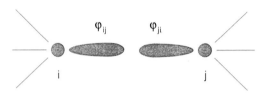

Figure 1-7. Pair of sp³ hybrids involved in one bond, as defined in Eq. (1-69).

atomic matrix elements of the Hamiltonian are given by

$$\langle \varphi_{ij} | H | \varphi_{ji} \rangle = -\beta \qquad \beta > 0 \qquad (1\text{-}70)$$

while the diagonal elements $\langle \varphi_{ij} | H | \varphi_{ji} \rangle$ are equal to the average sp³ energies of the atoms, which we denote E_A for atom A and E_B for atom B. In the first step we neglect all other matrix elements. The problem is then equivalent to a set of identical diatomic molecules, each one leading to one bonding and one antibonding state of energy:

$$E_{\substack{b\\a}} = \frac{E_A + E_B}{2} \mp \sqrt{\left(\frac{E_A - E_B}{2}\right)^2 + \beta^2} \quad (1\text{-}71)$$

Of course these states will be strongly degenerate since their degeneracy is equal to the number N of bonds in the system. The wave function of the bonding and antibonding states will take the form, for a

bond ij connecting two neighbors i and j,

$$\psi_{b,ij} = \frac{\varphi_{ij} + \lambda\,\varphi_{ji}}{\sqrt{1 + \lambda^2}}$$

$$\psi_{a,ij} = \frac{\lambda\,\varphi_{ij} - \varphi_{ji}}{\sqrt{1 + \lambda^2}} \qquad (1\text{-}72)$$

In the following discussion we take the convention that i is an A atom and j a B atom.

As there are two electrons per bond, the ground state of the system corresponds to completely filled bonding states and empty antibonding states. This description defines the "molecular" or "bond orbital model" in which the bonding states give a rough account of the valence band and the antibonding states, of the conduction band. This model has been extremely successful in describing semiquantitatively the trends in several physical properties of these materials: ionicity, effective charges, dielectric susceptibilities, average optical gaps, and even cohesive properties (see Harrison (1980) for more details).

With such a simple starting point, the formation of the band structure is easy to describe. The inclusion of further interactions which were neglected in the molecular model will tend to lift the degeneracy of the bonding and antibonding states. Exactly the same arguments as those developed in Sec. 1.1 lead to the conclusion that there will be the formation of a bonding band from the bonding states. For this we write the wave function ψ_b as a combination of all $\psi_{b,ij}$:

$$\psi_b = \sum_{\substack{\text{pairs}\\ij}} a_{b,ij}\,\psi_{b,ij} \qquad (1\text{-}73)$$

The leading correction term in the Hamiltonian matrix will be the interaction between two adjacent bonds which takes one of two values, Δ_A or Δ_B, depending on the common atom. Projecting Schrödinger's equation on these states one gets the set of equations:

$$(E-E_b)\, a_{bij} = \Delta_A \sum_{j' \neq j} a_{bij'} + \Delta_B \sum_{i' \neq i} a_{bi'j} \tag{1-74}$$

where the sums are over adjacent bonds having an A or B atom in common. At this stage it is interesting to introduce the following sums

$$S_i = \sum_j a_{bij} \quad \text{and} \quad S_j = \sum_i a_{bij} \tag{1-75}$$

so that one can rewrite Eq. (1-74) as

$$(E-E_b + \Delta_A + \Delta_B)\, a_{bij} = \Delta_A S_i + \Delta_B S_j \tag{1-76}$$

Summing this either over j or i, one gets two equations:

$$(E-E_b - 3\Delta_A + \Delta_B)\, S_i = \Delta_B \sum_{j \in i} S_j$$
$$(E-E_b - \Delta_A + 3\Delta_B)\, S_j = \Delta_B \sum_{i \in j} S_i \tag{1-77}$$

where the sums are over the nearest neighbors of one given atom. Injecting the second Eq. (1-77) into the first one gives $\left(\text{with } \Delta = \dfrac{\Delta_A + \Delta_B}{2} \text{ and } \delta = \dfrac{\Delta_A - \Delta_B}{2}\right)$:

$$\{(E-E_b)^2 - 4\Delta(E-E_b) - 12\,\delta^2\}\, S_i =$$
$$= (\Delta^2 - \delta^2) \sum_{i_2} S_{i_2} \tag{1-78}$$

where now the sum is over the second nearest neighbors of atom i. This set of equations on sublattice A is just the same as what would be obtained for a tight binding s band on an f.c.c. lattice. This leads to the following solutions:

$$E = E_b + \tag{1-79}$$
$$+ 2\Delta \pm [4\Delta^2 + 12\,\delta^2 + (\Delta^2 - \delta^2)\,\varphi]^{1/2}$$

with

$$\varphi = 4\left(\cos\frac{k_x a}{2} \cos\frac{k_y a}{2} + \right.$$
$$\left. + \cos\frac{k_x a}{2} \cos\frac{k_z a}{2} + \cos\frac{k_y a}{2} \cos\frac{k_z a}{2} \right) \tag{1-80}$$

a being the lattice parameter, and k_x, k_y, and k_z the components of the wave vector along the cube axes. It is clear that φ varies continuously from 12 to -4. The extrema of the bands ($\varphi = 12$) are then given by:

$$E = E_b + 2\Delta \pm 4\Delta = E_b \begin{array}{c} +6\Delta \\ -2\Delta \end{array} \tag{1-81}$$

while for $\varphi = -4$ a gap is opened in the band, its limits being given by

$$E = E_b + 2\Delta \pm 4\,|\delta| \tag{1-82}$$

Equation (1-79) gives only two bands while we started with four bonding orbitals per atom of the A sublattice. The two bands which appear to be missing can be obtained by noting that Eq. (1-76) has a trivial solution for which all the S_i are zero, with non-zero a_{bij} if

$$E = E_b - 2\Delta \tag{1-83}$$

This is the equation of a twofold degenerate flat band with pure p character, since all S_i and S_j are zero.

To summarize the results, one can say that the interaction between bonding orbitals has broadened the bonding level into a valence band consisting, in the present model, of two broad bands and a twofold degenerate flat band. Exactly the same treatment can be applied to the antibonding states by changing E_b into E_a and defining Δ_A and Δ_B describing the interactions between antibonding states. This antibonding band will thus represent the conduction band. The resulting band structure is compared in Fig. 1-8 to a more sophisticated calculation from Chaney et al. (1971) (its parameters have been adjusted to give overall agreement). It can be seen that it already reproduces the essential features of the valence band.

The inclusion of further matrix elements left out from the previous simple picture will have the following qualitative effects:

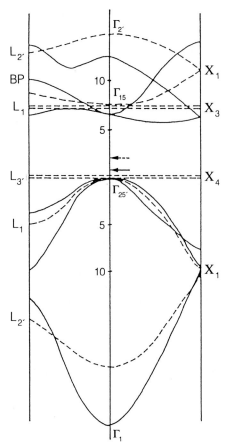

Figure 1-8. Comparison of the simple model description of diamond (- - -) with a more sophisticated calculation (———). The vertical axis corresponds to energies in eV.

i) the bonding-antibonding interactions lead to a slight repulsion between the valence and conduction bands, and ii) the inclusion of interactions between more distant bonds induces some dispersion into the flat bands.

The advantage of the simple picture we have just detailed is that it can be generalized, as we shall see, to a lot of other situations such as covalent systems with lower coordination (Sec. 1.4) or to non-crystalline and amorphous semiconductors. Furthermore, all tight binding descriptions with parametrized interactions give results

which are in good correspondence with those we have derived, with the minor corrections we have mentioned. One such empirical model is the $sp^3\ s^*$ description of Vogl et al. (1983) which leads, for GaAs, to the band structure of Fig. 1-9, which can be shown to be in good agreement with the experiments for the valence band and the lowest conduction band.

1.3.1.2 The Empirical Pseudopotential Method

A major improvement in the detailed description of the bands of tetrahedral semiconductors has been achieved with the use of empirical pseudopotentials. Let us then first discuss its application to purely covalent materials like silicon and germanium. The basis vectors of the direct zinc-blende lattice are $a/2$ (110), $a/2$ (011), and $a/2$ (101). The corresponding basis vectors of the reciprocal lattice are $2\pi/a$ (11$\bar{1}$), $2\pi/a$ ($\bar{1}$11), and $2\pi/a$ (1$\bar{1}$1). The reciprocal lattice vectors G which have the lowest square modulus are the following, in increasing order of magnitude:

$$\frac{a}{2\pi}\,G \qquad \left(\frac{a}{2\pi}\right)^2 G^2$$

000	0	
111	3	(1-84)
200	4	
220	8	
311	11	

For elemental materials like Si and Ge there is only one form factor $v(G)$ but we have seen in Eq. (1-67) that the matrix element of the potential involves a structure factor which is given here by:

$$S(G) = \cos G \cdot \tau \qquad (1-85)$$

where the origin of the unit cell has been taken at the center of a bond in the (111) direction and where τ is thus the vector

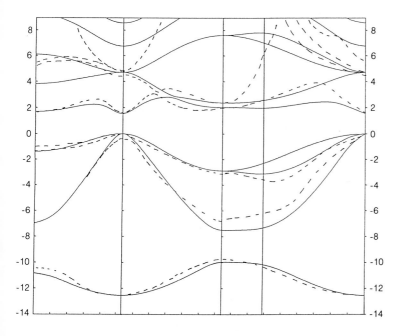

Figure 1-9. Comparison between the sp^3 s* band structure of GaAs (——) and the empirical pseudopotential one (- - -). Vertical scale: energies in eV, horizontal scale: k values.

$a/8$ (111). For local pseudopotentials this matrix element $\langle k+G|V|k+G'\rangle$ can be written $V(G)$ and is thus given by:

$$V(G) = v(G) \cos(G \cdot \tau) \tag{1-86}$$

The structure factor part is of importance since, among the lowest values of $|G|$ quoted in Eq. (1-84), it gives zero for $2\pi/a$ $(2,0,0)$. If one indexes $V(G)$ by the value taken by the quantity $(a/2\pi)^2 \ G^2$, then only the values V_3, V_8, and V_{11} are different from zero. It has been shown (in Cohen and Bergstresser, 1966) that the inclusion of these three parameters alone allows us to obtain a satisfactory description of the band structure of Si and Ge. This can be understood simply by the consideration of the free electron band structure of these materials which is obtained by neglecting the potential in the matrix elements Eq. (1-56) of the Hamiltonian between plane waves. The eigenvalues are thus the free electron energies $\hbar^2/2m \ |k+G|^2$ which, in the f.c.c. lattice, lead to the energy bands plotted in Fig. 1-10. The similarity is strik-

ing, showing that the free electron band structure provides a meaningful starting point.

The formation of gaps in this band structure can be easily understood at least in situations where only two free electron branches cross. To the lowest order in perturbation theory, one will have to solve the 2×2 matrix

$$\begin{bmatrix} \dfrac{\hbar^2}{2m}|k+G|^2 & V(G'-G) \\[2ex] V^*(G'-G) & \dfrac{\hbar^2}{2m}|k+G'|^2 \end{bmatrix} \tag{1-87}$$

The resulting eigenvalues are

$$E(k) = \frac{\hbar^2}{2m}\left[\frac{|k+G|^2+|k+G'|^2}{2}\right] \pm$$
$$\pm \left\{\left(\frac{\hbar^2}{2m} \cdot \frac{|k+G'|^2-|k+G|^2}{2}\right)^2 + \right.$$
$$\left. + |V(G'-G)|^2\right\}^{1/2} \tag{1-88}$$

whose behavior as a function of k is pictured in Fig. 1-11. The conclusion is that

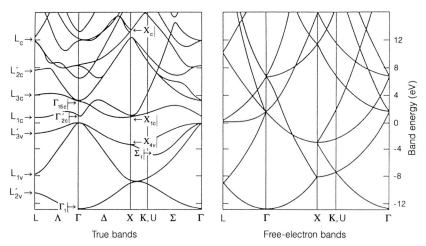

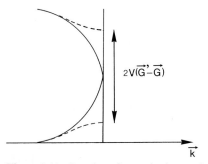

Figure 1-10. Correspondence between free-electron and empirical pseudopotential bands, showing how degeneracies are lifted by the pseudopotential.

there is the formation of a gap at the crossing point whose value is $2|V(G' - G)|$. Note that for this to occur the crossing point at $k = -\dfrac{G' + G}{2}$ must lie within the first Brillouin zone or at its boundaries. For points where several branches cross, one will have a higher order matrix to diagonalize but this will generally also result in the formation of gaps. This explains the differences between the free electron band structure and the actual one in Fig. 1-10.

The number of parameters required for fitting the band structures of compounds is

Figure 1-11. Opening of a gap in the nearly free electron method. The full line corresponds to the two free electron branches, the dashed lines to the two branches split by the potential Fourier component.

different in view of the fact that there are now two different atoms in the unit cell with form factors $v_A(G)$ and $v_B(G)$. The matrix elements $V(G)$ of the total pseudopotential will thus be expressed as

$$(1-89)$$

$$V(G) = V^s(G)\cos(G \cdot \tau) + i\, V^A(G)\sin(G \cdot \tau)$$

where V^s and V^A are equal to $(v_A + v_B)/2$ and $(v_A - v_B)/2$, respectively. The number of fitting parameters is then multiplied by 2, the symmetric components V_3^s, V_8^s, and V_{11}^s being close to those of the covalent materials and the antisymmetric components being V_3^A, V_4^A, and V_{11}^A since the antisymmetric part of V_8 vanishes.

1.3.2 The *k*-*p* Description and Effective Masses

We have already seen for hydrogenic impurity states (Sec. 1.1) that the concept of effective masses near a band extremum is very powerful. This will prove still more important for heterostructures which we discuss later. In any case it is desirable to provide a general framework in which to analyze this problem. This is obtained directly via the k-p method which we present in this section.

The basis of the method is to take advantage of the crystalline structure which allows us to express the eigenfunctions as Bloch functions and to write a Schrödinger-like equation for its periodic part. We start from

$$\left\{ \frac{p^2}{2m} + V \right\} e^{ik \cdot r} u_k(r) = E(k) e^{ik \cdot r} u_k(r) \tag{1-90}$$

where we have written the wave function in Bloch form. We can rewrite this in the following form:

$$\left\{ \frac{(p + \hbar k)^2}{2m} + V \right\} u_k(r) = E(k) u_k(r) \tag{1-91}$$

which is totally equivalent to the first form. To solve this we can expand the unknown periodic part $u_k(r)$ on the basis of the corresponding solutions at a given point k_0, which we label $u_{n k_0}(r)$:

$$u_k(r) = \sum_n c_n(k) u_{n, k_0}(r) \tag{1-92}$$

The corresponding solutions are the eigenvalues and eigenfunctions of the matrix with the general element

$$A_{nn'}(k) = \left\langle u_{n, k_0} \left| \frac{(p + \hbar k)^2}{2m} + V \right| u_{n', k_0} \right\rangle \tag{1-93}$$

We now use the fact that u_{n, k_0} is an eigenfunction of Eq. (1-91) for $k = k_0$, with energy $E_n(k_0)$. This allows us to rewrite the matrix element Eq. (1-93) in the simpler form:

$$A_{n, n'}(k) = \left\{ E_n(k_0) + \frac{\hbar^2}{2m}(k - k_0)^2 \right\} \delta_{n, n'} +$$
$$+ \frac{\hbar (k - k_0)}{m} p_{n, n'}(k_0) \tag{1-94}$$

with

$$p_{nn'}(k_0) = \langle u_{n, k_0} | p | u_{n', k_0} \rangle \tag{1-95}$$

Diagonalization of the matrix $A(k)$ given by Eq. (1-94) can give the exact band struc-

ture (an example of this is given in Gardona and Pollak, 1966). However, the power of the method is that it represents the most natural starting point for a perturbation expansion. Let us illustrate this first for the particular case of a single non-degenerate extremum. We thus consider a given non-degenerate energy branch $E_n(k)$ which has an extremum at $k = k_0$ and look at its values for k close to k_0. The last term in Eq. (1-94) can then be considered as a small perturbation and we determine the difference $E_n(k) - E_n(k_0)$ by second order perturbation theory applied to the matrix $A(k)$. This gives

$$E_n(k) = E_n(k_0) + \frac{\hbar^2}{2m}(k - k_0)^2 + \tag{1-96}$$
$$+ \frac{\hbar^2}{m^2} \sum_{n' \neq n} \frac{[(k - k_0) \cdot p_{nn'}][(k - k_0) \cdot p_{n'n}]}{E_n(k_0) - E_{n'}(k_0)}$$

which is the second order expansion near k_0 leading to the definition of the effective masses. The last term in Eq. (1-96) is a tensor. Calling 0α its principal axis, one gets the general expression for the effective masses m_α^*:

$$\frac{m}{m_\alpha^*} = 1 + \frac{2}{m} \sum_{n' \neq n} \frac{|(p_\alpha)_{nn'}|^2}{E_n(k_0) - E_{n'}(k_0)} \tag{1-97}$$

This shows that when the situation practically reduces to two interacting bands, the upper one has positive effective masses while the opposite is true for the lower one. This is what happens at the Γ point for GaAs, for instance.

Another very important situation is the case of a degenerate extremum, i.e., the top of the valence band in zinc blende materials which occurs at $k = 0$. We still have to diagonalize the matrix Eq. (1-94) taking $k_0 = 0$ and, for $k \approx 0$, the last term can still be treated by the second order perturbation theory. By letting i and j be two members of the degenerate set at $k = 0$ and l, any other state distant in energy, we now must apply the second order perturbation theory on a degenerate state. As shown in standard textbooks (Schiff, 1955) this leads to diagonalization of a matrix:

$$A_{ij}^{(2)}(\mathbf{k}) = \left[E_i(0) + \frac{\hbar^2}{2m} k^2 \right] \delta_{ij} +$$

$$+ \frac{\hbar^2}{m^2} \sum_l \frac{(\mathbf{k} \cdot \mathbf{p}_{il})(\mathbf{k} \cdot \mathbf{p}_{lj})}{E_i(0) - E_l(0)} \qquad (1\text{-}98)$$

The top of the valence band has three-fold degeneracy and its basis states behave like atomic p states in cubic symmetry (i.e., like the simple functions x, y, and z). The second order perturbation matrix is thus a 3×3 matrix built from the last term in Eq. (1-98) which, from symmetry, can be reduced to (Kane, 1956, 1957; Kittel and Mitchell, 1954; Dresselhaus et al., 1955)

$$\begin{bmatrix} Lk_x^2 + M(k_y^2 + k_z^2) & Nk_x k_y & Nk_x k_z \\ Nk_x k_y & Lk_y^2 + M(k_x^2 + k_z^2) & Nk_y k_z \\ Nk_x k_z & Nk_y k_z & Lk_z^2 + M(k_x^2 + k_y^2) \end{bmatrix} \qquad (1\text{-}99)$$

where L, M, and N are three real numbers, all of the form:

$$\frac{\hbar^2}{m^2} \sum_l \frac{(p_\alpha)_{il}(p_\beta)_{lj}}{E_i(0) - E_l(0)} \qquad (1\text{-}100)$$

It is this matrix plus the term $\dfrac{\hbar^2 k^2}{2m}$ on its diagonal which define the $h_{ij}(\mathbf{k})$ matrix of Sec. 1.1.4 to be applied in effective mass theory to a degenerate state. Up to this point we have not included spin effects and in particular spin-orbit coupling, which plays an important role in systems with heavier elements. If we add the spin variable, the degeneracy at the top of the valence band is double and the $\mathbf{k} \cdot \mathbf{p}$ matrix becomes a 6×6 matrix whose detailed form can be found in (Bassani et al., 1974; Altarelli, 1986; Bastard, 1988). One can slightly simplify its diagonalization when the spin orbit coupling becomes large,

from the fact that

$$\mathbf{L} \cdot \mathbf{S} = 1/2 (J^2 - L^2 - S^2) \qquad (1\text{-}101)$$

where $\mathbf{J} = \mathbf{L} + \mathbf{S}$. Because here $L = 1$ and $S = 1/2$, J can take two values $J = 3/2$ and $J = 1/2$. From Eq. (1-101) the $J = 3/2$ states will lie at higher energy than the $J = 1/2$ ones and, if the spin orbit coupling constant is large enough, these states can be treated separately. The top of the valence band will then be described by the $J = 3/2$ states leading to a 4×4 matrix whose equivalent Hamiltonian has been shown by Luttinger and Kohn (1955) to be:

$$H = \frac{\hbar^2}{m} \left\{ \left(\gamma_1 + \frac{5}{2} \gamma_2 \right) \frac{k^2}{2} - \gamma_2 \sum_\alpha k_\alpha^2 J_\alpha^2 - \right.$$

$$\left. - \gamma_3 \sum_{\alpha \neq \beta} k_\alpha k_\beta \frac{(J_\alpha J_\beta + J_\beta J_\alpha)}{2} \right\} \qquad (1\text{-}102)$$

where $\alpha, \beta = x, y$ or z.

Finally, as shown by Kane (1956, 1957), it can be interesting to treat the bottom of the conduction band and the top of the valence band at $\mathbf{k} = 0$ as a quasi-degenerate system, extending the above described method to a full 8×8 matrix which can be reduced to a 6×6 one if the spin orbit coupling is large enough to neglect the lower valence band.

1.3.3 Optical Properties and Excitons

One of the major sources of experimental information concerning the band structure of semiconductors is provided by optical experiments. In particular fine structures in the optical spectra might reflect characteristics of the band structure. To see this in more detail let us discuss the absorption of the light which is proportional to the imaginary part of the frequency dependent dielectric constant, i.e.,

$$\varepsilon_2(\omega) = \frac{\alpha}{\omega^2} \frac{2}{(2\pi)^3} \int |M_{vc}(\mathbf{k})|^2 \cdot$$
$$\cdot \delta(E_c - E_v - \hbar\omega) \, d^3\mathbf{k} \qquad (1\text{-}103)$$

where the optical matrix element is:

$$M_{vc} = \langle \psi_{c,\mathbf{k}'} | e^{i\boldsymbol{\varkappa} \cdot \mathbf{r}} \nabla_{\mathbf{r}} | \psi_{v,\mathbf{k}} \rangle \qquad (1\text{-}104)$$

$\boldsymbol{\varkappa}$ being the wave vector of light. From the Bloch theorem this matrix element is non-zero only if $\mathbf{k}'$ is equal to $\mathbf{k} + \boldsymbol{\varkappa} + \mathbf{G}$ ($\mathbf{G}$ is the reciprocal lattice vector). As $|\boldsymbol{\varkappa}|$ is much smaller than the dimensions of the first Brillouin zone, this means that the transition is vertical within the Brillouin zone (i.e., it occurs at fixed $\mathbf{k}$). Fine structures in $\varepsilon_2(\omega)$ due to allowed transitions can be studied by discarding the $\mathbf{k}$ dependence of the optical matrix element $M_{vc}(\mathbf{k})$. In a narrow energy range around such a structure, $\varepsilon_2(\omega)$ becomes proportional to the joint densities of states

$$(1\text{-}105)$$
$$J_{vc}(\omega) = \frac{2}{(2\pi)^3} \int \delta(E_c(\mathbf{k}) - E_v(\mathbf{k}) - \hbar\omega) \, d^3\mathbf{k}$$

which can be expressed as an integral over the surface $S(\hbar\omega)$ in $\mathbf{k}$ space such that $E_c(\mathbf{k}) - E_v(\mathbf{k}) = \hbar\omega$. This gives

$$(1\text{-}106)$$
$$J_{vc}(\omega) = \frac{2}{(2\pi)^3} \int\limits_{S(\omega)} \frac{dS}{|\nabla_{\mathbf{k}}(E_c(\mathbf{k}) - E_v(\mathbf{k}))|_s}$$

It is clear from this expression that important contributions will come from critical points where the denominator of Eq. (1-106) vanishes. These can originate from $\mathbf{k}$ points where one has the separate conditions

$$\nabla_{\mathbf{k}} E_c(\mathbf{k}) = 0, \quad \nabla_{\mathbf{k}} E_v(\mathbf{k}) = 0 \qquad (1\text{-}107)$$

Critical points of this kind only occur at high symmetry points of the Brillouin zone (such as $\mathbf{k} = 0$, for instance). Other critical points are given by

$$\nabla_{\mathbf{k}}(E_c(\mathbf{k}) - E_v(\mathbf{k})) = 0 \qquad (1\text{-}108)$$

The behavior near such Van Hove singularities can be discussed quite generally by expanding $E_c - E_v$ to the second order in $\mathbf{k}$, around the singular point $\mathbf{k}_0$. This formally gives

$$E_c - E_v = E_0 + \sum_{\alpha=1}^{3} a_\alpha (k_\alpha - k_{0\alpha})^2 \qquad (1\text{-}109)$$

and the qualitative behavior of the joint densities of states $J_{vc}(\omega)$ near such a point depends on the sign of the a_α. Typical results are shown in Fig. 1-12, showing that, from the shape of the absorption spectrum, one can infer a part of the characteristics in the band structures. As an illustration, we show in Fig. 1-13 the curves $\varepsilon_2(\omega)$ for silicon and germanium in comparison to the predicted curves (Greenaway and Harbeke, 1968).

Up to now we have only discussed optical transitions between one particle states in which an electron is excited into the conduction band, leaving a hole in the valence band. This is permissible if we treat the electron and hole as independent particles. However, these quasi-particles have opposite charges and attract each other via an

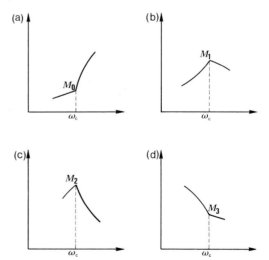

Figure 1-12. Schematic joint densities of states near the critical points for different situations (see text).

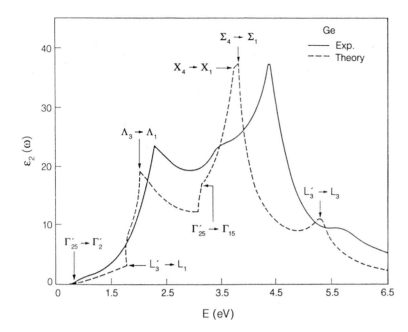

Figure 1-13. Experimental and predicted $\varepsilon_2\,(\omega)$ for Ge.

effective potential $-e^2/\varepsilon r$. This coulombic potential can give rise to localized gap states such as hydrogenic impurities, so that the absorption spectrum, instead of starting at $h\nu = E_g$, the energy gap, can show lines at $E_g - \dfrac{m^* e^4}{2\,\hbar^2\,\varepsilon^2}\dfrac{1}{n^2}$, where m^* is a suitable effective mass. Again the justification of this proceeds via the effective mass theory, but in a way slightly more complex than for impurities.

To find the excitonic wave function, we can write the total wave function of the excited states in the form (Knox, 1963; Kittel, 1963):

$$\psi_{exc} = \sum_{k_e, k_h} a\,(k_e, k_h)\,\Phi\,(k_e, k_h) \qquad (1\text{-}110)$$

where the Φ correspond to the excited states obtained from the ground state by exciting a valence band electron of wave vector k_h to a conduction band state k_e. To use the effective mass approximation we introduce a two particle envelope function

by the Fourier transform:

$$F\,(r_e, r_h) = \sum_{k_e, k_h} a\,(k_e, k_h)\,e^{\,i\,(k_e\cdot r_e - k_h\cdot r_h)} \qquad (1\text{-}111)$$

For simple band extrema with isotropic effective masses, it can be shown along the same lines as in Sec. 1.1 that $F\,(r_e, r_h)$ obeys the effective mass equation:

$$\left\{ E_g + \frac{p_e^2}{2\,m_e^*} + \frac{p_h^2}{2\,m_h^*} - \frac{e^2}{\varepsilon\,|r_e - r_h|} \right\} \cdot F\,(r_e, r_h) = E F\,(r_e, r_h) \qquad (1\text{-}112)$$

One can separate the center of mass and relative motion in this Hamiltonian in such a way that the total energy becomes:

$$E = E_g + \frac{\hbar^2\,k^2}{2\,M} - \frac{m^*\,e^4}{2\,\varepsilon^2\,\hbar^2}\frac{1}{n^2} \qquad (1\text{-}113)$$

where M and m^* are the total and reduced masses respectively and k is the wave vector for the center of mass motion. From this it is clear that the lowest excited states are those for $k = 0$ and these give rise to the hydrogenic lines.

It is interesting to determine the oscillator strength for exciton absorption for comparison with one particle transitions. We have seen before that, at a given frequency, the strength of the absorption is determined by the optical matrix element. For many electron states, this element is given by

$$M_{exc} = \langle \varphi_0 | \sum_i p_i | \psi_{exc} \rangle \qquad (1\text{-}114)$$

where $\sum_i p_i$ is the one electron sum of the individual momenta, φ_0 is the ground state, and ψ_{exc} is one of the exciton states whose general form is given in Eq. (1-110). One can expand ψ_{exc} and express φ_0 and $\varphi(k_h \to k_e)$ as Slater determinants, in which case M_{exc} becomes:

$$M_{exc} = \sum_{k_e, k_h} \langle k_h | p | k_e \rangle a(k_e, k_h) \qquad (1\text{-}115)$$

We have seen that the one particle matrix elements are non-zero for vertical transitions, i.e., only if $k_e = k_h = k$. This matrix element is identical to the $M_{vc}(k)$ defined above and we take it to be constant over the small range of k involved. We then get

$$M_{ext} = M_{vc} \sum_k a(k, k) \qquad (1\text{-}116)$$

From the definition of the envelope function Eq. (1-111) we see that $\sum_k a(k, k)$ is equal to $\int F(r, r) d^3 r$ where we take $r_e - r_h = r$. We thus obtain the final result:

$$|M_{exc}|^2 = |M_{vc}|^2 | \int F(r, r) d^3 r |^2 \qquad (1\text{-}117)$$

For the simple model we have just considered, the lowest exciton wave function is

$$F(r_e, r_h) = \frac{e^{i k \cdot R}}{\sqrt{V}} \frac{e^{-\frac{|r_e - r_h|}{a}}}{\sqrt{\pi a^3}} \qquad (1\text{-}118)$$

where V is the volume of the specimen, R is the center of mass position and a the

exciton Bohr radius, thus leading to

$$|M_{exc}|^2 = |M_{vc}|^2 \frac{V}{\pi a^3} \qquad (1\text{-}119)$$

This is to be compared to the one particle spectrum which is given by

$$|M(\omega)|^2 = \qquad (1\text{-}120)$$
$$= |M_{vc}|^2 \sum_k \delta \left(\hbar \omega - E_g - \frac{\hbar^2}{2 m^*} |k|^2 \right)$$

where m^* is the reduced mass. This sum reduces to the density of states of a 3D electron gas, so that

$$|M(\omega)|^2 = \qquad (1\text{-}121)$$
$$= |M_{vc}|^2 \frac{V}{4 \pi^2} \left(\frac{2 m^*}{\hbar^2} \right)^{3/2} (\hbar \omega) - E_g)^{1/2}$$

It is better for comparison with $|M_{exc}|^2$ to calculate the integrated $|M(\omega)|^2$ up to a frequency ω. One thus gets

$$\frac{|M_{exc}|^2}{|M|^2_{integr\,(\omega)}} = 6 \pi \left[\frac{\hbar^2}{2 m^* a^2 (\hbar \omega - E_g)} \right]^{3/2} =$$
$$= 6 \pi \left(\frac{E_{1s}}{\hbar \omega - E_g} \right)^{3/2} \qquad (1\text{-}122)$$

where E_{1s} is the exciton binding energy in the 1s state. Typical values are $E_{1s} \approx$ 10 meV, $\hbar \omega - E_g \approx 200$ meV in which case the ratio, Eq. (1-122), is of the order of 20%.

1.3.4 A Detailed Comparison with Experiments

We give an account here of some recent results comparing empirical pseudopotential bands with X-ray photo-emission and inverse photo-emission results (Chelikowsky et al., 1989). These experimental techniques combined with reflectivity data can yield nearly complete information concerning the occupied and empty states. A comparison of experimental data with theory will then provide a stringent test of the validity of the predictions.

The empirical pseudopotential used by Chelikowsky et al. (1989) and Cohen and Chelikowsky (1988), was built from a local and a non-local part. The local part was, as described before, restricting the non-zero components to be V_3, V_4, V_8, and V_{11}. As emphasized by Chelikowsky et al. (1989), this procedure yields an accurate description of the reflectivity and photo-emission spectra. However, for Ge, GaAs, and ZnSe, non-local corrections to the pseudopotential are necessary to produce similar accuracy. This is caused by d states within the ion core which modify the conduction band structure of these three materials and make the pseudopotential non-local. As discussed before, a simple correction for a specific l-dependent term can be written

$$V_{NL}(k, G - G') = \frac{4\pi}{\Omega_a}(2l+1) P_l(\cos\theta) \cdot$$

$$\cdot \int dr\, r^2\, V_l(r)\, j_l(\varkappa_r)\, j_l(\varkappa'_r) \qquad (1\text{-}123)$$

where $\varkappa = k + G$, $\cos\theta = \varkappa \cdot \varkappa'/\varkappa\varkappa'$, Ω_a is the atomic volume, P_l is a Legendre polynomial, j_l is a spherical Bessel function, and $V_l(r)$ is the non-local correction. In Chelikowsky et al. (1989) a simple Gaussian form has been used for V_l whose parameters are fitted to experiment. The local and non-local parameters are tabulated in Chelikowsky et al. (1989). The cut-off in the plane wave expansion was taken at an energy $E_1 \approx 8$ Ry and extra plane waves up to $E_2 \approx 13$ Ry were introduced by the perturbation theory. The corresponding results for Ge and GaAs are compared to experimental data in Figs. 1-14 and 1-15. The agreement is fairly good, especially if one notes that both photo-emission and optical data are reproduced with similar accuracy.

At this point it is also interesting to measure the accuracy of the first principles G-W

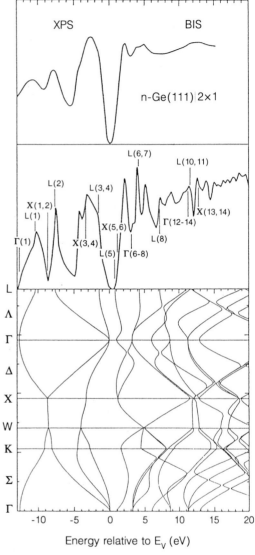

Figure 1-14. Comparison between photo-emission measurements and calculated densities of states for Ge: experimental intensities (top), theoretical density of states (middle), corresponding calculated bands (bottom) for which the vertical axis corresponds to the wave vector. The experimental data correspond to X-ray photo-emission spectroscopy (XPS) or to Bremsstrahlung isochromate spectroscopy (BIS).

calculations for these materials. This must be done keeping in mind that these calculations are performed starting from local density calculations which, as we have

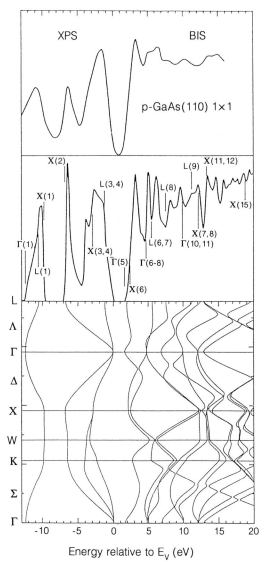

Figure 1-15. Comparison between photo-emission measurements and calculated densities of states for GaAs.

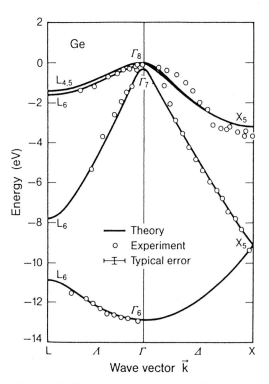

Figure 1-16. Comparison between angular resolved photo-emission and the G-W calculation of Hybertsen and Louie (1986) showing typical experimental error bars.

1.4 Other Crystalline Materials with Lower Symmetry

In this section other cases of crystalline semiconductors are examined. In these the bonding is more complex than that in zinc-blende materials which constitute, in a sense, the prototype of covalent or partly ionic bonding. We begin by generalizing the tight binding arguments discussed for tetrahedral systems to cases with lower co-ordination. We then specifically consider chain-like structures such as Se or Te for which we use the tight binding description, comparing the results with photo-emission data. We briefly discuss the case of lamellar materials. Finally, we consider a class of

seen, lead to large discrepancies in excitation energies. One result taken from Hybertsen and Louie (1986) compares the valence band structure with angular resolved photo-emission (see Fig. 1-16).

semiconductors with unconventional bonding, the Sb chalcogenides.

1.4.1 General Results for Covalent Materials with Coordination Lower than Four

Let us consider systems where each atom has N equivalent bonds to its neighbors. As with tetrahedral compounds, we want to build a molecular model which provides a simple basis for the understanding of their band structure. On each atom we build N equivalent orbitals which point exactly or approximately towards the nearest neighbors. We consider in all cases an sp minimal basis; this is always possible since one has four basis states from which one forms only N directed orbitals, with $N < 4$. The remaining atomic states are then chosen by taking into account the local symmetry (we shall later see specific examples of how this can be achieved).

Once this is done the basic electronic structure follows almost immediately. Again the directed states strongly couple in pairs as in diatomic molecules and form σ bonds with a σ bonding state and a σ^* antibonding state. It is this coupling which dominates the Hamiltonian matrix and the cohesive properties. To the lowest degree of approximation, all other states remain uncoupled at their atomic value. The resulting level scheme then consists of $N/2 \, \sigma$ bonding and $N/2 \, \sigma^*$ antibonding states plus $4 - N$ non-bonding atomic states per atom at energies which depend on the specific case under consideration. Of course, these levels are all strongly degenerate and, if the molecular model is meaningful, further interactions will lift this degeneracy to form well-defined and separate energy bands.

1.4.2 Chain-Like Structures Like Se and Te

One instructive example of a chain-like structure is pictured in Fig. 1-17, which represents an elemental system where each atom forms two equivalent bonds with an interbond angle of 90°. This situation does not exactly represent the crystalline structure of Se and Te but is very close and will help us in understanding the properties of these materials. We will discuss this case from the tight binding point of view, the same qualitative description of the bands being obtained with the pseudopotential approach.

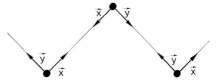

Figure 1-17. Simplified chain-like structure for Se and Te.

Let us begin the tight binding picture by starting from the molecular model. As discussed before we first have to build two directed states pointing towards the neighbors. With the local axes of Fig. 1-17, these are pure p states p_x and p_y with the positive lobe oriented towards the neighbors. Then, from symmetry, the other atomic states which do not participate in σ bonds will be the s state and the p_z state, perpendicular to the plane of the two bonds.

The parameters appropriate to Se and Te are such that roughly $E_p - E_s \approx 10$ eV, $H_{\sigma\sigma} \approx -2$ eV which from Eq. (1-64) means that $H_{s\sigma} \approx -1.3$ eV and $H_{\pi\pi} \approx -0.6$ eV. The results of the molecular model are pictured in Fig. 1-18 with the corresponding electron population per atom and the nature of the states. The upper valence band is the non-bonding p_z band corresponding to the well-known "lone pair" electrons. Comparison with the photo-emission re-

Figure 1-18. Molecular levels for Se and Te, the energy scale being of order 15 eV between the s and σ* states.

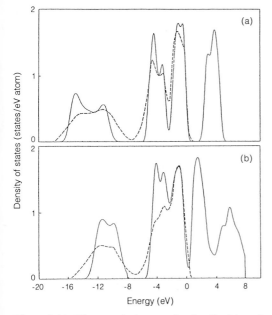

Figure 1-19. Photo-emission results for Se (a) and Te (b) compared to the calculated densities of states: - - - (experiment), —— (theory).

sults of Fig. 1-19 (Shevchik et al., 1973) shows that the molecular model already gives an essential account of the results.

As for covalent systems, we can now study the broadening of the molecular levels into bands. We treat each band separately, which is valid to the first order in perturbation theory. We make use of a nearest neighbors tight binding Hamilto-

nian as defined in Sec. 1.2.3.1 with four parameters H_{ss}, $H_{\sigma\sigma}$, $H_{s\sigma}$, and $H_{\pi\pi}$ (note that the $H_{\sigma\sigma}$ connecting the p_x, p_y functions involved in σ bonds are already included in the molecular model). The case of the s band is the simpler one, with an interaction H_{ss} between nearest neighbors. We are dealing with a system containing two atoms per unit cell (Fig. 1-17) and easily get the s band dispersion relation:

$$E_s(k) = E_s \pm 2|H_{ss}| \cos \frac{k\,a}{2} \qquad (1\text{-}124)$$

The p_z lone pair band can be treated in the same way since the p_z orbitals are coupled together only via the interaction $H_{\pi\pi}$. This gives:

$$E_z(k) = E_p \pm 2|H_{\pi\pi}| \cos \frac{k\,a}{2} \qquad (1\text{-}125)$$

The broadening of the σ and σ* bands is slightly more involved. Let us refer the bonding and antibonding orbitals to one sublattice whose atoms are labelled i, their neighbor in the cell being labelled i'. From Fig. 1-17 the bonding and antibonding states can be written: (1-126)

$$\psi_{bx_i} = \frac{\varphi_{xi} \pm \varphi_{xi'}}{\sqrt{2}} \qquad \psi_{by_i} = \frac{\varphi_{yi} \pm \varphi_{y(i-1)'}}{\sqrt{2}}$$

The energies of these states are $\pm|H_{\sigma\sigma}|$. With our tight binding Hamiltonian there is no interaction between adjacent bonds but only with second nearest neighbor bonds. Moreover, the x-like bonds do not couple with the y-like bonds giving rise to doubly degenerate bands. It is simple to show that only the $H_{\pi\pi}$ interaction is involved and one gets the following dispersion relations:

$$E_b(k) = -|H_{\sigma\sigma}| + |H_{\pi\pi}| \cos(k\,a)$$
$$E_a(k) = +|H_{\sigma\sigma}| + |H_{\pi\pi}| \cos(k\,a) \qquad (1\text{-}127)$$

Of course the linear chain model of Fig. 1-17 does not exactly correspond to the real structure of Se and Te. These materials are characterized by helicoidal chains with three atoms per unit cell (Hulin, 1966). The interbond angle is 100° for Se and 90° for Te. Under such conditions the basic features of the previous model still remain valid. When the bond angle is not exactly 90°, one builds on each atom two symmetrical p_x and p_y orbitals in the plane of its two bonds, pointing only approximately towards its neighbors. One then builds a p_z orbital perpendicular to the plane and, finally, the s orbital is left uncoupled. This leads to a molecular model exactly identical to the previous one except that the bonding and antibonding levels will now be at $\pm |H_{\sigma\sigma}| \sin^2(\theta/2)$, where θ is the bond angle. The dispersion relations will also be slightly modified with respect to those in the simple model, especially for the lone pair band since the p_z orbitals are no longer parallel. However, all the basic qualitative features will remain unchanged.

All conclusions of the tight binding description are confirmed by experiments (Shevchik et al., 1973) and also by empirical pseudopotential calculations (Schlüter et al., 1974). The main difference is that the weak interaction between chains induces some degree of three-dimensional characters which wash out to some extent the one-dimensional divergences of the density of states (Schlüter et al., 1974).

1.4.3 Layer Materials

It is not possible here to give a complete account of what has been done on layer materials. We thus restrict ourselves to the case of the crystalline germanium monochalcogenides Ge–Se and Ge–Te for which we can generalize the simple model discussed above for the Se and Te chains. In these materials both Ge and Se (or Te)

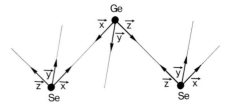

Figure 1-20. Simplified layer structure for GeSe.

atoms have threefold coordination with interbond angles close to 90°. Furthermore, they consist of two-dimensional layers, see Fig. 1-20.

We can generalize the molecular model to this kind of system in a straightforward manner. We idealize the situation by considering interbond angles of 90° and build local axes 0x, y, z, on each atom along these bonds. The natural basis states for the molecular model will be the s states and the corresponding p_x, p_y, p_z orbitals, each of these having its positive lobe pointing toward one neighbor. In the molecular model the s states remain uncoupled while all p orbitals couple by pairs, leading to σ bonding and σ* antibonding states. The resulting level structure is shown in Fig. 1-21. The number of electrons per GeSe unit is 10. As there are 2 "s" states, 3 σ, and 3 σ* states per unit, one concludes that all s and σ states are filled and these form the valence band. The fundamental gap then takes place between the σ bonding and σ* antibonding states.

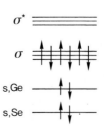

Figure 1-21. Molecular levels for GeSe, the energy scale between the s and σ* levels being of order 15 eV.

The level positions in Fig. 1-21 have been calculated using the parameter values given in Bergignat et al. (1988) for GeSe. The s level of Se is much lower, followed by the Te s level, and the σ and σ^* levels. The resulting valence band compares extremely well with the XPS measurements for crystalline GeSe, noting that the densities of states of the σ levels are 3 times larger than for s states.

1.4.4 New Classes of Materials: the Antimony Chalcogenides

One major interest of these materials is that they clearly show how band theory can clarify the understanding of the chemical bond in a class of semiconductors with fairly complex structure. This subsection is a summary of what can be found in Lefebvre et al. (1987).

Tin and antimony atoms have an electronic configuration Sn:[Kr] $4d^{10}5s^25p^2$, Sb:[Kr] $4d^{10}5s^25p^3$. Some chalcogenide compounds are insulators in which the bonds have strong covalent character while others are characterized by a lone pair, $5s^2$, which does not take part in the bonding but whose properties are directly correlated to the Sn or Sb coordination and to the structural packing (Gillespie and Nyholm, 1957) (for instance, such a correlation is the basis of chemical valence shell electron pair repulsion VSEPR theories (Gillespie, 1972)). These materials are characterized by a large range of electrical behaviors (insulator, semi-conductor, semimetal). Their general formula can be expressed as B_bX_x, $A_aB_bX_x$, or $A_aB_bX_xI_i$ where the atoms are alkaline or alkaline-earth or Tl, Pb for A, Sn or Sb for B, a chalcogen for X (S, Se, Te), and iodine for I.

Until recently, systematic analyses of this family of materials have consisted of a determination of their atomic structure and their electrical conductivity and also of Mössbauer experiments performed on ^{119}Sn and on ^{121}Sb. This has allowed Ibanez et al. (1986) to build a simple chemical bond picture using the concepts of "asymmetry" and "delocalization" of the $5s^2$ lone pair. It is thus interesting to analyze the exact meaning of such notions through the combined use of photo-emission (UPS and XPS) measurements and band structure calculations. We show in the following discussion that this allows us to obtain a coherent picture of the electronic properties of this class of materials and, furthermore, that one can derive a molecular approximation of the full band structure which allows a clear understanding of the physical nature of the distortion experienced by the $5s^2$ lone pair electron distribution. We consider five representative elements of the Sb family chosen to exhibit the whole range of electronic properties, i.e., SbI_3, Sb_2Te_3, $SbTeI$, $TlSbS_2$, and Tl_3SbS_3.

Let us first summarize the previous understanding of the electronic properties of these materials. First, X-ray diffraction studies have provided the bond lengths and interbond angles which have been connected to the bonding character (covalent, ionic, or Van der Waals), the distortion of the packing around antimony atoms being attributed to the stereochemical activity of the $5s^2$ lone pair $E(Sb)$. Mössbauer spectroscopy gives information on the electron distribution around the Sb atom through the isomer shift δ (directly connected to the 5s electron density at the nucleus) and the quadrupole splitting Δ which reflects the electric field gradient. Using these data plus the electrical properties (conductivity σ, band gap E_g), these materials have been classified into three groups on the basis of their $E(Sb)$ behavior:

i) E (Sb) is stereochemically inactive, localized around the Sb nuclei with strong $5s^2$ character. This corresponds to octahedral surrounding of the Sb atoms and to insulating behavior (ex: SbI_3).

ii) E (Sb) is stereochemically active as is seen by the distorted surrounding of the Sb atom and the corresponding reduction of the 5s character at the Sn nucleus, which can be attributed to s-p hybridization. These materials are semiconductors ($1.2\,eV \leq E_g \leq 2\,eV$) with weak conductivity ($\sigma \approx 10^{-6} - 10^{-8}\,\Omega^{-1}\,cm^{-1}$).

iii) E (Sb) is again stereochemically inactive, i.e., the Sb environment is again octahedral as in the first group but there is a 5s density loss at the nuclei and these compounds are semi-metals ($E_g \approx 0.1\,eV$, $\sigma \approx \approx 10^3\,\Omega^{-1}\,cm^{-1}$). The lone pair E (Sb) is then considered as partly delocalized (Ibanez et al., 1986).

On the basis of these known properties a tentative description of the band structure has been proposed (Ibanez et al., 1986) whose main features are schematized in Fig. 1-22. Fig. 1-22 (1) corresponds to case (i), (2) and (3) to case (ii) with sp hybridization, and (4) to the delocalized $5s^2$ lone pair. The important point is that such a description assumes that the "5s" band is moving from the bottom to the top of the

valence band which is relatively hard to understand on simple grounds.

A better understanding of these electronic properties requires a band structure calculation which is difficult to perform in view of the large numbers of atoms per unit cell (between 6 and 24 for the materials considered here). For this reason it is necessary to use empirical tight binding theory whose simplicity allows a full calculation. The tight binding description for these materials is based on the use of an "s-p" minimal basis set and, as usual, on the neglect of interatomic overlap terms. The Hamiltonian matrix elements are taken from Harrison's most recent set of empirical parameters (Harrison, 1981). However, the situation here is more complex since there are several close neighbor distances. To account for this, we apply Harrison's prescription to the interatomic matrix elements $H_{\alpha\beta}(R_1)$ for atoms which are at the nearest neighbor's distance R_1 and determine the other $H_{ab}(R)$ by the scaling law:

$$H_{\alpha\beta}(R) = H_{\alpha\beta}(R_1)\exp\left[-2.5\left(\frac{R}{R_1} - 1\right)\right]$$

$$(1-128)$$

valid for R lying between R_1 and a cut-off distance R_c chosen to correctly represent the crystal. The use of an exponential de-

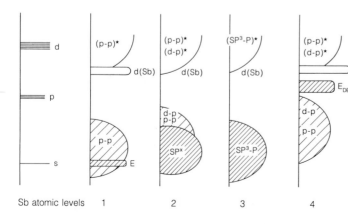

Figure 1-22. Initially proposed densities of states for the antimony chalcogenides, the energy scale being 10 to 15 eV between the lower and upper levels.

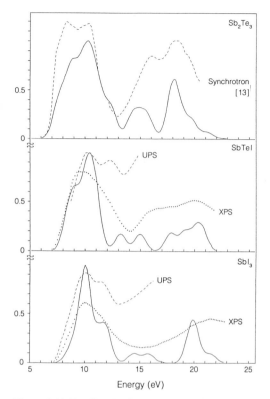

Figure 1-23. Predicted valence band densities of states for antimony chalcogenides compared to UPS and XPS spectra. The full lines correspond to theory. The vertical represents intensities or densities of states in arbitrary units.

Table 1-2. Predicted gaps (E_g, p) compared with experimental gaps (E_g, e) in eV.

Compound	E_g, p	E_g, e
SbI_3	2.40	2.30
Sb_2Te_3	0.14	0.21
$SbTeI$	1.32	1.45
$TlSbS_2$	1.73	1.77
Tl_3SbS_3	2.12	1.80

tion of valence bands but Table 1-2 shows that the predicted gaps also compare well with the experiments. One can then calculate the number N_S of 5s electrons on the Sb atom which is the basic quantity determining the Mössbauer isomer shift δ. Fig. 1-24 shows the linear correlation be-

pendence and of the parameter 2.5 is discussed in Allan and Lannoo (1983). Of course, any type of empirical theory has to be tested in several ways and, for this reason, we use XPS-UPS measurements to confirm the nature of the predictions for the valence band densities of states for which tight binding should give a reasonable description. Some characteristic results are reproduced in Fig. 1-23 where it can be seen that the empirical tight binding calculation correctly gives the number and position of the main peaks of the valence band density of states and also gives the correct valence band width. In principle, tight binding is better suited to the descrip-

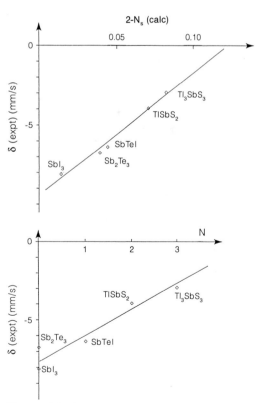

Figure 1-24. The Sb Mössbauer isomer shift δ versus 2-N_s (N_s being the number of Sb s electrons) and the number N of missing neighbors.

tween the measured δ and the computed values of N_S. This proportionality demonstrates that the band structure determination gives a completely coherent picture of the electronic structure of these compounds. A striking feature of these results is that the loss of 5s electrons is extremely small with a maximum value of about 0.1. The notion of strong or weak 5s character on Sb is thus quite relative as is the notion of delocalization since the band states are always delocalized over the whole crystal. However, if we sum up their contribution to the 5s population on the Sb atoms we get something that is always close to $N_S \approx 2$. A final comment that one can make concerning the band structure is that in all cases one finds the 5s Sb density of states at the bottom of the valence band in contradiction with the qualitative chemical picture of Fig. 1-22 that relates the $5s^2$ delocalization to a shift of this density of states towards higher energies.

At this level, a better understanding can only be obtained using a simple physical model. To do this we idealize the atomic structure by considering that a reasonable first order description of the immediate environment of an Sb atom consists of the perfect octahedron of Fig. 1-25. However, among the 6 possible sites $i = 1$ to 6, N are taken to be vacant. We then treat such a

Sb-M_{6-N} unit as a molecule for which we take as basis functions: φ_s, φ_x, φ_y, and φ_z which are the s and p states of the Sb atom and χ_i which are the "p" states of the existing M atoms that point towards the Sb atom (i.e., their positive lobe is in its direction). The "p" levels of the Sb and M atoms fall in the same energy range and, to simplify, we take them as degenerate while the s level of the Sb atom is about 10 eV lower. For this reason we first treat the coupling between the p states alone. For a given direction $\alpha = X$, Y, or Z two situations can occur (see Fig. 1-25 b):

i) There are two M atoms $i (=1, 2,$ or 3) and $j (=4, 5,$ or 6). In this case the Sb p state φ_α only couples to the corresponding antisymmetrical combination $(\chi_i - \chi_j)/\sqrt{2}$ giving rise to a σ bonding state and a σ^* antibonding state while the symmetrical $(\chi_i + \chi_j)/\sqrt{2}$ remains at the p energy.

ii) There is only one M atom i. In this case φ_α couples to χ_i to give again a σ^* and σ state but with a smaller splitting.

The resulting level structure is shown in Fig. 1-25 c. All these molecular levels have to be filled except for the σ^* antibonding levels. At this stage the 5s electron population on the Sb atoms is exactly $N_S = 2$. The only factor that allows for a reduction in N_S is the coupling of φ_S with the empty antibonding states σ^*. However, in case i)

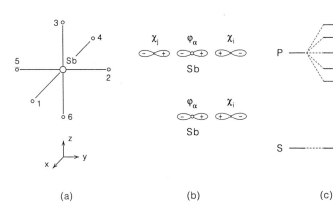

Figure 1-25. Simplified molecular model of the antimony chalcogenides: a) idealized octahedron around one Sb atom, b) the two possible situations along one axis with no or one missing neighbor, c) the resulting schematic level structure.

(a)   (b)   (c)

these are antisymmetrical so that the coupling vanishes by symmetry. This is not true in case ii) where such coupling will exist and will induce a reduction in N_S. We then arrive at the conclusion that the total reduction in N_S will be proportional to the number N of missing M atoms, if $0 \leq N \leq 3$. Fig. 1-24 b shows that the plot of the chemical shift δ versus N also gives a straight line but with a reduced accuracy as compared to Fig. 1-24 a. The constant factor relating $2 - N_S$ to N can be evaluated by the second order perturbation theory and gives a result comparable to Fig. 1-24 a. This means that the simple molecular model contains the essence of the behavior of the 5s lone pair.

In conclusion the combination of XPS measurements and band structure calculations leads to a fully coherent description of these materials allowing us, for instance, to understand the trends in the Mössbauer chemical shift related to the 5s electron density at the Sb nucleus. This gives a precise meaning to the empirical notions of asymmetry and delocalisation of the $5s^2$ lone pair which were previously used in solid state chemistry.

1.5 Non-Crystalline Semiconductors

In the preceding section we have discussed the properties of several crystalline covalently bonded systems with varying coordination numbers. These are usually determined in such a way that each atom satisfies its local valence requirements. In most cases this leads to the $8 - N$ or octet rule which links the coordination Z to atomic valence through the relation $Z = 8 - N$ for $N \geq 4$. However, there are a lot of exceptions to this rule, for example, the case of crystalline Ge–Se in which Z is

3 for both Ge and Se instead of being respectively 4 and 2. In all these covalently bonded materials the cohesive energy mainly results from the formation of the local bonds between nearest neighbors and not from incomplete filling of a broad band as in metals. This cohesive energy is much less sensitive to variations in interbond angles and to long range order than to stretching of the covalent bonds. This explains why, under appropriate preparation conditions, most of these materials can be found either in the amorphous or in the glassy state characterized by a loss of the long range order. However, as shown by the determination of their radial distribution function, these systems still possess a well-defined short range order similar to what is observed in the crystalline phases.

In this section we examine some features of non-crystalline semiconductors. We first consider some elemental amorphous systems like a-Si, a-As, and a-Se and examine possible modifications in the density of states. We then detail the properties of a typical intrinsic defect likely to be present in a-Si, i.e., the isolated dangling bond. Finally we make some comments on the electronic structure of more complex systems like SiO_x, Ge_xSe_{1-x} . . .

1.5.1 The Densities of States of Amorphous Semiconductors

From our qualitative discussion in Sec. 1.1 remember that it is likely that amorphous semiconductors will give rise to energy bands. In these cases, the pseudo-continuum of states arises not only because there are a large number of atoms but also because there is some disorder inherent to such structures which tends to spread the energy spectrum, leading to band tails.

To get a more precise feeling for what happens in the amorphous state, it is necessary to build idealized models which could be mathematically tractable and be considered as reference situations. It is for such reasons that continuous random networks have been developed to model systems like a-Si, a-Ge, at a-SiO$_2$. These are constructed by representing atoms and bonds as balls and sticks and connecting them together randomly without loose ends or dangling bonds. Usually such networks lead to a predicted radial distribution function relatively close to the experimental one. However, they remain idealized descriptions since the real material can contain clusters, dangling bonds, and other eventual deviations.

Assume for a moment that we are dealing with such an idealized lattice for a-Si. The first problem that arises concerns the general shape of the band structure: does it lead to a fundamental energy gap and to the same structures in the valence band as in the crystalline system? An interesting answer to such complex questions can be obtained via simplified Hamiltonians such as those offered by the empirical tight binding theory. This was achieved by Weaire and Thorpe (1971) on the basis of Hamiltonian first proposed by Leman and Friedel (1962). This was based on the use of sp^3 hybrids and included only two parameters: the intra-bond coupling between such hybrids and the coupling between any two hybrids centered on the same atom. Here we shall reproduce the same conclusions by using the more involved Hamiltonian described in Sec. 1.3.1.1 for crystalline Si. We again build bonding and antibonding orbitals which are exactly the same as for the crystal. We treat the broadening of the bonding and the antibonding band separately. Concentrating on the bonding band, we get the analog of Eq. (1-76),

$$(E - E_b + 2\varDelta)\, a_{bij} = \varDelta\, (S_i + S_j) \qquad (1\text{-}129)$$

in view of the fact that we are dealing with an elemental system ($\varDelta_A = \varDelta_B = \varDelta$). We directly sum this expression over j and get

$$(E - E_b - 2\varDelta)\, S_i = \varDelta \sum_j S_j \qquad (1\text{-}130)$$

where the sum over j extends over the nearest neighbors of atom i. Equations (1-130) are the same as those one would obtain for an s band on the same lattice with unit nearest neighbors interactions

$$\varepsilon\, S_i = \sum_j S_j \qquad (1\text{-}131)$$

The eigenvalues of the two matrices are thus related through

$$E = E_b + 2\varDelta + \varDelta\varepsilon \qquad (1\text{-}132)$$

The Hamiltonian matrix defined by Eq. (1-131) is known as the connectivity matrix and general topological theorems (Ziman, 1979) allow us to show that its eigenvalues ε must lie in the interval $[-4, +4]$ imposed by the coordination number. This means that the energies in the bonding band must lie in the range

$$\qquad (1\text{-}133)$$
$$E_b + 2\varDelta - 4|\varDelta| \leqq E \leqq E_b + 2\varDelta + 4|\varDelta|$$

This is the same result as for the crystalline case and thus shows that the bonding band for this idealized model of a-Si must be contained in the same energy interval as for c-Si. The same reasoning applies to the antibonding band. Thus the gap between these two bands is at least as large as for the crystal in the same model. Inclusion of the interactions between the bonding and antibonding states can only increase this gap by mutual repulsion of the two bands.

The first conclusion of the model is thus that the gap still exists in a-Si. The second question concerns the structures in the valence band. In the crystal some characteristic structures of the density of states are due to the Van Hove singularities which are a signature of long range order, as we

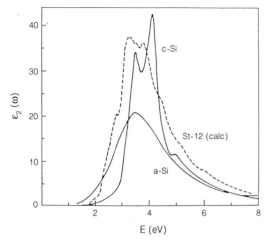

Figure 1-26. Comparison between the curves $\varepsilon_2(\omega)$ for c-Si and a-Si. The dashed line corresponding to the curve calculated for the ST12 structure.

discussed above. In the amorphous system these should be washed out as shown by the comparison of the density of states $n(\varepsilon)$ of the connectivity matrix Eq. (1-131) between the crystal and a Bethe-lattice with coordination 4 (Lannoo, 1973). This is effectively what the XPS and $\varepsilon_2(\omega)$ curves represented in Fig. 1-26 show for c-Si and a-Si.

The real atomic structure in a-Si is likely to deviate from the idealized one. To satisfy the local bonding constraints, distortions in bond angles and even in bond lengths are likely to occur. These will give rise to band tails. It is thus expected that the gap region will not be free of states since the conduction and valence band tails will overlap as shown schematically in Fig. 1-27. As regards the transport properties,

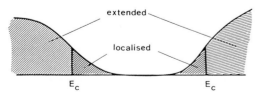

Figure 1-27. Schematic density of states in the gap region of amorphous semiconductors.

the localized or the delocalized nature of these states is of importance. A commonly adopted view (Ziman, 1979; Weaire, 1981) is that of a mobility edge separating the two types of states as shown in Fig. 1-27. Finally, the situation is complicated further by the existence of defects in the bonding, in particular the isolated dangling bonds which are treated in the following section.

A more complete review of amorphous or glassy semiconductors is given in Ziman (1979) and detailed consideration about localized and extended states can be found in Ziman (1979) and Weaire (1981).

1.5.2 Dangling Bonds

A lot of physical situations in tetrahedrally coordinated materials involve the rupture of bonds. The simplest well-documented case is the P_b center at the $Si-SiO_2$ interface which corresponds to a tricoordinated silicon atom, i.e., to the isolated dangling bond. Such defects are also likely to occur in a-Si. Another well-known situation is the vacancy in silicon (and to a less extent in compounds) where there are four interacting dangling bonds.

Let us first shortly recall the basic physical properties of dangling bonds. The simplest description comes from a tight binding picture based on an atomic basis consisting of sp^3 hybrid orbitals. The properties of the bulk material are dominated by the coupling between pairs of sp^3 hybrids involved in the same nearest neighbor's bond. This leads to bonding and antibonding states which are then broadened by weaker interbond interactions to give, respectively, the valence and conduction bands. In the bonding-antibonding picture, the rupture of a bond leaves an uncoupled or "dangling" sp^3 orbital whose energy is midway between the bonding and antibonding states. When one allows for interbond coupling, this results in a dangling bond state whose energy falls in the gap region and whose wave function is no longer of pure sp^3 character, but is somewhat delocalized along the backbonds.

Experimentally this isolated dangling bond situation is best realized for the P_b center, i.e., the tricoordinated silicon atom at the $Si-SiO_2$ interface (Poindexter and Caplan, 1983; Caplan et al., 1979; Johnson et al., 1983; Brower, 1983; Henderson, 1984) but it can also occur in amorphous silicon (Jackson, 1982; Street et al., 1983) as well as in grain boundaries or dislocations. It has been identified mainly through electron spin resonance (ESR) (Poindexter and Caplan, 1983), deep level transient spectroscopy (DLTS) (Johnson et al., 1983; Cohen and Lang, 1982) and capacitance measurements versus frequency and optical experiments (Jackson, 1982; Johnson et al., 1985). The following picture emerges: i) The isolated dangling bond can exist in three charge states: positive D^+, neutral D^0, and negative D^-. These respectively correspond to zero, one, and two electrons in the dangling bond state. ii) The effective Coulomb term U_{eff}, i.e., the difference in energy between the acceptor and donor levels, ranges from ≈ 0.2 to 0.3 eV in a-Si (Jackson, 1982) to about 0.6 eV at the $Si-SiO_2$ interface (Johnson et al., 1983).

The ESR measurements give information on the paramagnetic state D^0 through the g tensor and the hyperfine interaction. Their interpretation indicates that the effective "s" electron population on the trivalent atom is 7.6% and the "p" one 59.4%, which corresponds to a localization of the dangling bond state on this atom amounting to 67% and an "s" to "p" ratio of 13% instead of 25% in a pure sp^3 hybrid. This last feature shows a tendency towards a planar sp^2 hybridization.

Several calculations have been devoted to the isolated dangling bond. However, only two of them have dealt with the tricoordinated silicon atom embedded in an infinite system other than a Bethe lattice. The first one is a self-consistent local density calculation (Bar-Yam and Joannopoulos, 1986) which concludes that the purely electronic value of the Coulomb term (i.e., in the absence of atomic relaxation) is $U \approx 0.5$ eV. The second one is a tight-binding Green's function treatment in which the dangling bond levels are calculated by imposing local neutrality on the tricoordinated silicon (Petit et al., 1986). In this way the donor and acceptor levels are respectively $\varepsilon(0, +) = 0.05$ eV and $\varepsilon(0, -) = 0.7$ eV. Their difference corresponds to $U = 0.65$ eV, which is in good agreement with the local density result. Both values correspond to a dangling bond in a bulk system and can be understood simply in the following way: the purely intra-atomic Coulomb term is about 12 eV for a Si atom; it is first reduced by a factor of 2 since the dangling bond state is only localized at 70% on the trivalent atom; finally, dielectric screening reduces it by a further factor of $\varepsilon \approx 10$. The final result $6/\varepsilon$ gives the desired order of magnitude 0.6 eV. At the $Si-SiO_2$ interface, however, the situation becomes different because screening is less efficient. A very simple argument leads to the replacement of ε by $(\varepsilon+1)/2$ so that the electronic Coulomb term for the P_b center should be twice the previous value, i.e., $U(P_b) \approx 1.2$ eV.

An extremely important issue is the electron-lattice interaction. There is no reason for the tricoordinated atom to keep its tetrahedral position. A very simple tight binding model (Harrison, 1976) shows that this atom does indeed experience an axial force that depends on the population of the dangling bond state. This is confirmed by more sophisticated calculations (Bar-Yam and Joannopoulos, 1986). The net result is that, when the dangling bond state is empty (D^+) then the trivalent atom tends to be in the plane of its three neighbors (interbond angle 120°). On the other hand, when it is completely filled (D^-) it moves away to achieve a configuration with bond angles smaller than 109° as for pentavalent atoms. Finally, the situation for D^0 is obviously intermediate with a slight motion towards the plane of its neighbors.

For the three charge states D^+, D^0, and D^-, corresponding to occupation numbers $n = 0$, 1, and 2, respectively, one can then write the total energy in the form

$$E(n, u) = \qquad\qquad (1\text{-}134)$$
$$= n E_0 + (1/2) U n^2 - F(n) \cdot u + (1/2) k u^2$$

where u is the outward axial displacement of the tricoordinated silicon atom, $F(n)$ the occupation dependent force, U the electron-electron interaction, and k the corresponding spring constant which should show little sensitivity to n. We linearize $F(n)$

$$F(n) = F_0 + F_1(n - 1) \qquad (1\text{-}135)$$

and minimize $E(n, u)$ with respect to u to get $E_{\min}(n)$. The first order derivative of $E_{\min}(n)$ at $n = 1/2$ and $n = 3/2$ gives the levels $\varepsilon(0, +)$ and $\varepsilon(-, 0)$. The second order derivative gives the effective correlation energy:

$$U_{\text{eff}} = U - \frac{F_1^2}{k} \qquad (1\text{-}136)$$

Theoretical estimates (Bar-Yam and Joannopoulos, 1986; Harrison, 1976) give $F_1 \approx 1.6 \text{ eV} \text{Å}^{-1}$ and $k \approx 4 \text{ eV} (\text{Å}^2)^{-1}$ (Lannoo and Allan, 1982) so that F_1/k is of the order 0.65 eV. This has strong implications for the dangling bond in a-Si where U_{eff}

becomes slightly negative as concluded in Bar-Yam and Joannopoulos (1986) but this result should be sensitive to the local environment. On the other hand, with $U \approx 1.2$ eV, the P_b center at the $Si-SiO_2$ interface would correspond to $U_{\text{eff}} \approx$ ≈ 0.6 eV, in good agreement with the experiment.

The theoretical finding that U_{eff} is slightly negative for the dangling bond in a-Si leads to an inverted order for its levels, in which case the D^0 state could never be stable (Bourgoin and Lannoo, 1983). This does not agree with the experiment, in which an EPR spectrum which seems characteristic of D^0 has been observed. One possible reason for this discrepancy is the suggestion that a-Si may contain overcoordinated atoms (Pantelides, 1986) which might be responsible for the observed features. However, recent careful EPR measurements (Stutzmann and Biegelsen, 1989) seem to rule out this possibility, practically demonstrating that dangling bonds indeed exist and with a positive U_{eff}. This would mean that theoretical calculations have underestimated the electronic U for reasons which are still unclear.

1.5.3 The Case of SiO_x Glasses

We now give a simplified analysis of the valence band structure of these glasses based on an extension of the tight binding arguments developed before.

Let us first consider the case of SiO_2. The molecular model is essentially the one developed by Harrison and Pantelides (1976). The building $Si-O-Si$ unit is shown in Fig. 1-28. Again one builds sp³ hybrids on the Si atoms while on the oxygen atom one keeps the natural sp basis. The oxygen "s" state is by far the lowest in energy and, to the first order, its coupling with other states can be neglected. It will

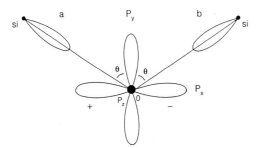

Figure 1-28. Si–O–Si unit for building the molecular model with the two sp³ hybrids of the Si atoms and the three p states of the O atom, p_z being perpendicular to the plane of the figure.

remain atomic-like at its atomic value $E_s(O)$. On the other hand, the oxygen p energy $E_p(O)$ is closer to the silicon sp³ energy $\bar{E}$ and the interaction of the corresponding states must be taken into account. The molecular states of the Si–O–Si unit of Fig. 1-28 are then built from the two sp³ hybrids a and b pointing towards the oxygen atom and the p_x, p_y, and p_z oxygen "p" states. It is clear that p_z, being perpendicular to the Si–O–Si plane, will remain uncoupled at this level of approximation giving one state at the atomic value $E_p(O)$. Thus the sp³ states a and b will only couple to p_x and p_y via the projection of these states along the axis of the corre-

sponding nearest neighbor direction. All interactions reduce to one parameter β_p defined as the interaction between an sp³ orbital and the p orbital along the corresponding bond. By symmetry p_x only interacts with $(a-b)/\sqrt{2}$ giving rise to strong bonding and antibonding states at energies

$$E_{S_A^B} = \frac{\bar{E} + E_p(O)}{2} \mp$$ (1-137)

$$\mp \sqrt{\left(\frac{\bar{E} - E_p(O)}{2}\right)^2 + \beta_p^2 \sin^2 \theta}$$

while p_y and $(a+b)/\sqrt{2}$ lead to weak bonding and antibonding states

$$E_{W_A^B} = \frac{\bar{E} + E_p(O)}{2} \mp$$ (1-138)

$$\mp \sqrt{\left(\frac{\bar{E} - E_p(O)}{2}\right)^2 + \beta_p^2 \cos^2 \theta}$$

where 2θ is the Si–O–Si bond angle.

The resulting valence band density of states per Si–O–Si unit is pictured in Fig. 1-29. It consists of delta functions at energies $E_s(O)$, E_{SB}, E_{WB}, and $E_p(O)$, the weight of each state per Si–O–Si unit being equal to unity. The influence of further interactions can now be analyzed as for pure Si. If we call Δ_S and Δ_W the interac-

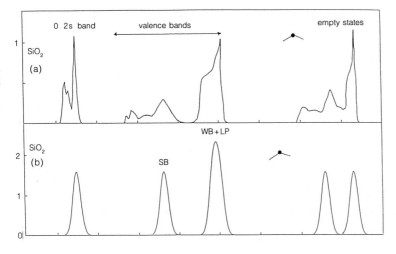

Figure 1-29. a) SiO₂ density of states calculated by Robertson et al. (1983). b) Density of states in the molecular model with a Gaussian broadening of 0.5 eV. – The vertical axis corresponds to densities of states in arbitrary units, the horizontal axis to energies, in units of 5 eV. SB, WB and LP denote strong bonding, weak bonding and lone pair states.

tion between strong and weak bonding states belonging to adjacent Si−O−Si units then we can repeat the treatment previously applied to Si simply by replacing Δ with Δ_s or Δ_w. This means that we get densities of states in the strong and weak bonding bands that have exactly the same shape as for Si, consisting of the superposition of a broad and a narrow, almost flat band. This behavior is apparent in the calculated density of states in Fig. 1-29 a. These results are in good qualitative correspondence with photo-emission data (Hollinger et al., 1977; Di Stefano and Eastman, 1972; Ibach and Rowe, 1974) and more sophisticated numerical calculations (Chelikowsky and Schlüter, 1977; O'Reilly and Robertson, 1983). Essential information provided by Eqs. (1-137) and (1-138) is that the splitting between the strong and weak bonding bands is a very sensitive function of the Si−O−Si angle 2θ. Any cause of randomness in θ such as the existence of a strained SiO_2 layer is then likely to induce some broadening of these two bands and partially fill the gap between them. Note that in an extreme situation where $\theta \approx 90°$, like in $GeSe_2$, the strong and weak bonding energies become identical, leading to a qualitative change in the shape of the density of states.

We are now in the position to discuss qualitatively the behavior of the SiO_x layer as a function of composition. Numerical calculations have been performed that are all based on a tight binding treatment combined with a more or less refined cluster Bethe lattice approximation (Lannoo and Allan, 1978; Martinez and Yndurain, 1981). Again the molecular model gives precious information that is confirmed by the full calculation. For this, let us consider the results of Lannoo and Allan (1978) which are pictured in Fig. 1-30. In the molecular model each Si−O−Si bond corresponds to

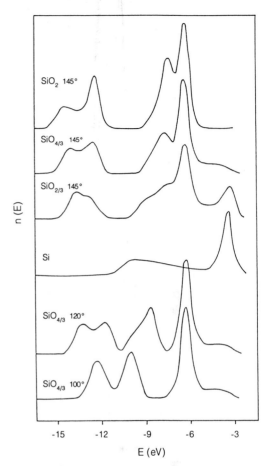

Figure 1-30. Theoretical density of valence band states for SiO_x systems for different compositions and different bond angles. The flat parts correspond to the zero of $n(E)$ in each case.

a density of states as given in Fig. 1-29 b, while each Si−Si bond corresponds to one bonding state which, for the parameters corresponding to Fig. 1-30, falls at an energy slightly higher than $E_p(O)$, the energy of the SiO_2 non-bonding state. If we start from the SiO_2 limit each Si−Si bond acts like an isolated defect giving rise to a defect level at the energy of the bonding state of Si, i.e., just above the SiO_2 valence band. When the concentration of Si−Si bonds increases this defective state will begin to broaden into a band with no defined structure at small concentrations. This corresponds to $SiO_{4/3}$ in Fig. 1-30. It is only when the concentration is high enough for Si chains to appear that the situation changes qualitatively. If, as before, Δ characterizes the interaction between two adjacent Si–Si bonding states, then the DOS of a Si–Si chain exhibits two divergences at $\pm 2\Delta$ from the bonding energy. At still higher Si concentrations there will be three and then four Si–Si bonds connected to each Si atom. For reasons discussed above such situations are characterized by a flat band at $+2\Delta$ and a broad band at lower energies. This regime will thus exhibit a peak of increasing height at the energy of the top of the pure Si valence band. This is exactly what happens in Fig. 1-30. Note also that the strong bonding band has exactly the reverse behavior: the height of its peak and its width both decrease. The last point that is clearly seen in the figure is that, for smaller θ, the separation between the strong bonding states and the weak and non-bonding states is smaller.

All these conclusions are in qualitative agreement with experimental observations concerning the valence band of the SiO_x systems and of the transition layer at the $Si–SiO_2$ interface.

1.6 Disordered Alloys

Compounds with a well-defined lattice but where there is substitutional disorder on the lattice sites compose a series of important electronic structure problems. This is, for instance, the case of pseudobinary semiconductor alloys like $Ga_xAl_{1-x}As$, $In_xGa_{1-x}As$.... Since such alloys are of much importance, here we give an account of recent work performed on these systems. However, before doing this we must introduce general methods like the virtual crys-

tal approximation (VCA), the average t-matrix approximation (ATA), and the coherent potential approximation (CPA). Again these methods, which correspond to complex systems, are applied within the framework of the tight binding approximation.

To perform such calculations on disordered systems one must use the Green's function formalism. For this we introduce the resolvent operator $G(\varepsilon)$ of the system, defined as

$$G(\varepsilon) = \lim_{\eta \to 0^+} (\varepsilon + i\eta - H)^{-1} \qquad (1\text{-}139)$$

where H is the Hamiltonian of the system and ε is the energy.

One major property of the resolvent operator is that the density of states $n(\varepsilon)$ can be expressed as

$$n(\varepsilon) = -\frac{1}{\pi} \, \text{Im Tr} \, G(\varepsilon) \qquad (1\text{-}140)$$

where Tr stands for the trace and Im means imaginary part. When considering a large disordered system, the fact that one takes the trace means that one performs an average over different local situations. One would get the same result by performing an ensemble average, i.e., by considering all configurations that the system could take and weight them by their probability. This means that $n(\varepsilon)$ can also be obtained by performing the average $\langle G(\varepsilon) \rangle$ of the resolvent operator and writing

$$n(\varepsilon) = -\frac{1}{\pi} \, \text{Tr Im} \, \langle G(\varepsilon) \rangle \qquad (1\text{-}141)$$

The advantage is that the quantity $\langle G(\varepsilon) \rangle$ becomes statistically homogeneous, i.e., if one has a random alloy on a given lattice, $\langle G(\varepsilon) \rangle$ acquires the lattice periodicity, while the original $G(\varepsilon)$ has not. It is to be noted that $\langle G(\varepsilon) \rangle$ is not equal to $(\varepsilon + i\eta - \langle H \rangle)^{-1}$; instead we define a self-energy operator $\Sigma(\varepsilon)$ such that one can write

$$\langle G(\varepsilon)\rangle = (\varepsilon + i\eta - \langle H\rangle - \Sigma(\varepsilon))^{-1} \quad (1\text{-}142)$$

The determination of this Σ is the aim of the following different approximations.

1.6.1 Definitions of the Different Approximations

We want to present different possible levels of approximation here and illustrate them on a tight binding model of a random alloy. This model will consist of a tight binding s band designed to treat a random alloy $A_x B_{1-x}$, the atomic sites forming a lattice with one site per unit cell. This tight binding Hamiltonian can thus be written:

$$H = \sum_l \varepsilon_l |l\rangle\langle l| + V\sum_{ll'} |l\rangle\langle l'| \quad (1\text{-}143)$$

where V is a nearest neighbor interaction taken to be independent of disorder while the on-site termes ε_l can take two values; ε_A for an A atom, ε_B for a B one.

Let us begin with the simplest type of approximation: the VCA (virtual crystal approximation). As indicated by its name, this consists of assuming that the average Hamiltonian $\langle H\rangle = \bar{H}$ gives a correct account of the electronic structure. This means that the exact $\langle G(\varepsilon)\rangle$ is replaced by an approximate expression $\bar{G}(\varepsilon)$:

$$\bar{G}(\varepsilon) = (\varepsilon + i\eta - \bar{H})^{-1} \quad (1\text{-}144)$$

From its definition, the expression of $\bar{H}$ in the simple tight binding model is given by:

$$\bar{H} = [x\,\varepsilon_A + (1-x)\,\varepsilon_B]\sum_l |l\rangle\langle l| +$$
$$+ V\sum_{ll'} |l\rangle\langle l'| \quad (1\text{-}145)$$

which can be solved by using the Bloch theorem.

We now want to go a step further and do the proper averaging of $\langle G\rangle$ over one site while the rest of the crystal is treated in an average field approximation. The first obvious thing to do is to start from the virtual crystal, determine $\bar{G}$, then consider the local fluctuations on a given site as perturbations and perform the average $\langle G_{ll}\rangle$ for that site, where G_{ll} is the diagonal element of G. Let us illustrate this with our specific model. We thus start from $\bar{H}$ defined in Eq. (1-145) and look at the possible fluctuations at site l. If site l is occupied by an A atom, the on site perturbation will be

$$W_A = \varepsilon_A - [x\,\varepsilon_A + (1-x)\,\varepsilon_B] =$$
$$= (1-x)(\varepsilon_A - \varepsilon_B) \quad (1\text{-}146)$$

while for a B atom it will be

$$W_B = \varepsilon_B - [x\,\varepsilon_A + (1-x)\,\varepsilon_B] = -x(\varepsilon_A - \varepsilon_B) \quad (1\text{-}147)$$

We treat both cases at the same time by assuming that there is a perturbation W_l on site l with possible values W_A or W_B. The diagonal matrix element G_{ll} can be obtained by applying Dyson's equation $(G = \bar{G} + \bar{G}WG)$,

$$G_{ll} = \bar{G}_{ll} + \bar{G}_{ll} W_l G_{ll} = \frac{\bar{G}_{ll}}{1 - W_l \bar{G}_{ll}}$$
$$= \frac{\bar{G}_{00}}{1 - W_l \bar{G}_{00}} \quad (1\text{-}148)$$

since $\bar{G}$ has translational periodicity. We now say that the average $\langle G_{ll}\rangle$ with G_{ll} given by Eq. (1-148) represents a good approximation. This defines the average matrix approximation which can be written

$$\langle G_{ll}\rangle_{ATA} = \left\langle \frac{\bar{G}_{00}}{1 - W_l \bar{G}_{00}} \right\rangle \quad (1\text{-}149)$$

On the other hand, $\langle G\rangle$ is quite generally related to $\Sigma(\varepsilon)$ through Eq. (1-142). To the same degree of approximation as before, one should thus write that $\langle G_{ll}\rangle_{ATA}$ is due to the on site perturbation Σ_{ATA} applied to site l. This leads to

$$\langle G_{ll}\rangle_{ATA} = \frac{\bar{G}_{00}}{1 - \Sigma_{ATA} \bar{G}_{00}} \quad (1\text{-}150)$$

A comparison between the two expressions allows us to express Σ_{ATA} in the form

$$\Sigma_{\text{ATA}} = \frac{\left\langle \dfrac{W_l}{1 - W_l \, \bar{G}_{00}} \right\rangle}{\left\langle \dfrac{1}{1 - W_l \, \bar{G}_{00}} \right\rangle} \tag{1-151}$$

It is customary to introduce the average t matrix $\bar{t}$, equal to the numerator in Eq. (1-151). In this way one gets the usual form of Σ_{ATA} (see Ziman, 1979, for details):

$$\Sigma_{\text{ATA}} = \frac{\bar{t}}{1 + \bar{t} \, \bar{G}_{00}} \tag{1-152}$$

This provides a considerable improvement on the VCA, but a definitely better approximation is given by the CPA (coherent potential approximation) which we will now discuss.

The CPA uses the same basic idea as the ATA, i.e., it is a single site approximation. However, instead of using an average medium corresponding to the average Hamiltonian $\bar{H}$, it is certainly much better to use $\bar{H} + \Sigma(\varepsilon)$, where the Σ is the unknown self-energy. Starting with this, the perturbation at site l is now given by $W_l - \Sigma(\varepsilon)$, W_l being defined as before. Thus the average $\langle G_{ll} \rangle$ is given by

$$\tag{1-153}$$

$$\langle G_{ll} \rangle_{\text{CPA}} = \left\langle \frac{\bar{G}_{00}(\varepsilon - \Sigma(\varepsilon))}{1 - [W_l - \Sigma(\varepsilon)] \, \bar{G}_{00}(\varepsilon - \Sigma(\varepsilon))} \right\rangle$$

But one must note that the self-energy is assumed to be the same on each lattice site so that the average crystal resolvent is simply $\bar{G}_{00}$, where ε is shifted by the coherent potential $\Sigma(\varepsilon)$. On the other hand, for Σ to be consistently defined, $\langle G_{ll} \rangle_{\text{CPA}}$ should be equal to $\bar{G}_{00}(\varepsilon - \Sigma(\varepsilon))$. Applying this to Eq. (1-153) one immediately gets the condition

$$\tag{1-154}$$

$$\left\langle \frac{[W_l - \Sigma(\varepsilon)] \, \bar{G}_{00}(\varepsilon - \Sigma(\varepsilon))}{1 - [W_l - \Sigma(\varepsilon)] \, \bar{G}_{00}(\varepsilon - \Sigma(\varepsilon))} \right\rangle = 0$$

which defines the expression of $\Sigma_{\text{CPA}}(\varepsilon)$ (Ziman, 1979).

1.6.2 The Case of Zinc Blende Pseudobinary Alloys

Here we summarize fairly recent calculations performed on these disordered alloys specifically, $In_{1-x}Ga_xAs$ and $ZnSe_xTe_{1-x}$ (Lempert et al., 1987). This work makes use of an extension of the CPA called the MCPA (molecular coherent potential approximation) which is particularly well adapted to these systems.

This case of alloys is of technological interest and it is important to be able to treat disorder effects accurately. This disorder can be conveniently divided into chemical and structural components. The former is related to the different atomic potentials of the two types of atoms, while the latter is associated with local lattice distortions, essentially due to differences in bond length. Such an effect was observed by Mikkelsen and Boyce (1982, 1983) in EXAFS (extended X-ray fine structure) measurements on $In_{1-x}Ga_xAs$. There the In–As and Ga–As bond lengths were found to vary by less than 2% from their limiting perfect crystal values, despite a 7% variation in the average X-ray lattice constant. This was also found in other zinc blende alloy systems (Mikkelsen, 1984).

The electronic structure of these materials is described in the tight binding approximation (extended to second nearest neighbors in the particular calculation of Lempert et al., 1987). The structural problem in an $A'_{1-x}A''_x B$ alloy causes a difficulty since it leads to local distortions with respect to the average zinc-blende structure. This is overcome by assuming that the atoms lie on the sites of the average crystal but scaling the dominant nearest neighbor interaction to the value appropriate to an

A'B or A"B bond. In such a way the Hamiltonian is defined on a perfectly regular lattice, but even in this case it is not appropriate to apply a simple CPA approximation in view of the existence of several orbitals per atom and of diagonal and non-diagonal disorder.

The possibility of performing an MCPA calculation is related to the choice of the particular unit cell shown in Fig. 1-31. As in the previous sections, use is made of a sp^3 energy ε_A, the coupling V_{1A} between two different sp^3 hybrids, and the intra bond coupling V_2^{AB}. As we have seen, such terms dominate the Hamiltonian for zinc-blende semiconductors and they are likely to give the dominant contribution of disorder effects.

As explained at the beginning of this section the disordered alloy Hamiltonian will be replaced by an effective Hamiltonian:

$$H_{\mathrm{eff}}(\varepsilon) = \langle H \rangle + \Sigma(\varepsilon) \qquad (1\text{-}155)$$

MCPA is based on the assumption that the self-energy $\Sigma(\varepsilon)$ is cell-diagonal (the cell being defined in Fig. 1-31). It is thus represented by an 8×8 matrix in the basis of the sp^3 orbitals. With this one can obtain exactly the same expressions as in ordinary CPA except that one must replace all quantities W_l, $\bar{G}_{00}$ by 8×8 matrices within the same sp^3 basis. The consistency condition (Eq. (1-153)) also applies but with matrix inversion and multiplication. We can rewrite it for our two component system (atoms A' and A") as:

$$\{[(W_{A'} - \Sigma)\,\bar{G}_{00}]^{-1} - 1\}^{-1} + \qquad (1\text{-}156)$$
$$+ \{[(W_{A''} - \Sigma)\,\bar{G}_{00}]^{-1} - 1\}^{-1} = 0$$

which can be transformed to:

$$(W_{A'} - \Sigma)^{-1} + (W_{B'} - \Sigma)^{-1} = 2\,\bar{G}_{00} \quad (1\text{-}157)$$

or, since $W_{A'} + W_{A''} = 0$, to:

$$\Sigma(\varepsilon) = -[W_{A'} - \Sigma(\varepsilon)] \cdot \qquad (1\text{-}158)$$
$$\cdot \bar{G}_{00}(\varepsilon - \Sigma(\varepsilon)) \cdot [W_{A''} - \Sigma(\varepsilon)]$$

which is the basic consistency condition allowing the determination of the complex self-energy matrix $\Sigma(\varepsilon)$. Equation (1-158) is then solved iteratively to get the unknown $\Sigma(\varepsilon)$ matrix.

In practice, advantage can be taken of the tetrahedral symmetry of the unit cell. The eight sp^3 orbitals of this cell can be transformed into basis states for the irreducible representations of the T_d symmetry group, i.e., A_1 and T_2 in the present case. On the A site this comes back to the s and p atomic orbitals, while on the neighbors this gives the same combinations of the four sp^3 orbitals as the s and p on the central atom. The matrix Eq. (1-158) is then split into four independent 2×2 blocks, which considerably reduces the difficulties.

It is interesting to give the explicit form of the perturbation matrices $W_{A'}$ and $W_{A''}$. We have seen that chemical disorder is modeled by three distinct matrix elements ε_A, V_{1A}, and V_{2AB} in Fig. 1-31, which for an atom A lead to an 8×8 matrix which we label h_A. The average crystal is then given

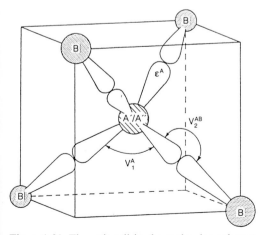

Figure 1-31. The unit cell in the molecular coherent potential approximation, defining the relevant interactions between sp^3 hybrids.

(for the compound $A'_{1-x}A''_xB$) by:

$$\bar{h}_A = (1-x)\,h_{A'} + x\,h_{A''} \qquad (1\text{-}159)$$

and the perturbation matrices $W_{A'}$ and $W_{A''}$ are:

$$W_{A'} = h_{A'} - \bar{h}_A = x\,(h_{A'} - h_{A''}) \qquad (1\text{-}160)$$

and

$$W_{A''} = h_{A''} - \bar{h}_A = -(1-x)\,(h_{A'} - h_{A''}) \qquad (1\text{-}161)$$

They are thus defined in terms of the differences $\varepsilon_{A'} - \varepsilon_{A''}$, $V_{1A'} - V_{1A''}$, and $V_{2A'B} - V_{2A''B}$.

We will not discuss here the details of the parametrization scheme used in ref. (Lempert et al., 1987), but instead summarize some of their results. Figures 1-32 and 1-33 give the densities of states, comparing the results of the VCA, the site CPA with chemical disorder only, and the MCPA.

Clearly the VCA is unsatisfactory for some of the spectral features, particularly for $ZnSe_{0.5}Te_{0.5}$. A fairly important physical quantity is the band gap bowing, i.e., the deviation from linearity with concentration of the alloy band gaps. The results are shown in Table 1-3 demonstrating that theory does not give a full account of the experimental value of the bowing effect. The difference could be attributed to several causes: uncertainties in the perturbative tight binding parameters, intercell disorder effects, clustering, etc. In any case, it must be realized that the bowing effect is very small, < 0.3 eV.

It is also of interest to briefly discuss other calculations performed on these alloys and compare them with the MCPA. The first of them is the bond-centered CPA of Chen and Sher (1981) which includes

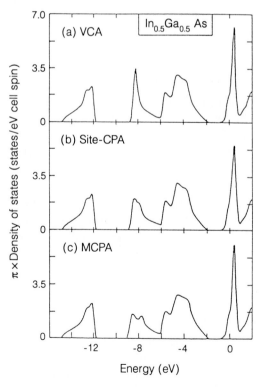

Figure 1-32. Density of states in VCA, CPA and MCPA for $In_{0.5}Ga_{0.5}As$. For a discussion see text.

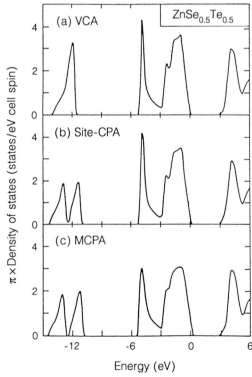

Figure 1-33. Density of states in VCA, CPA and MCPA for $ZnSe_{0.5}Te_{0.5}$. For a discussion see text.

Table 1-3. Comparison of experimental with theoretical values of band gap bowing.

	$In_{1-x}Ga_xAs$ E_{gap}	$ZnSe_xTe_{1-x}$ E_{gap}
Linear interpolation	$530-1030x$	$3040-330x$
Deviation from linearity ($x = 0.5$)		
VCA	-59	$+10$
Site CPA	-79	-5
MCPA	-66	-90
Supercell	-60	-49
Expt.	$-123, -133$	-270

disorder effects associated with each bond, i.e., in the form of a 2×2 matrix in the corresponding bonding-antibonding state basis. This includes structural disorder due to V_{1A} and should be, in principle inferior to the MCPA from that point of view. A second approach is the supercell method which models an $A'_{0.5}A''_{0.5}B$ zinc-blende pseudobinary alloy as a chalcopyrite crystal (Zunger and Jaffe, 1983). This allows them to perform self-consistent local density calculations. However, the relation to a disordered alloy is not trivial and, for this reason, a calculation has been performed in Lempert et al. (1987) with the same tight binding Hamiltonian in the supercell approach. We reproduce the results in Figs. 1-34 and 1-35 which demonstrate that the supercell results are much closer to the MCPA than are those of the bond-centered CPA. This shows that the main features of the density of states are primarily related to local bonding properties.

We should finally mention the recursion method (Haydock, 1980) which should, in principle, be well adapted to this kind of problem. It allows a direct determination of the local Green functions. This would be fairly efficient for getting the shape of the density of states, but much less efficient for the determination of band edges which require much more precision.

1.7 Systems with Lower Dimensionality

With the progress of growth techniques it is now possible to build semiconductor structures which have two-, one-, or even zero-dimensional character. Such structures are not only important for semiconductor devices but they also exhibit a lot of interesting physical properties. We shall shortly describe some basic aspects of these systems. We begin with their qualitative features, then analyse the use of the envelope function approximation, especially the problem of the correct boundary conditions. We then treat the case of quantum wells and superlattices. We also discuss hydrogenic impurities and excitons.

1.7.1 Qualitative Features

Molecular beam epitaxy and other growth techniques have allowed to deposit semiconductors layer by layer. The simplest system one can obtain in this way is the quantum well, corresponding, for example, to a layer of GaAs within bulk GaAlAs (Fig. 1-36). The GaAlAs gap is larger than the GaAs gap, and the conduction and valence band offsets are such that one gets attractive potentials both for electrons and holes. Assuming that one can treat both electrons and holes in an effective mass approximation, this leads to a standard quantum mechanical one-dimensional problem except for the free particle motion in directions parallel to the layer. More recently it has become possible to grow quantum boxes in which the particles are confined in all directions.

Let us concentrate on the electrons, for instance, and discuss the general features of the density of states in such systems. By definition this is given by:

$$n(\varepsilon) = \sum_{n,k} \delta(\varepsilon - \varepsilon_n - \varepsilon_k) \tag{1-162}$$

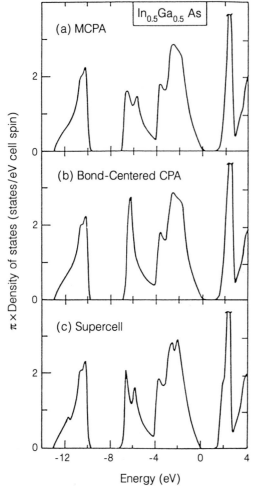

Figure 1-34. Density of states in MCPA, bond-centered CPA and the supercell approach for $In_{0.5}Ga_{0.5}As$.

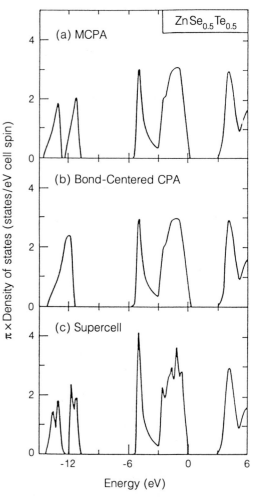

Figure 1-35. Density of states in MCPA, bond-centered CPA, and the supercell approach for $ZnSe_{0.5}Te_{0.5}$.

For a system confined in $3-d$ directions and free in the other d directions, ε_k is the dispersion relation in d dimensions (with k the corresponding wave vector), while ε_n are the levels resulting from the confinement. We can rewrite $n(\varepsilon)$ as

$$n(\varepsilon) = \sum_n \int d\varepsilon' \, \delta(\varepsilon - \varepsilon_n - \varepsilon') \sum_k \delta(\varepsilon' - \varepsilon_k) \tag{1-163}$$

The quantity $\sum_k \delta(\varepsilon' - \varepsilon_k)$ defines the density of states $\nu_d(\varepsilon')$ of the d dimensional free

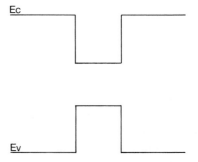

Figure 1-36. Conduction and valence band square well potentials for a layer of GaAs in $Ga_{1-x}Al_xAs$.

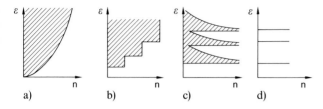

Figure 1-37. Schematic densities of states for: a) a bulk material, b) a quantum well, c) a quantum wire, d) a quantum dot.

electron gas (only defined for $\varepsilon' > 0$). This allows us to express $n(\varepsilon)$ in the form

$$n(\varepsilon) = \sum_n v_d(\varepsilon - \varepsilon_n) \qquad (1\text{-}164)$$

For a free electron gas in d dimensions, it is well known that $v(\varepsilon) \propto \varepsilon^{d/2-1}$ so that we get, quite generally,

$$n(\varepsilon) \propto \sum_n (\varepsilon - \varepsilon_n)^{d/2-1} \qquad (1\text{-}165)$$

This is plotted in Fig. 1-37 for bulk systems, quantum wells, wires, and dots, each showing remarkably distinct behaviors.

It is also possible to produce gradual transitions between 2D and 3D behavior, for instance. For this, one can increase the width of the layers in which case the ε_n will tend to form a pseudocontinuum as discussed in Sec. 1.1. One can also build a superlattice with equidistant quantum wells. When these are close enough, the states ε_n broaden into 1D bands, the steps in the density of states of Fig. 1-37 become smoother and $n(\varepsilon)$ tends to 3D behavior. All these cases represent nice examples of general considerations about band structure formation.

1.7.2 The Envelope Function Approximation

The electronic structure problem of these structures is more complex than for bulk materials, especially when space-charge effects are included, in which cases the potential has not only discontinuities but also macroscopic curvature (e.g., n-p heterojunctions). Fortunately, in regions where the potential is slowly varying and where the carriers are close to the band extrema, it is possible to simplify the treatment by using the effective mass approximation discussed previously in Sec. 1.1.4. For a single band extremum the wave function takes the form of Eq. (1-32), i.e., it is the product of the Bloch function at the extremum times a slowly varying envelope function $F(r)$ which is the solution of the effective mass equation, Eq. (1-30). This equation can be solved in regions where the potential has no discontinuity, but then one has to match the solutions from both sides of one discontinuity. Let us illustrate this at a simple interface where there is a conduction band discontinuity along the z direction; at $z = 0$ between materials A on the left and B on the right. Assuming isotropic effective masses m_A^* and m_B^*, the envelope function takes the form

$$F(r) = f(z)\, e^{i k_\parallel \cdot r_\parallel} \qquad (1\text{-}166)$$

where $\parallel$ stands for quantities parallel to the interface plane. The matching problem only concerns $f_A(z)$ and $f_B(z)$. For such a potential in a vacuum, the matching would require continuity of the wave function and of its derivative. However, here $f_A(z)$ and $f_B(z)$ are not the true wave functions but envelope functions in materials with different bulk properties. Normally it is the total wave functions which must satisfy the standard boundary conditions and not their envelope part. It is, however, customary to use the following boundary conditions: $f(z)$ continuous, $\dfrac{1}{m^*}\dfrac{df}{dz}$ continuous. These reduce to the normal conditions for materials with the same effective mass. They also have the advantage of leading to the conservation of the current probability across the interface, which is a constraint that must necessarily be fulfilled.

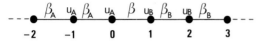

Figure 1-38. The A · B interface for a linear tight binding chain with the intraatomic and nearest neighbors' parameters.

An illustration of the problems raised when choosing the boundary conditions is provided by a 1D tight binding s band model with diagonal energies U_A or U_B and nearest neighbors interactions β_A or β_B depending on the material A or B, the coupling between A and B being taken as β (see Fig. 1-38). Writing the wave function $\sum_n C_n \varphi_n$, where φ_n is the "s" orbital of atom n and the C_n are unknown coefficients, one gets the following set of equations:

$$(E - U_A) C_n = \beta_A (C_{n-1} + C_{n+1}) \quad n \leq -1$$
$$(E - U_A) C_0 = \beta_A C_{-1} + \beta C_1$$
$$(E - U_B) C_1 = \beta_B C_2 + \beta C_0$$
$$(E - U_B) C_n = \beta_B (C_{n-1} + C_{n+1}) \quad n \geq 2$$

(1-167)

For bulk states of the infinite materials A or B, the Bloch theorem tells us that the coefficients C_n behave as $C_0 \exp(i k n a)$ where a is the lattice parameter and the energy bands are given by $U_{A,B} + 2\beta_{A,B} \cos(k a)$. The extrema of these bands occur at $k = 0$ or $k = \pi/a$ and the Bloch function at these extrema is such that $C_n = C_0 (\pm 1)^n$. To make use of the effective mass approximation for the interface problem we must consider energies falling near an extremum for both materials A and B; this extremum corresponds to $U_A \pm 2\beta_A$ and $U_B \pm 2\beta_B$, and the nature of

the extremum is not necessarily the same in both materials. In this case we write

$$C_{An} = C_{A0} \, \varepsilon_A^n \, f_{An}$$
$$C_{Bn} = C_{B1} \, \varepsilon_B^{n-1} \, f_{Bn}$$

(1-168)

where $\varepsilon_A = \pm 1$, $\varepsilon_B \pm 1$, the sign depending on the nature of the extremum. In this case f_{An} and f_{Bn} are expected to be weakly dependent on n and one can express them as the value taken by the continuous envelope functions $f_A(z)$ or $f_B(z)$ at $z = na$. These envelope functions are solutions of the bulk equations, i.e., the first and fourth Eq. (1-167), given by

(1-169)

$$(E - U_A) f_A(na) = \varepsilon_A \beta_A \{ f_A[(n-1)a] + f_A[(n+1)a] \} \quad n \leq -1$$

$$(E - U_B) f_B(na) = \varepsilon_B \beta_B \{ f_B[(n-1)a] + f_B[(n+1)a] \} \quad n \geq 2$$

The second and third equation can be written in a similar way:

(1-170)

$$(E - U_A) f_A(0) = \varepsilon_A \beta_A f_A(-a) + \beta f_B(+a)$$
$$(E - U_B) f_B(a) = \varepsilon_B \beta_B f_B(2a) + \beta f_A(0)$$

We can inject into these last two equations the analytic continuations $f_A(+a)$ and $f_B(0)$, solutions of Eq. (1-169). This leads directly to

$$f_A(+a) = \varepsilon_A \frac{\beta}{\beta_A} f_B(+a)$$

(1-171)

$$f_A(0) = \varepsilon_B \frac{\beta_B}{\beta} f_B(0)$$

As the functions $f_{A,B}(z)$ are slowly varying, the quantities $\frac{1}{2}[f_{A,B}(a) \pm f_{A,B}(0)]$ give: (1) for the "+" sign, the value of the functions $f_{A,B}$ at the interface, and (2) for the "−" sign, the value of $a \cdot f'_{A,B}$ where f' is the derivative. Considering f and f' as the components of a column vector, one can then formally write:

$$\begin{pmatrix} f_A \\ f_A' \end{pmatrix} = \mathbf{T} \begin{pmatrix} f_B \\ f_B' \end{pmatrix} \qquad (1\text{-}172)$$

where $\mathbf{T}$ is a 2×2 transfer matrix given by:

$$\mathbf{T} = \begin{bmatrix} \dfrac{1}{2}\left(\dfrac{\varepsilon_A \beta}{\beta_A} + \dfrac{\varepsilon_B \beta_B}{\beta} \right) & \dfrac{a}{2}\left(\dfrac{\varepsilon_A \beta}{\beta_A} - \dfrac{\varepsilon_B \beta_B}{\beta} \right) \\[2ex] \dfrac{1}{2a}\left(\dfrac{\varepsilon_A \beta}{\beta_A} - \dfrac{\varepsilon_B \beta_B}{\beta} \right) & \dfrac{1}{2}\left(\dfrac{\varepsilon_A \beta}{\beta_A} + \dfrac{\varepsilon_B \beta_B}{\beta} \right) \end{bmatrix}$$

$$(1\text{-}173)$$

This expression of the boundary conditions shows that the simple boundary conditions used in most calculations are not generally obeyed. In general, one should replace them by the transfer matrix formulation Eq. (1-172) which has been generalized in Ando et al. (1989); Ando and Akera (1989) to realistic band structures. It is interesting to discuss the situations for which the transfer matrix Eq. (1-173) leads to the usual boundary conditions. For instance, continuity of the envelope function ($f_A = f_B$) requires that $T_{11} = 1$ and $T_{12} = 0$. This last condition implies that $\beta^2 = \dfrac{\varepsilon_B}{\varepsilon_A} \beta_A \beta_B$. The signs of the different coupling parameters are obviously the same; this requires that $\varepsilon_B / \varepsilon_A = 1$, i.e., that one is dealing with extrema of the same nature on both sides. If this is realized, the $\mathbf{T}$ matrix becomes:

$$\mathbf{T} = \varepsilon_A \begin{bmatrix} \sqrt{\dfrac{\beta_B}{\beta_A}} & 0 \\[2ex] 0 & \sqrt{\dfrac{\beta_B}{\beta_A}} \end{bmatrix} \qquad (1\text{-}174)$$

The sign of ε_A can be absorbed as a change in phase of the function $f_B(z)$. In this case, continuity of the wave function is ensured only if $\beta_B / \beta_A = 1$ corresponding to equality of the effective masses for the two materi-

als. If this is so, continuity of the first derivative is automatically fulfilled. This is the only situation for which the usual continuity equations automatically apply. This seems to hold true for GaAlAs/GaAs systems as shown in Ando et al. (1989); Ando and Akera (1989).

1.7.3 Applications of the Envelope Function Approximation

The simplest application of the envelope function approximation corresponds to the isolated quantum well for single isotropic band extrema of the same nature in the two materials. This occurs for the conduction band of $GaAs-Al_xGa_{1-x}As$ systems in which the minimum is of Γ symmetry in both materials when $0 < x < 0.45$. Furthermore, the effective masses are of the same order of magnitude in the two materials. As we have seen, the well and barriers are in the GaAs and GaAlAs parts respectively. In such a situation the normal boundary conditions should apply and one has thus to solve a simple square well problem, the electron mass being replaced by the effective mass m^*. The superlattice case is treated as a Krönig-Penney-type model and leads to a similar broadening of the quantum well levels into bands.

The corresponding valence band problem under the same conditions is not as simple. As we have seen in the impurity case, it is necessary to solve a set of coupled differential equations which can more or less be simplified after some approximations as discussed in Altarelli (1986); Bastard (1988).

Another interesting problem concerns the behavior of hydrogenic impurities in quantum wells. Again the basic case concerns the isotropic band minimum with effective mass m^*. In three dimensions the binding energy is $R^* = \dfrac{m^* e^4}{2 \hbar^2 \varepsilon^2} \approx 5.8 \text{ meV}$

in GaAs. The same problem in two dimensions gives a binding energy equal to $4R^*$. This exact value is of interest in understanding the trends as a function of the quantum well thickness. The first calculation of this problem was performed variationally (Bastard, 1981) for a quantum well bounded by two infinite barriers. The variational wave function was written as

$$\psi(r) = \chi_1(z) \exp - \alpha \sqrt{\varrho^2 + (z - z_i)^2} \quad (1\text{-}175)$$

where z is the direction perpendicular to the layer, z_i the impurity position, ϱ the in plane distance from the impurity, and $\chi_1(z)$ the state of the ground quantum well subband. As expected, the resulting ground state binding energy increases from its 3D value as the thickness decreases. This method runs into difficulties in the small thickness limit where effective mass theory should not apply in the z-direction. In this case, it is more appropriate to perform a strict 2D application of effective mass theory where the variational function is taken as a product of the exact Bloch function at the bottom of the lowest sub-band times $\exp - \alpha \varrho$ (Priester et al., 1984). This gives the 2D limit exactly by construction and is valid as long as the binding energy is smaller than the intersub-band separation. Finally, the case of acceptor impurities is extremely complex and we do not discuss it here (references can be found in Altarelli, 1986; Bastard, 1988).

It is interesting to look at the exciton states and the corresponding optical absorption in systems with reduced dimensionality. Let us first consider a quantum well and the simple effective mass equation, Eq. (1-112), for the exciton envelope function. The problem is complicated by the fact that there is confinement in the z-direction. There can still be free motion of the center of mass in the directions x and y parallel to the layer. The ground state

obviously corresponds to no such motion and two variational wave functions for this state have been sought (Bastard, 1988):

$$F(z_e, z_h, \varrho) = N_1 \chi_{1,e}(z_e) \chi_{1,h}(z_h) \exp - \frac{\varrho}{\lambda}$$

or $\qquad\qquad\qquad\qquad\qquad\qquad (1\text{-}176)$

$$F(z_e, z_h, \varrho) = N_2 \chi_{1,e}(z_e) \chi_{1,h}(z_h) \cdot$$
$$\cdot \exp\left[-\frac{1}{\lambda} \sqrt{\varrho^2 + (z_e - z_h)^2} \right] \quad (1\text{-}177)$$

where $\varrho = \sqrt{(x_e - x_h)^2 + (y_e - y_h)^2}$, N_1 and N_2 are normalization constants, λ is the variational parameter and $\chi_{1e}(z_e)$ and $\chi_{1h}(z_h)$ are the lowest states in each quantum well in the absence of electron hole interaction. With these trial wave functions one gets the result that the binding energy increases when the width L of the well decreases to reach a limiting value of four times the bulk one, as for hydrogenic impurities. It is interesting to calculate the oscillator strength by using the same method as in Sec. 1.3.2. Using for χ_1 and χ_2 a $\sin(kz)$ form, one readily obtains from Eq. (1-117)

$$|M_{2,\text{exc.}}|^2 = |M_{vc}|^2 \frac{2s}{\pi \lambda^2} \quad (1\text{-}178)$$

Thus the relative strength of optical absorption between the quantum well and bulk exciton is, for the same volume $V = SL$ of material, given by

$$\left| \frac{M_{2,\text{exc.}}}{M_{3,\text{exc.}}} \right|^2 = \frac{2a^3}{L\lambda^2} \quad (1\text{-}179)$$

For strong confinement, λ tends to its 2D limiting value $a/2$ so that this ratio becomes equal to

$$\left| \frac{M_{2,\text{exc.}}}{M_{3,\text{exc.}}} \right|^2 = \frac{8a}{L} \quad (1\text{-}180)$$

where L is of the order of the interatomic spacing. This ratio can thus become, for

GaAs, as large as 300 showing that the 2D confined exciton has much more oscillator strength than the bulk one. This is confirmed by experimental observations (Dingle et al., 1974).

Let us finally try to extend this to the quantum box. We then consider a spherical box with infinite potential at the boundary $r = R$. We only consider the limit of strong confinement where the kinetic energy terms dominate the electron-hole attraction (they scale respectively as $1/R^2$ and $1/R$). The latter can thus be included in the first order perturbation theory. At the lowest order, the envelope function for the lowest excitonic state takes the form

$$F(r_e, r_h) = N \frac{\sin(k\, r_e)}{r_e} \frac{\sin(k\, r_h)}{r_h} \quad (1\text{-}181)$$

with $k = \pi/R$. One can now evaluate $|M_{0,\text{exc.}}|^2$ for this 0D case from the general expression Eq. (1-117). This gives

$$M_{0,\text{exc.}} = |M_{vc}|^2 \quad (1\text{-}182)$$

Thus the strength of this exciton relative to the bulk one for the same volume of material given here by $V = \frac{3}{4}\pi R^3$ is equal to

$$\left| \frac{M_{0,\text{exc.}}}{M_{3,\text{exc.}}} \right|^2 = \frac{3}{4}\left(\frac{a}{R}\right)^3 \quad (1\text{-}183)$$

which is the result of Kayanuma (1988).

Again the bulk exciton radius is $a \approx 100\,\text{Å}$ while in the limit of very strong confinement $R \approx 10\,\text{Å}$ or less. This means that the enhancement is still larger than in quantum wells.

1.8 References

Allan, G., Lannoo, M. (1983), *J. Phys. (Paris), 44,* 1355.

Altarelli, M. (1986), in: *Heterojunctions and Semiconductor Superlattices:* Allan, G., Bastard, G., Boccara, N., Lannoo, M., Voos, M. (Eds.). Heidelberg: Springer Verlag, p. 12.

Ando, T., Wakahara, S., Akera, H. (1989), *Phys. Rev. B 40,* 11609.

Ando, T., Akera, H. (1989), *Phys. Rev. B 40,* 11619.

Bachelet, G. B., Hamann, D. R., Schlüter, M. (1982), *Phys. Rev. B 26,* 4199.

Baraff, G. A., Schlüter, M. (1984), *Phys. Rev. B 30,* 3460.

Bar-Yam, Y., Joannopoulos, J. D. (1986), *Phys. Rev. Lett. 56,* 2203.

Bassani, F., Iadonisi, G., Preziosi, B. (1974), *Rep. Prog. Phys. 37,* 1099.

Bastard, G. (1981), *Phys. Rev. B 24,* 4714.

Bastard, G. (1988), *Wave Mechanics applied to Semiconductor Heterostructures.* Paris: Les Editions de Physique.

Bergignat, E., Hollinger, G., Chermette, H., Pertosa, P., Lohez, D., Lannoo, M., Bensoussan, M. (1988), *Phys. Rev. B 37,* 4506.

Bourgoin, J., Lannoo, M. (1983), *Point Defects in Semiconductors II, Springer Series in Solid State Science,* Vol. 35, Berlin, Heidelberg, New York: Springer Verlag.

Brower, K. L. (1983), *Appl. Phys. Lett., 43,* 1111.

Caplan, P. J., Poindexter, E. H., Deal, B. E., Razouk, R. R. (1979), *J. Appl. Phys., 50,* 5847.

Cardona, M., Pollak, F. H. (1966), *Phys. Rev. 142,* 530.

Chadi, D. J. (1977), *Phys. Rev. B 16,* 3572.

Chaney, R. C., Lin, C., Lafon, E. E. (1971), *Phys. Rev. B 3,* 459.

Chelikowski, J. R., Schlüter, M. (1977), *Phys. Rev. B 15,* 4020.

Chelikowski, J. R., Wagener, T. J., Weaver, J. H., Jin, A. (1989), *Phys. Rev. B 40,* 9644.

Chen, A. B., Sher, A. (1981), *Phys. Rev. B 23,* 5360.

Chen, A. B., Sher, A. (1982), *J. Vac. Sci. Technol., 21,* 138.

Cohen, M. L. (1983), *Proc. Enrico Fermi Summer School,* Varenna.

Cohen, M. L., Bergstresser, T. K. (1966), *Phys. Rev. 141,* 789.

Cohen, M. L., Chelikowski, J. R. (1988), *Electric Structure and Optical Properties of Solids.* New York: Springer Verlag.

Cohen, J. D., Lang, D. V. (1982), *Phys. Rev. B 25,* 5285.

Cohen, M. L., Heine, V., Weaire, D. (1970), *Solid State Physics:* Ehrenreich, H., Seitz, F., Turnbull, D. (Eds.), New York, London: Academic Press, Vol. 24.

Dingle, R., Wiegmann, W., Henry, C. H. (1974), *Phys. Rev. Lett., 33,* 827.

Di Stefano, T. H., Eastman, D. E. (1972), *Phys. Rev. Lett., 29,* 1088.

Dresselhaus, G., Kip, A. F., Kittel, C. (1955), *Phys. Rev. 98,* 368.

Euwema, R. N., Wilwhite, D. L., Suratt, G. T. (1973), *Phys. Rev. B 7,* 818.

Faulkner, R. A. (1968), *Phys. Rev., 175,* 991.

Faulkner, R. A. (1969), *Phys. Rev., 184,* 713.

Gillespie, R. J. (1972), *Molecular Geometry*. London: Van Nostrand Reinhold, Com. LTD, p. 6.

Gillespie, R. G., Nyholm, R. (1957), *Quart. Rev. Chem. Soc., 11,* 339.

Godby, R. W., Schlüter, M., Sham, L. J. (1986), *Phys. Rev. Lett., 56,* 2415, and (1987), *Phys. Rev. B 35,* 4170.

Greenaway, D. L., Harbeke, G. (1968), *Optical Properties and Band Structure of Semiconductors*. Oxford: Pergamon Press.

Hamann, D. R. (1979), *Phys. Rev. Lett., 42,* 662.

Hamann, D. R., Schlüter, M., Chiang, C. (1979), *Phys. Rev. Lett., 43,* 1494.

Harrison, W. A. (1973), *Phys. Rev. B 8,* 4487.

Harrison, W. A. (1976), *Surf. Sci., 55,* 1.

Harrison, W. A. (1980), *Electronic Structure and The Properties of Solids, The Physics of the Chemical Bond*. New York: Freeman.

Harrison, W. A. (1981), *Phys. Rev. B 24,* 5835.

Haydock, R. (1980), in: *Solid State Phys:* Ehrenreich, H., Seitz, F., Turnbull, D. (Eds.), New York: Academic Press, Vol. 35, p. 215.

Hedin, L., Lundquist, S. (1969), *Solid State Physics, 23,* 1.

Henderson, B. (1984), *Appl. Phys. Lett., 44,* 228.

Hohenberg, P., Kohn, W. (1964), *Phys. Rev., 136 B,* 864.

Hollinger, G., Jugnet, Y., Tran Minh Duc (1977), *Sol. State Comm., 22,* 277.

Hulin, M. (1966), *J. Phys. Chem. Sol., 27,* 441.

Hybertsen, M. S., Louie, S. G. (1985), *Phys. Rev. Lett., 55,* 1418.

Hybertsen, M. S., Louie, S. G. (1986), *Phys. Rev. B 34,* 5390.

Hybertsen, M. S., Louie, S. G. (1986), *Phys. Rev. B 34,* 5390.

Ibach, H., Rowe, J. E. (1974), *Phys. Rev. B 10,* 710.

Ibanez, A., Olivier-Fourcade, J., Jumas, J. C., Philippot, E., Maurin, M. (1986), *Z. Anorg. Allg. Chem. 6,* 540/541, 106.

Jackson, W. B. (1982), *Solid State Comm., 44,* 477.

Johnson, N. M., Biegelsen, D. K., Moyer, M. D., Chang, S. T., Poindexter, E. H., Caplan, P. J. (1983), *Appl. Phys. Lett., 43,* 563.

Johnson, N. M., Jackson, W. B., Moyer, M. D. (1985), *Phys. Rev. B 31,* 1194.

Kane, E. O. (1956), *J. Phys. Chem. Sol. 1,* 82.

Kane, E. O. (1957), *J. Phys. Chem. Sol. 1, 249.*

Kane, E. O. (1976), *Phys. Rev. B 13,* 3478.

Kayanuma, Y. (1988), *Phys. Rev. B 38,* 9797.

Kittel, C. (1963), *Quantum Theory of Solids*. New York: Wiley and Sons Inc.

Kittel, C., Mitchell, A. H. (1954), *Phys. Rev. 96,* 1488.

Knox, R. S. (1963), *Theory of Excitons, Sol. State Phys., Supplt. 5:* Seitz, F. and Turnbull, D. (Eds.) New York: Academic Press.

Kohn, W., Sham, L. J. (1965), *Phys. Rev. 140 A,* 1133.

Lannoo, M. (1973), *J. de Phys., 34,* 869.

Lannoo, M., Allan, G. (1978), *Sol. State Comm., 28,* 733.

Lannoo, M., Allan, G. (1982), *Phys. Rev. B 25,* 4089.

Lannoo, M., Decarpigny, J. N. (1973), *Phys. Rev. B 8,* 5704.

Lannoo, M., Schlüter, M., Sham, L. J. (1985), *Phys. Rev. B 32,* 3890.

Leman, G., Friedel, J. (1962), *J. Appl. Phys. Supplt. 33,* 281.

Lefebvre, I., Lannoo, M., Allan, G., Ibanez, A., Fourcade, J., Jumas, J. C., Beaurepaire, E. (1987), *Phys. Rev. Lett., 59,* 2471.

Lempert, R. J., Hass, K. C., Ehrenreich, H. (1987), *Phys. Rev. B 36,* 1111.

Louie, S. G. (1980), *Phys. Rev. B 22,* 1933.

Luttinger, J. M., Kohn, W. (1955), *Phys. Rev. 97,* 869.

Luttinger, J. M. (1956), *Phys. Rev. 102,* 1030.

Mauger, A., Lannoo, M. (1977), *Phys. Rev. B 15,* 2324.

Martinez, E., Yndurain, F. (1981), *Phys. Rev. B 24,* 5718.

Mikkelsen, J. C. Jr., Boyce, J. B. (1982), *Phys. Rev. Lett., 49,* 1412.

Mikkelsen, J. C. Jr., Boyce, J. B. (1983), *Phys. Rev. B 28,* 7130.

Mikkelsen, J. C. Jr. (1984), *Bull. Am. Phys. Soc., 29,* 202.

Motta, N. et al. (1985), *State Comm., 53,* 509.

O'Reilly, E. P., Robertson, J. (1983), *Phys. Rev. B 27,* 3780.

Pantelides, S. T. (1978), *Rev. Mod. Phys., 50,* 797.

Pantelides, S. T. (1986), *Phys. Rev. Lett., 59,* 688.

Pantelides, S. T., Harrison, W. A. (1976), *Phys. Rev. B 13,* 2667.

Petit, J., Lannoo, M., Allan, G. (1986), *Sol. State Comm., 60,* 861.

Perdew, J., Levy, M. (1983), *Phys. Rev. Lett., 51,* 1884.

Philippot, E. (1981), *J. Solid State Chem., 38,* 26.

Poindexter, E. H., Caplan, P. J. (1983), *Prog. Surf. Science, 14,* 201.

Priester, C., to be published.

Priester, C., Allan, G., Lannoo, M. (1983), *Phys. Rev. B 28,* 7194.

Priester, C., Allan, G., Lannoo, M. (1984), *Phys. Rev. B 29,* 3408.

Resca, L., Resta, R. (1979), *Sol. State Comm., 29,* 275.

Resca, L., Resta, R. (1980), *Phys. Rev. Lett., 44,* 1340.

Robertson, J. (1983), *Advances in Phys., 32,* p. 383.

Sawyer, J. F. Gillespie, R. J. (1987), *Progress Inorg. Chem., 34,* 65. New York: InterSciences Publi. J. Wiley and Sons.

Schiff, L. I. (1955), *Quantum Mechanics*, 2nd ed. New York: McGraw-Hill, p. 158.

Schlüter, M. (1983), *Proc. Enrico Fermi Summer School, Varenna*.

Schlüter, M., Joannopoulos, J. D., Cohen, M. L. (1974), *Phys. Rev. Lett., 89,* 33.

Sham, L. J., Schlüter, M. (1983), *Phys. Rev. Lett., 51,* 1888.

Sham, L. J., Schlüter, M. (1985), *Phys. Rev. B 32*, 3883.

Shevchik, N. J., Tejeda, J., Cardona, M., Langer, D. W. (1973), *Sol. State Comm., 12*, 1285.

Slater, J. C. (1960), *Quantum Theory of Atomic Structure, Vol. 1.* New York: McGraw-Hill.

Slater, J. C., Koster, G. J. (1954), *Phys. Rev. 94*, 1498.

Street, R. A., Zesch, J., Thomson, M. J. (1983), *Appl. Phys. Lett., 43*, 672.

Stutzmann, M., Biegelsen, D. K. (1989), *Phys. Rev. B 40*, 9834.

Talwar, D. N., Ting, C. S. (1982), *Phys. Rev. B 25*, 2660.

Vogl, P., Hjalmarson, H. P., Dow, J. D. (1983), *J. Phys. Chem. Sol., 44*, 365.

Weaire, D. (1981), in: *Fundamental Physics of Amorphous Semiconductors,* Springer Series in Solid State Science, Yonezawa, F. (Ed.). New York: Springer Verlag, p. 155.

Weaire, D., Thorpe, M. F. (1971), *Phys. Rev. B 4*, 2508.

Ziman, J. M. (1979), *Models of Disorder.* Cambridge University Press.

Zunger, A., Jaffe, J. E. (1983), *Phys. Rev. Lett., 51*, 662.

2 Optical Properties and Charge Transport

Rainer G. Ulbrich

IV. Physikalisches Institut, Georg-August-Universität, Göttingen,
Federal Republic of Germany

List of Symbols and Abbreviations

A	vector potential
a_0	lattice constant
a_B^*	exciton Bohr radius
c	phase velocity
d	interatomic distance (bond length)
D	lattice force constant
$D(\hbar\omega)$	combined density of states
D_{e-h}	combined density of states of an e–h pair
$E(t), E$	electric field
$E(k)$	electron dispersion relation
δE	shift of the conduction band edge
ΔE	excitation energy
e	electron charge
E_c, E_v	conduction and value band edge
E_D	donor binding energy
E_F	Fermi energy
E_G, E_{gap}	gap energy
E_G'	reduced gap energy
$E_F^{(e)}, E_F^{(h)}$	quasi-Fermi levels
E_i, E_f	initial, final state energy
$E_n(k)$	energy band with index n
E_r	resonant electric field
$E_v(k)$	valence band energy
E_0	energy of the ground state
E_0	amplitude of the electric field
E_{ck}, E_{vk}	eigenstates of conduction and valence band
E_{DP}	deformation potential constant
E_{ex}	excited state energy
e_{14}	component of the piezo-electric tensor for zinc blende symmetry
E_{1s}	$= E_G - Ry^*$
$E_{kin}^{(e)}$	kinetic energy of the final electron state
$E_{kin}^{(h)}$	kinetic energy of the final hole state
e–h	electron-hole pair
$f(k)$	scalar distribution function
$f_e(k), f_h(k)$	scalar distribution function for electrons, holes
f_0	scalar distribution function in thermal equilibrium
f_{field}	scalar distribution part induced by electrical transport
f_W	Wigner phase-space distribution function
G	quantized conductance
g_{Spin}	spin degeneracy
$\hbar = h/2\pi$	Planck constant
$\mathcal{H}_0$	Hamilton operator
$\mathcal{H}'$	small perturbation (Hamilton operator)

J, j	total angular momentum
$\boldsymbol{k}, k, \boldsymbol{k}'$	wave vector (state)
k	Boltzmann constant
$\boldsymbol{K}, \boldsymbol{K}_{\text{pair}}$	$= \boldsymbol{k}_e - \boldsymbol{k}_h$, pair wave vector
k_c, k_v	conduction and valence band states
$\boldsymbol{k}_e, \boldsymbol{k}_h$	wave vector of an electron, hole
k_{F}	Fermi wave vector
$\boldsymbol{k}_0$	wave vector referring to the ground state
$\boldsymbol{k}_1, \boldsymbol{k}_2$	initial wave vectors of two colliding particles
$\boldsymbol{k}'_1, \boldsymbol{k}'_2$	final wave vectors of two colliding particles
k_{max}	$= \pi/a_0$
L, L	orbital angular momentum
$\mathscr{L}(E, \lambda)$	function which accounts for screening
L_{free}	mean free path
L_{typ}	typical confinement dimension
L_{s}	structural lenght parameter
m	magnetic quantum number
M	atomic mass
m_J	component of $\boldsymbol{J}$ in a given direction
m_0	free electron mass
m^*	effective mass
m_e^*, m_h^*	effective mass of an electron, hole
n	band index (quantum number)
n	density of free carriers
n_e, n_h	electron, hole density
N	number of electrons
N	occupation number
n_c	critical exciton density
$N_{\text{A}}, N_{\text{D}}$	acceptor, donor impurity concentrations
N_q	phonon occupation number
n_{pair}	e–h pair density
N_{PHOTON}	occupation number per photon mode
$\boldsymbol{p}$	induced polarization
$\boldsymbol{P}, P$	electronic polarization density, macroscopic dipole moment
$\mathscr{P}$	interband momentum matrix element
$\hat{\boldsymbol{p}}$	momentum operator
P_{eh}	electron-hole pair polarization
$\boldsymbol{q}, q$	phonon wave vector
q_{pe}	piezoelectric charge constant
$\boldsymbol{r}$	distance vector
r_{ee}, r_{hh}, r_{eh}	mean interparticle distances
r_{e-h}	characteristic separation of relative motion in an e–h pair
Ry^*	exciton Rydberg energy
s, p, d	quantum numbers of orbitals
$\boldsymbol{S}$	spin vector

S	energy flux density
t, t'	time
T	absolute temperature
T_{L}	lattice temperature
$u_k(r)$	cell-periodic part of the wave function
U_{eb}	bias voltage
v	group velocity
V	volume, normalization volume
δV	volume change
$V(r)$	interaction potential
$V_{\mathrm{c}}(r)$	periodic crystal field potential
v_{g}	group velocity
v_{s}	sound velocity
v_{DR}	mean drift velocity
W	probability of scattering per unit time
W_{cv}	induced transition probability
$W_{i \rightarrow f}$	probability of scattering from state E_i to state E_f
$x, \langle x \rangle$	amplitude, expectation value
$\hat{x}$	displacement operator
x_{cv}	displacement matrix element
$Y_{j,m}$	eigenfunctions belonging to the total angular momentum J
z	distance (between two tunnel contacts)
Γ	broadening parameter
Γ, L	points in the Brillouin zone
$\varepsilon(\omega)$	complex dielectric function
ε	dielectric constant
ε_0	permittivity constant
ε^*	value of $\varepsilon(\omega)$ at a frequency corresponding to $Ry^*/\hbar$
$\varkappa$	inverse bulk modulus
λ	wavelength
λ_{DH}	Debye-Hückel screening length
λ_{TF}	Thomas-Fermi screening length
μ	reduced mass of an excitation
μ	drift mobility (mean drift velocity per unit field)
μ_{II}	mobility due to ionized impurity scattering
μ_e, μ_h	drift mobility of an electron, hole
μ_{DP}	drift mobility due to deformation potential scattering
v	exponent in the energy dependence of the scattering time
ϱ	density of a crystal
$\sigma(\omega)$	ac conductivity
τ	scattering time
τ	mean free time (between collisions)
τ_{m}	momentum relaxation time
τ_{E}	energy relaxation time

τ_{life}	lifetime
τ_{rad}	spontaneous radiative lifetime
$\phi(\boldsymbol{k})$	wave function in $\boldsymbol{k}$-space
φ_{ext}	external perturbation potential
φ_{ind}	induced perturbation potential
φ_{tot}	total perturbation potential
$\Phi(r_e - r_h)$	envelope function of electron-hole motion
$\Phi_{1s}(r_{eh})$	envelope function of exciton ground state
$\chi(\omega)$	dielectric polarizability (or "response function")
χ_1	in-phase "dispersive" part of χ
χ_2	out-of-phase "absorptive" part of χ
$\chi^{(1)}, \chi^{(2)}$	coefficients in Taylor expansion of χ
$\psi_{n,k}(\boldsymbol{r})$	wave function of crystal electrons
$\psi(\boldsymbol{r}), \psi^*(\boldsymbol{r})$	wave function in $\boldsymbol{r}$-space
ω	frequency, excitation frequency
ω_{g}	frequency referring to the gap energy
ω_{s}	frequency of an acoustic phonon mode
ω_{LO}	frequency of an optical phonon
ω_0	natural frequency of a simple mass-spring system
ω_{vib}	lattice vibrational frequency
BTE	Boltzmann transport equation
B.Z.	Brillouin zone
CW	continuous wave
EMA	effective mass approximation
LA	longitudinal-acoustic phonon mode
MOS-FET	metal-oxide semiconductor field-effect transistor
RPA	random phase approximation
TA	transversal-acoustic phonon mode

2.1 Introduction

Interest in charge transport and optical properties of semiconductors dates back almost forty years. The experiments of Shockley and co-workers upon transport of photo-injected minority carriers in germanium and later silicon identified for the first time the three basic phenomena which are common to electrical transport in semiconductors (Shockley, 1950):

 (i) the effect of changing carrier energies,
 (ii) the effect of heating the lattice, and
 (iii) changes in the number of carriers.

They pointed out the fundamental importance of electronic band structure for charge transport. The work on Ge and Si initiated later investigations of III–V and II–VI compounds. Scientific progress in this field accelerated considerably when the enormous potential and far-reaching practical applications of semiconductor devices became evident in the late 1950s and early 1960s.

The last two decades have brought tremendous diversification to semiconductor physics. The steady evolution in the understanding of specific semiconductor properties in the broader context of solid state physics has been accompanied by an unparalleled development of sophisticated growth and processing techniques for semiconducting materials. The refinement of evaporation and lithographic procedures led ultimately to the fabrication of manmade microstructures on the atomic scale. At the same time, the fabrication of ultrapure and highly perfect bulk crystal material has been improved continuously, so that at present the two prototype semiconductors – silicon and gallium arsenide – are the best-understood crystalline solids.

The fabrication of more and more complex microstructures as well as the availability of well-defined semiconductor bulk materials have produced rich varieties of new and unforeseen phenomena. Quantum size effects in geometrically confined structures play the key role: Their generic wealth is manifested in the unique properties of so-called superlattices, quantum wires, and quantum dots. Low-dimensional carrier systems, with their unusual transport properties in electric and magnetic fields, have become topics of great practical interest. Spectacular new findings have come about in the context of localization and charge transport in semiconductors: the quantum Hall effect in 2-d confined carrier systems; strong correlations of carrier motion in such systems leading to the fractional quantum Hall effect; and finally the quantized conductance in laterally confined ("quantum wire") configurations.

A closely related development is the growing interest in optical properties of structured semiconductors and their applications in optoelectronic devices. The renaissance of the seemingly mature field of optics, which had started with the invention of the laser, continued with key contributions from semiconductor physics: injection lasers, efficient photon detectors in the near infrared, optical nonlinearities and bistability, ultrafast optical switching, and coherent pulse generation are just a few examples.

Electrical transport and optical properties of semiconductors have been treated accordingly in many review articles and textbooks. The aim of the present article is to give a comprehensive survey of some of the basic concepts, to illustrate their immediate application in specific situations, and to discuss some key experimental results. The level of presentation should help the unfamiliar reader to become aware of the state-of-the-art in the field and assist him in evaluating the contemporary primary literature.

Elementary coverage of band structures, optical transitions, carrier scattering, and electrical transport will be given to some extent, and emphasis is placed on direct application of each topic to practical problems and recent experiments. This article does not cover the aspects of magneto-optics and magneto-transport, but does contain some selected material on nonlinear transport and nonlinear optics in semiconductors. For further information on these topics, the reader is referred to the monographs and the review articles cited at the end of the chapter.

2.2 Electronic Band Structure and Intrinsic Optical Properties

The optical properties of crystalline semiconductors in the visible part of the spectrum are closely connected with their electronic band structure. All optical spectra reflect essentially the unperturbed energy spectrum $E(k)$ and underlying wave functions $\psi_{n,k}(r)$ of crystal electrons. To a certain degree, details of optical spectra in real crystals are modified by the presence of impurities and structural defects. At temperatures $T > 0$, the agitation of thermal lattice vibrations broadens and shifts the optical spectra.

Calculations of electronic band structures – at any level of sophistication – use input parameters which are derived from spectroscopic data. In retrospect, practically all important physical quantities related to semiconductor band structures were extracted from empirical spectra of absorption, emission, and reflectivity. The most relevant parameters are fundamental and higher band gaps, the magnitude of momentum matrix elements, symmetries of

wave functions at special points in the Brillouin zone, and joint density-of-states of the bands.

The common starting point for a discussion of electronic properties of crystalline solids is the periodic crystal potential $V_c(r)$: one assumes a priori that each crystal electron moves in the field of all nuclei and all other electrons, and that all electrons see effectively the *same* potential. More precisely, the potential $V_c(r)$ is said to be of Hartree–Fock type, i.e. a self-consistent mean field (Anderson, 1963). The use of $V_c(r)$ leads to the concept of the "one-electron" approximation: it is taken for granted that all solid state properties can be described in terms of products of one-electron wave functions. Despite the large valence electron density of approximately 3×10^{23} cm^{-3} in typical semiconductors, this concept works surprisingly well and is a good basis for all further treatments of semiconductor properties.

In the next paragraphs we will describe on an elementary level the connection between single-electron band structures $E(k)$ and fundamental optical spectra of semiconductors. Necessary extensions leading to excitonic effects in optical spectra, and screening effects connected with high $e–h$ pair densities will be discussed in Sec. 2.2.3.

2.2.1 Electron-Hole Pair Excitations

Filled valence bands, empty conduction bands, and a band gap of width $E_G \gg kT$ between them define the ground state of the N-electron-system "semiconductor" with energy E_0. Since all N electron charges may conduct in principle, the terms valence band and conduction band are not really useful in this context. What counts is the distribution of occupied and empty states in each energy band $E_n(k)$ with band index n. An electron in an otherwise empty band and a hole in a filled band are both charge "carriers". They are the electronic excitations which contribute to the response of the semiconductor to external electric fields; i.e. they are responsible for charge transport and optical properties.

An excited state with energy E_{ex} of the N-electron system may be created by taking an electron out of a valence band state, and putting it into an empty conduction band state (see Fig. 2-1). This procedure leads to the creation of an electron–hole (e–h) pair. The total energy change ΔE connected with this process is the sum of three terms:

(i) the potential energy of the electron with respect to $V_c(r)$. It is called "gap energy" E_G and corresponds to the change in band index n (the analog of the atomic orbital quantum number);
(ii) kinetic energy $E_{kin}^{(h)}$ of the initial electron state, or the final hole state;
(iii) kinetic energy $E_{kin}^{(e)}$ of the final electron state.

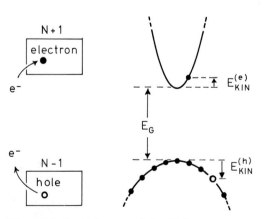

Figure 2-1. Excited states of the N-electron system (left). The energy–wave vector diagram (right) shows a valence band "hole" and a conduction band "electron" with kinetic energies $E_{kin}^{(h)}$ and $E_{kin}^{(e)}$. In addition, the potential energy E_G is included (see text).

Any rearrangement of the valence electron distribution due to the excitation of an e–h pair is neglected at this level of the one electron approximation.

We may therefore write

$$\Delta E = E_{ex} - E_0 =$$
$$= E_{kin}^{(e)} + E_{kin}^{(h)} + E_G - E_0 \qquad (2\text{-}1)$$

For wide enough bands and small kinetic energies of the carriers, the so-called effective mass approximation (EMA) applies (Cohen and Chelikowsky, 1988), and the kinetic energies are simply given by

$$E_{kin}^{(j)} = \frac{\hbar^2 k_j^2}{2 m_j^*}, \qquad j = e, h \qquad (2\text{-}2)$$

and are counted from the band edges E_c and E_v. The coefficients m_e^* and m_h^* represent the effective masses of electrons and holes.

Electrons and holes of given wave vectors k_e and k_h are distinguishable particles if their band index n is different. They can be described by fermion creation and annihilation operators, which act on the "Fermi sea" of filled valence band and empty conduction band states (Haken, 1973).

If we want to go further than counting numbers, we have to study wave functions in realistic potentials $V_c(r)$, and the discussion has to be more specific (Harrison, 1980). The prototype semiconductors which we have in mind are crystalline solids with tetrahedral coordination. This class includes the group IV element Ge, the group III–V compound GaAs, and the II–VI compound ZnSe. These three prototypes have their chemical homologues, which appear in the columns of the periodic table of elements: the monatomic C, Si, Ge family; the diatomic semiconductors with indium, InP, InAs, InSb...; and those with Cadmium, CdS, CdSe, ... etc.

It turns out that these materials, which all crystallize in either diamond, zinc blende, or wurtzite lattice structures, have many basic features in common: their non-bonding conduction band states at and around E_c have the symmetry of atomic $L=0$ or "s-like" orbitals (with even parity in the diamond case) and twofold degeneracy (for the electron spin). Their effective mass parameter m_e^* is usually low. The valence band edge at energy E_v has eigenfunctions which resemble "p-like" orbitals (with odd parity in the diamond structure). Its sixfold degeneracy is split by the spin-orbit interaction such that the $j=3/2$ terms lie usually above the $j=1/2$ split-off level (Kane, 1966). In the diatomic zinc blende and wurtzite lattice structures the potential $V_c(r)$, and hence the valence electron density, is quite different at the cation and the anion site. This asymmetry causes characteristic linear terms in the $E_n(k)$ relation at $k=0$ (Kane, 1957; Cardona et al., 1986).

Away from the zone center $k=0$ the valence band extremum is split by the $k \cdot p$ – interaction into heavy and light hole bands because of the spatial anisotropy of the sp^3 hybrid wave functions which form the tetrahedrally coordinated bonds (Parmenter, 1955; Harrison, 1980). For the same reason, the effective masses at $k=0$ and beyond are anisotropic, in general. With increasing k, the effective mass parameters usually become larger. In the wurtzite structure (with two different bond lengths), the valence bands are nondegenerate at $k=0$. The basic shape of cell-periodic wave functions and their nearest-neighbor overlap are shown qualitatively in Fig. 2-2. A simple 2-dimensional quadratic lattice (representing fourfold coordination in a monatomic crystal) has been chosen for clarity.

The proper eigenstates of both electrons and holes in the periodic crystal potential $V_c(r)$ are described by Bloch wave functions. They extend over the entire crystal volume, with equal probability density in each lattice cell:

$$\psi_{n,k}(r) = \frac{1}{\sqrt{V}} e^{i k \cdot r} u_k(r) \qquad (2\text{-}3)$$

Common basis functions for the symmetry classification of the cell-periodic part of the wave function, $u_k(r)$, are either the atomic orbitals "s", "p", "d", etc., in cartesian coordinates or the eigenfunctions $Y_{j,m}$ belonging to the total angular momentum $J = L + S$, with $J = 3/2$ or $1/2$. The overlap integrals between wave functions in adjacent cells for a given wave vector direction

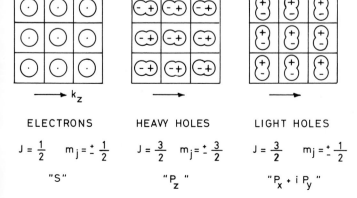

ELECTRONS

$J = \dfrac{1}{2} \qquad m_j = \pm\,\dfrac{1}{2}$

"S"

HEAVY HOLES

$J = \dfrac{3}{2} \qquad m_j = \pm\,\dfrac{3}{2}$

"P_z"

LIGHT HOLES

$J = \dfrac{3}{2} \qquad m_j = \pm\,\dfrac{1}{2}$

"$P_x + i\,P_y$"

Figure 2-2. Cell-periodic wave functions $u_k(r)$ in a cubic lattice for $k \parallel z$. Only the spatial part is shown; the spin parts $|\uparrow\rangle$ are chosen to be aligned in the z-direction (schematic).

(i.e., quantization axis) depend on the component m_j of J in that direction. This is the physical origin of heavy and light hole mass difference; it can be inferred qualitatively from Fig. 2-2. A lucid discussion of crystal symmetry and band degeneracy can be found in Tinkham (1964) and Bassani and Pastori Parravicini (1975).

2.2.2 Dielectric Polarization and Response Function

Optical studies of solids probe the dielectric polarizability. At sufficiently high frequencies the response of any solid is dominated by electronic transitions. When static or low-frequency electric fields act on a diatomic (i.e., polar) semiconductor, there is an additional effect: The two sublattices in zinc blende (or wurtzite) structure will move with respect to each other and give a lattice contribution to the induced polarization. In the following we will restrict ourselves to excitation frequencies ω that are high in comparison to the vibrational frequencies ω_{vib} of the lattice. The natural frequency ω_0 of a simple mass-spring system is in the infrared region of the spectrum for all common values of lattice force constants D and atomic masses M in semiconductors (see, e.g., Ashcroft and Mermin, 1976). D is directly related to the inverse bulk modulus $\varkappa$ and the interatomic distance (bond length) d through

$$\omega_0 = \sqrt{\frac{D}{M}} = \sqrt{\frac{d}{\varkappa \cdot M}} \qquad (2\text{-}4)$$

We find the range $\omega_{\text{vib}} \sim 0.32 \times 10^{14} \dots 2.4 \times 10^{14}$ rad/s when we go from the heaviest II–VI compound, CdTe, to the lightest group IV crystal, diamond (C). In the visible part of the spectrum ($\omega = 3 \times 10^{15}$ rad/s at 2 eV), $\omega \gg \omega_{\text{vib}}$ is fulfilled and we may safely neglect the contribution of lattice polarizability, because the nuclei cannot

follow the driving field. We will now discuss the dielectric properties of semiconductors as a function of incident photon energy $\hbar\omega$, or light frequency. The low-frequency limit, or "static" case, implies energies which are small compared to characteristic electronic excitation energies, but which are still well above ω_{vib}.

Let us consider as the driving force a spatially uniform electric field $E_0 \cdot \exp(-i\omega t)$ which interacts with all N electrons in the crystal. The induced polarization p of electronic charge in the directed bonds and (to a much lower degree) of inner shells of the atoms will generally result in a nonuniform electric field in the solid and hence reduce the effective field at the atomic sites. The importance of such local field corrections depends on the degree of ionicity and the localization of the bond charges. The reader is referred to the article by Lannoo in this volume for a thorough discussion (see also Resta, 1983). For simplicity we neglect here the local field effect altogether and keep in mind that the assumption of a uniform electric field means a rigid displacement of electronic charge distributions (see Fig. 2-3).

The electronic polarization density P within the crystal is the sum of all induced

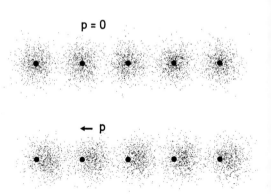

Figure 2-3. Charge density distribution in periodic crystal with a) external electric field $E = 0$; b) induced dipole moments $p = \chi \cdot E$.

polarizations p per volume:

$$P = \frac{1}{V} \cdot \sum_j p_j \qquad (2\text{-}5)$$

The crystal field due to the crystal potential V_c close to the nucleus is 10^8 V/cm. Provided that E_0 is much smaller than this, it is reasonable to assume that the induced P is proportional to E. This basic assumption defines the regime of "linear optics" in solids. For ordinary incoherent light sources, where power densities

$$S = \frac{1}{2} \cdot \frac{c}{4\pi} \cdot |E_0|^2 \qquad (2\text{-}6)$$

do not exceed a few kilowatts per cm^2, the condition is perfectly well fulfilled:

$$E_0 = 19 \text{ V/cm} \quad \text{for} \quad S = 1 \text{ W/cm}^2 \qquad (2\text{-}7)$$

Only laser light sources offer the possibility of entering the nonlinear regime. Focused and short-pulsed lasers are now routinely used to study the nonlinear optical response of solids (see Sec. 2.4 below).

The proportionality constant between P and E is the dielectric polarizability χ, or "response function", defined by

$$P = \chi(\omega) \cdot E \qquad (2\text{-}8)$$

in the frequency domain, with sinusoidally varying fields

$$E = E_0 \cdot e^{-i\omega t} \qquad (2\text{-}9)$$

Here, it is understood that only the real part of E represents the actual electric field. Summing over frequencies in Eq. (2-9) allows one to describe any time dependence of $E(t)$ without the loss of generality. $\chi(\omega)$ will generally be a complex function of ω

$$\chi(\omega) = \chi_1(\omega) + i\chi_2(\omega) \qquad (2\text{-}10)$$

with an in-phase "dispersive" term χ_1 and an out-of-phase "absorptive" term χ_2.

We can extract the dielectric response from the electronic band structure as fol-

lows: the driving field is treated as a small perturbation $\mathcal{H}'$ to the Hamilton operator $\mathcal{H}_0$. The latter contains V_c and stands for the unperturbed crystal. The vector potential is chosen such that

$$E = -\frac{1}{c} \cdot \frac{\partial A}{\partial t} = \frac{i\omega}{c} A \qquad (2\text{-}11)$$

for the plane wave of Eq. (2-9). We add the field momentum $e \cdot A/c$ to the momentum operator $\hat{p}$ in $\mathcal{H}_0$. The coupling term between a single electron and the light field becomes to lowest order in A

$$\mathcal{H}' = \frac{e}{m_0 c} \hat{p} \cdot A = -\frac{e\hbar}{m_0 \omega} E \cdot \nabla_k \qquad (2\text{-}12)$$

Here m_0 is the free electron mass. The next term is proportional to A^2 and is usually negligible (Bassani and Pastori Paravicini, 1975). The incorporation of Eq. (2-12) into time-dependent perturbation theory is straightforward. With the aid of Fermi's "golden rule" one obtains the induced current density in the crystal and relates its out-of-phase component to χ_2:

$$\chi_2(\omega) = \frac{\pi}{V}\left(\frac{e\hbar}{m_0\omega}\right)^2 \sum_{c,v} \left|\langle k, c| \frac{\partial}{\partial x} |k, v\rangle\right|^2 \cdot$$
$$\cdot \delta(E_c(k) - E_v(k) - \hbar\omega) \qquad (2\text{-}13)$$

V is the volume appropriate for normalization of ψ, the sum includes all wave vector pairs $(c, k; v, k)$ which fulfill the condition that the initial state is full and the final state is empty (with a plus sign), or vice versa (with a minus sign). The driving field is chosen along the x-axis, and P is taken to be parallel to E. Equations (2-8)–(2-13) are the simplest and qualitatively correct semiclassical approach to the linear optical response of semiconductors:

(i) The absorption of light is mediated by the linear coupling between E and single electron momenta p in Eq. (2-12);

(ii) the sum in Eq. (2-13) contains transitions from occupied valence to unoccupied conduction band states ("upward transitions") and their time-reversed counterparts ("downward transitions"). With the appropriate signs, Eq. (2-13) describes both fundamental optical processes of induced absorption and emission.

The extension toward a full quantum mechanical description is conceptually simple: the vector potential A is quantized and the elementary process of photon absorption (emission) occurs only if the two coupled states $(c, k; v, k)$ differ by $\hbar\omega$ in energy. Since the photon momentum $\hbar c q \sim 10^5\,\mathrm{cm}^{-1}$ in the visible part of the spectrum is small on the scale $k_{\mathrm{max}} = \pi/a_0 \sim 6 \times 10^7\,\mathrm{cm}^{-1}$ of electron quasi-momenta $\hbar k$, one speaks of "vertical" transitions in k space.

The matrix element of Eq. (2-13) may be rewritten by using the relation

$$\left| \langle c, k | \frac{\partial}{\partial x} | v, k \rangle \right| =$$
$$= \frac{m_0 [E_c(k) - E_v(k)]}{\hbar^2} |\langle c, k | \hat{x} | v, k \rangle| \quad (2\text{-}14)$$

which is valid for localized electron orbitals (Harrison, 1980). Equation (2-14) connects the operator for the dipole mo-

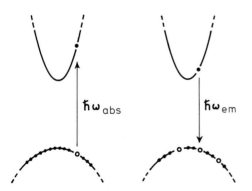

Figure 2-4. Light absorption from the crystal ground state, $\chi_2 > 0$ (left), and light emission from an excited state, $\chi_2 < 0$ (right). Both elementary processes are contained in the dielectric polarizability (Eq. (2-13), see text).

ment, $e \cdot \hat{x}$, with $\chi(\omega)$, and allows one to substantiate the intuitive interpretation of Fig. 2-3: the process of optical absorption may be seen as the time-proportional admixture of an "s-like" wave function amplitude to the initial "p-like" valence band wave function mediated by the dipole moment operator $\hat{x}$. The superposition of both eigenstates produces an oscillatory motion of the bond charges with amplitude x out of their equilibrium positions $x = 0$. Inspection of Eqs. (2-5) and (2-14) shows that the expectation value of x corresponding to one optically coupled e–h pair per site in the crystal is

$$\langle x \rangle = \frac{\pi \cdot e \cdot \hbar}{m_0^2 \, \omega^3} \left| \langle c | \frac{\partial}{\partial x} | v \rangle \right|^2 \cdot E_0 \cdot \cos(\omega t)$$
$$(2\text{-}15)$$

It is convenient to introduce the interband momentum matrix element P (Lawaetz, 1971):

$$\mathscr{P} = \frac{2\hbar^2}{m_0} \left| \langle c | \frac{\partial}{\partial x} | v \rangle \right|^2 \quad (2\text{-}16)$$

According to Eq. (2-15), we find that for a typical semiconductor with $\mathscr{P} = 30\,\mathrm{eV}$, $E_G = 1.5\,\mathrm{eV}$, the induced dipole moment is $\sim 1 \times 10^{-8}$ Ångstrom per V/cm and elementary charge.

Equation (2-13) reveals that all optical transitions are directly connected with momentum matrix elements – the same quantities that also determine the band gaps and band curvatures near band extrema in the framework of $k \cdot \hat{p}$ perturbation theory. This connection leads to a number of useful relations (so-called sum rules) which help to determine mass parameters and g factors from optical spectra, and vice versa (Kane, 1966; Lawaetz, 1971; Hermann and Weisbuch, 1977; Harrison, 1980).

Up to this point we have only discussed the dielectric polarization for resonant excitation with light, i.e., $\hbar\omega > E_G$. It follows from very general arguments, which are based only on the causality and linearity of the system, that $\chi_1(\omega)$ and $\chi_2(\omega)$ are closely connected with each other. One relation is as follows (Stern, 1963; Cardona, 1969):

$$\chi_1(\omega) = \frac{2}{\pi} \int_0^\infty \frac{\omega' \chi_2(\omega')}{\omega'^2 - \omega^2} \, d\omega' \qquad (2\text{-}17)$$

It gives an explicit expression for $\chi_1(\omega)$

$$\chi_1(\omega) = \frac{2\hbar}{V} \left(\frac{e}{m_0}\right)^2 \sum_{k,\,k'} \frac{1}{\omega_{kk'}} \frac{\left| \langle c, k | \frac{\partial}{\partial x} | v, k \rangle \right|^2}{\omega_{kk'}^2 - \omega^2}$$

$$(2\text{-}18)$$

The real part of $\chi(\omega)$ describes the response of all electrons weighted with a resonance denominator. The form of the dielectric susceptibility resembles very closely the behavior of an ensemble of massive charges on springs which are characterized by individual eigenfrequencies $\omega' = (E_{ck} - E_{vk})/\hbar$ and are driven by an external force field with frequency ω. This analogy with forced harmonic oscillations was recognized very early – in fact long before quantum mechanics – in the work of Drude and Lorentz. They interpreted the empirical results of their time, which are now described quantum-mechanically by Eqs. (2-13) and (2-18), in terms of electro-mechanical oscillators (Jackson, 1975).

For a given frequency ω one must sum over all possible resonances of Eq. (2-18), i.e., all e–h pair states with a total momentum of zero. The occupation of initial and final states is taken into account by summing only over filled valence and empty conduction band states (or vice versa, with the opposite sign). For the consistent description of matter in Maxwell's equations the complex dielectric function is defined:

$$\varepsilon(\omega) = 1 + 4\pi \chi(\omega) \qquad (2\text{-}19)$$

It describes the dielectric response of a semiconductor and is the quantum me-

chanical version of the response of a linear system to a harmonic time-dependent perturbation (Kubo, 1957).

An important extension of Eqs. (2-13)–(2-18) can be made in order to include the possibility of partially filled (or empty) states k_v and k_c. Each term of the k sum is multiplied with scalar distribution functions f_e, f_h obeying $0 \le f \le 1$. The crystal is assumed to be excited, either in thermal equilibrium ($f = f_0$) or with an arbitrary f, which characterizes nonequilibrium. This weighting procedure leads to an average optical response of the system; phase relations between the electrons and holes are not taken into account at this level of description (see Sec. 2.3 below).

Any wave vector or energy-dependent structure in $f_e(k)$ and $f_h(k)$ will obviously alter the response at a given frequency ω and show up in all optical spectra as dispersive (or absorptive) features. It turns out that transmission and emission spectra are essentially dominated by the imaginary part of χ, whereas reflectivity spectra are mainly controlled by the real part of χ (Wooten, 1972).

2.2.3 Correlated Electron-Hole Pairs: Excitons

The model of independent electrons moving in a strictly static crystal potential $V_c(r)$ works reasonably well in the context of the dielectric response $\chi(\omega)$. It provides a qualitatively correct description of the important features of optical spectra – band gaps, critical points, density-of-state peaks – for all common semiconductors. However, a closer look at empirical spectroscopic data and their quantitative comparison with the results of one-electron band structure calculations inserted into Eqs. (2-13) and (2-18) reveals a problem: at and around all critical points of $E(k)$, and especially vicinity of band edges, which are of greatest importance for all practical applications (see below), the independent particle scheme is grossly inaccurate. The dense electron gas with an average density of $n \sim 3 \times 10^{23}$ cm^{-3} *has* correlations; that is, the excited electrons and holes interact strongly with each other and polarize the surrounding electrons. In other words, a more realistic picture of interband transitions has to include the screened Coulomb interaction between electrons and holes.

In principle, we expect three different kinds of correlated motion when electron–hole pairs – or "excitons" – are excited in a semiconductor:

(i)   the excited electron (Eq. (2-12)) does not leave the atomic site of its origin and stays bound to it. The pair is said to be trapped (Toyozawa, 1980);

(ii) both electron and hole are mobile after the excitation, but stay bound together; the pair moves through the crystal but keeps a characteristic separation r_{e-h} of relative motion. The two limiting cases $r_{e-h} \sim a_0$ and $r_{e-h} \gg a_0$ (a_0 is the lattice constant) are known as Frenkel and Wannier–Mott excitons (Wannier, 1937; Mott, 1961; Elliott, 1957);

(iii) electron and hole are unbound and move away from each other (Dow, 1976; Stahl and Balslev, 1986).

The situation is controlled by several factors: kinetic energies of electrons and holes, the pair density with its influence on screening of the $e-h$ Coulomb interaction, and the strength of electron–phonon coupling. The most common case in group IV, III–V and II–VI semiconductors with band gaps in the range $1-3$ eV and moderate electron–phonon coupling is the extended exciton, case (ii). We will now briefly discuss its basic features (Dimmock,

1967; Elliott, 1963; Cho, 1979; Rashba and Sturge, 1982).

The natural length and energy scales of one interacting $e-h$ pair are the exciton Bohr radius a_B^* and the exciton Rydberg energy Ry^* (see Chap. 1):

$$a_B^* = \frac{\hbar^2 \, \varepsilon_0 \cdot \varepsilon^*}{\mu \cdot e^2} \qquad (2\text{-}20)$$

$$Ry^* = \frac{\mu \, e^4}{2\hbar^2 \, (\varepsilon_0 \cdot \varepsilon^*)^2} = \frac{e^2}{2\varepsilon_0 \cdot \varepsilon^* \, a_B^*} \qquad (2\text{-}21)$$

They are basic parameters which describe the $e-h$ pair motion in relative coordinates, $r_{e-h} = |r_e - r_h|$, in the presence of the long-range attractive Coulomb interaction modified by the dielectric function $\varepsilon(\omega)$. ε^* is the value of $\varepsilon(\omega)$ taken at a frequency corresponding to $Ry^*/\hbar$, and μ stands for the reduced effective mass of the two-particle problem "exciton", that is, the short form for a correlated $e-h$ pair. Typical values for Ry^* are $1-100$ meV, and $a_B^* = 10 - 100$ Ångstrom; in other words

$$Ry^* \ll E_G \quad \text{and} \quad a_B^* \gg a_0 \qquad (2\text{-}22)$$

The existence of bound $e-h$ pair states and the adjacent continuum of correlated pair states change the functional form of the dielectric response function $\chi(\omega)$ appreciably. Instead of the plane wave states, which were assumed in Sec. 2.2.2 to be the correct eigenstates of the crystal and had been used in the derivation of Eqs. (2-13) and (2-18), we must incorporate the actual envelope wave function for relative $e-h$ motion. The problem is tractable for two interacting carriers in effective mass approximation (Wannier, 1937; Cho, 1979).

2.2.3.1 Single-Particle versus Pair Excitations

The spatial correlation of electron and hole motion represents the simplest type of *collective* excitation in a solid: both quasi-particles attract each other by the potential $e^2/\varepsilon_0 \cdot \varepsilon^* |(r_e - r_h)|$ and move freely through the crystal. It is plausible to assume that for low e–h pair densities – say $n_{pair} \ll (a_B^*)^{-3}$ – the interactions between two different pairs can be ignored. This assumption is the concept for the elementary excitation "exciton". An exciton with given total momentum $\hbar K$ is represented by the product of a phase factor $\exp(iKr)$ with the envelope function $\Phi(r_e - r_h)$. The wave vector $K = k_e - k_h$ characterizes the center-of-mass motion of the pair, the envelope describes the relative e–h motion. This new "quasi-particle exciton" is composed of two particles with half-integral spins; its creation and annihilation operators obey commutation rules which are approximately those of bosons (Hopfield, 1969).

The sum of electron and hole momenta, k_e and k_h, is conserved in the exciton. Therefore the pair wave vector, $K_{pair} = k_e - k_h$, is a natural choice for its translational motion. Figure 2-5a summarizes the essentials of the single-particle description in terms of electron (or hole) momenta and

energies for a noninteracting pair. Figure 2-5b shows the corresponding spectrum with interaction in the so-called exciton (or pair) representation. The lowest excitation energy is smaller than the (single-particle) gap energy E_G. For $K=0$ we find $E_{1s} = E_G - Ry^*$ because of the attractive e–h Coulomb interaction.

Discrete exciton resonances with their energy spectrum ranging from E_{1s} to E_G and the continuum of correlated pair states above E_G transform the single-particle optical spectrum of Eq. (2-13) in a characteristic manner (Dow, 1976; Hanke, 1978). We will discuss this so-called Coulomb enhancement effect in detail in Sec. 2.2.4.5 below.

2.2.3.2 Exciton Screening

What is the influence of other free carriers on the e–h pair correlation? The screening of the e–h interaction must be inspected more closely in order to answer this question. Consider the mutual Coulomb interaction $V(r)$ of a given e–h pair in the presence of n_e and n_h free carriers:

a) both carriers repel or attract (i.e., polarize) the bound valence charges and the nuclei;

b) in addition, the e–h pair is surrounded by clouds of mobile extra electrons and holes in the conduction and valence band which screen the pair interaction at long distances.

In the static limit $\omega = 0$, assuming temperature $T=0$, $n_e = n_h = n$, and for simplicity $m_e = m_h$, which thus allows for one Fermi energy parameter E_F in both bands, the model of Thomas and Fermi for screening (Platzman and Wolff, 1973) gives a qualitative idea of the screened potential

$$V(r) = \frac{e}{4\pi\varepsilon_0 \cdot \varepsilon^* r} \exp\left(-\frac{r}{\lambda_{TF}}\right) \qquad (2\text{-}23)$$

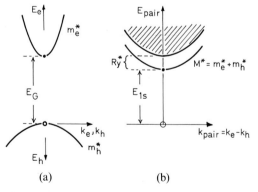

(a)   (b)

Figure 2-5. Single particle versus pair excitations: a) Dispersion relations $E_e(k_e)$ and $E_h(k_h)$ with gap energy E_G for non-interacting electrons and holes; b) Pair spectrum E_{Pair}. The ground state is at $E_{Pair} = 0$, $k_{Pair} = 0$, and one discrete bound pair state (the "1s exciton") is shown together with the continuum of unbound, but still correlated electron–hole pair states above (cross-hatched region).

with

$$\lambda_{TF} = \sqrt{\frac{E_F \cdot \varepsilon_0 \cdot \varepsilon^*}{6\pi n e^2}} \qquad (2\text{-}24)$$

Taking $n = 2 \times 10^{15}\,\text{cm}^{-3}$, $m_e^* = 0.067\,m_0$, we find that $E_F = 1\,\text{meV}$ and $\lambda_{TF} = 135\,\text{Å}$. More realistic treatments of bipolar screening of e–h pairs by electrons *and* holes are difficult. They have to take into account exchange and correlation effects in the plasma of carriers (Appel and Overhauser, 1982; Zimmermann, 1990). Authoritative reviews on high density e–h plasma effects were written by Rice (1977), Haug (1985), and Haug and Schmitt-Rink (1985).

In any case, the potential $V(r)$ in Eq. (2-23) contains the two distinct screening mechanisms mentioned above. On the one hand, ε^* stands for the quasi static screening via polarizable valence bond charges. This contribution can be traced back to the interband terms in Eq. (2-18) and may be referred to as "virtual" e–h pair excitations driven at frequency $\omega \sim 0$. On the other hand, the exponential function in Eq. (2-23) is the result of a balance between the attraction of mobile screening carriers by the screened electron (or hole) and the electrostatic repulsion in the screening cloud, both considered at frequency $\omega \sim 0$. In the model of Thomas and Fermi the radial dependence of the extra charge density around the screened carriers is calculated from an approximate (i.e., smoothed) solution of the Schrödinger wave equation with the inclusion of an electrostatic energy term (Platzman and Wolff, 1973).

If we deal with the same situation at finite temperature T, the thermal motion of the electrons adds to the repulsive action of the screening charges. Here the approach of Hückel and Debye leads in the limiting case of $kT \gg E_F$ to the "classical" version (Debye and Hückel, 1923) of the screening length, which has to be used in Eq. (2-23):

$$\lambda_{DH} = \sqrt{\frac{kT \cdot \varepsilon_0 \cdot \varepsilon^*}{4\pi n e^2}} \qquad (2\text{-}25)$$

Both expressions were derived for the static case; that is, the screened charge is at rest, $\omega = 0$. If the screened charge moves through the crystal, one must account for the dynamics of the screening carriers. We will return to the concept of dynamical screening below.

The length scale a_B^* which was introduced for the extended exciton states (Eq. (2-20)) separates two excitation density regimes: if the mean interparticle distances, i.e., r_{ee} and r_{hh}, exceed a_B^*, the picture of isolated bound pairs is adequate and we may think of a dilute gas of excitons. If the condition is not met, the situation is more complicated. Screening of the long range part of the Coulomb interaction between e and h will become increasingly important, leading eventually to a screening length shorter than a_B^*, so that individual pair states (Eqs. (2-20), (2-21)) are no longer a good basis for description. The term e–h "plasma" is used for such a dense system of electrons and holes in a semiconductor (Rice, 1977; Comte and Noziere, 1982). Two essential characteristics of the bipolar plasma are the existence of two distinct quasi-Fermi levels, $E_F^{(e)}$ and $E_F^{(h)}$, for both types of carriers, a reduced (renormalized) gap energy E_G', and acoustic plasmons (Appel and Overhauser, 1982). A second qualitative change is expected for still higher densities, n_e and n_h. It marks the transition toward true metallic behavior, and is accompanied by the closing of the semiconductor band gap (Combescot and Bok, 1983).

In the dilute case, the relative motion of e and h can be described in terms of plane wave states originating from unperturbed conduction and valence bands. Solutions of the wave equation for this problem of screened exciton motion have been extensively discussed in the last years (Zimmermann et al., 1978; Haug, 1985; Haug, 1988). We will skip a thoroughly detailed discussion here and instead sketch only plausible aspects of the problem.

Suppose the relative e–h motion of a single exciton is described by the envelope function

$$\Phi_{1s}(r_{eh}) = \frac{1}{\sqrt{V}} \cdot \exp\left(-\frac{r_{eh}}{a_B^*}\right) \qquad (2\text{-}26)$$

The expansion of this wave function into plane wave states $|k_e\rangle$ and $|k_h\rangle$ has Lorentzian shape, is centered around $k_e = k_h = 0$, and extends to roughly $(a_B^*)^{-1}$ in k space (Landau and Lifschitz, 1966). If we consider the available density of states in this wave vector range, we can estimate that approximately a density of $n_c \sim (a_B^*)^{-3}$ excitons can be created in a given crystal volume before phase space occupation reaches $f_e \sim f_h \sim 1$. A further increase in e–h pair density requires more kinetic energy, because of the Pauli principle, and increases the quasi-Fermi energies $E_F^{(e)}$ and $E_F^{(h)}$.

The interplay between the increase of kinetic energy and the lowering of the screened potential energy of the e–h pair motion controls the behavior of the system as a function of density. There is a relatively smooth transition from individual excitons at low density into a plasma of highly correlated e–h pairs around n_c, and finally into a dense metallic plasma with density much higher than n_c and a less pronounced long-range correlation (Rice, 1977; Haug, 1985).

2.2.4 Interband Absorption in Semiconductors

In this paragraph we discuss the application of Eqs. (2-13)–(2-18) to specific cases:

semiconductors with different band structures, dependence on dimensionality, and influence of external boundary conditions.

2.2.4.1 Direct versus Indirect Transitions

The conservation of crystal momentum k in Eq. (2-13) requires that only "vertical" transitions occur in the fundamental absorption and emission process. It is natural to classify band structures accordingly as "direct" and "indirect" with respect to their lowest (or any higher) gap. There is a large qualitative difference between the two cases, due to the symmetry of the wave functions at the band extrema. The upper valence band edge of all common semiconductors belongs to p-like wave functions, and is centered around $k = 0$. Figure 2-6 shows schematically the full wave functions $u_k(r)$ with phase factors $\exp(ik \cdot r)$ for a direct gap material such as InP and an indirect semiconductor such as Ge in reciprocal space (left) and direct space (right). In dipole approximation the small wave vector of light is neglected and the matrix element of Eq. (2-13) gives zero upon integration over all crystal unit cells in the indirect case: the transition is called "dipole-forbidden". Only second-order processes, e.g., the additional emission or absorption of a phonon (or the participation of an impurity) may cause light absorption. For intrinsic Ge and Si it has been shown that both variants of electron–phonon and hole–phonon coupling which are depicted in Fig. 2-6 can occur. They involve phonon modes of different symmetry, so that in high-resolution absorption and emission spectra both paths can be clearly distinguished (Rühle, 1981). Any perturbation of the symmetry of the crystal lattice, e.g., through the presence of impurities, interfaces or a free surface, or an electric field, will increase the probability of such forbid-

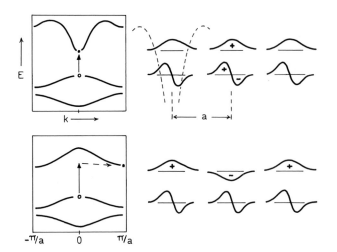

Figure 2-6. Dispersion relations (left) and full wave functions in periodic crystal potential (right). Top: In a "direct-gap" semiconductor the adjacent site wave functions are in phase and this leads to strong interaction with the driving light field. Bottom: The "indirect-gap" case shows alternating phase factors which lead to destructive interference; integration over the crystal volume gives vanishing interband transitions.

den transitions (Abstreiter, 1984; Abstreiter et al., 1984).

2.2.4.2 Radiative Recombination Rate

It is of basic interest to calculate the induced transition probability W_{cv} of one conduction band state (say at $k=0$) with the matching valence band state under the influence of a resonant electric field E_r. We obtain directly from Eqs. (2-13), (2-14), and (2-18) that

$$W_{cv} = \frac{e^2 \omega^2}{2\pi \hbar^2 c^3} |x_{cv}|^2 \cdot E_r^2 \qquad (2\text{-}27)$$

If we insert for E_r the amplitude of the electric field in one mode of the electromagnetic field at frequency $\hbar \omega_g$ with occupation number $N = 1/2$, we may identify Eq. (2-27) with the inverse spontaneous radiative lifetime τ_{rad} of the coupled e–h pair at $k=0$:

$$\tau_{rad} = \frac{2\pi \hbar c^3 \varepsilon_0 \cdot \varepsilon^*}{e^2 \omega^3 |x_{cv}|^2} \qquad (2\text{-}28)$$

We find with the parameters $\mathscr{P} = 30\,\text{eV}$, $E_g = 1.5\,\text{eV}$ and $\varepsilon^* = 12$ a value of $\tau_{rad} = 1.05\,\text{ns}$. This is in good agreement with empirical values of the interband radiative recombination time, which were deter-

mined from low-temperature bound-exciton decay times (Henry and Nassau, 1970) and also from direct band edge luminescence decay in connection with absolute quantum efficiency measurements (Hwang, 1972).

The e–h recombination rate of indirect gap materials, such as Si or Ge, is orders of magnitude smaller and reflects the strength of coupling between phonons and (i) zone-center valence band and (ii) zone-edge conduction band electronic states. Values of $\tau_{life} \sim 35\,\text{ms}$ for Si and $\sim 20\,\mu\text{s}$ for Ge have been found experimentally in high-purity intrinsic crystals (Thomas, 1979; Yablonovitch et al., 1986). Electron–phonon coupling is sensitive to small changes in the wave function $u_k(r)$, and it is not surprising that there is only qualitative agreement between the experimental data and ab initio calculations.

2.2.4.3 Restricted Dimensionality

In the past two decades, much effort in semiconductor materials engineering has been directed toward controlled carrier confinement along one ore more dimensions (Esaki, 1983). The terms quantum "well", "wire", and "dot" are now established abbreviated forms for semiconductor devices with so-called "mesoscopic" length scales in one, two, or three dimensions, that is, structural parameters L_s with a magnitude between the lattice constant a_0 of the atomic constituents, and the inverse of a relevant electron (or hole) wavenumber k:

$$a_0 < L_s < k^{-1} \qquad (2\text{-}29)$$

Here k may represent the magnitude of a thermal wave-vector, or the Fermi wave vector of the ensemble of free carriers which is confined in the structure. A large part of current work is aimed at novel optical properties of such devices (Haug, 1988).

Within certain limits we may adopt the simple approach of Eq. (2-13) to describe the optical properties of such microstructures. Provided that the energy spectrum E_n of the electrons in the structure is known, we can safely proceed along the lines of Sec. 2.2.2 and calculate the optical response: The sum over k space in Eq. (2-13) automatically takes care of any quantization effects, if we identify the "old" quantum number k (in the 3-dimensional case) with the "new" modified quantum numbers n of the structure with restricted degrees of freedom. The joint density of states is usually the eminent factor in the optical response $\varepsilon(\omega)$, just as in the 3-d case.

The effective Coulomb interaction between electrons and holes in a lavered structure with confined carrier motion is qualitatively different and leads to characteristic changes in the 2-d exciton spectrum. In the ideal 2-d case, the exciton binding energy is increased fourfold over the 3-d case, and screening is more efficient and has a density dependence which differs appreciably from that given in Eqs. (2-24) and (2-25) (Chemla and Miller, 1985).

The basic optical spectra of quasi-two-dimensional semiconductor structures with not too narrow well thicknesses are by now well understood (Dingle, 1975). It is not surprising to find a qualitatively different edge absorption spectrum: Instead of the familiar square-root dependence of the combined density of states $D(\hbar\omega)$ in 3-d, one observes a step-like, piecewise constant $D(\hbar\omega)$ in the 2-d case. Optical spectra of quasi 1-d structures are beginning to emerge and seem to follow the prediction of $D(\hbar\omega) \sim 1/\sqrt{(E - E_i)}$ (Kohl et al., 1989).

There are obvious limits beyond which the use of effective masses and factorized wave functions (atomic orbitals and simple phase factors) in very small structures becomes questionable (Bastard, 1988). The field is in rapid progress and the reader is referred to a recent survey of experimental work (Weisbuch and Vinter, 1990).

2.2.4.4 Angular Dependence of Interband Transitions

In full analogy with atomic spectroscopy, the orbital and spin parts of the initial and final electron wave functions ψ determine the angular characteristics of the interband optical transitions. We can therefore anticipate selection rules, that is, a pronounced angular dependence of the matrix element in Eq. (2-13). The wave functions at the valence band extremum in zinc blende structures are such that at $k = 0$ the transition to the conduction band creates both heavy- and light holes, and mixed spin states are produced in the process. In the lower (hexagonal) symmetry of the wurtzite lattice this degeneracy is lifted and we find two different transition energies with a strict selection rule at $k = 0$. An electric field vector $E \parallel c$ couples the heavy hole ($M_j = 3/2$), and $E \perp c$, the light hole band ($M_j = 1/2$) with the conduction band states. Away from $k = 0$ the symmetry is generally lower, and mixing of orbital p- and s-components occurs (Bassani and Pastori Parravicini, 1975).

Electron spins in zinc blende crystals may be partially aligned in the process of optical absorption by proper selection of levels and a suitable choice of photon angular momentum (Lampel, 1974; Zakharchenya and Meier, 1984). The method is analogous to optical pumping in atomic spectroscopy and has been applied to transitions at the Γ point of the Brillouin zone (Lampel, 1974) and also at the L point (Planel et al., 1977).

2.2.4.5 Fundamental Gap Spectra at Low Excitation

Under conditions of low temperature and low-excitation power, the direct band edge spectrum of intrinsic semiconductors is always dominated by sharp-excitonic line spectra.

In bulk crystals, where the sample dimensions are much larger than the exciton

Bohr radius a_B^*, the natural unit for e–h pair density is the reciprocal exciton volume. The pair density for complete volume filling with 1 s excitons is $n_c \sim 2 \times 10^{17} \, \text{cm}^{-1}$ in a typical semiconductor like GaAs with $a_B^* \sim 100 \, \text{Å}$. Taking the radiative recombination time of Eq. (2-28), and converting it to a generation rate, we find that at least $2 \times 10^4 \, \text{W/cm}^2$ of CW pump power is needed to maintain such a density in a layer of 1 µm thickness, which is roughly the absorption length.

At densities much lower than n_c, discrete bound e–h pair resonances below the gap energy E_G and a smooth continuum above give clear experimental evidence for the long-range attractive Coulomb interaction between electrons and holes. Figure 2-7 shows the absorption spectrum of a 4.2 µm thick high-purity GaAs crystal at low temperature, $T = 1.2$ K.

The material was grown by vapor-phase epitaxy and had residual shallow acceptor and donor impurity concentrations $N_A < N_D \sim 2 \times 10^{14} \, \text{cm}^{-3}$. The spectrum has been measured with a filtered incandescent CW light source, with approxi-

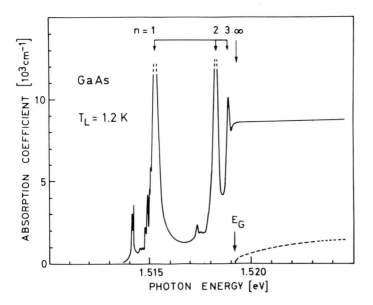

Figure 2-7. Band edge absorption spectrum of high-purity gallium arsenide at low temperature. The spectrum shows a Rydberg series of discrete bound pair states $n = 1, 2, 3, \ldots$ and the smooth continuum above E_G. Additional sharp lines below $n = 1$ and $n = 2$ are due to residual impurities and their localized bound pair states. A fictitious independent-particle absorption spectrum is shown for comparison (dashed line).

mately $10^{10}\,\mathrm{cm}^{-3}$ excited pairs, and the spectrometer behind the sample. A line-shape analysis of the two lowest reso-nances, $1s$ and $2s$, in the polariton frame-work (Hopfield, 1969; Weisbuch and Ulbrich, 1982) gave a broadening parame-ter of $\Gamma = 40\,\mu\mathrm{eV}$. This value is close to the intrinsic damping caused by scattering of excitons on piezoelectrically active TA and LA phonon modes (see Sec. 2.3.2 below). All measured parameters of the spectrum, i.e., the exciton Rydberg of 4.2 meV, the absolute value $8.5 \times 10^3\,\mathrm{cm}^{-1}$ of the ab-sorption coefficient at $\hbar\omega = E_G$, and the spectral shape above the gap, are consis-tent with the Wannier exciton picture in the regime of effective mass approximation (Elliott, 1963; Dow, 1976; Cho, 1979). This is remarkable, because of the valence band degeneracy and the strong mixing of light and heavy hole wave functions in the exci-ton problem. Nevertheless, a simple two-band description with average masses $m_e^* = 0.067\,m_0$ and $m_h^* = 0.32\,m_0$ gives a reasonable idea of the wave functions in-volved. In Figure 2-8 we have plotted the exciton envelope wave functions for four different energies in the optical spectrum of Fig. 2-7. Energies are given in units of the exciton Rydberg Ry^*, only states for angu-lar momentum $L = 0$ are shown, and nor-malization is carried out in a fixed volume (Ulbrich, 1988).

The strong correlation between electron and hole at small relative distances r is responsible for the excitonic peaks which are observed in all empirical spectra: the true optical absorption evolves from the single-particle spectrum displaying a red-shift $\chi_2(\omega)$ multiplied by the square of the exciton envelope wave function at $r=0$. This relation also holds above the gap. At the gap energy, and even several Rydberg energies above E_G, the persistent peak of $\psi(r)$ at $r=0$ (see Fig. 2-8) enhances the

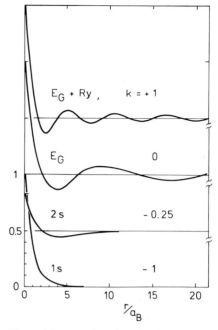

Figure 2-8. Wave functions for the relative motion of a single electron–hole pair. The parameter k indicates the pair energy relative to E_G, in units of the exciton Rydberg Ry^*. Note that there is strong correlation for small values of r even for the continuum states.

transition probability appreciably. The magnitude of this "Coulomb enhance-ment" effect is demonstrated in Fig. 2-7. The dashed curve is the calculated absorp-tion spectrum for a fictitious system with no Coulomb interaction between electron and hole, where the interband transition matrix elements are taken between plane wave states according to the single-particle description which led to Eq. (2-13).

2.3 Charge Transport and Scattering Processes

The transport properties of electrons in a crystalline semiconductor are deter-mined principally by their group velocity

$v_g = d\omega/dk = \hbar^{-1} \, grad \, E(k)$. An ensemble f of occupied stationary states with wave vectors k and density of states $D(k)$ will contribute to charge transport in a given direction with the current density

$$j = e/\hbar \int_{B.Z.} f(k) \, grad \, E(k) \, D(k) \, d^3k \qquad (2\text{-}30)$$

The distribution function f defines the number N of electrons in a given phase space volume $(k, k+dk; r, r+dr)$, that is, $N = f(k, r, t) \cdot d^3k \cdot d^3r$. Fermion statistics requires $0 \le f \le 1$. According to Eq. (2-30), a current which has once been established, say, by an electric field pulse, will persist infinitely: the electrical resistivity of electrons in the strictly periodic crystal potential $V_c(r)$ of Sec. 2.2 vanishes.

Finite resistivity is caused by spatial and temporal perturbations of $V_c(r)$. They couple a given electron to other excitations in the crystal and cause scattering processes which make the momentum of charge carriers non-stationary. In a typical semiconductor, the transport of charge over macroscopic distances is controlled by

- lattice vibrations
- impurities and lattice defects, and
- interactions between the mobile carriers.

Vibrations of the nuclei about their equilibrium positions induce changes in the energy of the electrons and constitute a time-dependent perturbation $\mathscr{H}'$ to the otherwise time-independent one-electron Hamiltonian $\mathscr{H}_0$, which is the starting point for the determination of the original electron dispersion relation $E(k)$ and the corresponding wave functions $\psi_{n,k}(r)$. $\mathscr{H}'$ is the adiabatic electron–phonon interaction term. In many cases it can be regarded as a small perturbation to $\mathscr{H}_0$, and a treatment in first-order perturbation theory is adequate. The total scattering-"out" rate from an initial state E_i into all energetically possible final states E_f is then given by the

familiar

$$W_{i \to f} = \frac{2\pi}{\hbar} \sum_f |\langle f | \mathscr{H}' | i \rangle|^2 \, \delta(E_f) \qquad (2\text{-}31\,a)$$

The probability per unit time for making the transition $k \to k'$ is written as $W = 1/\tau$, and the inverse quantity τ is called scattering time. It can be interpreted as the mean time between scattering events or "collisions". A direct analogy is possible with the classical concept of a collision time. The scattering time τ of Eq. (2-31 a) may be an explicit function of the wave vector k of the electron; i.e., it may be dependent on its initial direction, or may depend only on its energy, $\tau = \tau(E)$.

Since Eq. (2-31 a) is based on perturbation theory, there is an artifact concerning energy-conservation during the scattering event. Only in the limit $t \to \infty$ is the transfer of energy between electron and phonons strictly energy-conserving; i.e., it obeys the delta-function in Eq. (2-31 a). It can be shown that this is not a real problem as long as the relation

$$\frac{\hbar}{|M|^2} \frac{d}{dE} (|M|^2) < \tau(E) \qquad (2\text{-}31\,b)$$

is fulfilled (Peierls, 1955; Paige, 1964). M is the transition matrix element. Most scattering mechanisms fall in this category.

One implication of Eqs. (2-31) is that there is no meaningful concept for the duration of the scattering event. Obviously one would need a description beyond perturbation theory to define such a quantity, which will then depend on details of the interaction potential (see Sec. 2.3.3 below).

Nevertheless, the most common and highly successful approach to the transport problem is based on Eq. (2-31 a). According to this equation, the rate of change of occupation of a given electronic state k and a specific phonon mode q depends on all other occupancies and respective transition probabilities. The full dynamics of electron transport is then uniquely determined by two sets of equations like Eq. (2-31 a) – for electrons and phonons –, provided that the driving field and the initial values for f and N are known.

If there is no explicit time dependence in $f(k)$ and $N(q)$, i.e., if the mean charge and heat currents are stationary, then we deal with the important special case of dynamical equilibrium. To obtain stationary conditions, we must explicitly include the dissipation of energy into an external heat sink.

In order to carry out this process, a number of assumptions and definitions need to be made. In the foregoing section (Sec. 2.2), electrons and holes were treated as extended eigenstates (Bloch waves) in a perfect crystal lattice and characterized by their wave vector k and energy $E(k)$. This is appropriate for the weak coupling of light of wavelength $\lambda \gg a_0$ with electrons. In electrical transport two modifications of this view are useful:

(i) The electronic excitations are seen as more or less localized particles in ordinary space, and will be called charge carriers;

(ii) the potential $V_c(r)$ is explicitly time-dependent and hence influences the energy of a given carrier in any of the original bands $E_n(k)$.

The corresponding variations in $E_n(k)$ with time lead to electron transitions between the former eigenstates of $\mathcal{H}_0$. At this point the statistical ensemble of carriers is defined by assigning time-dependent probabilities f to the occupancy of the stationary electron states of the perfect lattice $V_c(r)$. The k- and r-dependent distribution function $f(k, r, t)$ describes the evolution of the carriers in phase-space. To distinguish between the different kinds of carriers, one defines $f_e = f$ for electrons in the conduction band, and $f_h = 1 - f$ for holes in the valence band.

A similar procedure is adopted for other excitations, such as phonons, impurities, plasmons, etc., which play a role in the transport process under consideration. The classical concept of phase–space distribution functions – taken from kinetic gas theory –, together with the quantum mechanical transition rates in first-order perturbation of Eq. (2-31), is the usual starting point for studies of electron and heat currents in solids (Chapman and Cowling, 1939; Ziman, 1963).

The connection between electrical transport and interband optical excitations in a semiconductor is shown in Fig. 2-9. It illustrates as a common aspect the dissipation of momentum and energy between the sub-systems of electrons, impurities, and phonons.

An external electric field with frequency ω is applied to the semiconductor and feeds momentum and energy into the system of carriers. The domain of "electrical transport" covers the frequency range $0 < \hbar \omega \ll E_G$, that is, dc to high frequencies of ~ Terahertz. The term "optical exci-

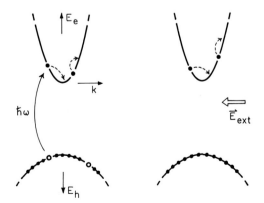

Figure 2-9. Dissipation of carrier momentum and energy in the process of interband optical excitation with $\hbar \omega > E_G$ (left) and in the case of charge transport induced by a slowly varying electric field, $\hbar \omega \ll E_G$.

tation" stands for $\hbar\omega > E_G$, that is, ac fields with frequencies high enough to induce interband transitions. The ac conductivity $\sigma(\omega)$ is directly related to the dielectric response function

$$\sigma(\omega) = \frac{i\omega}{4\pi}[1 - \varepsilon(\omega)] \qquad (2\text{-}32)$$

and describes the response of free carriers in a semiconductor to an external perturbation, that is, electric current in transport, and dielectric polarization in optics. The real part of σ contains, as discussed in Sec. 2.2, the dispersive (in-phase) component of j and p, the imaginary part all dissipative (out-of-phase) processes.

Energy-conserving (i.e., elastic) and inelastic carrier scattering mechanisms both contribute to σ.

Optical excitation of electron–hole pairs with narrow-band (i.e., more or less monoenergetic) laser light sources with $\hbar\omega - E_G > kT$ creates ab initio nonequilibrium distributions with large deviations δf from f_0, the equilibrium Fermi function at temperature T. More precisely, we have $\delta f(k) \gg f_0(k, T)$ even in the regime of linear optics with small external fields $|E| \ll |E_c|$ (see Sec. 2.2.2). Electrical transport, on the other hand, induces smoothly varying distributions, at least for those common scattering mechanisms which may be treated in a relaxation time approximation (see below), and which obey the smallness criterion $\delta f_{\text{field}}(k) \ll f(k)$. Both methods are insofar complementary; the combination of optical methods with traditional wire-bound techniques of electrical transport has emerged as a powerful analytic tool in the field of high-speed devices and circuitry testing (Auston et al., 1984; Fattinger and Grischkowsky, 1989).

Optical spectroscopy with its inherent energy selectivity (see Fig. 2-9) is able to extract very detailed information on specific scattering mechanisms in semiconductors. In fact, recent progress in time-resolved spectroscopy (see Sec. 2.4) has allowed qualitatively new insights and understanding of electron and phonon transport on a microscopic level.

2.3.1 Momentum and Energy Relaxation of Carriers

Let us investigate in some detail the basic processes which relax the initial momentum and energy of a charge carrier. Consider a single electron or hole in an otherwise empty band, and assume that all other bands are either filled or empty. All single-particle excitations of the crystal – electrons, phonons, plasmons, excitons, etc. –, may couple to the carrier under consideration, provided that momentum and energy conservation are fulfilled. If the kinetic energy E_{kin} of the carrier (measured from the nearest band extremum) satisfies

$$E_{\text{kin}} < E_{\text{gap}} \qquad (2\text{-}33)$$

we are in the regime of intraband scattering. The range of possible excitations is now restricted to relatively low energies. If Eq. (2-33) is not fulfilled, impact ionization processes (i.e., the interband Auger effect) and plasmon excitation with higher energies become possible (Platzman and Wolff, 1973). Under the influence of the external electric field, the carrier will steadily take up momentum and energy in the time interval δt

$$\delta E = e \cdot E \cdot v_G \cdot \delta t \qquad (2\text{-}34)$$

At the same time the carrier will undergo scattering processes, which mediate the transfer of momentum and energy into other, typically low lying excitations of the crystal with excitation energy ΔE. This transfer and the reverse mechanism determine the net momentum and energy relaxation rate of the given carrier, see Fig. 2-10.

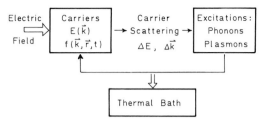

Figure 2-10. The driving electric field feeds momentum and energy into the system of free carriers, which in turn mediates the transfer into other, typically low-lying excitations of the crystal. To obtain stationary transport, dissipative processes via coupling to a thermal bath must be included.

The requirements of the continuity equation in phase space (k, r) and proper averaging over the actual carrier and phonon distributions leads us to the Boltzmann transport equation for $f(k, r, t)$

$$\frac{df}{dt} = v_G \cdot \nabla_r f - \frac{e}{\hbar} E \cdot \nabla_k f - \frac{\partial f}{\partial t}\bigg|_{\text{Coll}} \quad (2\text{-}35)$$

The left side is the total change of f with time. It may contain source and sink terms for particle generation and annihilation (see below) and it vanishes in equilibrium, resp. stationary situations. The last term on the right side contains the total scattering "in-" and "out-rates" of all active scattering mechanisms, and the two other terms describe drift in r and k space, in other words "ballistic transport" in real and velocity space.

The Boltzmann transport equation (BTE) is a first-order integro-differential equation which, in principle, can be integrated by elementary means if the starting distribution $f(k, r, t=0)$ is given and the following conditions are met: the source term $df/dt|_s$ is known for all times, and the collision term $\partial f/\partial t|_{\text{Coll}}$ is well-defined for all the distributions $f(t)$ which will eventually evolve in the time interval $(0, T)$ of interest.

For sufficiently simple configurations, the BTE has plausible, easy-to-find solutions. Let us first discuss the trivial case with no scattering at all, $\partial f/\partial t|_{\text{Coll}} = 0$, particle number conservation, $df/dt|_s = 0$, and spatial homogeneity, $\partial f/\partial r = 0$. Taking $f(t=0) = \delta(k - k_0)$ and with a little algebra involving the delta-function derivative gives

$$k(t) = k_0 + e/\hbar \cdot E \cdot t \quad (2\text{-}36)$$

as a solution. It is identical with the "ballistic" propagation of a constantly accelerated electron, which can be deduced directly from the Schrödinger equation for one electron in the crystal potential $V_c(r)$ with the static potential $e \cdot E \cdot r$ added to it. The solution is straightforward and gives the wave function

$$\psi(r, t) =$$
$$= (1/N) \cdot \exp\left(i k \cdot r - i/\hbar \int_0^t E(t') \, dt'\right) \quad (2\text{-}37)$$

with the time-dependent wave vector of Eq. (2-36), energy $E(t) = \hbar^2 k^2(t)/2m^*$, and effective mass $m^* = 1/\hbar^2 (d^2E/dk^2)^{-1}$.

Equally simple is the spatially inhomogeneous case, with no driving force, $E = 0$, and no scattering. Here we find that with $f(k, r, 0) = \delta(k_0) \cdot \delta(r_0)$ a solution of the BTE is

$$f(t) = \delta(k_0) \cdot \delta(r_0 + t \cdot \hbar k_0/m^*) \quad (2\text{-}38)$$

Less trivial situations require not only much more effort, but in practice heavy use of numerical methods is also necessary, either to solve for the integrals or to simulate the scattering terms with the help of Monte-Carlo (or cellular) methods, which provide a statistical sample of the configuration space (k, r). The reader is referred to the recent literature in this rapidly developing field (Reggiani, 1985).

To facilitate the search for solutions of the BTE, several simplifications have been devised in the past. One familiar procedure is the linearization of the BTE with respect to the driving field term E. It is based on the assumption that only scattering events with relatively small energy transfer $\delta E \ll E$ occur. Hence the scattering-in and scattering-out terms of Eq. (2-35) do not depend explicitly on the global form of $f(k)$, but are functions of k only. In other words, the correct dependence of $\partial f/\partial t|_{\text{Coll}}$ on the true distribution $f(k)$ is replaced by a local dependence on k (or only on the energy). The resulting scattering term has the much simpler structure

$$\partial f/\partial t|_{\text{Coll}} = (f(k) - f_0(k))/\tau(k) \qquad (2\text{-}39)$$

and defines the so-called relaxation-time approximation (Ziman, 1963). It can be shown that for all quasi-elastic scattering mechanisms, such as acoustic phonon scattering and impurity scattering, there is a well-behaved solution $f(k, r, t)$ of the BTE in the limit $|E| \to 0$, the "low-field limit", with mean drift velocity

$$v_{\text{DR}} = \int_{\text{B.Z.}} v \cdot f(k, r)\, \mathrm{d}^3 r\, \mathrm{d}^3 k \qquad (2\text{-}40)$$

By definition, v_{DR} satisfies the linear relation

$$v_{\text{DR}} = \mu \cdot E \qquad (2\text{-}41)$$

that is Ohm's law. μ is the mean drift velocity per unit field, known as drift mobility, so that the electrical conductivity is expressed (Eqs. (2-30) and (2-36)) in the familiar form

$$\sigma = e \cdot n \cdot \mu = n \cdot e^2 \langle \tau \rangle / m^* \qquad (2\text{-}42)$$

The brackets $\langle \cdots \rangle$ denote averaging in the sense of Eq. (2-40). The magnitude of the low-field drift mobility μ is a measure for the average strength of the scattering which relaxes the deviation $\delta f(k)$ of the actual distribution from the equilibrium distribution $f_0(k)$. The action of the driving field E on the distribution function in k

space and at a given temperature, T_1, together with the "restoring force" represented by momentum scattering, Δk, is shown schematically in Figs. 2-11a and b. The same system at a higher temperature, T_2, generally has a different value of $\delta f(k)$, and correspondingly μ, which "probes" larger k values because of the higher kinetic energies involved.

It is straightforward to show that the temperature dependence of μ can be expressed as

$$\mu(T) \sim T^{\nu} \qquad (2\text{-}43)$$

where ν stands for the exponent in the energy dependence of the scattering time $\tau = \tau(E) \sim E^{\nu}$ (Ziman, 1963).

The second factor in Eq. (2-42) is the density of free carriers in the semiconductor. Doping with shallow impurities of one

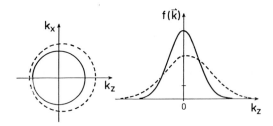

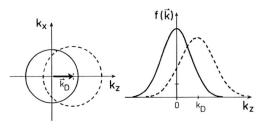

Figure 2-11. Carrier distribution functions in the non-degenerate case at a given lattice temperature (full curves: equilibrium distributions f_0). An electric field in the z-direction drives f out of equilibrium (dashed curves). (a) Acoustic phonon scattering leads to pronounced heating of carriers. (b) Strong inter-carrier scattering (via Coulomb interaction) gives larger drift momentum and less anisotropy of the distribution.

kind – donors or acceptors – allows for preferential population of one band at sufficiently high T with one type of mobile charge carrier, electrons or holes. The majority carrier concentration in a strongly compensated n-type semiconductor is (Blakemore, 1974)

$$n(T) \sim (N_A - N_D) \cdot$$
$$\cdot [1 + (N_A/N_C \cdot \beta) \cdot \exp(E_D/kT)]^{-1} \quad (2\text{-}44)$$

where E_D represents the donor binding energy, and N_D, N_A are the concentrations of impurities, $\beta \sim 1$, and $N_C \sim 4 \times 10^{17}$ $(T/300)^{3/2}$ cm^{-3} in gallium arsenide. At high enough temperatures $n(T) \sim N_D - N_A$.

Each ionized impurity is a localized charge with a long-range static electric field around it. The ensemble of randomly distributed ions causes a static random electric field throughout the crystal which has profound consequences for charge transport: it scatters the mobile carriers very efficiently, especially at low temperatures. The mechanism of ionized impurity scattering is closely related to carrier–carrier scattering (see below) and gives a mobility $\mu_{II} \sim T^{3/2}$ (Seeger, 1985; Blakemore, 1982).

The effective total mobility of carriers which are subject to several different scattering mechanisms can be obtained approximately by adding the reciprocal mobilities for each individual process (Ashcroft and Mermin, 1976). This procedure, known as Matthiessen's rule, is valid when quantum-mechanical interference effects between the scattering mechanisms do not play a role. It is of great practical interest to quote for a given crystal its combined total mobility curve $\mu(T)$. This function is characteristic for each semiconductor material and type of carrier (e, h), and is the basis for all device applications (Hess, 1980; Sze, 1981).

The presence of ionized impurities generally lowers the mobility, especially at low temperatures, and modifies the intrinsic $\mu(T)$ relations. They are tabulated for all common semiconductors (Landolt-Börnstein, 1982). Maximum values occur typically in the temperature range 20–100 K, and vary from low (100 cm^2/Vs) in p-type materials to high (10^6 cm^2/Vs) in n-type crystals with small effective masses, such as the III–V compound GaAs.

Undoped semiconductor crystals have thermally excited e–h pairs of density

$$n_0 \cong N_C \cdot \exp(-E_G/2kT) \quad (2\text{-}45)$$

where N_C has a value of 8×10^{17} cm^{-3} at $m^* = m_0$, and $T = 300$ K. For i-GaAs we find $n_0 \sim 10^6$ cm^{-3}.

The net current flow in this ambipolar case is composed of two counter-propagating particle currents: a hole current and an electron current. Application of Matthiessen's rule would yield the ambipolar mobility

$$\mu_{tot} = \mu_e \cdot \mu_h/(\mu_e + \mu_h) \quad (2\text{-}46)$$

but this overestimates μ_{tot} considerably because of the strong relative momentum relaxation mediated through the attractive e–h Coulomb scattering (Auston and Shank, 1974; Kuhn and Mahler, 1989).

A genuine disadvantage of volume doping is the reduced mobility, especially at low temperatures, i.e., $T < 100$ K. It has been circumvented by the elegant method of so-called "modulation doping" (Dingle et al., 1979; Störmer, 1984), that is, the spatial separation of impurity centers from the mobile carriers. The application of remote-doping techniques has steadily increased the practical mobility limit in n- and p-type GaAs–Ga$_{1-x}$Al$_x$As heterojunctions and quantum well structures from $\sim 10^5$–10^7 cm^2/Vs (Mendez and Wang, 1985; Pfeiffer et al., 1989).

At this point it is useful to extract some numbers from the expressions in Eqs. (2-41)–(2-46). An n-type bulk semiconductor, e.g., GaAs, with relatively small electron effective mass m^* has mobilities for the doping levels $n = 10^{14}$, 10^{16}, 10^{18} cm^{-3} as shown in Fig. 2-12.

At room temperature, $T = 300$ K, we find that $\mu_n = 10^4$ cm^2/Vs, with a mean scattering time $\tau \simeq 3 \times 10^{-13}$ s or a corresponding mean free path of $L \simeq 8$ µm. There is a trade-off between high mobility values and their practical realization in devices, which involves the onset of nonlinearities. As we discussed in the context of the transport equation, a large mean free path, i.e., a long scattering time, causes a much stronger dependence of the mean carrier energy on driving field. "Hot carrier" effects combined with a drastic lowering of the mobility may therefore occur in these materials at relatively low field strengths. Excess energies of $\Delta E \gg kT$ have been found at low temperatures with fields $|E| = 10$ V/cm in a two-dimensional electron gas (Shah et al., 1983).

As already mentioned, all concepts surveyed in this section were originally developed in the classical context of the kinetic theory of gases (Chapman and Cowling, 1939). The central ideas are the mean free path L and mean free time τ between collisions, and both quantities are simply reformulated for charge carriers in semiconductors. There are, however, limitations in the analogy. A plausible criterion for the applicability of the classical concept of τ is the magnitude of the collision rate compared with the rate of energy change divided by the typical excitation energy ΔE

$$\frac{1}{\Delta E} \cdot \frac{dE}{dt} = \frac{1}{\tau_E} \qquad (2\text{-}47)$$

If the collision rate is much smaller than the latter quantity, the idea of individual collisions with well-defined kinematic properties is valid. In the other limit one may expect qualitatively new behavior, so-called quantum transport. This behavior represents conditions which require explicit consideration of the wave-mechanical properties of the electrons and phonons involved in the scattering. Recent developments in microminiaturization of MOSFET devices with channel widths below 1 µm and high field transport in layered structures gave some evidence for such intracollisional effects and the necessity for modifications of the classical Boltzmann transport picture (Barker, 1980; Hu et al., 1989).

When the mean free path L and the confinement dimension L_{typ} become comparable, wave-mechanical size quantization becomes dominant and cannot be ignored. In

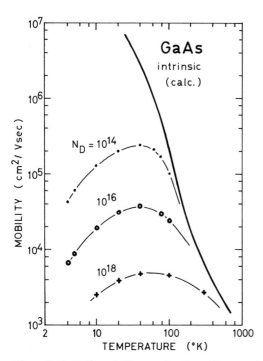

Figure 2-12. Drift mobility as a function of temperature for gallium arsenide under low-field conditions. Doping levels are indicated.

this case interference terms may show up between the scattering amplitudes into different channels (Eq. (2-31 a)), as well as final-state interactions, and the concept of the collision term in the Boltzmann transport equation has to be checked with care. We mention here two eminent examples of quantum transport in confined carrier systems: recent experimental findings on quantized conductance

$$G = g_{\text{Spin}} \cdot \frac{e^2}{h} \qquad (2\text{-}48)$$

in quasi-one-dimensional semiconductor structures (Beenacker and van Houten, 1988; Buettiker, 1988) and the quenching of the normal Hall-effect in thin-wire four-terminal configurations (Roukes et al., 1987 and 1990; see also Chapter 9).

On the other hand, a surprisingly broad class of experiments, even in very small structures, can be successfully described in terms of the simple wave-packet picture. This can be illustrated with the example of "hot carrier injection" through tunnel contacts into regions with a high potential gradient and a collecting electrode within a distance $z \sim L_{\text{free}}$ (Hayes et al., 1985; Heiblum and Fischetti, 1987).

During ballistic flight, see Fig. 2-13, the wave packets spread in space, and are eventually collected as a current in the back electrode. Measurements of the collection efficiency as a function of bias voltage U_{eb} can give direct information on the mean free path of carriers in the field region. Recent experiments in GaAs–GaAlAs heterojunctions at room temperature have reported values of several hundred Ångstroms, in good agreement with the above-mentioned scattering mechanisms. Closely related are recent experiments on ballistic transport of hot electrons through thin gold films via tunnel injection (Bell et al., 1990).

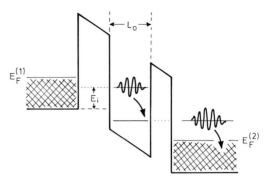

Figure 2-13. Ballistic transport of carrier wavepackets in a structured semiconductor with metal contacts and tunnel injection. The mean free path and the dominant energy relaxation processes can be extracted from geometry L_0 and current/voltage characteristics of the device.

A large part of the ongoing effort towards small devices is focussed at controlled ballistic transport in structures which can be fabricated and electrically controlled on a length scale $L \sim L_{\text{free}}$ (Kelly and Weisbuch, 1986; Capasso, 1989). The necessity to include quantum transport concepts is obvious and is a basic ingredient for further technological progress in micro-electronics.

2.3.2 Phonon Scattering

In this section the dominant carrier-phonon scattering mechanisms are discussed in some detail, so that the reader gets an idea of the relative importance of these mechanisms in different temperature ranges.

Acoustic waves in a solid are accompanied by strain. The central idea of the deformation-potential approach is that the electron-phonon coupling term $\mathscr{H}'$ of Eq. (2-31 a) is given approximately by the shift δE of the conduction band edge that is produced by a homogeneous strain in the crystal. It is assumed that the local strain which is set up by the phonon mode q in a given unit cell equals the strain for the $q=0$ mode. The approach is correct if (i) the wavelength $2\pi/q$ is much larger than the interatomic distance, so that the crystal can be treated as a continuous elastic medium, and (ii) the carrier kinetic energy is small compared with the total width of the energy band.

The deformation-potential mechanism occurs in all solids: any system of electrons and nuclei responds to enforced volume changes δV with changes δE in the total energy. The energy shift can be expressed as follows (Bardeen and Shockley, 1950; Shockley, 1950):

$$\delta E = E_{DP} \cdot \operatorname{div} \boldsymbol{u}(\boldsymbol{r}) \qquad (2\text{-}49)$$

It is obvious that only longitudinal acoustic waves with volume changes $\operatorname{div} \boldsymbol{u} \sim \boldsymbol{q} \cdot \boldsymbol{u}$ are accompanied by a non-vanishing deformation potential constant E_{DP}. Transverse acoustic modes (shear waves) produce no volume changes to first order in $\boldsymbol{u}$. The matrix element for deformation potential coupling is

$$\mathcal{H}' = E_{DP}\, \boldsymbol{q} \cdot \boldsymbol{u} = \qquad (2\text{-}50)$$
$$= E_{DP}\, q\, \frac{\hbar}{(2\varrho\, V \hbar\omega_s)^{1/2}} \cdot \sqrt{N_q + \tfrac{1}{2} \pm \tfrac{1}{2}}$$

Here ϱ is the density of the crystal, $\boldsymbol{q}$ the acoustic phonon mode with energy $\hbar\omega_s$ and the occupation number N_q, and V the normalization volume. If we assume for the moment that carriers and phonons are in equilibrium with each other, the mean free path can be evaluated relatively easily: If the carriers are non-degenerate, i.e., if they obey a Boltzmann distribution, and if the phonon ensemble is also in an equilibrium distribution of the same temperature T, and if the electron-phonon scattering can be treated as quasi-elastic, i.e., $\delta E \ll kT$, one finds for the mean free path of the electrons

$$L = 3\pi\hbar^4 \varrho\, v_s \cdot (4 m_e^* E_{DP}\, k\, T)^{-1} \qquad (2\text{-}51)$$

where v_s is the sound velocity, and m_e^* is the carrier effective mass. L is independent of the carrier energy. For a typical semiconductor with $m_e^* = 0.1\, m_0$, $E_{DP} = 10\,\text{eV}$, $\varrho = 5\,\text{g/cm}^{-3}$, $v_s = 3 \times 10^5\,\text{cm/s}$, and $T = 300\,\text{K}$, a value of $L = 2.3\,\mu\text{m}$ is obtained. From Eq. (2-51) it follows that the scattering time is also the momentum relaxation time and follows

$$\tau \sim E^{-1} \qquad (2\text{-}52)$$

For excess energies $E > k\, T_L$, that is, in the "hot electron" regime, the drift mobility μ_{DP} will therefore decrease with an increasing electric field.

In crystals with no inversion symmetry, there is an additional mechanism of acoustic phonon scattering caused by long-range electrostatic forces. An acoustic wave with wave vector $\boldsymbol{q}$ displaces both sublattices by different amounts and induces a macroscopic dipole moment $\boldsymbol{P}$, which in turn (see Sec. 2.2.2) produces an electric field (piezoelectric effect)

$$E = -(q_{pe}/\varepsilon_0) \cdot \operatorname{div} \boldsymbol{u} \qquad (2\text{-}53)$$

q_{pe} is the piezoelectric constant (Landolt-Börnstein, 1982). The electrostatic energy density of this field represents a change in carrier energy

$$\mathcal{H}' = \delta E = (e \cdot e_{14}/\varepsilon_0\, q) \cdot \operatorname{div} \boldsymbol{u} = $$
$$= (e \cdot e_{14}/\varepsilon_0\, q) \sqrt{k\, T_L/2 \cdot V \cdot v_s} \qquad (2\text{-}54)$$

Here e_{14} is the relevant component of the piezoelectric tensor for zinc blende symmetry, which connects the induced dielectric polarization with the applied mechanical stress.

The phases of both matrix elements (Eqs. (2-50) and (2-54)) with respect to the lattice wave $\boldsymbol{u}(\boldsymbol{r}, t)$ are different and differ by $\pi/2$. This produces an interference between both scattering mechanisms when they couple to the same phonon mode (Rode, 1975).

It is interesting to note that for kinematic reasons there is a threshold energy for carriers to interact with acoustic phonons: the quadratic electron dispersion relation $E(k)$ and the linear phonon dispersion relation intersect in two points only if the condition

$$E > E_{MIN} = m^* \cdot v_s^2/2 \approx 5 \,\mu eV \qquad (2\text{-}55)$$

is fulfilled (Levinson, 1977).

Optical phonon modes, i.e., lattice vibrations with large relative displacements in each unit cell, produce the same two types of interaction potentials. The short-range effect is called optical deformation potential scattering, and the long-range electrostatic part is the polar optical (or Fröhlich) scattering. Since optical phonon energies are typically large, $\hbar \omega_{LO} \sim 20-50 \,meV$, and the coupling is relatively strong, both processes provide a very efficient energy relaxation mechanism for carriers, provided their kinetic energy exceeds $\hbar \omega_{LO}$. Practically all optical and transport experiments with monoenergetic carrier injection show the characteristic threshold for optical phonon emission (Tsui, 1974; Heiblum and Fischetti, 1987; Becker et al., 1986).

The strong coupling of polar-optical phonons to electrons, the so-called polaron problem, requires more than first-order perturbation theory for an adequate treatment (Feynman, 1972; Devreese, 1976). The matrix element of the interaction operator is

$$\mathcal{H}' = e \cdot \sqrt{\frac{\hbar \omega_{LO} \cdot (\varepsilon_0 - \varepsilon_\infty)}{2 \cdot V \cdot \varepsilon_0 \cdot \varepsilon_\infty}} \cdot$$
$$\cdot (1/q) \cdot \sqrt{N_q + \tfrac{1}{2} \pm \tfrac{1}{2}} \qquad (2\text{-}56)$$

and favors small changes of carrier momentum. Since the energy transfer $\Delta E = \hbar \omega_{LO}$ is always high ($\Delta E > kT$ in all practical cases), a meaningful momentum

relaxation time in the sense of Eq. (2-49) cannot be defined at low temperatures. Both mechanisms of optical phonon coupling with their high threshold energy tend to dominate the total mobility curves $\mu(T)$ if T is comparable with, or larger than, the Debye temperature, which is $400-800 \,K$ for the common semiconductors.

In some cases the coupling constants for deformation potential and polar scattering were determined empirically from a variety of different experiments and have been checked against each other. The range of experiments includes electrical transport with quantitative mobility curves $\mu(T)$ (Rode, 1975), several methods of optical spectroscopy involving mechanical stress tunneling spectroscopy (Tsui, 1974), free-carrier scattering, and inelastic light scattering (Raman- and Brillouin scattering), with absolute cross sections related to the electron-phonon matrix elements (Grimsditch et al., 1979).

Despite this large number of independent experiments, there are still large ambiguities in the assignment of absolute values to certain coupling parameters. Even very basic quantities, e.g., the individual deformation potential constants E_{DP} for conduction and valence bands in common semiconductors, are only known with moderate precision. The example GaAs is discussed by Price (1985).

2.3.3 Carrier-Carrier Scattering

The scattering of an electron by the change in $V_c(r)$ due to the presence of *other* electrons is qualitatively different from phonon scattering. The latter process couples delocalized electronic states with lattice vibrations spread uniformly over the crystal volume, and the relatively weak interaction between electrons and ions occurs simultaneously in all unit cells. The former process, that is, the mutual interaction between mobile carriers – electrons, holes, or both – depends on the inverse relative distance between both scatterers and may therefore be deeply inelastic. As a consequence, it is difficult to describe the Coulomb scattering between carriers in a perturbative way.

A standard approach along the lines of the tractable two-body problem is to treat it as a simple binary collision process in relative coordinates. At all times only one nearest neighbor is included. Obviously this approach overestimates the total scattering rate. Higher-order effects, such as the presence of a third scatterer during the collision, as well as screening in the plasma of mobile carriers and the coupling with collective electron motions (plasmons), are disregarded altogether. Nevertheless, all practical efforts have used this approximation scheme, and reasonable agreement with experiments, at least to a limited extent in some cases, has been found (Fermi, 1948; Appel, 1961; Platzman and Wolff, 1973).

In the case of e–h scattering, the problem is closely connected with the spectrum of e–h pair (or "exciton") continuum states (see Sec. 2.2.3 above). For the sake of simplicity, transport theory usually starts ad hoc with plane wave states and tackles the problem in a perturbative way: the screened Coulomb interaction $V(r)$ between two carriers in real space is taken in the Born approximation in order to calculate the scattering rate. The rate of colisions between the two particles, with initial wave vectors k_1 and k_2, and final wave vectors k_1' and k_2', is then given by the following matrix element (Schiff, 1968):

$$\langle k_1' k_2' | e \cdot V(r) | k_1 k_2 \rangle =$$

$$= \frac{1}{V^2} \iint \exp[-i(k_1' r_1 + k_2' r_2)] \frac{e^2 \exp(-|r_1 - r_2|/\lambda)}{4\pi \varepsilon_0 \varepsilon |r_1 - r_2|} \exp[-i(k_1 r_1 + k_2 r_2)] \, dr_1 \, dr_2 \quad (2\text{-}57)$$

This expression is valid for two electrons in parabolic bands (i.e., in effective-mass-approximation) and only for moderately high energies, because of the static screening used. The collision rate in Eq. (2-57) depends only on the k-vector difference between both scatterers. The usual two-body transformation of coordinates r_1, r_2 and k_1, k_2, k_1', k_2' into the center-of-mass reference frame simplifies this expression into

$$\langle k_1' k_2' | e V(r) | k_1 k_2 \rangle =$$

$$= \frac{e^2}{\varepsilon \cdot V} \cdot \frac{1}{|k_{12}' - k_{12}|^2 + \lambda^{-2}} \quad (2\text{-}58)$$

and now shows the complete equivalence of the initial problem with that of a particle of mass μ (reduced mass) colliding with a fixed elementary charge, i.e., the familiar Rutherford scattering (Blatt, 1968). From the well-known scattering cross section for this configuration, one can easily extract *relative* momentum and energy relaxation rates (Landau, 1936). In the case of indistinguishable particles (two electrons, or two holes collide), the rates have to be slightly modified because of interference terms in the final state.

The scattering rate following from Eq. (2-58) has been evaluated in detail for two limiting cases:

(a) one "hot" carrier interacts with a sea of "cold" equilibrium carriers of given density n (Larkin, 1960).

(b) an initially monoenergetic distribution of carriers relaxes towards equilibrium (Hearn, 1980; Yoffa, 1980; Asche and Sarbej, 1984).

For case (a) the energy relaxation rate is

$$\frac{dE}{dt} = -\frac{n \cdot e^4}{4\pi\varepsilon^2 \sqrt{2m^* E}} \cdot \mathscr{L}(E, \lambda) \quad (2\text{-}59)$$

Here n is the carrier density, ε the static dielectric constant, m^* the carrier effective mass, and E the instantaneous kinetic energy of the incident carrier, presumed to be much larger than the thermal energy kT. $\mathscr{L}(E, \lambda)$ is a weakly energy-dependent and density-dependent function which accounts for screening. For low densities and in the range of practical interest ($n \sim 10^{14} - 10^{17}$ cm^{-3}) one finds $\mathscr{L} \simeq 1$. From Eq. (2-58) we obtain the relaxation rate τ^{-1} for the loss of relative momentum, i.e., the momentum loss in the center-of-mass coordinate system:

$$\tau_m^{-1} = - \frac{n \cdot e^4}{2 \cdot \sqrt{2} \cdot \pi \, (\varepsilon \cdot \varepsilon_0)^2 \sqrt{m_e^*} \, E^{3/2}} \cdot \mathscr{L}(E, \lambda) \tag{2-60}$$

Taking $\varepsilon = 10$, $E = 100$ meV, $m^* = 0.1 \, m_0$ we find that $\tau \sim 100$ ps at $n = 10^{17}$ cm^{-3}.

There are only a few direct experimental checks on the validity of this description. Transport coefficients have been calculated with explicit inclusion of e–e scattering in the BTE (Wingreen et al., 1986). The stopping power of fast electrons in metals and inelastic electron transmission through thin semiconductor crystals (e.g., in electron transmission microscopy) have been measured and compared to the rates discussed here (Elkomoss and Pape, 1982). The approach leading to Eq. (2-58) neglects all collective motions of carriers – plasmons and phonon/plasmon coupled modes – which may as a whole act as very efficient momentum and energy-relaxing scatterers; thus the total scattering rate will be underestimated (Petersen and Lyon, 1990). On the other hand, the model takes into account only the nearest scatterer (the deepest inelastic) and neglects the screening influence of all other electrons, which tend to reduce the scattering rate. Finally, the pair correlation of electrons and holes is disregarded completely. All three effects enter the final rate in qualitatively different ways. Obviously more quantitative studies are needed to understand the process of carrier–carrier scattering in optically excited semiconductors, e.g., in the context of ultrafast carrier dynamics (see below).

2.4 Nonlinear Optics and High Field Transport

In the limit of small driving fields, $E \rightarrow 0$, observation tells us that all transport and optical properties are independent of E. It has been shown that such a "linear response" behavior follows from very general grounds (Kubo, 1957). If one considers, say, the ubiquitous scattering mechanisms discussed in Sec. 2.3, the range of "small" electric fields in semiconductors extends from $\sim$ V/cm at helium temperature and low doping levels to several kV/cm at room temperature. When the driving field is no longer small and becomes a sizeable fraction of the internal crystal field of $\sim 10^8$ V/cm, which characterizes the ground state properties of the system, nonlinear response is to be expected.

At first glance, it is surprising that metals usually do not show deviations from Ohm's law, even under high field conditions. The large number of carriers involved in a typical metal – there are $D(E) \cdot dE \sim 10^{23}$ cm^{-3} electron states in the energy range $dE = kT$ around the Fermi energy E_F at room temperature – leads to an important consequence: the excess kinetic energy of the ensemble of carriers is tightly coupled to the lattice. This prevents the electrons from getting "hot" even for extremely high current densities up to the order of 10^{10} A/cm^2, which corresponds to a driving field of $\sim 10^4$ V/cm in good conductors such as Cu or Au

(Schoenlein et al., 1988; Heinrich and Jantsch, 1976).

Semiconductors, on the other hand, are generally prepared with much smaller densities of active carriers. Doping with suitable impurities controls the carrier density in the range $n_e = 10^{15} - 10^{18}$ cm^{-3}, i.e., approximately five to eight orders of magnitude less than in metals. The dissipation of momentum and energy has to be channeled through this "dilute" ensemble of carriers. At a given current density the mean carrier velocity – and hence the carrier kinetic energy E_{kin} – is higher by the same factor. This fact explains why "hot carrier" phenomena are easily observed in semiconductors for quite modest electric fields ranging from several V/cm at low temperatures to kV/cm at room temperature (Shockley, 1950; Conwell, 1967; Hess, 1980; Weisbuch, 1986).

If the same comparison is made for a fixed driving field strength, we find that the phonon system acts as the "bottle neck": in metals, all the $\sim 10^{23}$ cm^{-3} electrons feed energy into the lattice and produce Joule heating which – in the stationary case – must be transferred via phonon transport through the crystal boundaries into the heat sink. Under such conditions, phonon scattering controls the onset of nonlinearities. In a semiconductor the lattice-vibrational modes need to take much less total power at the same electric field, and it is the carriers themselves which determine the nonlinear properties. However, at extremely low temperatures, in the milli-Kelvin range, hot electron effects may also be observed in metals (Roukes et al., 1985).

For a more detailed discussion of nonlinear transport it is useful to expand the distribution functions f_e, f_h in a power series with the driving field as smallness parameter. The procedure is justified for the common scattering mechanisms which involve acoustic phonons. However, if only one highly inelastic process is present, e.g., optical phonon scattering (see above Sec. 2.3.3), the concept does not work (Peierls, 1955). An arbitrarily small electric field is now sufficient to drive the system into the nonlinear regime with $E_{kin} > \hbar \omega_{LO}$. This threshold marks the onset of real phonon emission, i.e., the process of "polaron stripping" (Feynman, 1972).

The treatment of nonlinear optical properties under nonresonant conditions, that is, in the transparent part of the optical spectrum, starts with a similar expansion: The induced response P is written as a power series in the driving field (Bloembergen, 1962). The description of nonlinear optical properties in terms of coefficients $\chi^{(1)}$, $\chi^{(2)}$, etc., has turned into an enormously successful concept for practical applications (Yariv, 1982; Levensohn, 1982).

The situation under resonant conditions, that is, for driving frequencies at and above the fundamental gap, is more complicated and much less well understood. An example is the edge spectrum of a direct gap semiconductor: It is dependent on the degree of excitation. The presence of electron–hole pairs or of phonons strongly influence the shape of $\chi(\omega)$ at and around $\hbar \omega_g$. The absorption of light, or heat, and also the presence of impurities and defects changes the energy spectrum $E_n(k)$, alters the occupation of electronic states in the crystal, and modifies its optical properties, especially at and below E_G.

Some of the optically coupled states may be driven into high occupancy or even saturation, and the concept of a series expansion of the susceptibility becomes questionable. Ab-initio treatments of this case have begun to emerge; they are closely connected with the above-mentioned high-field quantum transport phenomena (Haug,

1988; Combescot, 1988; Zimmermann, 1990).

A peculiar problem is the inclusion of wave-mechanical ingredients, such as the phase of the electron wave function, into a description with local distribution functions $f_e(k,r)$ and $f_h(k,r)$ (see Sec. 2.3). Wigner has introduced a natural generalization of these functions in the form of

$$f_W(k,r) = \frac{1}{\pi} \int \psi^*\left(r - \frac{R}{2}\right) \psi\left(r + \frac{R}{2}\right) \cdot$$
$$\cdot \exp(-i k \cdot R) \, d^3R \qquad (2\text{-}61)$$

Here $\psi(r)$ is the wave function of the particle under consideration. This so-called Wigner phase-space distribution function f_W yields $|\psi|^2$, when integrated with respect to k, and $|\phi(k)|^2$, when integrated with respect to r. It incorporates certain wave-like properties of the particle (electron, hole). However, its interpretation as a local probability is restricted, because it is in general not non-negative (Wigner, 1932).

The single-particle approximation described in Sec. 2.2 works well for static screening and for not overly abrupt spatial variations in the perturbing potential φ_{ext}. At densities around $n_c \sim a_B^{*-3}$, the screening of the e–h pair interaction leads to a screening length of the order of a_B^*. In this intermediate density regime, a more refined treatment of screening is necessary. There are extensions of the Hartree-Fock self-consistent field equations which include screening from the beginning and account for the response of the electron–hole plasma to a given charge carrier which is at rest or is kept in motion by an external perturbation φ_{ext} with frequency ω (Collet, 1989). The treatment given by Lindhard considers only those contributions of the induced screening charge density around the carrier which are linear in the total potential $\varphi_{tot} = \varphi_{ext} + \varphi_{ind}$. The plasma is assumed to be in an equilibrium state described by Fermi functions f_e, f_h. Application of first-order time-dependent perturbation theory leads to an effective dielectric function of the form (Wooten, 1972)

$$\varepsilon(\omega,q) = 1 + \frac{e^2}{\varepsilon_0 q^2} \cdot$$
$$\cdot \frac{1}{V} \sum_k \frac{f_0(k) - f_0(k+q)}{E(k+q) - E(k) - \hbar\omega + i\hbar\eta} \qquad (2\text{-}62)$$

Here ω is the frequency of the external perturbation, q is its wave vector, and the sum extends over all occupied single-electron levels $E(k)$ of the unperturbed system, and f_0 is the equilibrium distribution function at temperature T. The derivation assumes that the perturbation is stationary, and agrees for $q \to 0$ with the Thomas-Fermi screening discussed in Sec. 2.2.

A direct consequence of Eq. (2-62) is that for longer distances the screened Coulomb potential of a point charge in an e–h plasma has characteristic oscillations in space with wavelength $\lambda \sim \pi/k_F$ (Friedel, 1958). Treatments of screening beyond the framework of the so-called random phase approximation (RPA) consider the distribution functions f_e, f_h as dynamic variables. A self-consistency procedure leads to the concept of a dynamically screened dielectric function (Schäfer, 1988; Haug, 1988; Hu et al., 1989).

2.4.1 Carrier Distributions Far from Equilibrium

The injection of real e–h pairs into a semiconductor by means of short pulse optical excitation creates distributions which are generally far from equilibrium. If the band structure $E(k)$ and the corresponding joint density of states is known, the resulting initial distribution can be inferred directly from the light pulse spectrum.

In the case of a direct-gap material with $E_G = 1.5\,\text{eV}$, $m_e^* = 0.1\,m_0$, $m_h^* = 0.5\,m_0$, and with a 100 fs light pulse with a center energy of 2 eV, the combined density of states is $D_{e-h} \sim 10^{20}\,\mathrm{cm^{-3}\,eV^{-1}}$. With the known absorption coefficient of $5 \times 10^4\,\mathrm{cm^{-1}}$, this produces distributions of electrons $f_e \sim 0.05$ and holes $f_h \sim 0.01$ per Joule/cm^2 excitation energy. Saturation of the interband transition because of the Pauli principle occurs for these conditions at $n_c \sim 10^{17}\,\mathrm{cm^{-3}}$. This so-called "bleaching" or "blocking" of transitions has been observed recently in a series of elegant experiments (Chemla and Miller, 1985; Oudar et al., 1985; Knox et al., 1988 a, b).

The relaxation behavior of optically excited non-equilibrium pair distributions in the density regime $10^{15}-10^{18}\,\mathrm{cm^{-3}}$ is at present not well understood (Kash et al., 1989; Petersen et al., 1990). Strong Coulomb scattering causes an extremely rapid spread of the width of the distribution in energy even at low densities of $\sim 10^{16}\,\mathrm{cm^{-3}}$. The rate of increase in relative energy spread is of the order of 1 eV/ps for the parameters given above. The rates scale approximately with the initial excess energy dE of the distribution (Böhne et al., 1990).

Under certain conditions, high-field transport may also produce strong deviations from equilibrium. Selective population of subsidiary minima in the band structure (the Gunn-effect), and thresholds in electron–phonon coupling or inter-band Auger processes may lead to peaked carrier distributions with very similar relaxation behavior.

Qualitatively new phenomena are expected when the carrier distributions undergo rapid transients, either by application of a fast electric field pulse, or by optical injection with strong density gradients: the anisotropy in k space leads to different time constants of energy and momentum relaxation, the so-called "overshoot" effect (Seeger, 1985).

2.4.2 Carrier Dynamics on Ultrashort Time Scales

Recent advances in the design of mode-locked dye lasers have enormously expanded the range of available pulse durations. The production of ultrashort laser light pulses with less than 10 fs duration has been demonstrated (Becker et al., 1988) and spectroscopy with time-resolution in the 100 fs range is now performed routinely. The breakthrough was triggered by the optimization of dispersion compensation in the laser resonator, an improved understanding of self-phase modulation in saturable absorbers, and the generation of synchronized "white" continua with a comparably short pulse. These advances have opened up new possibilities for studying carrier dynamics directly in the time domain with a time resolution down to a few femtoseconds. This corresponds to a frequency bandwidth of more than ≈ 1000 Thz. Before the invention of femtosecond light pulses this frequency range was not directly accessible; wire-bound electronic circuits extend to approximately 10 Ghz. This limit is imposed by practical restriction in geometry and the unavoidable scaling of charge capacitance effects in small-scale devices (Kelly and Weisbuch, 1986).

Meanwhile, a number of classical experimental techniques – e.g., nonlinear optical transmission, reflectance, and emission spectroscopy – have been implemented with ultrashort light pulses and applied to semiconductors and metals (Grischkowsky, 1985; Lin et al., 1988; Schoenlein et al., 1987, 1988). The dynamics of carrier ensembles became directly accessible for the first time, and relevant scattering rates were deduced from time-resolved spectra (Haight and Silberman, 1989; Cho et al., 1990).

The aspect of mutual coherence between the driving light field, and the optically coupled electron–hole pairs has been worked out in a series of experiments on the optical (AC) Stark effect with $\hbar\omega < E_G$ (Froehlich et al., 1985; Mysyrowicz et al., 1986). Resonance effects, level shifts and polarization selection rules have recently been investigated in theory and discussed in some detail (Joffre et al., 1989; Schmitt-Rink, 1990).

A common guideline in the investigations of carrier dynamics is the concept of longitudinal and transverse relaxation times of electronic states. This concept had originally been developed for isolated two-level systems (spins, atoms) interacting with an external driving field (Bloembergen, 1962). However, it needs to be modified to be applicable to a dense ensemble of atoms (i.e., a solid) and its resonant interaction with light. What is needed is a consistent description of the macroscopic electrodynamical response of the crystalline solid which also incorporates the correct quantum mechanical "two-level" behavior on the atomic scale.

One approach starts with the so-called Maxwell–Bloch equations and introduces appropriate electron–hole pair variables to describe coherent pulse propagation and damping at resonance, $\hbar\omega > E_G$, in semiconductors (Stahl and Balslev, 1986). The term "optically coupled states" has been coined to stress the analogy with atomic physics. Up to now, however, most experimental observations have been interpreted simply in terms of phenomenological longitudinal and transverse relaxation times in the language of isolated two-level atoms (Oudar et al., 1985; Haug, 1988).

In the resonant case, $\hbar\omega > E_G$, intense coherent light pulses with occupation numbers $N_{PHOTON} \gg 1$ per photon mode create an initial electron–hole pair polarization P_{eh} (see Sec. 2.2) with well-defined temporal and spatial coherence. During and after build-up of P_{eh} the phase correlation between driving field and induced polarization is relaxed by carrier momentum scattering. This transition from coherently driven pair states into the "random motion" of carriers in an e–h plasma is not well understood at present. The former regime is related to ballistic transport, and special effects such as photon echo, optical nutation, and optically induced band gaps are associated with it. The latter condition is characteristic of traditional charge transport, with no phase correlation between the carriers.

2.5 Conclusion

One point to be noted in concluding is that the concepts discussed in this article were all developed in the context of ideal, bulk crystalline solids. Their immediate application to semiconductors, and especially to microstructures with mesoscopic length scales, necessarily requires caution and modifications. Some of the assumptions were made for the sake of simplicity and cannot represent realistic experimental conditions.

At present, the views on nonlinear charge transport in small semiconductor structures and their optical properties on the ultrashort time scale are not fully consistent. The concepts of ballistic transport of electrons on the one hand and coherent electron–hole pair dynamics on the other are closely related to each other but are usually presented in very different formalisms. There is still much to be done to fill this gap. It is thus safe to say that the field is far from being mature.

2.6 References

Abstreiter, G. (1984), in: *Advances in Solid State Physics Vol. 24:* Grosse, P. (Ed.). Braunschweig: Vieweg.

Abstreiter, G., Cardona, M., Pinczuk, A. (1984), in: *Topics in Applied Physics Vol. 54.* Berlin, Springer Verlag, p. 5.

Anderson, P. W. (1963), *Concepts in Solids.* Reading: Benjamin, Ch. 2.

Appel, J. (1961), *Phys. Rev. 122,* 1760.

Appel, J., Overhauser, A. W. (1982), *Phys. Rev. B 26,* 507.

Asche, M., Sarbej, O. G. (1984), *Phys. Stat. Sol. (b) 126,* 607.

Ashcroft, N. W., Mermin, N. D. (1976), *Solid State Physics.* Philadelphia: Holt, Rinehart and Winston, Chs. 22 and 23.

Auston, D. H., Shank, C. V. (1974), *Phys. Rev. Lett. 32,* 1120.

Auston, D. H., Cheung, K. P., Valdmanis, J. A., Kleinman, D. A. (1984), *Phys. Rev. Lett. 53,* 1555.

Bardeen, J., Shockley, W. (1950), *Phys. Rev. 80,* 72.

Barker, J. R. (1980), in: *Physics of Nonlinear Transport in Semiconductors, NASI Series B Vol. 52.* New York: Plenum Press, p. 126.

Bassani, F., Pastori Paravicini, G. (1975), *Electronic States and Optical Transitions in Solids.* Oxford: Pergamon Press, Chs. 1 and 5.

Bastard, G. (1988), *Wave Mechanics applied to Semiconductor Heterostructures.* Paris: Les Editions de Physique.

Becker, P. C., Fragnito, H. L., Brito Cruz, C. H., Fork, R. L., Cunningham, J. E., Henry, J. E., Shank, C. V. (1988), *Phys. Rev. Lett. 61,* 1647.

Becker, W., Gerlach, B., Hornung, T., Ulbrich, R. G. (1986), in: *Proc. 18th Int. Conf. Physics of Semiconductors, Stockholm, 1986:* Engstrom, O. (Ed.). Singapore: World Science Publishers, p. 1713.

Beenacker, C. W. J., van Houten, H. (1988), *Phys. Rev. Lett. 60,* 2406.

Bell, L. D., Hecht, M. H., Kaiser, W. J., Davis, L. C. (1990), *Phys. Rev. Lett. 64,* 2679.

Blakemore, J. (1974), *Solid State Physics.* Philadelphia: Saunders, Ch. 4.

Blakemore, J. (1982), *J. Appl. Phys. 53,* R123.

Blatt, F. J. (1968), *Physics of Electronic Conduction in Solids.* New York: McGraw-Hill.

Bloembergen, N. (1962), *Nonlinear Optics.* New York: Benjamin.

Böhne, G., Sure, T., Ulbrich, R. G., Schaefer, W. (1990), *Phys. Rev. B 41,* 7549.

Buettiker, M. (1988), *Phys. Rev. B 38,* 9375.

Capasso, F. (1989), in: *Physics of Quantum Electron Devices (Springer Series in Electronics and Photonics Vol. 28).* Berlin: Springer, Ch. 1.

Cardona, M. (1969), in: *Optical Properties of Solids:* Nudelman, S., Mitra, S. S. (Eds.). New York: Plenum, Ch. 6.

Cardona, M., Christensen, N. E., Fasol, G. (1986), *Phys. Rev. Lett. 56,* 2831.

Chapman, S., Cowling, T. G. (1939), *Mathematical Theory of Non-Uniform Gases.* London: Cambridge University Press.

Chemla, D., Miller, D. A. B. (1985), *J. Opt. Soc. Am. B 2,* 1155.

Cho, K. (1979), *Excitons,* in: *Springer Topics in Current Physics Vol. 14.* Berlin: Springer.

Cho, G. C., Kuett, W., Kurz, H. (1990), *Phys. Rev. Lett. 65,* 764.

Cohen, M. L., Chelikowski, J. R. (1988), *Electronic Structure and Optical Properties of Semiconductors.* Berlin: Springer, Ch. 1.

Collet, J. H. (1989), *Phys. Rev. B 39,* 7659.

Combescot, M. (1988), *Solid State Electron. 31,* 657.

Combescot, M., Bok, J. (1983), in: *Cohesive Properties of Semiconductors under Laser Irradiation, NATO ASI series Vol. E 69.* The Hague: Nijhoff Publishers, p. 289.

Comte, C., Nozieres, P. (1982), *J. Physique 43,* 1069.

Conwell, E. M. (1967), *High Field Transport in Semiconductors.* New York: Academic Press, Ch. III.

Debye, P., Hückel, E. (1923), *Phys. Z. 24, 185;* ibid. 305.

Devreese, J. T., Evrard, R. (1976), in: *Linear and Nonlinear Electronic Transport in Solids:* Devreese, J. T., van Doren, V. E. (Eds.). New York: Plenum Press, p. 91.

Dimmock, J. O. (1967), in: *Semiconductors and Semimetals Vol. 3:* Willardson, R. K., Beer, A. C. (Eds.). New York: Academic Press, Ch. 5.

Dingle, R. (1975), in: *Advances in Solid State Physics Vol. 15:* Queisser, H. J. (Ed.). Braunschweig: Vieweg, p. 21.

Dingle, R., Störmer, H. L., Gossard, A. C., Wiegmann, W. (1979), *Appl. Phys. Lett. 3,* 665.

Dow, J. (1976), in: *New Developments in the Optical Properties of Solids:* Seraphin, B. (Ed.). Amsterdam, North Holland, Ch. 2.

Elkomoss, S. G., Pape, A. (1982), *Phys. Rev. B 26,* 6739.

Elliott, R. J. (1957), *Phys. Rev. 108,* 1384.

Elliott, R. J. (1963), in: *Polarons and Excitons:* Kuper, C. G., Whitfield, G. D. (Eds.). Edinburgh: Oliver and Boyd, p. 269.

Esaki, L. (1983), in: *Recent Topics in Semiconductor Physics:* Kamimura, H., Toyozawa, Y. (Eds.). Singapore: World Scientific.

Fattinger, Ch., Grischkowsky, D. (1989), *Appl. Phys. Lett. 54,* 490.

Fermi, E. (1940), *Phys. Rev. 57,* 485.

Feynman, R. P. (1972), *Statistical Mechanics.* Reading: Benjamin, Ch. 2.

Friedel, J. (1958), *Nuovo Cimento Suppl. 7,* 287.

Froehlich, D., Noethe, A., Reimann, K. (1985), *Phys. Rev. Lett. 55,* 1335.

Grimsditch, M. H., Olego, D., Cardona, M. (1979), *Phys. Rev. B 20,* 1758.

Grischkowsky, D. R. (1985), *J. Opt. Soc. B2*, 582 and references therein.

Haight, R., Silberman, J. A. (1989), *Phys. Rev. Lett. 62*, 815.

Haken, H. (1973), *Quantum Field Theory of Solids.* Stuttgart: Teubner, Ch. 2.

Hanke, W. (1978), in: *Advances in Solid State Physics Vol. 19.* Braunschweig: Vieweg, p. 43.

Harrison, W. A. (1980), *Electronic Structure and the Properties of Solids.* San Francisco: W. H. Freeman, Ch. 9.

Haug, H. (1985), *J. Lumin. 30*, 171.

Haug, H. (1988), in: *Optical Nonlinearities and Instabilities in Semiconductors:* Haug, H. (Ed.). New York: Academic Press, Ch. 3.

Haug, H., Schmitt-Link, S. (1985), *J. Opt. Soc. Am. B2*, 1135.

Hayes, J. R., Levi, A. F. J., Wiegmann, W. (1985), *Phys. Rev. Lett. 54*, 1570.

Hearn, C. J. (1980), in: *Physics of Nonlinear Transport in Semiconductors, NASI Series B Vol. 52.* New York: Plenum Press, p. 153.

Heiblum, M., Fischetti, M. V. (1987), in: *Physics of Quantum Electron Devices (Springer Series in Electronics and Photonics Vol. 28).* Berlin: Springer.

Heinrich, H., Jantsch, W. (1976), *Phys. Letters 57A*, 485.

Henry, C. H., Nassau, K. (1970), *Phys. Rev. B1*, 1628.

Hermann, C., Weisbuch, C. (1977), *Phys. Rev. B15*, 823.

Hess, K. (1980), in: *Physics of Nonlinear Transport in Semiconductors, NASI Series B Vol. 52.* New York: Plenum Press, p. 32.

Hopfield, J. J. (1969), *Phys. Rev. 182*, 945.

Hu, Ben Yu-Kuang, Sarker, Sanjoy K., Wilkins, John W. (1989), *Phys. Rev. B39*, 8468.

Hwang, C. J. (1972), *J. Appl. Phys. B6*, 1355.

Jackson, J. D. (1975), *Classical Electrodynamics.* New York: John Wiley & Sons, Ch. 7.

Joffre, M., Hulin, D., Migus, A., Combescot, M. (1989), *Phys. Rev. Lett. 62*, 74.

Kane, E. O. (1957), *J. Phys. Chem. Solids 1*, 249.

Kane, E. O. (1966), in: *Semiconductors and Semimetals Vol. 1:* Willardson, R. K., Beer, A. C. (Eds.). New York: Academic Press, p. 75.

Kash, J. A., Ulbrich, R. G., Tsang, J. C. (1989), *Solid State Electron. 32*, 1277.

Kelly, M. J., Weisbuch, C. (1986), *The Physics and Fabrication of Microstructures and Microdevices*, in: *Springer Proceedings in Physics Vol. 13.* Berlin: Springer, Part 2.

Knox, W. H., Chemla, D. S., Livescu, G. (1988a), *Solid-State Electronics 31*, 425.

Knox, W. H., Chemla, D. S., Livescu, G., Cunningham, J. E., Henry, J. E. (1988b), *Phys. Rev. Lett. 11*, 1290.

Kohl, M., Heitmann, D., Grambow, P., Ploog, K. (1989), *Phys. Rev. Lett. 63*, 2124.

Kubo, R. (1957), *J. Phys. Soc. Japan 12*, 570.

Kuhn, T., Mahler, G. (1989), *Phys. Rev. B39*, 1194.

Lampel, G. (1974), in: *Proceedings of the 12th Int. Conf. on the Physics of Semiconductors.* Stuttgart: Teubner, p. 743.

Landau, L. D., Pomerantschuk, I. (1936), *Phys. Z. Sowjetunion 10*, 649.

Landau, L. D., Lifschitz, E. M. (1966), *Lehrbuch der Theoretischen Physik, Vol. III.* Berlin: Akademie-Verlag, p. 122ff.

Landolt-Börnstein (1982), *Numerical Data and Functional Relationships in Science and Technology, Group III, Vols. 17a, b.* Berlin: Springer.

Larkin, A. I. (1960), *Sov. Phys. JETP 37*, 186.

Lawaetz, P. (1971), *Phys. Rev. B4*, 3760.

Levenson, M. D. (1982), *Introduction to Nonlinear Laser Spectroscopy.* New York: Academic Press, Ch. 3.

Levinson, I. B. (1977), *Sov. Phys. Solid State 19*, 1500.

Lin, W. Z., Schoenlein, R. W., Fujimoto, J. G., Ippen, E. P. (1988), *IEEE J. Quantum Electronics 24*, 267.

Mendez, E. E., Wang, W. I. (1985), *Appl. Phys. Lett. 46*, 1159.

Mott, N. F. (1961), *Phil. Mag. 6*, 287.

Mysyrowicz, A., Gibbs, H. M., Peyghambarian, N., Masselink, W. T., Morkoc, H. (1986), in: *Ultrafast Phenomena V.* Berlin: Springer, p. 197.

Mysyrowicz, A., Hulin, D., Antonetti, A., Migus, A., Masselink, W. T., Morkoc, H. (1986), *Phys. Rev. Lett. 56*, 2748.

Nakajima, S., Toyozawa, Y., Abe, R. (1980), *The Physics of Elementary Excitations*, in: *Springer Series in Solid-State Sciences Vol. 12.* Berlin: Springer, Ch. 2.

Oudar, J. L., Hulin, D., Migus, A., Antonetti, A., Alexandre, F. (1985), *Phys. Rev. Lett. 55*, 2074.

Paige, E. G. S. (1964), *Prog. Semicond. 8*, 1.

Parmenter, R. H. (1955), *Phys. Rev. 100*, 573.

Peierls, R. E. (1955), *Quantum Theory of Solids.* London: Oxford University Press, Ch. 6.3.

Petersen, C. L., Lyon, S. A. (1990), *Phys. Rev. Lett. 65*, 760.

Pfeiffer, L., West, K. W., Störmer, H. L., Baldwin, K. W. (1989), *Appl. Phys. Lett. 55*, 1888.

Planel, R., Nawrocki, M., Benoit à la Guillaume, C. (1977), *Phys. Rev. B15*, 5869.

Platzman, P. M., Wolff, P. A. (1973), in: *Solid State Physics Vol. 13 (Waves and Interactions in Solid State Plasmas).* New York: Academic Press, Ch. 1.

Price, P. (1985), *Phys. Rev. B32*, 2643.

Rashba, E. I., Sturge, M. D. (1982), *Excitons.* Amsterdam: North-Holland.

Reggiani, L. (1985), *Hot-Electron Transport in Semiconductors (Springer Topics in Applied Physics Vol. 58).* Berlin: Springer, p. 66ff.

Resta, R. (1983), *Phys. Rev. B27*, 3620.

Rice, M. (1977), in: *Solid State Physics 32*, 1.

Rode, D. L. (1975), in: *Semiconductors and Semimetals Vol. 10:* Willardson, R. K., Beer, A. C. (Eds.). New York: Academic Press, Ch. 1.

Roukes, M. L., Freeman, M. R., Germain, R. S., Richardson, R. C., Ketchen, M. B. (1985), *Phys. Rev. Lett. 55,* 422.

Roukes, M. L., Scherer, A., Allen jr., S. J., Craighead, A. G., Ruthen, R. M., Beebe, E. D., Harbison, J. P. (1987), *Phys. Rev. Lett. 59,* 3011.

Roukes, M. L., Scherer, A., van der Gaag, B. P. (1990), *Phys. Rev. Lett. 64,* 992.

Rühle, W. (1981), *Phys.* Rev. B 22, 1255.

Schäfer, W. (1988), in: *Optical Nonlinearities and Instabilities in Semiconductors:* Haug, H. (Ed.). New York: Academic Press, Ch. 6.

Schiff, L. I. (1968), *Quantum Mechanics, 3rd Ed.* New York: McGraw-Hill, Ch. 10.

Schmitt-Rink, S. (1990), *Phys. Rev. B 41,* 2247.

Schoenlein, R. W., Lin, W. Z., Fujimoto, J. G., Eesley, G. L. (1987), *Phys. Rev. Lett. 58,* 1680.

Schoenlein, R. W., Lin, W. Z., Brorson, S. D., Ippen, E. P., Fujimoto, J. G. (1988), *Solid-State Electronics 31,* 443.

Schoenlein, R. W., Fujimoto, J. G., Eesley, G. L., Capehart, T. W. (1988), *Phys. Rev. Lett. 61,* 2596.

Seeger, K. (1985), *Semiconductor Physics.* Berlin: Springer, Ch. 6.

Shah, J., Pinczuk, A., Stoermer, H. L., Gossard, A. C., Wiegmann, W. (1983), *Appl. Phys. Lett. 42,* 55.

Shockley, W. (1950), *Electrons and Holes in Semiconductors.* Princeton: van Nostrand, Chs. 1 and 2.

Stahl, A., Balslev, I. (1986), *Electrodynamics of the Semiconductor Band Edge.* Berlin: Springer, Ch. 4.

Stern, F. (1963), in: *Solid State Physics Vol. 15:* Seitz, F., Turnbull, D. (Eds.). New York: Academic Press, p. 299.

Störmer, H. L. (1984), in: *Advances in Solid State Physics Vol. 24.* Braunschweig: Vieweg, p. 25.

Sze, S. M. (1981), *Physics of Semiconductor Devices.* New York: John Wiley & Sons, Ch. 3.

Thomas, G. A., Blount, E. I., Capizzi, M. (1979), *Phys. Rev. B 19,* 702.

Tinkham, M. (1964), *Group Theory and Quantum Mechanics.* New York: McGraw-Hill.

Toyozawa, Y. (1980), in: *The Physics of Elementary Excitations (Springer Series in Solid-State Sciences, Vol. 12).* Berlin: Springer, Ch. 7.

Tsui, D. C. (1974), *Phys. Rev. B 10,* 5088.

Ulbrich, R. G. (1985), in: *Advances in Solid State Physics Vol. 25:* Grosse, P. (Ed.). Braunschweig: Vieweg, p. 299.

Ulbrich, R. G. (1988), in: *Optical Nonlinearities and Instabilities in Semiconductors:* Haug, H. (Ed.). New York: Academic Press, Ch. 5.

Wannier, G. H. (1937), *Phys. Rev. 52,* 191.

Weisbuch, C. (1986), *Fundamental Properties of III–V Semiconductor 2-d Quantized Structures,* in: *Semiconductors and Semimetals Vol. 24.* New York: Academic Press.

Weisbuch, C., Ulbrich, R. G. (1982), in: *Topics in Applied Physics, Vol. 51.* Berlin: Springer, Ch. 7.

Weisbuch, C., Vinter, B. (1990), *Quantum Semiconductor Structures: Fundamentals and Applications.* Boston: Academic Press.

Wigner, E. P. (1932), *Phys. Rev. 40,* 749.

Wingreen, N. S., Stanton, C. J., Wilkins, J. W. (1986), *Phys. Rev. Lett. 57,* 1084.

Wooten, F. (1972), *Optical Properties of Solids.* New York: Academic Press, Ch. 5.

Yablonovitch, E., Allara, D. L., Chang, C. C., Gmitter, T., Bright, T. B. (1986), *Phys. Rev. Lett. 57,* 249.

Yariv, A. (1982), *Quantum Electronics, 2nd Ed.* New York: Wiley, Ch. 7.

Yoffa, Ellen J. (1980), *Phys. Rev. B 21,* 2415.

Zakharchenya, B. P., Meier, F. (1984), *Spin Orientation.* Amsterdam: North Holland.

Ziman, J. (1963), *Electrons and Phonons.* Oxford: Oxford University Press, Ch. 7.

Zimmermann, R. (1990), in: *Advances in Solid State Physics Vol. 30.* Braunschweig: Vieweg.

Zimmermann, R., Kilimann, K., Kraeft, W. D., Kremp, D., Roepke, G. (1978), *Phys. Stat. Sol. B 90,* 175.

3 Intrinsic Point Defects in Semiconductors

George D. Watkins

Department of Physics, Sherman Fairchild Laboratory for Solid State Studies 161, Lehigh University, Bethlehem, PA, U.S.A.

List of Symbols and Abbreviations

a_1	symmetric representation for a molecular orbital
A_1	symmetric representation for a many-electron state
C_{3v}	trigonal symmetry
D_{2d}	tetragonal symmetry
e	doubly degenerate representation for a molecular orbital
e^-	electron
e_D^-	electron bound to a shallow donor
E	double degenerate representation for a many-electron state
E_c	electron energy at the conduction band edge
E_{JT}	Jahn-Teller relaxation energy
E_v	electron energy at the valence band edge
$E(i/j)$	Fermi level position at which defect changes from charge state i to j
g	gyromagnetic ratio for a paramagnetic defect
h^+	hole
i	interstitial (subscript)
I	interstitial
k_B	Boltzmann constant
k_Q	spring force constant for distortion mode Q
s	substitutional (subscript)
S	electronic spin
T	absolute temperature (K)
t_2	triply degenerate representation for a molecular orbit
T_2	triply degenerate representation for a many-electron state
T_d	tetrahedral symmetry
U	Hubbard correlation energy
U_a	activation energy (barrier) for defect migration
V	group-V atom
V_Q	linear Jahn-Teller coupling coefficient to distortion mode Q
V	vacancy
W	defect formation energy
v	frequency of a rate limited process
π	polarization parallel to a defect symmetry axis
ϱ	resistivity
σ	polarization perpendicular to a defect symmetry axis
σ_e	capture cross section for electrons
σ_h	capture cross section for holes
τ	time constant for a rate limited process
CC	configurational coordinate
DLTS	deep level transient capacitance spectroscopy
ENDOR	electron-nuclear double resonance
EPR	electron paramagnetic resonance
LEC	liquid encapsulated

LVM   localized vibrational mode
MBE   molecular beam
ODENDOR optical detection of electron-nuclear double resonance
ODMR   optical detection of electron paramagnetic resonance
OMVPE   organometallic vapor phase

3.1 Introduction

In this chapter, we will define "intrinsic point defects" to mean the simplest structural defects that can be formed in a crystalline lattice by rearrangements of the host atoms, no impurities being involved. With this definition, there are three basic types: (1) a lattice *vacancy* (a missing host atom); (2) an *interstitial* (an extra host atom squeezed into the lattice); and (3) an *antisite* (atom A on site B, or vice versa, in an AB compound crystal lattice). These in turn can, of course, be considered the elementary "building blocks" for all other more complex structural defects – divacancies, trivacancies, voids; interstitial pairs, interstitial aggregates; etc. Even dislocations, grain boundaries and external surfaces can be considered to be constructed from arrays of these three fundamental units.

Vacancies, interstitials, and antisites are also in a sense the only true *intrinsic* defects in that a significant concentration of each is thermodynamically stable at any temperature. [With its formation energy W_j, the fractional concentration of defect j is given approximately by $\exp(-W_j/k_B T)$]. The binding energy between defects to form complexes is invariably significantly smaller than the individual defect formation energies so that the thermodynamic equilibrium concentration of complexes becomes vanishingly small as the size grows. Complexes can and do form, of course, in the growth and processing of materials, but they must result from nonequilibrium processes such as aggregation and precipitation of the quenched-in intrinsic defects, etc.

The fact that vacancies and interstitials are thermodynamically stable has important consequences in high temperature processing of materials in that they supply the principal mechanisms for mass transport and diffusion. With an activation barrier U_a for the migration of the defect, its contribution to diffusion is proportional to $\exp[-(W+U_a)/k_B T]$. The migration barrier U_a and the formation energy W are therefore important properties that we would like to establish for each defect.

In a semiconductor, the intrinsic defects are of interest for a number of other important reasons as well. Because a defect perturbs the periodicity of the perfect lattice, its presence can introduce electronic states in the forbidden gap of the material. We therefore need to understand its electronic structure as well – its available charge states, the local lattice relaxations around it, and how these affect its electrical and optical properties, its stability and mobility, and its interactions with other defects.

And so the study of intrinsic defects in semiconductors is a very rich, complex, and often frustrating enterprise. Fragments of information and clues have been assembled in all of the elemental and compound semiconductors after many years of study. However, in many cases they are only clues and the interpretations are often controversial. Even when a consensus seems to have occurred, our previous experiences in such matters urge caution, for it could well be wrong. The principal problem is *identification* of the defect being studied.

In the case of vacancies and interstitials, therefore, the first question that has had to be addressed experimentally is how to produce them and know that it is their properties that are being studied. In semiconductors, attempts to quench them in by rapid cooling from elevated temperatures, often successful in metals, has invariably failed due to the combination of high diffusional mobility of the defect and its low concentration near the melt, plus the ubiquitous presence of fast diffusing trace impurities from the ambient, or already present in the material, that tend to dominate the material property changes. The only clean, direct, and unambiguous way that has been found to produce the defects is by high energy ($\gtrsim 1$ MeV) electron irradiation at cryogenic temperatures (Corbett, 1966). In this way, the simple interstitial and vacancy produced by a Rutherford "collision" of electron with a host atom nucleus can be frozen into the lattice for study. The scenario of such an event is shown in Fig. 3-1. As the temperature is subsequently raised, one of the species will become mobile first. At this stage, some of the pairs will recombine, while for others, the mobile species will escape and diffuse through the lattice until trapped by other defects to form new pair defects. At a higher temperature stage, the remaining partner becomes mobile and is trapped also by other defects.

Secondly, identification requires experimental observation techniques that reveal the structure of the defect at the atomic level. The only techniques that fully satisfy this requirement are the magnetic resonance ones. These include electron paramagnetic resonance (EPR) (Watkins,

1975a), electron-nuclear double resonance (ENDOR) (Ammerlaan et al., 1985), and optical detection of EPR (ODMR) (Cavenett, 1981) and ENDOR (ODENDOR) (Spaeth, 1986). The unique information provided by these techniques, when they work, is resolved hyperfine interaction with the atomic nuclei in the core and neighborhood of the defect which allows a detailed mapping of the atoms involved and their structural arrangements. Once the defect is identified, then electrical, optical, and other measurements can be performed in close correlation with the magnetic resonance experiments, to further establish its properties.

In this chapter we therefore plan to deal only with those intrinsic defects for which magnetic resonance studies have supplied relatively firm identifications. The best documented system in this regard is *silicon*. Here, the isolated vacancy has been observed, the electrical and diffusional properties of several of its charge states determined, and a host of its interactions with impurities and other defects identified. The isolated interstitial has not been directly observed. However, a microscopic identification of its interactions with other defects has served to reveal many of its properties. In this case, theory has served also to fill in some of the important gaps.

The next best understood materials are the II–VI semiconductors. In ZnSe, for instance, isolated interstitials and vacancies and close Frenkel pairs on the metal sublattice have been identified and studied in substantial detail. In several of the other II–VI compounds, metal vacancies with similar characteristics have been detected and clues to the properties of the chalcogen vacancy have also been obtained.

Finally, there are the III–V compounds. Here the important role of the antisites has clearly been established by magnetic reso-

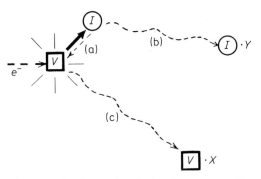

Figure 3-1. An electron irradiation damage event. Recovery occurs by a) vacancy-interstitial recombination, b) interstitial migration until trapped, c) vacancy migration until trapped.

nance studies although there remain many controversies about their properties and their interactions with other defects. There are fewer hard facts about vacancies and interstitials, however. The isolated gallium vacancy in GaP appears to have been identified and magnetic resonance of a few defects believed to be associated with gallium interstitials have been reported in GaP, and AlGaAs. Otherwise, there are no firm identifications.

In what follows, therefore, we will restrict ourselves to what has been learned in these three systems, with major emphasis on silicon and ZnSe, the best documented of the three. What we have learned from the elemental semiconductor silicon and the substantially ionic II–VI compound semiconductors can hopefully serve as a guide to interpreting results in the more covalent III–V compounds which they bracket and for which magnetic resonance methods have so far been less successful.

3.2 Silicon

3.2.1 Defect Production

A most surprising and unexpected result is found when p-type silicon is irradiated with 1.5–3.0 MeV electrons at 4.2 K. EPR spectra of *isolated vacancies* are observed but none that can be identified with interstitials or close Frenkel pairs. Instead spectra of defects identified with interstitials trapped by impurities are observed with approximately 1:1 intensity to those of the isolated vacancies. The unavoidable conclusion is that silicon interstitials must be mobile, migrating long distances even at 4.2 K! (Watkins, 1964). In terms of the expected scenario of Fig. 3-1, processes (a) and (b) are occurring even during the 4.2 K irradiation and only isolated vacancies and trapped interstitials remain. We

will return to this surprising result later when we discuss the interstitial in more detail. Let us at this point accept the fact and concentrate on what has been learned about the isolated vacancy.

3.2.2 The Silicon Vacancy

3.2.2.1 Electronic Structure

The vacancy is observed to take on five charge states in the forbidden gap ($V^=$, V^-, V^0, V^+, V^{++}). Two of these (V^+ and V^-), having an odd number of electrons, are paramagnetic and have been identified directly by EPR (Watkins, 1964). Using these two spectra as probes, it has been possible to develop a simple model which accounts in remarkable detail for the electronic structure of all the charge states (Watkins, 1975b, 1986). This is illustrated in Fig. 3-2. Here the electrons are depicted as occupying highly localized molecular orbital states made up primarily as linear combinations of the broken bond orbitals (a, b, c, d) of the four silicon atoms surrounding the vacant site.

An undistorted vacancy has full tetrahedral T_d symmetry and simple group theory arguments require that a singlet (a_1) and triplet (t_2) set are formed. The fully bonding a_1 state will be the lowest in energy, as shown, due to the attractive Coulomb interaction of the electrons with the positive cores of the four silicon neighbors (the "crystal field"). For V^{++}, Fig. 3-2a, two electrons occupy the a_1 state, spins paired. The defect is diamagnetic and no EPR is observed.

For V^+, Fig. 3-2b, the third electron goes into the t_2 orbital. Because of the degeneracy associated with the orbital, a tetragonal Jahn-Teller distortion occurs (Jahn and Teller, 1937; Sturge, 1967), lowering the total energy of the defect with atoms a and d, and atoms b and c, pulling

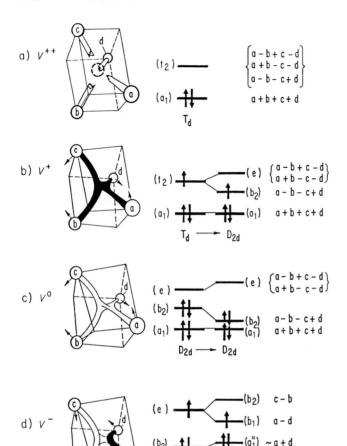

Figure 3-2. Simple one-electron molecular orbital model for the vacancy in silicon (Watkins, 1983).

together by pairs, as shown. The resulting orbital (b$_2$ in reduced D$_{2d}$ symmetry) contains the unpaired electron which is spread equally over the four silicon atoms, as has been established in the EPR studies from resolved ^{29}Si hyperfine interactions at each site. The existence of the tetragonal distortion is evidenced by small tilts in the dangling bond directions which are detected in the hyperfine interactions, as well as by anisotropy in the g-values of the center.

In forming V^0, Fig. 3-2 c, the next electron goes also into the b$_2$ orbital, spin paired, further enhancing the tetragonal Jahn-Teller distortion because now the energy gain of two electrons is involved. In

this state the defect is diamagnetic and no EPR is observed.

For V^-, Fig. 3-2d, the fifth electron goes into the degenerate e orbital and an additional Jahn-Teller distortion occurs. This distortion is of b$_2$ symmetry (in D$_{2d}$) with atoms b and c further pulling together and a and d separating slightly. As a result, the unpaired electron ends up on atoms a and d, again as established by EPR.

For $V^=$, the absence of EPR suggests again a diamagnetic state, the sixth electron perhaps going paired off into the a−d orbital. (This is not shown in the figure, however, because so far no direct or indirect evidence of its symmetry has been obtained. It cannot be ruled out that a completely different set of relaxations set in for this charge state.)

For the two paramagnetic charges states (V^+ and V^-), analysis of the ^{29}Si hyperfine interactions indicates that $\sim 60–65\%$ of the wavefunction is accounted for in these dangling orbitals. In the case of V^-, ENDOR studies have resolved weaker ^{29}Si hyperfine interactions with an additional 26 inequivalent shells of neighboring silicon atom sites accounting for most of the remaining wavefunction (Sprenger et al., 1983). The true wavefunctions therefore of the states derived from the t_2 orbitals can be considered to have $\sim 60–65\%$ of their density highly localized in these dangling orbitals with a weak tail, extended over many shells of neighbors, accounting for the remaining $\sim 35–40\%$.

The remarkable success of these simple one-electron models reveals that the electron-electron interactions that tend to favor parallel spin coupling (Hund's rules for atoms, etc.) are small. We are, in effect, in the strong crystal field regime. We fill each level before going to the next. When degeneracy occurs, a Jahn-Teller distortion occurs which imposes a new crystal field, decoupling the electrons again. It is by no

means obvious that this necessarily should be the case. There has been much controversy in the theoretical literature through the years on this point, with no clear consensus (Stoneham, 1975; Lannoo and Bourgoin, 1981; Lannoo et al., 1981; Malvido and Whitten, 1982; Lannoo, 1983). And so it is really only through experiment that we know that this simple case applies in silicon. We must be prepared for other possibilities in other systems, as we shall see in the compound II–VI and possibly the III–V semiconductors.

3.2.2.2 Effects of Lattice Relaxations – General Considerations

When lattice relaxations such as these Jahn-Teller distortions occur at a defect, significant consequences may result in terms of the defect's electrical and optical properties, and its stability. Since this is a feature applicable to many of the intrinsic defects that will be discussed in this chapter, not just the silicon vacancy, let us now develop some of the general concepts. In Fig. 3-3, we show a "configurational coordinate" (CC) diagram for a hypothetical defect D which undergoes a distortion (mode Q) in capturing an electron to become neutral, D^0. Shown are three total energy surfaces, one for the initial D^+ state, a second displaced by the band gap E_g to represent D^+ plus a free electron (e^-) and hole (h^+), and a third when the defect traps an electron, $D^0 + h^+$, and distorts.

As illustrated in the figure, the electrical level position measured with respect to the conduction band edge is the energy difference between the neutral charge state $D^0 + h^+$ and the ionized state $D^+ + e^- + h^+$ each determined in its *fully relaxed state*. Jahn-Teller distortions therefore directly alter the electrical level posi-

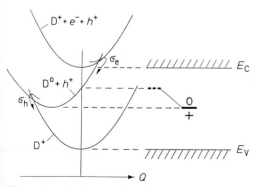

Figure 3-3. Configurate coordinate (CC) model for a defect with large lattice relaxation in a semiconductor (Watkins, 1983).

tions in the gap. Secondly, they cause Stokes shifts for optical ionizing transitions which are "vertical" on the diagram, causing differences between ionization energies determined optically and electrically. (The true electrical level position corresponds to the "zero phonon line" of the optical transition.) Third, the intersecting energy surfaces supply a "multiphonon" mechanism for electron (σ_e) and hole (σ_h) capture processes, which are otherwise difficult for a deep level. When the distortion is large, as depicted in the figure, they provide in addition a radiationless recombination path for electrons and holes, the neutral state providing a "short circuit" across the gap by alternate electron and hole capture. Finally, carrier capture means entry into a high vibrational state of the new charged state and this burst of energy may serve to assist the defect over its diffusional barrier, a phenomenon called *recombination-enhanced migration* (Kimerling, 1978; Bourgoin and Corbett, 1978).

3.2.2.3 Electrical Level Positions of the Vacancy

Fig. 3-4 summarizes what is presently known about the level positions for the vacancy (Watkins, 1986). The levels in Fig. 3-4a represent a consensus of recent theoretical estimates for the single particle t_2 and a_1 levels for the undistorted neutral vacancy (Baraff and Schlüter, 1979; Bernholc et al., 1980). These can be considered to correspond to the t_2 and a_1 orbitals of Fig. 3-2. In Fig. 3-4b we have included corrections to the single particle neutral t_2 gap level positions vs. charge state to account for the Coulomb repulsion energy between the electrons as the occupancy of the t_2 level increases. The level separation U (the Hubbard correlation energy), taken

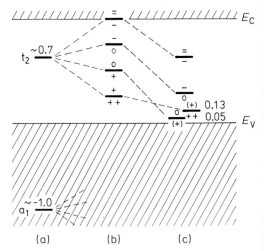

Figure 3-4. Electrical level positions of the vacancy in silicon: a) Calculated single particle levels for unrelaxed V^0. b) Corresponding estimates for the level positions. c) Experimental results, after Jahn-Teller relaxations (Watkins, 1983). (Level positions in eV from the valence band edge, E_V).

in the figure to be ~ 0.3 eV, is a typical value found experimentally for deep levels in silicon and also agrees well with theoretical estimates for the vacancy (Baraff et al., 1980a).

In Fig. 3-4c, we show the results after the Jahn-Teller distortions. Each level drops in the gap because with each added electron the Jahn-Teller distortions increase, Fig. 3-2. The first $(0/+)$ and second $(+/++)$ donor level positions have been measured directly in EPR and correlated deep level transient capacitance spectroscopy (DLTS) studies (Watkins and Troxell, 1980; Newton et al., 1983; Watkins, 1986), and are indicated with respect to the valence band edge in Fig. 3-4c.

We note here a remarkable result: The levels are reversed, in *negative-U* ordering (Watkins, 1984), the first donor state $(0/+)$ being deeper than the second $(+/++)$. The gain in Jahn-Teller energy in going from V^+ to V^0 is apparently great enough to overcome the Coulomb repulsion energy between the electrons so that the ionization energy for $V^0 \rightarrow V^+ + e^-$ is 0.08 eV greater than that for the removal of the second electron $V^+ \rightarrow V^{++} + e^-$. (Alternatively, the 0.13 eV ionization energy to remove the first hole $V^{++} \rightarrow V^+ + h^+$ is greater than the 0.05 eV required for the second hole $V^+ \rightarrow V^0 + h^+$.) There is in effect a net *attraction* between the two electrons (holes) at the defect.

This effect was first predicted theoretically for the vacancy by Baraff et al. (1979, 1980a, b) and subsequentially confirmed in detail by experiment. It can be understood in a very simple way. We outline this below because with it, and the experimental level positions, a direct estimate of the magnitude of the Jahn-Teller energies is possible (Baraff et al., 1980b).

For $n = 0$, 1, or 2 electrons in the b_2 orbital of Fig. 3-2b and c, the relaxation energy of the corresponding charge state (V^{++}, V^+, or V^0, respectively) can be approximated as

$$E(n) = -n V_Q Q + (1/2) k_Q Q^2 \quad (3\text{-}1)$$

where V_Q is the linear Jahn-Teller coupling coefficient for the b_2 orbital to the tetragonal distortion mode, Q is it amplitude and k_Q is the force constant describing the elastic restoring forces on the atoms for this mode. ($-V_Q Q$ is therefore the lowering of the b_2 orbital which in this simple independent electron model is independent of occupancy). Minimizing Eq. (3-1) with respect to Q gives for the Jahn-Teller stabilization energy

$$E_{JT}(n) = -n^2 V_Q^2 / 2 k_Q \quad (3\text{-}2)$$

The $(+/++)$ level is therefore lowered from its relaxed state by $E_{JT}(0) - E_{JT}(1) = V_Q^2 / 2 k_Q$ and the $(0/+)$ level is lowered by $E_{JT}(1) - E_{JT}(2) = 3 V_Q^2 / 2 k_Q$. From this,

$$U_{eff} = E(0/+) - E(+/++) = U - V_Q^2 / k_Q \quad (3\text{-}3)$$

which is negative if $V_Q^2 / k_Q > U$, the criterion originally derived by Anderson (1975).

Setting U_{eff} to the experimental value of -0.08 eV and with $U = {\sim}0.3$ eV, Eq. (3-3) gives $V_Q^2 / k_Q = {\sim}0.38$ eV. With Eq. (3-2), the Jahn-Teller relaxation energies for V^+ and V^0 become ${\sim}0.19$ eV and ${\sim}0.76$ eV, respectively. The Jahn-Teller distortions of Fig. 3-2 are therefore not just interesting curiosities that describes subtle features of the electronic structure. They are large, being comparable to the band gap with serious consequences for the electrical level positions as well. We will see that they also have important bearing on the diffusional mobility of the vacancy.

No information is presently available on the positions of the acceptor levels $(-/0)$ and $(=/-)$. The failure to detect DLTS levels associated with them in n-type material has been interpreted as indicating that they must be greater than ${\sim}0.17$ eV below the conduction band edge (Troxell, 1979).

3.2.2.4 Activation Energy for Vacancy Diffusion

As the temperature is raised, the vacancy EPR spectra disappear and a large assortment of new spectra emerge which have been identified as vacancies paired off with impurities. The annealing therefore is the result of long range migration of the vacancy. Detailed study of the kinetics via EPR and DLTS have given the following results:

In low resistivity n-type silicon, where the vacancy is in the $V^=$ charge state, the activation energy has been determined to be 0.18 ± 0.02 eV (Watkins, 1975b). In low resistivity p-type material with the vacancy in the V^{++} state, the energy is 0.32 ± 0.02 eV (Watkins et al., 1979). Under reverse bias in DLTS studies, the activation energy for the annealing process was

Table 3-1. Activation energies for the silicon vacancy migration.

Conductivity type	Charge state	U_a(eV)
Low ϱ n-type	$V^=$	0.18 ± 0.02
High ϱ p-type (reverse bias)	V^0	0.45 ± 0.04
Low ϱ p-type	V^{++}	0.32 ± 0.02

found to be 0.45 ± 0.04 eV, presumably the property of V^0 (Watkins et al., 1979).

These results are summarized in Table 3-1. Such low activation energies for vacancy migration are indeed surprising for a material which doesn't melt until 1450 °C and for which activation energies of self diffusion are $\sim 4-5$ eV. The logical conclusion is that the major part of the diffusion energy $W + U_a$ must be the formation energy W. This, however, is in conflict with the traditional interpretations of high temperature diffusion experiments where W and U_a have been estimated to be comparable in energy (Tan and Gösele, 1985). Recently *ab initio* local density calculations have concluded that vacancy formation energies are indeed ~ 4 eV (Car et al., 1985; Bar-Yam and Joannopoulos, 1985) and high temperature quantum molecular dynamics calculations of the same sophistication are matching the low ~ 0.4 eV barriers (Smargiassi and Car, 1990). At this juncture, it looks therefore as though high temperature diffusion measurements will have to be reinterpreted in terms of these values.

It has also been established that vacancies can be made to migrate through the lattice at temperatures well below those required for thermally activated diffusion, under conditions of minority carrier injection or optical excitation. Detailed studies have been performed in both n- and p-type silicon which demonstrate unambiguously

athermal (not thermally activated) migration at 4.2 K under such conditions and have served to estimate the efficiency of the process (Watkins et al., 1983; Watkins, 1986). As pointed out in Section 3.2.2.2, and in Fig. 3-3, Jahn-Teller distortions provide a natural mechanism for such processes where the energy associated with carrier capture or "vertical" optical excitation is funneled into local lattice vibrations which assist the defect over its migration barrier. In the case of the vacancy where the estimated Jahn-Teller energies exceed the small diffusional barriers, no additional thermal energy is required and the process is athermal.

3.2.2.5 Vacancy Interactions with Other Defects

Some of the simple vacancy-defect pairs which have been identified to form when the vacancy anneals are illustrated in Fig. 3-5. In Fig. 3-5a and b, the vacancy is trapped next to substitutional germanium. Both $(V \cdot Ge)^+$ and $(V \cdot Ge)^-$ have been studied by EPR (Watkins, 1969) and the electronic structure is found to be only slightly perturbed from that for V^+ and V^-, the presence of germanium being established by additional hyperfine lines for the ^{73}Ge $(I = 9/2, 7.6\%$ abundant) isotope. The Jahn-Teller distortions and level positions (including possible negative-U) appear very similar to those for the isolated vacancy.

Substitutional oxygen, Fig. 3-5c, results from the trapping of a vacancy by interstitial oxygen, a common impurity in silicon. On-center substitutional oxygen would normally be expected to be a double donor. However, EPR experiments show that it moves off-center, bonds to two of the silicon atoms and the defect becomes instead a single acceptor with its (−/0) level at

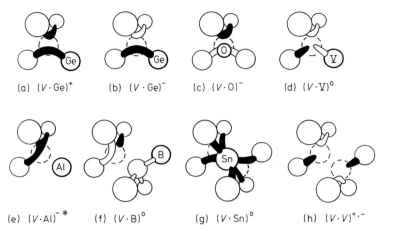

(a) $(V \cdot Ge)^+$ (b) $(V \cdot Ge)^-$ (c) $(V \cdot O)^-$ (d) $(V \cdot V)^0$

(e) $(V \cdot Al)^{-*}$ (f) $(V \cdot B)^0$ (g) $(V \cdot Sn)^0$ (h) $(V \cdot V)^{+,-}$

Figure 3-5. Vacancy-defect pairs identified in silicon. Jahn-Teller distortions are evidenced by the bond reconstruction by pairs. The localization of the unpaired electron observed by EPR is indicated in black (Watkins, 1986).

$E_c - 0.17$ eV (Watkins and Corbett, 1961; Bemski, 1959; Brower, 1971, 1972).

Vacancies trapped by the substitutional group-V atoms P (Watkins and Corbett, 1964), As, or Sb (Elkin and Watkins, 1968) take on the configuration shown in Fig. 3-5d, introducing a single acceptor level at $E_c - 0.43$, $E_c - 0.47$, or $E_c - 0.44$ eV, respectively (Troxell, 1979). The aluminum-vacancy pair has been studied only in a photo-excited state which has the configuration shown in Fig. 3-5e (Watkins, 1967). A level at $E_v + 0.52$ eV has been tentatively associated with it (Troxell, 1979). The boron-vacancy pair differs from the others in that the boron atom appears to prefer the next-nearest-neighbor position to the vacancy, Fig. 3-5f (Watkins, 1976; Sprenger et al., 1985). The level positions for it have not been established.

The tin-vacancy pair, Fig. 3-5g, takes on the interesting configuration in which the tin atom resides halfway between the two atom sites, bonding partially with all six resulting dangling bonds (Watkins, 1975c). Levels at $E_v + 0.32$ eV and $E_v + 0.07$ eV have been identified as its first $(0/+)$ and second $(+/++)$ donor levels, respectively (Troxell, 1979). The divacancy results by pairing of two vacancies during

vacancy anneal and its structure is shown in Fig. 3-5h (Watkins and Corbett, 1965; deWit et al., 1976; Sieverts et al., 1978). [It has also been observed to form directly as a primary event in the irradiation with a somewhat higher threshold energy than that for single vacancy formation (Corbett and Watkins, 1965)]. It takes on four charge states (VV^+, VV^0, VV^-, $VV^=$) and has been observed by EPR in the single plus and minus charge states. Its second acceptor level ($=/-$) has been determined to be at $E_c - 0.23$ eV, its first acceptor level $(-/0)$ at $E_c - 0.39$ eV, and its single donor state $(0/+)$ at $E_v + 0.20$ eV (Kimerling, 1977).

For most of these defects again simple one-electron molecular orbital models similar to that of Fig. 3-2 for the vacancy (but with reduced symmetry due to the nearby defect) provide a satisfactory description of their structures. For the ground states, the levels are populated according to the large crystal field regime and when degeneracy remains, Jahn-Teller distortions occur, decoupling the electrons leading to minimum multiplicity ($S = \frac{1}{2}$ or 0) as for the isolated vacancy. The only exception is the tin-vacancy pair, which has an $S = 1$ ground state. In the figure, Jahn-Teller distortions are evidenced by the bond reconstruction by pairs. The localization of the unpaired electron observed by EPR is indicated in black. As for the isolated vacancy, hyperfine interactions again indicate $\sim 65\%$ localization on the atoms adjacent to the vacancy.

In the case of the divacancy (Cheng et al., 1966) and the group-V atom-vacancy pairs (Watkins, 1989), related optical absorption bands have also been reported. These have been interpreted in terms of transitions between occupied and unoccupied molecular orbital states for the centers, providing further confirmation for the molecular orbital models. In addition, sharp local vibrational mode absorption spectra have been observed for both charge states of the substitutional oxygen defect (Corbett et al., 1961; Bean and Newman, 1971).

In Fig. 3-6, the stability of the various vacancy and first generation vacancy-defect pairs is summarized schematically as would be observed in ~ 15 min isochronal annealing studies.

3.2.3 The Silicon Interstitial

3.2.3.1 Migration During a 4.2 K Irradiation

As mentioned in Section 3.2.1, electron irradiation of p-type silicon at 4.2 K produces isolated vacancies and interstitial silicon atoms which are trapped at impurities. The unavoidable conclusion is that interstitial silicon atoms must be highly mobile at 4.2 K at least under the conditions of irradiation in p-type material, and migrate until trapped by impurities. This has been verified directly by EPR for the trapping by substitutional boron (Watkins, 1975d) and aluminum (Watkins, 1964), and indirectly for gallium (Watkins, 1964). At somewhat higher temperatures (~ 100 K), trapping by substitutional carbon also becomes important (Watkins and Brower, 1976). These processes can also be monitored by local vibrational mode (LVM) spectroscopy of the light boron (Tipping and Newman 1987) and carbon (Chappell and Newman, 1987) atoms, these techniques also revealing similar trapping by interstitial oxygen (Brelot and Charlemagne, 1971).

In n-type silicon, the situation is less clear. On the one hand, evidence has been cited that the interstitial silicon atom is more stable, not migrating long distances until ~ 150–175 K (Harris and Watkins, 1984). For example, it is in this temperature region that the carbon interstitials are observed to emerge in n-type material after annealing from a 4.2 K irradiation. Also EPR studies reveal that most divacancies produced at 4.2 K or 20.4 K in n-type ma-

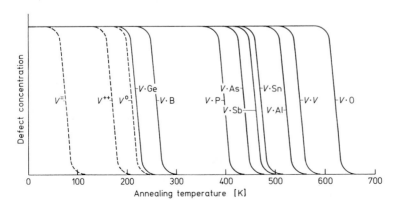

Figure 3-6. Schematic of vacancy and vacancy-defect pair annealing stages in silicon (~ 15 min isochronal) (Watkins, 1986).

terial are perturbed by some other defect nearby. These divacancies disappear in a broad annealing stage beginning at ~ 100 K, consistent with identifying the perturbing defects as the interstitials produced in the primary event and frozen into the lattice nearby, which recombine with their divacancy upon annealing.

On the other hand, interstitial boron atoms are observed to be produced at 4.2 K in n-type silicon partially compensated by boron, albeit at an order of magnitude lower rate than in p-type material (Watkins, 1975 d). This argues that long range motion is still occurring. Also the emergence of interstitial carbon coincides with the disappearance of two unidentified EPR centers, the annealing kinetics of the dominant one giving (Harris and Watkins, 1984).

$$\tau^{-1} = 2.8 \times 10^{13} \exp\left(-0.57\,\text{eV}/k_{\text{B}}T\right) \text{s}^{-1} \tag{3-4}$$

The preexponential factor is characteristic of a single jump process suggesting that these precurser defects may also be trapped interstitials with the kinetics reflecting the rerelease of the interstitials.

So, whether significant long range motion of the silicon interstitial is occurring in n-type silicon during electron irradiation at cryogenic temperatures or not is still unclear. We know, however, that the subsequent emergence of interstitial carbon at ~ 150–175 K signals its arrival at the trace substitutional carbon impurities. The 0.57 eV of Eq. (3-4) is therefore an upper limit to its diffusion activation energy in the n-type material.

3.2.3.2 Trapped Interstitials

The structures deduced from EPR studies of some of the trapped interstitial configurations are summarized in Fig. 3-7. In Fig. 3-7a, the interstitial silicon atom has traded places with a substitutional aluminum atom, ejecting it into the interstitial site. In the Al_i^{++} state seen by EPR, it resides on center in the empty tetrahedral interstitial site of the lattice (Watkins, 1964; Brower, 1974; Niklas et al., 1985). In Fig. 3-7b, boron is also displaced from its substitutional site but it prefers to nestle into a bonding configuration between two

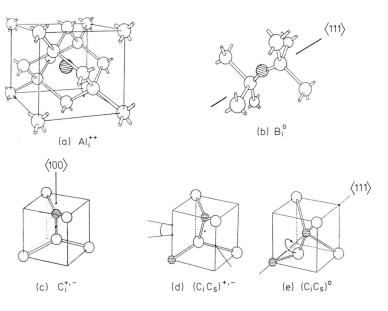

(a) Al_i^{++} (b) B_i° (c) $C_i^{+,-}$ (d) $(C_i C_s)^{+,-}$ (e) $(C_i C_s)^{\circ}$

Figure 3-7. Structures identified by EPR that are produced when interstitial silicon is trapped by substitutional (a) aluminum, (b) boron, and (c) carbon. Subsequent anneal of C_i produces a $C_i C_s$ pair with the two nearly energetic configurations shown in (d) and (e). The impurity atoms are shaded.

nearest neighbor substitutional silicon atoms, at least as deduced from EPR studies of its neutral B_i^0 state (Watkins, 1975 d). The carbon atom, on the other hand, Fig. 3-7 c, tends to share a single substitutional site with the extra silicon atom, as observed by EPR in both its positive and negative charge states (Watkins and Brower, 1976; Song and Watkins, 1990).

Also shown in Figs. 3-7 d and e are two configurations that have been determined for an interstitial carbon-substitutional carbon pair $(C_i C_s)$ which is produced when interstitial carbon begins to diffuse through the lattice at $\sim 350\,^\circ$C and is trapped by another substitutional carbon impurity. These two configurations are almost equally energetic and the stable configuration flips between the two depending upon the defect charge state (Song et al., 1988).

Taken together, the configurations of Fig. 3-7 are revealing because they demonstrate the rich variety of configurations available for an interstitial atom. In a sense they provide the most direct experimental clues that we have about what the isolated interstitial silicon atom might look like. Also their electrical and diffusional properties are interesting in this regard, as we now describe.

Interstitial aluminum is a double donor and its second donor state $(+/++)$ has been located at $E_v + 0.17$ eV (Troxell et al., 1979). Its diffusional activation energy has been determined in p-type material to be 1.2 eV. However, under minority carrier injection, the diffusion is greatly enhanced with a remaining activation energy of only 0.3 eV. Apparently 0.9 eV is being effectively dumped into the defect during an electron-hole recombination event. No experimental information is available concerning the other charge states of the defect but the arguments of Section 3.2.2.2

suggest large lattice relaxational changes associated with them in order to account for the recombination-enhanced migration.

Interstitial boron also displays strong evidence of large lattice relaxational rearrangements with charge state change. Its thermal activation energy for diffusion has been measured to be ~ 0.6 eV but under injection conditions it migrates athermally (Troxell and Watkins, 1980). In addition, it is a negative-U defect, its acceptor level $(-/0)$ at $E_c - 0.35$ eV being below its single donor level $(0/+)$ at $E_c - 0.13$ eV (Harris et al., 1987). Interstitial carbon on the other hand appears well behaved with an acceptor state at $E_c - 0.1$ eV and a donor state at $E_v + 0.27$ eV, in normal ordering (Kimerling et al, 1989). It has been reported to display recombination-enhanced migration near room temperature, but a detailed study has not been performed.

Finally, in Fig. 3-8, the stability of the various first generation interstitial-impurity pairs is summarized schematically as would be observed in ~ 15 min isochronal annealing studies (Kimerling et al., 1989). When C_i begins to migrate, $C_i C_s$ (Song et al., 1988), $C_i O_i$ (Trombetta and Watkins, 1987), and C_i-group-V atom pairs (Zhan and Watkins, 1990) emerge and their stabilities are also indicated. [In addition to these simple C_i-related centers identified by EPR, optical studies have revealed a rich variety of other centers which have been successfully modeled as C_i and C_i-related defects serving as nucleation centers for interstitial silicon (Davies, 1988; Davies et al., 1987)]. Similarly, when Al_i migrates, it forms $Al_i Al_s$ pairs (Watkins, 1964; Niklas et al., 1985). The fate of B_i has not been established by EPR.

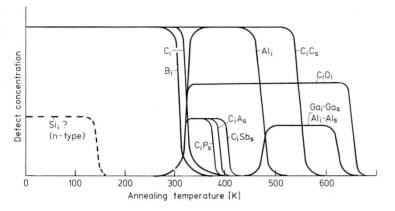

Figure 3-8. Schematic of the annealing of interstitial-related defects in silicon (~ 15 min isochronal).

3.2.3.3 Theory of the Silicon Interstitial

In the absence of direct experimental observation of the isolated interstitial, we turn to theory. Two independent groups have recently modeled the silicon interstitial using *ab initio* pseudo-potential local density quantum mechanical calculations and the results are remarkably consistent with our clues above (Car et al., 1985; Bar-Yam and Joannopoulos, 1984, 1985). In the Si_i^{++} state, the tetrahedral site is predicted to be the most stable, like Al_i^{++} in Fig. 3-7 a, being ~ 0.5–0.7 eV more stable than the bond-centered site, Fig. 3-7 b or e, the next stable one. However, in the Si^+ and Si^0 states, the tetrahedral site is no longer the minimum-energy position and the interstitial moves to one of several possible other positions which appear to be very close in energy. These include the bond-centered site, Fig. 3-7 b or e, another bonding configuration having a split $-\langle 110 \rangle$ character, and the hexagonal site (the open lattice site halfway between two tetrahedral positions). These form a set of rather flat energy surface paths through the crystal so that in these charged states the interstitial silicon atom should have a very low activation energy for migration. In addition, alternate cycling back to the Si^{++} tetrahedral site with hole capture, and return with electron capture, provides an effective athermal migration mechanism which can explain its long range migration under electron irradiation conditions at 4.2 K.

An interesting prediction also of the calculations is that interstitial silicon is a negative-U center. In the calculations of Car et al. (1985), the tetrahedral interstitial Si^{++} can only bind electrons in shallow Coulombically bound effective-mass like states. Upon trapping one electron it flips over to one of the other sites – in their calculations the bond-centered site – where the electron can be bound deeply, with a net total energy gain of ~ 0.4 eV. A second electron can now be bound by ~ 0.8 eV. The first donor state $(0/+)$ is predicted therefore to be at $E_c - 0.8$ eV, with the second $(+/++)$ at $E_c - 0.4$ eV, in negative-U ordering.

One question that must be addressed is how accurate are such calculations? Can we trust them? At present, the consensus seems to be that they are remarkably good – perhaps much better than we have good theoretical basis to expect. We have already discussed their successes for the vacancy. Here again for the interstitial they appear to be coming up with reasonable results that explain its high mobility and its configurations and properties as evidenced by its trapped states. Another impressive example in this regard is a similar computational study for interstitial aluminum where by considering the diffusive motion from tetrahedal to hexagonal sites, the experimental activation energies for both thermal and recombination-enhanced mechanisms could be matched extremely well to experiment (Baraff and Schlüter, 1984). Finally, the estimates for interstitial formation energies are ~ 4–5 eV, which, with the negligible barrier for diffusion, agree well with high temperature self-diffusion results. (As for the vacancy contribution to diffusion, traditional interpretations of the interstitial component have often involved larger contributions from its migration barrier with smaller formation energies. Again it begins to look rather convincing that these interpretations were incorrect and a fresh interpretation is required.)

And so, combining theory with our indirect experimental information, a rather complete picture of the isolated self interstitial and its properties seems to have emerged.

3.3 II–VI Semiconductors

3.3.1 ZnSe

3.3.1.1 The Zinc Vacancy

Electron irradiation (1–3 MeV) at or below room temperature produces isolated zinc vacancies which have been studied in considerable detail by EPR (Watkins, 1971, 1975e, f, 1977) and ODMR (Lee, 1983; Lee et al., 1980, 1981; Jeon, 1988; Jeon et al., 1986; Watkins, 1990a). The zinc vacancy produces a double acceptor ($=/-$) in the lower half of the ZnSe band gap, and in normally n-type material it is

doubly negatively charged ($V_{Zn}^{=}$) and is not paramagnetic. Illumination with ~ 500 nm light serves to photoionize it producing V_{Zn}^{-} which is observed by EPR.

The structure of V_{Zn}^{-} deduced from these studies is shown in Fig. 3-9. The hole (unpaired spin) is highly localized on only one of the four nearest Se neighbors and is highly p-like pointing toward the vacancy. This "small polaron" structure follows naturally from a simple one-electron mole-

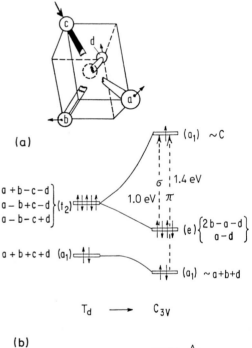

(a)

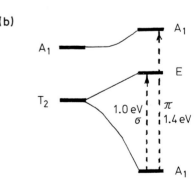

(b)

Figure 3-9. a) Simple one-electron molecular orbital model for V_{Zn}^{-} in ZnSe. b) The corresponding many-electron states (Watkins, 1990a).

cule orbital model for the defect, similar to that of Fig. 3-2 developed for the silicon vacancy. As for the silicon vacancy, the a_1 and t_2 orbitals in the undistorted T_d symmetry are made up as linear combinations of the p orbitals pointing into the vacancy on each of the four Se (a, b, c, d) neighbors. For $V_{Zn}^=$, its eight electrons completely fill the orbitals, no degeneracy remains, the many electron $a_1^2 t_2^6$ state is of A_1 symmetry, and no Jahn-Teller distortion occurs.

V_{Zn}^-, however, has degeneracy associated with a missing electron in the t_2 orbital, is of T_2 symmetry, and is therefore unstable to a Jahn-Teller distortion, which in this case is trigonal, as shown. In Fig. 3-9 b, the energies of the corresponding ground and excited many electron states (sum of the one-electron energies of Fig. 3-9 a) are also shown. Also shown are optical absorption transitions predicted by the model. These have been observed experimentally as broad absorption bands centered at 1.4 eV (π, polarized parallel to the trigonal distortion axis) and at 1.0 eV (σ, polarized perpendicular to the axis) and confirmed unambiguously to arise from the zinc vacancy by ODMR studies.

The presence of the zinc vacancy also produces luminescence at 1.72 eV (720 nm) which has been established by the ODMR studies to arise from electron transfer from a distant shallow donor to the $(=/-)$ acceptor level of the vacancy. Combining all of this optical information, a configurational coordinate diagram has been constructed for the double acceptor state of the defect. This is shown in Fig. 3-10, where Q is the amplitude of the trigonal distortion. The lowest energy curve is for the $V_{Zn}^=$ charge state which is stable in the undistorted position, $Q=0$. Shown also at $E_g = 2.82$ eV higher in energy is the curve for $V_{Zn}^=$ plus a free electron and hole. The remaining curves are for the ground and excited states of $V_{Zn}^- + e^-$, as given in Fig. 3-9 b. (The CC diagram is for the total energy, the curvature upward for each state reflecting the elastic energy stored in the lattice due to the distortion, which must be added to the electronic energies of Fig. 3-9 b.)

The curves shown in Fig. 3-10 for each state have been derived from a simple linear Jahn-Teller coupling theory with a single elastic restoring force constant for all states (Schirmer and Schnadt, 1976). They are completely determined by matching to the four observed optical transitions – the 1.0 eV and 1.4 eV absorption bands of V_{Zn}^-, the donor-acceptor pair luminescence at 1.72 eV and the photoionization and luminescence excitation band at 2.51 eV. That this is a good representation of the defect is confirmed by other experimental observations: (1) The splitting of the T_2 level vs. Q matches well an estimate of the Jahn-Teller coupling coefficient from uniaxial stress studies of the V_{Zn}^- spectrum;

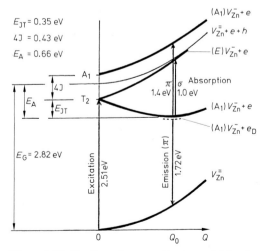

Figure 3-10. Configuration coordinate model for the second acceptor state $(=/-)$ of V_{Zn} in ZnSe (Watkins, 1990 a). e_D^- denotes an electron trapped on a shallow donor.

(2) Fig. 3-10 predicts the double acceptor level $(=/-)$ to be at $E_v + 0.66$ eV (the energy difference between $V_{Zn}^- + e^-$ and $V_{Zn}^= + e^- + h^+$, each determined in its relaxed state). This agrees with a value of $\sim E_v + 0.7$ eV estimated from the kinetics of hole release from V_{Zn}^- observed by EPR. The CC diagram also contains other important information: It reveals the magnitude of the Jahn-Teller energy to be 0.35 eV, the $A_1 - T_2$ "crystal field" splitting to be 0.43 eV, and the characteristic phonon frequency associated with the trigonal distortion to be ~ 11 meV (from the curvature of the V_{Zn}^- curve) a reasonable value, being of the order of characteristic TA phonons in ZnSe.

The zinc vacancy in ZnSe is therefore an extremely well characterized center as far as its electrical and optical properties, and its microscopic electronic and lattice structures are concerned.

3.3.1.2 Zinc Vacancy Diffusion and Interaction with Other Defects

Fig. 3-11 shows schematically the result of annealing after a 4.2 K or 20.4 K electron irradiation (Watkins, 1977). The vacancy is stable up to ~ 400 K at which point it disappears and new defects emerge that have been identified as vacancy-impurity pairs. This establishes that the annealing is a result of long range migration of the vacancy to be trapped by the impurities. The kinetics of the process have been studied giving for the zinc vacancy migration barrier

$$U_a = 1.26 \text{ eV} \qquad (3-5)$$

The zinc vacancy-impurity pairs that have been identified include zinc vacancies adjacent to the isoelectronic impurities S and Te, and to substitutional chemical donors (Holton et al., 1966). In all cases, the electronic structure is very similar to that of the isolated zinc vacancy, the hole again highly localized on one of the four vacancy near neighbors but with the symmetry of the center being lowered by the presence of the impurity. The donor-vacancy pairs have also been established to be important defects in as-grown material giving rise to the characteristic orange-yellow "self-activated" luminescence in ZnSe (Nicholls et al., 1978; Lee et al., 1980, 1981).

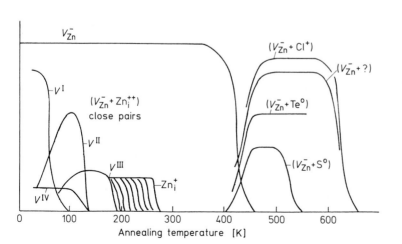

Figure 3-11. Schematic of the annealing stages (15 min isochronal) for the V_{Zn}^--related defects observed by EPR after 1.5 MeV electron irradiation at 20.4 or 4.2 K (Watkins, 1990 a).

3.3.1.3 Interstitial Zinc and Close Frenkel Pairs

In addition to the zinc vacancy, zinc interstitials and several distinct zinc-vacancy–zinc-interstitial close Frenkel pairs have been observed to be frozen into the lattice after 1–3 MeV electron irradiation at $T \leq 20.4$ K. In this case, as opposed to silicon, the expected scenario of Fig. 3-1 is indeed observed.

Four distinct close pairs have been observed directly by EPR (Watkins, 1974, 1975e, f, 1977, 1990a). Labeled I–IV, they anneal in discrete steps between 60 K and 180 K as illustrated in Fig. 3-11. Their electronic structure, deduced from the EPR studies, remains, like that of the vacancy-impurity pairs above, essentially that of a V_{Zn}^- perturbed only slightly by the presence of interstitial Zn_i^{++} in its different nearby lattice positions. In addition, many more distinct close pairs, as well as the isolated interstitial zinc atom, have been detected by ODMR in luminescence bands in the visible and near infrared produced by the irradiation (Rong and Watkins, 1986a, b, 1987; Rong et al., 1988; Watkins et al., 1988). Well resolved spectra for at least 17 distinguishable close pairs (different separation distances) have been observed. They anneal in a series of discrete steps, the closest ones first, the more distant ones at progressively higher temperatures, with the isolated interstitial zinc last at ~ 260 K, as indicated also in the figure. No significant increase in the isolated vacancy EPR intensity is observed through the annealing stages indicating that the primary annealing process for the pairs is annihilation due to recombination (process (a) in Fig. 3-1).

In these ODMR studies, the interstitial is observed in its Zn_i^+ paramagnetic state, either when isolated, or exchange-coupled with its paramagnetic V_{Zn}^- partner in the close pairs. From a scientific viewpoint, the pairs are particularly interesting because they provide a unique system where the exchange and Coulomb interactions, the radiative lifetimes, and the overall combined wavefunctions can be probed directly vs. lattice separation. The interested reader is encouraged to refer to the original literature. Here we will concentrate only on what has been learned about the isolated interstitial.

For the isolated Zn_i^+, hyperfine interactions have been resolved for the central zinc and two shells of neighboring selenium neighbors. The results establish that the Zn_i^+ ion is on-center in the tetrahedral site surrounded by four Se atoms (not in the other available tetrahedral site surrounded by four zinc atoms). The electronic structure is that of an electron in a highly localized S-orbital bound to the Zn_i^{++} closed shell core. By analysis of the luminescence wavelength shift vs. close pair distance and the Bohr radius deduced from hyperfine analysis for the isolated Zn_i^+, it has been possible to estimate the level position for the second donor state $(+/++)$ of the interstitial to be at $\sim E_c - 0.9$ eV.

Detailed kinetic studies of the low temperature annealing stages involving the interstitial in Fig. 3-11 have not been performed. However, assuming a characteristic single jump rate

$$\nu \sim 10^{13} \exp\left(-U_a/k_B T\right) \text{s}^{-1} \qquad (3\text{-}6)$$

and $10^2 - 10^5$ jumps required before being trapped or annihilated in 15 min at $T \sim 260$ K for the distant pairs and for isolated zinc leads to an estimate for the interstitial migration barrier of

$$U_a \sim 0.60 - 0.70 \text{ eV} \qquad (3\text{-}7)$$

3.3.1.4 Defects on the Selenium Sublattice

No direct spectroscopic evidence is available concerning the selenium vacancy or selenium interstitial in ZnSe. In the next section we will briefly review what has been learned about intrinsic defects in the other II–VI materials. There, limited information has been obtained concerning the chalcogen vacancy which may be applicable to ZnSe as well.

3.3.2 Other II–VI Materials

3.3.2.1 The Metal Vacancy

The metal vacancy has also been observed by EPR in its single-negative charge state in ZnS (Watkins, 1975f; Shono, 1979), CdS (Taylor et al., 1971), BeO (Hervé and Maffeo, 1970; Maffeo et al., 1972; Maffeo and Hervé, 1976), and ZnO (Taylor et al., 1970; Galland and Hervé, 1970, 1974). The electronic structure in each case is very similar to that in ZnSe, the hole being highly localized in a p-orbital on a single chalcogen neighbor.

In the case of ZnS, correlative optical (Watkins, 1975f) and ODMR (Lee et al., 1982) studies have identified an optical absorption band at 3.26 eV associated with $V_{Zn}^{=}$, one at 1.65 eV associated with V_{Zn}^{-}, and shallow donor to V_{Zn}^{-} recombination luminescence at 2.18 eV. From this, a CC diagram has been constructed and is shown in Fig. 3-12. This predicts the double acceptor level $(=/-)$ to be at $\sim E_v + 1.1$ eV and the Jahn-Teller relaxation energy to be ~ 0.5 eV.

Annealing kinetic studies have been performed for the disappearance of the V_{Zn}^{-} spectrum in ZnS and give an activation energy of 1.04 ± 0.07 eV (Watkins, 1977) (see Table 3-2). The preexponential factor for the decay is consistent with long range migration. The activation energy has

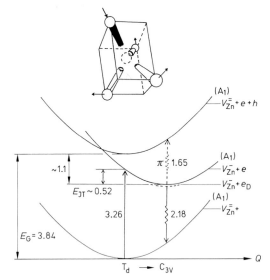

Figure 3-12. Configuration coordinate model for the second acceptor state $(=/-)$ of V_{Zn} in ZnS (Lee et al., 1982) (energies in eV).

therefore been tentatively identified with the migration activation energy for the vacancy. It must be considered tentative, however, because vacancy-impurity pairs were not observed to grow in during the process. Vacancy-donor pairs have been observed in as-grown material and, again, their structures are very similar to those of the corresponding pairs in ZnSe (Holton et al., 1966).

In the case of CdS, BeO, and ZnO, no optical or annealing information is available, except that the vacancies are stable at 300 K. An interesting additional bit of information does come, however, from stud-

Table 3-2. Level positions and migration energies (in eV), determined for intrinsic defects on the metal sublattice in ZnSe and ZnS.

		ZnSe	ZnS
V_{Zn}	$E(=/-)$	$E_v + 0.66$	$E_v + 1.1$
	U_a	1.26	1.04
Zn_i	$E(+/++)$	$\sim E_c - 0.9$	
	U_a	0.6–0.7	

ies in BeO (Maffeo et al., 1970) and ZnO (Galland and Hervé, 1970). There, a deep *neutral* state is also stable in the gap and has been observed by EPR. The defect has an $S = 1$ ground state, the electronic structure being that of two bound holes, one on each of two neighbors. In terms of our simple one-electron molecular orbital model, Fig. 3-9, this can only occur if the many-electron effects are greater than the Jahn-Teller-produced crystal field energies, forcing Hund's rules (maximum spin). And so, at least for the oxides this represents a significant departure from what has been learned for the vacancy in silicon.

3.3.2.2 The Chalcogen Vacancy

The chalcogen vacancy has been observed by EPR and ENDOR in its singly positive charge state in ZnS (Schneider and Rauber, 1967), ZnO (Smith and Vehse, 1970; Gonzalez et al., 1975), and BeO (Du-Varney et al., 1969; Garrison and DuVarney, 1973). For all three, the centers are similar to F-centers in the ionic NaCl crystal structure lattices (F^0 in the alkali halides, F^+ in the alkaline earth oxides), with an unpaired electron highly localized but symmetrically distributed primarily over the four nearest neighbors of the metal vacancy. Optical transitions have been associated with the chalcogen vacancies in some cases, indicating that the second donor levels $(+/++)$ are deep, being ~ 2–3 eV below the conduction band. No systematic studies of the migrational properties have been reported.

3.3.2.3 The Other Intrinsic Defects

No evidence exists for antisites (atom A on site B, etc.) in the II–VI materials. This is as expected considering the large chemi-

cal difference between the group II and group VI constituents.

Finally, interstitial chalcogen atoms are being produced in a radiation damage event and must therefore be present, in isolated or trapped form. They also presumably play an important role in high temperature transport processes. No direct EPR or other convincing spectroscopic evidence of them or their role has been reported, however.

3.4 III–V Semiconductors

3.4.1 Antisites

In the III–V semiconductors, a class of intrinsic defects not considered for the elemental group-IV and the compound II–VI materials becomes important. It is the antisite, where a group-V atom occupies a group-III atom site (denoted V_{III}) or vice versa (III_V). The first, the anion antisite, should, in the usual chemical impurity sense, be a double donor. The second, the cation antisite, should be a double acceptor. These become important because the chemical difference between the group III and V atoms is sufficiently small that each has a finite solubility on the other sublattice. They are therefore present in as-grown materials. They can also be produced by electron irradiation and by plastic deformation.

3.4.1.1 The Anion Antisite V_{III}

The presence of V_{III} antisites has been established directly by EPR, ODMR, and ODENDOR in GaP(P_{Ga}) (Kaufmann et al., 1976, 1981; O'Donnell et al., 1982; Killoran et al., 1982; Spaeth, 1990), GaAs(As_{Ga}) (Weber and Omling, 1985; Meyer et al., 1987), and InP(P_{In}) (Kennedy and Wilsey, 1984; Deiri et al., 1984; Kana-ah et al., 1985; Jeon et al., 1987). As expected they appear to be deep double donors, the magnetic resonance studies being performed on the singly positively charged paramagnetic $(S = \frac{1}{2})$ state. In this state, the group-V atom is located on-center in the group-III atom site and the paramagnetic electron is strongly localized in a Coulombically bound S-like orbital (a_1) on the atom. As in the case of interstitial Zn_i^+ in ZnSe (Section 3.3.1.3), this is evidenced in the EPR and ODMR studies by an isotropic $g \sim 2$ value and a large isotropic hyperfine interaction with the central group-V atom.

However, because the defects are being observed in an S-state, the ODMR and EPR spectra are relatively insensitive to whether the defect is truly isolated or whether it is complexed with some other defect nearby. In GaP, the ODMR and EPR resolution is sufficient to resolve hyperfine interactions with the nearest neighbor phosphorus atoms. In this case two different antisite-related centers can be distinguished in both EPR and ODMR studies, one with four P neighbors (PP_4), as expected for an isolated antisite, and one with only three [PP_3X in EPR (Kennedy and Wilsey, 1978), PP_3Y in ODMR (O'Donnell et al., 1982; Killoran et al., 1982), which may or may not be the same defect]. The latter presumably have an impurity or a vacancy replacing one of the P nearest neighbors but otherwise no firm identification is yet available. But even in the case of PP_4 one cannot be sure that a defect in one of the next shells is not an integrable part of the center.

In the case of PP_4 in GaP, this question is still unsettled. Preliminary higher resolution ODENDOR results of the four nearest phosphorus atoms and the next shells of neighbors indicate that at least two distinct antisite structures may contribute to the ODMR signal and that neither may actually be the isolated one (Spaeth, 1990). In the case of InP, slight shifts in the EPR spectra between n- and p-type materials provided an early hint also of such effects which have been confirmed subsequently in ODENDOR studies (Gislason et al., 1990). There, two distinct P_{In} ODENDOR spectra have been revealed, one of which appears to be consistent with an isolated antisite, but the other is not. In GaAs, evidence that more than one antisite-related defect may be involved comes from EPR studies where the spin-lattice relaxation times differ greatly for as-grown defects vs. those produced by plastic deformation or radiation damage (Hoinkis and Weber, 1989). ODENDOR studies also indicate more than one but so far only one As_{Ga} ENDOR spectrum has been studied in any detail, presumably the one for the defect with the longer relaxation times. The ENDOR spectrum was originally analyzed as the isolated antisite because the signals from the neighbor shells could be analyzed as regular. Closer inspection, however, revealed an extra signal which was subsequently assigned to a nearby arsenic interstitial (Meyer et al., 1987).

This points up the extreme difficulty that may occur in making this determination from the magnetic resonance study of an S-state system. In effect, the presence of a nearby defect often makes little difference to the electronic structure of the antisite. On the other hand, it is of extreme importance in understanding the electrical and optical properties of the total defect, its stability, etc.

An interesting and much studied example of a V_{III} antisite-related defect with this ambiguity is a center labeled *EL2 in GaAs* (Martin and Markram-Ebeid, 1986). The importance of this native defect is the fact that, being deep, it serves to make semi-insulating GaAs, desirable for device substrates. The scientifically intriguing feature of the defect is that it displays metastability. This is illustrated by the CC diagram of Fig. 3-13 a. In its stable configuration (A), it is a deep donor at $E_c - 0.75$ eV which can be detected by DLTS and by an optical absorption band at 1.15 eV. Optical excitation into this band converts the defect into a metastable configuration (B) for which no optical absorption is observed, the vertical optical ionizing transition now being above the band gap, as shown. Conversion back to the stable configuration occurs with an activation energy barrier of ∼0.3 eV.

EPR, ODMR and ODENDOR studies have provided strong arguments that EL2 involves the As_{Ga} antisite (Weber and Omling, 1985; Meyer et al., 1987). The ODENDOR results, as mentioned above, have been interpreted to indicate that the defect is actually an $As_{Ga}-As_i$ pair, the metastability presumably arising from the interstitial arsenic which hops from one separation distance to another, as shown in Fig. 3-13 b (Von Bardeleben et al., 1986). On the other hand, optical studies of a zero-phonon line at 1.04 eV, believed to be associated with the 1.15 eV photoexcitation band, have provided strong evidence that the center has full tetrahedral symmetry, suggesting that the metastability is an intrinsic property of the *isolated* As_{Ga} antisite (Kaminska et al., 1985; Trautman et al., 1990).

Theoretical studies have provided additional arguments for the isolated antisite model (Dabrowski and Scheffler, 1989; Chadi and Chang, 1988; Kaxiras and Pandey, 1989). They have predicted that the metastable state can result from the ejection of the As_{Ga} atom into an interstitial position leaving a gallium vacancy behind (Fig. 3-13 c). The energetics that are calculated agree closely with the experimental observation for EL2. At the present writing, there is no consensus on the correct model.

Still, in some cases, the perturbation of a nearby defect has been sufficient to detect directly in EPR or ODMR. We have already mentioned PP_3X and PP_3Y in GaP. Additional P_{Ga}-defect complexes have been detected by these techniques in electron-irradiated (Kennedy and Wilsey, 1981) and impurity-diffused (Godlewski et al., 1989; Chen et al., 1989) GaP. In InP, a lower symmetry P_{In}-related center has also been detected by ODMR (Kennedy et al., 1986).

3.4.1.2 The Cation Antisite III_V

No direct magnetic resonance identification of a III_V antisite exists at present for

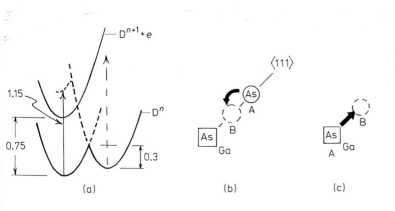

Figure 3-13. CC diagram for EL2 in GaAs (energies in eV). Two current models are: b) an As_{Ga} antisite-As_i interstitial pair, and c) an isolated As_{Ga} antisite (Watkins, 1990 b).

any III–V semiconductor. On the other hand, an *impurity* group-III atom cation antisite – *boron* on the As site in GaAs – has been identified by local vibrational mode (LVM) spectroscopy. In this case, resolved structure due to the two gallium isotopes for each of the four neighbors is observed providing a direct confirmation (Gledhill et al., 1984). This provides strong evidence that the intrinsic III$_V$ antisite should also be present and therefore play an important role in the properties of the material.

In Ga-rich GaAs, a photoluminescence peak at 1.441 eV and an electrical acceptor level at $E_V + 0.078$ eV have been suggested to arise from the Ga$_{As}$ antisite (Yu, 1983; Elliot et al., 1982; Roos et al., 1989). This appears to be a reasonable deduction, confirmed somewhat by the fact that B$_{As}$ also appears to have an acceptor level near the valence band as indicated by the observation that the local mode disappears abruptly when the Fermi level goes below $\sim E_v + 0.2$ eV (Woodhead et al., 1983).

Theory has again made some interesting predictions for the III$_V$ antisite. In the first place, in one set of calculations, it has been predicted to be the dominant defect in Ga-rich GaAs, and when on-center in the As site, to be a double acceptor with its first acceptor level (–/0) at $\sim E_v + 0.25$ eV (Baraff and Schlüter, 1985). A more recent calculation has led to the interesting conclusion that the Ga atom is on-center only in the Ga$_{As}^=$ state. Upon two hole capture, it is predicted to break its bond with one Ga neighbor and eject toward the interstitial position, forming what can be considered a closely bound Ga$_i$ + V$_{As}$ pair (Zhang and Chadi, 1990). It is predicted to be a negative-U center with the $(=/0)$ occupation level at $\sim E_v + 0.17$ eV. (The coincidence of this level position with that for which the LVM B$_{As}$ band disappears is

suggestive, as pointed out by these authors.)

In conclusion, we have in GaAs some tentative experimental indications for the properties of the intrinsic III$_V$ antisite and an intriguing suggestion of structural instabilities for it from theory. The lack of magnetic resonance identification in this or any of the III–V materials, however, still makes this only speculation at present.

3.4.2 Group-III Atom Vacancies

The group-III atom vacancy has been identified only in GaP. The identification came first from EPR studies of room temperature electron-irradiated GaP where the characteristic hyperfine structure of the four phosphorus neighbors is resolved (Kennedy and Wilsey, 1981; Kennedy et al., 1983). Subsequent ENDOR studies have confirmed the identification (Hage et al., 1986). In terms of the simple molecular orbital model used for silicon (Fig. 3-2) and for ZnSe (Fig. 3-9), the center is observed in the neutral state which has three electrons in the t$_2$ orbital. It is therefore isoelectronic with V_{Si}^-. Contrary to V_{Si}^-, however, the total spin is $S = 3/2$ and no Jahn-Teller distortion occurs. Assuming that this is the ground state, the reduction of the Coulomb repulsive interactions between the three electrons achieved by placing them in a spatially antisymmetric combination of the three orthogonal t$_2$ orbitals apparently exceeds the gain available from a Jahn-Teller distortion, and the high spin many-electron $^4A_2$ state dominates. (As pointed out in Section 3.2.2.1 for the silicon vacancy, the balance between the two competing coupling schemes may be a delicate one that can vary from one system to another.)

Annealing studies reveal that V_{Ga} in GaP disappears in a 15 min isochronal anneal at $\sim 350\,^{\circ}\mathrm{C}$. Detailed kinetic studies have not been performed. However, using again the simple arguments of Section 3.3.1.3, Eq. (3-6), we can make a rough estimate for its migration barrier of ~ 1.5 eV. A level at $E_v + 0.64$ eV has been tentatively identified with a single donor level $(0/+)$ of V_{Ga} by correlation of its production and annealing with the EPR results (Mooney and Kennedy, 1984).

Theoretical calculations have been performed for V_{Ga} in GaAs which come to a most surprising conclusion (Baraff and Schlüter, 1985; Bar-Yam and Joannopoulos, 1986). They predict that the gallium vacancy should exhibit instability as a neighbor As atom moves into the gallium site producing an As_{Ga} antisite with an adjacent arsenic vacancy. Associated with this conversion is a strong negative-U property, the defect converting from $V_{Ga}^{=}$ to $(As_{Ga} + V_{As})^{3+}$ at $\sim E_v + 0.5$ eV. All intermediate charge states are predicted to be thermodynamically unstable although they might persist as excited metastable states.

The corresponding calculations have not been performed for GaP but it is interesting to speculate as to whether something similar might be happening there also. The neutral $^4A_2$ $S = 3/2$ V_{Ga} state seen by EPR and ENDOR in that material has only been seen after prolonged optical excitation and it is conceivable therefore that it is a metastable state. This brings up the further question as to whether the PP_3X or PP_3Y P_{Ga}-related centers discussed in Section 3.4.1.1 could possibly be the analogous $P_{Ga}V_P$ configuration for the gallium vacancy (Beall et al., 1984). Clearly further discussions concerning the metal vacancies in III–V materials is premature at this juncture.

3.4.3 Metal Interstitials

An $S = \frac{1}{2}$ ODMR spectrum has been reported in as-grown liquid encapsulated (LEC) GaP that has been attributed to an isolated gallium interstitial (Ga_i^{++}) (Lee, 1988). Strong isotropic hyperfine interactions with the two naturally abundant ^{69}Ga and ^{71}Ga isotopes are partially resolved as expected for a tightly bound S-state (a_1) of Ga_i^{++} on-center in the tetrahedral interstitial site. The alternative possibility of a Ga_{As} antisite could be ruled out on theoretical grounds because in a paramagnetic $S = \frac{1}{2}$ state it would have a partially filled t_2 level, with a node at the gallium nucleus. The identification as involving Ga_i^{++} is therefore probably correct but we must remember, as in the case of the V_{III} antisites (Section 3.4.1.1) that an S-state is very insensitive to a nearby defect and it cannot be ruled out that the interstitial is involved in a complex with a nearby impurity or defect.

Two triplet $(S = 1)$ low symmetry centers have also been reported in ODMR studies of luminescence in GaP that have been assigned to Ga_i-related centers. One, labeled here as Ga_iX was originally attributed to an oxygen impurity (Gal et al., 1979), but has more recently been assigned to a Ga_i-defect complex (Lee, 1988; Godlewski and Monemar, 1988). Another, labeled Ga_iY, has been observed in Cu- and Li-doped GaP (Chen and Monemar, 1989). Both reveal the characteristic hyperfine of the gallium nucleus expected for an excited $S = 1$ center involving the gallium interstitial.

In molecular beam (MBE) and organometallic vapor phase (OMVPE) epitaxially grown $Al_xGa_{1-x}As$, an $S = \frac{1}{2}$ ODMR spectrum detected in luminescence has also been reported that has been attributed to Ga_i^{++} (Kennedy et al., 1988). Here the partially resolved gallium hyperfine interaction displays anisotropy which could result simply from the alloy disorder or could also be evidence for a nearby defect. The spectral dependence of the luminescence associated with the center has led to a tentative assignment of its second donor state $(+/++)$, if isolated, at $\sim E_v + 0.5$ eV.

The apparent observation of Ga_i-related defects in as-grown samples of GaP and $Al_xGa_{1-x}As$ suggests that interstitial gallium is either relatively immobile in the lattice, or is paired off with some other immobile defect or impurity. Consistent with this, after annealing the $Al_xGa_{1-x}As$ samples at 850 °C, some of the ODMR signals attributed to Ga_i still remain. No reports of metal interstitials exist for the other III–V materials.

3.4.4 Defects on the Group-V Sublattice

No direct magnetic resonance *identification* of any group-V vacancy or interstitial exists for a III–V semiconductor.

On the other hand, these defects are definitely being produced by electron irradiation. In GaAs (Pons and Bourgoin, 1981) and InP (Massarani and Bourgoin, 1986), for example, the production of defects by displacements on the group-V atom sublattice have been established by electrical measurements. This has been accomplished by DLTS studies using bombarding electron energies very near the displacement threshold where the unique anisotropy of defect production expected for the group-V atom sublattice is observed.

In GaAs, arguments have been presented that most of the several DLTS peaks are associated with the primary $V_{As} + As_i$ Frenkel pairs of different lattice separations that are frozen into the lattice and stable up to ~ 500 K where they anneal primarily by recombining and annihilating

themselves (Pons and Bourgoin, 1985; Bourgoin et al., 1988). Evidence of charge state effects on the annealing rates and recombination-enhanced processes for the pairs have been presented (Stievenard and Bourgoin, 1986; Stievenard et al., 1986). A broad almost structureless EPR signal has been observed that anneals in the same temperature region as these DLTS levels which could possibly be associated therefore with these pairs. However, in the absence of resolved hyperfine interactions, this cannot be considered an identification.

In the modeling of the annealing processes in GaAs, it has been suggested that it is the As_i that is probably the mobile species. A spectroscopic piece of evidence that may be consistent with this is the observation in infrared absorption studies that room temperature electron irradiation produces new LVM spectra in both GaAs and GaP that can be identified as substitutional B_{Ga} and C_V impurities which have trapped a nearby defect (Newman and Woodhead, 1984). The vibrational frequencies suggest that the defect is interstitial. Identification with the group-V interstitial requires its separation from the Frenkel pair and migration at room temperature. It has been suggested that this could occur by recombination-enhanced processes accompanying the ionization associated with the long irradiation required for these studies or perhaps by an enhanced mobility in p-type material.

Again, in the absence of specific benchmark identifications by magnetic resonance, this scenario for defects on the group-V sublattice remains speculative.

3.5 Summary and Overview

It has been a difficult task to write a review of a rapidly evolving field such as this for a series intended to serve as a useful reference for years to come. It is this sobering fact that has dictated the decision here to concentrate primarily on what is believed to be well established. In doing this, we have left out many exciting and interesting models that have been postulated for intrinsic defects and their reaction processes. This is particularly true for the III–V semiconductors where there has been a great deal of current research activity but where magnetic resonance clarification has been somewhat less successful so far. There are two comprehensive recent reviews that deal with this subject for GaAs, the most thoroughly studied system. These are recommended to the interested reader who would like to explore further this fascinating subject (Pons and Bourgoin, 1985; Bourgoin et al., 1988).

Still, even though we have limited ourselves only to the better established defects, it is clear that we have begun to accumulate valuable insight into the properties of the intrinsic defects in semiconductors. Vacancies, interstitials, and/or antisites have been identified and studied in a few representative cases throughout the semiconductor series from covalent silicon, through the III–V's to the substantially ionic II–VI's. In this final section, we will first summarize briefly these cases, as presented in this chapter. We will then attempt an overview by pointing out a useful systematic pattern that they appear to fall into. This hopefully may serve as a guide to understanding, or predicting, the properties of all of the intrinsic defects – including those not yet identified – throughout the semiconductor series.

3.5.1 Summary

3.5.1.1 Silicon

The isolated vacancy can take on five charge states in the forbidden gap ($V^=$,

V^-, V^0, V^+, V^{++}). A simple one-electron molecular orbital model for the occupancy of the a_1 and t_2 orbitals for the undistorted vacancy gives a good description of the electronic structure of the defect, the Jahn-Teller lattice relaxations involved, and the electrical level position for each of its charge states. The activation energy for migration is small, ranging from 0.18–0.45 eV depending upon charge state, and it migrates athermally under electronic excitation associated with e–h pair recombination at the defect. The vacancy has negative-U properties, the first donor level $(0/+)$ at $E_v + 0.05$ eV, the second $(+/++)$ at $E_v + 0.13$ eV. This plus the recombination-enhanced migrational properties follow as a natural consequence of the large Jahn-Teller distortions.

Interstitial silicon has not been observed directly by magnetic resonance techniques but a great deal is believed to be known about it from study of its trapped configurations and from theory. It is highly mobile (athermal) even at 4.2 K under electron irradiation conditions in p-type material. In n-type material its activation energy for migration must be ≤ 0.57 eV. Several nearly energetic interstitial configurations (in the interstices and in bonding arrangements) appear to exist for the interstitial. Theory predicts a conversion between them vs. charge state and negative-U properties between Si_i^{++} which is stable in the tetrahedral site and Si_i^0, which prefers a bonded configuration. Alternation between the two sites accompanying e^- and h^+ capture provides an explanation for the athermal migration during electron irradiation.

3.5.1.2 II–VI Materials

The metal vacancy produces a double acceptor level $(=/-)$ in most II–VI materials (ZnS, ZnSe, CdS, BeO and ZnO) and in its single negatively charged state undergoes a large static trigonal distortion. In ZnS and ZnSe, complete CC diagrams have been constructed that describe the electrical and optical properties of the defects and identify the distortion as Jahn-Teller in origin. For ZnS and ZnSe, the second acceptor levels $(=/-)$ are at $E_v + 1.1$ eV and $E_v + 0.66$ eV and the migration energy barriers are 1.04 eV and 1.25 eV, respectively.

Interstitial zinc in ZnSe is on-center in the tetrahedral interstitial site surrounded by four Se atoms. It is a double donor with its second donor state $(+/++)$ deep at $\sim E_c - 0.9$ eV. Its migration energy is $\sim 0.6-0.7$ eV.

The chalcogen vacancy is also a deep donor, its singly ionized V^+ state being S-like, spread equally over its four group-II neighbors. It is stable at room temperature.

3.5.1.3 III–V Materials

The V_{III} antisite is a double donor which is present in as-grown materials. It has been observed by magnetic resonance techniques in GaP, GaAs, and InP in its singly positively charged state. In this charge state, the group-V atom is on-center in the group-III atom site with the unpaired electron highly localized in an S-like orbital centered on the atom.

Interstitial Ga_i^{++} also appears to have been detected in as-grown GaP and $Al_xGa_{1-x}As$. For it the unpaired electron is similarly in a deep S-orbital centered on the atom.

For both the V_{III} antisite and interstitial Ga_i^{++} the isotropic character of their S-electronic states makes it difficult to distinguish whether the defects being studied are truly isolated or paired with another impu-

rity or defect. Although this distinction is important in identifying the defect, the chemistry of its formation, and its electrical properties, the insensitivity to complexing assures us on the other hand that the *electronic structure* of the isolated defect is being correctly deduced.

The neutral group-III atom vacancy V_{III} has been observed in electron-irradiated GaP in a high spin $S = 3/2$ state. In this state the defect displays the full tetrahedral (T_d) symmetry of the undistorted lattice. It is observed after prolonged optical excitation at cryogenic temperatures, however, so there remains an uncertainty whether this is the true ground state of the neutral defect. The vacancy is stable only to $\sim 350\,°C$.

Theory predicts interesting structural instabilities for many of the intrinsic defects. These include the V_{III} and III_V antisites and the metal vacancy, V_{III}, which have been predicted to convert into intrinsic defect pairs:

$$V_{III} \rightleftharpoons V_i + V_{III} \tag{3-8}$$

$$III_V \rightleftharpoons III_i + V_V \tag{3-9}$$

$$V_{III} \rightleftharpoons V_{III} + V_V \tag{3-10}$$

These calculations have been for GaAs. For the As_{Ga} antisite conversion, Eq. (3-8), the pairs on the right have been predicted to be metastable and to account for the much studied EL2 defect in GaAs. The Ga_{As} antisite, Eq. (3-9) and the gallium vacancy, Eq. (3-10), are predicted to be bistable, the isolated symmetric defect on the left being stable in n-type material but converting to the pairs in p-type material with strong negative-U character. These instabilities have not been confirmed experimentally, however, at the date of this writing.

3.5.2 Overview

In the case of the silicon vacancy and the metal vacancy in II–VI materials, it was found instructive to construct a singly degenerate (a_1) and a triply degenerate (t_2) molecular orbital for the defect in its undistorted T_d lattice arrangement which could be populated by the appropriate number of electrons for each defect charge state. This served to account for the cases where large symmetry lowering lattice relaxations are encountered as being the result of electronic degeneracy and a consequent Jahn-Teller distortion. These distortions in turn were found to have important consequences for the electrical level positions, the optical transitions, and the migrational properties of the defect.

Let us attempt here therefore to extend this approach to all of the intrinsic defects throughout the semiconductor series to see what insight this might provide. For a vacancy on either sublattice, one can always construct a symmetric combination of the ruptured bonds on the nearest neighbors of the other sublattice to form an a_1 orbital, and three equally energetic combinations (antisymmetric by pairs) to form the t_2 orbitals, higher in energy, as was described first for the silicon vacancy in Section 3.2.2.1. This is reproduced in Fig. 3-14a. This can also be done for any interstitial as well, where the a_1 state now represents the outer s- and the t_2 states the outer p-orbitals of the interstitial atom. In this simple approach, therefore, the only qualitative difference between a vacancy on either sublattice or an interstitial in any semiconductor is the electron occupancy and the positions of the occupied and unoccupied a_1 and t_2 orbitals with respect to the band gap.

3.5.2.1 Vacancies

In Table 3-3 we summarize the a_1 and t_2 level occupancies for the various charge states of vacancies in the group IV, III–V, and II–VI semiconductors. We have included also the ionic I–VII alkali halides for the further insight that they can provide. The configuration in each case is denoted $a_1^m t_2^n$ where m is the number of electrons in the a_1 orbital and n is the number in the t_2 orbital. We have assumed here that the strong crystal field limit applies (the $a_1 - t_2$ splitting is greater than the electron-electron interactions) and we fill the a_1 level before occupying the t_2 levels.

Table 3-3. Electron occupancies ($a_1^m t_2^n$) for the various charge states of vacancies in the group IV, III–V, II–VI, and I–VII materials.

m	n	I	II	III	IV	V	VI	VII	
2	6	−	=	≡					
2	5	**0**	−	=					⎫
2	4		**0**	−	=				⎪
2	3			**0**	−				⎬ Jahn-Teller
2	2				**0**				⎪
2	1				+	**0**			⎪
2	0				++	+	**0**		⎭
1	0						+	**0**	
0	0						++	+	

Experimentally observed occupancies are shown in bold print.

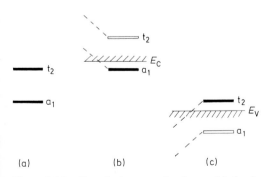

Figure 3-14. One-electron molecular orbitals in undistorted (T_d) symmetry for a) vacancy or host interstitial, b) V_{III} antisite, c) III_V antisite.

Orbital degeneracy occurs only for the vacancies which have partially filled t_2 shells, i.e., $n = 1$ to 5. For the others, Jahn-Teller symmetry lowering distortions should not occur and we anticipate full T_d symmetry for the defect. This has been confirmed directly for V_{VI}^+ in the II–VI materials and V_{VII}^0 [the F center (Klick, 1972)] in the I–VII alkali halides and inferred indirectly for $V_{II}^=$ in II–VI's (Fig. 3-9) and V_{IV}^{++} in silicon (Fig. 3-2). Accordingly, we expect full T_d symmetry also for V_{III}^{-3}, V_V^+, and $V_{VI}^{0,+,++}$ in the semiconductors.

On the other hand, we can anticipate symmetry lowering Jahn-Teller distortions for all of the other cases shown in the table where partial occupancy of the t_2 levels occurs. This has already been confirmed for the metal vacancies in II–VI's and the vacancy in silicon. The alkali halide extreme also confirms this for V_I^0 [V_F center (Kabler, 1972)]. Consider, therefore the group-III vacancy in a III–V semiconductor. Table III predicts that V_{III}^{-3} is stable in full T_d symmetry, but that ionization to $V_{III}^=$ and to the less negatively charged states should produce Jahn-Teller distortions which increase progressively vs. ionization as it goes from $a_1^2 t_2^5$ to $a_1^2 t_2^4$, etc. This provides a clear physical origin for the theoretical predictions for the gallium vacancy in GaAs, Eq. (3-10). An inward trigonal distortion of one of the As neighbors occurs for $V_{III}^=$ just as for V_{II}^- (both $a_1^2 t_2^5$). The distortion increases with further ionization, the As atom propelling toward the Ga site leaving an adjacent arsenic vacancy behind.

Contrary to this simple picture, the neutral gallium vacancy in GaP has been observed in its high spin $S = 3/2$ state with full T_d symmetry even though its configuration is $a_1^2 t_2^3$. There appear to be two possibilities: either as discussed in Section 3.4.2, the electron-electron interactions exceed the Jahn-Teller distortion energies in this case, or an excited metastable state is being observed. The large lattice relaxational energies that have been calculated in the GaAs case provide a strong hint that the latter explanation may be the correct one.

3.5.2.2 Interstitials

In Table 3-4 we present the corresponding a_1 (s) and t_2 (p) occupancies predicted for host interstitials. Again as predicted, the two observed cases, Zn_i^+ and Ga_i^{++}, with $a_1^1 t_2^0$ configurations, have the full T_d symmetry. The only other experimentally observed case is the interstitial halogen atom in alkali halides [H-center (Kabler, 1972)] and, as predicted for its $a_1^2 t_2^5$ configuration, it is strongly distorted.

This molecular orbital approach also provides a logical explanation for the theoretically predicted properties for the silicon interstitial. The predicted on-center tetrahedral site for Si_i^{++} follows naturally from its $a_1^2 t_2^0$ configuration. Its movement off this site in the Si^+ and Si^0 states in turn follows naturally from the partial t_2 occu-

Table 3-4. Electron occupancies ($s^m p^n$) for the charge states of host interstitial atoms in the group IV, III–V, II–VI, and I–VII materials.

m	n	I	II	III	IV	V	VI	VII	
2	6								
2	5							**0**	
2	4						0		
2	3				−	0	+		} Jahn-Teller
2	2				0	+	++		
2	1			0	+	++			
2	0		0	+	++				
1	0	0	**+**	**++**					
0	0	+	++						

Experimentally observed occupancies are shown in bold print.

pancy, the negative-U properties reflecting the much stronger Jahn-Teller driving force (X2) for the distortion in Si_i^0 with $a_1^2 t_2^2$. From Table 3-4, we are led to anticipate similar interesting effects for the Group-V interstitials in III–V materials and also the Group-VI interstitials in II–VI's.

3.5.2.3 Antisites

The antisites can also be cast into this molecular orbital picture, with the aid of Figs. 3-14b and c. Here the antisites are being viewed as simple substitutional double donors (or acceptors) and treated therefore as host atoms but with an increase (or decrease) in nuclear charge of $2\,e$. As normal host atoms, their s and p orbitals interact, respectively, with the a_1 and t_2 vacancy orbitals of the neighbors forming bonding states pushed downward into the valence band and antibonding states shifted upward into the conduction band leaving the gap empty (i.e., "healing" the vacancy). Increasing the nuclear charge as in the V_{III} antisite serves to pull these antibonding states down, the a_1 state returning into the gap as depicted in Fig. 3-14b. Decreasing the nuclear charge, as for the III_V antisite, permits the bonding t_2 states to rise back into the gap as shown in Fig. 3-14c.

Referring to Fig. 3-14c, the theoretically predicted instability for the III_V antisite, Eq. (3-9), again follows naturally. For the double minus state, the t_2 level is filled and the group III atom resides on-center in the group-V atom site. The singly ionized state III_V^- is t_2^5 and Jahn-Teller distorted, the group-III atom displacing trigonally toward the interstitial site. In the neutral state III_V^0 (t_2^4) the Jahn-Teller coupling is doubled, the atom displacing farther toward the interstitial position. Again as for the silicon vacancy, Eqs. (3-1) to (3-3) in

Section 3.2.2.3, negative-U can result between the neutral and doubly negative states.

Referring to Fig. 3-14b, the V_{III} antisite should remain undistorted for its three charge states: V_{III}^0 (a_1^2), V_{III}^+ (a_1^1), and V_{III}^{++} (a_1^0). This again has been confirmed directly for the paramagnetic V_{III}^+ state studied by EPR, ODMR and ODENDOR. The figure also povides an explanation for the theoretically predicted metastable EL2 behavior for the isolated V_{III} antisite as described in Section 3.4.1.1. Doubly occupying the t_2 level in the conduction band provides an excited state for the neutral defect ($a_1^0 t_2^2$) which can Jahn-Teller distort in a similar fashion to that for III_V above, the now central V atom injecting into the interstitial position as shown in Fig. 3-13c. In this case, the Jahn-Teller relaxation energy is not quite sufficient to overcome the $a_1 - t_2$ promotion energy and a long-lived metastable neutral state is formed.

3.5.2.4 Other Intrinsic Defects

These simple a_1 and t_2 molecular orbital models clearly supply therefore a framework which accounts with remarkable success for the electronic structure and lattice relaxational properties of all of the experimentally documented intrinsic defects discussed in this chapter as well as those for which recent *ab initio* local density calculations have made as yet untested predictions of meta- and bi-stability, negative-U, etc.

They should also therefore provide hints as to the properties of the intrinsic defects which have not yet been experimentally identified. Consider, for example, the interstitials, Table 3-4. The as yet undetected group IV, V, and VI interstitials all have partially filled t_2 states for most of their expected charge states. This suggests, as predicted for the silicon interstitial, large lattice relaxational changes vs. charge state, which can contribute to charge dependent and recombination-enhanced migrational properties and possible negative-U. As described in Section 3.4.4, evidence for charge dependent and/or recombination-enhanced migration of arsenic interstitials in GaAs has indeed been cited. The failure so far to detect any of these interstitials by magnetic resonance could be explained by negative-U properties which tend to make the paramagnetic odd-electron charge states thermodynamically unstable.

From Table 3-3, it is the group II, III, and IV vacancies for which partial t_2 occupancy dominates the expected charge states. The importance of Jahn-Teller effects for the group-II and group-IV vacancies has been clearly established. Table 3-3 suggests therefore that we should expect the same for the group-III vacancies. As mentioned earlier, this implies that V_{Ga}^0 in GaP, the only group-III vacancy detected so far, may actually have been observed in an excited metastable state, its undistorted (T_d) $S = 3/2$ state not representative of the ground state which would be strongly Jahn-Teller distorted. Again, negative-U properties for the group-III vacancies would contribute to the difficulties of magnetic resonance detection.

3.5.2.5 Migration Barriers

Migration barrier energies for the intrinsic defects in silicon have been found to be small (~ 0–0.6 eV) and charge state dependent. Under recombination, migration for both is athermal. For the compound semiconductors, we expect somewhat higher barriers, motion on one sublattice being hindered by the presence of atoms on the other. This is borne out for metal vacancies in ZnS (1.04 eV), ZnSe (1.26 eV) and GaP (~ 1.5 eV) and also for the zinc interstitial in ZnSe (~ 0.6–0.7 eV). In all cases where large lattice relaxational changes are predicted vs. charge state, we may anticipate strong charge state dependence and recombination-enhancement contributions to the diffusional mechanisms for the defects.

3.6 Acknowledgements

This review was made possible by support from National Science Foundation grant DMR-89-02572 and Office of Naval Research Electronics and Solid State Program grant N000014-90-J-1264.

3.7 References

Ammerlaan, C. A. J., Sprenger, M., Van Kemp, R., Van Wezep, D. A. (1985), in: *Microscopic Identification of Electronic Defects in Semiconductors:* Johnson, N. M., Bishop, S. G., Watkins, G. D. (Eds.). Pittsburgh: Mat. Res. Soc., Vol. 46, p. 227.

Anderson, P. W. (1975), *Phys. Rev. Lett. 34,* 953.

Baraff, G. A., Kane, E. O., Schlüter, M. (1979), *Phys. Rev. Lett. 43,* 956.

Baraff, G., Schlüter, M. (1979), *Phys. Rev. B19,* 4965.

Baraff, G. A., Kane, E. O., Schlüter, M. (1980a), *Phys. Rev. B21,* 5662.

Baraff, G. A., Kane, E. O., Schlüter, M. (1980b), *Phys. Rev. B21,* 3563.

Baraff, G. A., Schlüter, M. (1984), *Phys. Rev. B30,* 3460.

Baraff, G. A., Schlüter, M. (1985), *Phys. Rev. Lett. 55,* 1327.

Bar-Yam, Y., Joannopoulos, J. D. (1984), *Phys. Rev. B30,* 2216.

Bar-Yam, Y., Joannopoulos, J. D. (1985), in: *Thirteenth International Conference on Semiconductors:* Kimerling, L. C., Parsey, J. M. Jr. (Eds.). Warrendale: A.I.M.E., Vol. 14a, p. 261.

Bar-Yam, Y., Joannopoulos, J. D. (1986), *Mat. Sci. Forum 10–12,* 19.

Beall, R. B., Newman, R. C., Whitehouse, J. E., Woodhead, J. (1984), *J. Phys. C. 17,* L963.

Bean, A. R., Newman, R. C. (1971), *Solid State Commun. 9,* 271.

Bemski, G. (1959), *J. Appl. Phys. 30,* 1195.

Bernholc, J., Lipari, N. O., Pantelides, S. T. (1980), *Phys. Rev. B21,* 3545.

Bourgoin, J. C., Corbett, J. W. (1978), *Rad. Eff. 36*, 157.

Brelot, A., Charlemagne, J. (1971), in: *Radiation Effects in Semiconductors:* Corbett, J. W., Watkins, G. D. (Eds.). London: Gordon and Breach, p. 161.

Bourgoin, J. C., von Bardeleben, H. J., Stievenard, D. (1988), *J. Appl. Phys. 64*, R65.

Brower, K. L. (1971), *Phys. Rev. B4*, 1968.

Brower, K. L. (1972), *Phys. Rev. B5*, 4274.

Brower, K. L. (1974), *Phys. Rev. B9*, 2607.

Car, R., Kelly, P. J., Oshiyama, A., Pantelides, S. T. (1985), in: *Thirteenth International Conference on Defects in Semiconductors:* Kimerling, L. C., Parsey, J. M. Jr. (Eds.). Warrendale: A.I.M.E., p. 269.

Cavenett, B. C. (1981), *Adv. Phys. 30*, 475.

Chadi, D. J., Chang, K. J. (1988), *Phys. Rev. Lett. 60*, 2187.

Chappell, S. P., Newman, R. C. (1987), *Semicond. Sci. Technol. 2*, 691.

Chen, W. M., Monemar, B. (1989), *Phys. Rev. B40*, 1365.

Chen, W. M., Monemar, B., Pistol, M. E. (1989), *Mat. Sci. Forum 38–41*, 899.

Cheng, L. J., Corelli, J. C., Corbett, J. W., Watkins, G. D. (1966), *Phys. Rev. 152*, 761.

Corbett, J. W., Watkins, G. D., Chrenko, R. M., McDonald, R. S. (1961), *Phys. Rev. 121*, 1015.

Corbett, J. W., Watkins, G. D. (1965), *Phys. Rev. 138*, A555.

Corbett, J. W. (1966), *Electron Radiation Damage in Semiconductors and Metals.* New York: Academic Press.

Dabrowski, J., Scheffler, M. (1989), *Phys. Rev. B40*, 10391.

Davies, G. (1988), *Mat. Sci. Forum 38–41*, 151.

Davies, G., Lightowlers, E. C., Newman, R. C., Oates, A. C. (1987), *Semicond. Sci. Technol. 2*, 524.

Deiri, M., Kana-ah, A., Cavenett, B. C., Kennedy, T. A., Wilsey, N. D. (1984), *J. Phys. C 17*, L793.

deWit, J. G., Sieverts, E. G., Ammerlaan, C. A. J. (1976), *Phys. Rev. B14*, 3494.

DuVarney, R. C., Garrison, A. K., Thorland, R. H. (1969), *Phys. Rev. 188*, 657.

Elkin, E. L., Watkins, G. D. (1968), *Phys. Rev. 174*, 881.

Elliot, K. R., Holmes, D. E., Chen, R. T., Kirkpatrick, C. G. (1982), *App. Phys. Lett. 40*, 898.

Gal, M., Cavenett, B. C., Smith, P. (1979), *Phys. Rev. Lett. 43*, 1611.

Galland, D., Hervé, A. (1970), *Phys. Lett. A 33*, 1.

Galland, D., Hervé, A. (1974), *Solid State Commun. 14*, 953.

Garrison, A. K., DuVarney, R. C. (1973), *Phys. Rev. B7*, 4689.

Gislason, H. P., Sun, H., Rong, F., Watkins, G. D. (1990), in: *The Physics of Semiconductors:* Anastassakis, E. M., Joannopoulos, J. D. (Eds.). Singapore: World Scientific, Vol. I, p. 666.

Gledhill, G. A., Newman, R. C., Woodhead, J. (1984), *J. Phys. C 17*, L301.

Godlewski, M., Monemar, B. (1988), *Phys. Rev. B37*, 2752.

Godlewski, M., Chen, W. M., Monemar, B. (1989), *Defect and Diffusion Forum 62/63*, 107.

Gonzalez, C., Galland, D., Herve, A. (1975), *Phys. Stat. Sol. B72*, 309.

Hage, J., Niklas, J. R., Spaeth, J.-M. (1986), *Materials Science Forum 10–12*, 259.

Harris, R. D., Watkins, G. D. (1984), *J. Elec. Mater. 14*, 799.

Harris, R. D., Newton, J. L., Watkins, G. D. (1987), *Phys. Rev. B36*, 1094.

Hervé, A., Maffeo, B. (1970), *Phys. Lett. A 32*, 247.

Hoinkis, M., Weber, E. R. (1989), *Phys. Rev. B40*, 3872.

Holton, W. C., deWitt, M., Estle, T. L. (1966), in: *International Symposium on Luminescence:* Riehl, N., Kallmann, H. (Eds.). Munich: Verlag Karl Thiemig KG, p. 454.

Jahn, H. A., Teller, E. (1937), *Proc. Roy. Soc. A161*, 220.

Jeon, D. Y., Gislason, H. P., Watkins, G. D. (1986), *Mat. Sci. Forum 10–12*, 851.

Jeon, D. Y., Gislason, H. P., Donegan, J. F., Watkins, G. D. (1987), *Phys. Rev. B36*, 1324.

Jeon, D. Y. (1988), Ph. D. dissertation, Lehigh University, unpublished.

Kabler, M. N. (1972), in: *Point Defects in Solids*, Vol I: *General and Ionic Crystals:* Crawford, J. H. Jr., Slifkin, L. M. (Eds.). New York: Plenum, Ch. 6.

Kaminska, M., Skowronski, M., Kuszko, W. (1985), *Phys. Rev. Lett. 55*, 2204.

Kana-ah, A., Deiri, M., Cavenett, B. C., Wilsey, N. D., Kennedy, T. A. (1985), *J. Phys. C 18*, L619.

Kaufmann, U., Schneider, J., Räuber, A. (1976), *App. Phys. Lett. 29*, 312.

Kaufmann, U., Schneider, J., Worner, R., Kennedy, T. A., Wilsey, N. D. (1981), *J. Phys. C 14*, L951.

Kaxiras, E., Pandey, K. C. (1989), *Phys. Rev. B40*, 8020.

Kennedy, T. A., Wilsey, N. D. (1978), in: *Defects and Radiation Effects in Semiconductors:* Albany, J. H. (Ed.). London: Inst. Phys., Conf. Se. No. 46, 375.

Kennedy, T. A., Wilsey, N. D. (1981), *Phys. Rev. B23*, 6585.

Kennedy, T. A., Wilsey, N. D., Krebs, J. J., Stauss, G. H. (1983), *Phys. Rev. Lett. 50*, 1281.

Kennedy, T. A., Wilsey, N. D. (1984), *App. Phys. Lett. 44*, 1089.

Kennedy, T. A., Wilsey, N. D., Klein, P. B., Henry, R. L. (1986), *Mat. Sci. Forum 10–12*, 271.

Kennedy, T. A., Magno, R., Spencer, M G. (1988), *Phys. Rev. B37*, 6325.

Killoran, N., Cavenett, B. C., Godlewski, M., Kennedy, T. A., Wilsey, N. D. (1982), *J. Phys. C 15*, L723.

Kimerling, L. C. (1977), in: *Radiation Effects in Semi-*

conductors: Urli, N. B., Corbett, J. W. (Eds.). London: Inst. Phys., Conf. Se. No. 46, p. 221.

Kimerling, L. C. (1978), *Solid State Elec. 21,* 1391.

Kimerling, L. C., Asom, M. T., Benton, J. L., Devrinsky, P. J., Caefer, C. E. (1989), *Mat. Sci. Forum 38–41,* 141.

Klick, C. C. (1972), in: *Point Defects in Solids, Vol 1 General and Ionic Crystals:* Crawford, J. H. Jr., Slifkin, L. M. (Eds.). New York: Plenum, Ch. 5.

Lannoo, M., Bourgoin, J. (1981), *Point Defects in Semiconductors I.* Berlin: Springer-Verlag.

Lannoo, M., Baraff, G., Schlüter, M. (1981), *Phys. Rev. B24,* 945.

Lannoo, M. (1983), *Phys. Rev. B28,* 2403.

Lee, K. M., LeSi, D., Watkins, G. D. (1980), *Solid State Commun. 35,* 527.

Lee, K. M., LeSi, D., Watkins, G. D. (1981), in: *Defects and Irradiation Effects in Semiconductors 1980:* Hasiguti, R. R. (Ed.). London: Inst. Phys., Conf. Se. No. 59, 353.

Lee, K. M., O'Donnell, K. P., Watkins, G. D. (1982), *Solid State Commun. 91,* 881.

Lee, K. M. (1983), Ph. D. dissertation, Lehigh University, unpublished.

Lee, K. M. (1988), in: *Defects in Electronic Materials:* Stavola, M., Pearton, S. J., Davies, G. (Eds.). Pittsburgh: Mat. Res. Soc., Vol. 104, 449.

Maffeo, B., Hervé, A., Cox, R. (1970), *Solid State Commun. 8,* 1205

Maffeo, B., Hervé, A., Rius, G., Santier, C., Picard, R. (1972), *Solid State Commun. 10,* 1205.

Maffeo, B., Hervé, A. (1976), *Phys Rev. B13,* 1940.

Malvido, J. C., Whitten, J. L. (1982), *Phys. Rev. B26,* 4458.

Martin, G. M., Markram-Ebeid, S. (1986), in: *Deep Centers in Semiconductors:* Pantelides, S. T. (Ed.). New York: Gordon and Breach, Ch. 6.

Massarani, B., Bourgoin, J. C. (1986), *Phys. Rev. B34,* 2470.

Meyer, B. K., Hofmann, D. M., Niklas, J. R., Spaeth, J. M. (1987), *Phys. Rev. B36,* 1332.

Mooney, P.M., Kennedy, T. A. (1984), *J. Phys. C 17,* 6277.

Newman, R. C., Woodhead, J. (1984), *J. Phys. C 17,* 1405.

Newton, J. L., Chatterjee, A. P., Harris, R. D., Watkins, G. D. (1983), *Physica 116B,* 219.

Nicholls, J. E., Dunstan, D. J., Davies, J. J. (1978), *Semicond. Insul. 4,* 119.

Niklas, J. R., Spaeth, J. M., Watkins, G. D. (1985), in: *Microscopic Identification of Electronic Defects in Semiconductors:* Johnson, N. M., Bishop, S. G., Watkins, G. D. (Eds.). Pittsburgh: Mat. Res. Soc., Vol. 46, p. 237.

O'Donnell, K. P., Lee, K. M., Watkins, G. D. (1982), *Solid. State Commun. 44,* 1015.

Pons, D., Bourgoin, J. C. (1981), *Phys. Rev. Lett. 18,* 1293.

Pons, D., Bourgoin, J. C. (1985), *J. Phys. C 18,* 3839.

Rong, F., Watkins, G. D. (1986a), *Phys. Rev. Lett 56,* 2310.

Rong, F., Watkins, G. D. (1986b), *Mat. Sci. Forum 10–12,* 827.

Rong, F., Watkins, G. D. (1987), *Phys. Rev. Lett. 58,* 1486.

Rong, F., Barry, W. A., Donegan, J. F., Watkins, G. D. (1988), *Phys. Rev. B37,* 4329.

Roos, G., Schoner, A., Penal, G., Krambrock, K., Meyer, B. K., Spaeth, J. M., Wagner, J. (1989), *Mat. Sci. Forum 38–41,* 951.

Schirmer, O. F., Schnadt, R. (1976), *Solid State Commun. 18,* 1345.

Schneider, J., Räuber, A. (1967), *Solid State Commun. 5,* 779.

Shono, Y. (1979), *J. Phys. Soc. Jpn. 47,* 590.

Sieverts, E. G., Muller, S. H., Ammerlaan, C. A. J. (1978), *Phys. Rev. B18,* 6834.

Smargiassi, E., Car, R. (1990), *Bull. Am. Phys. Soc. 35,* 621.

Smith, J. M., Vehse, W. H. (1970), *Phys. Lett. A 31,* 147.

Song, L. W., Zhan, X. D., Benson, B. W., Watkins, G. D. (1988), *Phys. Rev. B42,* 5765.

Song, L. W., Watkins, G. D. (1990), *Phys. Rev. B42,* 5759.

Spaeth, J. M. (1986), *Mat. Sci. Forum 10–12,* 505.

Spaeth, J. M. (1990), private communication.

Sprenger, M., Muller, S. H., Ammerlaan, C. A. J. (1983), *Physica 116B,* 224.

Sprenger, M., van Kemp, R., Sieverts, E. G., Ammerlaan, C. A. J. (1985), in: Thirteenth International Conference on Defects Semiconductors: Kimerling, L. C., Parsey, J. M. Jr. (Eds.). Warrendale, A.I.M.E., Vol. 14a, 815.

Stievenard, D., Bourgoin, J. C. (1986), *Phys. Rev. B33,* 8410.

Stievenard, D., Boddaert, X., Bourgoin, J. C. (1986), *Phys. Rev. B34,* 4048.

Stoneham, A. M. (1975), *Theory of Defects in Solids.* Oxford: Clarendon Press, Ch. 27.

Sturge, M. D. (1967), in: *Solid State Physics, Vol. 20:* Seitz, F., Turnbull, D., Ehrenreich, H. (Eds.). New York: Academic Press, pp. 91–211.

Tan, T. Y., Gösele, U. (1985), *Appl. Phys. A37,* 1.

Taylor, A. L., Filipovich, G., Lindeberg, G. K. (1970), *Solid State Commun. 8,* 1359.

Taylor, A. L., Filipovich, G., Lindberg, G. K. (1971), *Solid State Commun. 9,* 945.

Tipping, A. K., Newman, R. C. (1987), *Semicond. Sci. Technol. 2,* 389.

Trautman, P., Walczak, J. P., Baranowski, J. M. (1990), *Phys. Rev. B41,* 3074.

Trombetta, J. M., Watkins, G. D. (1987), *Appl. Phys. Lett. 51,* 1103.

Troxell, J. R., Chatterjee, A. P., Watkins, G. D., Kimerling, L. C. (1979), *Phys. Rev. B19,* 5336.

Troxell, J. R. (1979), Ph. D. dissertation, Lehigh University, unpublished.

Troxell, J. R., Watkins, G. D. (1980), *Phys. Rev. B22,* 921.

Von Bardeleben, H. J., Bourgoin, J. C. (1986), *Phys. Rev. B33,* 2890.

Von Bardeleben, H. J., Stievenard, D., Deresmes, D., Huber, A., Bourgoin, J. C. (1986), *Phys. Rev. B34,* 7192.

Watkins, G. D., Corbett, J. W. (1961), *Phys. Rev. 121,* 1001.

Watkins, G. D. (1964), in: *Radiation Damage in Semiconductors,* Paris: Dunod, 97.

Watkins, G. D., Corbett, J. W. (1964), *Phys. Rev. 134,* A 1359.

Watkins, G. D., Corbett, J. W. (1965), *Phys. Rev. 138,* A 543.

Watkins, G. D. (1967), *Phys. Rev. 155,* 802.

Watkins, G. D. (1969), *IEEE Trans. Nucl. Sci. NS-16,* 13.

Watkins, G. D. (1971), in: *Radiation Effects in Semiconductors:* Corbett, J. W., Watkins, G. D. (Eds.). New York: Gordon and Breach, 301.

Watkins, G. D. (1974). *Phys., Rev. Lett. 33,* 223.

Watkins, G. D. (1975a), in: *Point Defects in Solids,* Vol 2: *Semiconductors and Molecular Crystals:* Crawford, J. H. Jr., Slifkin, L. M. (Eds.) New York: Plenum, pp. 333–392.

Watkins, G. D. (1975b), in: *Lattice Defects in Semiconductors 1974:* Huntley, F. A. (Ed.). London: Inst. Phys. Conf. Se. No. 23, 1.

Watkins, G. D. (1975c), *Phys. Rev. B12,* 4383.

Watkins, G. D. (1975d), *Phys. Rev. B12,* 5824.

Watkins, G. D. (1975e), in: *Lattice Defects in Semiconductors 1974:* Huntley, F. A. (Ed.). London: Inst. Phys., Conf. Se. No. 23, 338.

Watkins, G. D. (1975f), *Intrinsic Defects in II–VI Compounds* (ARL TR 75-0011). Clearinghouse, Springfield, VA 22151 USA: National Technical Information Services.

Watkins, G. D. (1976), *Phys. Rev. B13,* 2511.

Watkins, G. D., Brower, K. L. (1976), *Phys. Rev. Lett 36,* 1329.

Watkins, G. D. (1977), in: *Radiation Effects in Semiconductors 1976:* Urli, N. B., Corbett, J. W. (Eds.). London: Inst. Phys., Conf. Se. No. 31, 95.

Watkins, G. D., Troxell, J. R., Chatterjee, A. P. (1979), in: *Defects and Radiation Effects in Semiconductors 1978:* Albany, J. H. (Ed.). London: Inst. Phys., Conf. Se. No. 46, p. 16.

Watkins, G. D., Troxell, J. R. (1980), *Phys. Rev. Lett. 44,* 593.

Watkins, G. D., Chatterjee, A. P., Harris, R. D., Troxell, J. R. (1983), *Semicond. Insul. 5,* 321.

Watkins, G. D. (1983), *Physica 117B–118B,* 9.

Watkins, G. D. (1984), in: *Festkörperprobleme XXIV:* Grosse, P. (Ed.). Braunschweig: Vieweg, pp. 163–189.

Watkins, G. D. (1986), in: *Deep Centers in Semiconductors,* Pantelides, S. (Ed.). New York: Gordon and Breach, pp. 147–183.

Watkins, G. D., Rong, F., Barry, W. A., Donegan, J. F. (1988), in: *Defects in Electronic Materials:* Stavola, M., Pearton, S. J., Davies, G. (Eds.). Pittsburgh: Mat. Res. Soc., Vol. 104, 3.

Watkins, G. D. (1989), *Rad. Eff. and Defects in Sol. 111 & 112,* 487.

Watkins, G. D. (1990a), in: *Defect Control in Semiconductors:* Sumino, K. (Ed.). Amsterdam: North Holland, Vol. I, pp. 933–941.

Watkins, G. D. (1990b), in: *Atomic Processes Induced by Electronic Excitation in Non-Metallic Solids:* Itoh, N., Fowler, W. B. (Eds.). Singapore: World Scientific, pp. 149–166.

Weber, E. R., Omling, P. (1985), in: *Festkörperprobleme XXV:* Grosse, P. (Ed.). Braunschweig: Vieweg, p. 623.

Woodhead, J., Newman, R. C., Grant, I., Rumsby, D., Ware, R. M. (1983), *J. Phys. C 16,* 5523.

Yu, P. W. (1983), *Phys. Rev. B27,* 7779.

Zhan, X. D., Watkins, G. D. (1991), *Appl. Phys. Lett.,* in press.

Zhang, S. B., Chadi, D. J. (1990), *Phys. Rev. Lett. 64,* 1789.

General Reading

Bourgoin, J. C., von Bardeleben, H. J., Stievenard, D. (1988), "Native defects in gallium arsenide", *J. Appl. Phys. 64,* R65–R91.

Pons, D., Bourgoin, J. C. (1985), "Irradiation-induced defects in GaAs", *J. Phys. C: Solid State Phys. 18,* 3839–3871.

Watkins, G. D., "EPR Studies of Lattice Defects in Semiconductors", in: *Defects and their Structures in Nonmetallic Solids (1976):* Henderson, B., Hughes, A. E. (Eds.). New York: Plenum, pp. 203–220.

Watkins, G. D. (1983), "Deep Levels in Semiconductors", *Physics 117B–118B,* 9–15.

Watkins, G. D. (1984), "Negative-U Properties for Defects in Solids", in: *Festkörperprobleme XXIV:* Grosse, P. (Ed.). Braunschweig: Vieweg, pp. 163–189.

Watkins, G. D. (1986), "The Lattice Vacancy in Silicon", in: *Deep Centers in Semiconductors:* Pantelides, S. (Ed.). New York: Gordon and Breach, pp. 147–183.

Watkins, G. D. (1990a), "Intrinsic Defects on the Metal Sublattice in ZnSe", in: *Defect Control in Semiconductors:* Sumino, K. (Ed.). Amsterdam: North Holland, Vol. I, pp. 933–941.

4 Deep Centers in Semiconductors

Helmut Feichtinger

Institut für Experimentalphysik der Karl-Franzens-Universität, Graz, Austria

List of Symbols and Abbreviations

A	hyperfine parameter
$\boldsymbol{B}$	magnetic field
D^0, D^+	neutral and ionized donors
E_a, E_e	energy of absorption and emission
E_b	pair binding energy
E_B	barrier height for capture rate
E_C, E_V	energy of the conduction and valence band edges
E_D	binding energy of shallow donors
E_f	Fermi energy, related to either E_C or E_V
E_m	migration energy
E_n	phonon energy of state n
E_{opt}, E_{th}	optical and thermal ionization energy
$E^{q/q+1}$	occupancy level
$E_{q,q+1}$	energy change upon adding one further electron
E_R	lattice relaxation energy
E_{ZP}	zero phonon transition energy
$\Delta E^{(ij)}$	transition energy
e	electron
G	Gibbs free enthalpy
$\Delta G^{q/q+1}$	standard chemical potential for ionization
g	spectroscopic splitting factor
g_q	degeneracy factor
$H(q)$	enthalpy
$\Delta H^{q/q+1}$	ionization enthalpy
h	Planck constant
$\hbar$	reduced Planck constant
$\boldsymbol{I}$	spin of nucleus
$\boldsymbol{J}$	total angular momentum
k	Boltzmann's constant
$\boldsymbol{k}$	wave vector
$\boldsymbol{L}$	orbital angular momentum
M, m	orbital and nuclear quantum number
N_A, N_D	acceptor and donor concentration
$N_{C,V}$	effective density of states
p	pressure
p	hole concentration
Q	configurational coordinate
q	charge state of the defect
S	Huang-Rhys factor
$S(q)$	entropy
$\boldsymbol{S}$	total electron spin
$\Delta S^{q/q+1}$	ionization entropy
T	temperature

U	Mott-Hubbard correlation energy
V	crystal volume
$V(r)$	defect potential
v	anion-cation transfer matrix element
$\langle v \rangle_{n,p}$	mean thermal velocity
x	mole fraction
Z	valency number
η	ratio of the radiative recombination rate to the total recombination rate
μ_e	chemical potential of an electron
ν	frequency
$\sigma_{n,p}$	thermal-capture cross section for electrons (n) or holes (p)
$\sigma_{n,p}^o$	optical cross section
τ_r, τ_{nr}	radiative or non-radiative lifetimes
ϕ	photon flux
ω	angular frequency
DLTS	deep level transient spectroscopy
DRAM	dynamic random access memory
EMT	effective mass theory
ENDOR	electron nuclear double resonance
EPR	electron paramagnetic resonance
EXAFS	extended X-ray absorption fine structure spectrocopy
FIR	far infrared spectroscopy
FTIR	Fourier transform infrared transmission spectroscopy
LDOS	local density of states
LED	light-emitting diode
LLR	large lattice relaxation
LMTO-ASA	linearized-muffin-tin-orbital in atomic-spheres approximation
MOD-FET	modulation field effect transistor
MOS	metal oxide semiconductor
NAA	neutron activation analysis
ODMR	optically detected magnetic resonance
OSB	Ourmazd-Schröter-Bourret model
PTIS	photo-thermal ionization spectroscopy
TD	thermal donors
VLSI	very large scale integration
YLID	Y-shaped thermal donor core

4.1 Introduction

4.1.1 Shallow and Deep Impurities: Technological and Physical Relevance

Among defects in semiconductors, those impurities that are widely used as dopants to control type and resistivity are well known as "shallow" donors or acceptors. Their ionization energies are very small compared to the fundamental gap of a given semiconductor (Fig. 4-1). Hence, at room temperature and well below, shallow donors are depleted from their electrons, and acceptors, from their holes.

Such centers are well understood in terms of effective mass theory (EMT) (Kohn, 1957, Pantelides, 1978). By adopting two basic assumptions, this reduces the bound-state equation for shallow defects to a hydrogenic Schrödinger equation with the potential $V(r)$ being a screened Coulomb potential, and the free-electron mass replaced by an effective mass depending on the energy band structure of the host matrix (see Chapter 1 in this volume), where the isolated defect is embedded.

In principle, the two assumptions require 1) the binding potential to be weak and to vary very slowly between two lattice points at a distance from the defect where static dielectric screening is meaningful and 2) the particle wavefunction to be localized in k-space. Thus the bound states are derived exclusively from nearest-band-edge states. For a shallow substitutional impurity, EMT then gives a hydrogen-like Rydberg series, terminated at the energy of the conduction band edge (for donors) or the valence band edge (for acceptors).

The binding energy for a donor has the form

$$E_D = \frac{Z^2 e^4 m_e^*}{(4\pi\varepsilon\varepsilon_0)^2 2h^2} \frac{1}{n^2} \qquad (4-1)$$

with a corresponding Bohr radius that extends over many lattice constants even for the 1s ground state. According to the simplified picture of Eq. (1-1) for a fixed host, Rydberg series of different shallow centers should be of similar structure.

Referring to the "deep" level of Fig. 4-1, it is easy to see that under steady-state conditions the electrical activity of the deep center depends on shallow doping and temperature. As an example, if the impurity visualized in Fig. 4-1 of concentration N_D acts as a deep donor in p-type material, it will compensate the shallow acceptors to $N_A - N_D = p$, the free hole concentration at room temperature. If the same impurity is incorporated into n-type material, it would be of practically no influence unless the shallow background doping is almost negligible compared to the deep donor concentration. Otherwise, the impurity will tend to increase the free-electron concentration, depending more or less on temperature.

Similar arguments hold if the deep impurity acts as an acceptor in the lower half of the band gap. Thus it becomes clear that the resistivity, or even the type of semiconductor in extreme cases, may be influenced by deep centers.

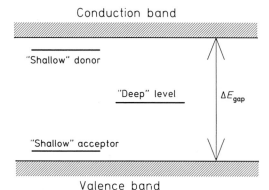

Figure 4-1. Simplified energy level scheme and ionization energies.

However, if these effects are unwanted, they may be avoided by keeping deep impurity contamination well below the shallow background doping. But even small amounts of deep impurities present in a semiconductor may greatly affect the carrier lifetime, which is an important parameter in all semiconductor devices. This is because non-radiative recombination proceeds most easily via deep centers since they can exchange carriers with both the conduction band and the valence band. But these effects of deep impurities such as gold and platinum are the ones exploited to reduce minority carrier lifetime in fast switching devices like thyristors (Lisiak and Milnes, 1975). Also, deep centers are used for extrinsic photoexcitation in photoconductors and infrared detectors (Sze, 1981; Sclar, 1981). For light-emitting diodes (LED) the quantum efficiency in an indirect gap material like GaP is strongly enhanced by the substitutional impurity nitrogen which is isoelectronic to phosphorus. On the other hand the ratio η of the radiative recombination rate to the total recombination rate is given by

$$\eta = \frac{\tau_{nr}^{-1}}{(\tau_r^{-1} + \tau_{nr}^{-1})} \tag{4-2}$$

where the subscripts nr and r stand for non-radiative and radiative lifetimes. In indirect gap materials, it often appears that $\tau_r \gg \tau_{nr}$, with the major LED deficiency in the case of GaP coming from the non-radiative recombination path via two not totally identified deep centers (Peaker and Hamilton, 1986).

In modern VLSI technology deep impurities may cause serious device problems, as they may act as generation-recombination centers in depleted-undepleted regions, respectively. Thus a series of device parameters in bipolar and MOS devices is degraded (Keenan and Larrabee, 1983).

Apart from harmful effects due to the electrical activity of deep centers, they are often also reponsible (even in electrically inactive form) for limiting the diffusion length, increasing leakage, or reducing and softening reverse breakdown voltages when precipitated in active layers. Recently, it was shown, for example, that gate oxide degradation in silicon MOS DRAM processing results to a large extent from the interaction of grown-in heavy metal contamination (preferably fast-diffusing species like Cu, Pd, Ni) (Bergholz et al., 1989). Still, increasing large scale integration in device manufacturing makes it highly desirable to keep metals in the starting silicon material well below concentrations of 10^{12} cm^{-3}.

It will follow from the few examples given above that physical understanding of deep impurities is urgently needed to support those scientists and engineers who are developing semiconductor materials and devices.

The physical nature of deep centers is reflected essentially, but by far not completely, by their deep ionization levels. Nevertheless, an initial approach may be to start again with shallow centers where the short-range central cell potential is responsible only for a chemical shift, especially of ground state terms, and hence for the chemical trends in the rather small ionization energies. As the energy levels become deeper, this should be a consequence of an increased short-range potential which tends to confine the particle wave function around the defect site. Finally, one may arrive at a situation, where a particle can be bound, at least in its ground state, by the short-range potential alone, whereas the Coulomb part of the defect potential plays only an almost negligible role. In other words, contrary to shallow centers, the bound state of a deep center is a spatially compact state, hence delocalized in *k*-space, and therefore is not exclusively derived from the nearest-band-edge states (Hjalmarson et al., 1980). This statement is manifested in Fig. 4-2 showing how the binding energy (ionization energy, see Sec. 4.2.1) depends on central cell potentials for different impurity-host systems (Vogl, 1984). Binding energies increase only slightly for rather small attractive central cell potentials (shallow centers) but steeply for large potentials (deep centers), the threshold separating the regions where either the long-range Coulomb tail or the short-range central cell potential binds an electron.

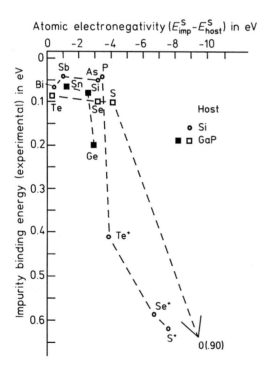

Figure 4-2. Experimental ionization energies of donors in Si and GaP (■: Cation site; □: Anion site) versus the central cell impurity potential strength, the latter given in differences between the s-orbital energy of the impurity atom (E_{imp}^s) and the replaced host atom (E_{host}^s). (After Vogl, 1984).

4.1.2 The Identification Problem and the Localization-Delocalization Puzzle

Deep levels are induced by substitutional impurities (where a host atom is replaced) that do not belong to the groups in the periodic table lying closest to that (or those, in the case of compound semiconductors) of the host crystal. But interstitial impurities and defect associates may also create deep levels. Here, transition metals play a key role in understanding deep impurity phenomena. Generally, this is also true for intrinsic defects like vacancies and self-interstitials. Therefore, intrinsic defects, their electronic structure, and first-generation interactions with other impurities are treated separately in the preceding chapter by G. D. Watkins.

One may literally say that shallow centers are rather exceptional, as can be seen from compiled data (e.g. Sze, 1981, p. 27). The vast number of impurity or defect systems gives rise to serious identification difficulties. These systems may differ in atomic structure, in host sites and bonding symmetries, in solid solubilities, in diffusivities, and in reactivities for interaction with other intrinsic or extrinsic defects present in the host crystal. This therefore results in very different sensitivities to various experimental techniques. A part of this problem is already inherent in the available experimental techniques, most of which are treated in a separate volume of this series (Volume 2: Characterization of Materials) and in various review articles on deep levels (Bourgouin and Lannoo, 1983; Kaufmann and Schneider, 1983; Clerjaud, 1985; Landolt-Börnstein, 1989).

Some of these techniques, such as mass spectroscopy, Mößbauer (spectroscopy, neutron activation analysis (NAA), and extended X-ray absorption fine structure spectroscopy (EXAFS), identify the chem-

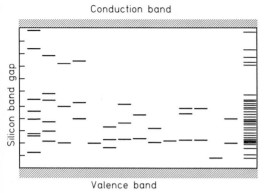

Conduction band

Silicon band gap

Valence band

Figure 4-3. Representation of experimental ionization energies (energy levels) ascribed to iron in silicon
and demonstrating the identification problem. (After
Graff and Pieper, 1981).

ical nature of an impurity in a host material and measure its total content but in general do not distinguish between various
configurations (isolated, associated, substitutional, interstitial, charge state, etc.) of
the defect species. Others, such as electron
paramagnetic resonance (EPR) and electron nuclear double resonance (ENDOR),
additionally identify different charge states
and lattice configurations under certain
circumstances. Absorption and luminescence techniques, optically detected magnetic resonance (ODMR), photo-thermal
ionization spectroscopy (PTIS), various
steady state techniques, such as the Hall
effect, or transient techniques based on
thermal activation (e.g., deep level transient spectroscopy (DLTS)) all measure
transitions between different electronic
configurations of deep impurities, but often fail to determine the chemical nature or
the constituents of a defect.

The best way to overcome the problem
is to relate "fingerprint" methods to methods sensitive to the chemical nature of a
defect species by applying them to identical semiconductor samples. In this context,
hybrid techniques like ODMR (Cavenett,

1981), photo EPR (Godlewski, 1985), or
spin-dependent emission monitored by
DLTS (Chen and Lang, 1983) are of valuable help, because they simultaneously
cross-link different responses from a defect
configuration. For example, Fig. 4-3
shows the picturesque "zoo" of ionization
energies ascribed to iron in silicon (Graff
and Pieper, 1981). However, the energy
level of interstitial iron was definitely
shown to be located at 0.375 eV above the
valence band, by simply combining the
EPR signal from neutral interstitial iron
(for the chemical nature) with the Hall-effect data (for the ionization energy) on
identical samples (Feichtinger et al., 1978).

As pointed out above, the behavior of
deep impurities is essentially influenced by
the local interaction of such a center with
the surrounding host atoms. For this interaction, experimental data yield a rather
paradoxical picture, which is especially
manifested in the 3d transition metals series.

1) An atomically localized model (Ludwig
   and Woodbury, 1962) is strongly favored by EPR data, showing that the net
   spin associated with 3d impurities obeys
   Hund's rule as in the case of free ions.
   This model is also supported by the multiplets of internal transitions, which resemble atomic multiplets, with ranges of
   excitation energies similar to free ions
   (Kaufmann and Schneider, 1983; Landolt-Börnstein, 1982).

2) A model suggesting delocalized states
   strongly interacting with the host follows from the fact that there is a substantial reduction in the slope of ionization energies with atomic number, relative to free ions (Fig. 4-4). The occurrence of multilevel ionization resulting
   from different charge states that are energetically confined within the small gap
   region (of the order of 1 eV) of various

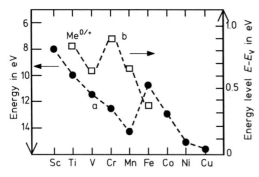

Figure 4-4. Experimental 3d ionization energies of neutral free 3d atoms (a) and energy levels for interstitial 3d donors in silicon (b). (After Feichtinger et al., 1984).

semiconductors points to an effective screening mechanism and hence also strongly favors a delocalized model.

Only the most striking features of experimental data supporting controversial models have been given here, but the puzzle can readily be outlined in more detail (Zunger, 1986).

Our general understanding of the physical nature of deep impurities comes from making experimental efforts with respect to the identification problem and from successfully resolving the above-cited paradoxical behavior of deep centers within electronic-structure theory.

Accordingly, after introducing some frequently used graphical representations like level schemes and configurational coordinate diagrams, the central purpose of the present article is to figure out different approaches to the electronic structure problem insofar as this is necessary for global understanding, and to emphasize what various models have in common.

The remainder of the contribution is devoted to highlighting special features of selected deep center systems via their electrical, optical or kinetic properties. Examples include excited states (chalcogens in silicon), large-relaxation effects (DX centers in semiconductor alloys), metastability and trends of deep donor-shallow acceptor pairs in silicon, and oxygen-related complexes (thermal donors). Some overlap with present structure models should be achieved from these results although long-lasting controversies – as in the case of thermal donors in silicon – and many other problems in deep-level physics can not yet be resolved. Finally, the role of hydrogen in the passivation of shallow acceptors and other point defects is sketched.

4.2 Deep Centers: Electronic Transitions and Concepts

4.2.1 Ionization at Thermal Equilibrium

As mentioned above, deep centers often show several charge states readily accessible to experiments that induce one or more ionizing transitions on the same defect. In thermal equilibrium, the fraction of centers being in a specified formal charge state relative to an adjacent charge state is given by a statistical relation (Shockley and Last, 1957):

$$\frac{N_q}{N_{q+1}} = \frac{g_{q+1}}{g_q} \exp\left(\frac{E_f - E_{q,q+1}}{kT}\right) \qquad (4\text{-}3)$$

Since thermal equilibrium in most experimental cases is maintained at fixed temperature and pressure, the quantities $E_{q,q+1}$ and E_f in the exponent of Eq. (4-3) can be interpreted as $-E_{q,q+1} = \Delta G^{q/q+1}$ and $-E_f = \mu_e$, where $\Delta G^{q/q+1}$ denotes the increase in Gibbs free enthalpy for the thermodynamical system upon ionization, and μ_e refers to the chemical potential representing the energy needed for transferring an electron from the Fermi level into the conduction band. The formal charge state of the defect, q, changes to $q+1$, thus making the defect more positive. Electronic degeneracy has been factorized out from ΔG and is separately contained within the g factors (Van Vechten, 1982) in Eq. (4-3). Entropy terms still included in ΔG (which is actually the standard chemical potential for the ionization process) arise from ionization-induced changes in the number of phonons populating the vibronic level systems due to nonlinear electron phonon coupling. At temperatures well above 500 °C, ΔG may drastically change upon anharmonicity effects in the local lattice vibrations, whereby the contribution from entropy to ΔG strongly increases (Gilles, 1990).

One may define ΔG according to the well-known thermodynamic relation

$$\Delta G^{q/q+1} = \Delta H^{q/q+1} - T \Delta S^{q/q+1} \qquad (4-4)$$

and separate the entropy terms ($S(q)$, $S(q+1)$) from enthalpy terms $H(q)$, $H(q+1)$. Then by definition of the "occupancy levels" $E^{q/q+1}$, with $E^{q/q+1} = H(q) - H(q+1) = -\Delta H^{q/q+1}$ (Baraff et al., 1980) one can return to Eq. (4-3), which now reads

$$\frac{N_q}{N_{q+1}} = \frac{g_{q+1}}{g_q} \exp \left[\frac{S(q) - S(q+1)}{k} \right] .$$
$$\cdot \exp \left[\frac{E_f - E^{q/q+1}}{kT} \right] \qquad (4-5)$$

where (for electron emission) E_f and $E^{q/q+1}$ are defined as negative energies. Occupancy levels are obtained experimentally in principle by measuring thermally activated quantities such as DLTS emission rates. These are of the form

$$e(T) = A(T) \exp \left(-\frac{\Delta G}{kT} \right) \qquad (4-6)$$

The ionization enthalpy ΔH can be obtained from this equation by use of an Arrhenius plot ($\log e(T)/A(T)$ over $1/T$

and Eq. (4-4). The factor $A(T)$ in Eq. (4-6) is given by

$$A(T) = \sigma_{n,p} \langle v \rangle_{n,p} N_{c,v} \qquad (4-7)$$

and contains $\sigma_{n,p}$ the thermal-capture cross section for electrons (n) or holes (p), the mean thermal velocity $\langle v \rangle_{n,p}$, and the effective density of states $N_{c,v}$ at either the bottom of the conduction band (c) or the top of the valence band (v). Since $\langle v \rangle_{n,p} \sim \sim T^{1/2}$ and $N_{c,v} \sim T^{3/2}$, the temperature dependence of both is $\langle v \rangle_{n,p} N_{c,v} \sim T^2$; but $\sigma_{n,p}$ may also depend on temperature (see Sec. 4.2.2).

An energy level scheme now can be constructed by relating each ΔH either to the conduction band edge ($E_c - \Delta H^{q/q+1}$), appropriate for levels in the upper half of the semiconductor band gap, or to the valence band edge ($E_v + \Delta H^{q+1/q}$), for levels residing in the lower half of the gap. Actually, this notation depends on whether an electron or a hole has been thermally activated from the impurity into the conduction band or the valence band, respectively. Figure 4-5 depicts the situation occurring

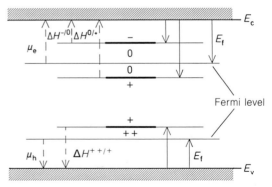

Figure 4-5. Level scheme for interstitial manganese in silicon (Si:Mn$_i$). Dashed arrows indicate the thermally activated processes with corresponding ionization enthalpies. Solid arrows denote occupancy levels $E^{q/q+1}$ (Mn$^{-/0}$: $E_c - 0.13$ eV, Mn$^{0/+}$: $E_c - 0.46$ eV, M$^{+/++}$: $E_v + 0.24$ eV). Note that for hole emission, E_f and $E^{q/q+1}$ are positive quantities. (Values are taken from Czaputa et al., 1983).

in the multilevel system of interstitial manganese in silicon (Czaputa et al., 1983).

Apart from degeneracies and entropies, it can be stated that when the Fermi level is about midway between the two adjacent $Mn^{-/0}$ acceptor and $Mn^{0/+}$ donor levels (Fig. 4-5, above), the impurity will be mainly in the neutral charge state. On the other hand, if the Fermi level crosses the $Mn^{+/++}$ level toward the valence band, for instance (Fig. 4-5), the double positive charge state of interstitial manganese will be dominant. Details concerning single level and multilevel systems have been extensively treated within the framework of semiconductor statistics (Blakemore, 1962; Milnes, 1973; Landsberg, 1982).

4.2.2 Franck-Condon Transitions and Relaxation

According to the requirement of thermal equilibrium in the energy level scheme defined above, any level involves a transition between two totally relaxed defect configurations. There is, however, another class of transitions that correspond to centers that have been excited or ionized optically.

Excitation by incident light may result, for example, in an internal transition between two localized electronic defect states leaving the formal charge state of the center unchanged. If electron-lattice coupling is sufficiently strong, the lattice atoms near the defect site may be rearranged in response to the changed charge distribution at the defect. But the corresponding relaxation process occurs only after the excitation energy has been absorbed by the defect's many-electron system, since lattice relaxations are considered to proceed at a rate three orders of magnitude smaller than that of the involved electronic system. Figure 4-6 shows the relations between op-

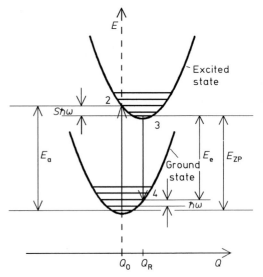

Figure 4-6. Franck-Condon transitions for an absorption-emission cycle resulting in a Stokes shift, $\Delta E_{Stokes} = 2 S \hbar \omega$. S is the Huang-Rhys factor, which is a measure of the number of phonons effectively involved in a transition and therefore also accounts for electron-lattice coupling strength.

tically induced transitions and lattice relaxation in a so-called configurational coordinate diagram, which helps one to keep track of important terms even in complicated processes. In Fig. 4-6, only the two interesting quantum states are shown by their adiabatic potential energy surfaces.

The underlying physical idea in such a diagram is the Born-Oppenheimer approximation (Born and Oppenheimer, 1927; Adler, 1982), which separates electronic motion from that of the nuclei or ion cores (if only outer electrons have to be considered for interaction). Adiabatic then refers to an evolution of a system of electrons and nuclei (cores), where electron eigenstates do not change suddenly on ion core displacements Q but provide the potential energy for ion core motion. The corresponding vibrational states are then harmonic oscillators with phonon energies $E_n = (n + 1/2) \hbar \omega$. Each electron quantum state is represented by its own parabola derived in principle by expanding total electronic energies to second order in the configurational coordinate Q at the stable positions (Q_0 and Q_R in Fig. 4-6). But it is only for linear electron-phonon coupling that parabolas are of the same shape as that found in Fig. 4-6.

Beyond linear coupling, spring constants my be altered in an excitation leading to a shift in mean transition energies with temperature (thereby inducing changes in enthalpies and entropies in the case of ionization) and to unsymmetrical line shapes and mean energies for absorption (E_a) and emission (E_e), relative to zero-phonon transitions (E_{ZP}). Mean transition energies corresponding to the most intense lines or to the center of a smoothed-out line shape (which depends on the Huang-Rhys factors S and is Gaussian in the high-coupling and high-temperature limit) depend on the overlap of vibrational wave functions in the initial and final states (Stoneham, 1975, Bourgoin and Lannoo, 1983).

According to the Franck-Condon principle, most probable absorption transitions occur between states centered at configuration Q_0 (Point 1 in Fig. 4-6) for the ground state and states which have their wave functions concentrated near the classical turning points of the same configuration for the excited states (Point 2). When the system has relaxed non-radiatively by phonon emission to the stable configuration at Q_R, the situation is reversed. Emission of light (luminescence) then proceeds again at constant configuration (Points 3 and 4) followed by non-radiative relaxation of the electronic ground state to the Q_0 configuration thus terminating the cycle described by Fig. 4-6. This simple picture will work surprisingly well in relating energy shifts, line shapes, and spatial extensions of wave functions to electron-phonon interactions, as long as non-degenerate electronic states can be coupled linearly to phonons of average frequency ω by a single representative lattice mode Q. The reason for this success emerges from the fact that strong restrictions exist on mean square displacements and mean square velocities of atoms in a harmonic solid irrespective of strongly temperature-dependent details in lattice dynamics (Johnson and Kassman, 1969; Housely and Hess, 1966).

Absorption of light can cause the photoionization of a deep center. The only difference to the charge-state-conserving excitations cited above lies in the fact that either the initial states or the final states are not localized near the defect. Thus, from experimental absorption studies of optical ionizations, thresholds as well as spectra may be obtained in terms of optical cross sections that can be fitted using as parameters the Franck-Condon shift $S\hbar\omega$, the spatial extension of the wave function associated with the localized level, and the respective oscillator strengths of transitions toward different extrema of a given band structure (e.g., Bourgoin and Lannoo, 1983). As mentioned before, in the occurrence of a Franck-Condon shift, the equilibrium position of the lattice embedding a defect may vary with its different formal charge states, depending on electron-lattice coupling.

Again, such a situation is closely tracked by a configuration diagram. Figure 4-7 depicts a hypothetical donor-type defect which undergoes a distortion in a generalized configurational mode Q, when in its neutral state. In order to include the band gap, the defect in its positive charge state and the completely filled valence band are chosen as an undistorted reference configuration. Therefore, three adiabatic energy

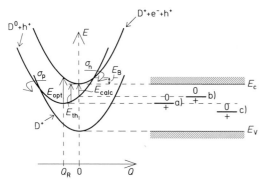

Figure 4-7. Configurational coordinate model for a deep level. (After Watkins, 1983). For the level scheme, see text.

surfaces (vibronic levels are not shown) are accounted for, whereby the two undistorted ones at $Q = 0$ are separated by the total energy needed for the creation of an electron-hole pair. This energy does not depend on Q since the process has practically no influence on the configuration of the atoms near the defect. The arrows at the intersection points (marked σ_n, σ_p) indicate capture and emission processes which have their physical origin in nonadiabatic transitions left out of the Born-Oppenheimer approximation. Transitions between two vibronic states that differ in electronic energy but have the same total energy are very important in non-radiative trapping and recombination processes. E_B in Fig. 4-6 denotes the barrier height by which capture rates appear thermally activated. If lattice relaxation occurs, thermal capture cross sections (Eqs. (4-6), (4-7)) become temperature dependent (e.g., Bourgoin and Lannoo, 1983). In the high-temperature limit, $\sigma_{n,p}$ takes the simple exponential form ($\sigma_{n,p} \sim \exp(-E_B/kT)$). The shifted parabola at Q_R represents the deep center in its distorted neutral charge state after having trapped an electron from the conduction band.

One may deduce a level scheme for comparison in three ways according to the previous section: (a) ionization via thermal activation, (b) calculation of the transition energy within a rigid, undistorted lattice, and (c) ionization via the absorption of light. It is clear from Fig. 4-7 that the position of the electrical level is given correctly only by the difference in total energies of the stable configurations $D^0 + h^+$ and $D^+ + h^+ + e^-$, with energies taken at $Q = Q_R$ and $Q = 0$, respectively. If relaxation effects are neglected, the ionization energy can be calculated from the difference in electron energies of the neutral and positive charge state of the deep center, both total energies being taken from the undistorted lattice configuration at $Q = 0$. Finally, if the ionization energy has been obtained from an optical experiment, the initial state will be defined by the relaxed configuration of the neutral defect state at $Q = Q_R$ while the final state is reached by a Franck-Condon transition at constant configuration. The three ionization energies from Fig. 4-7 appear therefore in a relation $\Delta E_{calc} < \Delta E_{therm} = \Delta H^{q/q+1} < \Delta E_{opt}$. One should note, however, that by the definition of the chemical potential

$$\mu = \left(\frac{\partial G}{\partial N}\right)_{p,T} = (\Delta G^{q/q+1})_{p,T} =$$

$$= \left(\frac{\partial E}{\partial N}\right)_{S,V} = (\Delta E^{q/q+1})_{S,V} \qquad (4\text{-}8)$$

and its application to an ionization process for an optical transition, the crystal volume V is kept constant by virtue of the Franck-Condon principle, whereas entropy S is conserved at the change of internal energy E if no phonons are involved. Therefore, at low temperatures $(\Delta H^{q/q+1})_{p,T} \simeq (\Delta E^{q/q+1})_{S,V}$. Zero-phonon energies that can be obtained by fitting the spectral dependence of optical cross sections to an appropriate model (e. g., Lucovsky, 1965), can be directly compared to the thermal activation energies $\Delta H^{q/q+1}$. Optical cross sections $\sigma_{n,p}^0$ are usually derived experimentally from optical emission rates

$$e_{n,p}^0 = \sigma_{n,p}^0 \, \Phi \tag{4-9}$$

where Φ is the photon flux applied. The dependence of ionization energies on symmetric (breathing mode) or symmetry-lowering distortions (e. g., Jahn-Teller distortions, induced by electronically degenerate defect states with partial occupation (see Chapter 3; Sturge, 1967)) may result in a reversed level ordering in extreme cases, where, for example, in a multilevel system like that shown in Fig. 4-5, the first donor level $(0/+)$ and the acceptor level $(-/0)$ appear in reversed order. This corresponds to a "negative U" situation envisioned by Anderson (1975), where for the difference of two subsequent occupancy levels, a relation

$$U = E^{q-1/q} - E^{q/q+1} < 0 \tag{4-10}$$

holds. In such a series, q never represents a dominant intermediate charge state contrary to the neutral state in the normal, "positive U" level ordering of Fig. 4-5. The role of relaxation is pointed out by splitting Eq. (4-8) into

$$U = U_0 + \Delta E_R \tag{4-11}$$

where $U_0 = E_0^{q-1/q} - E_0^{q/q+1}$ is the vertical (Franck-Condon) Mott-Hubbard correlation energy calculated at an undistorted $(Q=0)$ configuration. U_0 results from the repulsive Coulomb interaction of the electrons occupying a single-particle state and in the absence of lattice relaxation, is responsible for the separation of first donor $(0/+)$ levels and acceptor $(-/0)$ levels in the level scheme of Fig. 4-5. The contributions of lattice relaxation, which may be considerably different for two subsequent

ionizations, are given by $\Delta E_R = E_R^{q-1/q} - E_R^{q/q+1}$. This can best be understood with the help of diagrams such as Fig. 4-7. In cases where $\Delta E_R < 0$, the positive U_0 may be outweighed to promote negative-U behavior. Well-known examples for negative-U systems are the vacancy in silicon (Baraff et al., 1979, 1980; Watkins and Troxell, 1980) and interstitial boron in silicon (Harris et al., 1982).

4.3 Phenomenological Models and Electronic Structure

In the light of more recent theoretical results, a deep center is characterized by a series of bound states and resonances produced by localized perturbation. Nevertheless, it is often useful to introduce symmetry via point charges representing ligands at an impurity site, or to qualitatively select eigenvalues or band resonances representing states that are important for modeling various interactions of the defect with the host crystal. In order to circumvent the rather opaque mathematics in presenting results of modern electronic-structure theory, two models, each in favor of either a localized or a delocalized electronic configuration of a defect, will be introduced below and used throughout the following discussion.

4.3.1 The Point-Ion Crystal Field Model

The pioneering work of Ludwig and Woodbury (1962), based on EPR studies, led to a semiempirical model for the electronic structure of transition metals in silicon that successfully accounts for most of the experimental findings to date. Electron paramagnetic resonance (see, e.g., Slichter, 1980) exploits the manifold of interactions that unpaired electrons undergo in a solid.

These include interactions of orbital motion and spin with each other, with an applied magnetic field and with the magnetic moments of the nuclei. Electrostatic and covalent interaction reflects the symmetry of the crystalline environment. EPR spectra involve the lowest-lying states of a paramagnetic ion, usually considered a rather isolated system, which can be described in terms of a so-called spin-Hamiltonian

$$\mathscr{H} = g\,\beta\,\boldsymbol{J}\cdot\boldsymbol{B} + \frac{a}{6}(J_x^4 + J_y^4 + J_z^4) +$$

$$+ A\,\boldsymbol{J}\cdot\boldsymbol{I} - (\gamma+R)\,\beta_N\,\boldsymbol{B}\cdot\boldsymbol{I} \qquad (4\text{-}12)$$

Eq. (4-12) is appropriate for tetrahedral symmetry and effective angular momentum $J \geq 2$. The leading term describes the Zeeman splitting while the second accounts for zero-field splitting due to the crystal field; the remaining terms describe indirect and direct magnetic hyperfine interactions with a nucleus of spin I (Low, 1961; Abragam and Bleaney, 1970).

For $J < 2$, the second term in Eq. (4-12) vanishes and the microwave energy absorbed in a resonance transition from the state $(M-1,m)$ to the state (M,m) is

$$h\,v = g\,\beta\,B + A\,m + [I(I+1) - m^2 +$$

$$+ m(2M-1)]\,A^2/2\,h\,v \qquad (4\text{-}13)$$

Thus the spectrum consists of $2I+1$ hyperfine components due to the range of the nuclear quantum number $m\,(+I \geq m \geq -I)$, each of which is split into $2J$ fine structure components due to the range of the electronic quantum number $M\,(J \geq M \geq -J)$ at fixed m. Figure 4-8 shows schematically the corresponding simple example for EPR and ENDOR (fixed M) transitions. In a well-resolved EPR spectrum, the effective angular momentum J, with components S (total electron spin) and L' (effective orbital angular momentum), and nuclear spin I may be

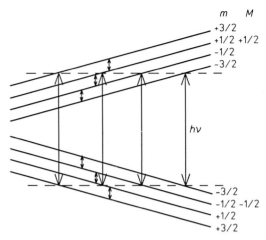

Figure 4-8. Allowed EPR transitions (long arrows) and ENDOR transitions (short arrows) for a paramagnetic center with $J=1/2$ and $I=3/2$. (After Ludwig and Woodbury, 1962).

determined by counting the lines; the electronic g factor and the hyperfine parameter A can be determined from microwave frequency and the magnetic field B, taken at the resonance transitions.

The fivefold orbital degeneracy of an atomic 3d transition metal is lifted by the interaction with the neighbors in the host crystal. In the most simple crystal field picture, neighboring host atoms are represented by point charges (point-ion approximation). Wave functions of a point defect in tetrahedral symmetry must transform according to one of the five irreducible representations of the T_d-point group: a_1, a_2, e, t_1, and t_2 (c.f., Ballhausen, 1962; Figgis, 1967). The tetrahedral crystal field of the four nearest silicon neighbors splits the 3d orbitals into a triplet of representation t_2 and a doublet of representation e, where the t_2 symmetric and e symmetric wave functions are of the form d_{xy} (d_{yz}, d_{zy}) and $d_{x^2-y^2}$ ($d_{3z^2-r^2}$), respectively (Fig. 4-9). Symmetry considerations do not reveal which of both orbitals (t_2 or e) is lower in energy. Only the t_2 symmetric (d_{xy}-like) or-

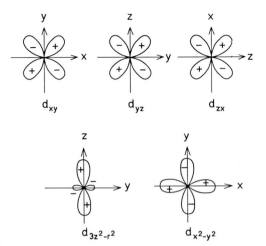

Figure 4-9. Angular dependency and relative signs of the set of five orthogonal basis functions forming the t_2 triplet and e doublet. Directions are given with respect to the cubic axes of the silicon unit cell (Fig. 4-10).

bitals point to the nearest neighbors for both substitutional and interstitial 3d impurities, which can be seen upon comparing Fig. 4-9 with Fig. 4-10. Thus it can be argued that for substitutional 3d-atoms, the t_2 orbitals interact with the negative charge of the dangling bonds residing on the four first neighbors leading to an in-

crease of t_2 orbital energy; hence, the level ordering is e below t_2, the standard arrangement for the tetrahedral crystal field. (Here and in the following discussion "level" should not be confused with the ionization levels in Sec. 4.2.1, since it denotes orbital energies and not transitions between different orbitals or between single-particle states and host band states.)

For interstitial impurities, nearly all of the charge can be thought concentrated on the 3d ion since these form only very weak directional bonds with their immediate neighbors. On the other hand, the positive cores of the neighboring atoms are only incompletely screened in the direction to the interstitial atom as the sp^3 bonds point off that direction. In this case the sign of the crystal field is reversed and the energy of the t_2 orbital is lowered to give the level ordering t_2 below e.

In detail, Ludwig and Woodbury (1962) established the following model from their EPR studies on transition metals in silicon.

i) The d shell of a transition metal incorporated into silicon contains all atomic s and d electrons not needed for bonding.

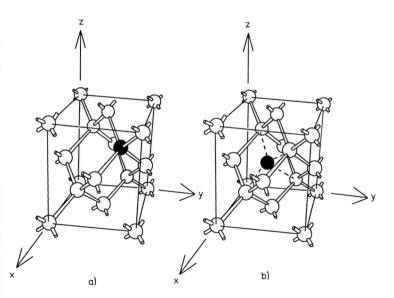

a)   b)

Figure 4-10. Substitutional (a) and interstitial (b) sites of tetrahedral symmetry in the diamond-like or zinc-blende structure.

ii) The e and t_2 levels are filled according to Hund's rule as in the case of free atoms.

The experimental findings interpreted within the given model, are shown in Fig. 4-11 a, where the total angular momentum of the ground state differs from the given spin values only for the $3d^6$ ($J = 1$) and the $3d^7$ ($J = 1/2$) configuration. For all other configurations, the effective orbital angular momentum L is zero so that $J = S$.

For substitutional atoms the number of electrons to be distributed over the e, t_2 levels is reduced by four to fill the sp^3 bonding orbitals.

The high-spin-like level arrangement manifested throughout the 3d series (as long as it is observed in EPR) seems to indicate in particular that the transition metal induced states are essentially impurity-like and sufficiently localized so that the perturbing crystal field is too weak to produce an $e - t_2$ splitting large enough to overcome the effect of spin polarization (Fig. 4-11 b). It must be noted, however, that for interstitial 3d atoms, considerable hybridization of t_2 orbitals with the nearest neighbors was made responsible for the partial or nearby complete quenching of the orbital contribution in the g factors for ions with non-vanishing effective angular momentum L. This was found to be consistent with the presence of many charge states (e.g., in manganese) within the gap (Ludwig and Woodbury, 1962).

4.3.2 The Defect Molecule Picture

4.3.2.1 Example: Nitrogen in Gallium Phosphide

Nitrogen in gallium phosphide constitutes an "isoelectronic" electron trap when it replaces a host phosphorus atom. Isoelectronic impurities are defined as substitu-

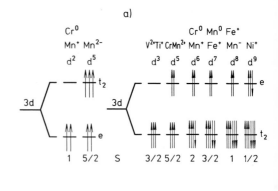

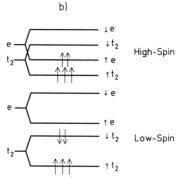

Figure 4-11. The Ludwig-Woodbury model for substitutional (left) and interstitial (right) 3d transition metals in silicon (a). Schematic representation of high-spin and low-spin configurations in terms of e-t_2 splitting and spin polarization (b).

tional atoms which have the same valence electron structure as the atom they replace.

Their special importance has been revealed in luminescence studies, where it was shown that they can bind excitons (weakly bound electron-hole pairs). Bound excitons give rise to an enhancement of the quantum efficiency in LEDs (light emitting diodes); nitrogen in GaP was extensively studied for this purpose (Dean, 1977).

Actually, the conduction band-related activation energy for an electron trapped at the N level in GaP is only 10 meV (Bergh and Dean, 1976), but this small energy may be seen as a consequence of the short-range potential alone, since a Coulomb tail does not exist (Hsu et al., 1977; Wolford et al., 1976). Therefore, qualitative physics of deep levels is contained in a simple molecular picture, borrowed from LCAO theory (linear combination of atomic orbitals) and focusing on sp^3 chemical bonding of the impurity with the host.

Figure 4-12 depicts a two-level model for N in GaP featuring a semi-empirical tight-binding approach that exploits chemical trends in the data of deep level impurities (Hjalmarson et al., 1980; Vogl, 1981). If a host atom is removed from the crystal, one fully symmetric a_1 state (s-like) and three t_2-symmetric (p-like) states can be formed from the four remaining "dangling-bond" hybrids, which essentially constitute the electronic structure of the vacancy (Bernholc et al., 1978; Lannoo and Bourgoin, 1981; Lindefelt and Zunger, 1981). Figure 4-12 concentrates on the s-like atomic orbitals for P (modeling the Ga-P interaction) and N (impurity-Ga interaction). The Ga atomic level of energy $E_{Ga} = (E_s^{Ga} + 3E_p^{Ga})/4$ lies above that of the s orbital for P (Fischer, 1972; Clementi and Roetti, 1974) and is the progenitor of the conduction band in the two-level model. If Ga and P are brought together to form a molecule,

this leads to a bonding-antibonding splitting proportional to $v^2/(E_{Ga} - E_s^P)$, where v is the anion-cation transfer-matrix element depending solely on the host bond length (Harrison, 1980).

A similar molecule can be constructed by substituting the host P by the isoelectronic impurity N. Since E_s^N is smaller than E_s^P, the resulting resonance splitting, based on unchanged bond length (omitting the relaxation around the N atom), is reduced. Thus the Ga-like antibonding level appears bound relative to the GaP antibonding level, whereas the occupied bonding N-like state is located below or deeply within the valence band. The latter is represented through the occupied P-like bonding state of the pure host.

Inherent in this simple model, the essential features of which have been expanded also to the interaction between p-impurity states and t_2-symmetric host states and worked out quantitatively within the framework of the tight binding method (Vogl et al., 1983; Vogl, 1984), are the following remarkable properties of the a_1-symmetric deep levels:

i) The deep level, acting as an electron trap by an attractive central cell potential, $E_s^{impurity} - E_s^{host} < 0$, is a host-like antibonding state, in contrast with the hyperdeep, electrically inactive impurity-like bonding state (Fig. 4-12). An electron is therefore trapped at the ligand site rather than at the impurity site, since the corresponding wave function will be excluded from the impurity sphere by the orthogonality constraint.

ii) With increasing attractive central cell potential (nominal donors), simulated in the two-state model by decreasing $E_s^{impurity}$, the antibonding host-like level appears asymptotically pinned to the dangling bond energy of Ga, E_{Ga}. This explains qualitatively, why deep levels

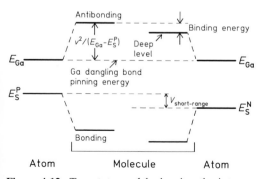

Figure 4-12. Two-state model, showing the interaction of s-like atomic orbitals for sp-bonded impurities in terms of bonding-antibonding splitting for the pure host (GaP) and for the impurity N, substituting the anion site (P). (After Vogl, 1984).

reflect the slope in central cell potential only in a strongly attenuated fashion (Fig. 4-2).

In a real semiconductor, the "bands" would broaden as the interaction between the defect molecule and the rest of the host is turned on. Therefore, the defect potential must exceed a certain threshold before a deep antibonding state can be "pulled" from the conduction band. In the case of a repulsive central cell potential $E^{impurity}$ $- E^{host} > 0$, the deep level is again host-like, but it shows some bonding character. Pinning to the dangling bond energy of the host occurs at increasing $E^{impurity}$, but this time the deep level enters the gap according to a valence band threshold. No hyper-deep level exists in the case of nominal acceptors.

For a large number of sp-bonded impurities that substitute the cation or the anion site and the zincblende hosts (Si, Ge, GaP, GaAs, InP, InAs, InSb, AlP, AlAs, ZnSe, ZnTe), predictions were made for the relative ordering of ionization energies of a_1 and t_2 states from the semi-empirical tight binding method. For alloys (e.g., $GaAs_{1-x}P_x$), deep trap levels were predicted and experimentally confirmed to follow the pinning energy of the Ga dangling bond (see Fig. 4-12) rather than the Γ_1 or X_1 points of band edges upon varying the composition (Vogl, 1984).

4.3.2.2 Transition Metals

In the two-state defect molecule introduced above, the bonding-antibonding repulsion between the impurity s or p orbitals and the a_1- and t_2-symmetric dangling hybrids of the host were identified to be of crucial importance for the creation of deep levels. If a 3d transition metal atom is put into the dangling bond system of the vacancy, the interaction now involves 3d

and 4s orbitals. As stated in Sec. 4.3.1, the 3d orbitals split into a t_2-symmetric (d_{xy}-like) triplet and into an e-symmetric ($d_{x^2-y^2}$-like) doublet under tetrahedral symmetry. For the d states of the impurity, it is obvious from Fig. 4-9 that only the t_2-symmetric d_{xy}-like states can couple effectively to the dangling hybrids of the host, whereas the e-symmetric states retain their atomic character (or, in other words, only the t_2-symmetric states can form σ-bonding states).

Apart from bound states (gap levels), defect-induced changes on the electronic structure of the host bands (resonances) as well as band states of the pure host that are important for host-impurity interaction, often are given in terms of the "spectral local density" of states (in the form of more or less pronounced peak structures). One may substitute the most important peaks by weighted delta functions in order to get an approximation for the defect molecule

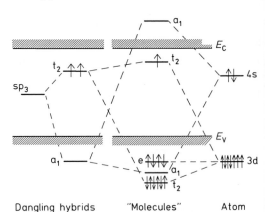

Dangling hybrids   "Molecules"   Atom

Figure 4-13. Simplified defect molecule picture (a further a_1 resonance from the backbonds of nearest neighbors (Scheffler et al., 1985) is not shown), reflecting qualitatively the interactions of s and d states of substitutional 3d transition metal impurities with the dangling bond system of the host. The single particle levels (spin polarization omitted) are shown as bound states in the gap, the nonbonding e levels and the bonding t_2 resonances are populated as an example according to neutral cobalt in silicon. (After Beeler et al., 1990)

picture. Figure 4-13 displays a standard situation as occurring for the heavier elements in the 3d series in silicon according to the results (see next section) of quantitative theories on transition elements in sp^3-bonded semiconductors. Again the a_1-symmetric (s-like) and the t_2-symmetric (p-like) states formed from the dangling bonds and the atomic s and d_{t_2} states of the impurity interact according to their relative energetic positions. Thus the antibonding impurity-like a_1 level (resonant with the conduction band) and the antibonding host-like t_2 state constituting the deep level in the gap are created. The resonant a_1 level in the conduction band remains unoccupied; however, in addition to the two electrons residing originally on the t_2 gap level of the neutral dangling bond system, four electrons from the impurity are needed for filling the bonding t_2 level deep within the valence band.

This confirms the findings of Ludwig and Woodbury presented in Sec. 4.3.1 suggesting a transfer of s electrons into the d shell and of four electrons into the sp^3 bonds. The nonbonding e level is not markedly affected by host interaction since the vacancy states have practically no e character (which is the actual justification for the simple picture given in Fig. 4-13).

However, for interstitial 3d impurities the situation is different. An interstitial atom is embedded in a host environment where all bonds are complete. Thus hybridization is expected only with the tails of sp^3 hybrid orbitals on the surrounding atoms. If one compares Fig. 4-9 and Fig. 4-10 for the interstitial site, one realizes that the lobes of d_{xy}-like (t_2) orbitals point to the four nearest silicon neighbors, whereas $d_{x^2-y^2}$-like (e) orbitals point to the six next-nearest neighbors. These directions are imposed by T_d symmetry for forming σ-type bonds between host p

states of e, t_2 representation, and d states of an impurity at the interstitial site. Local densities of states (LDOS) of the pure silicon crystal, projected onto the interstitial site (Zunger, 1986; Beeler et al., 1990) show the availability of both t_2 states from the first (nearest) coordination sphere and e states from the second (next-nearest) coordination sphere. In principle, this is also confirmed by cluster calculations (De Leo et al., 1981) in the form of discrete d-like levels of a $Si_{10}H_{16}$ cluster. Figure 4-14 depicts the interaction between d-like host states (as before approximated by discrete levels) and the atomic d_{e,t_2} orbitals of the transition metal at the T_d interstitial site. Since symmetric host a_1 states are also available at the interstitial site, by their interaction with the atomic s states a resonant antibonding impurity-like a_1 state is formed within the conduction band and remains unoccupied.

Once again, this is in formal agreement with the Ludwig-Woodbury model, suggesting a transfer of s electrons into the d shell for interstitials as well. The p–d hybridization leads to bound states of e and t_2 symmetry in the gap, showing antibonding character, and to their counterpart in the form of bonding resonances deep within the valence band.

Contrary to substitutional 3d atoms, the gap levels must accommodate all valence electrons of the impurity.

An interesting feature of the covalently delocalized models presented so far concerns the relative ordering of the e, t_2 gap levels: For substitutional 3d impurities, the effective crystal field splitting, with the order being t_2 above e, results from the hybridization of only the t_2-symmetric states with dangling bond host states of the same symmetry, in contrast to the e-symmetric impurity states which remain nonbonding and resemble atomic states. For interstitial

3d impurities, both e- and t_2-symmetric states couple to the relevant host states, but the host-like, antibonding t_2 gap level is repelled by antibonding conduction band states of t_2 character, while the e gap level is not; hence the order is e above t_2 (Fig. 4-14).

4.3.3 Transition Metals: Results of Quantitative Calculations

As pointed out above, the relative ordering of the crystal field splitting is also accounted for by the covalently delocalized model of a deep center. However, many more physical properties are covered by that point of view as has been revealed by quantitative calculations. In order to emphasize the most striking results, in the following sections the most important consequences are divided into several parts which, in reality, are interrelated.

4.3.3.1 Gap Levels and High Spin–Low Spin Ordering

The magnitude of the crystal field splitting (ΔE_{CF}) manifests itself directly in the trends of the lowest excitation energy of a 3d impurity when multiplet corrections are of minor importance and indirectly when competition with the exchange splitting ΔE_{Ex} for spin-up and spin-down states results in a high-spin or low-spin electronic configuration (see Fig. 4-11 b). In Fig. 4-15 the e–t_2 splitting for silicon is shown as calculated from density-functional theory in conjunction with the local-density approximation. Green's function technique within the frame work of LMTO-ASA (Linearized-Muffin-Tin-Orbital in Atomic-Spheres-Approximation) was used for solving the single particle equations (Beeler et al., 1990). Results were obtained for a rigid (unrelaxed) lattice, neglecting spin po-

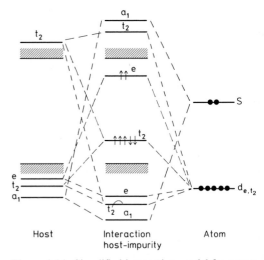

Figure 4-14. Simplified interaction model for a transition metal occupying an interstitial site of tetrahedral symmetry. Gap levels are populated according to neutral manganese in silicon.

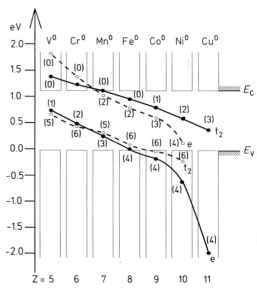

Figure 4-15. Gap levels and apparent crystal field splitting through the 3d transition metal series in silicon for neutral substitutional (solid lines) and interstitial (dotted lines) impurities. Z denotes the effective core charge and is equal to the total number of 3d and 4s valence electrons forming the many-electron system of the deep center. The numbers in parentheses give the population of the levels of different symmetry in the single-particle picture, neglecting spin polarization. (After Beeler et al., 1990).

larization (Spin-restricted calculation: The occupancy of all spin-up states is assumed to be equal to the occupancy of all corresponding spin-down states). In the main trend along the 3d series, there is also agreement with calculations based on other methods, such as the X_α cluster calculations in Si and GaAs for substitutional impurities (Cartling, 1975; Hemstreet, 1980; Fazzio and Leite, 1980), and in Si for interstitial impurities (De Leo et al., 1981), the empirical tight-binding method for interstitials in Si (Pecheur and Toussaint, 1983), and in local-density Green's function calculations for interstitial and substitutional 3d atoms in silicon (Zunger and Lindefelt, 1982). A critical discussion of various approaches regarding the choice of an appropriate Hamiltonian and the technique of quantitatively calculating states and energy levels that define a deep center was given by Pantelides (1986).

For substitutional transition metals, the trends of the e and t_2 levels with atomic number can be easily understood by the help of Fig. 4-13. The impurity-like (d-like) levels of either symmetry will reflect the decrease of atomic d_{e,t_2} orbital energy with increasing Z, but the antibonding t_2-symmetric states due to hybridization and pinning to the host dangling bond system show this trend only in a strongly attenuated fashion as compared to the non-bonding e-symmetric states. Thus the one-electron crystal field splitting is largest at the high-Z limit.

Interstitial 3d states will behave in a similar manner, but in this case both e and t_2 states hybridize with the host. With decreasing atomic number, as the impurity-host interaction increases, the antibonding t_2-symmetric states are increasingly repelled by antibonding conduction band states (Fig. 4-14) hence the $e-t_2$ splitting is largest at the low Z limit (Fig. 4-15).

Interstitial copper does exist only in the single positive charge state, since the completely filled atomic d shell leads to full occupancy of the antibonding e, t_2 levels, and the impurity-like a_1 level is resonant with the conduction band (Fig. 4-14); the remaining s electron may be bound by the positive Coulomb potential of interstitial copper resulting in shallow donor behavior. By separating anisotropic many-electron effects, Zunger (1986) has fitted an average crystal field energy Δ_{eff} to internal transitions ($e^n t_2^m \rightarrow e^{n'} t_2^{m'}$ where $n+m = = n' + m'$) determined experimentally from excitation spectra in III–V and II–IV compound semiconductors; the trend in Δ_{eff} behaves like that discussed for substitutional 3d elements in the one-electron picture (Fig. 4-15), at least from the heavier atoms down to manganese.

In calculations within a single-particle model all electrons populating a degenerate one-electron level (like the e or t_2 states of Fig. 4-15) feel the same average potential. Although average many-electron effects are contained in the various one-electron approaches by including exchange terms in effective potentials or energy functionals, this symmetrization in the case of partially occupied states does not allow for stabilizing effects of spatial correlation which may arise from the electrons occupying spatially separated and variationally independent orbitals (like that shown in Fig. 4-9). High-spin states are simply the result of spatial correlation, since alignment of the spins in strongly localized wave functions (for example, in e and t_2 orbitals) keeps the electrons apart and minimizes electrostatic repulsion despite the higher energy of the second orbital (e or t_2 for interstitial or substitutional 3d states, respectively). The effect of anisotropic contributions (multiplet corrections) to ground state and transition energies in heteropolar semiconductors was demonstrated in calculations of Fazzio et al. (1984).

The spin part of the correlation energy can be included within the one-electron model by carrying out spin-unrestricted calculations, i.e., allowing electrons of different spin to experience a different potential. The available electrons may then be distributed to the resulting levels in all possible ways. From such states, the various multiplet energies can be obtained by perturbation theory, taking into account residual electronic interactions (e.g., Sugano et al., 1970). Possible multiplets for transition metals have also been worked out by Hemstreet and Dimmock (1979) in combining results of one-electron calculation with ligand field theory. However, determining the lowest-energy configuration (the ground state multiplet) is often hampered by the fact that orbitally degenerated multiplets may lead to spontaneous symmetry-lowering Jahn-Teller distortions, thereby stabilizing a low-spin configuration against a non-degenerate, high-spin correlated multiplet originally lower in energy. The question as to the final result of such a competition usually has to be settled by experiment, as has been done for the singly negative vacancy in silicon (see Chapter 3 by Watkins).

EPR-results on total spin of 3d impurities in most cases are confirmed by spin-polarized calculations with respect to a rigid lattice made from first principles. This indicates that Jahn-Teller effects are presumably suppressed by substantial exchange splitting of the gap states. But even though lattice relaxation is neglected, such calculations do not predict an overall high-spin tendency throughout the 3d series and its various charge states (Beeler et al., 1985a, Katayama-Yoshida and Zunger, 1985). Fig. 4-16 and Fig. 4-17 show the

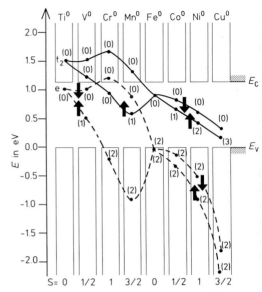

Figure 4-16. One-electron energies (spin-unrestricted) of the ground states of neutral substitutional 3d transition metals in silicon. The numbers in parentheses indicate the occupancy of gap levels and resonant states. (After Beeler et al., 1985a).

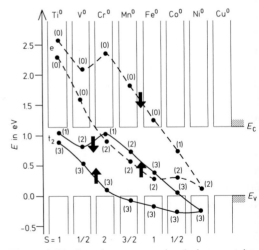

Figure 4-17. One-electron energies (spin-unrestricted) of the ground states of neutral interstitial 3d transition metals in silicon. (After Beeler et al., 1985a).

spin-polarized one-electron levels in the ground state configuration for neutral, substitutional and interstitial, 3d metals in silicon as obtained from local spin density calculations (Beeler et al., 1985a). The spin splitting depends on spin density and the localization of the defect states. As for the localization, the same qualitative arguments that were used in discussing the trends of Fig. 4-15 hold here as well.

For substitutional impurities, the non-bonding impurity-like e states remain strongly localized throughout the series; when they enter the gap, thereby becoming partially occupied, the spin splitting is largest. The antibonding host-like t_2 states are strongly delocalized at the high Z limit, due to their largely vacancy-like character, and they become increasingly localized with decreasing atomic number as they are shifted upward toward the conduction band. This causes the e-t_2 splitting to decrease (see Fig. 4-15) and the e↓ state is then pushed above the t_2↓ state for Mn^0 and Cr^0 near the center of the series, resulting in a high-spin configuration for both impurities. For Co^0, Ni^0, and Cu^0, however, the nearly constant spin splitting (the increasing magnetization is compensated by increasing delocalization in that series) is markedly overwhelmed by the e-t_2 splitting, the latter being largest at high Z. Therefore, from Fig. 4-16 a low-spin configuration is predicted for neutral substitutional Co, Ni, and Cu in silicon. Unfortunately, experimental evidence is lacking for that prediction, whereas high-spin states for neutral substitutional Cr and Mn are confirmed by EPR (Fig. 4-11 a).

Similar arguments hold for interstitial 3d impurities. In Fig. 4-17 low-spin configurations are indicated at the low- and high-Z limits, and only the transition elements at the center of the series should form high-spin states. Essentially, this is caused by the stronger localization of the e states compared to the t_2 states and the increase in e-t_2 splitting with decreasing atomic number (Fig. 4-15). Neutral interstitial Ni is nonmagnetic due to fully occupied

states. Therefore, its spin-polarized one-electron configuration is equivalent to that in Fig. 4-15. (This is also true for substitutional Fe^0, where the e levels are completely filled, and the t_2 levels are empty.) Since silicon 3d states dissolve mainly on interstitial sites, all high-spin states could be identified by EPR (Fig. 4-11 a); again no experimental information exists on the predicted low-spin ground states. For Ti and V on interstitial sites, the singly positive (Van Wezep and Ammerlaan, 1985) and the doubly positive charge states (Fig. 4-11 a) have been identified by EPR, but for these charge states high-spin and low-spin ordering coincide. (This is easily realized by the use of Fig. 4-17.)

Another interesting fact is revealed by the calculations: The local spin density at the impurity site appears more localized than the gap states themselves; this indicates that an essential part of the local magnetization originates from the impurity-induced valence band resonances, although they (like core states) do not contribute to the total spin. A substantial part of the spin density in many cases resides outside the impurity sphere as suggested by ENDOR data. For interstitial Fe^+ in silicon, experimental evidence indicates that about 80% (Greulich-Weber et al., 1984) or 72% (Van Wezep et al., 1985) of the spin density resides inside the impurity first neighbor shell; this value is reduced to about 60% for interstitial Ti^+ in silicon (Van Wezep et al., 1985). For comparison, theory yields 73% in the former case and 42% in the latter case (Beeler et al., 1985a). All these findings are also consistent with reduced spin-orbit splitting which is obvious from absorption, luminescence, and photo-EPR experiments on internal transitions (Kaufmann and Schneider, 1983; Clerjaud, 1985) and the quenching of the orbital angular momentum in the g values of EPR spectra mentioned earlier. The origin these effects share, is again the host-impurity interaction via p-d hybridization.

Beeler and Scheffler (1989) have also performed spin-unrestricted self-consistent LMTO Green's function calculations of the electronic structure of 4d transition metals in silicon. Figure 4-18 shows the results for substitutional impurities in the neutral charge state. In principle, all conclusions drawn from the interaction model of Fig. 4-13 which are applied to explain the trends depicted in Figs. 4-15, 4-16, and 4-17 remain valid. The only difference concerns the delocalization of both the e and t_2 states, which is substantially stronger in the 4d series than in the 3d series (This is even more true for the 5d shell.) Consequently, from the calculations represented by Fig. 4-18, low-spin ground states are predicted for substitutional 4d ions. Furthermore, at the high Z end, the t_2 states again resemble pure dangling bond states. This was already found to be true for the heaviest 3d elements, Cu and Ni.

This remarkable property, already inherent in the simplest two-state model of Fig. 4-12 by the existence of a dangling bond pinning energy, is exploited in the so-called vacancy model (Watkins, 1983). According to this model, the electronic structure of the "heavy" substitutional transition elements can be described in terms of a closed noninteracting d^{10} shell, with the remaining electrons residing at the dangling bond orbitals of the vacancy. One may arrive at this picture by steadily increasing the attractive impurity potential and thus lowering the atomic impurity d_{e,t_2} orbital in the defect molecule model of Fig. 4-13.

Finally, the nonbonding e states and the now also strongly localized impurity-like t_2 states, both fully occupied, form the practically noninteracting closed-shell (Krypton core) configuration. By now, the host-like gap levels of t_2 symmetry can be seen to have practically pure vacancy character and no d character. (Hemstreet, 1977). For example, Ni^-, Pd^-, Pt^- correspond to a configuration $3d^{10}+V^-$, $4d^{10}+V^-$, $5d^{10}+V^-$, respectively (V: vacancy). Thus a total spin $S=1/2$, observed in EPR studies for the cited elements and charge states (Pd^- and Pt^- in silicon, Ni^- in germanium), can be readily explained. Additionally, there is evidence that the characteristic Jahn-Teller induced dihedral distortion of V^- is also manifested in the EPR spectra of Pt^- and Pd^- (Ludwig and Woodbury, 1962).

Recently, the analogous EPR spectrum for Ni in silicon was detected (Vlasenko et al., 1987). However, Van Oosten et al. (1989) reproduced all these EPR spectra in silicon and argued that the large anisotropy in the electronic g factor for Pt^- points to a non-vanishing effective angular orbital momentum, contrary to what is expected from the vacancy model (Watkins,

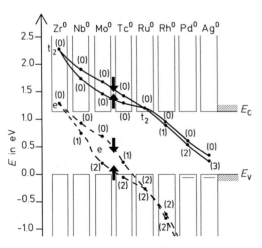

Figure 4-18. One-electron energies (spin-unrestricted) of the ground states of neutral 4d transition metals at the tetrahedral substitutional site. (After Beeler and Scheffler, 1989).

1983). They propose a dihedral electronic structure consisting of an open d^9 shell and two electrons bonded to only two silicon atoms, the two other dangling Si hybrids forming a reconstructed bond. This model should also be applicable to Pd^- and Ni^- where an open-shell orbital momentum possibly is reflected by strong anisotropy in the nuclear g factors.

4.3.3.2 Coulomb Induced Nonlinear Screening and Self-Regulating Response

The preceding sections were intended to demonstrate qualitatively why most transition metals can support bound states in the gap of a covalent semiconductor: Large variations in the energies of the atomic d states affect the gap levels only weakly via bonding-antibonding splitting.

The coexistence of various charge states of a given impurity in the semiconductor gap provides another puzzling fact if one bears in mind that energies in subsequent ionizations of the d shell differ by about 20 eV for a free-space atom. For example, for Mn the d^6 to d^3 ionization energies are 14.2, 33.7, 51.2, and 72.4 eV, respectively (Moore, 1959; Corliss and Sugar, 1977). The counterpart of this energy for a deep impurity is the Mott-Hubbard correlation energy U (U_0 for unrelaxed lattice), which in silicon is of the order of 0.3 eV (Eqs. (4-10), (4-11) and Fig. 4-5). This small value of U points to a nonlinear screening of the intra-d-shell repulsion, since a linear response would give a Mott-Hubbard U reduced only by one order of magnitude.

A self-consistent dielectric function, calculated by Zunger and Lindefelt (1983) for various transition metals in silicon, reveals that the impurity-induced perturbation is screened within an atomic distance. Consequently, a neutral 3d atom in silicon, substitutional or interstitial, already attains local charge neutrality at the central cell boundary and turns out to be very stable against the build-up of an ionic charge upon ionization. This remarkable dynamic aspect of nonlinear screening was first envisioned by Haldane and Anderson (1976), who concluded that several charge states in the gap occur as a result of rehybridization between transition metal d orbitals and host sp orbitals if the gap level occupation changes upon ionization. The details of this self-regulating mechanism have been worked out extensively by Zunger and Lindefelt (1983), Singh and Zunger (1985), and Zunger (1986).

Due to the antibonding character and hence rather extended nature of the gap states, their exclusive contribution to the total charge of the impurity-host system cannot effect the localization of the impurity-induced charge perturbation to the central cell. The difference is compensated by a rearrangement of valence band charge in response to the impurity-induced perturbation. Figure 4-19 shows the contributions of gap states and valence band states (resonances) to the net impurity charge. It can be seen that the gap states only at large distances from the impurity site produce all of the center's net charge (normalized to unity at the neutrality limit). In addition, the valence band resumes its unperturbed density at precisely these distances. Within the central cell region, the screening response from the valence band amounts to a net charge contribution of up to 50% in the case of neutral interstitial Mn in silicon. For substitutional Mn, a "screening overshoot" occurs ($Q_{net} > 1$) very close to the impurity site, arising from the strong bonding resonances.

Ionizing the impurity means that an electron is removed from a gap level of a neutral center. This relieves the valence band resonances (Fig. 4-13, 4-14) from

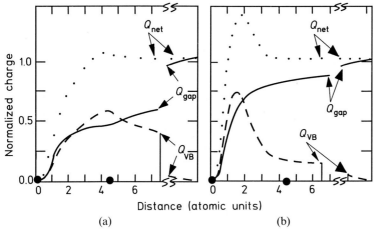

(a)   (b)

Figure 4-19. Decomposition of the net impurity charge Q_{net} into gap level contribution Q_{gap} and the valence band contribution Q_{VB} für interstitial (a) and substitutional (b) manganese in silicon. Local charge neutrality is reached at $Q_{net} = 1$, since Q_{net} is normalized by ΔZ, the difference in the number of valence electrons between impurity and host. In the cases (a) and (b) displayed in the figure, $\Delta Z = 7$ and $\Delta Z = 3$, respectively. Distances from the substitutional or interstitial impurity placed at the origin are given in atomic units (1 a.u. $\simeq 0.53$ Å $\simeq 0.053$ nm). (After Zunger and Lindefelt, 1982).

their Coulomb repulsion. As a consequence, they become lower in energy, with their coupling to the host being weakened.

Dehybridization now strengthens the (localized) d character and weakens the (delocalized) p character of the impurity-induced resonances. Consequently, the resonances increasingly localize in the central cell region thus making up for practically all of the charge that has been lost in the ionization process. Inversely, if an electron is added to a gap level, the mechanism just described would lead to an increase in hybridization, where the d content of the resonances markedly decreases, giving them more p character at the same time. The impurity-induced resonances therefore delocalize, and charge leaks out from the central cell, leaving Q_{net} nearly as close to neutrality as before. Therefore it appears that by this self-regulating response of the impurity-host system, an ionization-induced change of charge takes place at the ligands rather than the impurity site.

This remarkable stability of 3d impurities in semiconductors against the build-up of an ionic charge upon ionization has been confirmed by other calculations (Beeler et al., 1985a, Vogl and Baranowski, 1985) and explains why several charge states of a 3d metal may occur within the rather small energy limits of a semiconductor gap. It is noteworthy, however, that no comparable mechanism exists for the local magnetic moments (Zunger, 1986), so that exchange-induced screening is not very effective. Apparently, the main effect on the magnetic polarization of valence band resonances is the change in the net spin of the gap states upon ionization, whereas the varying localization of the resonances is of minor importance. Finally, it should be clear that nonlinear Coulomb screening only works in highly covalent semiconductors. It decreases in efficiency with decreasing covalency and vanishes for ionic crystals.

4.3.4 Ionization Energies and Trends

4.3.4.1 Transition Metals in Silicon

Ionization energies as well as excitation energies (Sec. 4.2) correspond to differences in the total energy of the initial and final states of a defect and must therefore include all contributions from electronic and lattice relaxation.

Nevertheless, in a rigid lattice, one may approximate the ionization energy of, for instance, a deep donor, within the local density formalism by the self-consistent calculation of Slater's transition state. This means removing half an electron from the one-electron level to be ionized (e.g., from a $t_2\uparrow$ gap level) and transferring it into the conduction band. A comparison with experimentally determined energy levels can then indicate the relative importance of lattice relaxation effects.

Figure 4-20 gives a comparison of calculated (Beeler et al., 1985a) and measured energy levels (Graff and Pieper, 1981a) of interstitial 3d transition metals in silicon. There is excellent agreement in most cases, but some drawbacks are also evident from Fig. 4-20. While theory yields a double donor for chromium, such behavior has been excluded by experiment (Feichtinger and Czaputa, 1981). Theory has failed to reproduce the correct ground states in the high-spin configuration for Cr_i^0 and Cr_i^+.

Apart from reasons originating from peculiarities of the applied formalism, a key to understanding the missing double-donor activity of chromium is the $Cr_i^{0/+}$ transition. In this transition, the electron is taken from the $t_2\downarrow$ level whereas the next electron for the $Cr_i^{+/++}$ ionization step has to come from the $e\uparrow$-single-particle gap level (see Fig. 4-17). Lattice relaxation might cause a general lowering of the calculated $e\uparrow$ level in this case, to give the observed high-spin ground states for C_i^0 and Cr_i^+. If this lowering would amount to about 0.3 eV, the $e\uparrow$ level would become resonant with the valence band due to electronic relaxation upon transferring one of its two electrons into the conduction band. Thus double-donor behavior of Cr_i is suppressed without seriously affecting the single-donor energy.

Unfortunately, for interstitial cobalt, the behavior of the $e\downarrow$ levels as to its sensitivity to lattice relaxation cannot be tested so far, since the very high diffusivity of interstitial cobalt (e.g., Utzig, 1989) in silicon causes the thermal stability of that species to be too low to observe any predicted

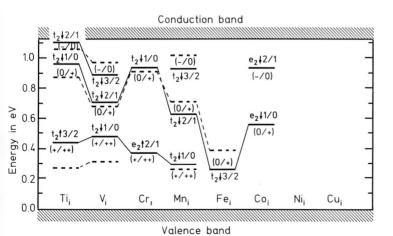

Figure 4-20. Calculated (solid lines) and experimental values (dashed lines) acceptor and donor levels for interstitial 3d transition metals in silicon. For each donor or acceptor level the corresponding change in the occupancy of the single-particle state involved is indicated. Calculations include spin polarization (spin-unrestricted theory). (After Beeler et al., 1985a).

energy levels. Copper has no deep levels, as mentioned before, and should be stable only in its positive charge state. For similar reasons, nickel in silicon should only be present in the neutral charge state due to its closed d shell.

The experimental single-donor levels shown in Fig. 4-20 particularly reflect the free-atom 3d ionization energies (Fig. 4-4) including the characteristic exchange-induced jump ("Hund's point") between Mn and Fe (d^5 and d^6 in the free atom, respectively). If neutral interstitial vanadium (V_i^0) has spin $S = 5/2$, the spin-related jump now should occur between the $V^{0/+}$ and the $Cr^{0/+}$ level and indeed does so. In case V_i^0 has spin $S = 1/2$ (see Fig. 4-15), this characteristic jump should be related to the increase of the spin splitting of the t_2 level caused by the change from low-spin to high-spin ground states.

For substitutional 3d impurities in silicon, with the exception of $Mn_s^{0/+}$ (Czaputa et al., 1985), no energy levels have definitely been identified until now. 3d metals in silicon dissolve mainly on interstitial sites (Weber, 1983) and often can be transferred to the substitutional site only by codiffusion of a highly soluble and fast-precipitating species such as Cu (for precipitation effects, see Chapter 11 of this Volume).

For $Mn_s^{0/+}$ theory (Beeler et al., 1985a) predicts an energy level at $E_v + 0.49$ eV, which is in good agreement with the experimental value of $E_v + 0.39$ eV. A few charge states, for example, Cr_s^0, Mn_s^+, Mn_s^{--} (Fig. 4-11), and Fe_s^+ (Muller et al., 1982), have been identified by EPR and were found to have high-spin configuration. With the exception of the trigonal center Fe_s^+, all the other ones show cubic symmetry. According to theory, neither Fe_s^+ nor Fe_s^- should be a stable configuration. Therefore, Fe_s is predicted to produce no deep level in the gap. But all this is based on the theoretical result that the low-spin configuration for Fe_s^0 ($S = 0$) is the correct ground state. A high-spin ground state ($S = 2$) would support both the existence of Fe_s^+ with spin $S = 3/2$ and a single-donor level.

Theory predicts 4d metals to be dissolved in silicon, preferentially on substitutional sites (Beeler and Scheffler, 1989). Calculated total energies show for the group IB ions (Cu-Ag-Au) representing the 3d, 4d, and 5d series that the stability of the substitutional site increases significantly relative to the interstitial site from Cu to Au because of the increasingly delocalized defect states. Experimental information on the energy levels of 4d metals is rather indecisive; only for Rh is there agreement with theory as to the amphoteric character and level positions. Predictions have been made for the double-donor to double-acceptor activity of substitutional Pd in silicon and for single-donor to triple-acceptor activity of substitutional Ag in silicon (Beeler and Scheffler, 1989). Table 4-1 shows amphoteric behavior for Rh (Czaputa, 1989), Pd (Landolt-Börnstein, 1989), and Ag (Baber et al., 1987) but no multilevel systems. The tabulated levels are assumed to be related to metals at the substitutional site, a fact which often is supported only by the high thermal stability of the species investigated. It seems clear that for the heavier 4d metals and particularly for the 5d series, Jahn-Teller distortions are no longer suppressed by exchange splitting, due to considerable delocalization of the defect states. Symmetry-lowering distortions in the resulting low-spin configurations are most naturally expected for states compatible with the above-mentioned vacancy model. In any case, there is no doubt that for the 4d and 5d series, lattice relaxation should be included in a calculation of level energies to meet the experimental values as closely as possible.

Fazzio et al. (1985) have studied the theoretical aspects of the substitutional IB series in silicon. They found that these impurities form a two-level, three-charge-state amphoteric system, where both the donor and acceptor transitions emerge from the antibonding t_2 gap state (compare Fig. 4-18 for Ag and Fig. 4-16 for Cu). But concomitant with the substantial delocalization in this system, Jahn-Teller distortions will occur. Therefore, the expected spin states for the neutral impurities will not be those derived from the unrelaxed lattice (e.g., $S=3/2$ for Ag in Fig. 4-18) but instead will follow from the occupancy of the lattice-distorted split-off t_2 states. These should give a doubly occupied spin-paired a_1-like level and a singly occupied b_1-like level to give $S=1/2$ (see again the vacancy model given by Watkins (1983)).

From Table 4-1, an effective Hubbard U (Eq. 4-10) of about 0.3 eV can be derived for Ag and Au. Fazzio et al. (1985) have compared their calculated values for the Franck-Condon U_0 (Eq. 4-11) with the experimental values for U. They found contributions from lattice relaxation of about 0.15 to 0.25 eV for both donor and acceptor transitions, which tend to cancel each other to give an effective U close to the ("vertical") Franck-Condon value.

Table 4-1. Energy levels (experimental) of substitutional 4d and 5d transition metals in silicon (References, see text).

Metal		Donor level (eV)	Acceptor level (eV)
4d	Rh	$E_c-0.57$	$E_c-0.31$
	Pd	$E_v+0.33$	$E_c-0.23$
	Ag	$E_v+0.34$	$E_c-0.54$
5d	Pt	$E_v+0.32$	$E_c-0.23$
	Au	$E_v+0.35$	$E_c-0.55$

The situation should be similar in the case of Pd and Pt. For these metals, the effective U appears to be increased to about 0.5 eV (Table 4-1), and the negative charge states should have spin $S=1/2$, which is in accordance with the experiment. However, there is still the problem of the correct identification of the isolated substitutional site of transition metals of the 4d and 5d series. For example, it has been discussed whether the donor and the acceptor transitions for gold in silicon evolve from the same center or not (Lang et al., 1980; Ledebo and Wang, 1983; Utzig and Schröter, 1984) although gold in silicon has been one of the most-studied systems over the past 25 years. One reason might be that no definite EPR signal for Au^0 ($S=1/2$) has been found so far, and that Au^+ and Au^- should not have any unpaired spins if the model just discussed is assumed to be adequate. On the other hand, misleading irregularities in determining gold concentrations from the gold donor or the gold acceptor can naturally be accounted for by a capture model of Feenstra and Pantelides (1985).

4.3.4.2 Compound Semiconductors and Bulk References

Transition metals substituting the cation site in compound semiconductors were studied theoretically by spin-polarized calculations within the framework of the semi-empirical tight-binding method (Vogl and Baranowski, 1985). Donors, acceptors, and double acceptors for the III−V compounds GaAs and GaP and for the II−VI compounds ZnS and ZnSe were predicted to form high-spin ground states. Their calculated energy levels were found to fit closely with the existing experimental data. (For example, properties of iron and chromium in III−V compounds have been discussed very thoroughly by Bishop (1986) and Allen (1986)). In Fig. 4-21, an example is given for the case of GaAs, from which (this time for the acceptors) the free-ion trend for the energy levels can again be seen in its attenuated form. Note that the characteristic spin-related jump between Mn and Fe arises again from effective d^5 (Mn^-) and d^6 (Fe^-) configurations constituting the initial states for the ionization of acceptors in GaAs. The same will be true for transition metal donors in II–VI compounds (see Fig. 4-22 below).

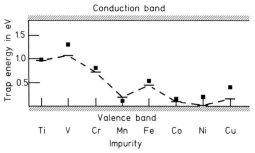

Figure 4-21. Calculated (dashed line) and experimental (squares) acceptor levels of 3d transition metals in GaAs. (After Vogl and Baranowski, 1985).

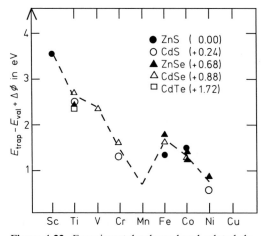

Figure 4-22. Experimental valence band related donor levels of the 3d series in III–V and II–VI compounds, corrected by the corresponding photothreshold $\Delta\Phi$. For ZnS, taken as a reference system, $\Delta\Phi$ is assumed to be zero. (After Vogl and Baranowski, 1985).

As long as high-spin states are involved, this jump will depend both on the formal charge states of the substitutional impurity and on the number of valence electrons of the atoms being substituted.

In a further step, this trend can be generalized and unified for a whole class of semiconductors. Level energies (ionization enthalpies) can be measured relative to the valence band edge, the changes of which, at least within certain classes of semiconductor materials (e.g., II–VI or III–V), are known to be given approximately by the differences in the experimental photothresholds (Harrison, 1980). In Fig. 4-22 experimental valence-band-related energy levels are shown, corrected for each solid by the corresponding photo-threshold (taken relative to ZnS as a reference host). Evidently, this procedure is very successful in clearly exhibiting the universal trend through the 3d transition metal series for different semiconductors (Vogl and Baranowski, 1985) and its close relation to that observed for free ions (Fig. 4-4a). Furthermore, this universal trend points to the existence of a certain reference level to which a given transition metal impurity appears pinned if one changes the host crystal.

There is experimental evidence that valence-band offsets in heterojunctions are simply the difference in the energy level positions of a given transition metal in the two compounds that form the heterojunction (Langer and Heinrich, 1985; Delerue et al., 1988). From these findings, an internal (bulk) reference level may be established, circumventing the contribution from the semiconductor surface to the (external) photo-threshold energy (Zunger, 1985; Tersoff and Harrison, 1987).

Caldas et al. (1984) suggested this reference level to be related to the antibonding cationic t_2-symmetric state (vacuum pinning). There are arguments, however, for the average self-energy of cation and anion dangling bonds constituting the reference in the band alignment of heterojunctions (Lefebvre et al., 1987). Both versions continue to be discussed, but for III–V compounds, the latter average gives the closer pinning and might be more appropriate at least for this class of semiconductors (Langer et al., 1989). Once such a reference is established, and a deep level of a given metal is characterized in one semiconductor, the energy levels in other crystals within the same class may be predicted to a certain degree of accuracy.

Further general trends have been discovered in isothermal pressure coefficients of donor or acceptor levels. Deep levels as a rule show pressure coefficients two orders of magnitude larger than those of shallow levels and may therefore be used to define deep centers experimentally (Jantsch et al., 1983). Nolte et al. (1987) linked pressure coefficients to the deformation potentials of the band edges by choosing deep centers with a vacancy-like level structure as a reference. Finally, pressure coefficients of deep levels were interpreted in terms of an isothermal change of crystal volume upon ionization to obtain a measure of symmetric (breathing mode) distortions around substitutional and interstitial deep centers (Samara and Barnes, 1986, Weider et al., 1989; Feichtinger and Prescha 1989).

4.3.5 Excited States

4.3.5.1 Internal Transitions

In Sec. 4.2.2, interelectronic transitions at a fixed charge state of both the defect and the accompanying electron-phonon interaction have been discussed. For transition metals, such excitations are better known as internal $d \rightarrow d^*$ transitions ac-

cording to a change of total energy of the many-electron system

$$\Delta E^{(ij)} = E^{(j)}(e^{m'} t_2^{n'}) - E^{(i)}(e^m t_2^n) \qquad (4\text{-}14)$$

where ij denotes the multiplet representation indices, and $e^{m'} t_2^{n'}$, $e^m t_2^n$ denote the predominant one-electron configurations (Singh and Zunger, 1985). In Eq. (4-14) an excitonic electron-hole contribution is contained, since both the excited electron and the remaining hole are still partially bound. Apart from ground states, inter-configuration mixing may be considerable, and therefore the one-electron level description denotes the configuration that mainly contributes to higher multiplet terms. Numerous internal transitions have been observed via absorption and luminescence spectroscopy (e.g., Landolt-Börnstein, 1989). Both can give valuable information on electronic structure and symmetries of a deep center, although luminescence may be masked by other competing centers or suppressed by strong electron-phonon coupling, leading to effective radiationless recombination. Zunger (1986) has compiled existing data on internal transitions studied for transition metals in various semiconductors.

As a rule, only one "oxidation" state of a given species is observed. This state d^N is determined by the total number N of "active" electrons (e.g., d^7 for Co^{2+}, or d^9 for Ni^+) and must not be confused with the charge state of an impurity relative to the given host used so far. For example, an "oxidation" state of Co^{2+} will correspond to the formal charge state Co^- in III–V compounds, to Co^0 in II–VI compounds when substituting the cation site, and to Co^{2-} in silicon. For the interstitial site, oxidation state and formal charge state coincide. In order to relate internal transitions to ionization, Fig. 4-23 represents an example, combining experimentally ob-

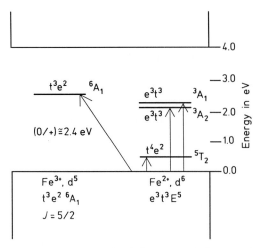

Figure 4-23. Energy level diagram, relating experimentally observed internal transitions to the single donor level of Fe in ZnS. The insert shows the parameters (including spin $J = S$ for the positive formal charge state) of the ground state configurations participating in the ionization (left) and internal excitations (right). (After Zunger, 1986).

served single donor activity ($Fe^{0/+}$) and the charge state conserving (Fe^0) transitions into excited states of substitutional Fe in ZnS. For all states involved, the many-electron multiplets $^{(2S+1)}\Gamma$ are denoted by spin multiplicity ($2S+1$) and orbital symmetry (Γ) in the tetrahedral point group. The oxidation state changes from Fe^{2+} to Fe^{3+} upon ionization. But transitions to higher excited states of a d^N system, occurring as resonances in the conduction band, have also been observed (Baranowski and Langer, 1971).

The occurrence of a single fixed oxidation state in the d→d* spectra (a few transitions have been observed additionally from neighboring states (Zunger, 1986)) and the fact that spectra from internal transitions of a given impurity resemble each other in different semiconductors are not very surprising in light of the electronic structure of deep centers discussed in the present contribution. Since the charge

state (or the oxidation state) is conserved, the dynamic aspect of screening described in Sec. 4.3.3.2 is lacking. Consequently, transition energies $\Delta E^{(ij)}$ (Eq. (4-14)) cover roughly the same range as found for free ions, suggesting a localized electronic structure (Sec. 4.2.1). Nevertheless, whenever an electron is excited from a non-bonding e level to a hybridized t_2 level, it is in fact transferred from the impurity site to the ligands. The optical cross section for the $Fe^{-/0}$ transition in GaAs was found to be strikingly smaller than that of similar ionizations for Cr and Cu (Kleverman et al., 1983), reflecting the different spatial extension of the initial states (e for Fe, t_2 for Cr and Cu).

It is precisely this difference in orbital character that makes it difficult to treat the multiplet effects of transition metals within empirical ligand field theory. This was pointed out recently by Watanabe and Kamimura (1989) who critically reviewed earlier attempts to include multiplet effects in first-principle approaches. Figure 4-24 demonstrates the state-of-the-art accuracy of such calculations compared to experi-

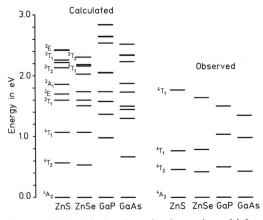

Figure 4-24. Calculated and observed multiplet structures of Co^{2+} (oxidation state) in III−V compounds. Theoretical values were calculated by Watanabe and Kamimura (1989).

mentally detected multiplet structures. Experimental values are taken from Weakliem (1962) and Koidl et al. (1973) for ZnS, Weber et al. (1980) for GaP, Hennel and Uba (1978) for GaAs and Baranowski et al. (1967) for ZnSe. From the upper $^4A_2 \rightarrow {^4}T_1$ transition a weak chemical trend with the increasing covalency from ZnS to GaAs becomes evident and is qualitatively reproduced by the calculations.

Multiplet structures of interstitial 3d transition metals in silicon were calculated combining the X_α-cluster method in a spin-unrestricted form with the above-cited approach of Hemstreet and Dimmock to account for space- and spin-induced correlations (DeLeo et al., 1982), but none of the corresponding $d \rightarrow d^*$ transitions has been detected so far.

4.3.5.2 Rydberg-Like States

In the discussion of the basic physics of a deep center in Sec. 4.1.1 and Fig. 4-2 it was mentioned, that the ground state energy of a deep center is almost exclusively determined by the short-range central cell potential. But whenever a neutral deep center is ionized by emitting either an electron into the conduction band or a hole into the valence band, the screened Coulomb potential due to the charged center should, in principle, give rise to a series of shallow excited states for which effective mass theory (EMT) should be applicable. These closely spaced energy levels near the relevant band edges play a role in the so-called cascade capture. In this process, a carrier is first captured in a highly excited state. It then goes down the ladder by emitting one or very few phonons each time. Finally, the carrier may drop from a lowest-lying cascade state (at low temperatures, re-emission from this state is negligible) into the deep-lying ground state by

multi-phonon emission or by a light-emitting transition. The cascade process was originally developed for shallow centers (Lax, 1960); in theory, the temperature dependence of the process is expressed in the form T^{-n}, where the value of n is usually $1 < n < 4$ (Grimmeis et al., 1980a, b).

Krag and Zeiger (1962) studied sulfur-doped silicon via absorption and obtained from their well-resolved line spectra the binding energies for the ground state and the excited states of a deep center, the latter being in perfect agreement with EMT. More recently, the chalcogen series in silicon was studied by applying different spectroscopic methods (see Sec. 4.4.1) and a He-like level structure was revealed. For transition metals in silicon, Rydberg-like series were observed from both absorption and photo-thermal ionization spectroscopy (PTIS) and interpreted in terms of EMT (Grimmeiss and Kleverman, 1989). For studying species of deep centers with low solubility, PTIS may be a very valuable tool, since the method was shown to be essentially independent of defect concentration (Kogan and Lifshits, 1977). PTIS operates by a two-step process: an optical excitation from the ground state into excited states followed by thermal ionization of the excited center. Since the PTIS signal is derived from the induced photocurrent or photocapacitance (in junction techniques), the sample studied acts as a photon detector itself. Under certain circumstances concerning the temperature range, the method yields a discrete line spectrum that can be obtained at a very high resolution of about 0.03 meV.

Figure 4-25b shows PTIS and FTIR (Fourier Transform Infrared Transmission Spectroscopy) spectra obtained from neutral interstitial iron (Fe_i^0) in silicon. The identification of the lines in the Fe_i^0 transmittance spectrum is based on three overlapping EMT series (Olajos et al., 1988), each in accordance with the EMT scheme pointed out in Fig. 4-25 a.

Whereas p states in a Coulomb potential are split by the anisotropic effective mass, s states are split by multi-valley interaction. For Fe_i the symmetry is tetrahedral, hence the s-like states should be split into an A_1 singlet, a T_2 triplet and an E doublet due to the six equivalent conduction-band minima of silicon, where the strongest splitting occurs for the 1s states.

The degeneracy of the 1s (T_2) and 1s (E) states may sometimes be lifted by spin-valley splitting (Aggarwal and Ramdas, 1965), Clearly, the true ground state is not determined by the Coulomb tail of the potential, since it constitutes the deep level. But even higher 2s and 3s states are systematically lower in energy than that expected from EMT, since they still feel the attractive localized central cell potential. Besides, the three series in Fig. 4-25 b involve EMT-forbidden transitions, where, on the other hand, transitions to the split 1s (T_2) states are missing. Thus the formal charge state for a deep center is easily deduced from the spacing of the p-like states in a well-resolved Rydberg series (which is proportional to Z^2), whereas for the symmetry of a center, perturbation spectroscopy (e.g., FTIR under uniaxial stress) may sometimes be more helpful (Krag et al., 1986). The occurrence of three EMT series in the transmittance spectrum shown in Fig. 4-25 b may be accounted for by spin-orbit splitting in the final states, since the Fe_i^+ center left behind upon excitation into the shallow states has an orbitally degenerate ground state ($^4T_2$).

In addition to the line spectrum of the shallow states, which is essentially similar to that found by FTIR, the PTIS spectrum of Fig. 4-25 b reveals a clearly resolved structure at higher energies. The final

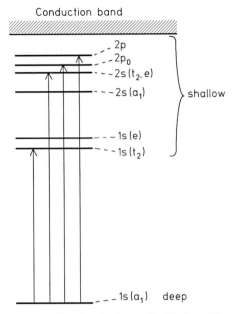

Figure 4-25 a. Level scheme for Rydberg-like states in silicon (see text).

states of the corresponding transitions must lie well within the conduction band continuum. The lines result from a higher-order absorption process, where a direct optical no-phonon ionization becomes resonant with a transition from the ground state to a bound excited state accompanied by the emission of a bulk phonon. Therefore, details of the PTIS spectrum are repeated at higher energies as they are just shifted by the energy of the participating phonons. These so-called phonon-assisted Fano resonances (Janzén et al., 1985) may be very helpful in deciding whether the studied center has donor character (electron excitation) or acceptor character (hole excitation), since in silicon, only intervalley phonons (Harrison, 1956) participate in the process for donors, and the zone-center phonon (Watkins and Fowler, 1977), in the process for acceptors.

Rydberg-like spectra were also observed for gold and platinum acceptors (Klever-

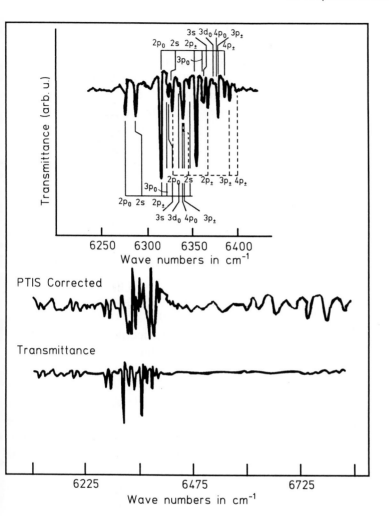

Figure 4-25b. Excitation spectra of interstitial iron (Fe_i^0) obtained from photothermal ionization (upper curve) and transmittance (lower curve and inset). (After Olajos et al., 1988).

man et al., 1987, 1988; Armelles et al., 1985) and were found to resemble those of shallow centers like indium in silicon, so that the compatibility of EMT with shallow excited states of deep centers can be seen as well established.

4.4 Properties of Selected Systems

4.4.1 Chalcogens in Silicon

Sulfur, selenium, and tellurium constitute very interesting impurities in silicon for practical and physical reasons. First,

these elements can be used in infrared detectors (Sclar, 1981; Migliorato and Elliott, 1978). Second, this group, which is isocoric to silicon (in the sense of always having ten outer core electrons), provides an electronic structure that shows both deeply bound ground states and excited Rydberg-like states. Besides, it represents an example in which the question as to the substitutional or interstitial site was definitely decided by theory (Beeler et al., 1985b). Oxygen, one of the most important impurities in device fabrication due to its strong involvement in gettering, precipitation effects, and the formation of the so-called thermal donors (TD) (Claeys and Vanhellemont, 1989; for TD's, see also Sec. 4.4.4), behaves quite different and does not occupy T_d symmetric lattice sites. As for other semiconductors, the role of oxygen in radiative recombination, especially in GaP, has been thoroughly discussed by Dean (1986).

4.4.1.1 Sulfur, Selenium, and Tellurium in Silicon

The electronic structure of the single-particle ground states for substitutional chalcogens was calculated by Bernholc et al. (1982), Singh et al. (1983), and Beeler et al. (1985b) via first-principles Green's function methods and by the empirical tight-binding method (Vogl, 1981). The calculations agree that the one-electron ground state exists in an a_1-symmetric host-like state, but they differ as to the order of binding energies (Singh et al.) compared to free ions. Figure 4-26 gives a simplified interaction scheme which should also hold for other sp-bonded impurities (compare also the introduction in-

to such schemes given by the two-state model in Sec. 4.3.2.1). Since the t_2-symmetric dangling bond state is pushed into the conduction band by the interaction with the atomic p state, and four of the six impurity electrons are needed to refill the bonding t_2 states in the valence band, the remaining two must populate the a_1-symmetric gap level, which can accommodate only two electrons of different spin. Accordingly, chalcogens can only act as double donors, a fact verified by experiment.

By monitoring thermal emission rates (Eq. (4-6)) via DLTS (deep level transient spectroscopy) in Si doped with S, Se, and Te, two energy levels in the upper half of the band gap could be established (Grimmeiss et al. 1980a, 1980b; Grimmeiss et al., 1981). The detailed electronic structure of these centers was revealed via absorption studies and PTIS lines, as shown in the example of Fig. 4-27 for the deeper of the two donor levels ascribed to the $Te^{+/++}$ transition (absorption is therefore studied for the positive charge state of Tellurium). The spectral dependence of the photo-ionization cross section (Eq. (4-9)) shows a clearly resolved structure due to photothermal ionization (see previous section), which may be expanded and found to agree with the absorption lines (upper part of Fig. 4-27) within 0.2 meV. With the assignments for the $2p_0$ and $2p_\pm$ Rydberg-like states for Te^+ from an overall spectrum, the spacing for the 2p lines is that expected from EMT (Faulkner, 1969) for a doubly ionized impurity ($Z=2$). From EMT the $2p_\pm$ state for a singly positive center is located 6.4 meV below the conduction band. The optical ionization energy at 5 K from Fig. 4-27 for the second donor is therefore 385 meV + 4 × 6.4 meV, which gives $E_{opt} = 410.8$ meV. This is in excellent agreement with the ionization energy taken from the PTIS lines.

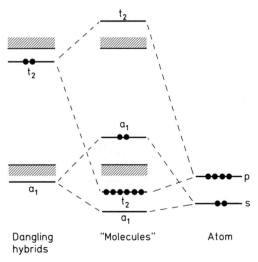

Figure 4-26. Simplified interaction model for neutral chalcogens at the substitutional site in silicon.

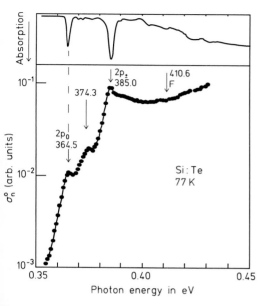

Figure 4-27. Photo-ionization cross section for Te$^+$ and PTIS structure just below the onset of direct photo-ionization, marked F (lower curve). $1s(A_1) \rightarrow 2p_0$ and $1s(A_1) \rightarrow 2p_\pm$ absorption lines of Te$^+$ in silicon (upper curve). (After Grimmeiss et al., 1981).

Similar studies were made for S and Se, which confirmed the picture of chalcogen double donors discussed so far (see the review of Grimmeiss and Janzén, 1986). Table 4-2 serves to illustrate the level structure accessible to optical excitation in the case of selenium. The atomic s orbital energies for Te, Se, and S are -19.06 eV, -22.77 eV, and -23.92 eV, respectively (Fischer, 1972; Clementi and Roetti, 1974). If those values are related to the interaction model of Fig. 4-26, the s orbital goes

down in energy in the series Te, Se, S as does the antibonding a_1 gap level, due to decreasing interaction between the s electrons of the impurity and the a_1-symmetric orbital of the vacancy. The chemical trend in the ionization energies of the series 199 meV, 306 meV, and 318 meV for the first donor and, more distinctly, 411 meV, 593 meV, and 613 meV for the second donor confirms the model, in which the results of the cited calculations are contained qualitatively.

In order to complete the overall agreement with the given electronic-structure model, an EPR spectrum of Te$^+$ is shown in Fig. 4-28, whose parameters in principle

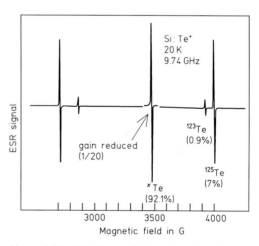

Figure 4-28. EPR lines of Te in Si. The off-center lines of much lower intensity are split by hyperfine interaction with the non-vanishing nuclear spin I of Te isotopes (I and relative abundances are inserted). (After Grimmeiss et al., 1981).

Table 4-2. Ground state related excitation energies of Se^0 and Se$^+$ in Silicon (all energies in meV). (After Janzen et al., 1984).

	$1s(T_2)$	$2p_0$	$2s(T_2)$	$2p_\pm$	$3p_0$	$3p_\pm$	Cond. band
Se^0	272.2	295.1	297.4	300.3	301.2	303.5	306.6
Se$^+$	427.3						
	429.5	547.2	553.9	567.6	571.6	581.0	593.2

are consistent with the spin-Hamiltonian of Eq. (4-12), with spin $S = 1/2$. One must note, however, that none of the experiments on chalcogens in silicon (including EPR and ENDOR, Niklas and Spaeth, 1983; Greulich-Weber et al., 1984) was able to decide definitely whether these atoms occupy the substitutional or the interstitial tetrahedral site. That question thus relies solely on the consistency with the theoretical results on electronic structure for substitutional atoms.

This remaining piece of the nearly completed puzzle was put in place by total energy calculations for substitutional and interstitial chalcogens (Beeler et al., 1985b). It was found that the difference in the energies of solution for both species (of the order of several eV) is strongly in favor of the substitutional site.

4.4.1.2 Oxygen and Nitrogen in Silicon

Both oxygen and nitrogen in silicon should be describable by the scheme of Fig. 4-26, which excludes Jahn-Teller distortions because the A_1 ground states are not orbitally degenerated; yet neither oxygen (Watkins and Corbett, 1961; Corbett et al., 1961) nor nitrogen (Brower, 1980, 1982) occupies T_d-symmetric sites. Instead, oxygen forms the so-called A center when trapped by a vacancy. Structurally, this consists of an off-centered oxygen atom bound to two silicon atoms, the remaining two forming a reconstructed bond. Contrary to the electrical activity of the other chalcogens, this center provides a single acceptor level 0.17 eV below the conduction band edge (Brotherton et al., 1983).

Oxygen may interrupt a silicon bond to form a nonlinear bridging molecule. In this rather interstitial position (Hrostowski and Adler, 1960), oxygen is electrically inactive (Corbett et al., 1964) but in super-saturated solution, it will be mobile at higher temperatures to form a variety of oxygen-related defects. The most intriguing of them will be discussed briefly in Sec. 4.4.4.

4.4.2 DX Centers in $Al_xGa_{1-x}As$

Detailed studies of donors in III–V ternary alloys have revealed a steadily increasing complexity in their electronic structures. The optical ionization energy for the deep levels was found to be drastically larger than that for thermal ionization, indicating a large lattice relaxation (Lang and Logan, 1977; Lang et al., 1979). Deep centers were thought to consist of the donor D (e.g., Si) and an unknown defect X, hence the name DX centers. The large lattice relaxation (LLR) was confirmed recently by very sensitive optical ionization studies (Mooney et al., 1988a; Legros et al., 1987). The view that all exciting physical features of the DX center (e.g., shallow-deep metastability, persistent photoconductivity, or hot-electron trapping) are simply properties of the isolated donor is now receiving increasing acceptance. Nevertheless, a generally confirmed microscopic model remains elusive – a breakthrough might be anticipated by a recent microscopic model (see Sec. 4.4.2.2).

Technologically, DX centers were made responsible for device instabilities in MODFETs (modulation field effect transistor) (Theis and Parker, 1987) and are suggested to limit the free-carrier density by self-compensation (Theis et al., 1988).

4.4.2.1 Large Lattice Relaxation and Metastability

As discussed in Sec. 4.2.2, moderate lattice relaxation most certainly influences the positions of various donor or acceptor levels in the gap (Fig. 4-7), sometimes to

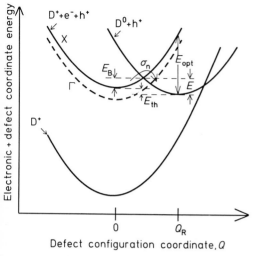

Figure 4-29. Conventional configuration coordinate diagram modeling large lattice relaxation effects of a DX Center. The shifted parabola at Q_R represents the defect in its stable position, giving rise to the deep level. The dashed-line parabola describes a situation where the deep level is resonant with the conduction band at the Γ symmetry point (see text and Fig. 4-31). (After Lang, 1986).

the extent of negative-U level ordering. In the case of very large relaxation, additional effects can occur. This is best demonstrated again by a configuration coordinate diagram (Fig. 4-29) that is typical for the DX family. At a first glance, no spectacular deviation might be seen if one compares the LLR diagram to that given in Fig. 4-7, but there are two important differences (to begin with, the dashed line parabola is ignored):

1) In Fig. 4-29, the lattice relaxation energy $S\hbar\omega$ (see Fig. 4-6), measured by the difference of the optical ionization energy, E_{opt}, and the thermal ionization energy, E_{th}, is much greater than the difference in total energies of the relaxed configurations measured by E_{th}. This is apparently not the case in the weak coupling regime of Fig. 4-7 ($E_{opt} - E_{th} < < E_{th}$).

2) The capture barrier E_B has increased, leading to a small and strongly thermally activated σ_n. But even more remarkable is the barrier E to be overcome for the thermal emission of an electron from the deep center into the conduction band, which now largely surmounts E_{th} ($E = E_{th} + E_B$).

Condition (1) represents a regime where extrinsic self-trapping may occur (Toyozawa, 1980, 1983). By this phenomenon, an isolated defect, which can bind a charge carrier only by its long-range Coulomb potential, is able to form a deep bound state. This mechanism is based on the interaction of strong electron-phonon coupling with the weak central-cell potential of the center. Thus the same donor species should give rise to both shallow and deep level behavior, the two levels being separated by an energy barrier.

Figure 4-30 displays experimental evidence for such a shallow-deep bistability. A slightly Si doped $Al_{0.29} Ga_{0.71} As$ sample (an Al mole fraction $x = 0.29$ makes the

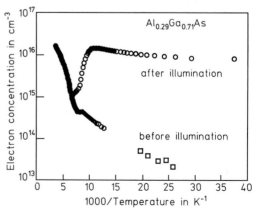

Figure 4-30. Free-carrier concentration in lightly Si-doped $Al_{0.29}Ga_{0.71}As$, obtained from Hall-effect measurements (After Theis, 1991). The true activation energy of the hydrogenic level cannot be inferred from the plot, after low-temperature illumination, since a metal-insulator transition might have occurred (Katsumoto et al., 1987).

alloy a direct semiconductor since the conduction band minimum at the Γ point is lowest in energy, see Fig. 4-31) was cooled in the dark and the Hall free-carrier concentration was determined as it was gradually warmed. During the cooling process, a cooling-rate-dependent temperature can be reached below which the electrons in the conduction band cannot longer equilibrate with the deep level due to both relaxation-induced barriers (condition (2) above). Thus most of the DX centers will be occupied in the deep-level configuration at Q_R (Fig. 4-29), but a small percentage of the electrons will have been frozen-out into the shallow level, which may also be represented for simplicity by the upper parabola (solid line) at $Q=0$ in Fig. 4-29. Actually, the parabola $D^0 + h^+$ representing the occupied shallow state should be placed slightly below that denoted by $D^+ + + e^- + h^+$ but also centered at $Q=0$, since the shallow level is mainly derived from X conduction-band states.

Warming the sample without any illumination will thermally activate electrons from the shallow level into the conduction band; the ionization energy will be reflected by the low temperature regime of Fig. 4-30. At higher temperatures, when the barrier E can be overcome, the deep level participates in the ionization, and the slope of the Hall curve changes drastically. Finally, nearly all centers will be unoccupied, assuming the configuration at $Q=0$. So far, from Fig. 4-30 (before illumination) an amphoteric center (compensated) is also imaginable, but low-temperature illumination results in persistent photoconductivity, which simply means that the free-carrier concentration reaches a high level under illumination and only changes slightly after the light has been turned off. In terms of Fig. 4-29, this means that by optical ionization (related to E_{opt}), the DX center has

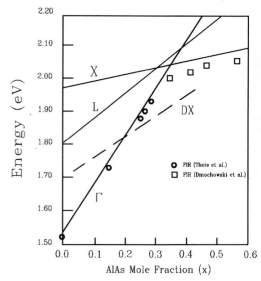

Figure 4-31. Hydrogenic level ionization energies in $Al_xGa_{1-x}As$ derived from far-infrared transitions. (Data from Theis et al. (1985) and Dmochowski et al. (1988)). The dashed line indicates the position of the deep DX level, with varying Al mole fraction.

been switched from the deep-level configuration to the shallow-level configuration. Fig. 4-31 shows data from far-infrared transitions where hydrogenic energy levels have been established for direct-gap material ($x<0.35$) (Theis et al., 1985) as well as indirect-gap material ($x>0.35$) (Dmochowski et al., 1988).

Condition (2) implies another property typical for DX centers. The deep level (Si doping) is located 160 meV below the conduction band edge at the point where the direct gap (Γ) becomes indirect (X), at $x_{crossing}=0.35$. Altering the composition of the semiconductor alloy toward lower Al mole fractions causes the ionization energy (E_{th}) to decrease. Finally, the deep level enters the Γ conduction band, represented by the dashed parabola in Fig. 4-29, at $x\approx0.2$ (see the dashed line labeled DX in Fig. 4-31). But even at such conditions, the center retains its compact bound state since it was shown experimentally that neither relaxation nor the thermal emission barrier (E in Fig. 4-29) depends on the host band structure (Mooney et al., 1986). In the conventional model, the stability of the barrier E against varying of the alloy composition requires the existence of selection rules by which thermal emission is linked mainly to L conduction band states. The reason for these rules is that the deep level seems to follow the L band minimum (see Fig. 4-31). Thus, while thermal trapping and persistent photoconductivity at the resonant deep level tend to vanish for $x < 0.22$ (because the capture barrier E_B increases), the level can be persistently populated by a hot-electron process (Theis and Parker, 1987). At very high doping levels, where the Fermi level is well above the conduction band edge, thermal trapping, even in GaAs, has also been observed (Mooney et al., 1988 b).

4.4.2.2 A Microscopic Model for DX Centers

While all energy barriers modeled by an LLR diagram like that of Fig. 4-29 can be established experimentally, the fundamental question as to the origin of the stabilizing relaxation remains. If the DX center is a complex, then changing the electronic structure upon ionization may result in either a molecular rearrangement (see the well-understood case of carbon-carbon pairs ($C_i C_s$) in Chapter 3 by Watkins) or in a change in one of the constituents of the complex to a more distant lattice site (see the case of donor-acceptor pairs discussed in the next section).

Most of the experiments on DX centers indicate, however, that the DX center concentration roughly equals that of the donor dopant. This was especially manifested in the hydrostatic pressure experiments of Mizuta et al. (1985) and Maude et al. (1987) on the donor-related level in GaAs. On the other hand, the majority of donors in GaAs occupy substitutional sites (Maguire et al., 1987). Morgan (1986) suggested an interstitial displacement of the donor atom. Chadi and Chang (1988) propose a similar model, where the lattice relaxation involves either a group IV donor (e.g., Si) moving from the substitutional Ga site into a threefold coordinated interstitial site or a Ga (or Al) atom moving into an interstitial site adjacent to a group VI donor (e.g., Te) in the As lattice. But these large distortions, with a vacancy (V_{Ga}) left behind, should be stabilized by trapping an extra electron whereby negative U behavior is effected (Khachaturyan et al., 1989). Photo-ionization of Si_{Ga} means in this case that the now negatively charged Si atom in the interstitial position passes through the unstable intermediate (neutral) state (as before, this state can be reached thermally in surmounting the barrier E) and relaxes back to the undistorted Ga site upon releasing a further electron. Inversely, by capturing an electron, the positive center passes again its intermediate state and relaxes into the interstitial position upon capturing another electron. Again, the capture barrier is E_B. An appropriate configurational coordinate diagram would now require a third parabola for the intermediate neutral state to be inserted between the parabola for the positive charge state (at $Q = 0$) and that for the negatively charged center (at $Q = Q_R$). Only very small barriers (compared to E_B and E) would be set up against the neighboring configurations. Remarkably, no selection rules need to be specified within this model in order to explain a composition-independent emission barrier, E, as long as both the intermediate single-electron state and the negatively charged two-electron state (at $Q = Q_R$) do not depend on the host band structure.

It has recently been shown that the famous EL2 center in GaAs (see Chapter 3) could consist of an As_{Ga} antisite defect which is metastable with a $V_{Ga} As_i$ pair (Dabrowski and Scheffler, 1989). This is very similar to the DX system described above, if one replaces As_{Ga} by Si_{Ga}^-.

4.4.3 Deep Transition Metal Donor–Shallow Acceptor Pairs in Silicon

Pairing with a shallow background dopant of proper charge state provides a possible path for the entropy-driven decay of a supersaturated solid solution. Such a supersaturation can be built up by diffusing an interstitial 3d metal (e.g., iron) into a silicon sample at high temperatures ($>1000\,^\circ$C) and then quenching the sample to room temperature. Interstitial 3d metals such as Cr_i, Mn_i, and Fe_i are still mobile at room temperature or slightly above, but in any case they are not lost to higher-order precipitates very quickly during the quenching process, as observed for Co or Cu in Si.

At room temperature, the above interstitial series in p-type material is positively charged (compare the position of the donor levels in Fig. 4-20), whereas practically all substitutional shallow acceptors (B, Al, Ga, In) are occupied and exist in the negative charge state. Thus the pairing is strongly catalyzed by the Coulomb-correlated reaction cross section and may finally be complete at room temperature with no isolated interstitials left. On the other hand, the pair binding energy turns out to be largely determined by Coulomb interaction and amounts to about 0.5 eV at the nearest substitutional-interstitial distance. This leads to a low barrier for pair dissociation and for the reaction

$$Fe_i^+ + B^- \Leftrightarrow (Fe_i^+ \, B^-)^0 \qquad (4\text{-}15)$$

equilibrium can actually be established between 30 °C and 150 °C (Kimerling et al., 1981; Graff and Pieper, 1981). At higher temperatures, the iron atoms diffuse irreversibly to other sinks and may form larger precipitates. Therefore, if pairs of this type are present in high-power semiconductor components (e.g., thyristors), they may cause thermal instability and degradation.

In the electronic structure of a shallow-deep pair configuration, the deep impurity retains its localized character, so that the well-known trends for isolated 3d impurities are again reproduced in the energy levels of the pairs. However, the perturbation induced by the shallow impurity makes the pair an interesting example for charge-state-controlled metastability. Thirty iron-related complexes have been identified by EPR including those about to be discussed (Ammerlaan, 1989). Thus one may get an impression of the problems of defect identification in cases where no defect-specific fingerprints are available from experiment.

4.4.3.1 Electronic Structure and Trends

In bringing together the isolated constituents Fe_i^0 and Al_s^0, the interaction between their electronic states can be modeled after theoretical results from the local density method by Scheffler (1989). In Fig. 4-32, Fe_i^0 is represented by its spin-split one-electron levels displayed in Fig. 4-17.

The single-particle t_2-symmetric ground state of neutral substitutional aluminum in silicon can be placed approximately 70 meV above the valence band, which value is found experimentally for the ionization energy of the shallow-acceptor level (this is feasible because the electronic relaxation is small due to the extended nature of the shallow state). When the interaction is initiated, the iron-like levels are shifted to higher energies because of the repulsive dielectrically screened Al potential (in this case covalent interactions are of minor importance, since strongly differing spatial extension allows for almost no overlap between the two wave functions). The t_2 levels are additionally split by the superimposed symmetry-lowering perturbation in the $\langle 111 \rangle$ pair (trigonal symmetry). One electron is transferred from the iron-like levels to the shallow t_2 state, which remains essentially unaffected for the reasons noted above. In general, this is not true for pairs formed exclusively from deep centers, for example, for the AuFe complex (Assali et al., 1985). Figure 4-32 shows that in contrast to isolated Fe_i, the pair is likely to have two iron-like ionization transitions in the gap: $(Fe_iAl)^{0/+}$ corresponding to $Fe_i^{+/++}$ and $(Fe_iAl)^{-/0}$, related to $Fe_i^{0/+}$.

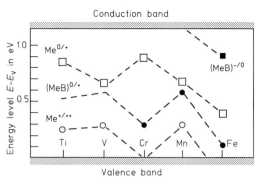

Figure 4-33. Trends in experimental energy levels of interstitial transition metal (Me_i) boron pairs in silicon. Data points have been interconnected and extrapolated. The donor and double-donor levels are shown for comparison. (After Feichtinger et al., 1984).

In Fig. 4-33, energy levels of nearest-neighbor $\langle 111 \rangle$ pairs are shown, together with donor and double-donor levels of the isolated interstitial impurities ($Me_i = Ti$, V, Cr, Mn, Fe). Only Me_iB pairs have been used in the figure, but their trend persists for the other pairs involving Al and Ga, a fact consistent with the interaction scheme of Fig. 4-32. The main trend is derived from the double-donor levels $Me^{+/++}$ of the isolated 3d metals since the shallow acceptors play a rather passive role. This can be explained by the iron-like one-electron-level structure in Fig. 4-32 and by simple tight-binding arguments (Feichtinger et al., 1984). The spin-related jump in the single-donor-level series (see 4.3.4) has now shifted and occurs between Cr and Mn. There is at least indirect evidence from pair-level trends that for Cr_i a second donor level ($Cr_i^{+/++}$) does not exist, which is in agreement with more direct experiments (Feichtinger and Czaputa, 1981).

4.4.3.2 Charge State Controlled Metastability

As to the pair binding energy the electronic structure of pairs discussed in the

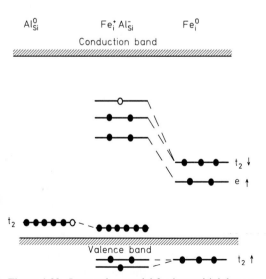

Figure 4-32. Interaction model for interstitial donor-shallow acceptor pairs in silicon. The $e\downarrow$ levels for both the isolated 3d impurity and the pair are resonant with the conduction band and are not shown. (After Scheffler, 1989).

previous sections suggests a simple ionic model. From such a model, the additional energy required to produce a separation from a first to a second nearest neighbor for the 3d interstitial in a positive charge state is

$$\Delta E_b = \frac{e^2}{4\pi\varepsilon\varepsilon_0}\left[\frac{1}{r_1}-\frac{1}{r_2}\right] \qquad (4\text{-}16)$$

where r_1 and r_2 are the distances for pairs aligned in the $\langle 111 \rangle$ direction and the $\langle 100 \rangle$ direction, respectively (ε is the static dielectric constant for silicon). Therefore, in thermal equilibrium one would expect a fraction

$$f \sim \exp\left[-\Delta E_b/kT\right] \qquad (4\text{-}17)$$

of pairs aligned along a $\langle 100 \rangle$ direction, with the metal sitting in the next nearest interstitial position adjacent to the substitutional acceptor. If one assumes a double positive charge for the interstitial, which also becomes stable for Fe_i and Cr_i when it is in a pair configuration (Mn_i^{++} does exist already for the isolated atom, see Fig. 4-33), then ΔE_b should be twice as large as before. Hence the charge state of the 3d constituent of a pair decides whether an appreciable fraction of $\langle 100 \rangle$ pairs can be found or not.

Further consequences may easily be seen from a configuration-coordinate diagram (Fig. 4-34), especially for the ionization energies, for which the relation $E_2 = E_1 - \Delta E_b$ should hold (note that no lattice relaxation has been included in the diagram). In their comprehensive study, Chantre and Bois (1985) not only found the energy levels related to the two pair configurations, but could establish overall agreement with the diagram in the case of Fe_iAl pairs. In their DLTS experiments, they could show that cooling of their samples (Schottky diodes) under reverse bias yields a fraction of pairs in the $\langle 100 \rangle$ con-

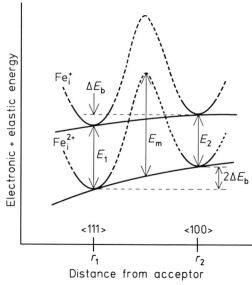

Figure 4-34. Configurational coordinate diagram for iron-acceptor pairs in silicon, based on a purely ionic model. E_m is the barrier to atomic motion (migration energy) from one configuration to the other. E_1 and E_2 denote the ionization transitions, e.g., for $(Fe_iAl)^{0/+}$ $\langle 111 \rangle$ and $(Fe_iAl)^{0/+}$ $\langle 100 \rangle$. (After Chantre and Bois, 1985).

figuration, and two energy levels, differing in intensity (at $E_v + 0.20$ eV and $E_v + 0.13$ eV), can be detected. Cooling with no reverse bias applied results in an almost entire suppression of the $E_v + 0.13$ eV level. Bias off corresponds to a situation where free holes are present in the junction to occupy the pair states in favor of $(Fe_iAl)^+$ (related to Fe_i^{++}), whereas bias on depletes the barrier from free carriers leaving the pairs in their neutral state (related to Fe_i^+) thus allowing for partial reorientation. Both pair configurations, of trigonal symmetry ($\langle 111 \rangle$ pairs) and orthorhombic symmetry ($\langle 100 \rangle$ pairs) have been identified in the case of Fe_iAl (Van Kooten et al., 1984; Gehlhoff et al., 1988).

This special type of configurational metastability apparently involves no lattice relaxation and should be observed for all similar systems composed of deep interstitial donors and shallow acceptors, but depends on whether the interstitials are mobile enough to allow for reorientation at relatively low temperatures. On the other hand, pairing at different pair distances and in different orientations might be the only way to get experimental data on the electronic structure for an ultrafast diffusing species like cobalt in silicon (Bergholz, 1982).

4.4.4 Thermal Donors in Silicon

Oxygen may well be the most-studied impurity found in silicon. A large fraction of all research efforts has been spent on the process of thermal donor (TD) formation, which works upon heat treatment in the 350–500 °C temperature range and generates complex defects which act as shallow double donors. Following the original idea of Kaiser et al. (1958), it was widely assumed that TD's are small oxygen aggregates in an early stage of oxygen precipitation. TD's are now believed possibly to consist of an oxygen-containing core around which Si self-interstitals agglomerate. It is of technological importance to clarify any facet of oxygen in Czochralski-grown silicon crystals (CZ), as this may play an important role in silicon device technology (Patel, 1977).

Thermal donors have been studied by electrical measurements, such as resistivity, Hall effect, DLTS, as well as by optical absorption, EPR, and ENDOR. But the task of synthesizing the vast amount of detailed information into a consistent whole has still not been brought to an end where any discussion has been settled. A successful model would have to cover the kinetic and electronic properties of a microscopic structure that changes steadily upon heat treatment.

Electrical measurements (Hall effect, resistivity) reveal the double-donor character of TD's with the apparent energy level positions shifted toward the conduction band upon prolonged annealing at about 450 °C (Kaiser et al., 1958; Gaworzewski et al., 1979). The initial rate of donor formation was found to be proportional to the fourth power of the initial oxygen concentration (Kaiser, 1957), whereas the maximum donor concentration attainable at 450 °C is approximately proportional to the cube of the initial oxygen concentration. The donor activity of TD's is destroyed by annealing at temperatures higher than 500 °C. Electron microscopy suggests that rodlike clusters grow during TD formation. They are aligned along $\langle 110 \rangle$ directions and should consist of hexagonal covalent Si (Bourret, 1987), contrary to an earlier interpretation of these clusters as ribbons of coesite, a high-pressure form of SiO_2 (Bourret et al., 1984).

The kinetics of TD's, which is determined from the electrical activity, is closely related to infrared (IR) absorption studies (Oeder and Wagner, 1983; Pajot et al., 1983). IR absorption spectra show clearly resolved Rydberg-like states (Sec. 4.3.5.2) of at least nine distinct double donors (Fig. 4-35). The maximum intensity within the donor series is gradually shifted to the shallower species upon heat treatment. This may well reflect the precipitation of impurities such as oxygen successively occurring at the core of the defect and thereby resulting in an increasingly repulsive perturbation. Figure 4-35 shows a distribution of the intensities (related to the concentration) of five different TD species; Table 4-3 lists the ground state (A_1) related ionization energies of all nine donors. A correlation study between DLTS and IR spectroscopy additionally links electrical

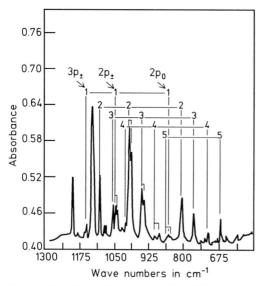

Figure 4-35. Absorption spectra of five different species of TD^+ in silicon, observed after 10 minutes pre-annealing at 770 °C, followed by 2 hours at 450 °C. (After Oeder and Wagner, 1983).

and optical properties of TD's, showing that both arise from the same defect (Benton et al., 1985).

EPR seemed to be the best instrument for shedding light on the initial core structure of TD, and two prominent EPR spectra associated with TD formation were published labeled NL8 and NL10 (Muller et al., 1978). As expected for a shallow center, the spectra show only small anisotropy in the g values (orthorhombic symmetry), which are close to that typically found for conduction band electrons. Unfortunately, no hyperfine interaction with oxygen could be resolved, since the line width of EPR evidently obscures the hyperfine structure (the samples were enriched with ^{17}O, which has a nuclear spin of $I = 5/2$).

Thus without detailed knowledge of the microscopic nature of the TD core, a number of models have been proposed. They are compatible with the kinetics and the orthorhombic symmetry of the center. A first group of models is based on the agglomeration of oxygen atoms, either of molecular oxygen to form O_4 complexes (Gösele and Tan, 1982) or of substitutional oxygen upon vacancy diffusion (Keller, 1984). The second group involves interstitial oxygen (e.g., the YLID configuration, Stavola and Snyder, 1983) as the saddle-point configuration of interstitial diffusion (Fig. 4-36a). The Ourmazd-Schröter-Bourret (OSB) model proposes an oxygen cluster of five atoms containing a silicon atom in its center, which is pushed into a near interstitial position along the $\langle 001 \rangle$ direction. According to this model the cluster subsequently grows by the stepwise addition of interstitial oxygen. In this model, for example, the electrical activity results from the broken bonds of the central silicon and ceases to exist by ejection of that atom, that is, by the emission of a self-interstitial upon a stress-relieving relaxation (Fig. 4-36b). A similar model has been proposed by Borenstein et al. (1986).

A third group of models relies on the agglomeration of silicon self-interstitials (Newman, 1985; Mathiot, 1987). Accordingly, the TD core is formed by two self-interstitials in the bond-centered position on

Table 4-3. First and second ionization energies ($TD^{0/+}$ and $TD^{+/++}$) of thermal donor species in silicon (After Pajot et al., 1983).

Species	A	B	C	D	E	F	G	H	I
$TD^0 \rightarrow TD^+$	69.1	66.7	64.4	62.1	60.1	58.0	56.2	54.3	52.9
$TD^+ \rightarrow TD^{2+}$	156.3	150.0	144.2	138.5	133.1	128.5	124	121	118

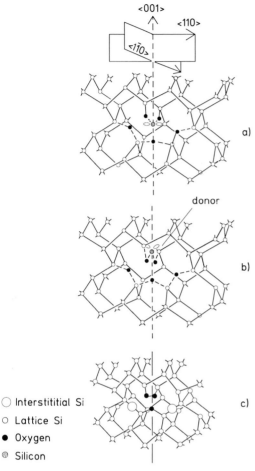

○ Interstititial Si
○ Lattice Si
● Oxygen
◉ Silicon

Figure 4-36. Models for thermal donor cores: YLID (a), OSB (b) and interstitial (Mathiot) model (c). For an explanation, see text. (After Claeys and Vanhellemont, 1989).

ment in sensitivity of three orders of magnitude. From their ^{17}O ENDOR on NL8, Michel et al. (1988) confirmed the involvement of oxygen in the central structure of a TD and proposed a model formed by four oxygen atoms in a vacancy, oriented in $\langle 111 \rangle$ directions from the center of the core. And the NL8-spectrum has been directly identified with the singly ionized TD^+ state by perturbation spectroscopy (infrared absorption under uniaxial stress, Lee et al., 1985). The structure for the NL10 spectrum has also been successfully resolved by ENDOR. The proposed microscopical model involves two oxygen atoms, which take the usual puckered, bond-centered interstitial position in one of the (100) planes, and a dopant aluminum atom. A vacancy is generated in the center of the defect to account for stress relaxation upon oxygen agglomeration (Gregorkiewicz et al., 1988). The NL 10 center has been claimed to represent the TD^- state of the thermal donor (Gregorkiewicz et al., 1989). However, it was concluded by comparison of ENDOR spectra of NL8 centers in boron- and aluminum-doped silicon that no dopant impurity is present in the TD^+ structure, while Al is involved in the NL10 centers. Also no ^{10}B or ^{11}B-ENDOR was found for NL10 in boron-doped silicon (Michel et al., 1989). So it seems that the microscopic core structure of NL8 given above would encompass all nine distinct donor species partially seen in Fig. 4-35. But if this is true, neither kinetic model discussed earlier can be entirely correct, and discussions on the "real model" of the TD will therefore certainly continue.

4.4.5 Hydrogen Passivation

A good deal of interest in the investigation of hydrogen as an impurity in semi-

a O_3 complex and can grow by adding further self-interstitials along the $\langle 110 \rangle$ direction (Fig. 4-36c). The advantage of these interstitial models is that they can account for the high mobility of the precipitating species.

A breakthrough in the microscopic identification of TD's, from which, however, many new questions arise, was achieved by recent ENDOR (Electron Nuclear Double Resonance (Fig. 4-8)) studies. This method typically can offer an improve-

conductors (for hydrogen in a-Si, see Chapter 10 of this Volume) comes from the important role hydrogen plays in passivating electrically active defects (Sah et al., 1983; Pankove et al., 1983). Hydrogen can be incorporated by a large number of device processing and operation steps. Among them are wafer polishing in alkali-based chemi-mechanical solutions, boiling in water, reactive ion etching and wet etching, sputter deposition of contact metals and others (Pearton et al., 1987). Passivation occurs for both shallow acceptors and donors in Si, GaAs, AlGaAs, and acceptors in Ge, InP, CdTe, and ZnTe (Pearton et al., 1989; Chevallier and Aucouturier, 1988). Even in as-grown GaAs (Shinar et al., 1986) or in InP and GaP (Clerjaud et al., 1987) these effects of hydrogen were detected. In Fig. 4-37, an example is given for acceptor and donor passivation in Si. Hydrogenation was performed by exposing both sides of a silicon wafer to hydrogen plasma for three hours at about 180 °C. The passivation effect shows different stability depending on the semiconductor and the species passivated; for H-B in Si, the initial resistivity can be restored at

temperatures around 150 °C. For donors in GaAs, restoration occurs after annealing at 400 °C, suggesting that a hydrogen-donor bond may be formed.

Thus, it is of special interest (as it was in the case of deep donor-shallow acceptor pairs, Sec. 4.4.3) to gain information initially on charge states and the preferential lattice sites of isolated hydrogen.

Van de Walle et al. (1988) concluded from their calculations that for H^+ and H^0 a bond-centered (BC) site along a $\langle 111 \rangle$ direction in the silicon lattice is lowest in energy, but for H^-, the interstitial tetrahedral (T) site is the most stable one. Therefore, H in Si may be another system showing charge-state-dependent metastability, at least theoretically (Watkins, 1989). The results of these calculations, like similar ones in other semiconductors (reviewed by Pearton, 1989) in the past, could only be compared to resonance experiments on muonium, the pseudo-isotope of hydrogen (1/9 proton mass and 2.2 μs lifetime). These experiments show that muonium may exist in a semiconductor in the two stable configurations just given before (e.g., Kiefl et al., 1988). Direct EPR observation of bond-centered hydrogen has only recently been reported (Gordeev et al., 1987; Gorelkinskii et al., 1990).

The models for passivation have also concentrated on these lattice sites but were controversial with regard to the mechanisms. Pankove et al. (1983), DeLeo and Fowler (1985) and DeLeo et al. (1988) proposed that the hydrogen atom sits at the BC site of an acceptor-silicon bond and forms a covalent bond with the adjacent Si atom. Assali and Leite (1986) placed the hydrogen atom in the interstitial antibonding site (AB) (this was the formerly labeled T). Total energy calculations, including lattice relaxation by use of the supercell method, show the BC site model to be

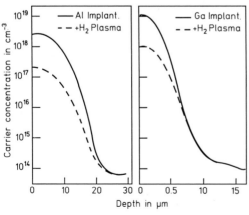

Figure 4-37. Spreading resistance profiles before (solid line) and after (dashed line) hydrogenation for acceptor passivation in silicon. (After Stavola et al., 1987).

more favorable (Sasaki and Katayama-Yoshida, 1989).

From tight binding calculations, a model was proposed to account for donor passivation in Si (Johnson et al., 1986). According to that model, the hydrogen atom is attached to one of the donor's nearest neighbors in an AB site along the $\langle 111 \rangle$ axis.

Figure 4-38 presents schematic models of a hydrogen-acceptor complex and a hydrogen-donor complex. Both centers were also studied by vibrational absorption under uniaxial stress (Bergman et al., 1988) and were found to be of trigonal symmetry. The B-H center showed a tendency to move off the $\langle 111 \rangle$ axis under compression.

Passivation should not be seen in the simple ionic picture of $(H^+ A^-)^0$ or $(H^- D^+)^0$ since covalency may play an important role in the hydrogen-acceptor or hydrogen-donor complex. The stable configurations and corresponding charge states of isolated hydrogen, however, suggest that the models displayed in Fig. 4-38 may be close to reality.

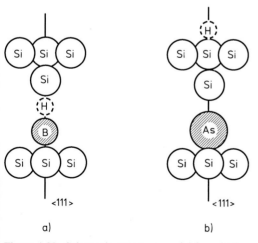

Figure 4-38. Schematic structure model for a hydrogen-acceptor complex (a) and a hydrogen-donor complex (b). (After Bergman et al., 1988).

4.5 References

Abragam, A., Bleaney, B. (1970), *Electron Paramagnetic Resonance of Transition Metals, The International Series of Monographs on Physics:* Marshall, W., Wilkinson, D. H. (Eds.). Oxford: Clarendon Press.

Adler, D. (1982), in: *Handbook on Semiconductors, Vol. 1:* Paul, W. (Ed.). Amsterdam, New York, Oxford: North-Holland.

Aggarwal, R. L., Ramdas, A. K. (1965), *Phys. Rev. 140,* A1246.

Allen, J. W. (1986), in: *Deep Centers in Semiconductors:* Pantelides, S. T. (Ed.). New York: Gordon and Breach, pp. 627–690.

Ammerlaan, C. A. J. (1989), *Solid State Phenomena Vols.* 6 and 7, pp. 591–602.

Anderson, P. W. (1975), *Phys. Rev. Lett. 37,* 953.

Armelles, G., Barrau, J., Brousseau, M., Pajot, B., Naud, C. (1985), *Solid State Commun. 56,* 303.

Assali, L. V. C., Leite, J. R., Fazzio, A. (1985) *Phys. Rev. B 32,* 8085.

Assali, L. V. C., Leite, J. R. (1986) *Phys. Rev. Lett. 55,* 980.

Baber, N., Grimmeiss, H. G., Kleverman, M., Omling, P. (1987), *J. Appl. Phys. 62,* 2853.

Ballhausen, C. J. (1962), *Ligand Field Theory,* New York: Mc Graw-Hill.

Baraff, G. A., Kane, O. E., Schlüter, M. (1979), *Phys. Rev. Lett. 43,* 956.

Baraff, G. A., Kane, O. E., Schlüter, M. (1980), *Phys. Rev. Lett. 21,* 5662.

Baranowski, J. M., Allen, S. W., Pearson, L. (1967), *Phys. Rev. 160,* 627.

Beeler, F., Scheffler, M. (1989), in: *Proceedings of the 15^t International Conference on Defect in Semiconductors, Budapest; Material Science Forum, Vols. 38–41:* Ferenczi, G. (Ed.). Switzerland: Trans Tech Publications, p. 257.

Beeler, F., Andersen, O. K., Scheffler, M. (1985a), *Phys. Rev. Lett. 55,* 1498.

Beeler, F., Scheffler, M., Jepsen, O., Gunnarson, O. (1985b), *Phys. Rev. Lett. 54,* 2525.

Beeler, F., Andersen, O. K., Scheffler, M. (1990), *Phys. Rev. B 41,* 1603.

Benton, J. L., Lee, K. M., Freeland, P. E., Kimerling, L. C. (1985) in: *Proceedings of the 13th International Conference on Defects in Semiconductors:* Kimerling, L. C., Parsey, J. M., Jr. (Eds.). N. Y.: The Metallurgical Society of AIME, p. 647.

Bergh, A. A., Dean, P. J. (1976), *Light-Emitting Diodes,* London, New York: Clarendon.

Bergholz, W. (1982), *Physika 116 B,* 195.

Bergholz, W., Mohr, W., Wendt, H., Dreves, W. (1989), *Materials Science and Engineering,* B 4.

Bergman, K., Stavola, M., Pearton, S. J., Hayes, T. (1988), *Phys. Rev. B 38,* 9643.

Bernholc, J., Lipari, N. O., Pantelides, S. T. (1978), *Phys. Rev. Lett. 41,* 895.

Bernholc, J., Lipari, N. O., Pantelides, S. T., Scheffler, M. (1982), *Phys. Rev. B 26*, 5706.

Bishop, S. G. (1986), in: *Deep Centers in Semiconductors:* Pantelides, S. T. (Ed.); New York: Gordon and Breach, pp. 541–626.

Blakemore, L. S. (1962), *Semiconductor Statistics*, Oxford: Pergamon.

Borenstein, J. T., Peak, D., Corbett, J. W. (1986), *J. Mat. Res. 1*, 527.

Born, M., Oppenheimer, J. R. (1927), *Ann. Phys. 84*, 457.

Bourgoin J., Lannoo, M. (1983), *Point Defects in Semiconductors II*, Berlin, Heidelberg, New York: Springer-Verlag; p. 109.

Bourret, A. (1987), *Inst. Phys. Conf. Ser. 87*, 39.

Bourret, A., Thibault-Desseaux, S., Seidman, D. N. (1984), *J. Appl. Phys. 55*, 825.

Brotherton, S. D., Parker, G. J., Gill., A. (1983), *J. Appl. Phys. 54*, 5112.

Brower, K. L. (1980), *Phys. Rev. Lett. 44*, 1627.

Brower, K. L. (1982), *Phys. Rev. B 26*, 6040.

Caldas, M. J., Fazzio, A., Zunger, A. (1984), *Appl. Phys. Lett. 45*, 671.

Cartling, B. G. (1975), *J. Phys. C 8*, 3171, 3183.

Cavenett, B. C. (1981), *Adv. in Phys. 30*, 475.

Chadi, D., Chang, K. J. (1988), *Phys. Rev. Lett. 61*, 873.

Chantre, A., Bois, D. (1985), *Phys. Rev. B 31*, 7979.

Chen, M. C., Lang, D. V. (1983), *Phys. Rev. Lett. 51*, 427.

Chevallier, J., Aucouturier, M. (1988), *Am. Rev. Mater. Sci. 18*, 265.

Claeys, C., Vanhellemont, J. (1989), *Solid State Phenomena Vols* 6 and 7, pp. 21–32.

Clementi, E., Roetti, C. (1974), *At. Data Nucl. Data Tables 14*, 177.

Clerjaud, B. (1985), *J. Phys. C 18*, 3615.

Clerjaud, B., Cote, D., Nand, C. (1987), *Phys. Rev. Lett. 58*, 1755.

Corbett, J. W., Watkins, G. D., Chrenko, R. M., McDonald, R. S. (1961), *Phys. Rev. 121*, 1015.

Corbett, J. W., McDonald, R. S., Watkins, G. D. (1964), *J. Phys. Chem. Solids 25*, 873.

Corliss, C., Sugar, J. (1977), *J. Phys. Chem. Ref. Data 6*, 1253.

Czaputa, R. (1989), *Appl. Phys. A 49*, 431.

Czaputa, R., Feichtinger, H., Oswald, J. (1983), *Solid State Commun. 47*, 223.

Czaputa, R., Feichtinger, H., Oswald, J., Sitter, H., Haider, M. (1985), *Phys. Rev. Lett. 55*, 758.

Dabrowski, J., Scheffler, M. (1989), *Phys. Rev. B 40*, 391.

Dean, P. J. (1977), in: *Topics in Applied Physics, Vol. 17, Electroluminescence*, Berlin, Heidelberg, New York: Springer-Verlag, p. 63.

Dean, P. J. (1986), in: *Deep Centers in Semiconductors:* Pantelides, S. T. (Ed.); New York: Gordon and Breach, pp. 185–348.

De Leo, G. G., Fowler, W. B. (1985), *Phys. Rev. B 31*, 6861.

De Leo, G. G., Watkins, G. D., Fowler, W. B. (1981), *Phys. Rev. 23*, 1851.

De Leo, G. G., Watkins, G. D., Fowler, W. B. (1982), *Phys. Rev. B 24*, 4962.

De Leo, G. G., Dorogi, M. S., Fowler, W. B. (1988), *Phys. Rev. B 38*, 7520.

Delerue, C., Lannoo, M., Langer, J. M. (1988), *Phys. Rev. Lett. 61*, 199.

Dmochowski, E., Langer, J., Raczynska J., Jantsch, W. (1988), *Phys. Rev. B 38*, 3276.

Faulkner, R. A. (1969), *Phys. Rev. 184*, 713.

Fazzio, A., Leite, J. R. (1980), *Phys. Rev. B 21*, 4710.

Fazzio, A., Caldas, M. J., Zunger, A. (1984), *Phys. Rev. B 30*, 3430.

Fazzio, A., Caldas, M. J., Zunger, A. (1985), *Phys. Rev. B 32*, 934.

Feenstra, R. M., Pantelides, S. T. (1985), *Phys. Rev. B 31*, 4083.

Feichtinger, H., Czaputa, R. (1981), *Appl. Phys. Lett. 39*, 706.

Feichtinger, H., Prescha, T. (1989), in: *Proceedings of the 15th International Conference on Defect in Semiconductors, Budapest; Material Science Forum*, Vols. 38–41: Ferenczi, G. (Ed.). Switzerland: Trans Tech Publications, pp. 427–432.

Feichtinger, H., Waltl, J., Geschwandtner, A. (1978), *Solid State Commun. 27*, 867.

Feichtinger, H., Oswald, J., Czaputa, R., Vogl, P. and Wünstel, K. (1984), *Proceedings of the 13th International Conference on Defects in Semiconductors:* Kimerling, L. C. and Parsey, J. M. (Eds.). The Metallurgical Society of AIME, p. 855.

Figgis, B. N. (1967), *Introduction to Ligand Fields*, New York: Jan Wiley.

Fischer, C. E. (1972), *At. Data 4*, 301.

Gaworzewski, P., Schmalz, K. (1979), *Phys. Stat. Sol. (a) 55*, 699.

Gehlhoff, W., Irmscher, K., Kreissl, J. (1988), in: *New Developments in Semiconductor Physics;* Berlin: Springer Verlag, p. 262.

Gilles, D., Schröter, W., Bergholz, W. (1990), *Phys. Rev. B 41*, 5770.

Godlewski, M. (1985), *Phys. States Solidi 90 a*, 11.

Gösele, U., Tan, T. Y. (1982), *Appl. Phys. A. 18*, 79.

Gordeev, V. A., Gorelkinskii, Y. V., Konopleva, R. F. Nevinnyi, N. M. N., Obukhov Y. V., Firsov, V. G. (1987), *"Anomalous States of Muonium and Implanted Hydrogen in Silicon"*, Leningrad: Academy of Sciences of the USSR. preprint 1340.

Gorelkinskii, Yu. V., Nevinnyi, N. N. (1990), in: *Proceedings of the 6th Trieste Semiconductor Symposium on: Hydrogen in Semiconductors: Bulk and Surface Properties* (to be published).

Graff, K., Pieper, H. (1981 a) in: *Semiconductor Silicon:* Huff, H. R., Krieger, R. J., Takeishi, Y. (Eds.). New Jersey: The Electrochemical Society, p. 331.

Graff, K., Pieper, H. (1981 b), *J. Electrochem. Soc. 128*, 669.

Gregorkiewicz, T., van Wezep, D. A., Bekman, H. H. P. Th., Ammerlaan, C. A. S. (1988), *Phys. Ref. B 38*.

Gregorkiewicz, T., Bekman, H. H. P. Th., Ammerlaan, C. A. J. (1989), in: *Defects in Semiconductors 15, Budapest; Materials Science Forum, Vols. 38–41:* Ferenczi, G. (Ed.). Switzerland: Trans Tech Publications.

Greulich-Weber, S., Niklas, J. R., Spaeth, J. M. (1984), *J. Phys. C 17*, L 911.

Grimmeiss H. G., Kleverman, M. (1989), *Solid State Phenomena, Vols. 6 and 7*, 277–288.

Grimmeiss, H. G., Janzén, E., Skarstam, B. (1980a), *J. Appl. Phys. 51*, 3740.

Grimmeiss, H. G., Janzén, E., Skarstam, B. (1980b), *J. Appl. Phys. 51*, 4212.

Grimmeiss, H. G., Janzén, E., Ennen, H., Schirmer, O., Schneider, Wörner, R., Holm, C., Sirtl, E., Wagner, P. (1981), *Phys. Rev. B24*, 4571.

Grimmeis, H. G., Janzén, E., Pantelides, S. T. (1986), in: *Deep Centers in Semiconductors:* Pantelides, S. T. (Ed.). New York: Gordon and Breach, pp. 1–86.

Haldane, F. D. M., Anderson, P. W. (1976), *Phys. Rev. B 13*, 2553.

Harris, R. D., Newton, J. L., Watkins, G. D. (1982), *Phys. Rev. Lett. 48*, 1271.

Harrison, W. (1956), *Phys. Rev. 104*, 1281.

Harrison, W. A. (1980), *Electronic Structure and the Properties of Solids*, San Francisco: Freeman.

Hemstreet, L. A. (1977), *Phys. Rev. B 15*, 834.

Hemstreet, L. A. (1980), *Phys. Rev. B 22*, 4590.

Hemstreet, L. A., Dimmock, J. P. (1979), *Phys. Rev. B 20*, 1527.

Hennel, A. M., Uba, S. M. (1978), *J. Phys. C 11*, 4565.

Hjalmarson, H. P., Vogl, P., Wolford, D. J., Dow. J. D. (1980), *Phys. Rev. Lett. 44*, 810.

Hohenberg, P., Kohn, W. (1964), *Phys. Rev. 136*, 864.

Housely, R. M., Hess, F. (1966), *Phys. Rev. 146*, 517.

Hrostowski, H. S., Adler, B. J. (1960), *J. Chem. Phys. 33*, 980.

Hsu, W. Y., Dow, J. D., Wolford, J. D., Streetman, B. G. (1977), *Phys. Rev. B 16*, 1597.

Huang, K., Rhys, A. (1950), *Proc. R. S. 204*, 406.

Jantsch, W., Wünstel, K., Kumagai, O., Vogl, P. (1983), *Physica 117 and 118 B*, 188.

Janzén, E., Stedman, R., Grossman G., Grimmeiss, H. G. (1984) *Phys. Rev. B 29*, 1907.

Janzén E., Grossman, G., Stedman, R., Grimmeiss, H. G. (1985), *Phys. Rev. 31*, 8000.

Johnson, D. P., Kassman, A. J. (1969), *Phys. Rev. 188*, 1385.

Johnson, N. M., Herring C., Chadi, D. J. (1986), *Phys. Rev. Lett. 56*, 769.

Kaiser, W. (1957), *Phys. Rev. 105*, 1751.

Kaiser, W., Frisch, H. L., Reiss, H. (1958), *Phys. Rev. 112*, 1546.

Katayama-Yoshida, H., Zunger, A. (1985), *Phys. Rev. B 31*, 8317.

Katsumoto, S., Komori, S., Kobayashi S. (1987), *J. Phys. Soc. Japan 56*, 2259.

Kaufmann, U., Schneider, J. (1983), *Adv. Electron. Phys. 58*, 81.

Keenan, J. A., Larrabee, G. B. (1983), in: *VLSI Electronics, Microstructure Science:* Einspruch, N. G., Larrabee, G. B. (Eds.). New York: Academic Press, Inc.

Keller, W. W. (1984), *J. Appl. Phys. 55*, 3471.

Khachaturyan, K., Weber, E. R., Kaminska, M. (1989), *Materials Science Forum, Vols. 38–41:* Ferenczi, G. (Ed.). Switzerland: Trans Tech Publications, pp. 1067.

Kiefl, R. R., Celio, M., Estle, T. L., Kreitzman, S. R., Luke, G. M., Riseman, T. M., Ansaldo, E. J. (1988), *Phys. Rev. Lett. 60*, 224.

Kimerling, L. C., Benton, J. L., Rubin, J. S. (1981), *Inst. Phys. Conf. Ser. 59*, 217.

Kleverman, M., Omling, P., Lebedo, L. A., Grimmeiss, H. G. (1983), *J. Appl. Phys. 54*, 814.

Kleverman, M., Olajos, J., Grimmeiss, H. G. (1987), *Phys. Rev. B 35*, 4093.

Kleverman, M., Olajos, J., Grimmeiss, H. G. (1988), *Phys. Rev. B 37*, 2613.

Kogan, M., Lifshits, T. M. (1977), *Phys. Stat. Sol. (a) 39*, 11.

Kohn, W. (1957), *Solid State Physics 5*, 257.

Kohn, W., Sham, J. L. (1965), *Phys. Rev. 140*, A 1133.

Koidl, P., Schirmer, O. F., Kaufmann, U. (1973), *Phys. Rev. B 8*, 4926.

Krag, W. E., Zeiger, H. S. (1962), *Phys. Rev. Lett. 8*, 485.

Krag, W. E., Kleiner, W. H., Zeiger, H. J. (1986), *Phys. Rev. B 33*, 8304.

Landolt-Börnstein (1989), *Numerical Data and Functional Relationships in Science and Technology, Vol. III/22b:* Madelung, O., Schulz, M. (Eds.). Berlin: Springer-Verlag.

Landolt-Börnstein (1982), *Numerical Data and Functional Relationships in Science and Technology, Vol. 17a, 17b.* Madelung, O. (Ed.). Berlin: Springer-Verlag.

Landsberg, P. T. (1982), in: *Handbook on Semiconductors Vol. 1:* Paul, W. (ed.). Amsterdam, New York, Oxford: North-Holland.

Lang, D. V., Grimmeiss, H. G., Mejer, E., Jaros, M. (1980), *Phys. Rev. B 22*, 3917.

Lang, D. V. (1986), in: *Deep Centers in Semiconductors:* Pantelides, S. T. (Ed.). New York: Gordon and Breach, p. 489.

Lang, D. V., Logan, R. A. (1977), *Phys. Rev. Lett 39*, 635.

Lang, D. V., Logan, R. A., Jaros, M. (1979), *Phys. Rev. B 19*, 1015.

Langer, J. M., Heinrich, H. (1985), *Phys. Rev. Lett. 55*, 1414.

Langer, J. M., Delerue, C., Lannoo, M. (1989), in: *Proceedings of the 15th International Conference on Defects in Semiconductors, Budapest; Materials Science Forum, Vols. 38–41:* Ferenczi, G. (Ed.), Switzerland: Trans Tech Publications, pp. 275–280.

Lannoo, M., Bourgoin, J. (1981), *Point Defects in Semiconductors I.*, Berlin, Heidelberg, New York: Springer-Verlag, p. 87.

Lax, M. (1960), *Phys. Rev. 119*, 1502.

Lee, K. M., Trombetta, J. M., Watkins, G. D. (1985), in: *Microscopic Identification of Electronic Defects in Semiconductors:* Johnson, N. M., Bishop, S. G., Watkins, G. D. (Eds.). Pittsburgh, Penn: Materials Research Society, p. 263.

Ledebo, L. A., Wang, Z.-G. (1983), *Appl. Phys. Lett. 42,* 680.

Lefebvre, I., Lannoo, M., Priester, C., Allau, G., Delerue, C. (1987), *Phys. Rev. B 33,* 1336.

Legros, R., Mooney, P. M., Wright, S. L. (1987), *Phys. Rev. B 35,* 7505.

Lindefelt, U., Zunger, A. (1981), *Phys. Rev. B 24,* 5913.

Lindefeldt, U., Zunger, A. (1982), *Phys. Rev. B 26,* 5989.

Lisiak, K. P., Milnes, A. G. (1975), *J. App. Phys. 46,* 5229.

Low, W. (1961), *Paramagnetic Resonance in Solids.* New York, London: Academic Press.

Lucovsky, G. (1965), *Solid State Commun. 3,* 299.

Ludwig, G. W., Woodbury, H. H. (1962), in: *Solid State Physics, Vol. 13:* Seitz, F., Turnbull, D. (Eds.). New York: Academic Press, pp. 223.

Maguire, J., Murray, R., Newman, R. C. (1987), *Appl. Phys. Lett. 52,* 126.

Mathiot, D. (1987), *Appl. Phys. Lett. 51,* 904.

Maude, D. K., Postal, S. C., Dmowski, L., Foster, T., Eaves, L., Nathan, M., Heilblum, M., Harris, J. J., Beall, R. B. (1987), *Phys. Rev. Lett. 59,* 815.

Michel, S., Niklas, S. R., Spaeth J.-M. (1988), in: *Defects in Electronic Materials:* Stavola, M., Pearton, S. J., Davies, G. (Eds). Pittsburgh: Mater. Res. Soc., p. 185.

Michel, J., Meilwes, N., Spaeth, J.-M. (1989), in: *Proceedings of the 15th International Conference on Defects in Semiconductors, Budapest;* Material Science Forum, Vols. 38–41: G. Ferenczi (Ed); Switzerland: Trans Tech Publications.

Milnes, A. G. (1973), *Deep impurities in Semiconductors,* New York, London: John Wiley and Sons.

Migliorato, P., Elliott, C. T. (1978), *Solid State Electronics 21,* 443.

Mizuta, M., Tachikawa, M., Kukimoto, H., Minomura, S. (1985), *Jpn. J. Appl. Phys. 24,* L 143.

Mooney, P. M., Calleja, E., Wright, S. L., Heilblum, M. (1986), in: *Defects in Semiconductors:* Bardeleben, H. J. (Ed.) Materials Science Forum Vols. 10–12. Switzerland: Trans Tech Publications, pp. 417–422.

Mooney, P. M., Northrup, G. A., Morgan, T. N., Grimmeiss, H. G. (1988a), *Phys. Rev. B 37,* 8298.

Mooney, P. M., Theis, T. N., Wright, S. L. (1988b), *Inst. Phys. Conf. Ser. 91,* 359.

Moore, C. E. (1959), *Atomic Energy Levels, Nat. Bur. Standards Cir. 467,* Vol. 3.

Morgan, T. N. (1986), *Phys. Rev. B 34,* 2664.

Muller, S. H., Tuynman, G. M., Sieverts, E. G., Ammerlaan, C. A. J. (1982), *Phys. Rev. B 25,* 25.

Muller, S. H., Sprenger, M., Sieverts, E. G., Ammerlaan, C. A. J. (1978), *Solid State Commun 25,* 987.

Newman, R. C. (1985), *J. Phys. C. 18,* L967.

Niklas, J. R., Spaeth, J. M. (1983), *Solid State Commun. 46,* 121.

Nolte, D. D., Walukiewicz, W., Haller, E. E. (1987), *Phys. Rev. B 36,* 9374.

Oeder, R., Wagner, P. (1983), in: *Defects in Semiconductors II:* Mahajan, S., Corbett, J. W. (Eds.). New York: North Holland, p. 171.

Olajos, J., Bech-Nielsen, M., Kleverman, M., Omling, P., Emmannuelson P., Grimmeiss, H. G. (1988), *Appl. Phys. Lett. 53,* 2507.

Ourmazd, A., Schröter, W., Bourret, A. (1984), *J. Appl. Phys. 56,* 1670.

Pajot, B., Compain, H., Lerouelle, J., Clerjaud, B. (1983), *Physica 117B and 118B,* 110.

Pankove, J. I., Carlson, D. E., Berkeyheiser, J. E., Wance, R. O. (1983), *Phys. Rev. Lett. 51,* 2224.

Pantelides, S. T. (1978), *Rev. Mod. Phys. 50,* 797.

Pantelides, S. T. (1986), in: *Deep Centers in Semiconductors:* Pantelides, S. T. (Ed.). New York: Gordon and Breach, pp. 1–86.

Patel, J. R. (1977), in: *Semiconductor Silicon:* Huff, H. R., Sirtl, E. (Eds.). Pennington, N. J.: The Electrochemical Society.

Peaker, A. R., Hamilton, B. (1986) in: *Deep Centers in Semiconductors:* Pantelides, S. T. (Ed.), New York: Gordon and Breach.

Pearton, S. J., Corbett, J. W., Shi, T. S. (1987), *Appl. Phys. A 43,* 153.

Pearton, S. J., Stavola, M., Corbett, S. W. (1989), in: *Defects in Semiconductors 15, Budapest; Materials Science Forum, Vols. 38–41:* G. Ferenczi (Ed). Switzerland: Trans Tech Publications.

Pecheur, P., Toussaint, G. (1983), *Physica 116 B,* 112.

Sah, C. T., Sun, S. Y.-C., Tzou J. J.-T. (1983), *Appl. Phys. Lett. 43,* 204.

Sasaki, T., Katayama-Yoshida, H. (1989), in: *Defects in Semiconductors 15, Materials Science Forum, Vols. 38–41:* Ferenczi, G. (Ed). Switzerland: Trans Tech Publications.

Scheffler, M. (1982), in: *Festkörperprobleme XXII;* P. Grosse (Ed). Braunschweig: Vieweg, p. 115.

Scheffler, M. (1989), in: *Festkörperprobleme XXII;* U. Rössler (Ed). Braunschweig: Vieweg, pp. 1–20.

Scheffler, M., Beeler, F., Jepsen, O., Gunnarsson, O., Andersen, O. K., Bachelet, G. B. (1985), *J. Electron. Mater. 14,* 45.

Sclar, N. (1981), *J. Appl. Phys. 52,* 5207.

Seibt, M. (1990), in: *Semiconductors Silicon,* Huff, H. R., Barraclough, K. G., Chikawa, J. I. (Eds.). Pennington N. J. L., The Electrochemical Society, p. 663–674.

Shinar, J., Kuna-ah, A., Cavenett, B. C., Kennedy, T. A., Wilsey, N. (1986), *Solid State Commun. 59,* 653.

Shockley, W., Last, J. T. (1957), *Phys. Rev. 107,* 392.

Singh, V., Zunger, A. (1985), *Phys. Rev. B 31,* 3729.

Slichter, C. P. (1980), *Principles of Magnetic Resonance, Springer Series in Solid State Sciences:* Cardona, M., Fulde, P., Queisser, H. S. (Eds.), Berlin, Heidelberg, New York: Springer Verlag.

Stavola, M., Snyder, L. C. (1983), in: *Defects in Silicon:* Bullis, W. M., Kimerling, L. C. (Eds.), Pennington, N. J., The Electrochem. Soc. Softbound Ser., p. 6.

Stavola, M. Pearton, S. J., Lopata, S., Dautremont-Smith, W. C. (1987), *Appl. Phys. Lett. 50,* 1086.

Stoneham, A. M. (1975), *Theory of Defects in Solids,* Chapter 10, Oxford: Clarendon Press.

Sturge, M. D. (1967), Solid State Physics 20, 91.

Sugano, S., Tanabe, Y., Kamimura, H. (1970), *Multiplets of Transition-Metal Ions in Crystals,* New York: Academic Press.

Sze, S. M. (1981), *Physics of Semiconductor Devices.* New York: John Wiley and Sons.

Tersoff, J., Harrison, W. A. (1987), *Phys. Rev. Lett. 58,* 2367.

Theis, T. N. (1991), *Inst. Phys. Conf. Ser. 95,* in press.

Theis, T. N., Parker, B. D. (1987), *Applied Surface Science 30,* 52.

Theis, T. N., Kuech, T. F., Palmateer, L. F., Mooney, P. M. (1985), *Inst. Phys. Conf. Ser. 74,* 241.

Theis, T. N., Mooney, P. M., Wright, S. L. (1988), *Phys. Rev. Lett 60,* 361.

Toyozawa, Y. (1980), in: *Relaxation of Elementary Excitations:* R. Kubo (Ed.). Berlin: Springer-Verlag, p. 3.

Toyozawa, Y. (1983), *Physica 116 B,* 7.

Utzig, J. (1989), *Appl. Phys. 65,* 3868

Utzig, J., Schröter, W. (1984), *Appl. Phys. Lett. 45,* 761.

Van de Walle, C. G., Bar-Yam Y., Pantelides, S. T. (1988), *Phys. Rev. Lett. 60,* 2761.

Van Kooten, J. J., Weller, A. G., Ammerlaan, C. A. J. (1984), *Phys. Rev. B 30,* 4564.

Van Oosten, A. B., Son, N. T., Vlasenko, L. S., Ammerlaan, C. A. J. (1989), in: *Proceedings of the 15th Conference on Defect in Semiconductors, Budapest;* Materials Science Forum, Vols. 38–41: Ferenczi, G. (Ed.). Switzerland: Trans Tech Publications, p. 355

Van Vechten, J. A. (1982), in: *Handbook on Semiconductors 3:* Keller, S. P. (Ed.). Amsterdam, New York, Oxford: North-Holland Publishing Company.

Van Wezep, Ḋ. A., Ammerlaan, C. A. J. (1985), *J. Electron. Mater. 14 a,* 863.

Van Wezep, D. A., Van Kemp, R., Sievers, E. G., Ammerlaan, C. A. J. (1985), *Phys. Rev. B 32,* 7129.

Vlasenko, L. S., Lebedev, A. A., Taptygov, E. S., Khramtsov, V. A. (1987), *Soviet Technical Physics Letters 13,* 1322.

Vogl, P. (1981), in: *Adv. in Solid State Physics XXI:* Treusch, J. (Ed.). Braunschweig: Vieweg, p. 191.

Vogl, P. (1984), in: *Advances in Electronics and Electron Physics 62,* New York: Academic Press, Inc., p. 101.

Vogl, P., Baranowski, J. M. (1985), *Acta Physica Polonica A 67,* 133.

Vogl, P., Hjalmarson, H. P., Dow, J. D. (1983), J. Phys. Chem. Solids 44, 365.

Watanabe, S., Kamimura, H. (1989), *Mat. Sci. Eng. B 3,* 313.

Watkins, G. D. (1975), in: *Point Defects in Solids,* Vol. 2: Crawford, J. H., Silfkin, L. M. (Eds.). New York: Plenum.

Watkins, G. D. (1983), *Physica 117/118 B,* 9.

Watkins, G. D. (1989) in: *Defects in Semiconductors 15, Materials Science Forum,* Vols. 38–41: G. Ferenczi, (Ed.), Switzerland: Trans Tech Publications.

Watkins, G. D., Corbett, J. W. (1961), *Phys. Rev. 121,* 1001.

Watkins, G. D., Fowler, W. B. (1977), *Phys. Rev. B 16,* 4524.

Watkins, G. D., Troxell, J. R. (1980), *Phys. Rev. Lett. 44,* 593.

Wolford, D. J., Streetman, B. G., Hsu, W. Y., Dow, J. D., Nelson, R. J., Holonyak, N., Jr. (1976), *Phys. Rev. Lett. 36,* 1400.

Weakliem, H. A. (1962), *J. Chem. Phys. 36,* 2117.

Weber, E. R. (1983), *Appl. Phys. A 30,* 1.

Weber, J., Ennen, H., Kaufmann, U., Schneider, J. (1980), *Phys. Rev. B 21,* 2394.

Weber, S., Ennen, H., Kaufmann, U., Schneider, J. (1980), *Phys. Rev. B 38,* 7520.

Weider, D., Scheffler, M., Scherz, U. (1989), in: *Proceedings of the 15th International Conference on Defects in Semiconductors, Budapest; Material Science Forum,* Vols. 38–41: Ferenczi, G. (Ed.). Switzerland: Trans Tech Publications, pp. 299–303.

Wolford, D. J., Streetman, B. G., Hsu, W. Y., Dow, J. D., Nelson, R. J., Holonyak, N., Jr. (1976), *Phys. Rev. Lett. 36,* 1400.

Zunger, A. (1985), *Ann. Rev. Mater. Sci. 15,* 411.

Zunger, A. (1986), in: *Solid State Physics,* Vol. 89, New York: Academic Press, Inc., pp. 275–464.

Zunger, A., Lindefelt, U. (1982), *Phys. Rev. B 26,* 5989.

Zunger, A., Lindefelt, U. (1983), *Phys. Rev. B 27,* 1191.

General Reading

Bourgoin, J., Lannoo, M. (1983), *Point Defects in Semiconductors II.* Heidelberg: Springer-Verlag.

Lannoo, M., Bourgoin, J. (1981), *Point Defects in Semiconductors I.* Heidelberg: Springer-Verlag.

Pantelides, S. T. (Ed.) (1986), *Deep Centers in Semiconductors.* New York: Gordon and Breach.

Scheffler, M. (1989), in: *Festkörperprobleme XXII:* Rössler, U. (Ed.). Braunschweig: Vieweg, pp. 1–20.

Stoneham, A. M. (1975), *Theory of Defects in Solids.* Oxford: Clarendon Press.

Vogl, P. (1981), in: *Advances in Solid State Physics XXI:* Treusch, J. (Ed.). Braunschweig: Vieweg, p. 191.

Zunger, A. (1986), in: *Solid State Physics* Vol. 89. New York: Academic Press, pp. 275–464.

5 Equilibria, Nonequilibria, Diffusion, and Precipitation

Ulrich M. Gösele and Teh Y. Tan

Department of Mechanical Engineering and Materials Science, Duke University, Durham, NC, U.S.A.

List of Symbols and Abbreviations

A, B	species of atoms
$\bar{A}$	stacking fault area per atom
A_i	atom of species A in interstitial site
A_s	atom of species A in substitutional site
C	concentration in atomic fractions, dimensionless
C_D	concentration of substitutional dopant atoms
C_D^s	concentration of electrically active dopant atoms
C_I	nonequilibrium concentration of self-interstitials
C_I^{eq}	equilibrium concentration of self-interstitials
C_s	nonequilibrium concentration of atoms in substitutional sites
C_s	carbon atoms in substitutional sites
C_s^{eq}	equilibrium concentration of atoms in substitutional sites
C_i	nonequilibrium concentration of foreign atoms in interstitial sites
C_i^{eq}	equilibrium concentration of foreign atoms in interstitial sites
C_X	nonequilibrium concentration of a point defect X
$C_{X^r}, C_{X^r}^{eq}$	nonequilibrium/equilibrium concentration of a point defect of charge state r
c	volume concentration
c_a, c_d	volume concentration of acceptors/donors
c_k	volume concentration of species k
c_{net}	difference in the concentrations of all donors and acceptors
c_s	concentration at the surface
CI	carbon-self interstitial complex
D	diffusivity
D_C	diffusivity of carbon
D_{eff}^{front}	effective diffusivity at front part of diffusion profile
$D_{eff}^{(I)}$	effective diffusivity of foreign atoms if their in-diffusion following kick-out mechanism is so quick that supersaturation of self-interstitials occurs
$D_{eff}^{(I, V)}$	effective diffusivity of foreign atoms if their in-diffusion is possible by both the kick-out and the Frank-Turnbull mechanism and is so quick that a supersaturation of self-interstitials and an undersaturation of vacancies occurs
$D_{eff}^{(i)}$	effective diffusivity of foreign atoms if their in-diffusion following kick-out mechanism is slow enough to allow the generated self-interstitials to migrate to surface
D_{eff}^{tail}	effective diffusivity at tail part of diffusion profile
$D_{eff}^{(V)}$	effective diffusivity of foreign atoms controlled by in-diffusion of vacancies from surface
D_I^s, D_V^s	diffusivity of substitutional atoms where diffusion is due to transport by self-interstitials/vacancies
$D_{I, V}^{eff}$	effective diffusivity of a perturbation in the concentrations of self-interstitials and vacancies
D_i	diffusivity of interstitial point defects
$\mathbf{D}_k$	diffusion tensor of species k
D_k	diffusion coefficient of species k

D_{per}^s	non-equilibrium diffusivity of dopants
D_r^s	diffusivity of dopant of charge state r, e.g., D_0^s, D_-^s, D_+^s, ...
D_S	diffusivity at surface
D_{X^r}	diffusivity of a point defect X^r
$D_{X^r}^{eff}$	effective diffusivity of a point defect X^r
D^s	total diffusivity of substitutional atoms
D^{SD}	self-diffusion coefficient
D^T	tracer self-diffusion coefficient
D^0	pre-exponential factor in fit function of diffusivity
E_A	energy of acceptor level
E_F	energy of Fermi level
$\mathbf{F}_l$	l^{th} force acting on species k
f_I	self-interstitial correlation factor
f_V	vacancy correlation factor
G_k	term in Eq. (5-20) accounting for generation of species k
G_X^F	Gibbs free energy of formation of a point defect X
g	spin degeneraty factor
H_X^F	enthalpy of formation of a point defect X
h	field enhancement factor
h^+	hole
I	self-interstitial
I_A, I_B	self-interstitials of species A/B in sublattice of compound semiconductors
j	charge number
j_k	flux of species k
k	species of atoms
k	charge number
k_B	Boltzmann's constant
L_k	term in Eq. (5-20) accounting for loss of species k
m	species of atoms
m	charge number
n	electron concentration
n_i	intrinsic carrier concentration
n_s	electron concentration
P_{As_q}	arsenic vapor pressure
$P_{As_q}^0$	reference arsenic vapor pressure
p	hole concentration
Q	activation enthalpy
q	electric charge
q	stoichiometric coefficient
r	charge state of a point defect
r_{crit}	critical radius
r_{SF}	radius of stacking fault
r_s	radius of dopant
S_X^F	entropy of formation of a point defect X
s_I	self-interstitial supersaturation ratio

s_V	vacancy supersaturation ratio
T	temperature
t	time
V	vacancy
V_A, V_B	vacancy in sublattice of species A/B in compound semiconductors
X	point defect
X^r	point defect of charge state r, e.g., X^0, X^+, ...
X^{r-}	negatively charged point defect, e.g., X^-, X^{2-}, ...
x	spatial coordinate
Z	number of different energetically equivalent defect locations and configurations per lattice site

α_{eff}	factor of proportionality in Eq. (5-84)
β	stoichiometric factor in compounds of slightly varying composition
γ	parameter describing concentration dependence of diffusivity
γ_{SF}	extrinsic stacking fault energy
Δ^s_{per}	normalized diffusivity enhancement
$\Delta^s_{per}(min)$	largest observed retardation of diffusion
ϱ	dislocation density
σ	interface energy
Φ_I	fractional interstitialcy diffusion component
Ω	volume per lattice atom

EPR	electron paramagnetic resonance
MBE	molecular beam epitaxy
MOCVD	metal-organic chemical vapor deposition
OED	oxidation-enhanced diffusion
ORD	oxidation-retarded diffusion
OSF	oxidation-induced stacking faults

5.1 Introduction

Silicon and gallium arsenide are presently the most important semiconductors for fabricating electronic and optoelectronic devices. Crystal growth and device processing involve temperatures high enough to allow intrinsic point defects such as vacancies and self-interstitials to migrate rapidly and to interact with extrinsic point defects such as dopants and impurities. Some of the typical point defects occurring in a silicon crystal are schematically shown in Fig. 5-1.

Intrinsic point defect concentrations under thermal equilibrium conditions will be covered in Section 5.2. In contrast to the case of metals, charge effects play a major role for point defects in semiconductors and may strongly influence their thermal equilibrium concentrations. In compound semiconductors, the thermal equilibrium concentration of intrinsic point defects also depends on the vapor pressure of the more volatile component. Frequently crystal growth or device processing introduces nonequilibrium concentrations of intrinsic point defects. Under most conditions, the nonequilibrium point defect concentrations may be treated in terms of deviations from their thermal equilibrium values. Various methods of generating nonequilibrium concentrations of point defects will be covered in Section 5.3.

Investigation of intrinsic point defects by spectroscopic techniques such as electron paramagnetic resonance (EPR), described in Chapter 3 of this volume, is not feasible at typical processing temperatures. Since intrinsic point defects are involved in self-diffusion and most other impurity diffusion processes, information on intrinsic point defects at high temperatures have been obtained by diffusion studies. Contrary to spectroscopic methods, diffusion

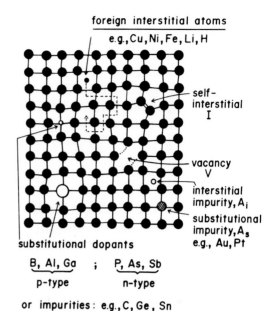

Figure 5-1. Schematic two-dimensional representation of various types of intrinsic and extrinsic point defects in an elemental semiconductor crystal such as silicon. I is a possible configuration of the self interstitial.

studies cannot give detailed *structural* information on point defects, but they can give information on point defect *charge states*. Diffusion experiments involving *equilibrium* or *nonequilibrium* point defects also allow one to identify which type of point defect (vacancy or self-interstitial) dominates a certain diffusion process. By these means long-standing controversies concerning the intrinsic point defects dominating diffusion processes in silicon and gallium arsenide have been solved.

A short introduction to the phenomenological description of diffusion processes and to atomistic diffusion mechanisms in semiconductors will be given in Sections 5.4 and 5.5. Sections 5.6–5.8 are devoted to diffusion in silicon, germanium and gallium arsenide. Finally, Section 5.9 will cover agglomeration phenomena of intrinsic point defects and the involvement of intrinsic point defects in precipitation phenomena associated with volume changes such as oxygen or carbon precipitation in silicon. This section will also contain a short note on the technologically important subject of gettering detrimental metallic impurities to predetermined gettering sites removed from the electrically active device region.

5.2 Intrinsic Point Defects Under Thermal Equilibrium Conditions

Intrinsic point defects are defects of atomic dimensions not involving any foreign atoms. The most basic and simple intrinsic point defects are vacancies (V) and self-interstitials (I) as indicated schematically in Fig. 5-1 for an elemental crystal such as silicon or germanium. III–V compound semiconductors (AB) consist of two sublattices for the Group III and the Group V atoms. In this case the basic intrinsic point defects are vacancies in the two sublattices (V_A and V_B), self-interstitials involving the two types of atoms (I_A, I_B), and anti-site defects which are atoms of one sublattice located on substitutional sites of the other sublattice (A_B, B_A), see Fig. 5.2.

Extrinsic point defects involve foreign atoms which may be incorporated into either *interstitial* sites such as Li, Cu, or Fe in

silicon or *substitutional* lattice sites such as the Group III elements (e.g., B, Al, Ga) or the Group V elements (P, As, Sb) commonly used as acceptors or donors in silicon devices. In silicon, the isoelectronic Group IV elements C, Ge, or Sn are substitutionally dissolved in a neutral charge state. Some elements, A, such as Au or Pt in silicon, Cu in germanium, or Zn in GaAs may exist in both *interstitial* (A_i) and *substitutional* sites (A_s), giving rise to peculiar diffusion mechanisms, discussed in Section 5.5.3.

The simple intrinsic and extrinsic point defects may form complexes, which increase the number of possible point defect configurations appreciably. That is, most substitutional dopants accomplish diffusion via complexes involving intrinsic point defects. Complexes of intrinsic point defects such as di-vacancies (two vacancies on adjacent lattice sites) are also quite common.

By introducing a thermal equilibrium concentration C_X^{eq} of intrinsic point defects ($X = V, I ...$), the Gibbs free energy of the crystal is minimized. In semiconductors, intrinsic point defects may occur in various charge states. Which charge state is actually realized cannot be reliably predicted based on simple principles or even elaborate quantum mechanical calculations. In the case of electrically neutral point defects X^0 the thermal equilibrium concentration is given by

$$C_{X^0}^{eq} = Z \exp(-G_X^F/k_B T) \qquad (5\text{-}1)$$

where

$$G_X^F = H_X^F - T S_X^F \qquad (5\text{-}2)$$

and H_X^F and S_X^F are the corresponding enthalpy and entropy of formation, respectively, T denotes the absolute temperature, and k_B is Boltzmann's constant. In Eq. (5-1) the dimensionless quantity Z characterizes

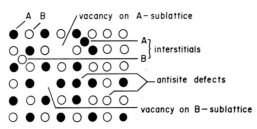

Figure 5-2. Various types of intrinsic point defects in a compound AB crystal.

the number of different possible energetically equivalent defect locations and configurations per lattice site. For simple vacancies, Z is unity. Throughout this chapter capital "C" will be used for dimensionless concentrations in atomic fractions and lower case "c" for volume concentrations. C and c are related via $c = C/\Omega$ where Ω is the volume per lattice atom (e.g., $\Omega = 2 \times 10^{-29}$ m^{-3} for silicon). The superscript "eq" indicates thermal equilibrium values. As with different charge states, in most semiconductors H_X^F and S_X^F are generally not well known either from clear-cut experimental results or from first principle calculations. However, results based on first principle calculations have become increasingly reliable and may serve as guidelines (Car et al., 1985; Baraff and Schlüter, 1986; Nichols et al., 1989).

The thermal equilibrium concentration of neutral point defects is independent of the position of the Fermi level, whereas the concentration of negatively charged point defects (X^-, X^{2-}) is increased by n-doping, and that of positively charged point defects ($X^+, X^{++}, \ldots$), by p-doping ("Fermi level effect"). For example, the thermal equilibrium concentration of negatively charged point defects in non-degenerate semiconductors is related to $C_{X^0}^{eq}$ via

$$C_{X^-}^{eq} = g\, C_{X^0}^{eq} \exp\left[(E_F - E_A)/k_B T\right] \tag{5-3}$$

E_F and E_A are the positions of the Fermi level and the point defect acceptor level, respectively (Shockley and Moll, 1960; Seeger and Chik, 1968). The dimensionless quantity g is the spin degeneracy factor, which, depending on the electron spin configuration of the charged and uncharged point defect, is $1/2$ or 2. Similar expressions hold for multiple or positively charged point defects. Frequently, the absolute concentration value is of less importance than its change with electron concentra-

tion n. The thermal equilibrium concentration $C_{X^r}^{eq}(n)$ of a point defect X^r compared to its value for intrinsic carrier concentration n_i is given by

$$C_{X^r}^{eq}(n)/C_{X^r}^{eq}(n_i) = (n/n_i)^{-r} \tag{5-4}$$

for a non-degenerate semiconductor. In Eq. (5-4) $r = 0$ denotes uncharged point defects, $r = 1, 2, 3, \ldots$ donor charge states, and $r = -1, -2, -3$ acceptor charge states of point defects. For example, for a triply negatively charged vacancy, V^{3-}, this ratio is

$$C_{V^{3-}}^{eq}(n)/C_{V^{3-}}^{eq}(n_i) = (n/n_i)^3 \tag{5-5}$$

For positively charged point defects and p-doped material, it is more convenient to express Eq. (5-4) in terms of the hole concentration p which is related to n via $np = n_i^2$. Since n_i strongly depends on temperature, the ratio in Eq. (5-5) for a given dopant (or electron) concentration is also strongly temperature dependent. Values of

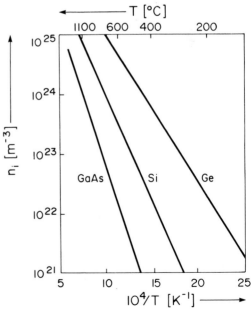

Figure 5-3. Intrinsic carrier concentrations of Ge, Si, and GaAs as a function of reciprocal temperature. (Adapted from Sze, 1985.)

n_i as a function of reciprocal absolute temperature are given in Fig. 5-3 for silicon, germanium, and gallium arsenide. The total intrinsic point defect concentration, including all charge states, generally depends on the electron concentration (or hole concentration or Fermi level) via

$$C_X^{eq}(n) = \sum_r C_{X^r}^{eq}(n_i)(n/n_i)^{-r} \qquad (5\text{-}6)$$

Each charge state of the intrinsic point defects may be associated with a different diffusivity D_{X^r}. These diffusivities are generally not individually known. Since the electronic equilibrium between different charge states is accomplished on a much shorter time scale than any diffusion process, only an effective intrinsic point defect diffusivity D_X^{eff} acts as a function of the electron concentration via

$$D_X^{eff} = \left(\sum_r D_{X^r} C_{X^r}^{eq}(n)\right)/C_X^{eq}(n) \qquad (5\text{-}7)$$

For III–V compounds, a thermodynamic range of non-stoichiometry exists according to the specific compound and the temperature. Within this range the thermal equilibrium concentration of intrinsic point defects in both sublattices depends on the vapor pressure of the more volatile component, e.g., on the arsenic vapor pressure in the case of GaAs (Kröger, 1973/74). For GaAs the dominating arsenic vapor species is either As_4 or As_2, Fig. 5-4 (Arthur, 1967), and the corresponding pressure dependencies of the concentrations of the point defects I_{As}, V_{As}, I_{Ga}, and V_{Ga}, are given by

$$\qquad (5\text{-}8)$$

$$C_{I_{As}}^{eq} \propto 1/C_{V_{As}}^{eq} \propto 1/C_{I_{Ga}}^{eq} \propto C_{V_{Ga}}^{eq} \propto (P_{As_q})^{1/q}$$

where P_{As_q} is the arsenic vapor pressure of the dominating As_q species with $q = 2$ or 4 (Kendall, 1968; Casey, 1973). Equation (5-8) is based on a reaction at the surface of the type

$$(1/q)\,As_q(g) \rightleftarrows I_{As} \qquad (5\text{-}9)$$

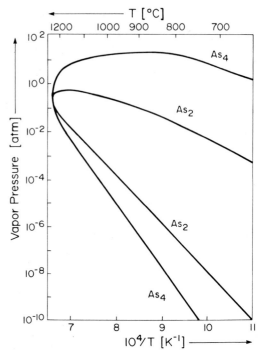

Figure 5-4. Partial pressures of the arsenic vapor species in equilibrium with most-gallium-rich (lower curves) or most-arsenic-rich GaAs (upper curves). (Arthur, 1967.)

Analogous relations hold for the other species of intrinsic point defects. The pressure dependence expressed in Eq. (5-8) presents a complication concerning performing experiments under defined conditions. However, in many cases the pressure dependencies of certain properties like diffusivities help to distinguish between different atomistic diffusion mechanisms.

5.3 Intrinsic Point Defects Under Nonequilibrium Conditions

Nonequilibrium concentrations of intrinsic point defects occur during crystal growth and during many of the steps for fabricating semiconductor devices. Crystal growth requires the cooling down of crystals from the melting point to room temperature. Since C_X^{eq} values decrease with decreasing temperature, the C_X^{eq} value attained at the solidification temperatures has to be decreased during cooling, which may be accomplished by diffusion to the surface or to dislocations. For large diameter dislocation-free crystals and sufficiently large growth speeds, point defect supersaturations may develop, which can be reduced by the formation of point defect agglomerates. Examples of crystal-growth induced point defect agglomerates in silicon are "swirl defects" termed after their swirl-like patterns due to the rotation of the crystal during crystal growth. Both interstitial-type and vacancy-type swirl defects have been observed in silicon (de Kock, 1981; Abe and Harada, 1983).

Ion implantation is one of the main methods of introducing p- and n-type dopants in devices. Depending on temperature, dose, and energy, the implanting process leads to a high nonequilibrium concentration of vacancy-self-interstitial pairs ("Frenkel pairs") which may form point defect agglomerates and dislocation loops or even may lead to amorphization of the crystal. At sufficiently high temperatures, vacancy-self-interstitial recombination and diffusion to the surface reduces the point defect concentrations quickly to their thermal equilibrium values. Therefore, implantation-induced nonequilibrium point defects do not play an important role for drive-in diffusions of dopants to a depth of many microns. In contrast, these nonequilibrium point defects are essential for both present day and future submicron devices especially if rapid thermal annealing techniques instead of conventional furnace annealing treatments are used. We will not discuss this important subject further in this Chapter and refer for more information to the review article by Fair (1989) and the references therein.

Surface reactions such as thermal oxidation, nitridation or silicidation of silicon may also lead to the injection of nonequilibrium intrinsic point defects (Fahey et al., 1989a). Contrary to the case of implantation, in which both vacancies and self-interstitials are generated in supersaturation, these surface reactions usually lead to a supersaturation of only one type of point defect (e.g., self-interstitials), whereas the other type of point defect (e.g., vacancies) is depleted according to the reaction

$$I + V \rightleftarrows 0 \qquad (5\text{-}10)$$

where "0" stands for the undisturbed lattice. In the presence of the injection of one type of point defect, reaction (5-10) leads approximately to local equilibrium between vacancies and self-interstitials in silicon via

$$C_I C_V = C_I^{eq} C_V^{eq} \qquad (5\text{-}11)$$

where C_I and C_V denote the perturbed self-interstitial and vacancy concentration, respectively. These nonequilibrium point defects affect dopant diffusion as well as the nucleation, growth or shrinkage of dislocation loops and are therefore of major importance for modern process simulation programs for silicon devices (Antoniadis, 1983, 1985; Fichtner, 1985; Kump and Dutton, 1988). Oxidation and nitridation-perturbed dopant diffusion will be dealt with in Section 5.6.4.2. These perturbed diffusion phenomena are not only of technical relevance, but properly applied, also allow one to distinguish between different atomistic diffusion mechanisms of dopant atoms (Frank et al., 1984; Fahey et al., 1989a).

Another source of nonequilibrium concentrations of intrinsic point defects are high-flux in-diffusion processes of foreign atoms dissolved on substitutional sites. Examples are high-concentration phosphorus diffusion into silicon leading to the so-called "anomalous" diffusion phenomena discussed in Section 5.6.4.3 or the in-diffusion of high-concentration zinc into GaAs (Sec. 5.8.4). It is much less recognized that high-concentration *out-diffusion* processes may also generate nonequilibrium concentrations of intrinsic point defects.

Impurity precipitation associated with volume changes may also lead to nonequilibrium concentrations of point defects. The best investigated example is the precipitation of oxygen in silicon, which may generate a supersaturation of self-interstitials in silicon due to the volume expansion associated with the formation of SiO_2 precipitates.

In an elemental semiconductor or within one sublattice of a compound semiconductor, a local point defect equilibrium according to Eq. (5-11) may be established based on the Frenkel pair generation/annihilation reaction Eq. (5-10). Global thermodynamic equilibrium $(C_I = C_I^{eq}, C_V = C_V^{eq})$ may be reached in elemental crystals by diffusion of the point defects from or to the surface, or from or to dislocations. For self-interstitials, dislocation climb leads to a change of C_I according to

$$(dC_I/dt)_{climb} \propto -D_I \varrho \, (C_I - C_I^{eq}) \qquad (5\text{-}12)$$

provided any effects due to dislocation line tension or possible stacking fault energies are neglected. In Eq. (5-12) ϱ is the dislocation density. A more detailed discussion of these climb processes will be presented in Section 5.9. A sufficiently high dislocation density generally guarantees intrinsic point defect concentrations close to their thermal equilibrium values. In the case of compound semiconductors, global point defect equilibrium normally requires diffusion of the nonequilibrium point defects to or from the surface. Since dislocation climb in binary compounds involves point defects from both sub-lattices, dislocation climb

enables only the establishment of local equilibrium between the two sublattices, e.g., according to

$$C_{I_A} C_{V_B} = C_{I_A}^{eq} C_{V_B}^{eq} \qquad (5\text{-}13)$$

In contrast to elemental semiconductors and contrary to popular opinion, a high dislocation density does not guarantee that the intrinsic point defect concentrations in compound semiconductors are close to their equilibrium values.

5.4 Phenomenological Description of Diffusion Processes

The flux j_k of a species of atoms, k, or point defects in a crystalline solid may be described by

$$j_k = -\mathbf{D}_k \nabla c_k + (\mathbf{D}_k/k_B T) \, c_k \sum_l F_l \qquad (5\text{-}14)$$

In Eq. (5-14) c_k is the volume concentration of this species, $\mathbf{D}_k$ the diffusion tensor governing its diffusion properties, and F_l the lth force acting on species k. The forces F_l may be due to internal or external electric fields for charged species or gradients in elastic fields, as detected near dislocations or close to mask edges during some device processing. The terms associated with the forces F_l are called *drift* or *field* terms, and the first term on the right-hand side is known as a *diffusion* term. Equation (5-14) has already been simplified from the general case of interacting particles by using concentrations in the diffusion term instead of the more general thermodynamic activities.

Most diffusion processes in technologically important semiconductors are isotropic, and the diffusion tensor $\mathbf{D}_k$ may therefore be replaced by a scalar diffusion coefficient D_k. In general, this diffusion coefficient can depend on temperature, and, indirectly via the Fermi level or directly, on its own concentration or on the concentration of other species m of concentration c_m:

$$D_k = D_k(T, n/n_i, c_k, c_m) \quad (5\text{-}15)$$

In Eq. (5-15) n/n_i represents the influence of the Fermi level. The effect of space charge regions is usually negligible at typical device-processing temperatures and the electron concentration n follows in good approximation from local charge neutrality which leads to

$$n = (c_{net} + \sqrt{c_{net}^2 + 4n_i^2})/2 \quad (5\text{-}16)$$

where c_{net} is the difference between the concentrations of all (singly) ionized donors and acceptors, c_d and c_a, respectively.

Of all the possible drift effects, usually only the one associated with the influence of the internal electric field is considered, which in terms of n/n_i and the electric charge q leads to

$$\quad (5\text{-}17)$$
$$j_k = -D_k \nabla c_k - (q/e) D_k c_k \nabla [\ln(n/n_i)]$$

In Eq. (5-17) e is the absolute value of the elementary charge, so that (q/e) is $+1$ for singly ionized donors and -1 for singly ionized acceptors. The basic structure of Eq. (5-17) is the same as that for electrons and holes in the absence of an external electric field. If the concentration of a charged diffusing species determines the Fermi level, then n in Eq. (5-17) depends only on the concentration of this species $c_k = c_{net}$ and on n_i via Eq. (5-16). Eq. (5-17) may then be simplified to

$$j_k = -h D_k \nabla c_k \quad (5\text{-}18)$$

where the so-called *field enhancement factor h* is given by

$$h = 1 + [1 + 4(n_i/c_k)^2]^{-1/2} \quad (5\text{-}19)$$

for singly ionized donors or acceptors. The value of h varies between 1 for $c_k \ll n_i$ and 2 for $c_k \gg n_i$. The use of Eq. (5-18) is limited to the diffusion of a dopant that solely determines the position of the Fermi level.

For calculating the time and spatial development of c_k the modified diffusion equation

$$\partial c_k/\partial t = -\text{div} j_k + G_k - L_k \quad (5\text{-}20)$$

has to be solved in which L_k is a term accounting for a possible *loss* of species k, e.g., due to precipitation or complexing. The term G_k accounts for the generation of species k, e.g., due to the dissolution of precipitates or complexes containing this species. The numerical solution of a set of such differential equations in one, two, or three dimensions with the appropriate boundary and initial conditions is one of the tasks of process simulation (Fichtner, 1985). In this Chapter we will deal with basic effects influencing diffusion processes, especially if they are related to intrinsic point defects. For this purpose, it is sufficient to deal solely with the one-dimensional version of Eq. (5-17).

As a point of reference, the simplest example of Eq. (5-20) occurs with only one species with a constant diffusivity D in one dimension in the absence of an internal electric field and without loss or generation terms. These conditions give Fick's second law:

$$\frac{\partial c}{\partial t} = D \frac{\partial^2 c}{\partial x^2} \quad (5\text{-}21)$$

Solutions of the simple diffusion equation (5-21) as well as its extension to two and three dimensions may be found in all standard text books on diffusion (Crank, 1957; Borg and Dienes, 1988) or in the mathematically equivalent problem of heat conduction (Carslaw and Jaeger, 1959). In the case of the in-diffusion of a species from the surface of a semi-infinite solid at $x = 0$ with the surface concentration kept at a constant value c_s, the solution is described in terms of a complementary error function:

$$c(x, t) = c_s \, \text{erfc} [x/\sqrt{4Dt}] \quad (5\text{-}22)$$

In many cases, diffusion in semiconductors depends strongly on the concentration of the diffusing species. For dopants this concentration dependence arises from the influence of the dopant concentration on the Fermi level and the associated influence of the Fermi level on the concentrations of charged point defects, which in turn act as vehicles of diffusion for the dopant. In other cases more complex reasons exist for the concentration dependence. Let us consider phenomenologically what kind of in-diffusion concentration profiles may be expected for a concentration-dependent diffusivity of the form

$$D = D_s (c/c_s)^\gamma \qquad (5\text{-}23)$$

when the concentration is kept constant (c_s) at the surface. D_s is the diffusivity value at the surface, and γ a parameter describing the concentration dependence. In Fig. 5-5 concentration profiles corresponding to the cases $\gamma = 0, 1, 2, 3$, and -2 are plotted in the normalized form $\log(C/C_s)$ versus $x/\sqrt{4 D_s t}$. The case $\gamma = 0$ represents the usual erfc-solution (cf. Eq. (5-22)) for constant diffusivity. For $\gamma > 0$ the diffusivity decreases as the concentration decreases. Therefore, increasingly box-like concentration profiles result for increasing γ. Although in reality the diffusivity of dopants is usually composed of several such terms with different values of γ (Sec. 5.6.4), $\gamma = 1$ is approximately realized for high-concentration diffusion of boron and arsenic in silicon, and $\gamma = 2$ both for high-concentration phosphorus diffusion in silicon and for the diffusion of zinc in GaAs. In the case of $\gamma = -2$ the *diffusivity increases* with *decreasing concentration* which leads to a concave profile shape (in the semi-logarithmic plot of Fig. 5-5). Such concave concentration profile shapes have been observed for Au, Pt, and Zn in silicon (Sec. 5.6.3) and for many elements in III−V compounds

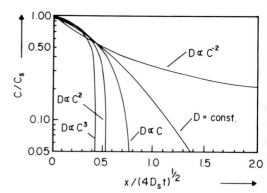

Figure 5-5. Normalized diffusion profiles for different concentration dependencies as indicated. (Adapted from Weisberg and Blanc, 1963.)

such as chromium in GaAs (Tuck, 1988). The concentration dependence of D may be determined from measured concentration profiles by means of a Boltzmann-Matano analysis as described in standard treatments of diffusion (Tsai, 1983). In the following section we will deal with the atomistic diffusion mechanisms, which are the underlying cause of concentration-dependent diffusivities.

5.5 Atomistic Diffusion Mechanisms

5.5.1 Diffusion Without Involvement of Intrinsic Point Defects

Interstitially dissolved atoms may diffuse by jumping from interstitial site to interstitial site (Fig. 5-1). In this *direct interstitial* diffusion mechanism no intrinsic defects are involved and the diffusivities are generally very high compared to those of substitutionally dissolved atoms as can be seen in Fig. 5-6 which shows a compilation of diffusion data for different elements in silicon. Examples are the diffusion of Li, Fe, and Cu in silicon. Oxygen in silicon also diffuses via interstitial sites, but in this case the oxygen interstitial is bound to two silicon atoms so that a diffusion jump requires the breaking of bonds. Therefore, the interstitial oxygen diffusivity is much lower than normal interstitial diffusivities but still much higher than the diffusivity of substitutionally dissolved dopants or self-diffusion.

Recently it has been suggested that silicon self-diffusion as well as the diffusion of Group-III and Group-IV dopants can also be accomplished without the involvement of intrinsic point defects via the *concerted exchange* mechanism (Pandey, 1986). Although it is impossible to prove that this mechanism does *not* contribute to self- and

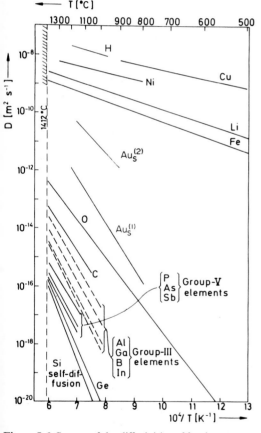

Figure 5-6. Survey of the diffusivities of foreign atoms in silicon and of silicon self-diffusion. The lines labeled with $Au_s^{(1)}$ and $Au_s^{(2)}$ correspond to different effective diffusivities of substitutional gold in silicon, as discussed in Section 5.6.3. (Adapted from Frank et al., 1984.)

dopant diffusion in silicon, it certainly cannot substitute for the diffusion mechanisms involving intrinsic point defects. Aside from other evidence, this assumption can be explained by the inability of the concerted exchange mechanism to carry out dislocation climb processes, which require the transport of a net amount of atoms.

5.5.2 Simple Vacancy Exchange and Interstitialcy Mechanisms

The diffusion of substitutionally dissolved atoms is facilitated by the presence of a neighboring intrinsic point defect. In a simple *vacancy exchange mechanism*, the substitutionally dissolved atom jumps into a vacancy on a nearest-neighbor site of the lattice or sublattice under consideration. In *interstitialcy* or *indirect interstitial* mechanisms, the substitutionally dissolved atom is first replaced by a self-interstitial and pushed into an interstitial position. From this location it changes over to a neighboring lattice site by pushing out that lattice atom. In self-diffusion no pair formation occurs between the lattice atom and the intrinsic defect involved, but substitutional impurities (or dopants) normally form complexes or pairs with the intrinsic point defects. These point-defect-impurity complexes explain the generally higher values for dopant diffusivities compared to those for self-diffusion (see Fig. 5-6 for silicon).

Within the simple vacancy exchange mechanism, the diffusivity D_V^s of substitutionally dissolved atoms is proportional to the available thermal equilibrium vacancy concentration,

$$D_V^s \propto C_V^{eq} \qquad (5-24)$$

Similarly, for substitutionally dissolved atoms using self-interstitials as vehicles of diffusion the diffusivity D_I^s is given by

$$D_I^s \propto C_I^{eq} \qquad (5-25)$$

Since, in principle, both vacancies and self-interstitials can contribute to the total diffusivity D^s, and since the intrinsic point defect may occur in various charge states, X^r, D^s as a function of the electron concentration n may be written as

$$D^s(n) = \sum_r \sum_X D^s_{X^r}(n_i)(n/n_i)^{-r} \quad (5\text{-}26)$$

The summation over the intrinsic point defects X for a given charge state r,

$$\sum_X D^s_{X^r}(n_i) = D^s_{I^r}(n_i) + D^s_{V^r}(n_i) \quad (5\text{-}27)$$

describes the self-interstitial and vacancy contributions to the diffusivity under intrinsic and thermal equilibrium conditions. If the concentrations C_X of intrinsic point defects X differ from their equilibrium concentrations C_X^{eq} due to an external perturbation, $D^s(n)$ changes to a perturbed diffusivity

$$D^s_{per}(n) = \quad (5\text{-}28)$$

$$= \sum_X \left(C_X(n)/C_X^{eq}(n) \sum_r D_{X^r}(n_i)(n/n_i)^{-r} \right)$$

Equation (5-28) has already taken into account that the ratio

$$C_{X^r}(n)/C_{X^r}^{eq}(n) = C_X(n)/C_X^{eq}(n) \quad (5\text{-}29)$$

is independent of the charge state r. We will apply and discuss this equation in the context of oxidation and nitridation-perturbed diffusion of dopants in silicon in Section 5.6.4. For compound semiconductors Eq. (5-28) has to be extended to account for the dependence of $D^s(n)$ on the pressure of the more volatile component as expressed in Eq. (5-8) (Casey, 1973).

5.5.3 Interstitial-Substitutional Mechanisms

5.5.3.1 Uncharged Species

An appreciable number of impurities A in semiconductors are mainly dissolved on substitutional sites A_s but accomplish dif-

fusion by switching over to an interstitial configuration A_i in which their diffusivity D_i is extremely high. Examples are Au, Pt, and Zn in silicon, Cu in Ge, and Zn, Be, Mn, Cr, and Fe in GaAs. The changeover from interstitial to substitutional sites and vice versa requires the involvement of intrinsic point defects. For uncharged species the two basic forms of changeover suggested are the *kick-out* mechanism (Gösele et al., 1980; Frank et al., 1984),

$$A_i \rightleftarrows A_s + I \quad (5\text{-}30)$$

involving self-interstitials, and the earlier-proposed *Frank-Turnbull* mechanism (Frank and Turnbull, 1956),

$$A_i + V \rightleftarrows A_s \quad (5\text{-}31)$$

involving vacancies. Both mechanisms are schematically shown in Fig. 5-7. It is worth mentioning that the kick-out mechanism is just the dynamic form of the Watkins replacement mechanism (Watkins, 1975) in which a self-interstitial generated by low-temperature electron irradiation pushed a substitutional atom A_s into an interstitial

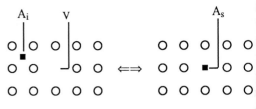

(a)   FRANK-TURNBULL MECHANISM

(b)   KICK-OUT MECHANISM

Figure 5-7. Schematic representation of the Frank-Turnbull or Longini mechanism (a) and of the kick-out mechanism (b).

position A_i. The kick-out mechanism is closely related to the interstitialcy diffusion mechanism. The main difference is that the foreign atom, once in an interstitial position, remains there for only one step in the interstitialcy mechanism and for many steps in the kick-out mechanism. In contrast, the Frank-Turnbull mechanism and the vacancy exchange mechanism are qualitatively different. Within the vacancy exchange mechanism an increase in vacancy concentration enhances the diffusivity, whereas within the Frank-Turnbull mechanism an increase in vacancy concentration decreases the effective diffusivity of the substitutional species.

A detailed description of the diffusion behavior of atoms moving via the kick-out or the Frank-Turnbull mechanism generally requires the solution of a coupled system of three differential equations describing diffusion and reaction of A_i, A_s, and the intrinsic point defect involved (V or I). Detailed discussions and methods of solutions may be found in the literature (Frank et al., 1984; Tuck, 1988; Morooka and Yoshida, 1989). In this Chapter, the effective diffusion coefficient D_{eff} of A_s via the kick-out mechanism is derived in a simplified manner for in-diffusion from the surface.

We assume for simplicity that the mass action law for the concentrations C_i of A_i, C_s of A_s, and C_I of I, in accordance with reaction (5-30), is fulfilled, so that

$$C_s C_I/C_i = C_s^{eq} C_I^{eq}/C_i^{eq} \qquad (5-32)$$

where the superscript "eq" indicates the thermal equilibrium concentrations (solubilities) of the corresponding species. If the in-diffusion of A_i is slow enough to allow the self-interstitials generated via the kick-out mechanism to migrate out to the surface and maintain their thermal equilibrium concentration, then the effective

diffusivity of A_s is given by

$$D_{eff}^{(i)} = D_i C_i^{eq}/C_s^{eq} \qquad (5-33)$$

provided $C_s^{eq} \gg C_i^{eq}$, which is generally the case. If, on the other hand, the in-diffusion of A_i is so fast that the generated self-interstitials can *not* escape quickly enough to the surface (i.e., if $D_i C_i^{eq} \gg D_I C_I^{eq}$), a *supersaturation of self-interstitials* will develop and further incorporation of A_s will be limited by the out-diffusion of the generated self-interstitial to the surface. This leads to an effective A_s diffusivity $D_{eff}^{(I)}$ following from the approximately flux balance

$$D_{eff}^{(I)} \partial C_s/\partial x = -D_I \partial C_I/\partial x \qquad (5-34)$$

and the mass-action law (5-32) as

$$D_{eff}^{(I)} = (D_I C_I^{eq}/C_s^{eq})(C_s^{eq}/C_s)^2 \qquad (5-35)$$

Analogously, for the Frank-Turnbull mechanism sufficiently slow in-diffusion $(D_i C_i^{eq} \ll D_V C_V^{eq})$ leads to the same $D_{eff}^{(i)}$ as given by Eq. (5-33). An effective A_s diffusivity $D_{eff}^{(V)}$, controlled by the in-diffusion of vacancies from the surface, results if $D_i C_i^{eq} \gg D_V C_V^{eq}$ holds:

$$D_{eff}^{(V)} = D_V C_V^{eq}/C_s^{eq} \qquad (5-36)$$

The strongly concentration-dependent effective diffusivity $D_{eff}^{(I)}$ leads to the peculiar concentration profiles shown in Fig. 5-5 for $\gamma = -2$. These profiles can easily be distinguished from the erfc-type profiles which are associated with $D_{eff}^{(V)}$. This macroscopic difference allows one not only to recognize different atomistic diffusion mechanisms of the specific foreign atom involved but also to obtain information on the *mechanism of self-diffusion*. The effective diffusivities in Eqs. (5-35) and (5-36) have been derived under the assumption of dislocation-free crystals. The presence of a high density of dislocations in an elemental crystal maintains the equilibrium concentration of intrinsic point defects. Thus an erfc-type profile characterized by the constant diffusivity $D_{eff}^{(i)}$ of Eq. (5-33) will result even if $D_i C_i^{eq} \gg D_I C_I^{eq}$ holds. For compound semiconductors this statement does not hold in general, since the presence of dislocations does not necessarily guarantee that intrinsic point defects will reach their thermal equilibrium concentrations. If self-interstitials and vacancies co-exist, such as in the case of silicon, then the effective A_s diffusion coefficient in dislocation-free material for $D_i C_i^{eq} \gg (D_I C_I^{eq} + D_V C_V^{eq})$ is given by

$$D_{eff}^{(I,V)} = D_{eff}^{(I)} + D_{eff}^{(V)} \qquad (5-37)$$

5.5.3.2 Charged Species

For III–V compounds, even regular p-type dopants such as zinc, beryllium or magnesium diffuse via an interstitial-substitutional diffusion mechanism (Tuck, 1974; Kendall, 1968; Casey, 1973; Tuck, 1988). In these cases the charge states of the involved species have to be taken into account. In a generalized form the kick-out mechanism now reads

$$A_i^{j+} \rightleftarrows A_s^{m-} + I^{k+} + (m+j-k)\,h^+ \qquad (5-38)$$

where j, k, and m are integers characterizing the charge state of the species, and h^+ stands for holes (Gösele and Morehead, 1981; Gösele, 1988). The self-interstitial is assumed to consist of the atomic species which forms the sublattice on which A_s is dissolved, e.g., a gallium interstitial in the case of zinc acceptors substitutionally dissolved on the gallium sublattice in GaAs. The corresponding extension of the Frank-Turnbull mechanism, which is often called *Longini* mechanism in the case of III–V compounds (Longini, 1962), may be written as

$$A_i^{j+} + V^{k-} \rightleftarrows A_s^{m-} + (m+j-k)\,h^+ \qquad (5-39)$$

In general, the intrinsic point defects, as well as the interstitials A_i, may occur in more than one charge state.

For the generalized kick-out mechanism, the mass-action law for local equilibrium between the various species reads

$$C_i / (C_s C_I p^{m+j-k}) = \mathrm{const}\,(T) \qquad (5-40)$$

where p is the hole concentration. For completely ionized substitutional acceptors, i.e. impurities $(m > 0)$ with concentration above the intrinsic electron concentration n_i, p may be replaced by $m C_s$. For donor impurities $(m < 0)$, the electron concentration is analogously given by $|m| C_s$. For dislocation-free material, considerations similar to those for uncharged species lead to

$$\qquad (5-41)$$
$$D_{eff}^{(i)} = (|m| + 1)[D_i C_i^{eq}/C_s^{eq}](C_s/C_s^{eq})^{|m| \pm j}$$

if the supply of A_i^{j+} limits the incorporation rate. The positive sign in the exponent holds for substitutional acceptors, and the negative sign for substitutional donors. The factor $|m| + 1$ accounts for the electric field enhancement (see also Sec. 5.4). Equation (5-41) holds both for the generalized kick-out and the Frank-Turnbull mechanism and is independent of the charge state of the intrinsic point defects.

When the diffusion of self-interstitials to the surface limits the incorporation rate of A_s, a *supersaturation of self-interstitials* will develop, and the effective diffusion coefficient for the A_s atoms is then given by

$$D_{eff}^{(I)} = \qquad (5-42)$$
$$= (|m| + 1)[D_I C_I^{eq}/C_s^{eq}](C_s/C_s^{eq})^{\pm k - |m| - 2}$$

where $C_I^{eq} = C_I^{eq}(C_s^{eq})$.

When the supply of vacancies from the surface limits the incorporation of A_s, an undersaturation of vacancies develops leading to

$$D_{\text{eff}}^{(V)} = (|m| + 1)[D_V C_V^{\text{eq}} (C_s^{\text{eq}})/C_s^{\text{eq}}] \cdot$$
$$\cdot (C_s/C_s^{\text{eq}})^{\pm k - m} \qquad (5\text{-}43)$$

For both Eq. (5-42) and (5-43) the same sign convention holds as that in Eq. (5-41). Equations (5-40)–(5-43) reduce to (5-33), (5-35), and (5-36) if all species involved are uncharged.

The quantities $D_I C_I^{\text{eq}} (C_s^{\text{eq}})$ and $D_V C_V^{\text{eq}} (C_s^{\text{eq}})$ refer to the self-diffusion transport coefficients of I^{k+} and V^{k-} under the *doping conditions* $C_s = C_s^{\text{eq}}$ and do not necessarily refer to the *intrinsic self-diffusion coefficient*. Even for charged species, constant effective diffusivities may be obtained. For example, for singly charged acceptor dopants ($m = 1$), I^{3+} ($k = 3$) and V^- ($k = 1$) lead to constant effective diffusivities $D_{\text{eff}}^{(I)}$ and $D_{\text{eff}}^{(V)}$, respectively. Since the applicable effective diffusion coefficient may change with the depth of the profile, complicated concentration profiles may result, as frequently observed in III–V compounds (Tuck, 1988).

Examples of foreign atoms diffusing via one of the interstitial-substitutional mechanisms will be discussed in Sections 5.6–5.8, for silicon and germanium as well as gallium arsenide. Understanding interstitial-substitutional diffusion mechanisms is fundamentally important for determining whether self-diffusion in a given semiconductor material is limited by vacancies or self-interstitials. This question can hardly be conclusively answered by other means.

5.6 Diffusion in Silicon

5.6.1 General Remarks

Silicon is the most important electronic material presently used and is likely to remain so in the foreseeable future. Diffusion of dopants is an important step during device processing. For sufficiently deep junctions, diffusion is required for generating the desired dopant profile. For submicron devices the tail of the implantation profile is already in the submicron regime, so that diffusion occurring during the necesssary annealing out of implantation-induced lattice damage may already be an undesirable effect. Methods such as rapid thermal annealing by flash lamps have been investigated to gain tighter control over the time spent at high temperatures (Fair, 1989). Even shallower junctions will probably require closely controlled diffusion processes from well-defined sources, as, for example, from doped polysilicon used for certain bipolar devices. The interest in dopant diffusion will increase in this context.

Diffusion processes in silicon were first described in terms of vacancy-related mechanisms in analogy to metals in which vacancies are the predominant intrinsic point defects at thermal equilibrium. In 1968 Seeger and Chik suggested that in silicon self-interstitials and vacancies both contribute to self- and dopant diffusion processes. After a controversy over the dominant intrinsic point defects in silicon that lasted almost 20 years, it is now almost generally accepted that self-interstitials and vacancies *both* have to be taken into account in order to consistently understand self-diffusion and most other impurity diffusion processes. This point of view has been adopted for the present Section. For dissenting opinions the reader is referred to articles by Bourgoin (1985) and Van Vechten (1980, 1990). The main indications for the involvement of self-interstitials in diffusion processes in silicon came from diffusion experiments performed under nonequilibrium intrinsic point defect conditions, such as experiments on the influence of surface oxidations or nitridation on dopant diffusion. Investigations of the diffusion properties of atoms such as Au or Pt migrating via an interstitial-substitutional mechanism were also crucial in establishing the role of self-interstitials in self-diffusion in silicon. What is still missing is a quantitative determination of the diffusivity and the thermal equilibrium concentrations of self-interstitials and vacancies, to be discussed in Section 5.6.6.

5.6.2 Sillicon Self-Diffusion

The transport of silicon under thermal equilibrium conditions is governed by the uncorrelated self-diffusion coefficient

$$D^{SD} = D_I C_I^{eq} + D_V C_V^{eq} \qquad (5\text{-}44)$$

As mentioned in Section 5.2, intrinsic point defects may exist in several charge states. The observed doping dependence of Group-III and Group-V dopant diffusion (Section 5.6.4) indicates the presence of neutral, positively charged, negatively and doubly negatively charged intrinsic point defects. It is presently not known whether all these charge states occur for both self-interstitials and vacancies. Taking all observed charge states into account we may write $D_I C_I^{eq}$ as

$$D_I C_I^{eq} = D_{I^0} C_{I^0}^{eq} + D_{I^-} C_{I^-}^{eq} + \qquad (5\text{-}45)$$
$$+ D_{I^{2-}} C_{I^{2-}}^{eq} + D_{I^+} C_{I^+}^{eq}$$

An analogous expression holds for vacancies. The quantity C_I^{eq} comprises the sum of the concentrations of self-interstitials in the various charge states according to

$$C_I^{eq} = C_{I^0}^{eq} + C_{I^-}^{eq} + C_{I^{2-}}^{eq} + C_{I^+}^{eq} \qquad (5\text{-}46)$$

Therefore, the diffusivity D_I is actually an effective diffusion coefficient consisting of a weighted average of the diffusivities in the different charge states according to Eq.

(5-7). The same holds analogously for C_V^{eq} and D_V.

The most common way to investigate self-diffusion in silicon is to measure the diffusion of silicon tracer atoms in silicon. These tracer atoms are silicon isotopes which can be distinguished from the usual silicon isotopes of the crystal by various experimental techniques. The tracer self-diffusion coefficient D^T differs slightly from Eq. (5-44) since it contains geometrically defined dimensionless correlation factors f_I and f_V,

$$D^T = f_I D_I C_I^{eq} + f_V D_V C_V^{eq} \qquad (5\text{-}47)$$

The vacancy correlation factor f_V in the diamond lattice is 0.5. The corresponding quantity $f_I \leqq 1$ depends on the unknown

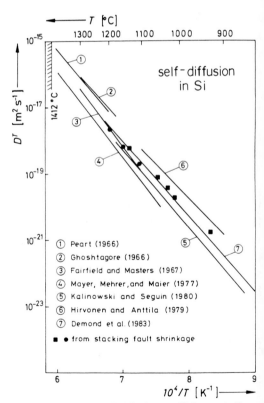

Figure 5-8. Tracer self-diffusion coefficients of silicon as a function of reciprocal absolute temperature. (Adapted from Frank et al., 1984.)

self-interstitial configuration. Measured results for D^T are shown in Fig. 5-8. Various results for D^T, which are usually fitted to an expression of the form

$$D^T = D^0 \exp(-Q/k_B T) \qquad (5\text{-}48)$$

are given in Table 5-1 in terms of the pre-exponential factor D^0 and the activation enthalpy Q.

Tracer measurements, including extensions measuring the doping dependence of D^T (Frank et al., 1984), do not allow the separation of self-interstitial and vacancy contributions to self-diffusion. Such a separation is possible by investigating the diffusion of Au, Pt, and Zn in silicon (as described in more detail in Section 5.6.3). These experiments allow a fairly accurate

determination of $D_I C_I^{eq}$ but only a crude estimate of $D_V C_V^{eq}$, derived from a combination of different types of experiments (Tan and Gösele, 1985). The resulting expressions shown in Fig. 5-9 are

$$D_I C_I^{eq} = 9.4 \times 10^{-2} \cdot \qquad (5\text{-}49)$$
$$\cdot \exp(-4.84 \text{ eV}/k_B T)\, \text{m}^2\, \text{s}^{-1}$$

$$D_V C_V^{eq} \approx 6 \times 10^{-5} \cdot \qquad (5\text{-}50)$$
$$\cdot \exp(-4.03 \text{ eV}/k_B T)\, \text{m}^2\, \text{s}^{-1}$$

Values of $D_I C_I^{eq}$, as determined by different groups are also given in Table 5-1 in the form of the pre-exponential factor D^0 and activation enthalpy Q. It is worth noting that $D_I C_I^{eq}$ coincides within experimental error with $D^T/2$ from tracer measurements. From the doping dependence of silicon

Table 5-1. Diffusivities of various elements including self-interstitials and vacancies in silicon fitted to $D = D^0 \exp(-Q/k_B T)$.

Diffusing species	Description of diffusivity	D^0 $[10^{-4}\,\text{m}^2\,\text{s}^{-1}]$	Q [eV]	References
		1800	4.77	Peart, 1966
		1200	4.72	Ghostagore, 1966
		9000	5.13	Fairfield and Masters, 1967
Si	D^T	1460	5.02	Mayer et al., 1977
		8	4.1	Hirvonen and Antilla, 1974
		154	4.65	Kalinowski and Seguin, 1980
		20	4.4	Demond et al., 1983
		914	4.84	Stolwijk et al., 1984
Si	$D_I C_I^{eq}$	320	4.80	Stolwijk et al., 1988
		2000	4.94	Hauber et al., 1989
		1400	5.01	Mantovani et al., 1986
Si	$D_V C_V^{eq}$	0.57	4.03	Tan and Gösele, 1985
		10^{-5}	0.4	Tan and Gösele, 1985
I	D_I	3.75×10^{-9}	0.13	Bronner and Plummer, 1985
		8.6×10^5	4.0	Taniguchi et al., 1983
V	D_V	0.1	2.0	Tan and Gösele, 1985
Ge	D^s	2500	4.97	Hettich et al., 1979
Sn	D^s	32	4.25	Teh et al., 1968
C_s	D^s	1.9	3.1	Newman and Wakefield, 1961
C_i	D_i	4.4	0.88	Tipping and Newman, 1987
O	D_i	0.07	2.44	Mikkelsen, 1986

self-diffusion (Frank et al., 1984), one can conclude that neutral as well as positively and negatively charged point defects are involved in self-diffusion. However, the data are not accurate enough to determine the individual terms of Eq. (5-45) or the analogous expression for vacancies. Since D^T as well as $D_I C_I^{eq}$ and $D_V C_V^{eq}$ each consist of various terms, their representation in terms of an expression of the type of Eq. (5-48) can only be an approximation that holds over a limited temperature range. In Section 5.6.6 we will discuss what is known about the individual factors D_I, C_I^{eq}, and D_V and C_V^{eq}.

5.6.3 Interstitial-Substitutional Diffusion: Au, Pt, and Zn in Si

Both gold and platinum can drastically reduce the minority carrier lifetime in silicon because their energy levels are close to the middle of the band gap. Both elements are used in some types of power devices to improve their frequency behavior. Generally, however, gold, and to a lesser extent platinum, are harmful contaminants in highly integrated circuits which should be avoided. For both reasons, the behavior of gold and platinum has been investigated extensively. Zinc is not a technologically important impurity in silicon but is of scientific interest as an element with a diffusion behavior in between substitutional dopants and gold and platinum in silicon.

Both gold and platinum in-diffusion profiles into dislocation-free silicon show the concave profile shape typical for the kick-out mechanism (Stolwijk et al., 1983, 1984; Frank et al., 1984; Hauber et al., 1989; Mantovani et al., 1986). Examples are shown in Fig. 5-10 for gold diffusion and in Fig. 5-11 for platinum diffusion. From profiles like these and from the measured solubility C_s^{eq} of Au_s and Pt_s in sili-

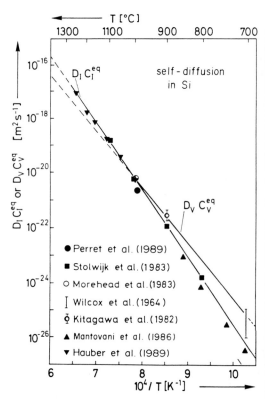

Figure 5-9. Comparison of the contributions $D_I C_I^{eq}$ and $D_V C_V^{eq}$ to the self-diffusion coefficient in silicon determined from the diffusion of Au (Wilcox et al., 1964; Stolwijk et al., 1983; Morehead et al., 1983), Pt (Mantovani et al., 1986; Hauber et al., 1989), Zn (Perret et al., 1989), and Ni (Kitagawa et al., 1982) in Si. Full symbols refer to $D_I C_I^{eq}$.

con, the values of $D_I C_I^{eq}$ shown in Fig. 5-9 have been determined. Diffusion of gold into thin silicon wafers leads to characteristic U-shaped profiles, even if the gold has been deposited on one side only. The increase of the gold concentration in the center of the wafer has also been used to determine $D_I C_I^{eq}$ (Frank et al., 1984).

In heavily dislocated silicon the dislocations act as efficient sinks for self-interstitials and keep C_I close to C_I^{eq} so that the constant effective diffusivity $D_{eff}^{(i)}$ from Eq. (5-33) governs the diffusion profile (Stolwijk et al., 1988). Analysis of the resulting

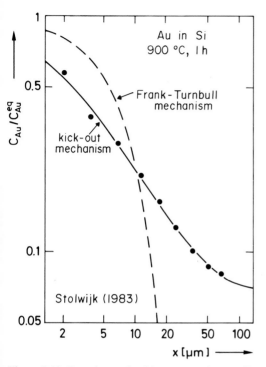

Figure 5-10. Experimental gold concentration profile in dislocation-free silicon (solid circles) compared with predictions of the Frank-Turnbull and the kick-out mechanism (Stolwijk et al., 1983).

erfc-profiles leads to

$$D_i C_i^{eq} \approx 6.4 \times 10^{-3} \cdot \qquad (5\text{-}51)$$
$$\cdot \exp\left(-3.93\,\text{eV}/k_B T\right) \text{m}^2\,\text{s}^{-1}$$

In Fig. 5-6, $D_i C_i^{eq}/C_s^{eq}$ (curve $\text{Au}_s^{(2)}$) is compared to $D_I C_I^{eq}/C_s^{eq}$ (curve $\text{Au}_s^{(1)}$). $D_i C_i^{eq}$ turns out to be much larger than $D_I C_I^{eq}$ from Eq. (5-49). This is consistent with the observation that gold concentration profiles are governed by $D_{eff}^{(I)}$ (according to Eq. (5-35)) in dislocation-free silicon. Zinc diffusion has also been investigated in highly dislocated and dislocation-free silicon (Perret et al., 1989). In highly dislocated material, an erfc-profile develops as expected (Fig. 5-12). In dislocation-free material only the profile part close to the surface shows the concave shape typical

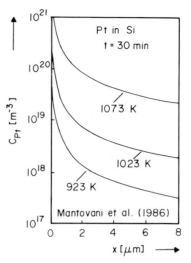

Figure 5-11. Platinum concentration profiles in dislocation-free silicon (Mantovani et al., 1986).

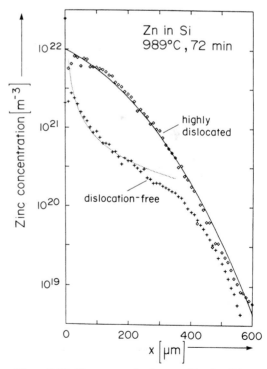

Figure 5-12. Zinc concentration profiles in dislocation-free and highly dislocated silicon. In highly dislocated silicon the results can be fitted by a complementary error function (solid line); in dislocation-free silicon the region close to the surface shows a kick-out profile (Perret et al., 1989).

for the kick-out diffusion mechanism (Fig. 5-12). This part can be used to determine $D_I C_I^{eq}$ values as indicated in Fig. 5-9. For lower zinc concentrations, a constant diffusivity takes over. The reason for this changeover from one profile type to another is as follows: In contrast to the case of gold, the $D_i C_i^{eq}$ value determined for zinc is not much higher than $D_I C_I^{eq}$ so that even in dislocation-free silicon only the profile close to the surface is governed by the term $D_{eff}^{(I)}$ of Eq. (5-35), which strongly increases with depth. For sufficiently large penetration depths $D_{eff}^{(I)}$ finally exceeds $D_{eff}^{(i)}$ and a constant effective diffusivity begins to determine the concentration profile, as shown in Fig. 5-12. A detailed analysis shows that the diffusivity in the tail region may be enhanced by the supersaturation of self-interstitials generated by the in-diffusion of zinc, which leads to an effective diffusivity in the tail region given by

$$D_{eff}^{tail} = D_{eff}^{(i)} (C_I/C_I^{eq}) \tag{5-52}$$

The changeover from a concave to an erfc-type profile has also been observed for the diffusion of gold either into very thick silicon samples (Huntley and Willoughby, 1973) or for short-time diffusions (Boit et al., 1990) into normal silicon wafers (300–800 µm).

The diffusion profile of gold in silicon is very sensitive to the presence of dislocations since dislocations may act as sinks for self-interstitials and thus enhance the local incorporation rate of Au_s. Even in dislocation-free silicon the self-interstitials created in supersaturation by the in-diffusion of gold may agglomerate and form interstitial-type dislocation loops which further absorb self-interstitials and lead to W-shaped profiles instead of the usual U-shaped profiles in gold-diffused silicon wafers (Hauber et al., 1986).

A detailed analysis of gold diffusion profiles at 1000 °C by Morehead et al. (1983) showed the presence of a small but noticeable vacancy contribution, which is consistent with the conclusion from dopant diffusion experiments that self-interstitials and vacancies are both present under thermal equilibrium conditions (see subsequent Section 5.6.4). Wilcox et al. observed in 1964 that the gold concentration profiles at 700 °C are characterized by a constant diffusivity, which indicates that at this temperature the kick-out mechanism is kinetically hampered whereas the Frank-Turnbull mechanism still operates. This appears to be the case for the incorporation of substitutional nickel also (Kitagawa et al., 1982; Frank et al., 1984). Attempts to repeat the 700 °C Au diffusion experiments have failed probably because of a much higher background concentration of grown-in vacancies or vacancy clusters present in the much larger diameter silicon crystals of today. Nevertheless, the 700 °C gold data of Wilcox et al. (1964) have been used to estimate $D_V C_V^{eq}$ at this temperature as indicated in Fig. 5-9.

5.6.4 Dopant Diffusion

5.6.4.1 Fermi Level Effect

Both n- and p-type regions in silicon devices are created by intentional doping with substitutionally dissolved Group V or Group III dopants which act as donors or acceptors, respectively. Technologically most important are the donors As, P, and Sb and the acceptors B, and to a lesser extent Al and Ga. Dopant diffusion has been studied extensively because of its importance in device fabrication. A detailed quantitative understanding of dopant diffusion is also a prerequisite for accurate and meaningful modeling in numerical process simulation programs. It is not our intention to compile all available data on dopant diffusion in silicon, which may conveniently be found elsewhere (Hu, 1973; Shaw, 1973, 1975; Tuck, 1974; Casey and Pearson, 1975; Fair, 1981 b; Tsai, 1983; Ghandi, 1983; Langheinrich, 1984). We will concentrate instead on the diffusion mechanisms and intrinsic point defects involved in dopant diffusion, on the effect of the Fermi level on dopant diffusion, and on nonequilibrium point defect phenomena induced by high-concentration in-diffusion of dopants.

The diffusivities D^s of all dopants in silicon depend on the Fermi level. The experimentally observed doping dependencies may be described in terms of the expression

$$D^s(n) = D_0^s + D_+^s(n_i/n) + D_-^s(n/n_i) + \\ + D_{2-}^s(n/n_i)^2 \qquad (5\text{-}53)$$

which for intrinsic conditions, $n = n_i$, reduces to

$$D^s(n_i) = D_0^s + D_+^s + D_-^s + D_{2-}^s \qquad (5\text{-}54)$$

Depending on the specific dopant, some of the quantities in Eq. (5-54) may be negligibly small. $D^s(n_i)$ is an exponential function of inverse temperature, as shown in Fig. 5-6. Values of these quantities in terms of pre-exponential factors and activation enthalpies are given in Table 5-2. Conflicting results exist on the doping dependence of antimony.

Table 5-2. Diffusion of various dopants fitted to Eq. (5-53) and values for activation enthalpies, Q, in eV. Each term fitted to $D^0 \exp(-Q/k_B T)$; D_i values in $10^{-2}\,\mathrm{m^1\,s^{-1}}$ (Fair, 1981; Ho et al., 1983).

Element	D_0^0	Q_0	D_+^0	Q_+	D_-^0	Q_-	D_{2-}^0	Q_{2-}
B	0.037	3.46	0.72	3.46	–	–	–	–
P	3.850	3.66	–	–	4.44	4.00	44.20	4.37
As	0.066	3.44	–	–	12.0	4.05	–	–
Sb	0.214	3.65	–	–	15.0	4.08	–	–

The higher diffusivities of all dopants relative to self-diffusion requires fast-moving complexes formed by the dopants and intrinsic point defects. The doping dependence of $D^s(n)$ is generally explained in terms of the various charge states of the intrinsic point defects carrying dopant diffusion as discussed in Section 5.5.2. Since self-interstitials and vacancies can both be involved in dopant diffusion, each of the terms in Eq. (5-54) generally consists of a self-interstitial and a vacancy-related contribution, such as

$$D_+^s = D_{I+}^s + D_{V+}^s \qquad (5\text{-}55)$$

following from Eq. (5-27). $D^s(n)$ may also be written in terms of a self-interstitial and a vacancy-related contribution as had first been suggested by Hu (1974),

$$D^s(n) = D_I^s(n) + D_V^s(n) \qquad (5\text{-}56)$$

with

$$D_I^s(n) = D_{I^0}^s + D_{I+}^s(n_i/n) + D_{I-}^s(n/n_i) + \\ + D_{I^{2-}}^s(n/n_i)^2 \qquad (5\text{-}57)$$

and an analogous expression for $D_V^s(n)$.

Contrary to popular opinion, although the observed doping dependence expressed in Eq. (5-53) shows that *charged point defects* are involved in the diffusion process, nothing can be learned about the relative contributions of self-interstitials and vacancies in the various charge states. Strictly speaking, in contrast to self-diffusion, the doping dependence of dopant diffusion does not necessarily prove the presence of *charged intrinsic point defects* but rather the presence of *charged point-defect dopant complexes*. In Section 5.6.4.2 we will describe a way to determine the relative contribution of self-interstitials and vacancies to dopant diffusion by measuring the effect of nonequilibrium concentrations of intrinsic point defects on dopant diffusion.

5.6.4.2 Influence of Surface Reactions

Thermal oxidation is a standard process in the fabrication of silicon devices and is used, e.g., in forming field or gate oxides or oxides protecting certain device regions from ion implantation. The oxidation process leads to the injection of self-interstitials which can either *enhance* the diffusivity of dopants using mainly self-interstitials as vehicles of diffusion or *retard* diffusion of dopants which diffuse mainly via a vacancy mechanism. Oxidation-enhanced diffusion (OED) has been observed for the dopants B, Al, Ga, P, and As and oxida-

tion-retarded diffusion (ORD) for Sb (Frank et al., 1984; Fahey et al., 1989a). The influence of surface oxidation on dopant diffusion is schematically shown in Fig. 5-13. The retarded diffusion of Sb is explained in terms of the recombination reaction (5-10) leading to a vacancy undersaturation in the presence of a self-interstitial supersaturation. The oxidation-induced self-interstitials may also nucleate and form interstitial-type dislocation loops on (111) planes which usually contain a stacking fault and are therefore termed oxididation-induced stacking faults (OSF). The growth and shrinkage kinetics of

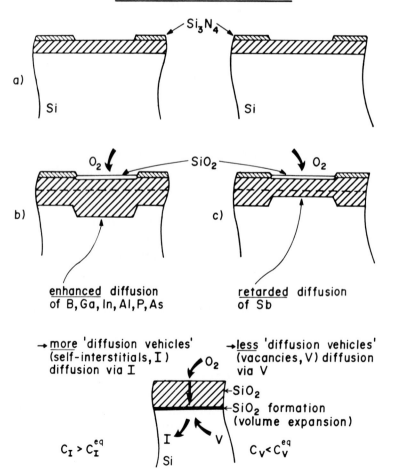

Oxidation-influenced diffusion

enhanced diffusion
of B, Ga, In, Al, P, As

retarded diffusion
of Sb

→ more 'diffusion vehicles'
(self-interstitials, I)
diffusion via I

→ less 'diffusion vehicles'
(vacancies, V) diffusion
via V

$\rightarrow SiO_2$ formation
(volume expansion)

$c_I > c_I^{eq}$

$c_V < c_V^{eq}$

Figure 5-13. Influence of surface oxidation on dopant diffusion in silicon. (a) Cross section of a silicon wafer doped near the surface with B, Ga, In, Al, P, As (left-hand side) or Sb (right-hand side) before oxidation. (b) Same cross section after surface oxidation indicating enhanced diffusion. (c) Retarded diffusion. For details, see text.

OSFs will be dealt with in Section 5.9, which covers precipitation phenomena.

The physical reason for the point defect injection during surface oxidation is simple (see Fig. 5-14). Oxidation occurs by the diffusion of oxygen through the oxide layer and reaction with silicon at the SiO_2/Si interface. The oxidation reaction is associated with a volume expansion of a factor of approximately two, which is mostly accommodated by viscoelastic flow of the oxide. It is also partly compensated by the injection of silicon self-interstitials into the silicon crystal, which leads to a supersaturation of these point defects. The detailed reactions occurring at the interface have been the subject of numerous publications (Fahey et al., 1989 a).

Oxidation can also cause vacancy injection provided the oxidation occurs at sufficiently high temperatures (typically 1150 °C or higher) and the oxide is thick enough. Under these circumstances silicon, probably in the form of SiO (Tan and Gösele, 1982; Celler, 1988), diffuses from the interface and reacts with the oxygen in the oxide away from the interface (Fig. 5-14 b). The resulting supersaturation of vacancies associated with an undersaturation of self-interstitials gives rise to retarded boron and phosphorous diffusion (Francis and Dobson, 1979) and enhanced antimony diffusion (Tan and Ginsberg, 1983). Thermal nitridation of silicon surfaces also causes a supersaturation of vacancies coupled with an undersaturation of self-interstitials, whereas oxynitridation (nitridation of oxides) behaves more like normal oxidation. Silicidation reactions have recently also been found to inject intrinsic point defects and to cause enhanced dopant diffusion (Hu, 1987; Fahey et al., 1989 a).

A simple quantitative formulation of oxidation- and nitridation-influenced diffu-

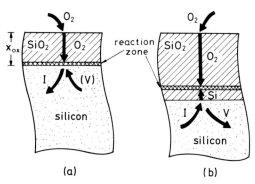

Figure 5-14. Schematic illustration of the injection or absorption of intrinsic point defects induced by surface oxidation of silicon according to Francis and Dobson (1979) and Tan and Gösele (1982). (a) Thin oxide layer and/or moderate temperature, (b) thick oxide layer and/or high temperature.

sion is based on Eq. (5-56) which changes for perturbed intrinsic point defect concentrations C_I and C_V approximately to

$$D_{per}^s (n) = D_I^s (n) \frac{C_I}{C_I^{eq} (n)} + D_V^s (n) \frac{C_V}{C_V^{eq} (n)} \tag{5-58}$$

For long enough times and sufficiently high temperatures (e.g., one hour at 1100 °C), local dynamic equilibrium between vacancies and self-interstitials according to Eq. (5-11) is established, and Eq. (5-58) may be reformulated in terms of C_I/C_I^{eq}. Defining a normalized diffusivity enhancement,

$$\Delta_{per}^s = [D_{per}^s (n) - D^s (n)]/D^s (n) \tag{5-59}$$

a fractional interstitialcy diffusion component,

$$\Phi_I (n) = D_I^s (n)/D^s (n) \tag{5-60}$$

and a self-interstitial supersaturation ratio,

$$s_I (n) = [C_I - C_I^{eq} (n)]/C_I^{eq} (n) \tag{5-61}$$

we may rewrite Eq. (5-58) in the form

$$\Delta_{per}^s = [2\,\Phi_I (n) + \Phi_I (n)\, s_I - 1]\, s_I/(1 + s_I) \tag{5-62}$$

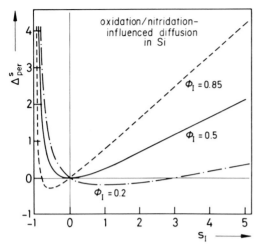

Figure 5-15. Normalized diffusion enhancement Δ^s_{per} versus self-interstitial supersaturation $s_I = (C_I - C_I^{eq})/C_I^{eq}$ for different values of Φ_I (Tan and Gösele, 1985).

provided Eq. (5-11) holds. Usually Eq. (5-62) is given for intrinsic conditions and the dependence of Φ_I on n is not indicated. Eq. (5-62) is plotted in Fig. 5-15 for Φ_I values of 0.85, 0.5 and 0.2.

The left-hand side of Fig. 5-15, where $s_I < 0$ (associated with a vacancy supersaturation), has been realized by high-temperature oxidation and thermal nitridation of silicon surfaces, as mentioned above. Another way to generate a vacancy supersaturation is by oxidation in an atmosphere containing HCl at sufficiently high temper-

atures and with a sufficiently high concentration of HCl (Tan and Gösele, 1985; Fair, 1989). As expected, $s_I < 0$ results in an enhancement of Sb diffusion and the retardation of P and B diffusion. Arsenic diffusion is enhanced as in the case of oxidation, which indicates that arsenic has appreciable components via vacancies and self-interstitials ($\Phi_I \approx 0.5$).

Several different procedures have been used to evaluate Φ_I for the different dopants, resulting in a wide range of conflicting published Φ_I values. With the capability of generating a self-interstitial supersaturation by oxidation ($s_I > 0$) and a vacancy supersaturation by thermal nitridation ($s_I < 0$), the most accurate procedure to determine Φ_I appears to be the following: The diffusion changes must be checked under oxidation and nitridation conditions. If for $s_I > 0$ the diffusion is enhanced, and for $s_I < 0$ it is retarded (as for P and B), then $\Phi_I > 0.5$ holds. Based on the largest observed retardation $\Delta^s_{per}(min)$ (which has a negative value), a lower limit of Φ_I may be estimated according to

$$\Phi_I > 0.5 + 0.5\,[1 - (1 + \Delta^s_{per}(min))^2]^{1/2} \tag{5-63}$$

Analogously, an upper limit for Φ_I may be estimated when retarded diffusion occurs for $s_I > 0$ and enhanced diffusion for $s_I < 0$, as in the case of Sb. A different procedure

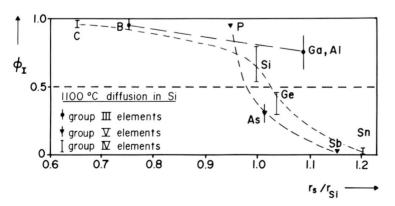

Figure 5-16. Interstitial-related fractional diffusion component Φ_I for Group III, IV and V elements versus their atomic radii in units of the atomic radius r_{Si} of silicon. The values for carbon and tin are derived from theoretical considerations and limited experimental results.

is required for elements with Φ_I values close to 0.5, such as arsenic.

In Fig. 5-16 values of Φ_I at 1100 °C are shown as a function of the atomic radius r_s of the various dopants for intrinsic doping conditions. Both the charge state (Group III/Group V dopants) and the atomic size influence Φ_I. Φ_I has a tendency to increase with increasing temperature. Oxidation and nitridation experiments and extrinsic conditions indicate that a decreasing value of Φ_I exists for phosphorus with increasing n-doping (see also subsequent Section), but that both phosphorus and boron still remain dominated by self-interstitials ($\Phi_I(n) > 0.5$). Inconsistencies in the determination of Φ_I by oxidation and nitridation experiments (Fahey et al., 1989 a) have led to speculations concerning the validity of the basic starting equation (Eq. (5-58)), and to more detailed approaches incorporating a diffusion contribution by the concerted exchange mechanism or the Frank-Turnbull mechanism (Cowern, 1988).

5.6.4.3 Dopant-Diffusion-Induced Nonequilibrium Effects

Nonequilibrium concentrations of intrinsic point defects may be induced not only by various surface reactions, as discussed in the previous section, but also by the in-diffusion of some dopants starting from a high surface concentration. These nonequilibrium effects are most pronounced for high-concentration phosphorus diffusion, but are also present for other dopants such as boron, and to a lesser extent aluminum and gallium. In high-concentration in-diffusion of phosphorus, the nonequilibrium concentrations of intrinsic point defects lead to a number of phenomena which had initially been termed "anomalous" (Willoughby, 1981), before a detailed understanding of these phenom-

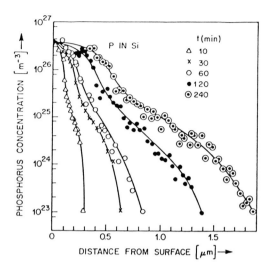

Figure 5-17. Concentration profiles of phosphorus diffused into silicon at 900 °C for the times t indicated (Yoshida et al., 1974).

ena had been acquired. We will only mention the most prominent of these phenomena. Phosphorus diffusion profiles (Fig. 5-17) show a *tail* in which the phosphorus diffusivity is much higher (up to a factor of 100 at 900 °C) than expected from isoconcentration studies. In n-p-n transistor structures in which high-concentration phosphorus is used for the emitter diffusion, the diffusion of the base dopant boron is similarly enhanced below the phosphorus diffused region. This so-called *emitter-push effect* is schematically shown in Fig. 5-18 a. The diffusion of B, P, or Ga in buried layers many microns away from the phosphorus-diffused area is also greatly enhanced (Fig. 5-18 b). In contrast, the diffusion of Sb in buried layers is retarded under the same conditions (Fig. 5-18 c). The enhanced and retarded diffusion phenomena are analogous to those occurring during surface oxidation. As has also been confirmed by dislocation-climb experiments (Strunk et al., 1979; Nishi and Antoniadis, 1986), all these phenomena are due to a supersaturation of silicon self-intersti-

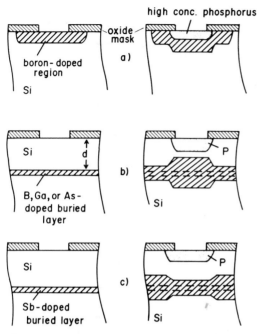

Figure 5-18. Anomalous diffusion effects induced by high-concentration phosphorus diffusion, (a) emitter-push effect of boron-doped base region, (b) enhanced diffusion of B, Ga, or As in buried layers and (c) retarded diffusion of Sb in buried layer (Gösele, 1989).

tials, associated with an undersaturation of vacancies, induced by high-concentration in-diffusion of phosphorus. The basic features of high-concentration phosphorus diffusion are schematically shown in Fig. 5-19, which also indicates the presence of electrically neutral precipitates at the phosphorus concentrations that exceed the solubility limit at the diffusion temperature. A much less pronounced supersaturation of self-interstitials is generated by boron starting from a high surface concentration as can be concluded from the boron profiles and from the growth of interstitial-type stacking faults induced by the boron diffusion (Claeys et al., 1978; Morehead and Lever, 1986).

Many qualitative and quantitative models have been proposed to explain the phe-

nomena associated with high-concentration phosphorus diffusion. The earlier models are mostly vacancy dominated and predict a phosphorus-induced vacancy supersaturation (Fair and Tsai, 1977; Yoshida, 1983; Mathiot and Pfister, 1984). These models contradict the experimental results that have been obtained in the meantime. Morehead and Lever (1986) presented a mathematical treatment of high-concentration dopant diffusion in which the point defect species dominates the diffusion of the dopant (e.g., self-interstitials for P and B and vacancies for Sb). The concentration of the other type of intrinsic point defect is assumed to be determined by the dominating point defect via the local equilibrium condition (cf. Eq. (5-11)). The dopant-induced self-interstitial supersaturation s_I may be estimated by the influx of dopants which release part of the self-interstitials involved in their diffusion process. These self-interstitials will diffuse both to the surface, where it is assumed that $C_I = C_I^{eq}$ holds, and into the material. Finally, a quasi-steady-state supersaturation of self-interstitials will develop in which the dopant-induced flux of injected self-interstitials cancels the flux of self-interstitials

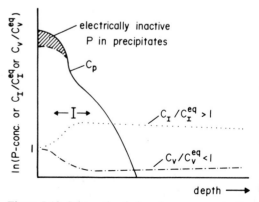

Figure 5-19. Schematic of phosphorus concentration profile (C_p) and of the normalized intrinsic point defect concentrations C_I/C_I^{eq} and C_V/C_V^{eq} (Gösele, 1989).

to the surface. The flux of self-interstitials into the bulk is considered small compared to the flux to the surface. This flux balance may be expressed in a similar manner to Eq. (5-34) in interstitial-substitutional diffusion:

$$\Phi_I \, h \, D^s(n) \frac{\partial C_D}{\partial x} = - D_I \frac{\partial C_I}{\partial x} \qquad (5\text{-}64)$$

where $1 \leq h \leq 2$ is the electric field enhancement factor, and C_D the substitutional dopant concentration. With a doping dependence of the dopant diffusivity in the simple form

$$D^s(n) \propto n^\gamma \qquad (5\text{-}65)$$

integration of Eq. (5-64) allows an estimate of the resulting supersaturation s_I of self-interstitials:

$$s_I = \frac{C_I - C_I^{eq}}{C_I^{eq}} = \frac{h \, \Phi \, D^s(n_s) \, C_D^s}{(\gamma + 1) \, D_I \, C_I^{eq}} \qquad (5\text{-}66)$$

For the derivation of Eq. (5-66) the fairly small doping dependence of $D_I \, C_I^{eq}$ has been neglected. In Eq. (5-66) n_s is the electron concentration (at the diffusion temperature), and C_D^s the concentration of electrically active dopants in dimensionless atomic fractions at the surface. For phosphorus, $\gamma = 2$ has to be used. By an analogous equation, a vacancy supersaturation may be estimated which may be induced by a dopant diffusing mainly via the vacancy exchange mechanism.

Let us briefly discuss the physical meaning of Eq. (5-66). The generation of a high supersaturation of intrinsic point defects requires not only a dopant diffusivity which is higher than self-diffusion (which holds for all dopants in silicon) but also a sufficiently high dopant solubility. Even further simplified, the condition for generating a high supersaturation of intrinsic point defects reads

$$D^s(n_s) \, C_D^s \gg D^{SD}(n) \qquad (5\text{-}67)$$

which is basically the same condition that has been used for generating a nonequilibrium concentration of intrinsic point defects by elements diffusing via the interstitial-substitutional mechanism (see Section 5.5.3). In short, diffusion-induced nonequilibrium concentrations of intrinsic point defects are generated if the effective flux of in-diffusing substitutional atoms (which either consume or generate intrinsic point defects) is larger than the flux of host atoms trying to re-establish thermal equilibrium concentrations of intrinsic point defects. We will use this principle again in the context of high-concentration zinc and beryllium diffusion in GaAs (Section 5.8).

Equation (5-66) predicts the proper high self-interstitial supersaturation for phosphorus, a factor of up to about eight for boron at 900 °C and negligible for Sb and As, in accordance with experimental results. Much more elaborate numerical models have recently been proposed for calculating diffusion-induced nonequilibrium point defect phenomena (Orlowski, 1988).

5.6.5 Diffusion of Group IV Elements

In Section 5.6.2 we dealt extensively with the diffusion of silicon in silicon. The other Group IV elements carbon, germanium and tin are also dissolved substitutionally but much less is known about their diffusion mechanisms. The diffusivities of C, Ge and Sn are given in Table 5-1 in terms of pre-exponential factors and activation enthalpies. The rate of germanium and tin diffusion is similar to silicon self-diffusion, whereas the rate of carbon diffusion is much faster (Fig. 5-6).

Based on volume considerations, it can be expected that the diffusion of carbon atoms, which are much smaller than silicon, involves mainly self-interstitials. This expectation is consistent with the experimental observation that self-interstitials injected by oxidation of high-concentration phosphorus diffusion enhance carbon diffusion (Ladd and Kalejs, 1986). Carbon diffusion is therefore most likely to be accomplished by a highly mobile carbon-self-interstitial complex (CI) according to

$$C_s + I \rightleftarrows (CI) \tag{5-68}$$

where C_s denotes substitutional carbon. The effective substitutional carbon diffusivity D_s may be expressed by $D_i C_i^{eq}/C_s^{eq}$, where D_i is the diffusivity of the fast-diffusing (CI) complex. C_i^{eq} and C_s^{eq} are the solubilities of the (CI) complex and substitutional carbon, respectively. Under the assumption that the (CI) complex is the same as that observed by Watkins and Brower (1976) and by Tipping and Newman (1987) after low-temperature electron irradiation with the diffusivity

$$D_i \approx 4.4 \times 10^{-4} \cdot$$
$$\cdot \exp(-0.88 \, eV/k_B T) \, m^2 \, s^{-1} \tag{5-69}$$

it may be estimated that

$$C_i^{eq}/C_s^{eq} \approx 5 \exp(-2.22 \, eV/k_B T) \tag{5-70}$$

which is in agreement with the assertion that carbon is almost exclusively dissolved on substitutional sites. The reaction (5-68) is basically the kick-out mechanism. Since $D_i C_i^{eq}$ of carbon is much smaller than $D_I C_I^{eq}$, it is expected that carbon diffusion is controlled by a constant diffusivity $[D_{eff}^{(i)}$ of Eq. (5-33)] in agreement with experimental observations (Newman and Wakefield, 1961; Rollert et al., 1989).

Germanium atoms are slightly larger than silicon atoms. Oxidation and nitridation experiments show a Φ_I value of Ge around 0.4 at 1100 °C (Fahey et al., 1989 b), which is slightly lower than that derived for silicon self-diffusion. For the much

larger tin atoms, it is expected that diffusion is almost entirely controlled by the vacancy exchanged mechanism, as in the case with the Group V dopant Sb. Consistent with this expectation, a nitridation-induced supersaturation of vacancies increases tin diffusion (Marioton and Gösele, 1989), but no quantitive determination of Φ_I is presently available for tin.

5.6.6 Diffusion of Self-Interstitials and Vacancies

Although the product $D_I C_I^{eq}$ is known, and estimates of $D_V C_V^{eq}$ are available, our knowledge of the individual factors D_I, D_V, C_I^{eq} and C_V^{eq} is astonishingly limited in spite of immense experimental efforts to determine these quantities. The individual diffusivities D_I and D_V are important quantities, which enter the most advanced numerical programs for simulating device processing. Their elusiveness has slowed down progress in this area (Kump and Dutton, 1988). The most direct way of measuring D_I is by injecting self-interstitials (e.g., via surface oxidation) at one location of the silicon crystal and then observing the effect on dopant diffusion or on growth or shrinkage of stacking faults at another location as a function of time and of distance between the two locations. That is, the two locations may be the front and the back side of a silicon wafer. Extensive experiments on the spread of oxidation-induced self-interstitials through wafers by Mizuo and Higuchi (1983) have shown that a supersaturation of self-interstitials arrives at about the same time as a corresponding undersaturation of vacancies. Therefore, these kinds of experiments at 1100 °C only give information on the *effective* diffusivity of a perturbation in the self-interstitial and vacancy concentrations. This effective diffusivity may be expressed approximately by

$$D_{I,V}^{eff} \approx (D_I C_I^{eq} + D_V C_V^{eq})/(C_I^{eq} + C_V^{eq}) \qquad (5\text{-}71)$$

and probably corresponds to the diffusivity values of about $3 \times 10^{-13}\, \mathrm{m^2\, s^{-1}}$ observed in the experiments of Mizuo and Higuchi at 1100 °C.

In most experiments aimed at determining D_I it has not been taken into account that self-interstitials may react with vacancies according to Eq. (5-10) to establish the local dynamical equilibrium described by Eq. (5-11). Based on experiments on oxidation-retarded diffusion of antimony (Antoniadis and Moskowitz, 1982; Fahey et al., 1989 a), it has been estimated that at 1100 °C an astonishingly long time (about one hour) is required to establish local dynamic equilibrium. This long recombination time indicates the presence of an energy or entropy barrier slowing down the recombination reaction. At lower temperatures much longer recombination times can be expected. These long recombination times hold for lightly doped material. There are indications that dopants or other foreign elements may act as "recombination centers" which can speed up the recombination reaction considerably. No solid data are available in this area. In order to demonstrate the state of affairs concerning D_I (and therefore indirectly also of C_I^{eq} via the known product $D_I C_I^{eq}$), the available D_I estimates, as a function of inverse absolute temperature, are shown in Fig. 5-20 (Taylor et al., 1989). At 800 °C the estimates differ by up to eight orders of magnitude.

The question of proper D_I values is further complicated by the observation that the measured effective diffusivity $D_{I,V}^{eff}$ depends on the type of silicon material used. In the experiments of Fahey et al. (1989 a) the transport of oxidation-induced self-interstitials through epitaxially grown silicon

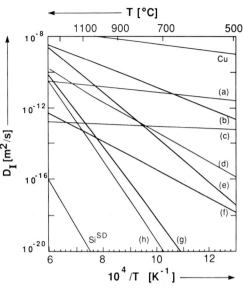

Figure 5-20. Diffusivity D_I of self-interstitials in silicon as a function of temperature as estimated by various authors (a–h) and compared to silicon self-diffusion and copper diffusion. (a) Tan and Gösele, 1985; (b) Morehead, 1988; (c) Bronner and Plummer, 1985; (d) Seeger et al., 1977; (e) Bronner and Plummer, 1987; (f) Griffin and Plummer, 1986; (g) Taniguchi et al., 1983; (h) Wada et al., 1983. (Adapted from Taylor et al., 1989.)

layers was much faster than through equally thick layers of as-grown float-zone or Czochralski silicon. This difference has been attributed to the presence of vacancy-type agglomerates left from the crystal growth process which might not be present in epitaxial silicon layers. These vacancy agglomerates would have to be consumed by the injected self-interstitials before further spread of interstitials can occur. Further experiments are definitely required to settle the question of self-interstitial and vacancy diffusivities as well as thermal equilibrium concentrations in silicon. We finally note that as yet it has not been possible to connect in a reasonable and consistent way the fairly high diffusivities of intrinsic point defects found after low-temperature electron irradiation with the much lower diffusivities which appear to be required to explain high-temperature diffusion experiments. Temperature-dependent configurations of intrinsic point defects have been suggested by Seeger and Chik (1968) and recently for interstitial manganese, iron and cobalt in silicon (Gilles et al., 1990).

5.6.7 Oxygen and Hydrogen Diffusion

Oxygen is the most important electrically inactive foreign element in silicon. In Czochralski-grown silicon, oxygen is incorporated from the quartz crucible and is usually present in concentrations of the order $10^{24}\,\mathrm{m}^{-3}$, which thus exceed the concentrations of electrically active dopants in certain device regions. Oxygen is interstitially dissolved and its diffusion requires the breaking of bonds. The diffusivity of interstitial oxygen, O_i, has been measured between about $300\,^{\circ}\mathrm{C}$ and the melting point of silicon and is in good approximation described by

$$D_i = 0.07 \exp(-2.44\,\mathrm{eV}/k_B T)\,\mathrm{m}^2\,\mathrm{s}^{-1} \quad (5\text{-}72)$$

as shown in Fig. 5-6 (Mikkelsen, 1986). The solubility C_i^{eq} of interstitial oxygen has been determined to be

$$(5\text{-}73)$$
$$C_i^{eq} = 1.53 \times 10^{27} \exp(-1.03\,\mathrm{eV}/k_B T)\,\mathrm{m}^{-3}$$

Since in most Czochralski-grown silicon crystals the grown-in oxygen concentration exceeds C_i^{eq} at typical processing temperatures, oxygen precipitation will occur, which will be dealt with in Section 5.9.3.

Around $450\,^{\circ}\mathrm{C}$ oxygen tends to form electrically active agglomerates ("thermal donors", Kaiser et al., 1958; Bourret, 1985). The formation kinetics of these agglomerates appears to require a fast-diffusing species, for which both self-interstitials (Newman, 1985) and molecular oxygen have been suggested (Gösele and Tan, 1982). The question of molecular oxygen in silicon has not yet been settled. In this context it is interesting to note that the presence of fast-diffusing nitrogen molecules in silicon has been demonstrated by Itoh and Abe (1988).

Hydrogen plays an increasingly important role in silicon device technology because of its capability to passivate electrically active defects. The passivation of dislocations and grain boundaries is especially important for inexpensive polycrystalline silicon used for solar cells. Both acceptors and donors can be passivated by hydrogen, which is usually supplied to the silicon from a plasma.

Hydrogen in silicon is assumed to diffuse as unbonded atomic hydrogen that may be present in a neutral or positively charged form. The diffusivity of hydrogen in silicon has been measured by Van Wieringen and Warmoltz (1956) in the temperature range of $970-1200\,^{\circ}\mathrm{C}$. These results are included in Fig. 5-6. Between room temperature and $600\,^{\circ}\mathrm{C}$, hydrogen diffusivities much lower than those extrapolated from the high-temperature data have been measured. Corbett and coworkers (Pearton et al., 1987) rationalized this observation by suggesting that atomic hydrogen may form interstitially dissolved, essentially immobile H_2 molecules. For a detailed understanding of the complex hydrogen concentration profiles, trapping at dopants must also be taken into account (Kalejs and Rajendran, 1990). As in the case of oxygen, the existence of hydrogen molecules has not been proven experimentally.

5.7 Diffusion in Germanium

Germanium lost its leading role for electronic devices more than two decades ago and is now mainly used either as a detector material or in Si/Ge superlattices. Therefore, only very few papers have recently been published on diffusion in germanium. Another reason might be that diffusion in germanium can consistently be explained in terms of vacancy-related mechanisms, and no self-interstitial contribution must be taken into account.

Fig. 5-21 shows the diffusivities of Group III and Group V dopants and of Ge in germanium as a function of inverse ab-

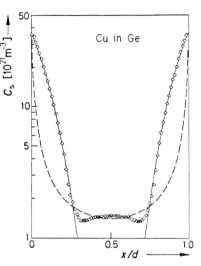

Figure 5-22. Concentration profiles of copper in a dislocation-free germanium wafer after diffusion for 15 min at 878 °C. The solid line represents the Frank-Turnbull mechanism, and the dashed line the kick-out mechanism (Stolwijk et al., 1985).

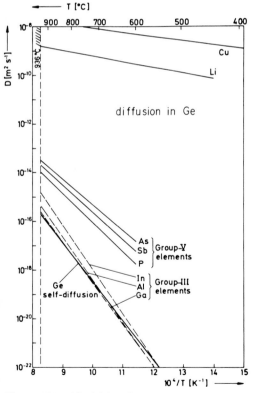

Figure 5-21. Diffusivities of various elements (including germanium) in germanium as a function of inverse absolute temperature. (Adapted from Frank et al., 1984.)

solute temperature for intrinsic conditions. The doping dependence of dopant diffusion can be explained by one kind of acceptor-type intrinsic point defect. These intrinsic point defects have been assumed to be vacancies since the earliest studies of diffusion in germanium (Seeger and Chik, 1968), but a convincing experimental proof was only given in 1985 by Stolwijk et al., based on the diffusion behavior of copper in germanium.

Copper diffuses in germanium via an interstitial-substitutional mechanism (Frank and Turnbull, 1956). In analogy to gold and platinum in silicon, the diffusion behavior of copper may be used to check diffusion profiles for any indication of a self-interstitial contribution via the kick-out mechanism. A concentration profile of copper diffusion into a germanium wafer is shown in Fig. 5-22 (Stolwijk et al., 1985). The dashed U-shaped profile which is typical for the kick-out mechanism obviously

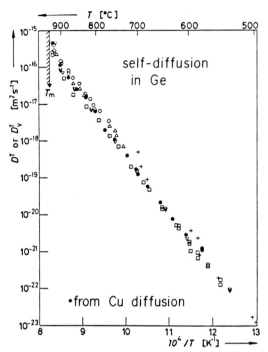

Figure 5-23. Tracer self-diffusion data compared to the vacancy contribution $D_V^T = 1/2\, D_V\, C_V^{eq}$ determined from copper diffusion in Ge (●) (Stolwijk et al., 1985).

does not fit the experimental data. In contrast, the experimental profiles may be described by the constant diffusivity $D_{eff}^{(V)}$ given by Eq. (5-36). In Fig. 5-23 values of the vacancy contribution to germanium self-diffusion,

$$D_V\, C_V^{eq} \approx 21.3 \times 10^{-4} \cdot \qquad (5\text{-}74)$$
$$\cdot \exp\left(-3.11\,\text{eV}/k_B T\right) \text{m}^2\,\text{s}^{-1}$$

as determined from copper diffusion profiles, are compared with corresponding tracer measurements of self-diffusion in germanium. The agreement is excellent, which shows that any kind of self-interstitial contribution is negligible and that vacancies appear to be entirely responsible for germanium self-diffusion. It is unclear why self-interstitials play such an important role in diffusion processes in silicon but no noticeable role in germanium.

5.8 Diffusion in Gallium Arsenide

5.8.1 General Remarks

Gallium arsenide is the most important compound semiconductor with applications ranging from fast electronic to optoelectronic devices such as light-emitting diodes and lasers. In combination with lattice-matched aluminum arsenide, gallium arsenide is also the main material for the fabrication of quantum well and superlattice structures. Although the diffusion of many elements in GaAs have been investigated (Tuck, 1988), most of the diffusion studies have concentrated on the main p-type dopants zinc and beryllium, on the main n-type dopants silicon and selenium, and on chromium, which is used for producing semi-insulating GaAs. Since zinc, beryllium, chromium, and a number of other elements diffuse via an interstitial-substitutional mechanism, this type of diffusion mechanism has historically received much more attention in GaAs than in elemental crystals. As with silicon and germanium diffusion, it had been assumed for a long time that only vacancies have to be taken into account to understand diffusion processes in gallium arsenide. This assumption has also been incorporated in process simulation programs (Deal et al., 1989).

Until recently, only a few studies of self-diffusion in GaAs had been available, but with the advances in GaAs/AlAs-type superlattices grown by Molecular Beam Epitaxy (MBE) or Metal-Organic Chemical Vapor Deposition (MOCVD), these kinds of studies have multiplied. The observation that high-concentration zinc diffusion into a GaAs/Al$_x$Ga$_{1-x}$As superlattice leads to a dramatic increase in the Ga/Al interdiffusion coefficient (Laidig et al., 1981) opened up the possibility of fabricating laterally

structured optoelectronic devices by locally disordering superlattices. Zinc-in-
duced superlattice disordering as observed
by Laidig et al. (1981) is shown in Fig. 5-24.

It turned out that this dopant-enhanced
superlattice disordering is a fairly general
phenomena which occurs for other p-type
dopants such as magnesium as well as for
n-type dopants such as silicon, selenium,
and tellurium (Deppe and Holonyak,
1988). Dopant-enhanced superlattice disordering is not only of technological importance but has also allowed scientists to
unravel the contributions of self-interstitials and vacancies to self-diffusion and dopant diffusion processes in the gallium sublattice. These superlattices with a typical
period of about 100 Å allow measurement
of Ga/Al interdiffusion coefficients (which
have turned out to be close to the gallium
self-diffusion coefficient) down to much
lower values than had been previously possible for gallium self-diffusion in bulk gallium arsenide with the use of radioactive
gallium tracer atoms. The dependence of
diffusion processes on the arsenic vapor
pressure, which is usually considered an
annoying feature of diffusion experiments
in GaAs, has helped in establishing the role
of self-interstitials and vacancies.

Presently, it appears that gallium vacancies and self-interstitials both have to be
taken into account in order to understand
self- and dopant diffusion processes in the
gallium sublattice of GaAs. Their relative
importance depends on the doping conditions ("Fermi level effect") and on the outside arsenic vapor pressure. Nonequilibrium concentrations of intrinsic point
defects may be induced by the in-diffusion
of dopants such as zinc starting from a
high surface concentration, in a similar
way to high-concentration phosphorus diffusion in silicon (Section 5.6.4.3). Since
much less is known on the diffusion mech-

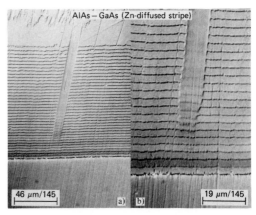

Figure 5-24. Shallow angle cross section of a portion
of an AlAs–GaAs superlattice wafer that, except for
a 10 μm stripe, has been masked with Si_3N_4 and has
been Zn diffused for 10 min at 575 °C. The shallow-
angle magnification is 145 in the vertical direction (no
horizontal magnification) and is skewed somewhat
relative to the orientation of the Zn-diffused stripe. In
the region of the Zn diffusion, the 40-period superlattice (45 Å/150 Å period) has become compositionally
disordered. Part (b) is a magnified view of part (a)
(Laidig et al., 1981).

anisms of atoms dissolved on the arsenic
sublattice, we will concentrate in the following on the diffusion of elements mainly
dissolved on the *gallium* sublattice. Compilation of diffusion data in GaAs in general
may be found elsewhere (Kendall, 1968;
Casey, 1973; Tuck, 1974, 1988; Jacob and
Müller, 1984).

5.8.2 Gallium Self-Diffusion and Superlattice Disordering

5.8.2.1 Intrinsic Gallium Arsenide

The self-diffusion coefficient of gallium
in intrinsic gallium arsenide $D_{Ga}(n_i)$ has
been measured by Goldstein (1961) and
Palfrey et al. (1981) with radioactive gallium tracer atoms (Fig. 5-25). This method
allows measurements of $D_{Ga}(n_i)$ down to
about $10^{-19} m^2/s$. Measurements of the
interdiffusion of Ga and Al in GaAs/
$Al_xGa_{1-x}As$ superlattices extended the

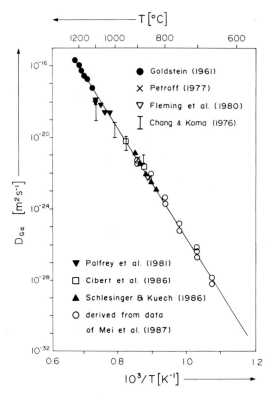

Figure 5-25. Plot of available data on gallium self-diffusion in GaAs and data on Ga/Al interdiffusion in GaAs/AlGaAs superlattices under intrinsic conditions together with D_{Ga}, derived from the data of Mei et al. (1987).

range to much lower values (Chang and Koma, 1976; Petroff, 1977; Fleming, et al., 1980; Cibert et al., 1986; Schlesinger and Kuech, 1986). The various data points have been approximately fitted to the expression

$$D_{Ga}(n_i) \approx 2.9 \times 10^4 \exp(-6\,eV/k_B T)\,m^2\,s^{-1} \tag{5-75}$$

by Tan and Gösele (1988). Eq. (5-75) does not fit exactly since the dependence of the experimental results on the different arsenic pressures have not been taken into account. The Ga/Al interdiffusion coefficient increases both for *high* and for *very low* arsenic vapor pressures (Furuya et al., 1987; Guido et al., 1987). This result indicates that

$D_{Ga}(n_i)$ is governed by gallium *vacancies* for sufficiently high arsenic vapor pressures and gallium *self-interstitials* for sufficiently low arsenic vapor pressures (Deppe and Holonyak, 1988). The role of gallium vacancies and self-interstitials becomes clearer when Ga diffusion in doped GaAs/$Al_xGa_{1-x}As$ superlattices is considered.

5.8.2.2 Doped Gallium Arsenide

No studies of gallium self-diffusion in doped bulk GaAs have been reported, but a wealth of data on Ga/Al interdiffusion in both n-type and p-type doped GaAs/AlAs superlattices is available. These interdiffusion experiments had been triggered by the observation of zinc diffusion enhanced superlattice disordering by Laidig et al. (1981). A number of disordering mechanisms have been proposed (Van Vechten, 1982, 1984; Laidig et al., 1981, Tatti et al., 1989) all of which are not general enough to account for the occurrence of an enhanced Ga/Al interdiffusion rate for other dopants. The observed dopant enhanced interdiffusion appears to be due to two main effects (Tan and Gösele, 1988):

i) The thermal equilibrium concentration of appropriately charged point defects is enhanced by doping (Fermi level effect, Section 5.2). In this case, which appears to hold for the n-type dopant silicon, only the *presence* of the dopant and *not its movement* is of importance. Compensation doping, e.g., with Si and Be, should not lead to enhanced Ga/Al interdiffusion which is in accordance with experimental results (Kawabe et al., 1985; Kobayashi et al., 1986).

ii) If one deals with a dopant with high diffusivity and solubility, so that the product $D^s C_D^s \gg D^{SD}(n)$ as expressed in Eq. (5-67) holds, then nonequilibrium intrinsic point defects are generated. Depending on whether a supersaturation or an undersatu-

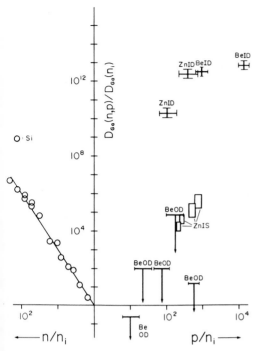

Figure 5-26. Plot of $D_{Ga}(n, p)/D_{Ga}(n_i)$ versus the ratio of extrinsic to intrinsic carrier concentration. The silicon doping data on the left are from Mei et al. (1987). On the right, the data for Zn in-diffusion (Zn, ID), for Be in-diffusion (Be, TD), Be out-diffusion (Be, OD) and for an implanted Zn source (Zn, IS) are taken from the literature. (Adapted from Tan and Gösele, 1988.)

ration of point defects develops, the disordering effect due to the Fermi level effect may be increased or decreased. Nonequilibrium point defects play an essential role in the diffusion of high-concentration zinc and beryllium in GaAs.

Let us first discuss Ga/Al interdiffusion in *n-type GaAs* and more specifically silicon-doped GaAs, where the type and the charge state of the intrinsic point defect dominating gallium diffusion in this lattice can be identified. On the left-hand side of Fig. 5-26 the enhanced Ga/Al interdiffusion coefficients as a function of the silicon n-doping level are plotted in a normalized form. These data, obtained by Mei et al. (1987),

show a clear doping dependence

$$D_{Ga}(n) \propto (n/n_i)^3 \tag{5-76}$$

which indicates the involvement of a triply negatively charged intrinsic point defect. Based on the pressure dependence of the interdiffusion coefficient of n-doped superlattices (Guido et al., 1987; Deppe and Holonyak, 1988) this defect has to be the *gallium vacancy* V_{Ga}^{3-}, as predicted by Baraff and Schlüter (1986). Values of $D_{Ga}(n_i)$ calculated from the data of Mei et al. and shown in Fig. 5-25 are in good agreement with values extrapolated from higher temperatures. Thus, including the arsenic vapor pressure dependence, we may write the gallium self-diffusion coefficient in n-type GaAs as

$$D_{Ga}(n) = D_{V_{Ga}^{3-}}(n_i, P_{As_q}^0)(n/n_i)^3(P_{As_q}/P_{As_q}^0)^{1/q} \tag{5-77}$$

where $P_{As_q}^0$ is a suitably chosen reference arsenic vapor pressure. For sufficiently high arsenic vapor pressure $q = 4$, and Eq. (5-77) can be extended to intrinsic GaAs for which

$$\tag{5-78}$$

$$D_{Ga}(n_i) = D_{V_{Ga}^{3-}}(n_i, P_{As_q}^0)(P_{As_q}/P_{As_q}^0)^{1/4}$$

should hold. Eq. (5-78) should be close to Eq. (5-75) which does not account for any specific arsenic pressure dependence.

Tellurium-doped GaAs based superlattices show a weaker dependence of the Ga/Al interdiffusion coefficient on the tellurium concentration than that expected from Eq. (5-76) (Mei et al., 1989). These results probably indicate that not all the tellurium is electrically active and contributes to the electron concentration.

The available Ga/Al interdiffusion data in *p-type GaAs* based superlattices are shown in a normalized form as a function of the Zn or Be induced p-doping concentration on the right-hand side of Fig. 5-25. The data obtained by various groups (Laidig et al., 1981; Lee and Laidig, 1984;

Kawabe et al., 1985; Myers et al., 1984; Hirayama et al., 1985; Ralston et al., 1986; Kamaba et al., (1987); Zucker et al., 1989) do not show a power law dependence as expected from the Fermi level effect. Instead, the data split into a group with an extremely high enhancement for experiments involving the presence of the outside doping source ("in-diffusion" conditions) and a group for which an enhancement could not be observed. These later experiments involved already grown-in dopants without an outside dopant source ("out-diffusion" conditions). Measurable, but considerably lower, enhancements than those observed under in-diffusion conditions have been described by Zucker et al. (1989) for implanted zinc ("implanted source").

The grossly different results for in- and out-diffusion conditions are due to nonequilibrium concentrations of intrinsic point defects induced by high-concentration diffusion of zinc or beryllium. Both zinc and beryllium diffuse via an interstitial-substitutional mechanism, as will be discussed in more detail in the next Section. Historically, most authors have considered the Frank-Turnbull or Longini mechanism (cf. Eq. (5-39)) for gallium vacancies as being applicable to the diffusion of p-type dopants (Kendall, 1968; Casey, 1973; Tuck, 1988). The superlattice disordering results indicate instead that the kick-out mechanism (cf. Eq. (5-38)) operates for these dopants and that Ga self-diffusion is governed by gallium self-interstitials. Within the framework of the kick-out mechanism the dopant in-diffusion generates a *supersaturation* of I_{Ga} (in analogy to the in-diffusion of gold or phosphorus in silicon) with a corresponding *increase* of dopant diffusion and the gallium self-diffusion component involving gallium self-interstitials. For grown-in dopants without

outside source, the kick-out mechanism involves the consumption of I_{Ga}, which leads to an *undersaturation* of I_{Ga} with a corresponding *decrease* in dopant diffusion (Kendall, 1968; Masu et al., 1980; Tuck and Houghton, 1981; Enquist et al., 1985; Enquist et al., 1988) and the gallium self-diffusion component involving gallium self-interstitials. The results of the superlattice disordering experiments are consistent with the expectations based on the kick-out mechanism. In contrast, the Frank-Turnbull mechanism predicts an undersaturation of vacancies for in-diffusion conditions and a supersaturation for out-diffusion conditions with a corresponding decrease and increase, respectively, of a vacancy-dominated gallium self-diffusion component. Since the predictions based on the Frank-Turnbull mechanism directly contradict the observed superlattice disordering results, it can be concluded that

i) zinc diffusion occurs via the kick-out mechanism, and

ii) gallium self-diffusion in p-type gallium arsenide is governed by gallium self-interstitials.

The pressure dependence of disordering of p-doped superlattices confirms the predominance of gallium self-interstitials in gallium self-diffusion (Deppe et al., 1988). The magnitude of the enhancement effect, its restriction to the dopant-diffused region and the implantation results of Zucker et al. (1989) indicate that a Fermi level effect has to be considered *in addition* to nonequilibrium point defects. The most likely k value of I_{Ga}^{k+} is 2 or 3 (Zucker et al., 1989; Winteler, 1971).

Combining the results for the p-type and the n-type dopant-induced disordering, including a self-interstitial supersaturation s_I defined according to Eq. (5-61), and a possible analogous vacancy supersaturation s_V, we may express the gallium self-diffusion coefficient approximately as

$$
\begin{aligned}
D_{\mathrm{Ga}}(n, p, P_{\mathrm{As}_q}) = &D_{\mathrm{I}^{k+}}(n_i, P^0_{\mathrm{As}_q})(p/n_i)^k \cdot \\
&\cdot (P_{\mathrm{As}_q}/P^0_{\mathrm{As}_q})^{-1/q}(1 + s_{\mathrm{I}}) + \\
&+ D_{\mathrm{V}^{3-}_{\mathrm{Ga}}}(n_i, P^0_{\mathrm{As}_q})(n/n_i)^3 \cdot \\
&\cdot (P_{\mathrm{As}_q}/P^0_{\mathrm{As}_q})^{1/q}(1 + s_{\mathrm{V}})
\end{aligned}
$$

Equation (5-79) describes all *presently known* effects on GaAs/AlAs superlattice disordering, but it cannot be excluded that Eq. (5-79) contains more than two terms (e.g., one for $k = 2$ one for $k = 3$, and possibly one for a neutral vacancy). For nonequilibrium gallium vacancies injected by a Si/As cap (Kavanagh et al., 1988), $s_{\mathrm{V}} > 0$ holds. In ion implantation, both $s_{\mathrm{I}} > 0$ and $s_{\mathrm{V}} > 0$ may hold, and both quantities will be time dependent. In diffusion-induced nonequilibrium point defects, the presence of dislocations will allow local equilibrium between intrinsic point defects to establish in the two sublattices according to Eq. (5-13). In this way a large supersaturation of I_{Ga} in the gallium sublattice may lead to an undersaturation of I_{As} or a supersaturation of V_{As} in the arsenic sublattice.

In contrast to the case of gallium diffusion very few data exist on arsenic diffusion in GaAs (Willoughby, 1983). The only experiment on the arsenic pressure dependence of arsenic diffusion indicates that arsenic vacancies are involved in arsenic diffusion in intrinsic GaAs (Palfrey et al., 1981). Disordering experiments in GaAs/GaP-type superlattices indicate that intrinsic point defects in the arsenic sublattice may also occur in various charge states. The question of whether point defect pairs (consisting, e.g., of one vacancy in each sublattice) may dominate self-diffusion under heavy n-doping conditions has not yet been sufficiently addressed.

5.8.3 Silicon Diffusion in Gallium Arsenide

Silicon is the main n-type dopant for GaAs based devices. Silicon is an amphoteric dopant which is mainly dissolved on the gallium sublattice but shows a high degree of self-compensation at high concentrations due to an increased solubility on the arsenic sublattice. The apparent concentration dependence of silicon diffusion has been modeled by a variety of mechanisms. Greiner and Gibbons (1985) propose that the silicon diffusion is predominantly carried out by $\mathrm{Si}_{\mathrm{As}} - \mathrm{Si}_{\mathrm{Ga}}$ pairs. Kavanagh et al. (1988) assume that the concentration dependence is due to a depth-dependent vacancy concentration generated by an Si/As type capping layer. More recently, it has been suggested that silicon diffusion is dominated by negatively charged gallium vacancies and that its apparent concentration dependence is actually a Fermi level effect (Tan and Gösele, 1988; Yu et al., 1989; Deppe et al., 1987, 1988). Results on silicon diffusion into n-type (Sn-doped) gallium arsenide confirm the Fermi level effect and contradict the Greiner-Gibbons pair-diffusion model. Deppe et al. (1988) suggest a charge state of -1 for the gallium vacancy. Yu et al. (1989) have used triply negatively charged gallium vacancies, in accordance with the species dominating superlattice disordering (Section 5.8.2), to fit the silicon diffusion profiles.

Let us finally comment on the generation of nonequilibrium point defects in silicon-doped GaAs. Even for high silicon concentrations, the condition in Eq. (5-67) for the generation of nonequilibrium point defects by in-diffusion processes is not fulfilled. Silicon diffusion in GaAs should therefore be essentially independent of whether it is an in- or an out-diffusion situation. This assertion is in agreement with experimental observations. In contrast, Enquist (1988) has reported that during the *growth* of highly Si-doped layers, buried layers of zinc diffuse excessively, which indicates a growth-induced supersaturation of gallium self-interstitials. This effect is technologically highly undesirable since it leads to the broadening of the zinc- or beryllium-doped base in hetero-bipolar transistors and consequently to a decrease in the maximally attainable speed. An explanation of this effect has been given by Deppe (1990) in terms of the well-known near-midgap Fermi level pinning at the newly grown surface region. As growth proceeds, the region changes from nearly intrinsic to n-type, and this leads to a higher concentration of charged gallium vacancies and self-interstitials in the region. The higher charged vacancy concentration is a thermal equilibrium requirement, but the concentration of Ga self-interstitials is a supersaturation which cannot be sustained. Thus, during growth, these self-interstitials flow to the surface and also into the bulk, which causes enhanced Zn or Be diffusion to occur. The phenomena of the enhanced base dopant diffusion induced by the growth of the n-type emitter layers resembles the emitter push effect in silicon (Section 5.6.4.3), but the physical origin is very different.

5.8.4 Interstitial-Substitutional Diffusion: Zn, Be, and Cr in GaAs

The main p-type dopants in GaAs based devices, zinc and beryllium, diffuse via an interstitial-substitutional mechanism in gallium arsenide as well as in many other III–V compounds. Since only the kick-out mechanism is consistent with the results of superlattice disordering (Section 5.8.2), we will base our discussion on this mechanism. In most papers, though, zinc, and

beryllium diffusion have been discussed in terms of the much earlier suggested Frank-Turnbull or Longini mechanism (Casey, 1973; Tuck, 1988).

In terms of the kick-out mechanism of p-type dopants, the condition for generating a supersaturation of self-interstitials during dopant in-diffusion is

$$D_i C_i^{eq} \gg D_{Ga}^{SD}(p) \tag{5-80}$$

as discussed in Section 5.5.3. The solubility C_i^{eq} of the zinc (or beryllium) interstitials will depend on the source conditions, such as the zinc vapor pressure if zinc is diffused from the vapor phase. For sufficiently low Zn or Be surface concentrations, the condition of Eq. (5-80) is not fulfilled, and intrinsic point defects can thus be considered to be in equilibrium. Under these conditions the effective diffusivity of substitutional p-type dopants is given by Eq. (5-41), which reads for $m = 1$ (single acceptors)

$$D_{eff}^{(i)} \propto p^{j+1} \propto C_s^{j+1} \tag{5-81}$$

with $j = 1$ for zinc (Tuck, 1988) and $j = 0$ for Be (Deal and Robinson, 1989). This concentration dependence which has been confirmed in isoconcentration experiments for zinc (Tuck, 1974) leads to the typical steep diffusion profiles shown for zinc in Fig. 5-27 (Cunnell and Gooch, 1960). For higher surface concentrations, when the condition of Eq. (5-80) becomes increasingly valid, a supersaturation of I_{Ga} develops. The out-diffusion of I_{Ga} to the surface will then determine the shape of the zinc concentration profile. Typical examples are shown in Fig. 5-28 from the pioneering work of Winteler (1971). The effective substitutional zinc diffusivity near the surface may then be obtained from Eq. (5-42) for $m = 1$:

$$\tag{5-82}$$
$$D_{eff}^{(I)} = 2[D_I C_I^{eq}(C_s^{eq})/C_s^{eq}](C_s/C_s^{eq})^{k-3}$$

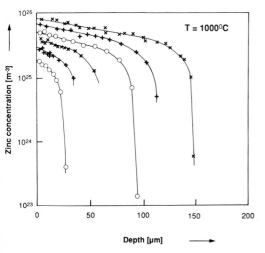

Figure 5-27. Zinc concentration profiles in GaAs for 10^4 min at $1000\,°C$ and different zinc vapor pressures (Cunnell and Gooch, 1960).

where $C_I^{eq}(C_s^{eq})$ is the thermal equilibrium concentration of $(k+)$ charged Ga self-interstitials at the surface. Based on an analysis of such profiles, Winteler (1971) arrived at the conclusion that positively charged gallium interstitials with $k = 2$ and 3 are involved in both zinc diffusion and gallium self-diffusion. For deeper parts of the profile, the conditions of Eq. (5-80) can no longer be applied. The front part of the profile becomes dominated again by $D_{eff}^{(i)}$ from Eq. (5-81), enhanced by a gallium self-

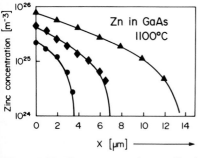

Figure 5-28. Zinc concentration profiles in GaAs for 15 min at $1100\,°C$ and different zinc vapor pressure (Winteler, 1971).

interstitial supersaturation that extends beyond the profile into the bulk,

$$D_{eff}^{front} = D_{eff}^{(i)}(C_{I_{Ga}}/C_{I_{Ga}}^{eq}) \qquad (5\text{-}83)$$

in analogy to Eq. (5-52). For even higher surface concentrations the profiles typically show a kink-and-tail structure (Fig. 5-29) and are associated with a high density of dislocations (Ball et al., 1981; Luysberg et al., 1989) and other defects generated by the zinc-diffusion-induced supersaturation of I_{Ga}. The most notable defects are voids (Winteler, 1971; Luysberg et al., 1989). The tail diffusivity again is given by Eq. (5-83). The intrinsic point defect agglomeration process in the two sublattices will briefly be discussed in Section 5.9.2.

For substitutional Zn or Be concentrations high enough to generate a *supersaturation* of I_{Ga} during *in-diffusion*, an *undersaturation* of I_{Ga} will develop for *grown-in* dopants or after taking away the outside dopant source, as indicated in Fig. 5-29. As a result, for these concentration conditions

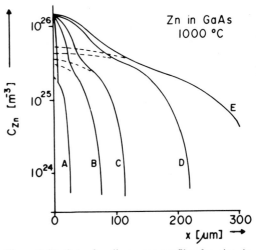

Figure 5-29. Set of radio-tracer profiles for zinc in GaAs at $1000\,°C$ for the following diffusion times A: 10 min; B: 30 min; C: 90 min; D: 9 h; E: 30 h (Tuck and Kadhim, 1972).

dopant "in-diffusion" will be much faster than dopant "out-diffusion", as has frequently been observed for zinc (Kendall, 1968; Tuck and Houghton, 1981, Enquist et al., 1988) and beryllium (Masu et al., 1980; Enquist et al., 1985). The supersaturation of gallium self-interstitials generated by in-diffusing zinc with an initially high surface concentration may enhance the diffusion of buried zinc or beryllium doped layers. This effect is analogous to the "emitter-push" effect in silicon. Although the effect is widely known in industrial laboratories concerned with contact diffusion of high-concentration zinc to p-doped layers in hetero-bipolar transistors or lasers, only one report is available in the open literature (Houston et al., 1988).

Chromium acts as a deep acceptor when substitutionally dissolved on gallium sites and is used for fabricating semi-insulating GaAs. As far as spatially uniform doping conditions are concerned, no charge effects have to be taken into account, and the interstitial-substitutional diffusion of chromium may be described in terms of the kick-out mechanism (cf. Eq. (5-30)) or the Frank-Turnbull mechanism (cf. Eq. (5-31)). In-diffusion profiles are fairly complex resembling a kick-out type profile near the surface and an erfc-type profile deeper in the material (Tuck, 1988; Deal and Stevenson, 1988). Out-diffusion profiles may be characterized by a constant diffusivity which is much lower than that of in-diffusion. Tuck (1988) and Deal and Stevenson (1988) have dealt with chromium diffusion in terms of the Frank-Turnbull mechanism. A satisfactory explanation of the diffusion behavior of chromium in intrinsic GaAs, however, probably has to include the co-existence of gallium vacancies and self-interstitials, the dependence of C_s^{eq} and C_i^{eq} on the outside chromium vapor pressure, and a dislocation-induced dynamic

equilibrium between the intrinsic point defects in the gallium and the arsenic sublattice. The chromium in-diffusion appears to be governed by the concentration-dependent $D_{eff}^{(I, V)}$ from Eq. (5-37) in the surface region, and by the much faster constant diffusivity D_{eff}^{tail} from Eq. (5-52) in the tail region. The observation that the increase of the chromium plateau in the dislocated bulk of the GaAs does *not* approach C_s^{eq} (which is in contrast to Au in Si) may be explained in terms of a build-up of non-equilibrium point defects in the *arsenic sublattice* by dislocation climb which leads to local point defect equilibrium according to Eq. (5-13) associated with a $C_s < C_s^{eq}$. In the case of out-diffusion the chromium vapor pressure is so low that a much lower diffusivity prevails, as in the out-diffusion of zinc. This slower out-diffusion appears to be dominated either by the constant vacancy component of $D_{eff}^{(I, V)}$ or the constant $D_{eff}^{(i)}$ which can be lower than $D_{eff}^{(I, V)}$ for low outside chromium vapor pressure.

5.8.5 Comparison to Diffusion in Other III–V Compounds

Gallium arsenide is certainly the one III–V compound in which self- and impurity diffusion processes have been studied most extensively. The available results on self-diffusion in III–V compounds have been summarized by Willoughby (1983). The Group III and the Group V diffusivities appear to be so close in some compounds that a common defect mechanism involving multiple intrinsic point defects appears likely, although no definite conclusion has been reached. There are hardly any experimental results available to draw conclusions on the type and charge state of the intrinsic point defects in self-diffusion processes. Zinc is an important p-type dopant for other III–V compounds as well, and its diffusion behavior also appears to be governed by an interstitial-substitutional mechanism. No information is available on whether the Frank-Turnbull mechanism or the kick-out mechanism operates in this case. It can be expected that dopant diffusion induced superlattice disordering may rapidly advance our understanding of diffusion mechanisms in other III–V compounds in a similar manner to what has been accomplished with GaAs. The state of understanding of diffusion mechanisms in II–VI compounds has recently been discussed by Shaw (1988).

5.9 Agglomeration, Precipitation, and Gettering

5.9.1 Agglomerates of Intrinsic Point Defects in Silicon

Nonequilibrium concentrations of intrinsic point defects occur in silicon during crystal growth, ion-implantation, and surface processes such as oxidation or nitridation. The nonequilibrium intrinsic point defects associated with crystal growth may agglomerate to generate various types of so-called swirl defects. "A-swirl" defects consist of interstitial-type dislocation loops resulting from a supersaturation of silicon self-interstitials. "B-swirl" defects are considered a precursor of A-swirl defects and probably consist of three-dimensional agglomerates of self-interstitials and carbon atoms (de Kock, 1981; Föll et al., 1981). Agglomerates of vacancies have been termed "D-swirl" defects (Abe and Harada, 1983). The agglomeration of these swirl defects depends on the growth speed and the temperature gradient along the crystal axis. A fully satisfactory quantitative description of swirl defect formation is not available although some attempts have

been made (Voronkov, 1982; Tan and Gösele, 1985).

In the following, we will deal with the much simpler case of the growth or shrinkage of dislocation loops containing a stacking fault on (111) planes. Such dislocation loops may be formed by the agglomeration of oxidation-induced self-interstitials and have been termed oxidation-induced stacking faults (OSF). These stacking faults may either nucleate at the surface (surface stacking faults) or in the bulk (bulk stacking faults). Approximating the shape of the stacking faults as semicircular at the surface with radius r_{SF} in the bulk, we may write their growth rate as

$$dr_{SF}/dt = \qquad (5\text{-}84)$$
$$= -\alpha_{eff}[D_I C_I^{eq} + D_V C_V^{eq})(\gamma_{SF}/k_B T) -$$
$$- D_I C_I^{eq} s_I + D_V C_V^{eq} s_V]\bar{A}/\Omega$$

In Eq. (5-84) α_{eff} is a dimensionless factor which can be approximated by about 0.5, γ_{SF} ($\approx 0.026\,\text{eV/atom}$) denotes the extrinsic stacking fault energy, $\bar{A}$ ($= 6.38 \times 10^{-20}\,\text{m}^2$) the stacking fault area per atom, and Ω ($= 2.0 \times 10^{-29}\,\text{m}^3$) the atomic volume. In the derivation of Eq. (5-84) it has been assumed that $\gamma_{SF}/k_B T \ll 1$, and that the line tension of the Frank partial dislocation surrounding the stacking fault may be neglected in comparison to the stacking fault energy. The first condition is always fulfilled (e.g., $\gamma_{SF}/k_B T \approx 0.2$ at 1300 K), the second for $r_{SF} \geq 1\,\mu\text{m}$. The quantities s_I and s_V denote self-interstitial and vacancy supersaturations, respectively, defined analogously to Eq. (5-61).

In an inert atmosphere, intrinsic point defect equilibrium is maintained ($s_I = 0$, $s_V = 0$), and Eq. (5-84) reduces to

$$(dr_{SF}/dt)_{in} = \qquad (5\text{-}85)$$
$$= -\alpha_{eff}(D_I C_I^{eq} + D_V C_V^{eq})(\gamma_{SF}/k_B T)\bar{A}/\Omega$$

which describes a linear shrinkage of stacking faults as has experimentally been observed (Fair, 1981a; Frank et al., 1984). From measured data of $(dr_{SF}/dt)_{in}$, the uncorrelated self-diffusion coefficient D^{SD} may be determined. The results are included in Fig. (5-8). Quantitative information on s_I has been extracted from the growth rate $(dr_{SF}/dt)_{ox}$ of OSF under oxidation conditions combined with the shrinkage rate $(dr_{SF}/dt)_{in}$ in an inert atmosphere at the same temperature:

$$s_I(t) \approx [1 - (dr_{SF}/dt)_{ox}/(dr_{SF}/dt)_{in}](\gamma_{SF}/k_B T) \tag{5-86}$$

Equation (5-86) yields for dry oxidation of a $\{100\}$ Si surface at temperatures in the vicinity of 1100 °C

$$s_I \approx 6.6 \times 10^{-9} t^{-1/4} \exp(2.52 \, eV/k_B T) \tag{5-87}$$

(Antoniadis, 1982; Tan and Gösele, 1982, 1985). The time t must be given in seconds. For $\{111\}$ surfaces the right-hand side of Eq. (5-87) must be multiplied by a factor of 0.6 to 0.7 (Leroy, 1986). For wet oxidation, multiplication factors larger than unity must be applied. The supersaturation ratios s_I, calculated based on Eq. (5-87), appear to overestimate s_I by 20–50%.

5.9.2 Void Formation During Zinc Diffusion in GaAs

In elemental crystals a supersaturation of intrinsic point defects may be eliminated by the nucleation and growth of dislocation loops. In this way C_X^{eq} may be established by dislocation climb processes. As discussed in Section 5.3, in compound semiconductors dislocation climb involves point defects in both sublattices. A supersaturation of I_{Ga}, as induced by high-concentration zinc diffusion into GaAs, can be reduced by dislocation-climb processes under the simultaneous generation of arsenic vacancies V_{As} (or the consumption of I_{As}).

Dislocation climb will stop when local point defect equilibrium according to Eq. (5-13) has been reached (Petroff and Kimerling, 1976; Marioton et al., 1989). Therefore, dislocation climb alone generally does not establish the thermal equilibrium concentration of intrinsic point defects. Thermal equilibrium concentrations in both sublattices may be reached if the arsenic vacancies generated in the arsenic sublattice via dislocation climb agglomerate and form voids. These voids will be half-filled with gallium atoms (in liquid form). The voids may act as sinks for more gallium interstitials. In the combination of the growth of interstitial-type dislocation loops and of voids partly filled with gallium atoms, supersaturations of I_{Ga} may be completely relieved and thermal equilibrium concentrations of intrinsic point defects in both sublattices may be established. Such a defect structure has in fact been observed in transmission electron microscopy studies of zinc-diffused GaAs (Luysberg et al., 1989).

5.9.3 Precipitation with Volume Changes in Silicon

Oxygen and carbon are the main electrically inactive impurities in silicon (Kolbesen and Mühlbauer, 1982; Kimerling and Patel, 1985). Oxygen is usually incorporated from the quartz crucible during the Czochralski growth process in the form of oxygen interstitials, O_i. The concentration of oxygen interstitials, C_i, at typical processing temperatures is higher than their solubility, C_i^{eq}, at these temperatures. Therefore, there exists a thermodynamic driving force for oxygen precipitation. A general description of precipitation, consisting of the nucleation and growth of precipitates may be found in many standard textbooks and also within this Series (Volume 5, Chapter 4) and will not be treated here. What is unusual about oxygen precipitation when compared to well-known precipitation phenomena in metals, is that a *volume change* associated with oxygen precipitate formation is combined with the presence of a *dislocation-free* matrix material. The volume increase of about a factor of 2 associated with oxygen precipitation leads to elastic deformation of the matrix. Further growth will be prevented by either the increase of elastic energy, unless the elastic strain is relieved by plastic deformation, or the emission or absorption of intrinsic point defects (Fig. 5-30). In the following, we will deal with the latter process. In a dislocation-free matrix this may cause a supersaturation or undersaturation of the appropriate intrinsic point defects which in turn may influence the nucleation and growth kinetics of precipitates. Let us first discuss the simple case where the intrinsic point defects relieve the elastic stress completely. In SiO_2 formation this requires

$$2\,O_i + (1+\beta)\,Si \rightleftarrows SiO_2 + \beta\,I \qquad (5\text{-}88)$$

where β is approximately 1. Assuming spherical SiO_2 nuclei and neglecting the influence of vacancies, we may write the critical radius r_{crit} above which precipitates grow as:

$$r_{crit} = \frac{\sigma\,\Omega}{2\,k_B T \ln[(C_i/C_i^{eq})(C_I^{eq}/C_I)^{1/2}]} \qquad (5\text{-}89)$$

(Gösele and Tan, 1982). In Eq. (5-89) σ is the SiO_2/Si interface energy, for which $0.09-0.5\,J/m^2$ have been reported, and Ω ($\approx 2 \times 10^{-29}\,m^{-3}$) is the volume of one silicon atom in the silicon lattice. For the derivation of Eq. (5-89) $\beta = 1$ has been used. During precipitation a supersaturation of self-interstitials will be produced ($C_I > C_I^{eq}$) which in return will increase the critical radius for further nucleation. This may also cause shrinkage of already-existing

precipitates if their radius is surpassed by the increasing r_{crit} (Ogino, 1982; Tan and Kung, 1986; Rogers et al., 1989).

After a sufficiently long time and for a sufficiently high supersaturation, the self-interstitials will nucleate interstitial-type dislocation loops (usually containing bulk stacking faults) which will reduce the self-interstitial concentration back to its thermal equilibrium value. For $C_I = C_I^{eq}$ Eq. (5-89) reduces to the classical expression for the critical radius. Vanhellemont and Claeys (1987) have given an expression for the critical radius in which not only self-interstitials but also vacancies and elastic stresses have been taken into account.

A more detailed look at oxygen precipitates shows that their shape, ranging from rod-like, to plate-like, to being almost spherical, depends on the detailed precipitation conditions. The different shapes can be explained by a balance between minimizing the elastic energy and the point defect supersaturation (Tiller et al., 1986). The growth of oxygen precipitates is limited by the diffusivity of D_i of oxygen interstitials given in Eq. (5-72). The growth kinetics of SiO_2 platelets has theoretically been dealt with by Hu (1986) and measured by Wada et al. (1983) and Livingston et al. (1984).

Carbon precipitation is associated with a *decrease* of about one silicon atomic volume for each carbon atom incorporated in an SiC precipitate. The same volume decrease holds for carbon agglomerates without compound formation. The volume requirements during carbon precipitation, which is just opposite to that during oxygen precipitation, may be fulfilled by the absorption of one self-interstitial for each carbon incorporated (Fig. 5-30).

If both carbon and self-interstitials are present in supersaturation, as is the case during crystal growth, co-precipitation is

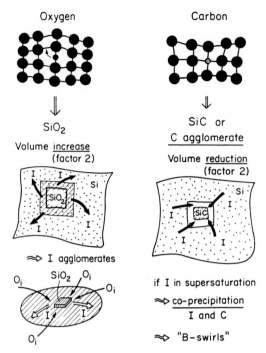

Figure 5-30. Schematic representation of volume changes during oxygen precipitation (left) and carbon precipitation (right) and the respective role of silicon self-interstitials, I (Gösele and Ast, 1983).

5.9.4 Gettering

During processing and handling, silicon wafer surfaces are frequently contaminated by fast-diffusing metallic impurities. The metallic impurities are then incorporated into the silicon bulk during high-temperature processing, where their solubility is high (Weber, 1983). Depending on the specific metallic atoms, they may stay in solution during cooling and drastically reduce the minority carrier lifetime, or they may precipitate out as silicides in the electrically active device regions close to the surface of the wafer. These precipitates may act as nuclei for oxidation-induced stacking faults (OSF) during later oxidation steps, and the partial dislocation surrounding the OSFs may in turn facilitate further silicide precipitation. In all these cases the device performance is severely degraded (Huff and Shimura, 1985; Shimura, 1989; Kolbesen and Strunk, 1985; Kolbesen et al., 1989).

There are a number of concepts of how device degradation and therefore low processing yield by metallic contamination may be reduced or eliminated. The ideal solution would be to avoid metallic contaminants from the very beginning by extremely careful and clean handling and processing of silicon wafers. Although this approach is technically feasible, the more common solution (for financial reasons) is to allow for a certain amount of contamination, and to apply procedures to *getter* away the undesirable metallic contaminants to regions where they cannot hinder the performance of the device. Although the idea of applying gettering in silicon dates back to Goetzberger and Shockley (1960), who showed that a coating of nickel around wafers increases the minority carrier lifetime, only much later was it used routinely in device processing.

likely to occur. B-swirl defects are thought to form in this way (Föll et al., 1981). A high supersaturation of self-interstitials can even trigger the co-precipitation of carbon atoms present in concentrations below their solid solubility limit.

If both carbon and oxygen are present simultaneously, it is obvious that co-precipitation of carbon and oxygen with a ratio of 1:2 will avoid stress and point defect generation and absorption. Co-precipitation of carbon and oxygen in this ratio has been observed by Zulehner (1983), Hahn et al. (1986), and Sun et al. (1990). Hahn et al. also showed that the silicon crystal remains essentially stress free in spite of a fairly large amount of co-precipitated carbon and oxygen.

One distinguishes frequently between *extrinsic* and *intrinsic* gettering techniques (Fair et al., 1985; Cerofolino and Maeda, 1988; Shimura, 1989). In extrinsic gettering, preferred nucleation sites for metal precipitates are usually offered at the back side of the wafers, by introducing a high density of dislocations. These dislocations may be generated by a number of means such as mechanical grinding, sand blasting, laser treatment or ion implantation. Another possibility involves offering highly n^+ or p^+ doped surface regions in which the solubility of negatively or positively charged point defects is enhanced. The best results so far have been obtained with high-concentration *phosphorus*, which is especially efficient in gettering elements, diffusing via an interstitial-substitutional mechanism such as gold (Lecrosnier et al., 1985) or platinum (Falster, 1985). The reason for this high gettering efficiency appears to be related to the high supersaturation of self-interstitials injected into the silicon by high-concentration phosphorus diffusion, which shifts the equilibrium between the substitutionally dissolved impurity atoms A_s to their fast-diffusing interstitial form via the reverse kick-out reaction

$$A_s + I \rightarrow A_i \qquad (5\text{-}90)$$

(Bronner and Plummer, 1987; Falster, 1985). In the case of predominantly interstitially dissolved impurities, for which Eq. (5-90) is not applicable, the effect of a supersaturation of self-interstitials appears to be related to the process of silicide formation (Ourmazd and Schröter, 1984). Oxidation processes are nowadays routinely performed in a chlorine containing atmosphere. This may also be considered an extrinsic gettering technique. Volatile metal chlorides as well as Si chlorides are formed at the surface, reducing both the amount of metal contaminants and OSF growth.

A definite disadvantage of extrinsic gettering techniques, in which the back side of the wafer is used, is that the metallic contaminants have to diffuse all the way through the wafer to be removed from the device regions. This is not the case for intrinsic gettering, in which oxygen precipitation is induced in the bulk of oxygen-containing Czochralski-grown silicon wafers. A *precipitate-free*, or *denuded*, zone is left in electrically active device regions close to the surface (Tan et al., 1977; Rozgonyi and Pearce, 1978; Fair et al., 1985). Oxygen precipitation in the denuded zone close to the surface is prevented by out-diffusion of oxygen interstitials to the surface during an appropriate high-temperature step. The efficiency of intrinsic gettering appears to be closely related to the high density of dislocations associated with oxygen precipitation. It has also been suggested that the self-interstitial generated during oxygen precipitation may play a major role (Bronner and Plummer, 1987; Ourmazd and Schröter, 1984). Intrinsic gettering is presently the standard gettering technique applied for integrated circuit processes. Intrinsic gettering is also one of the best examples of "defect engineering" in which an appropriate understanding of defect reactions facilitates a tailoring of these reactions for beneficial use in device processing (see also Chap. 11).

5.10 References

Abe, T., Harada, H. (1983), in: *Defects in Semiconductors II:* Mahajan, S., Corbetts, J. W. (Eds.). New York: North-Holland, pp. 1–17.

Antoniadis, D. A. (1982), *J. Electrochem. Soc. 129,* 1093–1097.

Antoniadis, D. A. (1983), in: *Process and Device Simulation for MOS-VLSI Circuits:* Antognetti, P., Antoniadis, D. A., Dutton, R. W., Oldham, W. G. (Eds.). Boston: Martinus Nijhoff, pp. 1–47.

Antoniadis, D. A. (1985), in: *VLSI Electronics, Vol. 12:* Einspruch, N. G., Huff, H. (Eds.). New York: Academic Press, pp. 271–300.

Antoniadis, D. A., Moskowitz, I. (1982), *J. Appl. Phys. 53,* 6788–6796.

Arthur, J. R. (1967), *J. Phys. Chem. Solids 28,* 2257–2267.

Ball, R. K., Hutchinson, P. W., Dobson, P. S. (1981), *Phil. Mag. 43,* 1299–1314.

Baraff, G. A., Schlüter, M. (1986), *Phys. Rev. Lett. 55,* 1327–1330.

Boit, C., Lau, F., Sittig, R. (1990), *Appl. Phys. A 50,* 197–205.

Borg, R. J., Dienes, G. J. (1988), *An Introduction to Solid State Diffusion.* San Diego: Academic Press.

Bourgoin, J. (1985), in: *Proc. 13th Int. Conf. Defects in Semiconductors:* Kimerling, L. C., Parsey, J. M. Jr. (Eds.). Warrendale: Metall. Soc. of AIME, pp. 167–171.

Bourret, A. (1985), in: *Proc. 13th Int. Conf. Defects in Semiconductors:* Kimerling, L. C., Parsey, J. M. Jr. (Eds.). Warrendale: Metall. Soc. of AIME, pp. 129–146.

Bronner, G. B., Plummer, J. D. (1985), *Appl. Phys. Lett. 46,* 510–512.

Bronner, G. B., Plummer, J. D. (1987), *J. Appl. Phys. 61,* 5286–5298.

Car, R., Kelly, P. J., Oshiyama, A., Pantelides, S. (1985), *Phys. Rev. Lett. 54,* 360–363.

Carslaw, H. S., Jaeger, J. C. (1959), *Conduction of Heat in Solids.* Oxford: Oxford Univ. Press.

Casey, H. C. (1973), in: *Atomic Diffusion in Semiconductors:* Shaw, D. (Ed.). New York: Plenum Press, pp. 351–429.

Casey, H. C., Pearson, G. L. (1975), in: *Point Defects in Solids, Vol. 2:* Crawford, J. H. Jr., Slifkin, L. M. (Eds.). New York: Plenum, pp. 163–253.

Celler, G. K., Trimble, L. E. (1988), *Appl. Phys. Lett. 53,* 2492–2494.

Cerofolino, G. F., Meda, L. (1988), *Int. Rev. Phys. Chem. 7,* 123–171.

Chang, L. L., Koma, A. (1976), *Appl. Phys. Lett. 29,* 138–141.

Cibert, J., Petroff, P. M., Werder, D. J., Pearton, S. J., Gossard, A. C., English, J. H. (1986), *Appl. Phys. Lett. 49,* 223–225.

Claeys, C. L., DeClerck, G. J., van Overstraeten, P. J. (1978), *Rév. Phys. Appliquée 13,* 797–801.

Cowern, N. E. B. (1988), *J. Apply. Phys. 64,* 4484–4490.

Crank, J. (1957), *The Mathematics of Diffusion.* London: Oxford Univ. Press.

Cunnel, F. A., Gooch, C. H. (1960), *Phys. Chem. Sol. 15,* 127–133.

Deal, M. D., Robinson, H. G. (1989), *Appl. Phys. Lett. 55,* 1990–1992.

Deal, M. D., Stevenson, D. A. (1988), *J. Appl. Phys. 59,* 2398–2407.

Deal, M. D., Hansen, S. E., Sigmon, T. W. (1989), *IEEE Trans. CAD 9,* 939–951.

de Kock, A. J. R. (1981), in: *Defects in Semiconductors:* Naryan, J., Tan, T. Y. (Eds.). New York: North Holland, pp. 309–316.

Demond, F. J., Kalbitzer, S., Mannsperger, H., Damjantschitsch, H. (1983), *Appl. Phys. Lett. 93 A,* 503–505.

Deppe, D. G. (1990), *Appl. Phys. Lett. 56,* 370–372.

Deppe, D. G., Holonyak, N. (1988), *J. Appl. Phys. 64,* R93–R113.

Deppe, D. G., Holonyak, N., Baker, J. E. (1987), *Appl. Phys. Lett. 52,* 129–131.

Enquist, P. (1988), *J. Cryst. Growth 93,* 637–645.

Enquist, P., Hutchby, J. H., Lyon, T. J. (1988), *J. Appl. Phys. 63,* 4485–4493.

Enquist, P., Wicks, G. W., Eastman, L. F., Hitzman, C. (1985), *J. Appl. Phys. 58,* 4130–4134.

Fahey, P., Griffin, P. B., Plummer, J. D. (1989a), *Rev. Mod. Phys. 61,* 289–384.

Fahey, P., Iyer, S. S., Scilla, G. J. (1989b), *Appl. Phys. Lett. 54,* 843–845.

Fair, R. B. (1981a), *J. Electrochem. Soc. 128,* 1360–1368.

Fair, R. B. (1981b), in: *Impurity Doping Processes in Silicon:* Wang, F. F. Y. (Ed.). New York: North-Holland, pp. 315–442.

Fair, R. B. (1989), *Advances in Chemistry Series, Vol. 221,* 265–323.

Fair, R. B., Tsai, J. C. C. (1977), *J. Electrochem. Soc. 124,* 1107–1118.

Fair, R. B., Pearce, C. W., Washburn, J. (Eds.) (1985), *Impurity Diffusion and Gettering in Silicon.* Pittsburgh: Mat. Res. Soc.

Fairfield, J. M., Masters, B. J. (1967), *J. Appl. Phys. 38,* 3148–3154.

Falster, R. J. (1985), *Appl. Phys. Lett. 46,* 737–739.

Fichtner, W. (1985), in: *Appl. Solid State Science, Suppl. 2, Part C:* Kahng, D. (Ed.). New York: Academic Press, pp. 119–336.

Fleming, R. M., McWhan, D. B., Gossard, A. C., Weigmann, W., Logan, R. A. (1980), *J. Appl. Phys. 51,* 357–363.

Föll, H., Gösele, U., Kolbesen, B. O. (1981), *J. Cryst. Growth 52,* 907–916.

Francis, R., Dobson, P. S. (1979), *J. Appl. Phys. 50,* 280–284.

Frank, F. C., Turnbull, D. (1956), *Phys. Rev. 104,* 617–618.

Frank, W., Gösele, U., Mehrer, H., Seeger, A. (1984), in: *Diffusion in Crystalline Solids:* Murch, G. E., Nowick, A. (Eds.). New York: Academic Press, pp. 31–142.

Furuya, A., Wada, O., Takamori, A., Hashimoto, H. (1987), *Jpn. J. Appl. Phys. 26,* L926–L928.

Ghandi, S. K. (1983), *VLSI Fabrication Principles.* New York: John Wiley & Sons, pp. 111–212.

Gilles, D., Schröter, W., Bergholz, W. (1990), *Phys. Rev. B. 43,* 6510–6519.

Goetzberger, A., Shockley, W. (1960), *J. Appl. Phys. 31,* 1821–1824.

Goldstein, B. (1961), *Phys. Rev. 121,* 1305–1311.

Gösele, U. (1988), *Ann. Rev. Mat. Science 18,* 257–282.

Gösele, U. (1989), in: *Microelectronic Materials and Processes:* Levy, R. A. (Ed.). Dordrecht: Kluwer Academic, pp. 588–634.

Gösele, U., Ast, D. G. (1983), *Energy Solar Technical Information Letter.* Office of Scientific and Techn. Inf.: U.S. Department of Energy, No. DOE/JPL/ 956046-83/9 (DE40009494).

Gösele, U., Morehead, F. (1981), *J. Appl. Phys. 54,* 4617–4619.

Gösele, U., Tan, T. Y. (1982), *Appl. Phys. A 28,* 79–92.

Gösele, U., Frank, W., Seeger, A. (1980), *Appl. Phys. 23,* 361–368.

Goshtagore, R. N. (1966), *Phys. Rev. Lett. 16,* 890–892.

Greiner, M. E., Gibbons, J. F. (1985), *J. Appl. Phys. 57,* 5181–5187.

Griffin, P. B., Plummer, J. D. (1986), *Tech. Digest Int. Electron. Device Meeting.* New York: IEEE, pp. 522–525.

Guido, L. J., Holonyak, Jr., N., Hsieh, K. C., Kalisiki, R. W., Plano, W. E., Burtham, P. D., Thornton, R. L., Epler, J. E., Paoli, T. L. (1987), *J. Appl. Phys. 61,* 1372–1379.

Hahn, S., Arst, M., Ritz, K. N., Shatas, S., Stein, H. J., Rek, Z. U., Tiller, W. A. (1986), *J. Appl. Phys. 64,* 849–855.

Hauber, J., Frank, W., Stolwijk, N. A. (1989), *Mat. Science Forum 38–41,* 707–712.

Hauber, J., Stolwijk, N. A., Tapfer, L., Mehrer, H., Frank, W. (1986), *J. Phys. C 19,* 5817–5836.

Hettich, G., Mehrer, H., Maier, K. (1979), *Inst. Phys. Conf. Sen. 46,* 500–507.

Hirayama, Y., Susuki, Y., Oleamoto, H. (1985), *Jpn. J. Appl. Phys. 24,* 1498–1502.

Hirvonen, J., Antilla, A. (1979), *Appl. Phys. Lett. 35,* 703–705.

Ho, Ch. P., Plummer, J. D., Hansen, S. E., Dutton, R. W. (1983), *IEEE Trans. Electron Devices ED-30,* 1438–1462.

Houston, P. A., Shephard, F. R., SpringThorpe, A. J., Mandeville, P., Margittai, A. (1988), *Appl. Phys. Lett. 52,* 1219–1221.

Hu, S. M. (1973), in: *Atomic Diffusion in Semiconductors:* Shaw, D. (Ed.). New York: Plenum, pp. 217–350.

Hu, S. M. (1974), *J. Appl. Phys. 45,* 1567–1573.

Hu, S. M. (1986), *Appl. Phys. Lett. 48,* 115–117.

Hu, S. M. (1987), *Appl. Phys. Lett. 51,* 308–310.

Huff, H. R., Shimura, F. (1985), *Solid State Technol. 3,* 103–118.

Huntley, F. A., Willoughby, A. F. W. (1973), *Phil. Mag. 28,* 1319–1340.

Itoh, T., Abe, T. (1988), *Appl. Phys. Lett. 53,* 39–41.

Jacob, H., Müller, G. (1984), in: *Landolt-Börnstein, Vol. III, 17 d:* Madelung, O., Schulz, M., Weiss, H. (Eds.). New York: Springer, pp. 12–34.

Kaiser, W., Frisch, H. L., Reiss, H. (1958), *Phys. Rev. 112,* 1546–1554.

Kalejs, I. P., Rajendran, S. (1990), *Appl. Phys. Lett. 55,* 2763–2765.

Kalinowski, L., Seguin, R. (1980), *Appl. Phys. Lett. 35,* 171–173.

Kamaba, N., Koboyashi, K., Endo, K., Sasudi, T., Misu, A. (1987), *Jpn. J. Appl. Phys. 26,* 1092–1096.

Kavanagh, K. L., Magee, C. W., Sheets, J., Mayer, J. W. (1988), *J. Appl. Phys. 64,* 1845–1854.

Kawabe, M., Shimizu, N., Hasegawa, F., Nannidi, Y. (1985), *Appl. Phys. Lett. 46,* 849–850.

Kendall, D. L. (1968), in: *Semiconductors and Semimetals, Vol. 4:* Willardson, R. K., Beer, A. C. (Eds.). New York: Academic Press, pp. 163–259.

Kimerling, L. C., Patel, J. R. (1985), in: *VLSI Electronics, Vol. 12:* Einspruch, N. G., Huff, H. (Eds.). New York: Academic Press, pp. 223–267.

Kitagawa, H., Hishimoto, K., Yoshida, M. (1982), *Jpn. J. Appl. Phys. 21,* 446–453.

Kobayashi, J., Nakajima, M., Fukunagon, T., Takamori, T., Ishida, K., Nakashima, H., Ishida, K. (1986), *Jpn. J. Appl. Phys. 25,* L736–L738.

Kolbesen, B. O., Mühlbauer, A. (1982), *Solid-State Electron. 25,* 759–775.

Kolbesen, B. O., Strunk, H. (1985), in: *VLSI Electronics, Vol. 12:* Einspruch, N. G., Huff, H. (Eds.). New York: Academic Press, pp. 144–222.

Kolbesen, B. O., Bergholz, W., Wendt, H. (1989), *Materials Science Forum 1–12,* 38–41.

Kröger, F. A. (1973, 1974), *The Chemistry of Imperfect Crystals.* Amsterdam: North-Holland.

Kump, M. R., Dutton, R. W. (1988), *IEEE Trans. Comp.- Aided Design 7,* 191–204.

Ladd, L. A., Kalejs, J. P. (1986), in: *Oxygen, Carbon, Hydrogen and Nitrogen in Crystalline Silicon:* Mikkelsen, Jr. J. C., Pearton, S. J., Corbett, J. W., Pennycook, S. J. (Eds.). Pittsburgh: Mat. Res. Soc., pp. 445–450.

Laidig, W. D., Holonyak, Jr., H., Camras, M. D., Hess, K., Coleman, J. J., Dapkus, P. D., Bardeen, J. (1981), *Appl. Phys. Lett. 38,* 776–778.

Langheinrich, W. (1984), in: *Landolt-Börnstein, Vol. III, 17 c:* Madelung, O., Schulz, M., Weiss, H. (Eds.). New York: Springer, pp. 118–149.

Lee, J. W., Laidig, W. D. (1984), *J. Electron. Mater. 13,* 147–165.

Lecrosnier, D., Paugam, J., Pelous, G., Richou, F., Salvi, M. (1985), *J. Appl. Phys. 52,* 5090–5097.

Leroy, B. (1986), in: *Instabilities in Silicon Devices:* Barbottin, G., Vapaille, A. (Eds.). Amsterdam: North-Holland, pp. 155–210.

Livingston, F. M., Messoloras, S., Newman, R. C., Pike, B. C., Stewart, R. J., Binns, M. J., Brown, W. P., Wilkes, J. G. (1984), *J. Phys. C. Solid State Phys. 17,* 6253–6276.

Longini, R. L. (1962), *Solid-State Electronics 5,* 127–130.

Luysberg, M., Jäger, W., Urban, K., Perret, M., Stolwijk, N., Mehrer, H. (1989), *Proc. 6th Oxford Conf. Microsc. Semicod. Mat.* Oxford: Inst. Phys. Conf. Ser. 100, pp. 309–415.

Mantovani, S., Nava, F., Nobili, C., Ottaviani, G. (1986), *Phys. Rev. B 33*, 5536–5544.

Marioton, B. P. R., Gösele, U. (1989), *Jpn. J. Appl. Phys. 28*, 1274–1275.

Marioton, B. P. R., Tan, T. Y., Gösele, U. (1989), *Appl. Phys. Lett. 54*, 849–851.

Masu, K., Konagai, M., Takahoshi, V. (1980), *Appl. Phys. Lett. 37*, 182–184.

Mathiot, D., Pfister, J. C. (1984), *J. Appl. Phys. 55*, 3518–3535.

Mayer, J. J., Mehrer, H., Maier, K. (1977), *Inst. Phys. Conf. Ser. 31*, 186–193.

Mei, P., Schwartz, S. A., Venkatesan, T., Schwartz, C. L., Colas, E. (1989), *J. Appl. Phys. 65*, 2165–2167.

Mei, P., Yoon, H. W., Venkatesan, T., Schwarz, S. A., Harbison, J. B. (1987), *Appl. Phys. Lett. 50*, 1823–1825.

Mikkelsen, Jr., J. C. (1986), in: *Oxygen, Carbon, Hydrogen and Nitrogen in Crystalline Silicon:* Mikkelsen, Jr., J. C., Pearton, S. J., Corbett, J. W., Pennycook, S. J. (Eds.). Pittsburgh: North-Holland, pp. 19–30.

Mizuo, S., Higuchi, H. (1983), *J. Electrochem. Soc. 130*, 1942–1947.

Morehead, F. (1988), in: *Defects in Electronic Materials:* Stavola, M., Pearton, S. J., Davies, G. (Eds.). Pittsburgh: Mat. Res. Soc., pp. 99–103.

Morehead, F. F., Lever, R. F. (1986), *Appl. Phys. Lett. 48*, 151–153.

Morehead, F., Stolwijk, N. A., Meyberg, W., Gösele, U. (1983), *Appl. Phys. Lett. 42*, 690–692.

Morooka, M., Yoshida, M. (1989), *Jpn. J. Appl. Phys. 28*, 457–463.

Myers, D. R., Biefeld, R. M., Fritz, I. J., Piccaux, S. T., Zipperian, T. E. (1984), *Appl. Phys. Lett. 44*, 1052–1054.

Newman, R. C. (1985), *J. Phys. C 18*, L967–L972.

Newman, R. C., Wakefield, J. (1961), *J. Phys. & Chem. Solids 19*, 230–234.

Nichols, C., Van de Walle, C. G., Pantelides, S. T. (1989), *Phys. Rev. Lett. 62*, 1049–1052.

Nishi, K., Antoniadis, D. A. (1986), *J. Appl. Phys. 59*, 1117–1124.

Ogino, M. (1982), *Appl. Phys. Lett. 41*, 847–849.

Orlowski, M. (1988), *Appl. Phys. Lett. 53*, 1323–1325.

Ourmazd, A., Schröter, W. (1984), *Appl. Phys. Lett. 45*, 781–783.

Palfrey, H. D., Brown, M., Willoughby, A. F. W. (1981), *J. Electrochem. Soc. 128*, 2224–2228.

Pandey, K. C. (1986), *Phys. Rev. Lett. 57*, 2287–2290.

Peart, R. F. (1966), *Phys. Stat. Sol. 15*, K119–K122.

Pearton, S. J., Corbett, J. W., Shi, T. S. (1987), *Appl. Phys. A 43*, 153–195.

Perret, M., Stolwijk, N. S., Cohausz, L. (1989), *J. Phys. Cond. Mater. 1*, 6347–6362.

Petroff, P. M. (1977), *J. Vac. Sci. Technol. 14*, 973–978.

Petroff, P. M., Kimerling, L. C. (1976), *Appl. Phys. Lett. 29*, 461–463.

Ralston, J., Wicks, G. W., Eastman, L. F., Deooman, B. C., Carter, C. B. (1986), *J. Appl. Phys. 59*, 120–123.

Rogers, N. B., Massond, H. Z., Fair, R. B., Gösele, U., Tan, T. Y., Rozgonyi, G. (1989), *J. Appl. Phys. 65*, 4215–4219.

Rollert, F., Stolwijk, N. A., Mehrer, H. (1989), *Materials Science Forum 38–41*, 753–758.

Rozgonyi, G. A., Pearce, C. W. (1978), *Appl. Phys. Lett. 32*, 747–749.

Schlesinger, T. E., Kuech, T. (1986), *Appl. Phys. Lett. 49*, 519–521.

Seeger, A., Chick, C. P. (1968), *Phys. Stat. Sol. 29*, 455–542.

Seeger, A., Föll, H., Frank, W. (1977), in: *Radiation Effects in Semiconductors 1976:* Urli, N. B., Corbett, J. W. (Eds.). Bristol: Inst. Physics, pp. 12–29.

Shaw, D. (1973) (Ed.), *Atomic Diffusion in Semiconductors.* New York: Plenum Press.

Shaw, D. (1975), *Phys. Stat. Sol. B 72*, 11–39.

Shaw, D. (1988), *J. Cryst. Growth 86*, 778–796.

Shimura, F. (1989), *Semiconductor Silicon Crystal Technology.* New York: Academic Press.

Shockley, W., Moll, J. L. (1960), *Phys. Rev. 119*, 1480–1482.

Stolwijk, N. A., Frank, W., Hölzl, J., Pearton, S. J., Haller, E. E. (1985), *J. Appl. Phys. 57*, 5211–5219.

Stolwijk, N. A., Perret, M., Mehrer, H. (1988), *Defect and Diffusion Forum 59*, 79–88.

Stolwijk, N. A., Schuster, B., Hölzl, J., Mehrer, H., Frank, W. (1983), *Physica 115 B*, 335–340.

Stolwijk, N. A., Schuster, B., Hölzl, J. (1984), *Appl. Phys. A 33*, 133–140.

Strunk, H., Gösele, U., Kolbesen, B. O. (1979), *Appl. Phys. Lett. 34*, 530–532.

Sun, O., Yao, K. H., Lagowski, J., Gatos, H. C. (1990), *J. Appl. Phys. 67*, 4313–4319.

Sze, S. M. (1985), *Semiconductor Devices, Physics and Technology.* New York: John Wiley & Sons, pp. 381–427.

Tan, T. Y., Ginsberg, B. J. (1983), *Appl. Phys. Lett. 42*, 448–450.

Tan, T. Y., Gösele, U. (1982), *Appl. Phys. Lett. 40*, 616–619.

Tan, T. Y., Gösele, U. (1985), *Appl. Phys. A 37*, 1–17.

Tan, T. Y., Gösele, U. (1988), *Mat. Science and Eng. B 1*, 47–65.

Tan, T. Y., Kung, C. Y. (1986), *J. Appl. Phys. 59*, 917–931.

Tan, T. Y., Gardner, E. E., Tice, W. K. (1977), *Appl. Phys. Lett. 30*, 175–176.

Taniguchi, K., Antoniadis, D. A., Matsushita, Y. (1983), *Appl. Phys. Lett. 42*, 961–963.

Tatti, S. R., Mitra, S., Stark, J. P. (1989), *J. Appl. Phys. 65*, 2547–2549.

Taylor, W., Marioton, B. P. R., Tan, T. Y., Gösele, U. (1989), *Rad. Eff. and Defects in Solids 111 u. 112*, 131–150.

Teh, T. H., Hu, S. M., Kastl, R. H. (1968), *J. Appl. Phys. 39*, 4266–4270.

Tiller, W. A., Hahn, S., Ponce, F. A. (1986), *J. Appl. Phys. 59*, 3255–3266.

Tipping, A. D., Newman, R. C. (1987), *Semicond. Sci. Technol. 2*, 315–317.

Tsai, J. C. C. (1983), in: *VLSI Technology:* Sze, S. M. (Ed.). New York: McGraw-Hill, pp. 169–218.

Tuck, B. (1974), *Introduction to Diffusion in Semiconductors*. Stevenage: Peter Peregrinus.

Tuck, B. (1988), *Atomic Diffusion in III–V Semiconductors*. Bristol: Adam Hilger.

Tuck, B., Houghton, A. J. N. (1981), *J. Phys. D 14*, 2147–2152.

Tuck, B., Kadhim, M. A. H. (1972), *J. Mater. Science 7*, 585–591.

Vanhellemont, J., Claeys, C. (1987), *J. Appl. Phys. 62*, 3960–3967.

Van Vechten, J. A. (1980), in: *Handbook of Semiconductors, Vol. 3:* Moss, T. S. (Ed.). Amsterdam: North-Holland, pp. 1–111.

Van Vechten, J. A. (1982), *J. Appl. Phys. 53*, 7082–7084.

Van Vechten, J. A. (1984), *J. Vac. Sci. Technol. B 2*, 569–572.

Van Vechten, J. A., Schmid, U., Myers, N. C. (1990), *Proc. Int. Conf. Science Technol. Defect Control in Semicond., Sept. 1989, Yokohama:* Sumino, K. (Ed.). Amsterdam: North-Holland, pp. 41–52.

Van Wieringen, A., Warmoltz, N. (1956), *Physica 22*, 849–865.

Voronkov, V. V. (1982), *J. Cryst. Growth 59*, 625–643.

Wada, K., Inone, N., Osaha, J. (1983), in: *Defects in Semiconductors II:* Mahajan, S., Corbett, J. W. (Eds.). New York: North-Holland, pp. 125–139.

Watkins, G. (1975), in: *Lattice Defects in Semiconductors 1974, Inst. of Phys. Conf. Sec. 23:* Huntley, I. A. (Ed.). London: Inst. Phys., pp. 1–22.

Watkins, G. C., Brower, K. L. (1976), *Phys. Rev. Lett. 36*, 1329–1332.

Weber, E. R. (1983), *Appl. Phys. A 30*, 1–22.

Weisberg, L. R., Blanc, J. (1963), *Phys. Rev. 131*, 1548–1552.

Willoughby, A. F. W. (1981), in: *Impurity Doping Process in Silicon:* Wang, F. F. Y. (Ed.). Amsterdam: North-Holland, pp. 1–53.

Willoughby, A. F. W. (1983), in: *Defects in Semiconductors:* Mahajan, S., Corbett, J. W. (Eds.). New York: North-Holland, pp. 237–252.

Wilcox, W. R., LaChapelle, T. J., Forbes, D. H. (1964), *J. Electrochem. Soc. 111*, 1377–1380.

Winteler, H. R. (1971), *Helv. Phys. Acta 44*, 451–486.

Yoshida, M. (1983), *Jpn. J. Appl. Phys. 22*, 1404–1413.

Yoshida, M., Arai, E., Nakamura, H., Terunuma, Y. (1974), *J. Appl. Phys. 45*, 1498–1506.

Yu, S., Gösele, U., Tan, T. Y. (1989), *J. Appl. Phys. 66*, 2952–2961.

Zucker, E. P., Hasimoto, A., Fukunaga, T., Watanabe, N. (1989), *Appl. Phys. Lett. 54*, 564–566.

Zulehner, H. W. (1983), in: *Aggregation Phenomena of Point Defects in Silicon:* Sirte, E., Goorissen, J., Wagner, P. (Eds.). Penningtan: Electrochem. Soc., pp. 89–110.

6 Dislocations

Helmut Alexander

II. Physikalisches Institut der Universität Köln, Köln, Federal Republic of Germany

Helmar Teichler

Institut für Metallphysik der Universität Göttingen,
Göttingen, Federal Republic of Germany

List of Symbols and Abbreviations

a	lattice parameter
$\boldsymbol{b}, b$	Burgers vector, Burgers modulus
D_k	kink diffusivity
D_n, D_{nn}	dipole-dipole interaction between nearest- and next-nearest-neighbor vacancies along the dislocation line
d, d'	dissociation widths on cross slip plane and primary glide plane
E_C, E_V	conduction and valence band edge
E_c	critical activation energy
E_{DK}^*	formation energy for a critical double kink
E_d	energy height of weak obstacles
E_{db}	dangling bond energy
E_F	Fermi energy
E^f	formation energy
E_0	dislocation level
e	electron charge
e^-	electron
e^*	effective charge
F_i	glide force
F_n	normal component of Peach-Köhler force
f	geometrical factor relating glide force difference and shear force
$J, \bar{J}$	nucleation rate and effective nucleation rate
k	Boltzmann constant
k_i	distribution coefficient
m	stress exponent of dislocation velocity
N_A, N_D	acceptor density, donor density
Q	activation energy of dislocation velocity
Q_0	activation energy of dislocation velocity at 1 MPa shear stress
Q'	activation enthalpy
q	charge
R_c	critical radius
R_i	friction stress
R^*	ratio of friction stress
S^*	critical length
T	temperature
T_m	absolute melting temperature
T_1	spin-lattice relaxation time
t_c	critical thickness
t_i, t_p	pulse length, pause length
t_i^*, t_p^*	critical pulse length, critical pause length
$\bar{t}_{DK}$	mean lifetime of a double kink
U	activation energy of the lower yield stress
U	intraatomic Coulomb integral
U_k	formation energy of a single kink

$v_k, \bar{v}_k$	kink velocity, mean kink velocity
W	activation enthalpy
W_c	activation energy
W_m	migration energy
γ	stacking fault energy
δ	lattice mismatch
$\varepsilon, \varepsilon_0$	relative and vacuum permittivity
$\dot{\varepsilon}$	strain rate
ε_{pl}	plastic strain
$\dot{\varepsilon}_{pl}$	plastic strain rate
μ	shear modulus
μ_i	mobility
ν_D	Debye frequency
ν_0	kink attempt frequency
σ	applied compression stress
σ_{el}	excess stress
τ	shear stress
τ_c	threshold stress
τ_{el}	yield stress of CRSS
ϕ_c	band bending energy in the dislocation core
$\phi(r)$	Coulomb potential of a screened dislocation

AD	antisite defect
CL	cathodo luminescence
CRSS	critical resolved shear stress
CS	constrictions
CZ	Czochralski
DK	double kink
DLTS	deep level transient spectroscopy
EBIC	electron-beam-induced current
EDSR	electric dipole spin resonance
EPR	electron paramagnetic resonance
FZ	floating zone grown
HFS	hyperfine structure
HREM	high-resolution electron microscope
L	luminescence
LCAO	linear combination of atomic orbitals
LEC	liquid encapsulation Czochralski
MWC	microwave conductivity
PD	point defect
PL	photoluminescence
PPE	photoplastic effect
RD	reconstruction defect
REDG	radiation-enhanced dislocation glide

SEM   scanning electron microscope
SF   stacking fault
SI   self-interstitial
TSCAP   thermally stimulated capacitance
V   vacancy

6.1 Introduction

At first glance the subject "dislocation in semiconductors" seems to be well defined. However, since dislocations in these materials are generated at elevated temperatures, their formation is always connected with a change in the number and distribution of point defects (intrinsic and extrinsic) in the crystal. Therefore, any change of a physical quantity measured at crystals with and without dislocations, respectively, has to be carefully analyzed as to which part of the change might be due to the dislocations themselves. The intricacy of this problem is illustrated by the current discussion of the extent to which clean dislocations are centers for carrier recombination. At present it is extremely difficult to arrive at a final answer to this question since methods for measuring the degree of decoration of dislocations with impurity atoms are lacking. In theory, the determination of the minimum energy core structure of straight partial dislocations has reached a state where fundamental changes are not to be expected. Regarding the electronic levels of straight dislocations, the theory indicates that no deep states exist (except perhaps for shallow levels due to extremely deformed bonds) for the so-called reconstructed variants of the dislocation cores, which seems to be the energetically stable configuration of the predominating partials in silicon. In the case of compounds, even more problems are unsolved than for the elemental semiconductors.

Nevertheless, during the last few years considerable progress has been made in dislocation research. Extended application of EPR spectroscopy to plastically deformed silicon has led to a clear distinction between point defects in the bulk and defects introduced into the core of dislocations when they move. Moreover, a thermal procedure could be worked out which strongly reduces the number of bulk defects after plastic deformation. As already mentioned, the core of straight partials in its ground state in silicon is reconstructed; after some motion it contains paramagnetic defects as (mainly) vacancies and possibly other singularities as kinks and jogs. Theoretical calculations indicate that deep states in the gap arise mainly from broken bond centers. Accordingly, the intensity of the DLTS signals of deformed silicon is also greatly reduced by moderate annealing.

Recently, shallow states in the gap connected to dislocations have come to the fore. Here, shallow is defined as being within 200 meV from the band edges. These states are responsible for near edge optical absorption, photoluminescence and transport phenomena such as microwave conductivity and combined resonance. Their nature is not yet clear and will be of central interest in the near future.

Dislocations are extended defects, i.e., their charge state may vary in wide limits. Consequently there is Coulomb interaction between charges confined to the dislocation which shifts the dislocation level with respect to the Fermi level (band bending around extended defects). The dependence of the EBIC contrast of dislocations on temperature and injection is analyzed along these lines. Careful analysis of the ionization of shallow donors (phosphorous) in plastically deformed n-type silicon by EPR clearly evidenced another source of band bending: The inhomogeneous distribution of deep acceptors (point defects) leads to inhomogeneous compensation of doping. Analysis of DLTS spectra under these conditions is not straightforward.

For semiconductor technology the most interesting properties of dislocations are their activity as a recombination center and as a sink for self-interstitials. Research is active with respect to both these effects.

In summary, it should be emphasized that in examining dislocations, the dislocation density does not satisfactorily characterize a crystal. Instead, its whole thermal history and the process which generated the dislocations have to be taken into account. Inhomogeneity after dislocation movement is unavoidable and may have a strong influence on physical properties. This message should be delivered before going into details.

In this chapter we comment on both mechanical and electronic properties of dislocations, which are often dealt with in separate reviews. We felt that there is encouraging convergence of theory and experiment but that the field is far from being well-rounded. Thus separate presentation of experimental and theoretical results seemed most helpful for the reader to form an unbiased opinion.

6.2 Geometry

The reader is assumed here to be familiar with the general conception of a dislocation (Friedel, 1964). As the most important semiconductors belong to the tetrahedrally coordinated crystals with diamond or sphalerite (cubic zinc blende) structure we have to deal primarily with dislocations in crystals with the face-centered-cubic (f.c.c.) Bravais lattice. As in f.c.c. metals, the Burgers vector of perfect dislocations is of the type $a/2 \langle 110 \rangle$. Also, the (close-packed) glide planes, determined by the Bravais lattice, correspond in diamond-like semiconductors as in f.c.c. metals to $\{111\}$ planes. However, the existence of two sublattices in the diamond as well as in the sphalerite structure brings about two

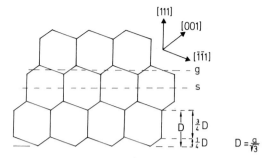

Figure 6-1. Projection of the diamond structure onto a (110) plane.

types of glide planes. Figure 6-1 shows a projection of the diamond structure onto a $\{110\}$ plane; obviously the $(1\bar{1}1)$ planes are arranged in pairs, the distance within one pair being three times shorter than between pairs. Thus at first glance the relative shift (shear) of one part of the crystal with respect to the other could proceed either in the wide space between two neighboring pairs of $(1\bar{1}1)$ planes or between the two planes of one pair. Since this shift is accomplished by the motion of dislocations those two possibilities result in dislocations with quite different core structures called "shuffle-set" and "glide-set" dislocations by Hirth and Lothe (1982).

In one of the first papers considering plastic deformation of germanium Seitz (1952) came to the conclusion that dislocations in diamond-like crystals should belong to the shuffle-set (Fig. 6-2). For a long period of time that opinion was generally accepted because it was in agreement with two principles: First, the stress needed for displacement of two neighboring lattice planes, one against the other, is generally smaller, the wider the distance is between these planes; second, in the case of covalent bonding, it is reasonable to assume that cutting one bond (per unit cell of the plane) would be easier than cutting three bonds.

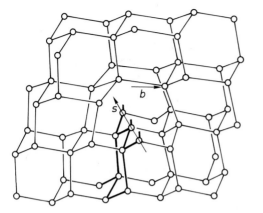

Figure 6-2. 60° shuffle-set dislocation.

also called dissociation, consists of a decomposition of the elementary step of shearing into two steps: an initial partial dislocation moves one part of the crystal against the other by a Burgers vector which is not a translation vector of the space lattice; this partial dislocation (abbreviated as "partial") leaves the stacking sequence disturbed; it is followed by a stacking fault ribbon. This region of wrong stacking sequence is closed by a second partial, whose Burgers vector completes the first Burgers vector to a space lattice vector.

In 1953 William Shockley gave a remarkable speech, documented as a short abstract (Shockley, 1953), in which he left open the question of whether dislocations in diamond-like crystals were of shuffle or glide type. He noticed that only in the second case is splitting of perfect dislocations into two (Shockley) partial dislocations possible, as in f.c.c. metals. This reaction,

$$a/2\,[011] \rightarrow a/6\,[121] + a/6\,[\bar{1}12]$$

Considering a model of the structure, it can easily be understood that only the two $(1\bar{1}1)$ planes constituting a narrow pair can be rebound by tetrahedral bonds after a relative displacement by $a/6\,[121]$. This means that only glide-set dislocations are able to dissociate (Fig. 6-3). Thus the clear

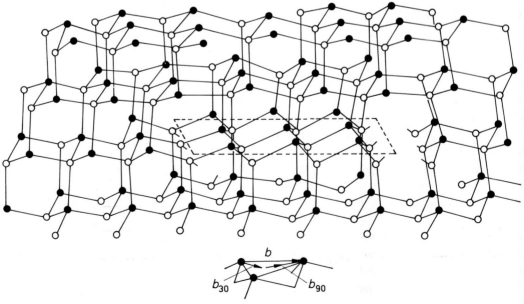

Figure 6-3. Dissociated 60° glide-set dislocation. A stacking fault ribbon is bound by the two partials: on the left side the 30° partial, on the right side the 90° partial. (Partials are not reconstructed).

proof of the then new weak-beam technique of transmission electron microscopy that glissile dislocations in silicon on most of their length are dissociated initiated a fundamental change of dislocation models and theory (Ray and Cockayne, 1970, 1971). Speculation about the association of a shuffle-set dislocation with a stacking fault ribbon in a neighboring narrow pair of planes (Haasen and Seeger, 1958) made clear that the only difference between a dissociated glide-set dislocation and a shuffle-set dislocation which is associated with a stacking fault bound by a dipole of partials is only one row of atoms (Fig. 6-4) (Alexander, 1974). EPR spectroscopy (Sec. 6.3.1) brings to light that when screw dislocations in silicon move, they introduce vacancies into the core of their partials. Thus an equilibrium between (dominant) glide segments and shuffle segments must be considered (Blanc, 1975). This equilibrium appears to be influenced by the tensor of stress acting on the dislocation as well as by the concentration of native point defects and possibly certain impurity species like carbon and oxygen (Kisielowski-Kemmerich, 1990).

One may ask why the former decision in favor of the shuffle-set dislocation was a wrong decision. This becomes clear from Fig. 6-3. Admittedly, the number of "dangling" bonds is larger in this structure than in Fig. 6-2 by a factor of three. But the arrangement of the orbitals containing the unpaired electrons is more suitable for pairwise rebonding ("reconstruction") in the partials of glide-set dislocations because the orbitals show much more overlap than in the core of shuffle-set dislocations (Fig. 6-2). Reconstruction of the core of the basic types of partials has been studied theoretically, as described in Sec. 6.4.3. The idea that this reconstruction is realized at least in silicon is in satisfactory agree-

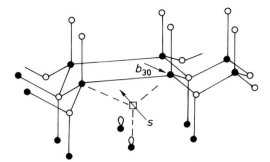

Figure 6-4. Replacing the row of atoms in the core of a 30° partial by vacancies generates the core of a shuffle-set 60° dislocation associated to a stacking fault between a neighboring close pair of (111) planes (from Alexander, 1974).

ment with the results of EPR spectroscopy which always came out with a much smaller number of unpaired electron spins than geometrically possible dangling bonds (Sec. 6.3.1). In summary, one may state that dissociated glide-set dislocations are favored over shuffle-set dislocations because in glide-set dislocations, both elastic energy and energy of unsaturated bonds are saved. For germanium, however, the situation is not as clear.

Proceeding from elements to compounds with sphalerite structure ($A^{III}B^{V}$ compounds and cubic $A^{II}B^{VI}$ compounds) an interesting complication arises: the two f.c.c. sublattices now are occupied by different atomic species. Nevertheless, in those compounds the overwhelming majority of mobile dislocations are also found dissociated into Shockley partials. Figure 6-3 can be used again, but the "black" and "white" atoms are now chemically different, and the bonding has an ionic contribution depending on the constituents. Obviously, all atoms which are seats of dangling bonds before reconstruction along a certain partial are of the same species. This makes reconstruction by pairwise rebonding more difficult; in fact, it is doubtful whether dislocations in com-

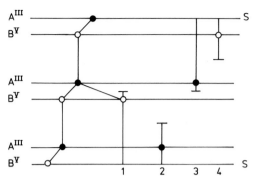

Figure 6-5. α and β (edge) dislocations in a cubic $A^{III}B^{V}$ crystal. S: surfaces of the crystal. 1,2: α dislocations. 3,4: β dislocations. 1,3: glide-set dislocations. 2,4: shuffle-set dislocations.

pounds are reconstructed – although without reconstruction, the preference for glide-set dislocations can no longer be easily understood.

A further consequence of the uniformity of atoms occupying the core sites of a partial is the doubling of the number of dislocation types when compared with elemental semiconductors: a 30° partial is conceivable with A atoms in its very core or with B atoms; the same holds for any dislocation. The naming of those chemical types is not uniform in the literature; there has been an attempt to remain independent from the decision whether the dislocations belong to the glide-set or to the shuffle-set. We take the view that dissocation into partials proves a dislocation to be of the glide-set type. Thus we call a dislocation with an extra half plane ending with A atoms (cations) β dislocation and its negative counterpart with anions in the center of the core α dislocation (Fig. 6-5). A more extended name would be A(g) and B(g), respectively. A B(s) dislocation would be of the same sign as A(g), but ending between widely spaced {111} planes. α-60° dislocations dissociate into an α-30° and an α-90° partial (Fig. 6-3), screw dislocations always consist of an α-30° and a β-30° partial.

The cubic sphalerite structure is assumed (at room temperature) by several $A^{II}B^{VI}$ compounds as well: cubic ZnS, ZnSe, ZnTe, CdTe. But a second group (ZnO, hexagonal ZnS, CdS, CdSe) belongs to the (hexagonal) wurtzite structure. This structure also is composed of tetrahedral groups of atoms, but the stacking sequence of those tetrahedra is ABAB ... instead of ABCABC ... Here the basal plane (0001) is the only close-packed glide plane. It is equivalent to the four (111) planes in the cubic structures. The Burgers vectors of perfect dislocations are of the type $b = (a/3) \langle \bar{2}110 \rangle$. Dislocations with these Burgers vectors may also glide on prismatic planes $\{10\bar{1}0\}$ and this secondary glide is indeed observed (Ossipyan et al., 1986). Since the $\{10\bar{1}0\}$ planes are chemically "mixed", there is no distinction between α and β dislocations on these planes. But there are two different distances between two $(10\bar{1}0)$ planes in analogy to shuffle and glide planes in the case of (111). Information describing which one is activated in wurtzite-type crystals is lacking (Ossipyan et al., 1986).

Stacking faults (SF) generated by dissociation of glissile dislocations on close-packed planes are of intrinsic type in all semiconductors investigated so far. An SF locally converts a thin layer from sphalerite into wurtzite structure and vice versa (Fig. 6-6). This is observed for ZnS, where

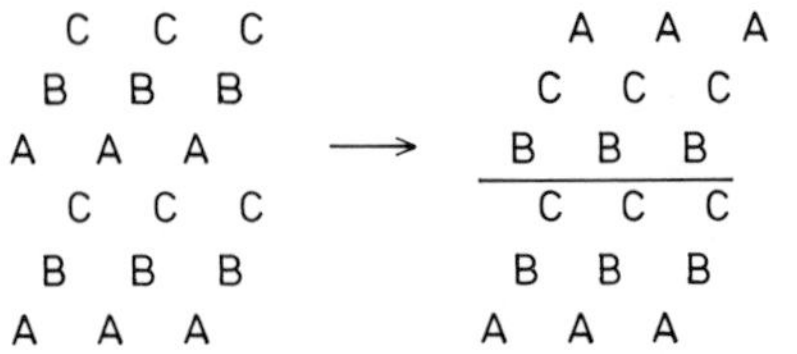

```
  C   C   C                    A   A   A
 B   B   B                      C   C   C
A   A   A         ⟶           B   B   B
 C   C   C                    ‾C‾ ‾C‾ ‾C‾
B   B   B                      B   B   B
A   A   A                     A   A   A
```

Figure 6-6. An intrinsic stacking fault generates a thin layer of hexagonal wurtzite (B C B C) in the cubic lattice.

one modification is transformed into the other by sweeping a partial over every second close-packed plane (Pirouz, 1989). Similarly, microtwinning of cubic crystals is equivalent to sweeping every plane by a Shockley partial (Pirouz, 1987). The stacking fault energy γ in $A^{III}B^{V}$ compounds decreases systematically with increasing ionicity of the compound (Gottschalk et al., 1978). This becomes clear as soon as γ is related to the area of a unit cell in the stacking fault plane and it can be understood from the local neighborhood of ions on both sides of the SF plane: here a 13th neighbor of opposite sign enters the shell of 12 next-nearest-neighbor ions, in this way reducing the Coulomb energy. Takeuchi et al. (1984) extended this consideration to $A^{II}B^{VI}$ compounds and showed good correlation of γ to the charge redistribution coefficient s, which accounts for the dependence of the effective ionic charge on the strain.

An interesting feature typical for dislocations in semiconductors are constric-tions (CS) where the dissociation into partials locally is withdrawn (Fig. 6-7). CS can be point-like on weak-beam micrographs (i.e., shorter than 1.5 nm) or segments on the dislocation line. It is well established that most CS are projections of jogs (Packeiser and Haasen, 1977; Tillmann, 1976). Considering density and distribution of CS in silicon after various deformation and annealing procedures we come to the conclusion that the majority of CS are products of climb events and not of dislocation cutting processes (Jebasinski, 1989). Point-like CS are jogs limiting a longer segment which redissociated after a climb on a new glide plane. Some of them may also be close pairs of jogs. Packeiser (1980) was able to measure the height of jogs in germanium and found them rather short (between 2 and 7 plane distances; elementary jogs were beyond the resolution of the technique). We recently measured the average distance L between two neighboring CS in p-type silicon and found that $L = 0.6\ \mu m$ for a strain of 1.6%, irrespec-

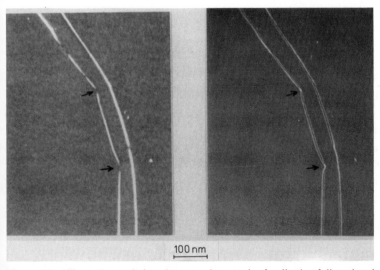

Figure 6-7. Silicon. Transmission electron micrograph of a dipole of dissociated edge dislocations with constrictions. Left image: stacking fault contrast ($g = (\overline{3}11)$). Right image: weak beam contrast ($022/0\overline{6}\overline{6}$) (from Jebasinski, 1989).

tive of whether the deformation was carried out at 650 °C or at 800 °C. Annealing after deformation leads to an increase of L for annealing below the deformation temperature, but to a decrease of L above T_{def}, exactly as was found previously for Ge (Haasen, 1979). But only above T_{def} ($=650$ °C) equilibrium was reached within the annealing time (16 h). It must be concluded that a fast process of annihilation of CS (climb of jogs along dislocations) is superimposed to a slow process of generation of new CS (net climb of the dislocation). It is noteworthy that Farber and Gottschalk (1991) in CZ-Si observed only very few CS.

There are extensive investigations of climbing of dissociated dislocations carried out with the help of high resolution electron microscopy (Thibault-Desseaux et al., 1989). The authors analyzed silicon bicrystals grown by the Czochralski technique and plastically deformed at 850 °C. Climb proceeds by nucleation of a perfect interstitial loop on the 90° partial (Fig. 6-8). These loops may or may not dissociate. It is interesting that the climb events are found at dislocations which are trapped by formation of a dipole or near the grain boundary. Slowly moving or resting dislocations are preferentially concerned when interaction with point defects is considered, irrespective of whether this involves impurity atoms (Sumino, 1989) or native defects. Thibault-Desseaux et al. (1989) estimate the concentration of interstitials necessary for the first step of climb as 10^{-4}–10^{-2} and believe that such a supersaturation (10^6) is the consequence of plastic deformation.

The above-mentioned mechanism of climb via constrictions is not excluded by Thibault-Desseaux et al. (1989), but it is not particularly suited for investigation by HREM. It would be worthwhile to investi-

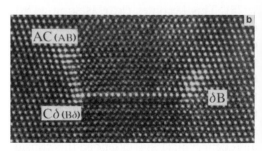

Figure 6-8. Silicon: Climb by formation of a complete dislocation (high-resolution TEM). (a) dissociated 60° dislocation (Aδ: 90° partial, δB: 30° partial); (b) The partial Aδ has decomposed into a complete dislocation AC (which has climbed by 7 atomic planes) and the partial Cδ (from Thibault-Desseaux et al., 1989).

gate whether climb in FZ-Si proceeds due to a supersaturation of vacancies, and in such a case CS would be prevalent.

6.3 Experimental Results on the Electronic Properties of Dislocations and Deformation-Induced Point Defects

Dislocations in semiconductors act as electrically activce defects: they can be "structural dopants" (acceptors and/or donors), recombination centers reducing the lifetime of minority carriers, or scattering centers. In the low-temperature region, dislocations are linear conductors. In additon to this direct influence on carrier density, lifetime, and mobility, there are indirect influences: electrically charged dislocations are surrounded by a screening space charge which causes local band bending and therefore may change the charge state of point defects in this region.

This multitude of electrical effects has attracted considerable research activity for a long time; however, on account of some special problems, understanding has developed rather slowly. The first problem, which had not been realized immediately, concerns the superposition of the electrical effects of point defects also produced by plastic deformation with that of dislocations (plastic deformation is the usual method to produce a sufficiently high density of well-defined dislocations). As revealed mainly by EPR spectroscopy, but also by Hall effect measurements, deformation produces a surprisingly high concentration of point defect clusters contributing to and sometimes dominating the electrical properties of deformed crystals.

The second problem arises from the character of dislocations as extended defects: While a point defect may change its charge state by one or two elementary charges, dislocations act as traps for majority carriers along their line; i.e., they may concentrate a considerable charge along their line. As any extra charge also this line charge q will be screened by a (cylinder of) space charge; in the low temperature regime this space charge consists of ionized dopant atoms, its radius R depending on the doping level: $R = (q/e\pi|N_D-N_A|)^{1/2}$ (Read, 1954). At a higher temperature or for small values of q free carriers are to be taken into account (Labusch and Schröter, 1980). Calculating the band bending, which represents the Coulomb interaction of the (screened) dislocation charge with free carriers, one finds remarkable shifts of the band edges for moderate line charge (Wilshaw and Fell, 1989). Thus, the position of the ener-

gy level of dopants and point defects relative to the Fermi level becomes locally inhomogeneous in the proximity of dislocations.

Finally, it is known that dislocations may change the spatial distribution of impurity atoms and that plastic deformation may introduce new impurities. Separation of the consequences of these processes from the effects of the ideal clean dislocation is extremely difficult.

It is not possible to give a comprehensive survey of all relevant papers in this field. The authors restrict themselves to a presentation of what they believe to be the most-important work, where the problems are understood in principle.

6.3.1 Electron Paramagnetic Resonance (EPR) Spectroscopy of Plastically Deformed Silicon

Where EPR spectroscopy is applicable, it yields more information on the defect under investigation than any other experimental method, because it reveals the symmetry of the defect. Via hyperfine structure (HFS) of the spectrum it also provides a hint as to the chemical species of atoms involved. Moreover, EPR spectroscopy can be calibrated to give numbers of defects.

Admittedly, EPR concerns only paramagnetic centers; this means that there may be electrically active defects not detectable by EPR. Other defects will be traced only in a certain charge state. The charge state may reveal something about the position of the defect in the energy gap by observing EPR spectra under illumination and of doped crystals.

Fortunately, silicon is one of the most suitable substances for EPR. Thus we will begin with a summary of what is known about the EPR of plastically deformed silicon crystals (Kisielowski-Kemmerich and Alexander, 1988).

So-called standard deformation by single slip ($T = 650\,^\circ$C, $\tau = 30$ MPa, $\Delta l / l \approx 5\%$) resulting in a dislocation density $N \approx 3 \times 10^9$ cm^{-2} produces about 10^{16} cm^{-3} paramagnetic centers. The majority (65–80%) are point defect clusters of high thermal stability. The remaining defects are related to the dislocation geometry by their anisotropy. A clear distinction between the two classes of defects can be made by several methods: (a) As mentioned before, the dislocation-related defects show anisotropy with respect to the (total) Burgers vector of the primary dislocations as a prominent axis. (b) Taking advantage of the fact that the spin-lattice relaxation time T_1 of these defects is up to four orders of magnitude shorter than that of the point defect clusters, Kisielowski-Kemmerich succeeded in separating the two parts of the spectrum completely (Fig. 6-9). On the other hand, using special passage conditions, one can detect both parts of the EPR spectrum at 15 K (they are normally seen at room temperature and at helium temperature, respectively). This disproves the assumption that the transition from the high-temperature spectrum to the low-temperature spectrum was due to a magnetic phase transition of the magnetic moments along the dislocation lines. (c) A two-step deformation interrupted by an annealing treatment suppresses to a large extent the production of the point defect spectrum during the second deformation step without changing the dislocation spectrum (Kisielowski-Kemmerich et al., 1986). (d) The attribution of the room temperature spectrum to point defect clusters is made more convincing by the production of a point defect cluster which is already well known from neutron irradiated

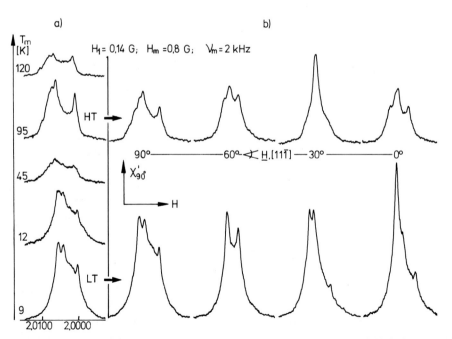

Figure 6-9. Silicon. EPR. (a) Dispersion spectrum at different temperatures. (b) Anisotropy of the spectrum of point defects (long spin-lattice relaxation time) and of the dislocation-related centers (short T_1) (from Kisielowski-Kemmerich et al., 1985).

Si. This, in fact, is possible by deformation at 390 °C (Brohl et al., 1987). (Radiation defects anneal out around 400 °C.) (e) Finally, the possibility to recharge the defects by doping is different for the two types of the paramagnetic defects.

EPR spectroscopy was applied to deformed silicon mainly by two groups for over more than two decades. Most of the experimental results were in perfect agreement. But in contrast to our interpretation given above and based on the anisotropy of the spectra and on variation of the deformation procedure, the Chernogolovka group insisted on ascribing the EPR spectra taken at room temperature and at low temperature to one and the same group of paramagnetic defects, located linearly along the dislocation cores (Ossypian, 1982). If all EPR centers are added, their number in fact corresponds to the number of dislocation sites. The change in the spectrum at around 60 K was ascribed to a magnetic phase transition which reduces the number of unpaired electrons because of the reconstruction of most of the dangling bonds. Admittedly, each EPR investigation on dislocations started with such a model in mind, but we feel the experimental results described above provide a convincing reason to abandon the model.

The most important conclusion to be drawn from the EPR spectrum is the following: most of the geometrically possible broken bonds in the core of partial dislocations are reconstructed or are at least lacking unpaired electrons. While most authors take that as proof for pairwise rebonding along the cores of partials, Pohoryles (1989) concluded from measuring photoconductivity of deformed germanium and silicon in helium gas under pressure that only helium atoms drive the reconstruction of the otherwise unreconstructed cores. He discusses negative U behavior as a reason for the EPR results.

According to our interpretation, about 3 % of the core sites of dislocations are occupied by unpaired electrons. It is worth noting here that the parameters describing the dislocation-related EPR spectrum are perfectly reproducible and do not depend on doping or on variation of the deformation conditions (the only exception being the number of centers). This is remarkable since plastic deformation causes significant strain in the lattice.

Evaluating now the anisotropy of the low-temperature (dislocation related) spectrum we can identify three different contributions: first, a wide line (10 G) similar to a certain extent to the EPR signal of amorphous silicon is apparent; the related center was called Si-Y by Suezawa et al. (1981). On top of this wide line, several narrow lines (1 G) stand out (Si-K 1), two for each activated slip system. Finally, a series of pairs of lines mark paramagnetic centers with spin $S \geq 1$ (Si-K 2). These lines also have a width of 10 G. It is possible to transform by light Si-K 1 centers into K 2 centers (Erdmann and Alexander, 1979).

Two pieces of information are most important for modeling the paramagnetic defects: The g tensor of Si-Y and of Si-K 2 is of orthorhombic I (C_{2v}) symmetry, with the axis where g is nearest to the free electron value g_e along [0$\bar{1}$1], therefore perpendicular to the (total) Burgers vector [011] of the primary dislocations. Hyperfine structure identifies Si-K 1 as a center of dangling bond type, the orbital being 22° from a $\langle 111 \rangle$ bond axis. (The two lines belonging to the primary dislocations are due to centers pointing "parallel" to [11$\bar{1}$] and [1$\bar{1}$1]. These two directions are perpendicular to b as well (Fig. 6-10); Weber and Alexander, 1979.) The line pairs Si-K 2 are most closely related to the Burgers vector: the axis of the fine structure tensor is exactly parallel to the total Burgers vector b (Bartelsen, 1977).

It is important to realize that the *total* Burgers vector of the dissociated dislocations does not influence the atomic neighborhood of any atom except constrictions (cf. Sec. 6.2) which, however, cannot be identified with the EPR centers because of their number. Rather, the atomic structure in the core of a partial dislocation is determined by the *partial* Burgers vector, being of the type $\langle 211 \rangle$. Thus we came to the conclusion that the distinction of the total Burgers vector in the spectra must mean that the related paramagnetic centers are located in the *screw* dislocations which run parallel to the Burgers vector (Weber and Alexander, 1979). Considering the core of $30°$ partials (Fig. 6-3) constituting screw dislocations, one notices that the broken bond of a reconstruction defect points along [011]. The dangling-bond-like orbitals Si-K 1 are nearly perpendicular to this direction. This observation suggested a vacancy in the core of such a $30°$ partial as the defect producing the Si-K 1 signal.

Quite recently, Kisielowski-Kemmerich (1989, 1990) started on this basis a group-theoretical analysis of the defect molecule consisting of a nearly planar group of 4 atoms corresponding to the arrangement of the innermost atoms in the core of a $30°$ partial (Fig. 6-10). In a first approximation, the defect has a threefold rotation axis imbedded into the crystal parallel to a twofold crystal axis. This situation is abnormal in solid state physics but occurs here because (on account of the stacking fault) one of the atoms (C in Fig. 6-10) is not in a regular lattice position. First, the reconstruction defect (7 valence electrons in the defect molecule) was considered; in agreement with the qualitative argument

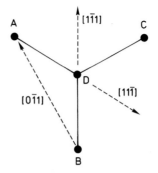

Figure 6-10. Model of the core of a $30°$ partial tackled by Kisielowski-Kemmerich (1990). Atom C belongs to the stacking fault. (All angles $120°$.)

above, it turned out that the g tensor (reflecting the local crystal field) must have its g_e axis perpendicular to the plane of the molecule.

The author then proceeded to a structure made up of three atoms (A, B, C) at the corners of an equilateral triangle (i.e., a vacancy in the core of the $30°$ partial). sp^2 hybridization is likely because the molecule is planar. From simple group-theoretical considerations it follows that 2 electrons will populate a low-lying a_1 level and the third electron will be in an electronically twofold degenerate e level. This situation is unstable against Jahn-Teller distortion. A reasonable assumption for the distortion is that one of the three sides of the triangle will become shorter or longer than the other two. As long as the three atoms A, B, and C are taken as equivalent, there are three possible orientations of a given side s*, i.e., three degenerate orientations of the defect or three EPR lines of equal type (g tensor, etc.). This contradicts the experimental situation (1 line Si-Y, 2 lines Si-K 1). In fact, the three atoms are not equivalent, since atom C is in a stacking fault position. If s* is opposite to atom C, one line arises; the center is of dangling-bond type parallel to [100] if s* is shorter than the other two sides. If s* is longer, the unpaired electron will be in an antibonding orbital parallel to the [0$\bar{1}$1] direction. Obviously, the latter case is in agreement with the character of Si-Y. At first glance, it is tempting to identify Si-K 1 with the other case where s* is adjacent to atom C, because here two equivalent orientations (lines) are to be expected. However, Si-K 1 is produced by an unpaired electron with 88% p character, as shown by hyperfine structure, and not 66% as expected for sp^2 orbitals. Consequently, Kisielowski-Kemmerich models Si-K 1 by a pyramidal defect: a vacancy next to a light impurity atom (carbon?) along the dislocation line ("fourfold coordinated vacancy"). The different widths of Si-Y and Si-K 1 correspond to other cases of three- and fourfold coordinated vacancies.

From the viewpoint of solid-state physics, the most-interesting EPR center in deformed Si is Si-K 2 because coupled spins are involved. We can either treat the spectrum as consisting of $S = 1$ line pairs ascribing the difference between the six pairs to different surroundings for the different pairs, or we can assume the coexistence of groups of coupled spins from $n = 2$ up to $n = 6$ (Bartelsen, 1977) (only one group of Si-K 2 is considered here with its maximum at 20 K). The latter model is accepted by Kisielowski-Kemmerich (1989), and it is interpreted as n threefold coordinated vacancies (Y centers) aligned along a dislocation core. The symmetry remains the same as for one Si-Y center (orthorhombic I = C$_{2v}$). The interaction between the electrons in their orbitals parallel to [0$\bar{1}$1] is of pure π character and is weak.

The total spin is $S = n \cdot \frac{1}{2}$. Compared with vacancy chains in irradiated silicon investigated by Lee and Corbett (1976), one should bear in mind that in that case the vacancies are arranged along zig-zag chains with *average* [011] direction, but in our case two vacancies are occupying next-nearest sites along a [011] line. Taking this into account, the interacton parameter D_n and D_{nn} as determined by Bartelsen (1977) show the right ratio[†]. Their absolute magnitude is smaller by about a factor of two, which may be due to a different form of the wave function of the unpaired electrons.

To summarize this section, we conclude that all EPR active centers which can be ascribed to dislocations belong to vacancies introduced into the core of the 30° partials forming screw dislocations. This introduction of vacancies corresponds to local transition from the glide-set structure into the shuffle-set structure (cf. Sec. 6.2). Having this model at hand, we can discuss the number of those defects.

First, we must stress that the paramagnetic centers are not intrinsic ingredients of (screw) dislocations; rather, they are produced during dislocation motion. Comparing different deformations under equal stress shows that the density of magnetic defects is proportional to the plastic strain, i.e., the area of glide planes that the dislocations have swept. Alternatively, comparing deformations under different stress shows that the production rate increases with stress. The centers can be annealed – with the exception of a small part of Si-Y – at or above 750 °C. A second deformation after this annealing starts with the high dislocation density reached at the end of the first deformation, but the density of magnetic defects now starts at zero and increases with ε_{p1} as during the first deformation (Kisielowski-Kemmerich et al., 1985).

[†] D_n and D_{nn} measure the (dipole-dipole) interaction between nearest- and next-nearest-neighbor vacancies along the dislocation line.

One of the great puzzles concerns the question of why only 30° partials give rise to the dislocation-related EPR spectra which are part of screw dislocations. Of course, there are 30° partials contained in 60° dislocation as well (depending on deformation mode, the ratio of length between the 30° partials in 60° dislocations and those in screws varies between 1 and 10).

Various assumptions can be made: (1) the stress direction during deformation may lead to preference of vacancies in screw dislocations, but the proximity of a 90° partial in a 60° dislocation may also prevent the existence of vacancies in the 30° partial nearby. (2) Since paramagnetism is restricted to a certain charge state, the position of the electronic states of screws and 60° dislocations relative to the Fermi level may be involved.

The close connection between generation of core vacancies and dislocation motion suggests kink generation or motion to be responsible for the vacancy generation. If we relate the number of EPR centers not to the entirety of dislocations, but only to the screws, the density of those centers becomes rather high: about 40% of the core sites are occupied by vacancies (Y centers) and 5% by K 1 centers after standard deformation. High-stress, low-temperature deformation ($\tau \geq 100$ MPa, 420°C) produces much fewer dislocation-related centers (1.5% of the screw sites), and at the same time, the mobility of screws compared to 60° dislocations is strongly reduced, as seen from the high proportion of screw dislocations (about 50%). This could indicate that the mobility of screws depends on the density of vacancies in their core. Louchet and Thibault-Desseaux (1987) actually discussed the possibility that kinks associated with a vacancy could be of higher mobility than

glide-set kinks. However, again it is not clear why this should be the case only for screws (90° partials turn out to be more mobile than 30° partials in Si and in Ge, so that the 30° partial should also determine the mobility of 60° dislocations).

The shear stress moving the dislocation drives the thermally activated introduction of vacancies into the core*. At 750°C the lifetime of those mobility-enhancing defects is too short; hence, they do not persist beyond the cooling period after deformation. At 420°C thermal fluctuations are too small to generate the optimal density of vacancies. An interesting observation in this connection is that internal dislocation loops (not reaching the crystal surface), after moving some 50 µm distance, take an extremely elongated shape parallel to the Burgers vector; this means that screws eventually become immobile when 60° segments continue to move.

Finally, it is of great interest that high-resolution electron microscopy (Bourret et al., 1983) shows the cores of 30° partials to be blurred in contrast to the cores of 90° partials. This can be interpreted as a high concentration of point defects in the 30° partials. Unfortunately, it has not been observed whether these 30° partials belonged to screw dislocations.

In summary, future research on dislocation motion should keep in mind that the production of intrinsic point defects may be involved. It should also be mentioned that the point defects (clusters) (HT EPR spectrum) which are produced by moving dislocations seem to influence the dislocation mobility. This was concluded by

---

* As shown by recent calculations (Teichler, 1990) the formation energy of a vacancy in the core of a 30° partial is strongly reduced – by 1.9 eV in Si and 1.5 eV in Ge – when compared with the bulk (cf. Sec. 6.4.3). Therefore, not only can vacancies be generated more easily in the dislocation core but also bulk vacancies will be bound to dislocations passing by.

Schröter et al. (1980) comparing the Hall effect and mechanical recovery of deformed germanium. In the case of silicon, annealing around 750 °C reduces strain hardening considerably as well.

Unfortunately, up to now EPR spectroscopy has yielded information on dislocations only in silicon. For germanium, the high nuclear spin of the isotope Ge^{73} and the strong spin-orbit interaction are impeding factors. Similar problems arise with III–V-compounds. Moreover, no method is known here to eliminate the spectra of point defects, which are more numerous than in Si.

6.3.2 Information on Dislocations and Point Defects from Electrical Measurements

Since the early days of semiconductor physics an impressive number of careful investigations have been devoted to the question of which defect levels in the energy gap are due to dislocations. Because dislocations interrupt the translational symmetry of the crystal, such levels are to be expected, and in 1953 Shockley proposed the broken bonds in the core of dislocations to act as acceptors. In fact, measurements of the Hall effect seemed to confirm this idea. Schröter (1978), investigating the Hall effect of deformed p-type germanium crystals in a wide temperature range, concluded that the dislocation states in the gap form a one-dimensional band which can compensate for shallow acceptors at low temperatures, and which can also accept electrons from the valence band at higher temperatures. The occupation limit of this amphoteric band when neutral was found to be 90 meV above of the valence band edge.

Ono and Sumino (1980, 1983) tried to evaluate along the same lines Hall data of p-type silicon crystals deformed at 750 °C. They concluded that this was impossible, because they were not able to fit the temperature dependence of the density of free holes. Instead, experimental results and theory could be brought into reasonable agreement under the assumption that plastic deformation produced point-like electrical centers. This means that no shift of the level of those defects in the gap by Coulomb interaction should be inferred, which is typical for extended defects. Quantitatively, it turned out that simultaneously with $5 \times 10^7 \, cm^{-2}$ dislocations (only etch pit densities have been determined) about $5 \times 10^{14} \, cm^{-3}$ acceptors and $7 \times 10^{13} \, cm^{-3}$ donors are produced. The energy levels of both types of point-like centers are approximately at the same position (0.3–0.4 eV above the valence band edge).

The decision in favor of point-like centers is natural because in view of the low doping of the material ($10^{14} \, cm^{-3}$), such a large density of acceptors in the dislocation lines would cause a band bending far exceeding the width of the band gap.

Thus, it can be stated that two methods as different as EPR spectroscopy and Hall effect measurement lead to the same important conclusion: at least in silicon, the consequences of plastic deformation for the electrical properties of a crystal are mainly due to point defects (PD) not located along dislocation lines. That does not exclude that other electron states do exist which are localized along dislocations (as are the EPR centers Y, K 1, and K 2). As shown by Wilshaw and Booker (1985), the temperature dependence of the *EBIC (electron-beam-induced current) contrast* of dislocations in deformed silicon can be accounted for by a certain number of rechargeable centers along the dislocations. These centers are subject to a shift in the gap equivalent to band bending by Coulomb interaction of the charges on the dislocation. (Obviously a charged dislocation assumes the character of a continuously charged line only if either the wave functions of the dislocation states overlap sufficiently to delocalize electrons along the line or if the actually charged (localized) states are less distant than the Debye screening length.)

For completeness, a third class of electronic states in the gap after plastic deformation should be mentioned: these states must be close to the edges of the valence band and the conduction band, and they form one-dimensional bands. Information on these shallow bands comes from microwave conductivity, electric dipole spin resonance, photoluminescence and optical absorption (cf. Sec. 6.3.3). Because most of these effects exhibit strong correlation with the dislocation geometry, these bands must be in the proximity of dislocation lines.

Before continuing to discuss further experiments on electrical effects we will give the reader at hand an approximate method for estimating the *dislocation charge and its screening* for given conditions. To be concrete (Fig. 6-11), we will consider an n-type doped material (N_D being the excess of chemical donors over chemical acceptors)

and a line of rechargeable centers, their density along the line being n_d. The level position of these centers when neutral will be E_0 above the valence band edge. The question is: How large will be the line charge in equilibrium at temperature T and how is the line charge screened, i.e., what is the potential around the line? A rigorous solution requires the solution of Poisson's equation, taking into account the ionized donors as well as free carriers; band bending by the Coulomb interaction of the line charge influences the distribution of free carriers, and the problem has to be solved self-consistently by iteration. Two approximations are widely used: the Read model, considering only ionized donors, and the Debye-Hückel model. Ferré et al. (1990) compare these approximations with a numerical solution of Poisson's equation and conclude that the Read model under certain conditions is the more adequate approximation. In this model the dislocation accepts electrons from the donors, in this way raising its position in the gap by an extra energy $e\phi_c$.

Whether or not the dislocation can be fully charged ($q = n_d \cdot e$) depends on the distance between the Fermi level and the dislocation level ($E_F - E_0$). In the latter case the maximum charge is reached when

$$e\,\phi_c = E_F - E_0$$

It is temperature dependent via $E_F(T)$.

The relation between line charge q and potential around the screened dislocation in the frame of the model is given by

$$\phi(r) = \frac{q}{\varepsilon\,\varepsilon_0\,\pi R^2}\left\{\frac{R^2 - r^2}{4} + \frac{R^2}{2}\ln\left(\frac{R}{r}\right)\right\}$$

R is the Read radius $R = (q/e\pi N_D)^{1/2}$ (Read, 1954).

To calculate the band bending ϕ_c in the dislocation core, we have to choose an in-

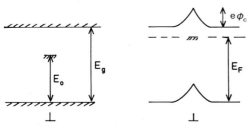

Figure 6-11. Band bending ($e\phi$) around a dislocation which acts as acceptor in n-type material. E_0: distance of the dislocation level from the edge of the valence band; E_F: Fermi level; E_g: width of the gap (left side undoped).

ner radius, e.g., $r = b$ (Burgers vector)

$$\phi_c = \phi(b) = \frac{q}{\varepsilon \varepsilon_0 \pi} \left\{ \frac{1 - \varrho^2}{4} + \frac{1}{2} \ln \varrho \right\}$$

with $\varrho = b/R$.

Since q and R are interrelated, the line charge determines ϕ_c. Of course, problems arise when the Read cylinders overlap.

Moreover, for applying the above calculation one has to be sure that N_D is the *effective* doping. Just in plastically deformed semiconductors, a great number of point defects may act as deep traps for majorities and thus compensate the chemical doping. Investigating the compensation of phosphorous as donors in deformed silicon by monitoring the number of neutral P atoms with EPR, we found that (at the same strain) the *percentage* of P atoms which are ionized is about the same, irrespective of the doping (from 5×10^{14} up to 10^{16} cm^{-3}) (Kisielowski et al., 1991). This must mean that only a certain portion of the *volume* (probably dislocation walls) contains deformation-induced acceptors (here mostly point defects). This gives a clear idea that not just effective doping but effective *local* doping has to be known for calculating dislocation charge and potential. This is one of the most serious problems in applying DLT spectroscopy to deformed crystals. (Other problems are discussed below.)

For silicon, we are now in a position to indicate a procedure which avoids those difficulties to a large extent: Deformation at 800 °C and annealing at the same temperature for 16 h greatly reduces the number of stable point defects. If subsequent deformation at lower temperatures is restricted to small strains, the number of deep point-like traps is small (Kisielowski-Kemmerich et al., 1986).

As soon as band bending is assumed to exist, another problem should be considered: If the dislocation essentially contains several electrical levels, filling of the lowest level by band bending may prevent higher levels from being filled.

On the basis of this model, Wilshaw and Fell (1989) could explain that the EBIC contrast of (internal) dislocation loops in n-type Si decreases with intensification of the electron beam which induces electron-hole pairs: the dislocation charge is reduced below the equilibrium value (EBIC contrast was shown to be proportional to the line charge q). With temperature, the EBIC contrast changes on account of changing q, too. From those experiments and the model, the authors deduce a density $n_d = 5 \times 10^7$ m^{-1} for the dislocation loops in Si. The level E_0 can only be limited to be deeper than 0.3 eV. For a (local) dislocation density of $N \approx 5 \times 10^6$ cm^{-2} the volume density of dislocation centers turns out to be 2×10^{12} cm^{-3}.

Such a density of recombination centers would never be detectable if it were not due to band bending. It has been questioned whether clean dislocations (i.e., free of impurity atoms) can cause an EBIC contrast at all (Kittler and Seifert, 1981). Such a contrast arises when the specimen contains recombination centers, whose effective capture cross section is to be compared with the square of the diffusion length of minorities. Assuming the cross section of the order of the core atom of a dislocation, the contrast would be undetectable with present means. However, as can be understood from the foregoing, a charged dislocation acts through the electric field in its screening space charge, the cross section of which is 10^5 to 10^6 times larger than an atom. Thus a continued effort should be made to distinguish EBIC contrasts from decorated dislocations and from clean dislocations. A direct demonstration of the electrical field in the depletion region around highly dislocated layers within silicon crystals is given by the bright EBIC contrast of those regions outside the Shottky contact. At the same time this phenomenon shows conductivity along such layers (Alexander et al., 1990).

Recently *DLTS (deep level transient spectroscopy)* has become the most effective technique to establish the density and position of levels in the gap. Its advantage before EPR is that the charge state of the defect to be investigated is not restricting. Defects, which are always sensitive to any annealing, are investigated using Schottky diodes on the crystal surface. In n-type material, electron traps in the upper half of the gap are detected, and in p-type crystals, the lower half of the gap is traced.

Unfortunately, there are severe problems when applying DLTS to deformed specimens:

(1) As we have seen, plastic deformation by some percentage produces 10^{15} to 10^{16} cm^{-3} deep traps (mainly point defect clusters). Doping should clearly exceed the number of traps, otherwise compensation must be taken into account. Therefore, weak deformation is optimal for DLTS; but then comparison with the less-sensitive EPR becomes difficult.

(2) Because of band bending, the position of levels near to extended defects will change during filling and emptying; moreover, this effect can be frequency dependent. Band bending by a small number of deep levels can obscure many shallow levels (Shikin and Shikina, 1988). In any case, superposition of band bending by the defect and by the depletion region of the Schottky barrier must be analyzed (Nitecki and Pohoryles, 1985).

(3) As shown first by Figielski (1978) and since then confirmed by experiment, the filling characteristics of extended defects are logarithmic in time. This makes

calibration of the number of traps often approximate.

(4) Finally, impurities can be confusing (Kronewitz and Schröter, 1987).

Because of all these problems, the results of DLTS for deformed crystals at the present time can be taken only as preliminary. Nevertheless, by comparing widespread literature, some levels can be identified. DLTS of deformed n-type silicon (10^{15} to 10^{16} cm^{-3} P) reveals three main peaks which each exhibit characteristic features.

E 0.27 eV* (= B line of Omling et al. (1985)) is one of the two dominant peaks, the estimated concentration of related electron traps after 1.6% strain being 10^{14} cm^{-3}. For both E 0.27 eV and E 0.55 eV (line D), it was shown that the filling behavior for very short pulses ($\leq 10^{-7}$ s) is exponential (as typical for point defects) and only later does it become logarithmic. This is a strong indication that it is in fact a point defect but is located in the space charge of dislocations growing with time. Another characteristic common to E 0.27 and to E 0.55 is that both line shapes can be fitted excellently to strain broadening (Omling et al., 1985).

E 0.55 is singular by its extremely temperature dependent filling behavior: the lower the temperature, the smaller the portion of E 0.55 centers which can be charged (Kisielowski-Kemmerich and Weber, 1991). Thus by TSCAP (filling pulse at 77 K) only a small part of E 0.55 can be detected. There are some indications that this part could be due to the same defects which produce the EPR spectrum Si-K 1/ K 2.

The number of E 0.55 centers depends strongly on deformation conditions and

---

* E 0.27 (eV): electron trap 0.27 eV from the edge of the conduction band. H 0.1 (eV): hole trap 0.1 eV from the edge of the valence band.

seems to dominate when (and where) dislocation interaction takes place (cutting processes?) (Kimerling et al., 1981).

Between E 0.27 eV and E 0.55 eV there is a third peak (C line), consisting of several overlapping lines (E 0.42 and E 0.48 eV). Neither line shape nor filling behavior are understood up to now. This is the thermally most stable DLTS peak in deformed silicon.

Deformed p-type silicon has been studied less extensively than n-type Si, in spite of the fact that the lower half of the gap contains about five times more traps than the upper half. The spectrum of the hole traps consists of a broad group of overlapping peaks (Kimerling and Patel, 1979). The total density of related centers ($\approx 3 \times 10^{15}$ cm^{-3}) is in good agreement with that of EPR centers with long spin lattice relaxation time (i.e., point defect clusters). After annealing, a broad peak at H 0.33 eV remains, which is ascribed to the dislocation by some authors. Photo-EPR showed that about 2.5×10^{15} cm^{-3} electrons can be excited from the energy range from 0.82 to 0.66 eV below the conduction band. Thus DLTS and EPR are in agreement as to the location of the number of hole traps (acceptors) in that energy range ($\varepsilon = 2.7\%$). If we take thermal stability as indicative for correlation of the particular defect with dislocations, the electron trap E 0.42/E 0.48 eV and the hole traps around $E_v + 0.33$ eV are most suspicious.

In this connection, it is worth having a look at the doping dependence of the intensity of the EPR spectra (Kisielowski et al., 1991). If we assume that plastic deformation generates the same types of defects in the doping range from 10^{17} cm^{-3} p type to 10^{17} cm^{-3} n type, we can determine the position of the Fermi level at which the defects enter or leave the paramagnetic charge state. Again there are

some problems limiting the accuracy of the method because of the high conductivity of the *doped* crystals at room temperature. The point defects, normally detected at room temperature in slow passage, have to be recorded in adiabatic fast passage at 24 K. The dislocation-related centers Si-Y and Si-K 1/K 2 are measured in adiabatic rapid passage at 8 K. The results are as follows: the spin density of both groups of paramagnetic centers decreases markedly beyond a critical doping of 3×10^{15} cm^{-3}, by phosphorous and by boron. Within those limits the behavior of point-like and dislocation-related defects is somewhat different:

The density of point defects in the paramagnetic charge state decreases from p doping over undoped material to n doping; in contrast, the density of the dislocation center Si-Y stays constant. The interpretation is as follows: All deformation-induced centers may assume at least three different charge states (Fig. 6-12); only when the Fermi level lies between their 0/+ and −/0 levels are they paramagnetic. Apparently, the Fermi level for most defects near a doping of +/− 3×10^{15} cm^{-3} passes through those levels. For the point defects the positions of those levels will be somewhat different depending on type; the decrease of the total number when E_F passes from the lower half of the gap into the

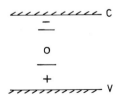

Figure 6-12. Energy levels of an amphoteric defect (schematic). If the Fermi level is in the middle range, the defect is neutral. The defect will be EPR active only in one charge state. C, V: edge of the conduction and valence band, respectively.

upper half reflects the fact that more types of point defect centers in fact have their characteristic levels in the lower half. Unfortunately, in deformed crystals the (local) Fermi level cannot be calculated from the doping but depends in a complicated manner on all the defects, including their spatial arrangement.

6.3.3 Phenomena Indicating Shallow Dislocation-Related States

6.3.3.1 Photoluminescence (PL)

In 1976 Drozdov et al. were the first to show that silicon crystals containing dislocations exhibit 4 PL lines (D 1 to D 4) with photon energies between 0.812 eV and 1.000 eV. Sauer et al. (1985), investigating the response of the spectrum to uniaxial stress, concluded that D 1 and D 2 in fact are to be ascribed to point defect centers with their (tetragonal) $\langle 100 \rangle$ axis in random orientation. D 3 and D 4 on the other hand appeared to be correlated to dislocations.

This is most convincingly shown by the modification that the line pair shows when the dislocation morphology is changed by high-stress–low-temperature deformation (Sauer et al., 1986). As mentioned earlier, this deformation procedure results in straight dislocations, parallel to three $\langle 110 \rangle$ directions; the percentage of screws is greatly increased and the dissociation width d of all dislocations is changed, i.e., partly increased and partly decreased ($3 \text{ nm} \leq d \leq 12 \text{ nm}$). Instead of the lines D 3 and D 4, crystals with this dislocation morphology show a new spectrum (D 5) consisting of a series of narrow lines with phonon replicas (Fig. 6-13). Very weak annealing (200 to 360 °C) transforms this spectrum back into D 4 (part of D 3 turning out to be a phonon replica of D 4). By this annealing, d is relaxed to its equilibri-

um value d_0 (5 nm). Consideration of the reaction kinetics of the new spectrum led Sauer et al. (1986) to assume donor-acceptor recombination as the actual type of PL. It could be shown that each of the lines of the D 5 series corresponds to a certain value of d in correspondence to the periodicity of the Peierls potential. Sauer et al. (1986) proposed transition between donors at one partial dislocation and acceptors at the other as the physical nature of the recombination process, the interaction between the two depending on the distance d between the two partials.

Later on, a similar spectrum was found in high-stress-deformed germanium (Lelikov et al., 1989). Using special compression axes recently, the dissociation width d of the dislocations could either be increased or decreased without moving the dislocation as a whole. In this way, it could be demonstrated that the lines fan out to the high-energy side when d decreases and vice versa (Izotov et al., 1990). This is confirmation for another model for the particular PL as recombination of one-dimensional excitons of the Mott type bound to the core of 90° partial dislocations. The energy of those excitons will be influenced by the strain field of the 30° partial completing a 60° dislocation. Lelikov et al. (1989) estimate the binding energy of electron and hole in deformation potential bands accompanying the dislocation line as 150 and 80 meV, respectively (germanium). Be that as it may, comparing the photon energies with the width of the band gap clearly prooves the existence of shallow states near or in the dislocations in silicon and germanium.

6.3.3.2 Optical Absorption

After plastic deformation, the optical absorption in front of the fundamental ab-

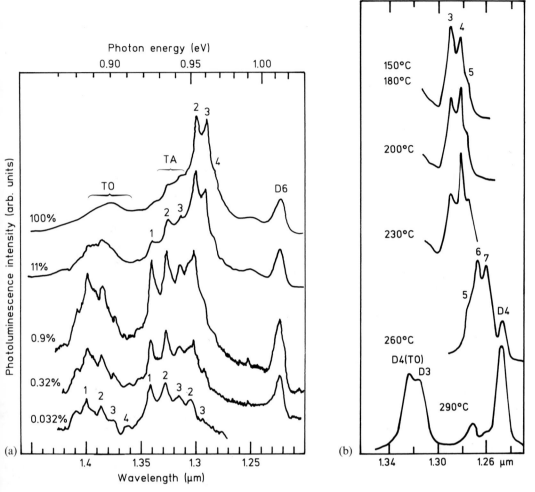

Figure 6-13. Silicon photoluminescence. The crystal is two-step deformed. (a) Excitation dependence; (b) change of the spectrum during isochronal annealing (1 h). T0, TA: phonon replica (from Sauer et al., 1986).

sorption edge is clearly increased for silicon and for gallium arsenide. Bazhenov and Krasilnikova (1986) calculate local band gap narrowing in the strain field of dislocations (deformation potential). The temperature dependence agrees well with the absorption spectrum observed with GaAs (gap narrowing by 200 meV), but for silicon this is not true before thermal annealing of the crystals at 800 °C. Then the gap is narrowed by 170 meV.

6.3.3.3 Microwave Conductivity (MWC)

One of the most attractive ideas in dislocation physics is to analyze (and possibly to use) dislocations as one-dimensional wires of high (metallic?) conductivity embedded into a matrix whose resistivity can be chosen and controlled. By the way, D. M. Lee et al. (1988) are realizing this idea by decorating straight and parallel misfit dislocations with nickel, but here the dislocations are only nucleation centers for a second phase.

We here focus on the ability of clean dislocations to carry current (dc or ac). A multitude of papers has appeared over the years in this field. The most recent work taking into account both the important role that point defects play and the possibility to completely change dislocation morphology is that by Brohl and co-workers concerning FZ silicon (Brohl and Alexander, 1989; Brohl, 1990; Brohl et al., 1990). The authors used pre-deformation at 800 °C ($\varepsilon = 1.6\%$) followed by an annealing step (16 h) at the same temperature. As shown by EPR and DLTS, no point defect (PD) centers can be found in those crystals. Subsequently the dislocations are straightened parallel to $\langle 110 \rangle$ by a short (30') deformation at 420 °C. This deformation step produces only a few PD's because the dislocations are moved only over a short distance.

An influence of these remaining PD's can be demonstrated by MWC under monochromatic illumination at a doping of 5×10^{14} cm^{-3}, but it is very weak at 4×10^{15} cm^{-3} ("effective doping").

The pronounced anisotropy of the dislocation morphology makes it easy now to demonstrate that MWC is parallel to the dislocations. This MWC (frequency 9 GHz) can be separated from the bulk contribution below 20 K, where it dominates by several orders of magnitude (Fig. 6-14).

Relaxing the high-stress morphology of the dislocations reduces the MWC and leads back to the predeformed state of both the anisotropy and the size of the effect.

Of key importance for understanding the MWC is the observation that a certain doping (n- or p-type) is necessary. Strictly speaking, MWC depends on effective dop-

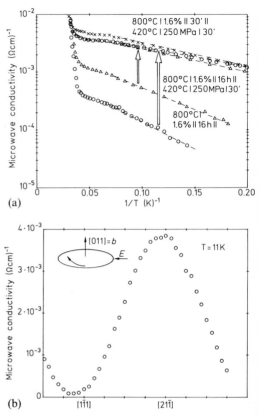

Figure 6-14. n-type (4.4×10^{15} cm^{-3} P) silicon, microwave conductivity. (a) Lower curves: only pre-deformed at 800 °C and annealed at 800 °C, 16 h. Upper curves: additionally deformed at 420 °C under high stress. o: electric field parallel to screw dislocations. △, ×: electric field parallel to edge dislocations. (b) Two step deformed crystal rotated around the Burgers vector of the dislocations. Maximum conductivity when the field is parallel to the glide plane (from Brohl, 1990).

ing, i.e., excess of chemical doping over the amphoteric effect of deep point defect levels. That (only) electrons or holes captured by the dislocation from dopants cause MWC had been shown before by Ossipyan (1985) in germanium by an elegant experiment: Neutron transformation doping produced (randomly distributed) dopant atoms – first gallium acceptors and then arsenic donors. Conductivity is developed exactly parallel with the actual effective doping. Discussing his results, Brohl (1990) ascribes MWC to band conductivity in shallow bands near to the conduction and valence band edges. The author first excludes hopping conductivity along an impurity band by a quantitative argument: even if all doping atoms were collected by dislocations, their mutual distance, would be much too big for hopping. Next, it is suggested that the dislocation may be connected with two bands; one is near the valence band and is full when the dislocation is neutral, and the other is near the conduction band and is empty. Conductivity is produced when the lower band accepts holes, or the upper band electrons. Considering the latter case, one may calculate the maximum line charge by assuming that all electrons which are lacking in the EPR of (neutral) phosphorous are accepted by dislocations. For a certain experiment q was $q \leq 1.4 \times 10^{-11}$ As m^{-1} (corresponding to 8.5×10^5 electrons per cm). The respective band bending is 120 meV. Because conductivity was observed, the Fermi level (coinciding with the phosphorous level $E_c - 45$ meV) must be inside the dislocation band. This means that band bending by at most 120 meV lifts the dislocation band onto the phosphorous level, so that the distance from the conduction band is at most 165 meV. A corresponding result is reached for the donor band near the valence band.

Influence of illumination by monochromatic light establishes the position of the deep levels competing with the dislocation band for the electrons from the phosphorous atoms; occupation limits of those states are 0.6 and 0.85 eV below the conduction band.

MWC by screw and 60° dislocations is of comparable magnitude. This, in our view, supports the idea that MWC is a matter of the elastic strain field and not of the dislocation core. Because dislocation segments of limited length are probably the conducting elements, it is extremely difficult to extract quantitative data on the conductivity of a single dislocation. A lower limit for the electron mobility of 100 cm^2 Vs^{-1} was estimated. The authors believe in the deformation potential of the strain field of the dislocations as attracting carriers to the dislocations. But also lateral confinement of the carriers by the potential wells accompanying charged dislocations can induce local hole bands at the top of the valence band and onedimensional resonant states below the conduction band (Fig. 6-11).

6.3.3.4 Electric Dipole Spin Resonance (EDSR)

Kveder et al. (1984, 1989), investigating deformed n-type silicon, found transitions between Zeeman terms induced by the *electrical* component of a microwave field (9 GHz). This type of combined resonance is due to spin-orbit interaction and also depends on high (band) mobility of electrons. A further requirement is a strong crystal field. In a cubic crystal the field is restricted to the core region of symmetry-breaking dislocations. That band mobility is under consideration can be seen from the fact that the effect is observed at 1.4 K. The EDSR is characterized by a special anisotropy from which the direction of the carrier motion can be deduced. In n-type silicon only one direction was found corresponding to the orientation of Lomer dislocations. The electrons were ascribed to a band 0.35 eV below the conduction band.

Quite recently, EDSR was demonstrated also in high-stress-deformed p-type silicon (Wattenbach et al., 1990). In contrast to n-type crystals, up to 9 lines are observed in this case. The analysis is not yet complete, but apparently dislocations of several orientations are active. Thus we suppose that the shallow bands established by MWC and connected with 60° and screw dislocations are responsible for the effect. This would mean that relatively widely extended wave functions (perpendicular to the dislocation line) feel enough spin orbit interaction to produce a very strong effect.

The Role of Contamination of Dislocations by Metal Atoms

It has been discussed repeatedly in the past whether the physical properties of dislocations in semiconductors might be influenced by unintentional decoration with impurity atoms. Those impurities are expected to be trapped in the strain field of the dislocation or precipitated as a new phase along the dislocation line. Quite recently first clear experimental results were published, mainly with respect to the influence of transition metal atoms on dislocation-related photoluminescence (PL) (Higgs et al., 1990a) and EBIC contrast (Higgs et al., 1991; Wilshaw, 1990) in silicon. The authors investigated dislocations in epitaxial layers and in crystals deformed under extremely pure conditions without detectable contamination ($< 10^{11}$ cm^{-3}) before and after contamination by back plating with Cu, Ni and Fe, respectively. In the "pure" state no PL nor EBIC contrast of dislocations could be detected. Materials with low levels of copper ($\approx 10^{13}$ cm^{-3}) showed both PL and EBIC contrast strongly. Interestingly, the three metals investigated exhibited almost identical effects. Considering TEM analysis there are two regimes of contamination: below and above one monolayer of metal atoms on the surface, respectively. Only in

the second regime can precipitates be seen at partial dislocations. Both PL and EBIC contrasts increase with contamination up to 0.1 monolayer. Further enhancement of decoration destroys PL (radiative recombination), while EBIC contrast persists and often increases up to dislocations with precipitates. The authors emphasize that various models for direct or indirect (via other dislocation-related point defects) effects of metallic impurities are conceivable.

Peaker et al. (1989) analyzed the effect of gold and platinum on oxygen-induced stacking faults (SF) surrounded by Frank partials. The basic material was vapor phase epitaxy silicon layers with an extremely low concentration of electrically active defects. After generating the SFs under clean conditions, the PL line D1 (cf. Sec. 5.3.1) and, by DLTS, a deep electron trap (activation enthalpy 0.415 eV) have been observed. After contamination with gold or platinum, D1 exhibited some modification of its shape and the trap position shifted gradually to midgap, irrespective of the fact that Au and Pt, respectively, produce as point defects quite different traps. At the same time the capture characteristics of the SF-related trap changed from being point-defect-like to logarithmic, which is typical for extended defects. From the *gradual* change of the trap position the authors conclude that the traps present at the clean Frank partials (415 meV) are modified by additional electrically active centers. The density of those centers is rather low (a few per 100 Å dislocation length).

These and other results raise the fundamental question whether clean dislocations in silicon are electrically active at all. Even such processes as dislocation motion and multiplication seem to be influenced by decoration, since Higgs et al. (1990b) found the generation of misfit dislocations in strained layers of Ge_xSi_{1-x} to be strongly influenced by the presence of copper.

Because extreme care is required to avoid any noticeable contamination of silicon by transition metals (Higgs et al., 1990a) one can assume that most experiments and applications concern at least lightly decorated dislocations. Thus, in order to obtain a better understanding of clean dislocations, further clarification of the role of low concentrations of impurities is of great importance. And, it can already be concluded now that some lack of reproducibility typical for measurements on dislocation properties may be traced back to these effects.

6.3.4 Germanium

The Hall data indicating the presence of a half-filled band 90 meV above the valence band in p-type germanium were mentioned at the beginning of this section. A DLTS study (Baumann and Schröter, 1983a) of deformed p-type germanium reveals four well-resolved peaks to be attributed to deformation induced defects. One of them, exhibiting logarithmic filling behavior and an activation enthalpy of the emission of holes of 0.075 eV, is tentatively ascribed to the dislocation band at $(E_v + 0.09 \text{ eV})$ deduced from Hall data. Two other DLTS lines (H 0.27 eV, H 0.19 eV) are saturated by the shortest filling pulse and are therefore interpreted as isolated point defect centers. The fourth DLTS line (H 0.39 eV) is only found when the deformation temperature was below $0.6 \, T_m$. Both this thermal instability and the filling characteristics proove that the center also is an isolated point defect. Its capture cross section is $\sigma_h = 10^{-19} \text{ cm}^2$, and its density was determined as $1.4 \times 10^{13} \text{ cm}^{-3}$ for a dislocation density $3.3 \times 10^7 \text{ cm}^{-2}$, which corresponds to a calculated strain $\approx 6 \times 10^{-3}$.

In *n-type* germanium (Baumann and Schröter, 1983b), a single broad and asymmetric DLTS peak exists after deformation at 420°C. Annealing at 580°C reduces the amplitude by an order of magnitude. The trap position is $(E_c - 0.20 \text{ eV})$ – coinciding with the position of the hole trap H 0.39 eV. After annealing, the filling behavior clearly shows this electron trap to be an isolated point defect; before annealing, the filling behavior is complicated, probably indicating that most of the defects here are under the influence of the barrier of charged dislocations. It is this part of the defects which anneals out.

Also, Hall effect data (Schröter et al., 1980) indicate that the density of holes suddenly increases markedly when the deformation temperature is lower than $0.6 \, T_m$ (of course, the critical temperature depends somewhat on the cooling procedure applied after deformation).

On the whole, the findings on the defect spectrum after deformation for germanium are similar to that of silicon: several deep traps are present, with more in the lower half of the gap. Most of them – if not all – are not directly dislocation related. Above 60% of the (absolute) melting temperature, a large portion of the point defect centers becomes unstable. Dislocations give rise to a shallow band.

6.3.5 Gallium Arsenide

One of the main problems for applying semi-insulating GaAs for high integration of semiconductor devices is spatial inhomogeneity of electrical parameters over the wafer, often correlated with the distribution of dislocations. Whether this is due to interaction between dislocations and point defects or to electrical properties of the dislocations themselves, it has induced strong research activity on dislocations in plastically deformed gallium arsenide.

Proceeding from elemental semiconductors to compounds, there is an increase in the number of material parameters which have to be known if results from different authors are to be compared: the crystals are grown by different techniques, producing quite different combinations of defects; there are indeed more *types* of defects, the most prominent among the new ones being the antisite (AD) defects As_{Ga} and Ga_{As}. Also, the number of dislocation types has doubled (Comp. Sect. 6.2).

A critical review of the literature gives the impression that all authors agree that the most striking effect of plastic deformation is a strong decrease in the density of free electrons in n-type GaAs (Gerthsen, 1986; Suezawa and Sumino, 1986; Skowronski et al., 1987; Wosinski and Figielski, 1989), which is proportional to the plastic strain ($\Delta n = -(2-6 \times 10^{16} \text{cm}^{-3})$ for $\varepsilon = 3\%$).

Optical DLTS locates these acceptors at about $E_v + 0.45$ eV (Skowronski et al., 1987). Other authors using other methods arrive at $E_v + 0.38$ eV (Gerthsen, 1986) and $E_v + 0.37$ eV (Wosinski, 1989), respectively. The acceptors compensate shallow donors in n-type material, but they are too deep to significantly influence the density of free holes in deformed p-type GaAs. In semi-insulating (si) GaAs they lower somewhat the Fermi level, transforming part of the EL 2 double donor into the paramagnetic charge state $(El\ 2)^+$. In spite of enormous efforts, the atomistic structure of EL 2 is not yet established, but according to common belief, it contains an anion AD As_{Ga} which in turn produces a characteristic four-line EPR spectrum. The increase of this spectrum with strain ε was previously explained as proof for the generation of EL 2 defects by moving dislocations. Later on, some doubt arose, because the additional defects did not exhibit the quench-

ability by light which is characteristic for EL 2 (Omling et al., 1986). Thus the present view is that the shift of the Fermi level caused by the deformation-induced acceptors changes only the charge state of part of the EL 2 defects present before the deformation. However, this does not seem to be the complete answer either: Fanelsa (1989) recently deformed highly n-doped $(1.8 \times 10^{17} \text{cm}^{-3}$ Si) GaAs which, as prooved by DLTS and known by others (Lagowski and Gatos, 1982), is free of EL 2. After plastic deformation by 3% at 400°C, DLTS showed $1.3 \times 10^{15} \text{cm}^{-3}$ EL 2 defects. In view of the inhomogeneous compensation, this is a lower limit. So we think that both effects – ionization and generation – can be responsible for the increase of the AD spectrum, and each case is to be analyzed separately.

No (other) donors are found after 3% deformation by Skowronski et al. (1987). Wosinski (Wosinski and Figielski, 1989; Wosinski, 1990), also using DLTS, was able to isolate two new traps occurring only after plastic deformation: an electron trap ED 1 ($E_c - 0.68$ eV, $\sigma = 10^{-14} \text{cm}^2$) and a hole trap HD 1 ($E_v + 0.37$ eV, $\sigma = 2 \times 10^{-14} \text{cm}^2$) (Fig. 6-15). Both traps have in common a logarithmic filling characteristic with variation of the filling pulse length over a range of six orders of magnitude. This is the reason why only a lower limit for the trap density can be given: 2.5×10^{15} and $2 \times 10^{15} \text{cm}^{-3}$ for a – calculated – dislocation density of $8.5 \times 10^8 \text{cm}^{-2}$. The authors conclude from the rigorously logarithmic filling and from the proportionality of the trap and dislocation densities that these traps belong to dislocation cores (α and β dislocations?). It has been shown that ED 1 acts as recombination center for photo-excited carriers.

Two optical effects are assumed to be strongly related to dislocations. The first

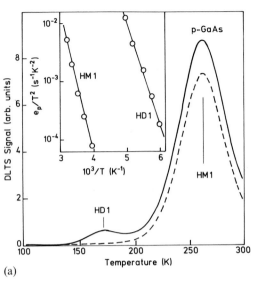

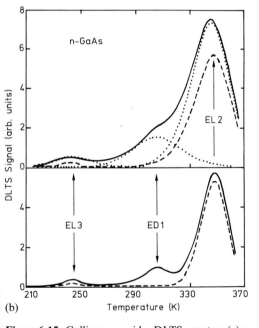

(a)

(b)

Figure 6-15. Gallium arsenide, DLTS spectra. (a) p-type GaAs dashed: undeformed; solid: plastic strain 2% (rate window: 17 s⁻¹, filling pulse: 1 ms (Wosinski, 1990)). (b) n-type GaAs (LEC). dashed: undeformed; solid: plastic strain 2.8%. Rate window: 5 s⁻¹, filling pulse: 5 ms (from Wosinski, 1990).

effect involves the absorption tail near the fundamental absorption edge, after subtraction of the contribution by intracenter transitions in EL 2 centers. This absorption tail is found by several authors (Bazhenov and Krasilnikova, 1986; Farvacque et al., 1989; Skowronski et al., 1987) and is explained by gap narrowing due to elastic strain or electric fields (Franz-Keldysh effect) near charged dislocations. From comparison of this tail in n- and p-type GaAs, Farvacque et al. (1989) deduced the existence of two dislocation "bands", one near the valence band, the other at midgap. Extending the consideration to the photoplastic effect (REDG: radiation-enhanced dislocation glide), the authors hesitate to think of true (delocalized) band states, since the recombination process contributing part of the activation energy of dislocation motion should occur at a certain local center. But the existence of a space charge with electrical field is not restricted to a continuous line charge. The second optical effect is a photoluminescence band at 1.13 eV photon energy detected by Suezawa and Sumino (1986) and Farvacque et al. (1989). But the authors make contradictory remarks on the necessary doping.

Finally, we should mention that plastic deformation of GaAs brings into existence a second four-line EPR spectrum starting at $\varepsilon \approx 4\%$ (as the AD As$_{Ga}$ spectrum) and then increasing linearly with strain to about 10^{17} cm^{-3} at 10% strain (Wattenbach et al., 1989). The anisotropy of the center is much more complicated than that of As$_{Ga}$. Christoffel et al. (1990) attribute the spectrum to two trigonal arsenic interstitial complexes. We found the same spectrum in plastically deformed GaP, also (Palm et al., 1991). Because Ga is the only species (in high enough concentration) with nuclear spin 3/2 we feel sure that we are observing the spectrum of the AD Ga_{As}. The low symmetry may be due to Jahn-Teller distortion (Krüger and Alexander, 1991). To summarize: in plastically deformed GaAs comparable concentrations of both types of antisite defects are present.

6.3.6 $A^{II}B^{VI}$ Compounds

The compounds of Zn and Cd with anions from group VI of the periodic table (O, S, Se, Te)[†] are unique among the semiconductors with respect to electrical effects of plastic deformation. The crystal structures (cubic zinc blende = sphalerite and hexagonal wurtzite), both are generated from tetrahedral groups of atoms with pure heterocoordination. The dislocation geometry is the same as in $A^{III}B^{V}$ compounds with the only exception that in wurtzite-type crystals, only one close-packed glide plane (0001) exists instead of four in zinc blende (111). But dislocations with an edge component in $A^{II}B^{VI}$ compound crystals carry large electrical charges (up to one elementary charge per lattice plane).

This leads to a number of closely related effects. The following survey is mainly based on the recent review by Ossipyan et al. (1986).

Before some of these effects are described, the origin of the dislocation charge will be discussed. The first idea concerns the (partly) ionic bonding of the compounds: in the core of any (perfect or partial) dislocation, a row of ions of the same type will carry a net charge, although not necessarily of the same magnitude as a row of the same ions in the undisturbed bulk because of possible reconstruction. (For the latter, charged crystal surfaces

give some hints.) Detailed consideration shows that perfect 60° glide set dislocations carry charges of $\pm(3e^*/4b)$ per lattice plane when e^* is the effective charge of an ion in the lattice. With $0.28 \leq e^*/e \leq 0.53$ (Phillips and van Vechten, 1969), this results in inherent dislocation charges of $(0.2-0.4)$ e/b for the series of compounds. The experimentally determined line charges for moving dislocations in the dark vary from 0.12 up to 0.7 e/b and are negative in n-type crystals throughout and positive in the only p-type material investigated so far (ZnTe). Thus the inherent ionic charge of the atoms in the dislocation core must be taken into account, but it cannot be a determinant for the dislocation effect. This is clearly demonstrated by the dependence of the line charge q on illumination.

The next assumption could be that charged point defects are collected by the dislocation when moving. This is excluded from consideration, the main reason also being the role of illumination.

Free carriers swept along by the electric field of the moving dislocation can be neglected, since no screening of the line charge was observed. (It should be noted that Hutson (1983) takes a different view here.) Finally, carriers remain as a source of the dislocation line charge which are captured into deep trap levels connected with the dislocation core. In this case the sign of the charge is not necessarily related to the ionic type of the particular dislocation; rather, as shown by experimental findings, the dislocation seems to act as a trap for majority carriers, as is usual. Ossipyan et al. (1986), using accepted parameters, calculated the charge to be expected on the basis of the Read model for a perfect Se(g) 60° dislocation in ZnSe with 10^{15} cm^{-3} shallow donors (-0.6×10^{-10} C/m) and found it to be smaller than

---

[†] There are only seven such compounds, since CdO is an NaCl type compound.

in experiments $(-2 \times 10^{-10} \text{ C/m})$. From this observation as well as others, the authors concluded that moving dislocations in $\text{A}^{\text{II}}\text{B}^{\text{VI}}$ compounds are not in thermal equilibrium with the electronic subsystem as far as the occupation of the electronic states of the dislocations is concerned. They outline a model of a charged dislocation (the charge at the beginning of motion being the ionic charge) which on its way interacts with charged point defects. The dislocation may capture electrons (or holes) from those point defects, and on the other hand, electrons may (thermally assisted) tunnel through the barrier around the dislocation into the conduction band. During its path, the dislocation will assume a dynamical equilibrium charge (Kirichenko et al., 1978; Petrenko and Whitworth, 1980). It amounts to $q = 1.67 \times 10^{-10} \text{ C/m}$ for the case mentioned above. Comparing capture rate and emission rate, a logarithmic increase of q with dislocation velocity v is obtained; in fact, such a relation between q and strain rate is found for strain rates exceeding 10^{-6} s^{-1}. For smaller strain rates the theory is not applicable because here the screening charge (made from ionized shallow centers) is believed to move with the dislocation.

Experiments

The experiments which bring to light the dislocation charge (some of them indeed spectacular) can be divided into two groups: Either an outer influence (illumination, electric field, carrier injection) changes the flow stress of a deforming crystal, or during deformation electrical effects are induced as charge flow from one side face of the crystal to the opposite side face ("dislocation current"), change of the electrical conductivity or luminescence.

We restrict ourselves here to a description of one experiment from each group. Charge flow connected with dislocations (Ossipyan and Petrenko, 1975) can be observed by macroscopic methods only if there are differences between the net charge transported by positive and negative dislocations. Since the sign of the charge is the same for the two types in $\text{A}^{\text{II}}\text{B}^{\text{VI}}$ compounds, only the difference of the size of the charge, or of the product from dislocation number times distance travelled by the average dislocation remains. Fortunately, the mobility of A(g) and B(g) dislocations mostly is extremely different, so that one type practically determines the charge flow. In this manner, it was possible to measure dislocation charges. Care must be taken in the case of low-resistivity samples where screening by mobile carriers can reduce the effective charge transported. On the basis of dislocation current, it can be easily understood that immersing the deforming crystals into mercury will decrease the flow stress (by avoiding surface charges), and that application of an electric field to the side faces also influences the crystals' flow stress (electroplastic effect).

The second key experiment to be discussed is the sudden and reversible change of the flow stress induced by illumination during deformation (photoplastic effect PPE (Ossipyan and Savchenko, 1968)). In most cases, the flow stress increases (up to a factor of two) – positive PPE – but there are also cases of negative PPE. Takeuchi et al. (1983) gave an extended review on the PPE in $\text{A}^{\text{II}}\text{B}^{\text{VI}}$ compounds. The nature of the PPE is by no means clear, and we will come back to it in Sec. 6.5.5. However, there is no reasonable doubt that illumination changes the dislocation charge. It was shown that the spectral dependence of both the dislocation charge q and the PPE are virtually identical; maximum effects are produced by photons just below the respective band gap energy. Moreover, there is a linear relation between q and the flow stress (Petrenko and Whitworth, 1980).

Deformation-Induced Luminescence

CdS and CdSe, in contrast to other semiconductors, are ductile down to liquid helium temperature. This makes those compounds very attractive for optical in-situ analysis of the deformation process. Weak thermal fluctuations prolong the lifetime of primary products of dislocation activity and allow to better separate point defect-related from dislocation-related effects. Tarbaev et al. (1988) applied very little strain (of the order of 10^{-4}) to CdS and CdSe crystals at low temperature, and measured optical absorption and luminescence spectra; by use of a scanning electron microscope (SEM) in cathodoluminescence (CL) mode, luminescence activity could be attributed directly to slip bands: the activity was found not only around dislocation etch pits but clearly also behind dislocations in their wake.

The authors found new peaks at a corresponding wavelength in both absorption and luminescence spectra. They attribute them to optical transitions between electronic states of one particular type of point defect complex. Piezospectroscopic investigation reveals C_s symmetry for the electronic states and probably also for the defects. All optical modifications of the crystal caused by plastic deformation disappear by storage of the crystal at room temperature. ·

Following the luminescence (L) activity for higher deformation degrees, it turns out that nonradiative centers must be destroyed by moving dislocations, while the new L centers, mentioned above, are produced. As a result the radiative efficiency of the crystal increases by deformation. The PL left behind by moving dislocations in CdS have been imaged; a domain structure of the polarization direction of the emitted light demonstrates collective behavior of the recombination centers. A model of those centers is outlined (Ossipyan et al., 1987).

For CdTe a number of investigations have been carried out to detect possible deep dislocation core states and to determine their position. Thermopower and Hall effect data from Müller (1982) and Haasen et al. (1983) for n- and p-CdTe are interpreted to give evidence for a defect level at $E_v + (0.3 \ldots 0.4 \text{ eV})$ in CdTe due to plastic deformation. DLTS measurements at n-CdTe by Gelsdorf and Schröter (1984) revealed a mid-gap line at $E_c - 0.72 \text{ eV}$ after plastic deformation. Since the trap density of this line turned out one order of magnitude larger than that of possible dislocation core states, the level was interpreted to be due to deformation-induced point defects. In deformed p-CdTe Zoth (1986) detected a DLTS line at $E_v + 0.7 \text{ eV}$ with trap density about one tenth of possible dislocation core states which might be associated with the latter. The defect levels observed by Zoth and Gelsdorf recently were confirmed by Nitecki and Labusch (1988) by means of photocapacity investigations, where the logarithmic time dependence of the filling factor was also reconfirmed for both levels. Zoth's results are in agreement with theoretical investigations (Teichler and Gröhlich, 1987) (comp. Sec. 6.4) which predict core states at Cd (g) partial dislocations with levels up to 0.68 eV below E_c (and levels at Te (g) partials up to 0.15 eV above E_v).

6.4 Theoretical Investigations about Electronic Levels of Dislocations

Many theoretical investigations have been made in the last decade to determine the position of possible deep electronic levels introduced by the dislocations in the band gap of semiconductors. The level position depends strongly on the arrangement of atoms and bonds in the dislocation cores, and therefore the field of level calculations is intimately woven together with the field of core structure simulations. Questions of importance in this context concern the detailed atomic arrangement in the dislocation cores; the deep electron level structure of a given arrangement; the relative stability of different core configurations of the same dislocation; and the formation energy, atomic structure, and electronic level position of local imperfections in otherwise perfect dislocations. The theoretical developments in this field are strongly influenced by the fact that the introduction of dislocations in a lattice changes the topology of the whole system. This prevents one from being able to characterize dislocations simply by local perturbations in a more or less perfect lattice, which is usually possible with point defects or even surfaces.

6.4.1 Core Structure Calculations

Present knowledge regarding the atomic arrangement in dislocation cores primarily comes from computer simulations. The core structure is determined through numerical minimization of the total energy as a function of the atomic positions, making use of suitable models for the structure-dependent part of the energy. As in the first application to the dislocation problem by Marklund (1980), these simulations take advantage of the fact that the arrangement is periodic along the dislocation line, and that the atomic positions are known sufficiently far away from the core in terms of the elastic deformation field. Because of this, the whole atomic pattern can be described by specifying the atomic positions within a flat cylindrical region around the dislocation (with a thickness of one or two atomic layers normal to the line), which typically contains from thirty to several hundred atoms embedded in an elastically deformed lattice.

Most of the core structure simulations are carried out by modeling the energy of the system within a "valence force field" approach. This was of significant influence on the notation used to characterize different core configurations. The valence force fields for the diamond structure rely on the existence of covalent bonds among neighboring atoms with the coordination number of the atoms fixed to four. They model the structural energy of an array of atoms in terms of deformations of these bonds, i.e., by bond-stretching and bond-bending contributions. In order to use these models to treat atoms with lower coordination numbers, the notation of dangling bonds must be introduced along with the concept of bond breaking. This concept implies a qualitative difference between "deformed" bonds and "broken" bonds, where deformed bonds suffer strong restoring forces with increasing deformations, but broken bonds exhibit no restoring force at all. A rather important parameter in this context is the bond-breaking energy, i.e., the asymptotic energy value ascribed to a broken bond, which measures whether bond deformation is energetically more favorable than bond breaking.

With regard to the distinction between broken bonds and deformed bonds, different core configurations are constructed for one and the same dislocation which differ in the density and distribution of broken bonds and dangling bonds. The so-called unreconstructed configurations contain as many dangling bonds as demanded by geometry when extending the elastic far field to the core region. From these variants, configurations of reduced density of dangling bonds are obtained by reconstructing neighboring pairs of dangling bonds under slight displacement of atoms, thereby gaining covalent bond energy on account of an increase in lattice deformation energy. The classification of the core configurations according to the density of dangling bonds has become of importance since the electron-theoretical calculations revealed that for the dislocations studied so far deep electron levels occur only in the case of dangling-bond-carrying cores. The theoretical studies concentrate on the glide-set partial dislocations, which according to weak-beam electron microscopy observations are considered to predominate. For Si and Ge in particular the 30° and the 90° partial dislocations are treated, resulting from dissociation of the perfect 60° and screw dislocations, where unreconstructed and completely reconstructed core configurations have been studied, as shown in Fig. 6-16. The reconstruction is accompanied by a change in symmetry, i.e., a doubling of the translation period along the dislocation in the 30° partials and breaking of the mirror symmetry normal to the dislocation in the 90° partials, which makes the different core configurations distinguishable according to symmetry, in addition to their density of broken bonds. Also, the 60° partial was investigated for Si and Ge where intermediate partially reconstructed configurations (Jones and Marklund, 1980; Veth and Teichler, 1984) are considered. In addition, unreconstructed 30° and 90° glide-set partials in GaAs

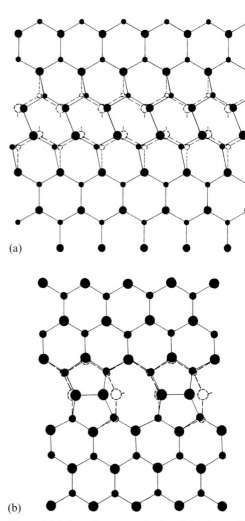

(a)

(b)

Figure 6-16. Atomic pattern in the glide plane of (a) the reconstructed 90° partial and (b) the reconstructed 30° partial dislocation in Si (dashed: unreconstructed configurations).

and CdTe have been investigated. In the following discussion we shall, however, concentrate on the elemental semiconductors and postpone the results concerning the compounds to the end of the section.

6.4.2 Deep Electron Levels at Dislocations

The theoretical calculations of the deep levels at dislocations extract the levels introduced by the dislocations from the comparison of calculated energy spectra for crystals with dislocations with those of ideal crystals. Such an approach seems indispensable, since, as mentioned above, the topological alterations brought about by dislocations prevents one from treating them in a perturbative way. The atomic positions in the crystal with dislocations are taken from independent studies such as the valence force field simulations described above. No significant effects on the level position were found when using simulations from different valence force fields. In order to model the Hamiltonian, either the tight-binding interpolation scheme was used (Jones, 1977; Marklund, 1979; Jones and Marklund, 1980; Veth and Teichler, 1984), or the more advanced LCAO approach (Northrup et al., 1981; Chelikowsky 1982; Teichler and Marheine, 1987; Wang and Teichler, 1989), or the simpler extended Hückel approximation (Alstrup and Marklund, 1977; Lodge et al., 1984). The geometrical models range from spherical clusters (Jones, 1977) of typically 700–1500 atoms containing one dislocation to supercell approximations (Marklund 1979; Northrup et al., 1981; Chelikowsky, 1982; Lodge et al., 1984; Wang and Teichler, 1989), where pairs of dislocations with an alternate Burgers vector embedded in a supercell (with cell size of about 50–200 atoms in the plane normal to the dislocation) are periodically repeated, up to treatments of isolated partials in an infinite crystal (Veth and Teichler, 1984; Teichler and Marheine, 1987). By means of the supercell approach, the overall translational symmetry of a periodic lattice is restored (although with an extremely large translation cell), which permits evaluation of the electronic level structure by application of the usual methods known for lattices. In the other investigations the continued-fraction recursion

method is applied to evaluate the level spectrum at the dislocations, as introduced in this field by Jones (1977) and developed further by Veth and Teichler (1984).

Because of the translational symmetry along the dislocation line, the bound states in the cores combine to one-dimensional Bloch wave-like states with the levels split into one-dimensional bands. The details of the theoretical studies reveal that reconstructed core configurations of the 30° glide-set partial in Si have no deep levels in the band gap (Chelikowsky, 1982; Veth and Teichler, 1984). The same holds for the corresponding 90° partial (Veth and Teichler, 1984; Lodge et al., 1989) – up to perhaps shallow levels near the band edges (Chelikowsky and Spence, 1984) – and for these dislocations in Ge (Veth and Teichler, 1984). For the unreconstructed 30° glide-set partials one band of bound states is predicted which covers the whole, or at least large parts, of the gap (Marklund, 1979; Veth and Teichler, 1984). For the unreconstructed 90° partial in Si two bands are deduced (Teichler and Marheine, 1987; Wang and Teichler, 1989), a lower band filled with electrons near the valence band edge E_v and an empty upper band reaching up to the conduction band. The two bands of the 90° partial reflect the two dangling bonds per periodicity length in this dislocation. The bands are separated by an energy gap of 0.05 eV width centered around 0.2 eV above E_v. (The earlier tight-binding treatments predicted for the 90° partial in Ge and Si two partially filled narrow bands of width 0.5 eV around E_v, but this result is due to an underestimation of the mutual interaction between neighboring dangling bonds in the tight-binding scheme.) For the 60° glide-set partial, a 'partially' reconstructed, dangling-bond-carrying configuration was found (Jones and Marklund, 1980; Veth and Teichler,

1984) with levels in the band gap in Ge and Si. The reconstructed configuration has no deep levels in the gap (Veth and Teichler, 1984). In addition, a 'weakly' reconstructed configuration with an extremely stretched bond was found to have an empty bound state split off from the conduction band in Si at about 0.8 eV above E_v (Jones and Marklund, 1980; Veth and Teichler, 1984), whereas in Ge this state turns into a conduction band resonance (Veth and Teichler, 1984).

6.4.3 Core Bond Reconstruction and Reconstruction Defects

The profound difference in the electron spectrum of dangling-bond-carrying and dangling bond-free configurations was realized in an early study by Marklund (1979). This observation immediately initiated attempts to estimate the energy difference between the different types of core configurations, which, within the framework of valence force field descriptions, particularly demands proper estimates of the bond-breaking energy. Besides the uncertainty in this value, additional difficulties arise from the fact that different valence force field models predict different lattice deformation energies, although the models yield rather similar atomic structures. Table 6-1 presents the energy difference between reconstructed and unreconstructed configurations, $\Delta E = E_{unrec} - E_{rec}$, deduced from various valence force fields, where the dangling-bond energy entering Table 6-1, E_{db}, is half the bond-breaking energy. The scatter of the results reflects the fact that the different valence force models are constructed to simulate effectively different features of deformed systems. Keating's (1966) model in its original version, along with its anharmonic generalization by Koizumi and Ninomiya (1978), is adapted to the elastic long-wavelength deformations but overestimates the energy of short-wavelength deformations

Table 6-1. Energy difference $\Delta E = E_{unrec} - E_{rec}$ (per line length b_0) between unreconstructed and reconstructed configurations.

Dislocation	ΔE (eV)		Structural energy model
	Si	Ge	
90° glide-set partial	$2 E_{db} - 1.7$	—	Orig. Keating[a]
	$2 E_{db} - 1.6$	$2 E_{db} - 1.5$	Orig. Keating[b]
	$2 E_{db} - 0.6$	—	Mod. Keating[c]
	$2 E_{db} - 1.7$	—	Lifson-Warshel[d]
	$2 E_{db} - 1.03$	$2 E_{db} - 0.98$	Bond charge[e]
	0.45	—	Tersoff[f]
30° glide-set partial	$E_{db} - 0.25$	—	Anh. Keating[a]
	$E_{db} - 0.17$	$E_{db} - 0.15$	Anh. Keating[b]
	$E_{db} - 0.1$	—	Mod. Keating[c]
	$E_{db} - 0.23$	$E_{db} - 0.22$	Bond charge[e]
	0.55	—	Tersoff[f]
60° glide-set partial[g]	$2 E_{db} - 1.5$	$2 E_{db} - 1.4$	Anh. Keating[b]

[a] Marklund (1980)   [b] Veth and Teichler (1984)   [c] Marklund (1981)
[d] Lodge et al. (1984)   [e] Trinczek (1990)   [f] Heggie and Jones (1987)
[g] The values refer to the difference between 'weakly' and 'completely' reconstructed core of the 60° partial.

in the dislocation cores. The modified potential (Baraff et al., 1980; Mauger et al., 1987) describes suitably well the energy of short-wavelength deformations but underestimates the energy in the elastic field. Applicable to both regions of deformations are, e.g., Weber's (1977) bond charge model or the Lifson-Warshel potential (Lodge et al., 1984) which thus provide the most reliable estimates for the lattice deformation energies.

The dangling-bond energy, E_{db}, has to be deduced from peripheral considerations, mainly from electron-theoretical arguments, since covalent bonding is of electronic nature. For Si the estimates of E_{db} range from the early value of 0.5 eV (Marklund, 1980) as a lower limit up to the more recent values of about 1.75 eV (Lodge et al., 1984) and 2.05 eV (Teichler, 1990), where, however, the latter estimates refer to bond-breaking energy values without residual interactions of the dangling bonds with their environment. The residual interactions depend on the geometry of the atomic structure and have to be determined for each defect separately. For the dangling bonds in the 90° glide-set partials of Si a reduction of about 0.7 eV was recently deduced by use of a quantum-chemical LCAO model (Teichler, 1989a) yielding an effective $E_{db} \approx 1.35$ eV. Estimates of E_{db} for Ge are not very frequent in the literature, but from comparison of the cohesive energies, a lower E_{db} in Ge than in Si seems reasonable. In accordance with this assumption, for the bond-breaking energy without residual interactions a value of 1.65 eV was estimated for Ge (Teichler, 1990). In the case of the dangling bonds in the 90° glide-set partials of Ge, this value may be reduced by about 0.5 eV (Teichler, 1989a) through residual interactions.

With regard to the lattice deformation energies in Table 6-1 from bond-charge-model and Lifson-Warshel-potential calculations (which, as mentioned, are more reliable than the Keating data), values for effective E_{db} of the order of 1 eV (or larger) mean that for all partial dislocations considered so far the reconstructed variants are energetically more favorable than the unreconstructed variants. This implies that the stable variants of the dislocations in their defect-free configuration do not provide any deep electron level in the band gap. In Si this case seems to be realized according to the estimates of E_{db} presented above and in accordance with the experimental observations (cf. Secs. 6.3.1, 6.3.2). For Ge the data at present are not clear enough to rule out the possibility that in this material glide-set partials with deep levels exist.

A treatment by Heggie and Jones (1987) for the 90° partial in Si (also included in Table 6-1) goes beyond the valence force field models and makes use of Tersoff's interatomic potential. This interatomic potential model was particularly constructed to compare structures with different coordination numbers like f.c.c. or s.c. silicon and it permits estimation of the total energies of reconstructed and unreconstructed variants without the concept of bond breaking. The deduced energies, however, have to be considered with caution since Tersoff's model has difficulties in simulating correctly the deformation energies in the elastic far field of the deformed system (Teichler, 1989b) and predicts an unreliable ranking of the different vacancy configurations (Heggie and Jones, 1987). Additional difficulties with the use of this (and similar) empirical Si models can be found from recent dislocation core studies by Duesbery et al. (1991). But taking all matters into consideration, we can say that Tersoff's model finds the deep-level-free reconstructed configuration of the 90°

glide-set partial in Si as a stable variant. This statement is in agreement with the valence force field treatments.

The theoretical treatments of the dislocation-induced deep electron levels discussed so far do not explicitly take into account effects of the finite intra-atomic Coulomb integral U. Finite U effects on the dislocation level spectrum have been considered by Usadel and Schröter (1978) in a phenomenological model designed to simulate the undissociated 60° shuffle-set dislocation. They found a Hubbard-type level splitting between the electron- and hole-like excitations in the narrow one-dimensional band of dislocation states, and they showed that most of the Hall effect data on deformed Ge could be fitted by assuming that $U \approx 0.3$ eV. (But there was also one set of experimental data where $U \approx 3.6$ eV.) Successful fitting of the Si data was not possible by this model. Besides this quantitative approach, a qualitative discussion of finite U effects on the excitation spectrum of dislocations was provided by Grazhulis (1979) assuming a Hubbard splitting between electron and hole excitations. Application of these ideas to the above-described quantitative one-electron results for the 30° and 90° glide-set partials reveals that significant effects of finite U may be expected in the case of the unreconstructed 90° partials with their small energy gap between empty and filled electron states. In the case of reconstructed partials the gap between full and empty states seems too large to speculate about finite U effects, and the strong instability of the unreconstructed 30° partials against reconstruction indicates that these variants may hardly be accessible to experimental verification.

The results presented on core bond reconstruction raise the question of whether dangling-bond-carrying defects on the re-

constructed dislocations account for the observed electronic activity of the latter. Possible defects are, e.g., local reconstruction defects (RD's) on the reconstructed partials (as shown in Fig. 6-17) and complexes of the RD's with other imperfections such as vacancies or kinks in the dislocation cores (Hirsch, 1980) sketched in Fig. 6-18, or vacancies in the cores. As discussed in Sec. 6.3.1, the vacancies as well as the RD's in complexes with vacancies and in vacancy clusters on the 30° partial have the right symmetry to account for the experimental EPR spectra and can explain the observed conversion between different EPR-active centers by formation and dissociation of these complexes (Kisielowski-Kemmerich et al., 1985; Kisielowski-Kemmerich, 1989; 1990).

Atomistic models of RD's and corresponding complexes can be constructed by computer simulation as in the case of straight dislocations, e.g., by embedding a cluster of atoms containing local defects into a crystal with dislocation or by studying a crystal with dislocations which has a periodic array of defects in the core. By such an approach, it was recently demonstrated within an LCAO model (Teichler, 1990) that vacancies may have a strong binding energy to the core of the 30° glide-

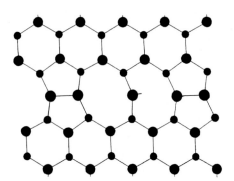

Figure 6-17. Reconstruction defect of the 30° partial dislocation.

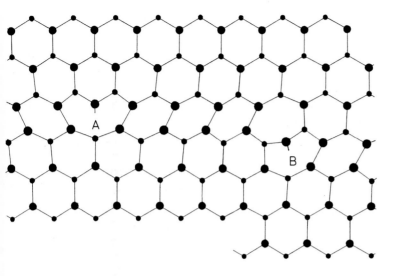

Figure 6-18. Atomic pattern in the glide plane of a reconstructed 90° partial (in Si) with reconstruction defect (A) and a complex of kink plus reconstruction defect (B).

set partials of about 1.9 eV in Si (1.5 eV in Ge), which might explain the large concentration of vacancies in the dislocation cores predicted from EPR measurements (cf. Sec. 6.3.1). Detailed investigations on the formation energy E^f were also carried

out for RD's, kinks and some of their complexes.

Table 6-2 presents estimates of E^f for these defects obtained from different valence force field models, which clearly indicate that a significant – or even the domi-

Table 6-2. Formation energies E^f of local defects in reconstructed partial dislocations and electron level positions E_D for computer-relaxed structures from tight-binding calculations. (All energies in eV.)[d]

Defect	Si		Ge		Model
	E^f (eV)	$(E_D - E_V)$ (eV)	E^f (eV)	$(E_D - E_V)$ (eV)	
Reconstruction defect					
30° glide partial	$E_{db} - 0.15$	0.15	$E_{db} - 0.14$	0.20	Orig. Keating[a]
90° glide partial	$E_{db} + 0.37$	0.02	$E_{db} + 0.33$	−0.02	Orig. Keating[a]
Kink on reconstructed					
90° glide partial	0.72	no	0.65	no	Anh. Kearing[a]
	0.14	–	0.16	–	Orig. Keating[b]
	0.13	–	0.13	–	Bond charge[b]
Kink pairs on reconstructed					
90° glide partial	0.37	–	–	–	Mod. Keating[c]
90° glide partial	0.24	–	–	–	Tersoff[c]
30° glide partial	>1.1	–	–	–	Mod. Keating[c]
Kink plus RD					
90° glide partial	$E_{db} + 0.95$	0.16	$E_{db} + 0.85$	0.12	Anh. Keating[a]
90° glide partial	$E_{db} + 0.36$	–	$E_{db} + 0.36$	–	Orig. Keating[b]
90° glide partial	$E_{db} + 0.26$	–	$E_{db} + 0.25$	–	Bond charge[b]

[a] Veth (1983)   [b] Trinczek (1990)   [c] Heggie and Jones (1987)
[d] Bars indicate that estimates are missing, 'no' means that no levels were found.

nant – part of E^f comes from the dangling-bond energy, E_{db}. In nearly all cases, E^f is larger than E_{db} since the local defects are accompanied by additional lattice deformation energy, with the exception of the RD's on the 30° glide-set partials, where local breaking of the reconstruction provides an energy gain due to lattice relaxation. Tight-binding calculation show the dangling-bond deep levels of these defects to be near E_v. The corresponding values are also included in Table 6-2, as obtained for the computer-relaxed structures. Regarding these values one must, however, take into account that changes in the atomic arrangement induce significant changes in level positions (Heggie and Jones, 1983). In addition, there are indications that on the reconstructed 90° glide-set partial in Si the RD level may be shifted midgap (Teichler, 1989a; Heggie et al., 1989) due to interactions of the dangling bond with its environment which are neglected in the tight-binding approach.

In contrast to the above-mentioned situations where dislocations introduce electronic levels in the band gap, there also exist cases where combinations of dislocations and otherwise electronically active centers may result in inactive complexes. A particular example of this kind is P in Si. According to a recent theoretical treatment by Heggie et al. (1989) in deformed Si an energetically favorable configuration occurs if P trapped in the core of a reconstructed 90° glide-set partial combines with an RD on the dislocation yielding P in threefold coordination. By this process, the dangling-bond level and the P level are eliminated from the band gap, whereby P, in additive, acts as a strong, local pinning center.

6.4.4 Shallow Dislocation Levels

So far, we have considered possible electron levels at dislocations and at defects in dislocations with states confined to the dislocation cores. Beyond this, theory predicts shallow levels corresponding to more extended states, either states associated with the stacking fault ribbon between the two dissociated partials or states of electrons and holes trapped in the elastic deformation field of the dislocations ('deformation potential states'). Stacking fault states were deduced for Si with levels up to 0.1 eV above the valence band edge (Marklund, 1981; Lodge et al., 1989) and thus cannot account for the deep levels observed by experiment. The deformation potential also induces shallow states where for Ge theory predicts electron states down to 0.1 eV below the conduction band and hole states up to about 0.02 eV above E_v (Celli et al., 1962; Claesson, 1979) for undissociated dislocations. In Si (Teichler, 1975) and for dissociated dislocations (Winter, 1978; Teichler, 1979) the corresponding levels are even closer to the band edges. Current photoluminescence measurements (cf. Sec. 6.3.3) at deformed Ge samples indeed show activation energies of 0.15 eV and 0.08 eV (Lelikov et al., 1989), which might be interpreted as shallow levels below the conduction band and above the valence band. For plastically deformed Si, mircowave conductivity investigations displayed thermal activation energies of 0.07 eV in p- and 0.08 eV in n-type material (Brohl and Alexander, 1989). The origin of these possible levels and their interrelationship with the hitherto studied deformation potential states is however, an open question.

6.4.5 Deep Dislocation Levels in Compounds

Regarding compounds, theoretical studies of deep electron levels are carried out for dislocation core states in GaAs and CdTe, where unreconstructed 30° and 90° glide-set partials are investigated. In AB compounds the defect-free unreconstructed partials are characterized by core rows of dangling-bond atoms of either A or B type (yielding a doubling of dislocation configurations compared to the elemental semiconductors). Since in compounds an electron transfer occurs between the cations and anions in the bulk (the direction depending upon whether the ionic or covalent bonding type dominates), rows of heavily charged atoms would appear in the dislocations if there were no alterations of charge distribution in the cores. Accordingly, for dislocations in compounds, one has the additional problem of modeling the charge distribution around the dislocations. In early investigations by Jones et al. (1981) on dislocations in GaAs, and by Öberg (1981) in CdTe, any effects of charge redistribution are neglected. In this case the atomic arrangement in the cores is simulated according to the valence force field approach of Martin using for all atoms their bulk effective charge. For evaluation of the electronic levels, the one-electron tight-binding approach for elemental semiconductors is applied. In the meantime, an improved method was deduced and applied to CdTe (Teichler and Gröhlich, 1987; Marheine, 1989) and GaAs (Myung, 1987) where the tight-binding scheme is extended beyond the independent-particle scheme taking into account Coulomb, exchange, and correlation corrections in a parameterized way. The effective charge on the atoms and the atomic positions are determined "self-con-

sistently" by making use of the molecular orbital approach to compounds. As in the case of elemental semiconductors, one band of deep electron levels results for the unreconstructed 30° glide-set partials and two bands for the 90° partials. For neutral dislocations with vanishing band bending, the dislocation bands at the As(g) and Te(g) partials (the notation refers to the type of dangling bond atoms in the cores) are situated around E_v with a width of about 0.5 eV or smaller, and the center of the band typically 0.1 eV above E_v for the As(g) and 0.1 eV below E_v for the Te(g) partials. In GaAs the levels of the Ga(g) dislocations are found about 0.9 eV above E_v, with the width being the same as for the As partials. For CdTe the self-consistent treatment gives a partially filled band about 0.5 eV below the conduction band. Here, the proper inclusion of the Coulomb effects seems of importance since neglect of these corrections, as in the earlier treatment, predicts for the neutral Cd(g) partials levels within the conduction bands which would lead to heavily charged dislocations in contrast to the underlying model assumptions of uncharged objects.

According to recent studies by Gröhlich (1987) and Marheine (1989), the defect-free Te(g) dislocations are unstable against electron capture. In intrinsic CdTe stable Te(g) dislocations carry a nominal charge of about 0.05 electrons per dangling bond (reduced according to the dielectric constant of 7 in CdTe) surrounded by a positive screening cloud (Marheine, 1989). The energetics involved in dangling-bond reconstruction for compounds and the question of possible core charge compensation by structural defects like vacancies on the lines have not yet been studied in detail. From the experiments (cf. Sec. 6.3.5), there are some indications for deep dislocation levels in GaAs in the lower part of the gap and at midgap (Wosinski and Figielski, 1989; Farvaque et al., 1989), whereas up to now the deep levels for CdTe found after plastic deformation (cf. Sec. 6.3.6) seem to be due to deformation-induced defects surrounding the dislocations.

6.5 Dislocation Motion

6.5.1 General

Notwithstanding the common glide geometry of diamond-like crystals and f.c.c. metals there is a fundamental difference in the mobility of dislocations in the two classes of materials. Whereas dislocations can move at the temperature of liquid helium in copper, temperatures of roughly half the (absolute) melting temperature are required to move dislocations over noticeable distances in elemental semiconductors. The reason for this is the localized and directed nature of covalent bonding. The disturbance of the lattice caused by the dislocation is concentrated into a much narrower range of the dislocation core, and this produces a pronounced variation of the core energy when the dislocation moves from one site in the lattice to the next one. There is a saddle point configuration in which the core energy is higher by a certain amount (the so-called Peierls potential) than in the equilibrium positions.

This holds for partial dislocations as well as for perfect dislocations, but the Peierls potential should be smaller for partials. This periodic profile of an important part of the dislocation energy depends on the direction of the dislocation in the glide plane; the energy minima are deepest for dislocations which are parallel to one of the three $\langle 110 \rangle$ lattice rows in a (111) plane. This can be demonstrated by deforming a crystal at a relatively low temperature (0.45 T_{m}) and with a high shear

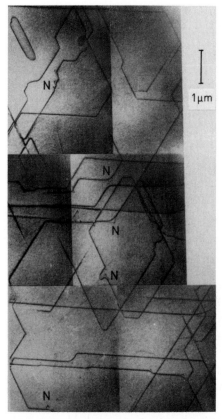

Figure 6-19. p-type silicon TEM: primary glide plane of a two-step deformed crystal (750 °C, $\tau = 12$ MPa; 420 °C, $\tau = 296$ MPa). N: so-called noses (extraordinarily wide dissociation).

stress (Fig. 6-19). The high stress allows the dislocations to turn with sharp bends from one direction to another. As Fig. 6-19 shows, dislocations follow $\langle 110 \rangle$ energy minima as far as possible. For this reason, 60° and screw dislocations are the basic types in these crystals and 30° and 90° partials, constituting these two types, are objects of many theoretical investigations (Sec. 6.4). At higher deformation temperatures and, therefore, smaller shear stress edge dislocations are prominent, consisting of two 60° partials. Examining lattice models, one notices that the core of a 60° partial is made from alternating elements of 30° and 90° partials (Hirsch, 1979). Applying high shear stress to a silicon crystal containing segments of edge character pinned by two constrictions (Sec. 6.2) transforms within a short period of time the edge dislocations into a triangle of two 60° dislocations. This happens at temperatures as low as 370 °C, where 60° and screw dislocations are about 50 times slower. This seems to confirm that edge dislocations can be considered as a dense array of kinks.

In what follows, we will summarize first the results of measurements of the velocity of perfect 60° and screw dislocations as revealed by various experimental methods (etch pitting at the crystal surface, X-ray live-topography, cathodo-luminescence). On account of space, we have to discard many details. There are several up-to-date reviews on the subject (Louchet and George, 1983; Alexander, 1986; George and Rabier, 1987). In the second part of this section, dissociation of perfect dislocations into partials is taken into account; it will be shown that 30° and 90° partials not only have different mobilities, but their position in front of or behind the stacking fault ribbon also influences the friction force to be overcome when moving. Those differences can explain some peculiarities of dislocation mobility under high stress. The authors believe that electron microscopic analysis of the morphology of single dislocations frozen in the state of motion will answer some of the questions still open after 30 years of investigating dislocation motion in semiconductors.

6.5.2 Measurements of the Velocity of Perfect Dislocations in Elemental Semiconductors

Since the first measurements of dislocation velocities in various semiconductors

by Patel and co-workers (Chaudhuri et al., 1962), most authors use expressions such as Eq. (6-1) to represent their results:

$$v = v_0 \left(\frac{\tau}{\text{MPa}} \right)^m \exp \left(- \frac{Q}{kT} \right) \qquad (6\text{-}1)$$

where v_0, m, and Q are quantities which first depend on the material under investigation, and second on the type of dislocation which moves. For small stresses ($\tau < 10$ MPa) Eq. (6-1) is not appropriate because there m depends not only on temperature but also on stress (George and Rabier, 1987). Actually, Q and m are found not to be independent constants, but rather Q increases with a lowering of the (shear) stress, indicating that an important portion of the stress dependence of dislocation velocity is due to a contribution by the applied stress to overcoming the activation barrier (Alexander et al., 1987).

From an experimental point of view, it should be mentioned that there is a controversy about the reliability of deducing the dislocation velocity from etch pitting the points where the dislocation penetrates the crystal surface before and after displacement. It is true that the dislocation half loops produced by scratching or hardness indentation show some irregularities near the surface (George and Champier, 1980; Küsters and Alexander, 1983). Thus it is absolutely necessary to use only the straight segments below those irregularities for measurements. Also, some pinning of dislocations can be inferred from the occurrence of a starting stress, i.e., a minimum stress for dislocation motion. Repeated loading of the same specimen must be avoided in these experiments. On the other hand, X-ray live topography suffers from the possibility that recombination of electron-hole pairs produced by irradiation influences the dislocation mobility (see below for the photoplastic effect). Using the double-etch-pitting technique with sufficient care yields very reproducible results from laboratory to laboratory at least in the range of lower temperature ($<0.6\ T_m$). Comparison of measurements of the dislocation velocity in the same material (n-type silicon) made by etch pitting and X-ray *in situ* topography showed agreement within 30% (George and Michot, 1982). Since the parameters Q and m extracted from macroscopic deformation tests with the help of the microdynamical theory (Alexander, 1986) are in satisfactory agreement with the values measured by etch pitting, there is little doubt that the latter technique gives information reflecting properties of dislocations in the bulk. However, this good correlation between macroscopic and microscopic sets (Q, m) holds only for elemental semiconductors (Si and Ge). In compounds, the large difference between different dislocation types obscures the connection between the activation enthalpy Q' of the state of optimal plasticity (lower yield stress or maximum creep rate) and the Q values of the dislocation movement.

Lower temperatures are to be preferred for measuring dislocation velocities not only because of slower gettering of impurities to dislocations but also for theoretical reasons: Although some authors doubt whether the Peierls potential is rate controlling for plasticity at high temperatures, it certainly is rate controlling below $0.6\ T_m$. Thus a comparison with the theory outlined in Sec. 6.6 should be made here.

For the "best values" of the parameters Q and m of Eq. (6-1) the most extensive data can be found for 60° dislocations in undoped FZ silicon. Critical discussion (Alexander et al., 1987) reveals Q to be weakly stress dependent in the region $4\ \text{MPa} \le \tau \le 200\ \text{MPa}$ corresponding to

(George et al., 1972)*:

$$Q(\tau) = Q_0 - E_1 \ln\left(\frac{\tau}{\text{MPa}}\right) =$$

$$= 2.6\ \text{eV} - 0.115\ \text{eV} \ln\left(\frac{\tau}{\text{MPa}}\right) \quad (6\text{-}2)$$

It is satisfactory that activation analysis applied to the yield stress of FZ silicon results in a barrier height $\Delta G_0 = 2.6\ \text{eV}$ (Omri et al., 1987) and in an activation volume $V = -\partial G/\partial\tau = E_1/\tau$ with $E_1 = 0.375\ \text{V}$ (Castaing et al., 1981). The stress dependence of the activation volume deduced from Eq. (6-2) is of the same type, but the absolute value is three times smaller. Comparing Eq. (6-2) with the empirical ansatz Eq. (6-1) allows the identification of Q with Q_0 and m with E_1/kT. In fact Kisielowski-Kemmerich (Kisielowski-Kemmerich, 1982; Alexander et al., 1983) extending measurement of the dislocation velocity in FZ Si to lower temperatures and higher stresses came to the conclusion that the stress exponent m consists of two independent parts:

$$m = m_0 + m_1 \quad (6\text{-}3)$$

where $m_1 = E_i/kT$ with E_i (screw) = 0.092 eV, and E_i (60° dislocations) = 0.122 eV (90/30) and 0.13 eV (30/90). The other (smaller) component m_0 is negative for compression tests and depends in a complicated manner on the deformation geometry (see below).

As mentioned above, it is not self-evident that the rate-controlling mechanism for dislocation motion stays the same in the entire temperature range up to the melting point. In fact Farber and co-workers found a sudden change of Q at about $0.75\ T_m$ both in silicon and germanium (both Czochralski-grown) (Farber and

---

* At $\tau = 300$ MPa Q is 1.8 eV (Küsters and Alexander, 1983) instead of 1.95 eV from Eq. (6-2).

Nikitenko, 1982; Farber et al., 1981). In silicon Q increases from 2.2 eV to 4 eV; nevertheless, because of a change of the prefactor, dislocations at higher temperatures are more mobile than is extrapolated from low temperatures. For germanium the situation is more complex: in the low-temperature regime ($< 600\,°C$), the activation energy Q is strongly stress dependent as already observed by Schaumburg (1972). Q increases from 1.6 eV ($\tau = 10$ MPa) to 3.5 eV ($\tau = 1$ MPa). Correspondingly, the stress exponent m below 10 MPa amounts to 3, while above that stress level it is 1.3. All these complications are absent above $600\,°C$: $Q = 1.8$ eV in $1 \le \tau \le 10$ MPa and $m = 1.3$.

The authors assume that point defects on the dislocation line act as obstacles for kink motion; at small stress, these obstacles cannot be overcome without thermal activation. As mentioned, the sudden increase of Q in the low-stress region is found also in undoped germanium. This may also be due to obstacles for kink motion. Möller (1978) explained it differently: Q should increase at a critical stress below which kink pairs on both partials have to generated simultaneously*.

---

* Recently Siethoff et al. (1991) emphasize an interesting parallelism between the activation energies of the lower yield point (U_{LY}) and of the first stage of dynamical recovery (U_{III}) of the stress strain curve. The authors calculate from U_{LY} the activation energy Q' of the motion of (a large collective of) dislocations, and from U_{III} the activation energy Q_D of the diffusion mechanism controlling the creep of edge-type dislocations. Comparing silicon and four $A^{III}B^V$ compounds the authors find the ratio Q'/Q_D to be about 2/3. This may mean that "an elementary diffusion step is involved in the process of kink formation and migration" (Siethoff and Behrensmeier, 1990), accounting for the remarks on the appearance of lots of point defects after dislocation motion (Sec. 6.3.1). It may, however, also indicate a common electronic process underlying both the dislocation motion and the particular diffusion process controlling climb in the related material.

Comparing screw dislocations and 60° dislocations shows that both types are qualitatively of equal mobility (within a factor of two) in silicon; in germanium, screws are more mobile than 60° dislocations (Schaumburg, 1972).

The ansatz (Eq. 6-1) implies that only the shear stress τ resolved to the glide system influences the dislocation velocity, and all other components of the tensor $\boldsymbol{\sigma}$ of applied stress are ineffective. This means that only the glide component of the Peach-Köhler force (Hirth and Lothe, 1982) is taken into account. Using hardness indentations, one is producing dislocation loops belonging to different glide systems simultaneously which move without interaction. Comparing those dislocation loops, we found (Alexander et al., 1983) that their mobilities at equal shear stress τ may be appreciably different. This must mean that other components of $\boldsymbol{\sigma}$ have to be taken into account besides τ (since Schmid's law states that only τ determines the yield stress of a crystal, it can be assumed that this law is violated in the case of silicon). However, efforts to correlate the mobility of dislocations belonging to different glide systems to the normal component of the Peach-Köhler force (i.e., climb force) were not fully successful (Weiß, 1980). The experimental results probably cannot be rationalized in terms of configurational forces on the dislocation as a quasiparticle but need calculation of the core structure of (partial) dislocations under the action of the full stress tensor.

The influence of the orientation of the glides system relative to the compression axis may be taken into account empirically by m_0 of Eq. (6-3):

$$m_0 = -7.2\,(1 - |\hat{\boldsymbol{b}} \cdot \hat{\boldsymbol{c}}|)^2 \, \frac{F_n}{b\,\sigma}$$

($\hat{b}$, $\hat{c}$: unit vectors parallel to Burgers vector and compression axis, respectively; F_n: normal component of the Peach-Köhler force; σ: applied compression stress).

6.5.3 Kink Formation and Kink Motion

As outlined in Sec. 6.6, both formation of kink pairs and thermally activated overcoming of a secondary Peierls potential and/or of localized obstacles by kinks may contribute to the apparent activation enthalpy Q of the dislocation motion. In order to come to a decision between various models proposed so far, it is of great importance to separate experimentally the two steps (nucleation and growth of kink pairs) from each other. There are three main approaches: first, one can try to analyze the microscopic displacement of partials at temperatures low enough that only a consumption of kinks which have been present before is possible. The second approach is to distribute the applied stress into a train of stress pulses of length t_i, separated by pauses of length t_p, and the third approach is to measure the internal friction.

1) The dissociation with d of straight dislocations can be extended by high shear stress up to about 100 Å and frozen in by cooling under stress (Wessel and Alexander, 1977). Heating such specimens in the electron microscope allows one to observe the relaxation of d to its stress free equilibrium value d_0. This was performed (with silicon) by Hirsch et al. (1981) and at even lower temperatures (100 to 140 °C) by Gottschalk et al. (1987). Hirsch et al. determined the migration energy W_m of kinks on 90° partials as 1.2 eV and Gottschalk et al. calculate for the same quantity (as an average for 30° and 90° partials) 1 to 1.2 eV. So W_m seems to contribute roughly one half to Q if one as-

sumes Q to represent the activation enthalpy of the uncorrelated motion of partials at high stress (the attraction between the partials by the surface tension of the stacking fault at the actual width d corresponds to a shear stress of the order of 200 MPa).

2) The idea of pulse deformation is as follows (Nikitenko et al., 1985; Farber et al., 1986): the distance moved by a dislocation during static loading is compared with the distance after periodic loading and unloading with the same total time of loading in both cases. Varying the pulse length t_i, one observes ceasing of the dislocation movement when the pulses become shorter than a certain critical length t_i^*. Similarly dislocations become immobile when the pauses become longer than about twice the pulse length ($t_p > t_p^*$). The authors explain these findings (for Si) in the frame of the Hirth-Lothe diffusion model (Sec. 6.6). The stress (7 MPa) should be so low that the kink density stays at its thermal equilibrium $C_k = (2/b)\exp(-U_k/kT)$ (U_k formation energy of a single kink). Under the stress τ, kink diffusion provides a drift velocity of kinks $v_k = D_k \tau b h / kT$. ($D_k = v_D \cdot \exp(W_m/kT)$ is the kink diffusivity, h height of a kink).

The dislocation velocity v under those conditions is $v = C_k v_k h$. In the model of Nikitenko et al. (1985), t_i^* marks the time in which a kink pair grows by diffusion to its critical length; on the other hand, in the time t_p^*, a kink pair vanishes by back diffusion which during the pulse length t_i has reached a length $2 v_k t_i$.

From the equation $\sqrt{4 D_k t_p^*} = \sqrt{4 D_k t_i^*} + 2 v_k (t_i - t_i^*)$, D_k can be calculated and from that $W_m = 1.58$ eV (for $\tau = 7$ MPa). After its publication, this simple explanation of the experimental results was criticized by the authors themselves (Nikitenko et al., 1987, 1989) and by Maeda (1988).

First of all, it cannot be understood why kink pairs should fully shrink by diffusion because the attraction between the two kinks is of short-range type. Second, as shown by simulations, the critical pulse length t_i^* is expected to be much shorter. Maeda (1988) repeated the experiments using α dislocations in GaAs and found qualitatively the same behavior as Nikitenko et al. Although there were some indications that the problem could be due to experimental imperfections, Nikitenko et al. (1987, 1989) were of the opinion that the simple model does not apply. They maintained that the deviation of the experimental results from the diffusion model is ascribable to point defects which change the energy of a kink pair and also produce the starting stress for dislocation motion. Therefore, they leave during the pauses a stress equal to the starting stress compensating for an effective back stress on the kinks, which can be explained by applying the theory of random forces to kink motion (Petukhov, 1988). In fact, now the shrinkage of kink pairs corresponds better to theory. The authors claim that the modification of the model does not change significantly the estimate of W_m (Farber, private communication). A comparison of starting stress and pulse deformation reveals that three quarters of W_m (1.2 eV) can be ascribed to the secondary Peierls potential, and one quarter to the effect of impurity atoms.

Maeda and Yamashita (1989) on the other hand extended the pulse deformation tests to low-doped (8.6×10^{16} cm^{-3} Si) GaAs. The authors found discrepancies between the diffusion model and experiment to be even more serious than with higher-doped crystals: now the critical pulse length t_i^* is more than ten times *longer* than the time for moving one Peierls distance during static loading! Maeda et al.

came to the conclusion that only locking of the dislocation by point obstacles and bowing out of dislocation segments can explain the particular type of intermittent loading characteristics they found. This would mean that the mere existence of an intermittent loading effect is no proof in favor of or against the kink diffusion model. In summary, no clear-cut conclusion can be drawn at this time from the intermittent loading effect revealed by the pulse loading experiments.

3) Jendrich and Haasen (1988) extended the analysis of internal friction of deformed germanium crystals to higher temperature. They found two peaks with activation enthalpies $H_1 = 1.11$ eV and $H_2 = 2.07$ eV. Within the framework of the kink diffusion theory and kink–kink annihilation (Sec. 6.6), the authors ascribe H_1 to the migration of kinks (consuming 70% of the activation energy Q of dislocation motion) and H_2 to the formation of kink pairs.

6.5.4 Experiments on the Mobility of Partial Dislocations

Application of an uniaxial compression stress σ to a crystal in general induces different glide forces F_i to the two partials of a perfect dislocation ($i = 1, 2$ numbers, the partial moving in front of and behind the stacking fault, respectively). For compression parallel to $\langle 213 \rangle$, $F_1 = 0.71 F_2$. Assuming that during steady state motion ($v_1 = v_2$) the effective force (made up from F_i, the attraction γ by the stacking fault and the repulsion A/d) is equal for both partials[†], one can calculate a stress-depen-

---

[†] $F_1 + \dfrac{A}{d} - \gamma = F_2 - \dfrac{A}{d} + \gamma.$

dent dissociation width (Wessel and Alexander, 1977):

$$d = \frac{d_0}{1+f\dfrac{b\tau}{2\gamma}} \qquad (6\text{-}4)$$

with $fb\tau = (F_2 - F_1)$.

Therefore, freezing in this high-stress state would lead one to expect the dislocations to be narrowed ($d < d_0$). But in fact the opposite is true (on average over a considerable scatter) for screws and those 60° dislocations which have the 90° partial moving in front (90/30 dislocations). Only 30/90 dislocations are narrowed as expected (the same is true for germanium) (Alexander et al., 1980). Wessel and Alexander (1977) explained those findings by introducing a lattice friction force R_i into the consideration leading to Eq. (6-4). Then, depending on the actual difference $(R_2 - R_1)$, widening as well as narrowing is possible:

$$d = \frac{d_0}{1+(f-\delta)\dfrac{b\tau}{2\gamma}} \qquad (6\text{-}4')$$

with

$$\delta = \frac{R_2 - R_1}{b\tau} = \frac{R_2 - R_1}{R_2 + R_1} = \frac{1 - R^*}{1 + R^*}$$

In this sense, the ratio $R^* = R_1/R_2$ of the friction stresses felt by the two partials has become a measurable quantity. A large number of measurements could not reduce the wide scatter of the measured width d; this scatter reveals the coexistence of dislocations in the whole range of R^* between zero and infinity. This fact and the mean values of R^* have been discussed by Alexander (1984). We will repeat here only the conclusions (for $T = 420\,°\text{C}$).

(1) It is not surprising to find the mobilities $\mu_i = (R_i)^{-1}$ of 30° and 90° partials to be different in view of their different core

structures (Sec. 6.2). Generally, 90° partials are more mobile than 30° partials.

(2) Statement 1 cannot explain that most screw dislocations consisting of two 30° partials are widened. In fact, the position of a partial before or behind the stacking fault also influences the mobility. For that reason, R^* (90/30) is not reciprocal to R^* (30/90).

(3) Examining whether R^* is a unique property of a given dislocation type W, Weiß (1980) measured R^* for five different glide systems. It turned out that the order of magnitude in fact stays constant ($R^*_{\text{screw}} \approx 0.5$, $R^*_{30/90} \approx 3$, $R^*_{90/30} \approx 0.35$), but the differences from one glide system to the other were well outside of the scatter of one system.

Grosbras et al. (1984) made similar experiments and proposed that the climb force acting on each partial – including its sign – should be responsible for the difference. Heister (1982) (cf. Alexander, 1984) tried to find the link between the velocity of a perfect dislocation and the friction forces acting on its partials both measured at 420°C and at high stress. He succeeded by assuming that the stress exponent m is the same for the partials as well as the total dislocation. Calling $v_0 \exp(-Q/kT)$ in Eq. (6-1) v_{0T}, he writes[†]:

$$\tau b = R_1 + R_2 = \left(\frac{v}{v_{0T}}\right)^{\frac{1}{m}} \text{MPa}\,b \qquad (6\text{-}5)$$

Using for each partial a similar ansatz $R_i = R_{0i}\left(\dfrac{v}{v_{0i}}\right)^{\frac{1}{m}}$; the left and the right side of

Eq. (6-5) fit together only if

$$m_1 = m_2 = m$$

[†] MPa is the dimensionality constant from Eq. (6-1).

Thus

$$R_{01}\left(\frac{v}{v_{01}}\right)^{\frac{1}{m}} + R_{02}\left(\frac{v}{v_{02}}\right)^{\frac{1}{m}} =$$

$$= \left(\frac{v}{v_{0T}}\right)^{\frac{1}{m}} \text{MPa}\, b \qquad (6\text{-}6)$$

The ratio of friction forces then is

$$R^* = \frac{R_1}{R_2} = \frac{R_{01}}{R_{02}}\left(\frac{v_{02}}{v_{01}}\right)^{\frac{1}{m}}$$

Using the exponents m as measured under the same conditions (Kisielowski-Kemmerich, 1982) as the R^*, one can separate (R_{01}/R_{02}) from (v_{02}/v_{01}) (Table 6-3). With these 6 numbers, Heister could fit 14 out of 15 measured dissociations d (3 dislocation types in 5 systems).

Moreover, the four parameters R_{01}, R_{02}, v_{01}, and v_{02} can be separated by putting absolute values of v into Eq. (6-6). Since for 30/90 dislocations velocities were measured in four systems, two systems provide control of the rather good fit. Table 6-4 presents the best values obtained for the four parameters of the three basic dislocation types in silicon. Calculating the combination $R_{0i}/(v_{0i})^{1/m}$ being characteristic for the (inverse) mobility of the related partial (Table 6-4), the result is fairly interesting: there is satisfactory agreement for the two 60° dislocations; the 90° partial is about three times more mobile than the 30° partial. However, the 30° partials of *screws* are clearly less mobile than the 30° partials belonging to 60° dislocations and, in addition, the trailing partial is appreciably less mobile than the leading partial. Closer inspection of Table 6-4 reveals the stress exponent m to be mainly responsible for the difference between 30° partials both in screw dislocations and in 60° dislocations.

For comparing different glide systems in Table 6-4, m is approximated by its

Table 6-3. Ratio of the parameters of lattice friction (R_0) and velocity (v_0) of Eq. (6-6) for the pairs of partials (1,2) forming the basic three dislocation types (silicon).

	Screw (30/30)	30/90	90/30
R_{01}/R_{02}	5	200	0.1
v_{02}/v_{01}	0.04	3.4×10^{-4}	10.7

Table 6-4. Parameters of Eq. (6-6) separated with the help of measured dislocation velocities and dissociation width (silicon).

	Screw (30/30) $m = 1.5$	30/90 $m = 1.9$	90/30 $m = 1.9$
R_{01} (N m^{-1})	1.6×10^{-2}	4.75×10^{-2}	4×10^{-3}
v_{01} (10^{-10} m s^{-1})	3.5	6.25	0.61
R_{02} (N m^{-1})	3.2×10^{-3}	2.4×10^{-4}	4×10^{-2}
v_{02} (10^{-10} m s^{-1})	0.14	2×10^{-3}	6.5
$R_{01}/(v_{01})^{m-1}$	3.2×10^{4}	3.3×10^{3}	9.5×10^{2}
$R_{02}/(v_{02})^{m-1}$	5.5×10^{4}	1.1×10^{3}	2.7×10^{3}

geometry-independent part $m_1 = E_i/kT$. The smaller mobility of 30° partials in screws, therefore, appears to correspond to E_1 (screw) $< E_1$ (60° dislocation). The special character of 30° partials out of screws is particularly interesting due to the fact that the dislocation related EPR active centers seem to be located in just these 30° partials (Sec. 6.3.1).

As shown by Rabier and Boivin (1990), the idea of a change in the dissociation width of screw dislocations under high shear stress is the key to understanding the plastic properties of GaAs (and probably other compounds as well) in the low temperature range (from 200 °C down to room temperature). Irrespective of whether α- or β-30° partial dislocations are less mobile, either positive or negative screw dislocations will be narrowed because each screw is composed from an α- and a β-30° partial.

Actually, in the temperature range just mentioned, stress-strain curves exhibit some pecularities when compared with medium temperatures (Rabier, 1989): (1) The yield point phenomenon is absent, and the curve is parabolic. This also holds for GaSb and InP. (2) The doping influence on the elastic limit is reversed. While at medium temperatures the yield stress of n-type GaAs is about five times larger than that of intrinsic and p-type GaAs, at room temperature for n-type GaAs it is smaller than that for the other two materials by about a factor of two. A similar trend has been found for InP. The authors (Rabier and Boivin, 1990) give convincing evidence that plastic deformation in the considered temperature (and stress) range is controlled by cross slip of screw dislocations. This is supported by TEM investigations. The threshold stress τ_c for cross slip can be calculated from Escaig's theory (1968), which assumes constrictions to be present and to provide nuclei for dissociation on the cross slip plane. The activation energy W_c for cross slip then depends on the ratio d/d' of the dissociation widths on cross slip plane and primary glide plane, respectively: the larger the rating d/d', the smaller the value of W_c. For the usual compression axis of the type $\langle 213 \rangle$, the cross slip plane is parallel to the compression axis, so that d is equal to the equilibrium width d_0. From Eq. 6-4' it follows:

$$\frac{d}{d'} = 1 + \frac{b\tau}{2\gamma}\left(f - \frac{1 - R^*}{1 + R^*}\right)$$

with $R^* = \mu_2/\mu_1$. Quantitative calculation for GaAs (Rabier and Boivin, 1990) showed that W_c at small stress (10 MPa) amounts to 6 eV and is practically independent from R^*. However, when τ approaches the high stress regime, W_c decreases reaching a certain value (e.g., 1.5 eV) at a lower stress for a higher value

of R^*. (Large R^* means $d < d_0$.) Therefore, cross slip is prominent for n-type GaAs at a lower stress than for intrinsic and p-type GaAs in agreement with what has been observed (item 2 above).

The particular shape of the stress-strain curve (item 1) also is explained by the decisive role of cross slip of screws. At the beginning of deformation, surface sources produce dislocation systems consisting of long parallel screw dipoles extended by fast-moving 60° dislocations of α type (Kesteloot, 1981). Several such loops emanate from each source, and the back stress from those loops eventually blocks the source, so that steep strain hardening results. But when the stress reaches a level where stress-induced narrowing of screws becomes important, cross-slip of screws out of the slip plane relaxes the pile-ups and the back stress on the sources. At the same time, spreading of glide onto new glide planes by double cross slip provides new dislocation sources. Both processes continuously reduce the slope of the stress-strain curve: dynamical recovery causes the deformation curve to take a parabolic shape.

Different behavior of the two types of 60° dislocations is observed also with respect to the *photoplastic effect* (PPE) (Küsters and Alexander, 1983). Here the 30/90 dislocations show strongly increased mobility when illuminated with bandgap light, while the 90/30 dislocations show little or no effect. This distinguishes the PPE from doping influence, which always concerns both types of 60° dislocations equally, and which may be superimposed to the PPE. The PPE can be rationalized as a decrease of the activation enthalpy Q by 0.68 eV (Si). The related effect was found by Maeda and Takeuchi (1981, 1985) in $A^{III}B^V$ compounds and was called radiation-enhanced dislocation glide (REDG).

Both groups interpret the enhancement of dislocation mobility by light (or irradiation with electrons) as recombination-enhanced kink motion. A different model is proposed by Belgavskii et al. (1985). The effect of electrical doping (Alexander, 1986) and of alloying (Sumino, 1989) on dislocation velocities has repeatedly been reviewed. Theoretical models are discussed in Sec. 6.6.2. In view of limited space, we refer to those references.

Note added in proof

Achievement of a better understanding of the interaction of oxygen atoms with moving dislocations in CZ-silicon by the work of Maroudas and Brown (1991) has to be mentioned. The authors calculated numerically the distribution of oxygen around moving 60° dislocations in the temperature range 600–1500 K for various dislocation velocities. The steady state is treated for the dislocation core being saturated. But the moving dislocation continuously has to dissociate from individual oxygen atoms which are replaced by new atoms. Therefore three components contribute to the drag stress to be overcome by the dislocation: the pure lattice (Eq. 6-1), the interaction with the oxygen cloud around the dislocation and the force necessary to dissociate from the oxygen in the core. The resulting dependence of the dislocation velocity v on shear stress and temperature agrees well with the experimental results of Imai and Sumino (1983).

6.5.5 Compounds

Dislocation velocities in compounds are mostly represented by relations of the type of Eq. 6-1. At equivalent temperatures (with respect to the absolute melting temperature T_m), dislocations are more mobile in $A^{III}B^{V}$ compounds than in elemental semiconductors. This fact is a source of problems for crystal growers. Because in compounds, α and β dislocations are distinguished by quite different values of the parameters v_0, m, and Q of Eq. 6-1, the activation energy of macroscopic plastic deformation (i.e., of the lower yield stress or maximum creep rate), is not simply related to the activation energies $Q_{\alpha, \beta}$ of dislocation motion. In comparing α and β dislocations, one generally finds that α dislocations (with anions in the core when belonging to the glide set, $B(g)$) are more mobile than β dislocations, but this difference strongly depends on doping: in p-type GaAs it may be even reversed. By high stress deformation, the order of mobilities of partial dislocations can be determined, as described above. It turns out that in undoped GaAs the mobility increases in the order $\mu(30\,\beta) < \mu(30\,\alpha) < \mu(90\,\alpha\beta)$, while in p-type GaAs the order of the two 30° partials is reversed (Androussi et al., 1987). The influence of irradiation on the mobility of dislocations in $A^{III}B^{V}$ compounds has been mentioned above. The literature on dislocation motion in GaAs has recently been reviewed (Alexander and Gottschalk, 1989).

Turning to $A^{II}B^{VI}$ compounds, Lu and Cockayne (1983) showed that also in those crystals dislocations stay dissociated when moving. Unfortunately, measurements of the velocity of single dislocations by classical methods are impossible because stationary dislocations become pinned (Ossipyan et al., 1986). Thus information on dislocation motion is gained from the electric current carried by groups of dislocation during cyclic deformation. In fact, the plastic strain rate $\dot{\varepsilon}$ is measured, and activation analysis is applied to $\dot{\varepsilon}(T, \tau)$. It is not clear if, and in which temperature range, a Peierls mechanism is active. Clearly, the electric charge carried by the dislocation is a determinant for the dislocation mobility. There are several mechanisms conceivable for this: The electrostatic energy of a kink in a charge line (Haasen, 1975), interaction of a charged dislocation with lattice rows of ions and/or with charged point defects (Ossipyan et al., 1986). $A^{II}B^{VI}$ crystals are plastically deformable down to much lower temperatures than other semiconductors and they exhibit a large positive photoplastic effect (PPE), i.e., during illumination the flow stress of a crystal drastically increases. This effect by most authors is interpreted as indicating reduced mobility of dislocations, but Takeuchi et al. (1983) claim to have proof of a reduced mean free path of dislocations. In CdS and CdTe there is also a negative PPE as in $A^{III}B^{V}$ compounds (Ossipyan and Shiksaidov, 1974; Gutmanas et al., 1979). In summary, dislocation motion in $A^{II}B^{VI}$ compounds shows many interesting properties, but it is far from being understood, even in the limited sense reached for elemental and $A^{III}B^{V}$ semiconductors.

6.6 Theory of Dislocation Motion

6.6.1 Dislocation Motion in Undoped Material

A number of theoretical models for dislocation dynamics have been derived to account for the experimental observations about the dislocation velocity discussed in the preceeding section. The development of these models is an interesting example of the interplay between theory and experiment, where the theory at present, however, has not reached its final form. There are some common features in the theories of dislocation dynamics which are based on rather general assumptions and there is a large number of facets which have

changed with time but nevertheless provide the framework of the current discussions. We will give a descriptive presentation of the actual state of the field by avoiding as far as possible detailed mathematical derivations. A review concerning the quantitative comparison between theory and experiment and including most of the necessary formulae was recently provided, e.g., by Alexander (1986). Three approaches to the theory of dislocation motion in undoped material will be presented here: the kink-diffusion model, as described by Hirth and Lothe (1982), the so-called 'weak-obstacle' theory of Celli, Kabler, Ninomiya, and Thomson (1963), and a modification of the weak-obstacle theory proposed by Möller (1978). The latter theory accounts for the fact that dislocations in semiconductors dissociate into partials, whereas the first two theories mentioned consider complete dislocations. In addition, theoretical approaches about the dislocation mobility in doped material will be discussed.

The theories of dislocation motion make use of the fact that in semiconductors the dislocations induced by deformations (except at high temperatures) are straight and aligned along a $\langle 110 \rangle$ direction of their $\{111\}$ slip plane, as revealed by Lang topographs and transmission electron microscopy observations. This behavior indicates a pronounced Peierls potential with Peierls valleys parallel to the $\langle 110 \rangle$ lattice rows. The dislocation velocity measurements are carried out with these straight screw and $60°$ dislocations (and their partials) forming hexagonal loops in the glide plane. The motion of such an "ideal" dislocation proceeds by kink propagation. The high Peierls potential (compared with the dislocation line energy) induces rather narrow kinks with kink width of the order of half the Burgers vector length b (b measures the fundamental translational period of the atomic arrangement along the dislocation line). Such a small kink width means that the dislocation passes from one Peierls valley to the next within a line segment of length b as indicated by the atomistic model in Fig. 6-18. Both features, the high Peierls barrier and the small kink width, are a consequence of the covalent bonding in the system. The fundamental elements of the theory of dislocation motion in this picture are the thermal nucleation of kinks, the propagation of the kinks along the dislocation, and, finally, the termination of the kink activity. Regarding the process of nucleation it commonly is accepted that by thermal fluctuations, kink pairs (called "double kinks", DK's) are generated with kink–kink distance larger then a critical length s^*. s^* is governed by the fact that for $s > s^*$, spreading of the DK's under the action of the local stress τ is energetically more favorable than regression, whereas for $s < s^*$ the increasing attractive kink interaction favors collapsing of the DK. With nucleation rate J for stable DK's, the dislocation velocity becomes

$$v = 2 h v_k J \bar{t}_{DK} \qquad (6\text{-}7)$$

(v_k: kink velocity, $\bar{t}_{DK}$: mean lifetime of a DK). Propagation of kinks may proceed as viscous flow or by thermally activated steps, the latter yielding a diffusive motion. Since kink movement in the semiconductors requires breaking and reconstruction of covalent bonds, which demand rather high energies compared to kT, this motion commonly is considered to be thermally assisted, giving for v_k under stress τ

$$v_k = \frac{\tau b h}{kT} a^2 v_D \exp(-W_m/kT) \qquad (6\text{-}8)$$

(a: diffusion step length, v_D: Debye frequency, W_m: activation enthalpy), the rela-

tionship already mentioned in Sec. 6.5.3. According to Hirth and Lothe (1982), the nucleation rate can be calculated under the condition of slow kink motion yielding

$$J = (v_k/b^2) \exp(-E_{DK}^*/kT) \qquad (6\text{-}9)$$

where the activation energy E_{DK}^* corresponds to the formation energy for a critical DK of width s^* and is slightly stress dependent. The proportionality of J with v_k reflects that the v_k scales the escape rate of DK's from the critical region around s^*. Within this Hirth-Lothe approach, it seems quite natural to identify W_m with the secondary Peierls potential, i.e., with the periodic variation of the kink self-energy along the dislocation due to the periodicity of the atomistic structure. Earlier internal friction experiments on Ge (Ohori and Sumino, 1972) have been interpreted as providing evidence for $W_m \approx 0.1$ eV. The more recent experiments, in particular those by Jendrich and Haasen (1988), support a different picture. Their internal friction measurements on Ge show two damping maxima, occurring after deformation only, with activation enthalpies of (1.108 ± 0.01) eV and (2.07 ± 0.2) eV, whereas the 0.1 eV peak of Ohori and Sumino (1972) was not found. With regard to a number of independent internal friction studies after 1972 which all failed in reproducing the 0.1 eV peak, Jendrich and Haasen (1988) came to the conclusion that this peak cannot be attributed to geometrical kink motion, because then it should have been detected by the later investigations, but may have been produced by point defects sensitive to purity and annealing treatment. They attribute their 1.108 eV peak to the movement of geometrical kinks on single partials, since this should be the process with lowest energy, giving $W_m = 1.108$ eV. The second peak is attributed to the formation of kink pairs, yielding $E_{DK}^* = 2.07$ eV. These conclusions are in accordance with computer simulation by Jones (1985) for Si, who deduced from an atomistic lattice model a value of $W_m = 1.3$ eV for kinks on the reconstructed 90° glide-set partials (1.4 to 1.9 eV for different variants of kinks on the 30° partials) in agreement with an early internal friction measurement by Southgate and Attard (1963).

The individual kinks contribute by their movement to the dislocation motion, until the kinks become immobile by reaching some impassable barrier ("strong obstacles", e.g., a node, a long jog, or possibly a sharp corner of the line) or until they are annihilated with an opposing kink from a neighboring DK. Depending upon whether immobilization (case I) or annihilation (case II) predominates, the mean active time of a DK, $\bar{t}_{DK}$, is determined either by $L/2v_k$ (with L the mean spreading width of a DK before immobilization, i.e., the average distance between the strong obstacles) or by L_{ann}/v_k (with $L_{ann} = 1/(J\bar{t}_{DK})$ the mean kink path before annihilation). For these situations v becomes

$$v = \begin{cases} h\,J\,L, & \text{(case I)} \\ 2h\sqrt{J\,v_k}, & \text{(case II)} \end{cases} \qquad (6\text{-}10)$$

In case I the velocity v scales with L and is thermally activated with enthalpy $Q = E_{DK}^* + W_m$. In case II the velocity is independent of L and has an activation enthalpy $Q = E_{DK}^*/2 + W_m$. v exhibits a lower increase with temperature in case II than in case I since the mean free path by which the kinks contribute to v is reduced with increasing density of kinks. According to experiments within the electron microscope (Louchet, 1981; Hirsch et al., 1981), the dislocation velocity in Si is proportional to the length L of the segments as long as L does not exceed some 0.2 μm. This indicates that $L_{ann} \approx 0.2$ μm. It implies that

under normal conditions (i.e., $L \geq 0.2$ μm) kink annihilation predominates in Si, and the measured activation enthalpy of v has to be interpreted as $Q = E_{DK}^*/2 + W_m$.

The Hirth-Lothe picture considered so far assumes that the kink motion is limited by W_m. For weak W_m Celli et al. (1963) argue that the motion of the kinks is significantly controlled by so-called "weak obstacles" (or dragging points), which are barriers distributed at random along the dislocation line with mean distance l and energy height E_d. l is considered to exhibit a temperature dependence like $l = l_0 \exp(-\varepsilon_1/kT)$ describing either a thermal instability of the weak obstacles or their finite binding energy to the dislocations. The kinks have to overcome these weak obstacles by thermal fluctuations where waiting in front of the obstacles has significant effects on the mean kink velocity $\bar{v}_k$ and on the effective DK nucleation rate, $\bar{J}$. It has to be taken into account for $\bar{v}_k$ that the time for traveling the distance l, $l/\bar{v}_k$, is given by the propagation time l/v_k plus the waiting time $v_0^{-1} \exp(E_d/kT)$ (with $v_0 \approx v_D$). For waiting times that are long compared with the propagation time this yields

$$\bar{v}_k = l\,v_0 \exp(-E_d/kT) \qquad (6\text{-}11)$$

The modification of the nucleation rate comes from the fact that stable DK's cannot be created too close to a weak obstacle, since a kink waiting in front of an obstacle tends to run backwards by fluctuations in its diffusion way and to be annihilated with its partner. Within this model (generalized by Rybin and Orlov, 1970), Celli et al. (1963) deduced the nucleation rate as follows:

$$\bar{J} = (v_0/b)\,(1 + E_d/(b\,h\,l\,\tau)) \cdot$$
$$\cdot \exp(-E_d/(b\,h\,l\,\tau) - E_{DK}^*/kT) \qquad (6\text{-}12)$$

Estimating the dislocation velocity from Eq. 6-10 with J and v_k substituted by $\bar{J}$ and $\bar{v}_k$ is the central point of the "weak-obstacle" theory of dislocation motion. The precise realization of the weak obstacles so far is not clear. Impurity atoms or atom clusters as well as lattice defects like vacancies or interstitials in the dislocation cores have been considered, as well as jogs and constrictions (see, e.g., Alexander, 1986). It seems that at present there are no convincing arguments particularly favoring one of these proposals, although there are arguments which make impurities or atom clusters and jogs or constrictions rather improbable candidates for weak obstacles in the case of freely moving straight dislocations with sufficiently large velocities (Alexander, 1986).

In the earlier investigations one of the main arguments in favor of the weak obstacle theory was its ability to describe the complicated stress dependence of the apparent activation energy $Q(\tau) := -d \ln v / d(1/kT)$ as deduced experimentally, e.g., from v data obtained by the etch pit technique (cf. Sec. 6.5.2). According to the recent discussion of the experimental situation by Sumino (1989), one has to be careful in using the dislocation velocity data from this technique. There are indications of a local pinning of dislocations in the surface regions sampled by this technique which are due to impurity clusters introduced from the surface when the crystal is kept at elevated temperatures to observe the dislocation motion. This pinning will be temperature dependent and particularly reduces the effective mobility of dislocations at low velocities, that is, at low stress τ. In highly pure Si, as discussed by Sumino (1989), the dislocation velocity is linear in τ with activation energy Q independent of τ (as long as the shape of the moving dislocation remains "regular"). This has

been confirmed in the stress range of 1 to 40 MPa/m^2 and temperatures between 600° and 800 °C with $Q = 2.20$ eV and 2.35 eV for 60° and screw dislocations in Si. Similar observations come from electron microscopy studies (Louchet, 1981; Hirsch et al., 1981).

The electron microscopy investigations revealed a continuous motion of dislocations without any waiting events down to the resolution limit of the method of 5 nm which is interpreted as an indication that weak obstacles, if they exist, have a mean distance of less than 5 nm. The electron microscopy observations have been analyzed (see Jones, 1983) in terms of the Hirth-Lothe picture and the weak obstacle model, yielding in the former case a secondary Peierls potential W_m of 1.2 and 1.35 eV. These estimates agree rather well with Jones's (1985) theoretical data from computer simulations mentioned above. This finding for Si, as well as the internal friction data of Jendrich and Haasen (1988) for Ge, might be considered to favor the Hirth-Lothe picture. As pointed out e.g. by Louchet and George (1983) for Si, the experimental values of v are larger than the predictions from this theory by some orders of magnitude, where the discrepancy was attributed to missing entropy terms. Following this idea Marklund (1985) succeeded in showing by computer simulations that the vibrational part of the kink migration entropy has, indeed, the right magnitude to account for the missing factor, which gives additional support to the correctness of the Hirth-Lothe theory.

An additional, necessary modification of the theory was introduced by Möller (1978) by considering that dislocations in Si and Ge are dissociated into partials with a stacking fault ribbon in between, where the partials are tightly bound to the $\langle 110 \rangle$ Peierls valleys. Since screw dislocations dissociate into two 30° partials, 60° dislocations in one 30° and one 90° partial, DK nucleation and kink motion on these partials must be considered to be the fundamental processes. Following a suggestion by Labusch, Möller (1978) took into account that the DK formation on the corresponding partials should be correlated below a critical stress τ_c, but that at stresses above τ_c uncorrelated nucleation occurs. τ_c turns out to be γ / d_0 (γ: stacking fault energy, d_0: dissociation width of the two partials) where $\tau_c \approx 1.0 \, \text{kg mm}^{-2}$ for 60° dislocations in Si, $1.6 \, \text{kg mm}^{-2}$ for screws (1.9 and $3.1 \, \text{kg mm}^{-2}$ in Ge). Möller (1978) compared his theory with etch pit measurements of the dislocation velocity in Ge (Schaumburg, 1972), Si (George et al., 1972), and GaAs (Choi et al., 1977). From this comparison he deduced estimates of the model parameters as compiled and critically considered by Alexander (1986). In the light of Sumino's (1989) remarks concerning the etch pit technique, the meaning of the parameters is, however, somewhat unclear, and a discussion of Möller's findings should be postponed until the controversy about this technique is resolved.

6.6.2 Dislocation Motion in Doped Semiconductors

So far, we have considered theoretical approaches concerning the dislocation velocity in undoped material. There are a number of theories about the effects of dopants on dislocation motion. The most significant feature in the doping dependence of the dislocation motion is the observation that v increases with doping in n- and p-Si as well as in n-Ge whereas it decreases with doping in p-Ge. The present theories consider as possible sources of this doping effect (Patel effect) a doping dependence of

the DK nucleation process or of the kink mobility. Particular examples of theories investigating the doping dependence of the DK formation are those by Patel and co-workers (Frisch and Patel, 1967; Patel et al., 1976), by Haasen (1975, 1979) and by Hirsch (1979). The doping dependence of kink mobility is studied by Kulkarni and Williams (1976), by Jones (1980, 1983), and by Jendrich and Haasen (1988). Since the earlier literature has been reviewed in detail (e.g., by Alexander, 1986) basic ideas of only some of these theories will be presented here in order to reflect the present discussion in this field.

Patel et al. (1976) proposed that the kinks are associated mainly with special charged dislocation sites, and that any mechanism that increases the electron concentration will increase the density of charged dislocation sites and consequently raise the kink concentration and dislocation velocity. In order to account for the observed effects, they assume that the dislocations in n- and p-type Ge as well as in n-Si introduce acceptor states whereas the dislocations are said to introduce donor-like states in p-Si. Patel and Testardi (1977a) succeeded in describing the relative change of the velocity of 60° dislocations in Ge by using a level position 0.13 eV below the conduction band. The best fit on screws in n-Si gave an acceptor level 0.6 eV above the valence band maximum E_v. Comparison with experiments on p-Si (at 450 °C) led to a donor level at the same position whereas velocity data for 550 °C result in a donor level about 0.75 eV above E_v. The approach was criticized by Schröter et al. (1977) since assuming centers of different type for dislocations in n- and p-Si yields severe inconsistencies if the doping is gradually lowered to the intrinsic range. Starting from n-doping, the theory predicts that the dislocations should be negatively charged in the intrinsic range. Starting from p-doping, it predicts a positive charge in this region. Patel and Testardi (1977 b) admitted this difficulty and claimed that their theory at least holds for Ge and n-Si.

A rather different approach was introduced by Haasen (1975, 1979). He assumed that the dislocations have partially filled electronic perturbation bands in the gap, as predicted by the microscopic theory for unreconstructed configurations (cf. Sec. 6.4.2), and that they are able to carry a net charge. As source of the Patel effect he considers a change in the DK formation enthalpy caused by a change in the effective charge on the dislocations due to doping. In his model the dislocation line displacement created by the DK reduces the electrostatic energy of the arrangement. The energy gain turns out proportional to the square of the line charge and acts as a doping-dependent reduction of the DK formation enthalpy. The possible charge on a dislocation is limited because of the Coulomb self-energy, which limits the possible gain of formation enthalpy. Regarding this it seems that the resulting energy reduction might be too small in most cases to solely account for the observed doping dependence of the dislocation velocity.

Hirsch (1979) in his theory treats dislocations which in their straight configuration have no deep energy levels in the band gap, as proposed by the microscopic theory for reconstructed core configurations (cf. Sec. 6.4.2). It is assumed that kinks have dangling bonds with deep donor and acceptor levels. Neutral and differently charged kinks are considered as independent thermodynamic species with individual equilibrium densities, where the density of charged kinks depends strongly on the relative position of the Fermi level compared to the the kink acceptor or donor level E_{kA}, E_{kD}. For the kink motion, Hirsch adopts the kink diffusion model with identical migration energy W_m for charged and uncharged kinks. Consequently, the doping dependence of the dislocation velocity goes with the doping dependence of the total density of charged plus uncharged kinks. For n-type material, negatively charged kinks are of importance where the ratio between charged and uncharged kink density is determined by the position of E_{kA}. The same seems to hold for p-Ge (which demands that $E_{kA} - E_v$ not be too large, and that the temperature be high enough), whereas for p-Si positively charged kink donor states play the main role in giving their electrons to the chemical acceptors. For Ge Hirsch arrives at $E_{kA} - E_v < 0.19$ eV. As reported by Jones (1983), Schröter has fitted earlier results in Si (George et al., 1972) to Hirsch's model yielding $E_{kA} - E_v = 0.67 \pm 0.04$ eV and $E_{kD} - E_v = 0.28 \pm 0.17$ eV.

According to the discussion by Jones (1980), it seems likely that the low-energy kinks in their equilibrium configuration on reconstructed dislocations have no dangling bonds and consequently no deep-donor or acceptor levels but provide shallow levels, in contrast to Hirsch's assumption. Regarding this observation, Jones (1980) deduced another model for the Patel effect. He assumes that in the process of migration, where in the saddle point configuration bonds may be stretched up to 30%, the shallow kink ground state levels deepen and hence, if charged, significantly lower the kink migration energy. In this picture, the kink density remains unaltered by doping, but the kink velocity changes. Within an atomistic model Heggie (1982) has shown that the kink levels in Si fall to $E_{kD} - E_v \approx 0.64$ eV, $E_{kA} - E_v \approx 0.87$ eV when the kinks reach their saddle point which indicates that considerable changes to the migration energy W_m are to be expected. Jones (1980) claims that the resulting expressions for the dislocation velocity in their formal structure agree with those of Hirsch (1979) although they have a different meaning and origin. Modified versions of the theory lateron are discussed by Jones (1983) now considering reconstruction defects (cf. Sec. 6.4.3), which he calls *solitons* or *antiphase defects,* on reconstructed dislocations as nucleation centers for DK's. The reconstruction defects represent local dangling-bond centers with deep electron levels and their density, of course, will change with doping, inducing a doping dependence of the DK nucleation rate. This picture is based on Jones's assumption (1983) of a sufficiently low formation enthalpy for reconstruction defects. Another estimate (Teichler, 1989a) deduces a rather high formation enthalpy and hence a low equilibrium density for these defects. Accordingly, the question about the equilibrium density of the reconstruction defects seems unsolved as far as theory is concerned. Experimentally (in Si), until now, no EPR signal of the isolated reconstruction defect has been detected despite intense efforts, which might be an indication that it is a rare configuration. In the same paper, Jones discusses that recombination of possibly highly mobile reconstruction defects with dangling-bond-carrying deep-level kinks, created during the DK nucleation process, may turn them into low energy, shallow-level kinks. The latter then exhibit a migration-energy reduction due to the deepening of the shallow levels in the saddle point configuration, as mentioned above.

An important experimental fact, initiating a further model, was provided by the internal friction measurements on deformed Sb-doped Ge by Jendrich and Haasen (1988). They showed that the acti-

vation enthalpy of the lower-energy process, reflecting the motion of geometrical kinks, is reduced by 0.3 to 0.4 eV under doping with (2 to 4) $\times 10^{17}$ cm^{-3} Sb. On the other hand, no doping dependence was observed in the activation enthalpy of the high-energy process attributed to kink pair formation. From a careful discussion of the available experimental and theoretical data, Jendrich and Haasen (1988) came to the conclusion that none of the models considered so far is able to provide a satisfying description of their experimental results. They propose that the doping dependence of the apparent W_m may reflect a change from low-mobile reconstructed kinks to highly mobile unreconstructed ones, where the latter have deep electron levels due to their dangling bonds, and the density of the latter increases under n-doping because of an energy gain when filling their levels with electrons.

Before closing this section, we should add one remark concerning the theories of dislocation velocity in undoped material. The theories sketched above are constructed to describe rather idealized situations where single dislocations move through a rather perfect crystal. Sumino's (1989) discussion about the pinning of near-surface dislocation segments at low stress by impurity clusters indicates that additional effects may be observed if the conditions are not as idealized as considered. A particular example of this is the effect of jogs introduced by climbing or by mutual cutting of dislocations from different glide systems. As proposed by Haasen (1979), these jogs may act as "weak obstacles" for kink motion or as strong pinning centers for the dislocations. They thus may introduce additional new features in the theoretical picture of dislocation mobility in the case of macroscopic plastic deformation of the samples.

6.7 Dislocation Generation and Plastic Deformation

6.7.1 Dislocation Nucleation

There are mainly three situations in which processors of semiconductors are faced with generation of dislocations: crystal growing, thermal processing, and growing epilayers on a substrate, leading to misfit dislocations. In the first part of this section, we will focus on the first two processes.

In contrast to intrinsic point defects, dislocations are never in thermal equilibrium, because of the small entropy of formation compared to the large enthalpy. Consequently, it is possible in principle to grow crystals of any substance dislocation free. Although in the case of ductile metals it is difficult to preserve such crystals without introducing dislocations by some surface damage, this is no problem with germanium and silicon, where dislocations are completely immobile at room temperature. Actually, up to now, it was not yet possible to grow dislocation free crystals of $A^{III}B^{V}$ compounds without high doping. This fact is connected to the higher mobility of dislocations in those materials and to some technical complications arising with evaporation of one component. The existence of such difficulties and, moreover, generation of dislocations in originally dislocation-free material make it worthwhile to study conditons under which dislocations may be generated (Alexander, 1989).

One idea involves nucleation of a dislocation loop by thermal fluctuations under the action of a mechanical stress. Obviously, introducing a dislocation means increasing the total energy of the crystal by the sum of strain and core energy of the loop. The driving force for nucleation of the loop is provided by the work done by the shear stress acting in the glide system of the loop; this work is proportional to the loop area. The resulting balance of energies goes through a maximum when the loop grows. Thus a critical radius R_c and an activation energy E_c are defined. Any reasonable estimate shows that E_c by far exceeds thermal fluctuations at any temperature. This means that in a perfect crystal (including equilibrium vacancies) dislocations cannot be generated by any stress.

We therefore must look for heterogeneous nucleation processes using some defects different from dislocations as nuclei. For semiconductors, the following are of importance:

(1) Surface damage;
(2) Agglomeration of native point defects;
(3) Punching of dislocation loops at precipitates of a second phase.

Before discussing these processes, it should be noted that Vanhellemont and Claeys (1988 a), when dealing with yielding (i.e., nucleation and multiplication of the source dislocations), considered only processes (1) and (3) above (together with multiplication of grown-in dislocations) to be heterogeneous and process (2) to be homogeneous (yielding).

1) Surface damage comprises all processes where large local stress applied to a thin surface layer of the crystal causes relative displacement of two regions of the crystal. This happens for instance with scratching, grinding, hardness indentations, impinging hard particles, etc. Hill and Rowcliffe (1974) analyzed hardness indentations on silicon surfaces and came to the conclusion that locally the theoretical shear strength is overcome followed by an out-of-register recombination of the two faces of the cut. Temperatures above of the brittle-ductile transition of the respective

substance are required to expand the dislocation loops produced, which are of the order of 10 μm diameter.

2) Intrinsic point defects in excess of the thermal equilibrium density are produced either by cooling from a high temperature, especially during crystal growth, or by precipitation of some impurity species (e.g., oxygen in silicon) producing a huge amount of self-interstitials (SI's). Admittedly the equilibrium density of vacancies (V's) and SI's in semiconductors are considerably lower than in metals, but any nonequilibrium concentration is hard to remove since annihilation of SI's with V's seems to be hindered by an energy barrier. Thus the common way to remove the excess point defects far from the crystal surface is agglomeration in spherical or – because of elastic strain energy – in platelike structures parallel to close-packed lattice planes. In the diamond structure, a double layer of V's or SI's embedded in the matrix is equivalent to an intrinsic or extrinsic stacking fault, respectively. We call this formation of an area of stacking fault (SF) step 1 of dislocation nucleation by agglomeration. In compounds, V's or SI's of the two sublattices should coprecipitate to form an ordinary SF. But this ideal case in general will not be realized. If, for example, in GaAs excess arsenic atoms are precipitating interstitially, a full layer of interstitial GaAs is formed by emission of the related number of Ga vacancies. (The subsequent complicated steps leading to precipitates of hexagonal arsenic are not of interest here (B. F. Lee et al., 1988).) The SF mentioned above is bound by a sessile Frank partial loop. If, by climb of this loop, the disc reaches a certain critical size, a Shockley partial dislocation may be spontaneously nucleated in its center, removing the SF and transforming eventually the Frank partial into a loop of a perfect

prismatic dislocation (unfaulting of the SF = step 2 of nucleation):

$$a/3\,[111] + a/6\,[\bar{2}11] \rightarrow a/2\,[011]\ (\text{in}\,(111))$$

The Burgers vector of the resultant dislocation makes suitable segments of the loop glissile in either the $(11\bar{1})$ or in the $(1\bar{1}1)$ plane (step 3).

Here, a fundamental principle of all heterogeneous nucleation mechanisms comes to light: An expanding dislocation loop needs to overcome the backstress which is due to interaction between all loop segments. A rough estimate for a circular loop shows that the critical radius for expansion under a shear stress τ is

$$R_c = \mu b/2\tau$$

($\mu b^2/2$ represents the line tension, μ being the appropriate shear modulus). For typical values (Si: $b = 3.85 \times 10^{-10}$ m, $\mu = 63.4$ MPa, $\tau = 10$ MPa) R_c turns out as 1.25 μm. Considerable stress concentration is necessary to expand dislocation loops smaller than one micrometer.

Thus the climb force (consisting of a chemical part due to the supersaturation of point defects and possibly a mechanical part) must be sufficiently large to increase the loop to the critical size so that glide motion and multiplication of a loop segment becomes possible. In this case, the second period of dislocation generation begins, namely, multiplication, which will be treated in the next section.

In a series of papers, Vanhellemont et al. analyzed the homogeneous nucleation of dislocations at the edge of a film (e.g., SiO_2, SiN_4, etc.) covering a silicon wafer (Vanhellemont et al., 1987; Vanhellemont and Claeys, 1988 a, b). From the calculated stress field of the film edge a particular dislocation is determined, which for a given orientation of substrate and film edge will grow fastest by climb and subsequently by glide. For CZ-grown substrates the point defects nucleating the dislocations are SI's produced by precipitating interstitial oxygen. The analysis is able to explain a great deal of various geometries and is confirmed by electron microscopical work. However, the origin of the first edge dislocation growing by climb is not clear. It is important to note that the stress field at the film edge promotes by mechanical climb force the precipitation of SI in this particular area and at particular dislocation types.

Föll and Kolbesen (1975) showed that A-swirls in "dislocation-free" silicon consist of perfect extrinsic dislocation loops which are also nucleated by agglomeration of SI's. Föll and Kolbesen found loops with SF's only if carbon (10^{17} cm^{-3}) and oxygen (10^{16} cm^{-3}) were present in high concentrations. Apparently, these impurities stabilized the stacking fault against unfaulting. The observed loops are rather large (0.5 to 1 µm). From their density (10^6-10^7 cm^{-3}) one may deduce a dislocation density of the investigated "dislocation-free" crystals of about 10^2-10^3 cm^{-2}. Only part of those dislocations becomes mobile by the release of a segment of the prismatic loop onto a glide plane. One may calculate from the data given in the paper that between 10^{13} and 5×10^{14} cm^{-3} of SI's are removed from the crystal by formation of A-swirls.

It should be stressed that the tendency of point defects to agglomerate in semiconductors is strongly correlated to their charge state and, therefore, to the actual position of the Fermi level. Convincing proof for that was given by Lagowski et al. (1984) in a paper treating the strong influence of electrical doping on the density N_0 of grown-in dislocations in GaAs crystals grown by the horizontal Bridgman technique. While N_0 for undoped GaAs is of

the order of 10^3 cm^{-3}, it decreases in n-type material (due to doping with Si or Se) to effectively zero and increases in p-type (Zn) crystals to 5×10^4 cm^{-3}. Actually, only the net doping ($N_D - N_A$) is of influence. Since the arsenic vacancy (V_{As}) is positively charged wherever the Fermi level is, it must be the V_{Ga} whose charge state determines the agglomeration of double layers of vacancies as the first step of dislocation generation under the low-stress conditions of the particular growth technique. The gallium vacancy is an acceptor with several states in the lower half of the energy gap. The authors claim that in a case where gallium vacancies dominate, arsenic antisite defects (As_{Ga}) are generated and not arsenic interstitials, to complete a double layer of vacancies. The reaction proceeds as follows:

$$V_{Ga}^{--} + As_{As} \rightarrow As_{Ga} + V_{As}^+ + 3\,e^-$$

The transition of an arsenic atom into a gallium vacancy (thereby providing the arsenic vacancy) is more promoted with decreasing numbers of free electrons. This means that the first step of dislocation nucleation by agglomeration of supersaturated vacancies in GaAs proceeds in proportion to $(n)^{-3}$, which is in good agreement with experiment. Generation of As_{Ga} antisite defects together with dislocations can be understood in this model as well.

3) Particles of a second phase (e.g., oxides) may produce dislocation loops in the surrounding matrix either by a volume misfit or by a difference of thermal expansion between particle and matrix. As was shown by Ashby and Johnson (1969), glide dislocation loops are generated around spherical particles in the glide plane of maximum shear stress. In many cases, the screw segments of those loops disappear afterwards by repeated cross slip, leaving prismatic dislocation loops. The transfor-

mation from a glide to a prismatic loop does not always take place: in dislocation-free FZ Si, a high shear stress (200 MPa) applied for some hours at 420 °C (after pretreatment at 700 °C) produces large glide loops in the plane of maximum shear stress (Krüchten v., 1984; Alexander et al., 1983). Calculating the critical radius for loop expansion under 200 MPa shear stress, one finds that the radius of the original loop, and therefore, of the particle, must be at least 70 nm.

Referring to crystal growth under liquid confinement (liquid encapsulation Czochralski technique), precipitations of one component of the compound or inclusions of the encapsulant glass may nucleate dislocations. For expansion of those nuclei of loops, the stress-temperature history of the crystal is decisive. This expansion and multiplication belongs to the second period of dislocation generation: growth and multiplication. Concluding this section, we should mention another type of dislocation nucleation which occurs frequently in metallic alloys: constitutional supercooling. Depending on the distribution coefficients k_i in the system under consideration, on the temperature gradient at the solidification interface, and on the growth rate, local supercooling may destabilize the planar solidification interface, and the freezing crystal will then be divided into cells separated by cell walls enriched in one of the components of the alloy. Those chemical inhomogeneities are connected with differences of the lattice constant and may eventually lead to small angle boundaries of misfit dislocations. It has been discussed whether the well-known cell structure of LEC-grown GaAs might be due to constitutional supercooling. But the observation that dislocation cells can be generated by after-growth anneal points to polygonization of otherwise produced dislocations.

6.7.2 Dislocation Multiplication (Plastic Deformation)

The dislocation content N_0 of as-grown crystals is often explained by plastic deformation under the action of thermal stress during the cooling period. This idea has its origin from the observation of dislocation etch pits on the crystal surface and on cross sections being arranged along slip lines, i.e., along the traces of slip planes. In Sec. 6.7.1 it was shown that this does not provide a complete explanation. First, some "source" dislocations have to be nucleated from defects of a different kind, whereby stress is helpful. Not before a critical density of such *mobile* dislocations (of the order of 10^3–10^4 cm^{-2}) is nucleated can motion and multiplication of dislocation – i.e. plastic deformation – take over the increase of the dislocation density (by several orders of magnitude). For improving growth methods, this distinction seems to be important, because it demonstrates that it may be more promising to control intrinsic point defects and precipitation of oxides than to remove fully thermal stress. Crystal cooling and its influence on plastic deformation is often tackled in the framework of the model of an elastic-plastic solid. Here it is assumed that any volume element of the crystal may accommodate elastically a certain part σ_{el} of the thermal stress, the excess stress being removed by plastic strain ε_{pl} with a linear relationship between ε_{pl} and $(\sigma - \sigma_{el})$. Calculating from the tensor of thermal stress the shear stress τ in the most-stressed glide system, τ_{el} is called "yield stress" or "critical resolved shear stress CRSS". The growth conditions are then adjusted so as not to reach the CRSS in any part of the growing crystal. This conception may be accepted for f.c.c. metals with "instantaneous" response of dislocations to the stress distribution. However, because of the thermally activated dislocation motion in semiconductors (cf. Sec. 6.5), the thermal history of the considered volume element plays an important role. Here the relation between shear stress and plastic strain is far from being unique. Moreover, the same $\dot{\varepsilon}_{pl}$ may be accommodated by a few fast-moving dislocations or by many slow ones. Thus the dislocation density that eventually appears depends on stress and temperature during the whole cooling history. The conception of CRSS therefore suffers mainly from its neglecting the dimension of time, not so much from the (not generally correct) assumption of a starting stress for dislocation activity. Semiquantitative analysis of density and distribution of grown-in dislocations in as-grown crystals of InP (LEC grown) on the basis of the dynamic properties of dislocations in that particular semiconductor has been carried out by Völkl (1988) and Völkl et al. (1987). Data on the properties of dislocations are obtained by standard deformation tests: single crystals are compressed uniaxially along an axis far from any highly symmetric direction. In this manner, mainly one glide system is activated: about 80% of the dislocations belong to one Burgers vector and glide plane (single slip). Beneficial for those tests are crystals with about 10^4 cm^{-2} grown-in dislocations so that nucleation of dislocations does not interfere. The compression test may be carried out with constant strain rate $\dot{\varepsilon}$ (the dynamical test resulting in a stress-strain curve) or with constant shear stress τ (creep test). An analysis of such deformation tests was carried out in the 1960s and is reviewed elsewhere (Alexander and Haasen, 1968; Alexander, 1986). Here we give just the essentials: Applying Eq. 6-1 for the dislocation velocity $v(\tau, T)$ to plastic deformation, i.e., simultaneous activity of very many dislocations, one has to replace the shear stress τ applied to the crystal by an effective stress

$$\tau_{eff} = \tau - A\sqrt{N} \qquad (6\text{-}13)$$

(N: actual dislocation density. For single slip, the term $A\sqrt{N}$ stems mainly from parallel dislocations of the primary slip system and can be calculated from the theory of elasticity). Equation (6-13) describes screening of the stress τ by the stress field of the other dislocations.

To calculate the development of the dislocation density N during deformation one has to know the law of dislocation multiplication. Experiments with Ge and Si revealed as a reliable approximation:

$$dN = N K \tau_{eff}^n v \, dt \qquad (6\text{-}14)$$

(n being 1 or 0). From Eq. (6-14), it becomes clear that multiplication proceeds by motion and not from fixed sources like the Frank-Read source. For compounds, this may be different because the segments of dislocation loops have extremely different mobility so that at the beginning of deformation only suitably oriented surface sources are active (Kesteloot, 1981). The nature of those sources is not well understood; it is possible that they are due to surface damage.

The stress exponent n in Eq. (6-14) depends on the dislocation density: for weakly deformed crystals, n is zero (no explicit dependence on stress); in a heavily deformed state, n becomes 1 (the extension of dipoles now dominates dislocation multiplication) (Alexander, 1986).

Combining Eqs. (6-1), (6-13), and (6-14) with the Orowan relation

$$\dot{\varepsilon}_{pl} = N b v \qquad (6\text{-}15)$$

which treats plastic strain rate as flux of mobile dislocations, one may calculate the dislocation density N and either stress τ

(for the dynamical test) or plastic strain $\varepsilon_{\mathrm{pl}}$ (creep test) as a function of time for a given temperature.

The Eqs. (6-1) and (6-13)–(6-15) offer an easy approach to the yield point phenomenon of dislocation-lean crystals (Johnston and Gilman, 1959). As long as N is small ($< 10^4$ cm^{-2}), the experimentally prescribed strain rate $\dot{\varepsilon}$ has to be provided mainly by elastic strain

$$\dot{\varepsilon} = \dot{\tau}/G + N\,b\,v$$

(G: effective shear modulus of the specimen and the machine).

Thus the stress increases with time quasielastically. Simultaneously, the dislocation density increases up to the upper yield point of the stress-strain curve where for the first time (Nbv) equals $\dot{\varepsilon}$. Continuing deformation further increases N, and stress can decrease. This causes the elastic term to become negative: the stress-strain curve adopts a negative slope. Eventually, at the lower yield point (Nbv) again equals $\dot{\varepsilon}$ but now with a dislocation density about three orders of magnitude bigger than at the upper yield point. From the lower yield point on, strain hardening dominates the stress, and the stress-strain curve follows the same three-stage scheme as with many metals.

The analysis of the first steps of plastic deformation as it was described above works so easily because most of the dislocations which become immobile during deformation are present as multipoles and therefore do not significantly contribute to the screening of stress, either. To calculate the (local) dislocation density that results from plastic deformation in a crystal cooling from the freezing temperature, we proceed as follows: first, a reasonable density of nucleated source dislocations of critical length is chosen, and then each volume element of the crystal is followed as it

moves through the particular temperature field of the particular apparatus for crystal growing. Of course, the growth rate introduces the dimension of time. Interestingly enough, not only slow growth may result in good crystals (because here the thermal stress may be better controlled) but also rather large growth rates again may be suitable: The time is too short for extensive dislocation multiplication before the considered volume element comes into cooler regions where dislocations move slowly. This variant is used for growing dislocation-free silicon, but for $A^{\mathrm{III}}B^{\mathrm{V}}$ compounds, with dislocations being more mobile, the necessary growth rate is too high to be realized at the present time.

6.7.3 Generation of Misfit Dislocations

One of the most frequent steps in device fabrication involves growing a layer of semiconducting material on top of a crystalline substrate with a different lattice parameter. It is clear from first principles that a critical thickness t_{c} of the layer will be present, beyond which the introduction of dislocations with an edge component into the interface region reduces the total energy of the system. t_{c} will depend on the lattice mismatch δ. However, experimental evidence shows that dislocation-free layers thicker than t_{c} often grow on a dislocation-free substrate. One must therefore conclude that the nucleation of misfit dislocations must overcome some barrier as well.

In fact, modern theories (Dodson and Tsao, 1987; Hull et al., 1989) do not consider an equilibrium situation but rather relaxation from a metastable dislocation-free state by nucleation and propagation of misfit dislocations. Applying the microdynamical theory as out-lined in 6.7.2 or similar models, these theories give calculations of the density of misfit dislocations to be expected for a given thermal history of the multilayer system. "A crucial factor determining strain relaxation is time at temperature as well as layer thickness" (Hull et al., 1989). Hull et al. (1989) describe for the system Ge_xSi_{1-x}/Si a method of measuring the material parameters needed for calculation, with respect to plasticity in the electron microscope. As in all computations along the lines of the microdynamical theory, the origin and the density of the first source dislocations must be known. One possible mechanism for the formation of such sources has been described by Hagen and Strunk (1978). Actually, there is no generally accepted model for the origin of the first misfit dislocations for those cases where the substrate does not provide threading source dislocations. (For a recent discussion stressing the nucleation of partial dislocations compare Hirsch, 1991.) For misfits exceeding 2% homogeneous nucleations of dislocation half loops at surface steps has been proposed (Eaglesham et al., 1989). For lower misfit, these authors quite recently found a new type of dislocation source. It consists of a particular type of stacking fault whose bounding partial can dissociate into a Shockley partial and a perfect dislocation. In fact, perfect dislocations on two different glide planes can be produced and the process is self-reproducing. These properties are deduced from electron microscopy of Ge_xSi_{1-x} layers on Si(100). It must be left for future research whether this type of source is frequent.

6.7.4 Gettering with the Help of Dislocations

Frequently during high-temperature processing of silicon devices, metal atoms precipitate, mostly as silicides; these precipitates have to be removed afterwards. As was shown by Ourmazd (1986), one may achieve that in principle by a two-step process: First, the precipitates are dissolved by a pulse of self-interstitials, and then the metal atoms are fixed (gettered) in a strained area far from the active region of the device. Strain-causing gettering mechanisms may be produced by surface damage (either on the front or on the back side), by phosphorous diffusion, by film deposition or by a defective layer in the bulk some 20 μm below the active zone. This latter technique is known as "intrinsic gettering". In fact, dislocation generation is induced by oxide precipitation. Before those oxides are formed, the oxygen has to be outdiffused from the active layer.

6.8 Acknowledgement

The authors wish to thank W. Schröter for critically reading the manuscript.

6.9 References

Alexander, H. (1974), *J. de Phys. 33*, C7-173.
Alexander, H. (1979), *J. de Phys. 40*, C6-1.
Alexander, H. (1984), In: *Dislocations 1984:* Veyssiere, P., Kubin, L., Castaing, J. (Eds.). Paris: CNRS 283.
Alexander, H. (1986), In: *Dislocations in Solids:* Nabarro, F. R. N. (Ed.). Amsterdam: North Holland, vol. 7, 113.
Alexander, H. (1989), *Rad. Eff. Def. Sol. 111–112,* 1.
Alexander, H., Gottschalk, H. (1989), *Inst. Phys. Conf. Ser. 104,* 281.
Alexander, H., Haasen, P. (1968), In: *Solid State Physics.* New York: Academic Press, Vol. 22, p. 27.
Alexander, H., Eppenstein, H., Gottschalk, H., Wendler, S. (1980), *J. Microsc. 118,* 13.
Alexander, H., Kisielowski-Kemmerich, C., Weber, E. R. (1983), *Physica 116B,* 583.
Alexander, H., Kisielowski-Kemmerich, C., Swalski, A. T. (1987), *Phys. Stat. Sol. (a) 104,* 183.
Alexander, H., Dietrich, S., Hühne, M., Kolbe, M., Weber, G. (1990), *Phys. Stat. Sol. (a) 117,* 417.
Alstrup, I., Marklund, S. (1977), *Phys. Stat. Sol. (b) 80,* 301.
Androussi, Y., Vanderschaeve, G., Lefebvre, A. (1987), *Inst. Phys. Conf. Ser. 87,* 291.

Ashby, M. F., Johnson, L. (1969), *Phil. Mag. 20*, 1009.

Baraff, G. A., Kane, E. O., Schlüter, M. (1980), *Phys. Rev. B21*, 5662.

Bartelsen, L. (1977), *Phys. Stat. Sol. (b) 81*, 471.

Baumann, F. H., Schröter, W. (1983a), *Phys. Stat. Sol. (a) 79*, K 123.

Baumann, F. H., Schröter, W. (1983b), *Phil. Mag. B48*, 55.

Bazhenov, A. V., Krasilnikova, L. L. (1986), *Sov. Phys. Sol. State 28*, 128.

Belgavskii, V. I., Darinskii, B. M., Sviridov, V. V. (1985), *Sov. Phys. Sol. State 27*, 658.

Blanc, J. (1975), *Phil. Mag. 32*, 1023.

Bourret, A., Desseaux-Thibault, J., Lancon, F. (1983), *J. de Phys. 44*, C4-15.

Brohl, M. (1990), Thesis Univ. Köln.

Brohl, M., Alexander, H. (1989), *Inst. Phys. Conf. Ser. 104*, 163.

Brohl, M., Kisielowski-Kemmerich, C., Alexander, H. (1987), *Appl. Phys. Lett. 50*, 1733.

Brohl, M., Dressel, M., Helberg, H. W., Alexander, H. (1990), *Phil. Mag. B61*, 97.

Castaing, J., Veyssiere, P., Kubin, L. P., Rabier, J. (1981), *Phil. Mag. A44*, 1407.

Celli, V., Gold, A., Thomson, R. (1962), *Phys. Rev. Lett. 8*, 96.

Celli, V., Kabler, M., Ninomiya, T., Thomson, R. (1963), *Phys. Rev. 131*, 58.

Chaudhuri, A. R., Patel, J. R., Rubin, J. (1962), *J. Appl. Phys. 33*, 2736.

Chelikovsky, J. R. (1982), *Phys. Rev. Lett. 49*, 1569.

Chelikovsky, J. R., Spence, J. C. H. (1984), *Phys. Rev. B30*, 694.

Choi, S. K., Mihara, M., Ninomiya, T. (1977), *Jap. J. Appl. Phys. 16*, 737.

Christoffel, E., Benchiguer, T., Goltzene, A., Schwab, C., Wang Guangyu, Wu Yu (1990), *Phys. Rev. B42*, 3461.

Claesson, A. (1979), *J. de Phys. 40 Suppl.*, C6-39.

Dodson, B. W., Tsao, J. Y. (1987), *Appl. Phys. Lett. 51*, 1325.

Drozdov, N. A., Patrin, A. A., Tkachev, V. D. (1976), *Sov. Phys. JETP Lett. 23*, 597.

Duesbery, M. S., Joos, B., Michel, D. J. (1991), *Phys. Rev. B43*, 5143.

Eaglesham, D. J., Maher, D. M., Kvam, E. P., Bean, J. C., Humphreys, C. J. (1989), *Phys. Rev. Lett. 62*, 187.

Erdmann, R., Alexander, H. (1979), *Phys. Stat. Sol. (a) 55*, 251.

Escaig, B. (1968), *J. de Phys. 29*, 225.

Fanelsa, A. (1989), *Diploma thesis Univ. Köln*.

Farber, B. Y., Gottschalk, H. (1991), in: *Polycrystalline Semiconductors, Schwäbisch Hall 1990*. Heidelberg: Springer, in press.

Farber, B. Y., Nikitenko, V. I. (1982), *Phys. Stat. Sol. (a) 73*, K14.

Farber, B. Y., Bondarenko, I. E., Nikitenko, V. I. (1981), *Sov. Phys. Sol. State 23*, 1285.

Farber, B. Y., Iunin, Y. L., Nikitenko, V. I. (1986), *Phys. Stat. Sol. (a) 97*, 469.

Farvacque, J. L., Vignaud, D., Depraetere, E., Sieber, B., Lefebvre, A. (1989), *Inst. Phys. Conf. Ser. 104*, 141.

Ferré, D., Diallo, A., Farvacque, J. L. (1990), *Rev. Phys. Appl. 25*, 177.

Figielski, T. (1978), *Sol. State Electron. 21*, 1403.

Föll, H., Kolbesen, B. O. (1975), *Appl. Phys. 8*, 319.

Friedel, J. (1964), Dislocations (Pergamon Press).

Frisch, H. L., Patel, J. R. (1967), *Phys. Rev. Lett. 18*, 784.

Gelsdorf, F., Schröter, W. (1984), *Phil. Mag. A49*, L35.

George, A., Champier, G. (1980), *Scripta Metall. 14*, 399.

George, A., Michot, G. (1982), *J. Appl. Cryst. 15*, 412.

George, A., Rabier, J. (1987), *Rev. Phys. Appl. 22*, 941.

George, A., Escaravage, C., Champier, G., Schröter, W. (1972), *Phys. Stat. Sol. (b) 53*, 483.

Gerthsen, D. (1986), *Phys. Stat. Sol. (a) 97*, 527.

Gottschalk, H., Patzer, G., Alexander, H. (1978), *Phys. Stat. Sol. (a) 45*, 207.

Gottschalk, H., Alexander, H., Dietz, V. (1987), *Inst. Phys. Conf. Ser. 87*, 339.

Grazhulis, V. A. (1979), *J. de Physique 40*, Suppl. C6-59.

Gröhlich, M. (1987), Thesis Univ. Göttingen.

Grosbras, P., Demenet, J. L., Garem, H., Desoyer, J. C. (1984), *Phys. Stat. Sol. (a) 84*, 481.

Gutmanas, E. Y., Travitzky, N., Haasen, P. (1979), *Phys. Stat. Sol. (a) 51*, 435.

Haasen, P. (1975), *Phys. Stat. Sol. (a) 28*, 145.

Haasen, P. (1979), *J. de Phys. 40*, C6-111.

Haasen, P., Seeger, A. (1958), *Halbleiterprobleme IV*: Schottky, W. (Ed.). Braunschweig: Vieweg, p. 68.

Haasen, P., Müller, H., Zoth, G. (1983), *J. de Phys. 44*, C4-365.

Hagen, W., Strunk, H. P. (1978), *Appl. Phys. 17*, 85.

Heggie, M. (1982), *Thesis Univ. Exeter*.

Heggie, M., Jones, R. (1983), *Phil. Mag. B48*, 365.

Heggie, M., Jones, R. (1987), *Inst. Phys. Conf. Ser. 87*, 367.

Heggie, M., Jones, R., Lister, G. M. S., Umerski, A. (1989), *Inst. Phys. Conf. Ser. 104*, 43.

Heister, E. (1982), *Diploma thesis Univ. Köln*.

Higgs, V., Lightowlers, E. C., Kightley, P. K. (1990a), *Mat. Res. Soc. Symp. Proc. 163*, 57

Higgs, V., Davies, G., Kubiak, R. (1990b), *Int. Conf. Shallow Impur. Semicond. London*.

Higgs, V., Norman, C. E., Lightowlers, E. C., Kightley, P. K. (1991), in: *Microscopic Semiconducting Materials*. Oxford: Oxford University Press.

Hill, M. J., Rowcliffe, D. J. (1974), *J. Mater. Sci. 9*, 1569.

Hirsch, P. B. (1979), *J. de Phys. 40*, C6-117.

Hirsch, P. B. (1980), *J. Microsc. 118*, 3.

Hirsch, P. B. (1991), in: *Polycrystalline Semiconductors, Schwäbisch Hall 1990*. Heidelberg: Springer, in press.

Hirsch, P. B., Ourmazd, A., Pirouz, P. (1981), *Inst. Phys. Conf. Ser. 60*, 29.

Hirth, J. P., Lothe, J. (1982), *Theory of Dislocations* (2. ed.). New York: Wiley.

Hull, R., Bean, J. C., Buescher, C. (1989), *J. Appl. Phys. 66*, 5837.

Hutson, A. R. (1983), *Physica 116B*, 650.

Imai, M., Sumino, K. (1983), *Philos. Mag. A47*, 599.

Izotov, A. N., Kolyubakin, A. I., Shevchenko, S. A., Steinman, E. A. (1990), *Defect Control in Semicond.:* Sumino, K. (Ed.), North Holland, 1447.

Jebasinski, R. (1989), *Diploma thesis Univ. Köln.*

Jendrich, U., Haasen, P. (1988), *Phys. Stat. Sol. (a) 108*, 553.

Johnston, W. G., Gilman, J. J. (1959), *J. Appl. Phys. 30*, 129.

Jones, R. (1977), *Phil. Mag. 35*, 57.

Jones, R. (1980), *Phil. Mag. B42*, 213.

Jones, R. (1983), *J. de Phys. 44*, C4-61.

Jones, R. (1985), In: *Dislocations in Solids:* Suzuki, H., Ninomiya, T., Sumino, K., Takeuchi, S. (Eds.). Univ. of Tokyo Press 343.

Jones, R., Marklund, S. (1980), *Phys. Stat. Sol. (b) 101*, 585.

Jones, R., Öberg, S., Marklund, S. (1981), *Phil. Mag. B43*, 839.

Keating, P. N. (1966), *Phys. Rev. 145*, 637.

Kesteloot, R. (1981), *Thesis Univ. Lille.*

Kimerling, L. C., Patel, J. R. (1979), *Appl. Phys. Lett. 34*, 73.

Kimerling, L. C., Patel, J. R., Benton, J. L., Freeland, P. E. (1981), *Inst. Phys. Conf. Ser. 59*, 401.

Kirichenko, L. G., Petrenko, V. F., Uimin, G. V. (1978), *Sov. Phys. JETP 47*, 389.

Kisielowski-Kemmerich, C. (1982), *Diploma thesis Univ. Köln.*

Kisielowski-Kemmerich, C. (1989), *Inst. Phys. Conf. Ser. 104*, 187.

Kisielowski-Kemmerich, C. (1990), *Phys. Stat. Sol. (b) 161*, 11.

Kisielowski-Kemmerich, C., Alexander, H. (1988), In: *Defects in Crystals:* Mizera, E. (Ed.), Singapore: World Scientific Publ., p. 290.

Kisielowski-Kemmerich, C., Weber, E. R. (1991), *Phys. Rev. B44*, 1600.

Kisielowski-Kemmerich, C., Weber, G., Alexander, H. (1985), *J. Electronic Mater. 14a*, 387.

Kisielowski-Kemmerich, C., Czaschke, J., Alexander, H. (1986), *Mater. Sci. Forum 10–12*, 745.

Kisielowski, C., Palm, J., Bollig, B., Alexander, H. (1991), *Phys. Rev. B44*, 1588.

Kittler, W., Seifert, W. (1981), *Phys. Stat. Sol. (a) 66*, 573.

Koizumi, H., Ninomiya, T. (1978), *J. Phys. Soc. Jap. 44*, 898.

Kronewitz, J., Schröter, W. (1987), *Izvest. Acad. Nauk USSR, Ser Fiz. 51*, 682.

Krüchten v., M. (1984), *Diploma thesis Univ. Köln.*

Krüger, J., Alexander, H. (1991), *16. Int. Conf. Def. Semic.*, to be publ.

Küsters, K.-H., Alexander, H. (1983), *Physica 116B*, 594.

Kulkarni, S. B., Williams, W. S. (1976), *J. Appl. Phys. 47*, 4318.

Kveder, V. V., Ossipyan, Y. A., Shalynin, A. I. (1984), *Sov. Phys. JETP Lett. 40*, 729.

Kveder, V. V., Koshelev, A. E., Mchedlidze, T. R., Ossipyan, Y. A., Shalynin, A. I. (1989), *Sov. Phys. JETP 68*, 1041.

Labusch, R., Schröter, W. (1980), In: *Dislocations in Solids:* Nabarro, F. R. N. (Ed.). Amsterdam: North Holland V. *5*, 129.

Lagowski, J., Gatos, H. C. (1982), *Appl. Phys. Lett. 40*, 342.

Lagowski, J., Gatos, H. C., Aoyama, T., Lin, D. G. (1984), *Appl. Phys. Lett. 45*, 680.

Lee, Y.-H., Corbett, J. W. (1976), *Phys. Rev. B13*, 2653.

Lee, D. M., Posthill, J. B., Shimura, F., Rozgonyi, G. A. (1988), *Appl. Phys. Lett. 53*, 370.

Lee, B. F., Gronsky, R., Bourret, E. D. (1988), *J. Appl. Phys. 64*, 114.

Lelikov, Y. S., Rebane, Y. T., Shreter, Y. G. (1989), *Inst. Phys. Conf. Ser. 104*, 119.

Lodge, K. W., Altmann, S. L., Lapiccirella, A., Tomassini, N. (1984), *Phil. Mag. B49*, 41.

Lodge, K. W., Lapiccirella, A., Battistoni, C., Tomassini, N., Altmann, S. (1989), *Phil. Mag. A 60*, 643.

Louchet, F. (1981), *Phil. Mag. 43*, 1289.

Louchet, F., George, A. (1983), *J. de Phys. 44*, C4-51.

Louchet, F., Thibault-Desseaux, J. (1987), *Rev. Phys. Appl. 22*, 207.

Lu, G., Cockayne, D. J. H. (1983), *Physica 116B*, 646.

Maeda, K. (1988), In: *Defects in Crystals:* Mizera, E. (Ed.). Singapore: World Scient. Publ., 153.

Maeda, K., Takeuchi, S. (1981), *Jap. J. Appl. Phys. 20*, L165.

Maeda, K., Takeuchi, S. (1985), In: *Dislocations in Solids:* Suzuki, H., Ninomiya, T., Sumino, K., Takeuchi, S. (Eds.). Tokyo: Univ. Press 433.

Maeda, K., Yamashita, Y. (1989), *Inst. Phys. Conf. Ser. 104*, 269.

Marheine, C. (1989), *Thesis Univ. Göttingen.*

Marklund, S. (1979), *Phys. Stat. Sol. (b) 92*, 83.

Marklund, S. (1980), *Phys. Stat. Sol. (b) 100*, 83.

Marklund, S. (1981), *Phys. Stat. Sol. (b) 108*, 97.

Marklund, S. (1983), *J. de Phys. 44*, C4-25.

Marklund, S. (1985), *Solid State Commun. 54*, 555.

Maroudas, D., Brown, R. A. (1991), *J. Appl. Phys. 69*, 3865.

Mauger, A., Bourgoin, J. C., Allan, G., Lannoo, M., Billard, I. (1987), *Phys. Rev. B35*, 1267.

Möller, H.-J. (1978), *Acta Metall. 26*, 963.

Müller, H. (1982), *Thesis Univ. Göttingen.*

Myung, H.-J. (1987), *Diploma thesis Univ. Göttingen.*

Nikitenko, V. I., Farber, B. Y., Iunin, Y. L. (1985), *JETP Lett. 41*, 124.

Nikitenko, V. I., Farber, B. Y., Iunin, Y. L. (1987), *Sov. Phys. JETP 66*, 738.

Nikitenko, V. I., Farber, B. Y. (1989), *Inst. Phys. Conf. Ser. 104*, 257.

Nitecki, R., Pohoryles, B. (1985), *Appl. Phys. A36*, 55.

Nitecki, R., Labusch, R. (1988), *Phil. Mag. B58*, 285.

Northrup, J. E., Cohen, M. L., Chelikowsky, J. R., Spence, J. C. H., Olsen, A. (1981), *Phys. Rev. B24*, 4623.

Öberg, S. (1981), *Phys. Stat. Sol. (b) 108*, 357.

Ohori, K., Sumino, K. (1972), *Phys. Stat. Sol. (a) 14*, 489.

Omling, P., Weber, E. R., Montelius, L., Alexander, H., Michel, J. (1985), *Phys. Rev. B32*, 6571.

Omling, P., Weber, E. R., Samuelson, L. (1986), *Phys. Rev. B33*, 5880.

Omri, M., Tete, C., Michel, J.-P., George, A. (1987), *Phil. Mag. A45*, 601.

Ono, K., Sumino, K. (1980), *Jap. J. Appl. Phys. 19*, L 629.

Ono, K., Sumino, K. (1983), *J. Appl. Phys. 54*, 4426.

Ossipyan, Y. A. (1982), *Sov. Scient. Rev. Sect. 4a:* Khalatnikov, J. M. (Ed.), 219.

Ossipyan, Y. A. (1985), In: *Dislocations in Solids:* Suzuki, H., Ninomiya, T., Sumino, K., Takeuchi, S. (Eds.). Tokyo: Univ. Press. 369.

Ossipyan, Y. A., Petrenko, V. F. (1975), *Sov. Phys. JETP 42*, 695.

Ossipyan, Y. A., Savchenko, I. B. (1968), *Sov. Phys. JETP Lett. 7*, 100.

Ossipyan, Y. A., Shiksaidov, M. S. (1974), *Sov. Phys. Sol. State 15*, 2475.

Ossipyan, Y. A., Petrenko, V. F., Zaretskii, A. V., Whitworth, R. W. (1986), *Adv. Phys. 35*, 115.

Ossipyan, Y. A., Negriy, V. D., Bulyonkov, N. A. (1987), *Izvest. Acad. Nauk USSR Ser Fiz. 51*, 1458.

Ourmazd, A. (1986), *Mat. Res. Soc. Symp. Proc. 59*, 331.

Packeiser, G. (1980), *Phil. Mag. A41*, 459.

Packeiser, G., Haasen, P. (1977), *Phil. Mag. 35*, 821.

Palm, J., Kisielowski-Kemmerich, C., Alexander, H. (1991), *Appl. Phys. Lett. 58*, 68.

Patel, J. R., Testardi, L. R. (1977a), *Appl. Phys. Lett. 30*, 3.

Patel, J. R., Testardi, L. R. (1977b), *Phys. Rev. B15*, 4124.

Patel, J. R., Testardi, L. R., Freeland, P. E. (1976), *Phys. Rev. B13*, 3548.

Peaker, A. R., Hamilton, B., Lahiji, G. R., Turc, I. E., Lorimer, G. (1989), *Mater. Sci. Eng. B4*, 123.

Petrenko, V. F., Whitworth, R. W. (1980), *Phil. Mag. A41*, 681.

Pethukov, P. V. (1988), *Sov. Phys. Sol. State 30*, 1669.

Phillips, J. C., van Vechten, J. A. (1969). *Phys. Rev. Lett. 23*, 1115.

Pirouz, P. (1987), *Scripta Metall. 21*, 1463.

Pirouz, P. (1989), *Scripta Metall. 23*, 401.

Pohoryles, B. (1989), *Inst. Phys. Conf. Ser. 104*, 175.

Rabier, J. (1989), *Inst. Phys. Conf. Ser. 104*, 327.

Rabier, J., Boivin, P. (1990), *Phil. Mag. A61*, 673.

Ray, I. L. F., Cockayne, D. J. H. (1970), *Phil. Mag. 22*, 853.

Ray, I. L. F., Cockayne, D. J. H. (1971), *Proc. Roy. Soc. London Ser. A325*, 543.

Read, W. T. Jr. (1954), *Phil. Mag. 45*, 775, 1119.

Rybin, V. V., Orlov, A. M. (1970), *Sov. Phys. Sol. State 11*, 2635.

Sauer, R., Weber, J., Stolz, J., Weber, E. R., Küsters, K.-H., Alexander, H. (1985), *Appl. Phys. A36*, 1.

Sauer, R., Kisielowski-Kemmerich, C., Alexander, H. (1986), *Phys. Rev. Lett. 57*, 1472.

Schaumburg, H. (1972), *Phil. Mag. 25*, 1429.

Schröter, W. (1978), *Inst. Phys. Conf. Ser. 46*, 114.

Schröter, W., Labusch, R., Haasen, P. (1977), *Phys. Rev. B15*, 4121.

Schröter, W., Scheibe, E., Schoen, H. (1980), *J. Microsc. 118*, 23.

Seitz, F. (1952), *Phys. Rev. 88*, 722.

Shikin, V. B., Shikina, N. I. (1988), *Phys. Stat. Sol. (a) 108*, 669.

Shockley, W. (1953), *Phys. Rev. 91*, 228.

Siethoff, H., Behrensmeier, R. (1990), *J. Appl. Phys. 67*, 3673.

Siethoff, H., Brion, H. G., Schröter, W. (1991), *Phys. Stat. Sol. (a) 125*, 191.

Skowronski, M., Lagowski, J., Milshtein, M., Kang, C. H., Dabkowski, F. P., Hennel, A., Gatos, H. C. (1987), *J. Appl. Phys. 62*, 3791.

Southgate, P. D., Attard, A. E. (1963), *J. Appl. Phys. 34*, 855.

Suezawa, M., Sumino, K. (1986), *Jap. J. Appl. Phys. 25*, 533.

Suezawa, M., Sumino, K., Iwaizumi, M. (1981), *Inst. Phys. Conf. Ser. 59*, 407.

Sumino, K. (1989), *Inst. Phys. Conf. Ser. 104*, 245.

Takeuchi, S., Maeda, K., Nakagawa, K. (1983), *MRS Symp. Proc. 14*, 461.

Takeuchi, S., Suzuki, K., Maeda, K., Iwanaga, H. (1984), *Phil. Mag. A50*, 171.

Tarbaev, N. I., Schreiber, J., Shepelski, G. A. (1988), *Phys. Stat. Sol. (a) 110*, 97.

Teichler, H. (1975), *Inst. Phys. Conf. Ser. 23*, 374.

Teichler, H. (1979), *J. de Phys. 40*, C6-43.

Teichler, H. (1989a), *Inst. Phys. Conf. Ser. 104*, 57.

Teichler, H. (1989b), In: *Polycrystalline Semicond.:* Möller, H. J., Strunk, H. P., Werner, J. H. (Eds.), Heidelberg: Springer, p. 25.

Teichler, H. (1990), In: *Proc. Polyse '90:* Werner, J. H., Strunk, H. P. (Eds.). Heidelberg: Springer, in press.

Teichler, H., Gröhlich, M. (1987), *Izv. Acad. Nauk USSR 51*, 657.

Teichler, H., Marheine, C. (1987), *Izv. Acad. Nauk USSR 51*, 663.

Thibault-Desseaux, J., Kirchner, H. O. K., Putaux, J. L. (1989), *Phil. Mag. A60*, 385.

Tillmann, J. (1976), *Thesis Univ. Köln.*

Trinczek, U. (1990), *Diploma thesis Univ. Göttingen.*

Usadel, K. D., Schröter, W. (1978), *Phil. Mag. 37*, 217.

Vanhellemont, J., Claeys, C. (1987), *J. Appl. Phys. 62,* 3960.

Vanhellemont, J., Claeys, C. (1988a), *J. Electrochem. Soc. 135,* 1509.

Vanhellemont, J., Claeys, C. (1988b), *J. Appl. Phys. 63,* 5703.

Vanhellemont, J., Amelinckx, S., Claeys, C. (1987), *J. Appl. Phys. 61,* 2170, 2176.

Veth, J. (1983), *Thesis Univ. Göttingen.*

Veth, H., Teichler, H. (1984), *Phil. Mag. B49,* 231.

Völkl, J. (1988), *Thesis Univ. Erlangen.*

Völkl, J., Müller, G., Blum, W. (1987), *J. Crystal Growth 83,* 383.

Wang, Y. L., Teichler, H. (1989), *Phys. Stat. Sol. (b) 154,* 649.

Wattenbach, M., Kisielowski-Kemmerich, C., Alexander, H. (1989), *Mater. Sci. Forum 38–41,* 73.

Wattenbach, M., Kisielowski-Kemmerich, C., Alexander, H., Kveder, V. V., Mchedlidze, T. R., Ossipyan, Y. A. (1990), *Phys. Stat. Sol. (b) 158,* K49.

Weber, W. (1977), *Phys. Rev. B15,* 4789.

Weber, E. R., Alexander, H. (1979), *J. de Phys. 40,* C6-101.

Weiß, W. (1980), *Diploma thesis Univ. Köln.*

Wessel, K., Alexander, H. (1977), *Phil. Mag. 35,* 1523.

Wilshaw, P. R. (1990), in: *Polycrystalline Semiconductors, Schwäbisch Hall 1990.* Heidelberg: Springer, in press.

Wilshaw, P. R., Booker, G. R. (1985), *Inst. Phys. Conf. Ser. 76,* 329.

Wilshaw, P. R., Fell, T. S. (1989), *Inst. Phys. Conf. Ser. 104,* 85.

Winter, S. (1978), *Phys. Stat. Sol. (b) 90,* 289.

Wosinski, T. (1990), *Defect Control in Semicond. Yokohama:* Sumino, K. (Ed.), North Holland, 1465.

Wosinski, T., Figielski, T. (1989), *Inst. Phys. Conf. Ser. 104,* 187.

Zoth, G. (1986), *Thesis Univ. Göttingen.*

General Reading

Alexander, H. (1986), "Dislocations in Covalent Crystals", in: *Dislocations in Solids,* Vol. 7: Nabarro, F. R. N. (Ed.). Amsterdam: North Holland, p. 113.

Alexander, H., Haasen, P. (1968), "Dislocations and Plastic Flow in the Diamond Structure", in: *Solid State Physics,* Vol. 22: Seitz, F., Turnbull, D., Eherenreich, H. (Eds.). New York: Academic Press, p. 27.

Benedek, G., Cavallini, A., Schröter, W. (Eds.) (1989), *Point and Extended Defects in Semiconductors.* New York: Plenum Press.

Philibert, J., Sieber, B., Zozime, A. (Eds.) (1983), "Propriétés et Structure des Dislocations dans les Semiconducteurs", *J. de Physique 44,* C4.

Roberts, S. G., Holt, D. B., Wilshaw, P. R. (Eds.) (1989), "Structure and Properties of Dislocations in Semiconductors", *Inst. Phys. Conf. Ser.* Vol. 104.

7 Grain Boundaries in Semiconductors

Jany Thibault, Jean L. Rouvière and Alain Bourret

Centre d'Etudes Nucléaires de Grenoble, Grenoble, France

List of Symbols and Abbreviations

A	rotation matrix
b, b_0, b_i	Burgers vector of the dislocation
b_g	glissile DSC dislocation
c_n	parameters which govern the dynamic properties
D	distance between two dislocations
e	electron charge
e_n	parameters which govern the characteristic frequency
E_0	shallow bulk level
$E_\mp$	Fermi energy
$E_\pm^{neutral}$	Fermi level in the neutral interface
E_C	energy of the conduction band
$E(\Theta)$	energy
ΔE_0	width of the exponential electron density decay
f	frequency
f	driving force
f_i	Fermi distribution function
h	projection parallel to the GB plane
h_0, h_i	projection on the GB normal of half the Burgers vector b_0, b_i
h, k, l	coordinates of the GB plane
h', k', l'	rotation axis
I	unit matrix
I_{CB}	current flow through the GB
K	coefficient dependent on the ratio of trapping and emission of the majority carriers
L	correlation length
m^*	effective mass of majority carrier
m_x	mass
n	normal of the boundary
N	Σ value
N	number of atoms
N	concentration of majority carrier
$N_i(E_i)$	arbitrary distribution of the interface states
Q	charge per unit surface
Q'	charge from the interface density of states
Q_i	charge increase
r	vector of the GB
r^o	O-lattice locus
R	generalized rotation tensor
s_1, s_2	displacement of the CSL
S	sum of the Burger vectors of the dislocation lines
t	rigid-body translation of one grain with respect to the other
u	unit vector of the rotation axis
U	potential energy

v	velocity of the GB
v_{th}	thermal velocity of majority carrier
V	voltage
w	variance of Gaussian potential fluctuations
x, y, z	coordinates
α	dislocation density tensor
$\bar{\alpha}$	tensor of the dislocation surface
β	distortion tensor
γ	energy contribution per surface area of a low angle tilt boundary
γ_d	elastic contribution to the energy of only one dislocation
ε	permutation tensor
Θ	orientation of the grains
μ	shear modulus
ν	Poisson's ratio
σ	conductivity
$\boldsymbol{\sigma}$	stress tensor
σ_{xy}	shear component of the stress tensor
Σ	coincidence index
τ	relaxation time
φ_n	function of n atomic positions
Φ_B	energy barrier height
ω	rotational component of $\boldsymbol{\beta}$
CSL	coincidence site lattice
dcp	dichromatic pattern
DIGM	diffusion-induced grain boundary migration
DLTS	deep level transient spectroscopy
DSC	discrete shift complete
EBIC	electron beam induced current
FIM	field ion microscopy
GB	grain boundary
GBD	grain boundary dislocation
HREM	high resolution electron microscopy
LBIC	light beam induced current
RBT	rigid-body translation
STEBIC	scanning transmission electron microscope
STM	scanning tunneling microscopy
SU	structural unit
TEM	transmission electron microscopy

7.1 Introduction

The idea that a grain boundary (GB) structure is a more or less crystallographically ordered zone is now well established. This said, however, an amorphous region can still appear, for example during the processing of ceramic materials, and the influence of these regions on the properties of the material cannot be ignored. In this chapter we will focus mainly on the former homogeneous type of GBs, although the latter more heterogeneous types will be included in the discussion where it is relevant.

This chapter will be divided into five parts. The first section will be devoted to the different descriptions of a grain boundary, although their historical development will not be pursued. Thus we shall start with the geometrical approach to GBs which is based essentially on the symmetry relationship between two lattices of the same material. This form of description provides a good introduction to the available structures of GBs and is a powerful means to describe the periodical interfaces, symmetry breaks and other associated defects within GBs, although it provides no information about stress fields or atomic positions. The description of a GB in terms of dislocations was in fact the first approach to fully understanding them, and the facility to enable the application of elastic theory was of great help in explaining a lot of GB behaviour (migration, interaction of GBs with other defects). This approach does not work for high-angle boundaries however, and a description in terms of structural units based on atomistic calculations would seem to go further towards explaining both the structure and behaviour of high angle GBs. There is a major limitation with this approach though, and this arises from the uncertainty regarding the interatomic potentials used in energetical simulations even though more reliable potentials are now available.

In the second section experimental results obtained regarding GB atomic structure will be presented. Many of these observations have been performed using high resolution electron microscopy (HREM) which in the field of elemental semiconductors at least, has been the major tool in the recent past providing much needed insight into the atomic structure of GBs. Whereas for years, in metals, most GB simulations have been calculated with only a few experimental HREM results being available for comparison, for the first time the reverse situation is now true for elemental semiconductors. Mention will also be made of new possibilities becoming available with the advent of scanning tunneling microscopy (STM) and atom probe field ion microscopy (FIM), although no details are provided concerning the experimental techniques themselves since a number of reference books covering these topics already exist.

The third section will be devoted to the electrical properties of GBs in semiconductors. These GBs may drastically affect the electronic response of semiconductor materials and special attention will be devoted to the additional defects or impurities segregated at the GB since these are most likely the origin of GB electrical activity.

The deformation of crystals containing GBs and the subsequent GB modifications will be described in section four. The strong interaction between point and linear defects will be shown to have major influence on the behaviour of GBs. Emphasis will be placed on some of the consequences of these events, such as GB migration, enhanced diffusion along migrated GBs, and GB recovery and cavitation.

Finally, several general references are provided to which the reader is invited to refer for completeness. They provide an overview of GBs in materials in general and in semiconductors in particular. Specially recommended however is the review paper written by Grovernor (1985) which provides a thorough grounding in GB properties in semiconductors.

7.2 Grain Boundary Descriptions

7.2.1 Definition and Descriptions

A grain boundary is a two dimensional defect which separates two grains of the same crystalline material that are related by a symmetry operation (Θ/t), a function that describes the relationship between the two grains in terms of orientation (Θ) and rigid body translation (t). In addition to the symmetry operation, the grain boundary is entirely defined if one knows the normal (n) of the boundary and the location of its interface with respect to a given origin. When n is perpendicular to the rotation axis the GB is called a tilt GB, whereas if n is parallel to the rotation axis, it would be known as a twist GB. Mixed GBs have both a tilt and a twist component and a symmetrical GB is one whose plane is in a symmetrical position with respect to both grains. Also, depending on the geometry of the interface, a GB can be flat, faceted, or curved although it has to be noted that these latter characterizations generally depend on the scale of observation.

The previous description is phenomenological, and GBs can also be described in terms of their physical properties such as diffusion, GB migration or sliding, and defects or impurities absorption; of these, it is the stress field induced by the GB and the GB defects on the one hand and the location of the atoms in the vicinity of the inter-face on the other hand that are of prime importance.

Since the discovery of GBs and the appreciation of their role in materials science, a major effort has been directed towards finding a suitable method to describe the deviation from perfect crystalline order induced by a GB which at the same time permits the explanation of its related properties. Several different approaches have been proposed:

i) the geometrical description,

ii) the dislocation description,

iii) the structural units description which in turn is linked to the energetical description.

These different approaches will all be considered in the following paragraphs, with emphasis on their capabilities and their limitations. However, in view of their increasing importance in fully understanding semiconductor properties, the different atomic potentials employed in the energetical simulations will be detailed in a separate, special paragraph.

7.2.2 Geometrical Descriptions

Basically, the geometrical approaches are built upon the premise that the two adjoining crystals have the same periodical structure and consequently, that a number of misorientation angles exist between the two crystals for which a superlattice common to both lattices can be described. This concept was first introduced by Friedel (1926) in order to describe twins. The geometrical descriptions themselves are thus derived from the use of the *coincidence site lattice* (CSL) idea, a theory whose validity was confirmed with the use of field ion microscopy by Brandon et al. (1964). This approach was first applied to cubic lattices and then extended to non-exact coincidence positions and to non-cubic structures in which the dimension of the common lattice might be reduced.

7.2.2.1 Coincidence Lattice, Discrete Shift Complete (DSC) Lattice, and Dichromatic Pattern

The so-called coincidence site lattice is the intersection of the lattices of two adjoining crystals. If the coincidence exists, it can be described by a integer called the coincidence index Σ which is defined as the ratio of the volumes of the unit cell of the CSL and of the crystal lattice. In other words, it represents the reciprocal density of coincidence lattice sites with respect to the original lattice. In the following, a GB will be denoted in the form:

$$(h\,k\,l)\langle h'\,k'\,l'\rangle \Sigma = N$$

where $(h\,k\,l)$ refers to the GB plane in the case of a symmetrical GB, the two adjoining planes being given in the case of an asymmetrical GB, $\langle h'\,k'\,l'\rangle$ represents the rotation axis and N is the Σ value (integer).

Tables of the CSLs in cubic systems have been established by Grimmer et al. (1974), and a checklist of the cubic coincidence lattice relation up to $\Sigma = 100$ has been provided by Mykura (1980). Bleris et al. (1982) studied hexagonal systems and generated tables where the c/a ratio was taken into account and a method to determine experimentally the exact relationship has been proposed (Bleris et al., 1981). A complete treatment of the CSL in hexagonal materials has been given by Grimmer (1989), and this was extended to rhombohedral and tetragonal lattices by Grimmer and Bonnet (1990).

Bollmann (1970) introduced the *DSC lattice (discrete shift complete)* in his consideration of cubic materials. The DSC lattice is the union of the two adjoining lattices and is useful in characterizing GB

defects induced by a small deviation from the exact coincidence position. The point is that a GB dislocation (GBD) with a Burgers vector belonging to the DSC lattice leaves the GB structure unchanged. A small deviation from the coincidence can be accommodated by a periodic array of such GB dislocations which, depending on the geometry of the CSL, might or might not be associated with a GB step. Grimmer and Bonnet (1990) applied the DSC approach in their treatment of rhombohedral and tetragonal materials.

Pond and Bollmann (1979) introduced the concept of *dichromatic pattern (dcp)* which is constructed using the two interpenetrating lattices, ignoring any interfacial planes. (The term dichromatic refers to the black and white dots conferred on the two lattices.) The most symmetrical pattern is called the *precursor*. Any modification induced by the introduction of, firstly an interface and, secondly an interfacial defect leads at each step to a loss of symmetry elements and consequently to the creation of crystallographically equivalent variants related by the lost symmetry operations. These variants may be found along the same interface. GB dislocations separating such energetically degenerated variants have been observed both by Pond and Smith (1976) and by Bacmann (1987). The Burgers vector of these dislocations does not belong to the DSC lattice and as such is called a partial DSC dislocation. Figure 7-1 shows a determination of the dcp, the CSL and the DSC lattice for the structure corresponding to $\Sigma = 5$ in a f.c.c. crystal.

Using group theory, these concepts have been extended to crystal coincidence by Gratias and Portier (1982) and by Kalonji and Cahn (1982). In addition, these authors showed that symmetry constraints might explain some interfacial properties such as the morphology of a crystal grow-

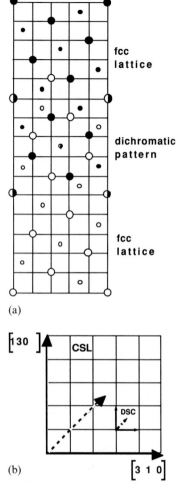

(a)

(b)

Figure 7-1. Black and white f.c.c. lattices, CSL and DSC lattices for [001] $\Sigma = 5$. The large circles are at the 0 level, and the small circles are at 1/2 level along the [001] axis. The dotted arrows have a 1/2 [001] component. The two interpenetrating lattices (Fm $\overline{3}$m) make up the dichromatic pattern with the space group (I 4/m m′m′) where "′" denotes an antisymmetry element, i.e., a change in color.

ing (during precipitation or solidification) in a crystalline environment. Pond (review paper 1989) extended the dichromatic pattern concept to crystals and introduced the dichromatic complex. This allowed him to propose a unified description of all the possible bulk and interfacial defects arising from the symmetry break of the bicrystal

complex. The result of applying these ideas to the semiconductors was described by Pond (1985). Figure 7-2 indicates the dichromatic complex corresponding to $\Sigma = 5$ in a cubic diamond structure.

Partial GB dislocations can also be used to account for junctions between GB facets (Pond, 1981) or for defect delimiting two GB areas whose rigid-body translation is different, a situation observed by Bacmann et al. (1981) in $\Sigma = 5$ and $\Sigma = 25$ Ge bicrystals.

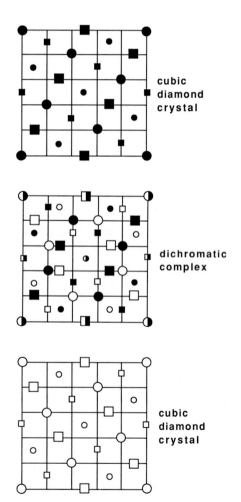

Figure 7-2. Dichromatic complexes for $\Sigma = 5$ related crystals with the diamond structure (F d $\overline{3}$ m). Its space group is (I 4_1/a m′ d′).

7.2.2.2 The O-Lattice

The *O-lattice concept* was first introduced by Bollmann (1970) in order to describe low angle boundaries in cubic materials. The O-lattice defines regions both where there is a good match between two adjoining lattices and regions where the match is at its worst. In contrast to the coincidence lattice, the O-lattice exists for any misorientation angle Θ, provided it is low: generally it ensures that the condition $\sin \Theta \simeq \Theta$ is fulfilled. The O-lattice is constructed using the sites of coincidence that occur between the two interpenetrating lattices. The site coordinates are the internal coordinates of the two lattices and consequently, they vary continuously with the misorientation angle. The positions of worst match give the locations of dislocations whose spacing also varies continuously with the misorientation angle.

The O-lattice locus is defined by r^{o} obeying the condition:

$$b = (I - A^{-1}) r^{o}$$

where b is the Burgers vector of the dislocation, A is the matrix defining the rotation of grain II with respect to grain I and I is the unit matrix. Bollmann (1970) introduced the concept of the b-net, which contains all the Burgers vectors for any possible boundary that can be chosen in a given O-lattice. As a result, the choice of GB plane determines completely the Burgers vectors which will be then attributed to the dislocation network of the boundary.

For high-angle boundaries, with a small deviation from an exact coincidence position, Bollmann (1970) introduced a O_2-lattice (second order O-lattice), which corresponds to the best match between the two DSC lattices, the reference DSC lattice being defined for the coincidence position. This lattice always exists, and varies con-

tinuously with the deviation angle from the coincidence position. The positions of worst match represent the GB dislocations distribution.

7.2.2.3 On the Utility of the Geometrical Description

The geometrical description provides the full set of crystallographic possibilities but says nothing either about the atomic relaxations in the interface or about the energy of the GBs. However, part of the understanding of GB properties is supported by a knowledge of the symmetry relationships between two crystals and of the appearance of GB defects arising from a break in the different levels of symmetry.

7.2.3 Dislocation Description

7.2.3.1 Low-Angle Grain Boundaries

Historically, the modelling of interfacial structure in terms of dislocations was developed before the geometrical approach was considered (Read and Shockley, 1952). The first experimental evidence of the validity of the dislocation model was given by Vogel et al. (1953) by imaging a low angle GB in Ge whose dislocations were revealed by chemical etching. The complete treatment was reviewed by Christian (1965). The dislocation distribution of a general low angle grain boundary was formulated by Frank (1950) as:

$$S = 2 \sin (\Theta / 2) u \times r \qquad (7\text{-}1)$$

where S is the sum of the Burgers vectors of the dislocation lines cut by an arbitrary vector r of the GB, u and Θ define the unit vector of the rotation axis and the rotation angle respectively.

An elegant way to derive this formula was developed by Kröner (1958 and 1981, which is an updated English version of the

1958 paper), who formulated mathematically the idea that the internal stresses are caused by dislocations whose 3-D distribution is given by the dislocation density tensor α. This continuous distribution of dislocations is one of the sources of distortion β such that:

$$\mathrm{rot}\,\beta = \alpha$$

and this leads to internal stresses through obeyment of Hook's law. Interestingly, this is equivalent to the 2-D treatment introduced by Bilby et al. (1955) who expressed the dislocations distribution on a surface delimiting two grains I and II with a discontinuity of distortion between them such that:

$$\varepsilon_{ijk}\,n_i\,\beta_{sj}^I - \varepsilon_{ijk}\,n_i\,\beta_{sj}^{II} = \bar{\alpha}_{ks} \qquad (7\text{-}2)$$

where $\bar{\alpha}$ is the tensor of the dislocation surface density, k refers to the dislocation Burgers vector and s to the dislocation line. n is the normal to the surface pointing from I to II and ε is the permutation tensor ($\varepsilon_{iij} = 0$). In the case of low angle GBs, one can consider only the rotational component ω (i.e. the antisymmetrical part) of the distortion β. ω can be deduced from the generalized rotation tensor R_{ij} via the formulae:

$$\omega_{ij} = R_{ij} - \delta_{ij} \qquad (7\text{-}3)$$

with

$$R_{ij} = u_i u_j + (\delta_{ij} - u_i u_j)\cos\Theta + \varepsilon_{ijk} u_k \sin\Theta$$

where u and Θ define the tilt axis and the rotation angle as above.

Eqs. (7-2) and (7-3) can be reformulated in matrix notation (Christian, 1965) to yield:

$$S = (R_I^{-1} - R_{II}^{-1})\,r$$

where S and r have the same meaning as in Eq. (7-1) and R_I and R_{II} are the rotation matrices defining the orientation of the two crystals with respect to a reference lattice. If one of the two crystals (crystal I for instance) is chosen as the reference the formula is simplified: R_I becomes the identity I and R_{II} is the rotation matrix giving the relative orientation of the two crystals. This corresponds to the O-lattice treatment given in Sec. 7.2.2.2. Regions of best fit are separated by regions of worst fit. A section created by the GB surface passing through this construction creates dislocations at the intersection of the GB with the zones of worst fit. This dislocation distribution is called the *primary dislocation network*.

Using elastic theory, the dislocation description enables one to estimate the energy contribution per surface area of a low angle tilt boundary. It varies with the orientation angle and can be expressed in terms of the function deduced by Read and Shockley (1952):

$$\gamma = \gamma_0\,\Theta\,\ln(\Theta_m/\Theta)$$

where $\gamma_0 = \mu b/4\pi(1-v)$ and $\Theta_m = e\alpha/2\pi$, b is the Burgers vector, e is the Neperian base, μ and v are the shear modulus and the Poisson's ratio respectively and α is about 1 or 2 depending on the dislocation core energy. When Θ tends to zero, the energy tends to zero, in accordance with the fact that the dislocation density tends to zero. On the other hand, the elastic contribution to the energy of only one dislocation whose expression is

$$\gamma_d = \gamma b/\Theta = b^2\mu/4\pi(1-v)\ln(1/\Theta)$$

decreases as Θ increases. What is happening here is that the compressive stress field of one dislocation annihilates the tensile stress field of the neighbouring dislocation. This holds not only for the dislocations but also for the regions on both sides of the interface. The extension of the stress field of an individual dislocation embedded in a low angle grain boundary is about half the spacing between the dislocations. Using the Kröner approach, the energy of more complex periodical arrays of dislocations has been calculated by Rey and Saada (1976) and Saada (1979).

7.2.3.2 High-Angle Grain Boundaries

In the case of high-angle grain boundaries, a description in terms of primary dislocations is not so straightforward. As the angle increases the distance between the dislocations reduces and the dislocation cores overlap. Thus, the elastic treatment in terms of Volterra dislocations is no longer valid and the above formulae do not hold. Nevertheless, the case of a small deviation from a coincidence position can be treated provided that the Brandon (1966) criterium is valid. This states that the maximum deviation from coincidence must be $\Theta_0(\Sigma)^{1/2}$ where Θ_0 is about $15°$. Thus for $\Sigma = 1$ (perfect crystal), Θ must be less than $15°$, which corresponds to the criterium for a low angle boundary.

Bollmann (1970) showed that this deviation can be accommodated by a periodic network of *secondary dislocations* whose distribution is superimposed on the coincidence GB and can be deduced either from the O_2-lattice (see Sec. 7.2.2) or from the equivalent Frank formula (7-1) applied to the DSC lattice. The Burgers vectors b of the secondary dislocations are DSC vectors. Generally, these GBDs are linked to the GB and cannot move in one or the other adjoining grain and in addition, they are generally, but not necessarily, associated with a GB step. The height h of the step associated with one particular GBD is geometrically determined by the CSL. The (b, h) pair can be determined using the method given by King and Smith (1980) in the following way. If s_1 and s_2 are the displacements of the CSL due to the GB de-

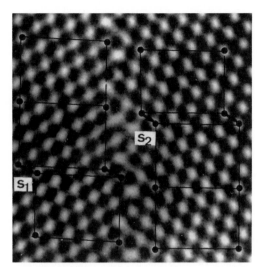

Figure 7-3. Determination of the (b, h) pair directly from a HREM micrograph. The case presented here is the one of a GBD in (122) [011] $\Sigma = 9$. $s_1 = a/2$ [011]$_I$ and $s_2 = a/4$ [21$\bar{1}$]$_{II}$; s_2 is the projection of $a/2$ [110]$_{II}$ or $a/2$ [10$\bar{1}$]$_{II}$, and $b = s_1 - s_2$ can be expressed simply in grain I as $a/6$ [$\bar{1}$21]$_I$ or $a/6$ [$\bar{1}\bar{1}$2]$_I$. h is the projection on the GB normal of $(s_1 + s_2)/2$ and is equal to the absolute value of $a/9$ [$\bar{1}$22]$_I = a/3$ (from Elkajbaji and Thibault-Desseaux, 1988).

fect (b, h) respectively in grain I and grain II, then b is given by the difference between s_1 and s_2 and h is the projection parallel to the GB plane of the half sum of s_1 and s_2. An example of such a determination directly on a HREM image of a DSC dislocation in (122)$\langle 011 \rangle \Sigma = 9$ is shown in Fig. 7-3.

7.2.3.3 Stress Field Associated with Grain Boundaries and Grain Boundary Defects

Predicting the stress field is of importance because of the possible stress induced attraction of lattice defects (point or linear defects) or impurities to the GB. This effect could be enhanced by the presence of GBDs through their long range stress field.

A general expression for the stress field of a low angle periodic tilt GB in an isotropic material was determined by

Hirth and Lothe (1968). The shear component σ_{xy} of the stress tensor which will act on external dislocations is defined by:

$$\sigma_{xy} = \frac{2\pi b \mu x}{(1-v) D^2} \exp\left(-\frac{2\pi x}{D}\right) \cos \frac{2\pi y}{D}$$

where x corresponds to the GB plane normal, y corresponds to the direction perpendicular to the tilt axis and contained in the GB plane, and D is the distance between two dislocations. The above expression is true for regions where $x \gg D/2\pi$. In these regions, σ is proportional to b and decreases exponentially from the GB plane with a typical distance of $D/2$. Near the GB, for $x < D/2$, the stresses can be considered to be induced by the first to the third nearest dislocations. The other stress components, in particular the diagonal component acting on a dilatation center such as a point defect or an impurity atom, behave as σ_{xy}.

In addition, this expression can be employed for a periodic set of secondary dislocations b' with a period D'. The closer a GB is to a coincidence GB, the larger is the period of secondary dislocations. As a consequence, the extension of the stress field is larger even if its amplitude becomes smaller with the decrease of the secondary dislocation Burgers vector b'.

This method of characterizing GB line defects has led to discussions (see for example a recent review by Priester, 1989) concerning the nature of extrinsic and intrinsic GBDs. Balluffi et al. (1972) emphasized that this type of dislocation network embedded in a coincidence boundary has an intrinsic character i.e. it forms part of the equilibrium of the GB. These dislocations have to be distinguished from additional ones known as extrinsic dislocations, which unlike the intrinsic type, give rise to long range stresses. Furthermore, the same authors pointed out that the response of the two types of dislocations to point defects absorption was quite different: Extrinsic GBDs with a large Burgers vector component normal to the GB strongly absorb point defects, the GB acting as a point defect sink due to the climbing possibility of the extrinsic GBD's.

It can be seen from the formula above, that a stress field exists whatever the GB dislocation type may be. The stress fields of secondary or primary dislocations at the same periodicity differ from the b value, and the stress field of intrinsic primary dislocations must be compared with that from the extrinsic dislocation considered as an isolated dislocation with the same b. This is expressed by the function:

$$\sigma_{xy} = \frac{\mu b x}{2\pi(1-v)} \frac{(x^2 - y^2)}{(x^2 + y^2)^2}$$

At a distance $x = D$, the stress field of GBDs is 0.06 times the stress field of the corresponding isolated dislocation.

The property of stress field segregation can be used for beneficial effects. The gettering process is a technique employed in semiconductor technology in order to concentrate metallic impurities in a given part of a device, precluding their undesired precipitation in other parts. Goetzberger and Shockley (1960) described the catastrophic effects of metallic precipitates on p-n junctions and proposed a method to remove these precipitates by means of gettering using glassy layer of oxide. At that time the mechanism responsible for destabilization of the precipitate remained unexplained. Queisser (1963) studied the diffusional and electrical properties of a twin boundary. The so-called coherent $\Sigma = 3$ GB in Si was found to act as the bulk whereas $\Sigma = 9$ GB had a very small tendency to precipitate metallic impurities but presented no enhancement of substitutional impurity diffusion.

It was thought that temperature could decrease the trapping efficiency of GBDs. As Sumino (1989) mentioned, provided the mean concentration of the impurities remains low enough (less than 10 ppm) and the temperature high enough to ensure that atomic diffusion is important, one can be sure that Cottrell atmosphere formation around the core of dislocations as a consequence of interaction between the elastic strain fields of the impurities and the dislocations is not the causal mechanism behind impurity gettering by dislocations. This proposal may be valid for GBDs.

7.2.4 Description in Terms of Structural Units

The geometrical theories as well as the dislocation descriptions of GBs have been very useful. However, they do not account for certain properties, such as electrical properties, which depend on atomic structure. Moreover, the core energy term is almost entirely dependent on the exact atomic structure which must now be considered as an intrinsic component itself.

7.2.4.1 Geometrical Structural Models

Hornstra developed the concept of structural units to describe dislocation cores in covalently tetracoordinated materials and extended this idea to the description of tilt GBs around the $\langle 011 \rangle$ axis (Hornstra, 1959) and the $\langle 001 \rangle$ axis (Hornstra, 1960). It is worth noticing how closely this first approach using stick and ball models concurs with the elaborate models more recently produced with the use for more sophisticated methods, and how well it reproduces the structures as they are actually found. In 1977, Krivanek, Isoda, and Kobayashi, obtained HREM images of an almost perfect $\Sigma = 9$ GB and using the Hornstra model they were able to deter-

mine the structure of the GB and showed that 5–7 atom rings accounted for the core of the Lomer dislocation in the bulk. Since then, d'Anterroches and Bourret (1984) and Vaudin et al. (1983) confirmed by HREM the Hornstra structure of respectively the second order twin ($\Sigma = 9$) in Ge and the second and the third order twins ($\Sigma = 27$) in Si. Further HREM observations detailed in Sec. 7.3 will show however, that more complicated structures predicted by Hornstra were not observed. Clearly, more elaborate models were necessary.

7.2.4.2 Energetical Structural Models

Followed the calculations of GB structure in Al performed by Hasson et al. (1972), Sutton and Vitek (1983a, b, c) extended the concept to describe the structure of GBs in metals. They showed that the structure of these GBs could be described in terms of clusters of a limited number of atoms called *structural units* (SU's). Furthermore, they showed that special GBs existed, with very low energy and comprised of only one type of SU, and these they termed *favoured* or delimiting GBs. Other GBs whose misorientation angle is between the angles of two favoured GBs can be described by employing a mixture of the SUs of the two favoured or delimiting GBs providing they share the same mean GB plane. However, as Sutton and Vitek (1983a) discussed, constitutive SUs are distorted with respect to the reference SU's and these distortions result in an increase in GB energy.

Using earlier geometrical work produced by Bishops and Chalmers (1968), Sutton et al. (1981), and then Sutton and Vitek (1983a, b, c) developed a new model, the SU/GBD model using energetical computations. This model was computed in f.c.c. and b.c.c. metals but could be extended without much difficulty to other structures such as cubic diamond structures. They established a direct relationship between their SU model and the secondary dislocation network and they extended its applicability far beyond dense CSL orientation. In fact, this description has two aspects; i) in a long period tilt GB, each minority unit (linked to one delimiting GB) can be considered as a GBD of the DSC lattice associated with the other delimiting GB (made up with the majority SUs) and ii) secondary dislocations can also be regarded as perturbations of the primary GB dislocations. Balluffi and Bristowe (1984) emphasized that the SU/GBD description of a GB between favoured GBs is not unique and established a hierarchy amongst the SU/GBD models. Moreover, they pointed out that the use of multiple delimiting GB allows one to minimize the distortions of the SUs.

A new approach has recently been proposed which links the SU description to a description of GB structure in terms of disclinations (Gertsman et al., 1989). As a GB is made up of a mixture of two different SUs, these SUs are deformed and this incompatibility between two different adjacent SUs can be accommodated by a disclination. Consequently, a GB might be considered as a distribution of disclination dipoles whose strain field can be formally estimated.

Recently, Sutton (1989) suggested that the SU model was severely restricted in its capacity to predict any GB structure and cited as examples of difficulties the case of a GB with high index rotation axis or a GB with multiple structure. Sutton and Balluffi (1990) therefore derived selection rules for combining SUs of delimiting GBs. The principal intention was to ensure compatibility between SUs in order to avoid long range stresses. Furthermore, they pointed

out that not all possible structures could be found using simulations because of the limitations ordained by the periodical conditions prerequisite in the computations.

This difficulty in predicting GB structure is further illustrated by high resolution electron microscopy observations mode of the multiple structure of tilt GBs as simple as $\Sigma = 11$ (233) [011] in Si (Putaux and Thibault, 1990) or $\Sigma = 13$ (520) [001] in Ge (Rouvière and Bourret, 1989). In both cases, the polymorphism found experimentally could not be theoretically predicted using the SU model although it could be justified a posteriori with the use of energetical computations.

Application of the SU energetical description to semiconductors will be discussed in detail in the next section.

7.2.5 Energetical Descriptions

It is only recently, within the last ten years, that computer-simulated energy minimization has been used to investigate semiconductor grain boundary structures. Grain-boundary computer simulations were first realized for rare gas or metallic grain boundaries, because in these materials interactions between atoms are, in a first approximation, satisfactorily described by simple two-body interaction laws. These numerical simulations have proved to be valuable tools in the study of detailed atomic configurations and have helped enormously in revealing important concepts such as:

– the possibility (in order to minimize the grain boundary energy) of a rigid-body translation (RBT) between to semi-infinite crystals,

– the usefulness of the notion of structural units in building and predicting grain boundary structures,

– the presence of deep cups in the Θ dependence of the energy.

GBs in covalent semiconductors (Si and Ge), as well as those in ionic semiconductors (up to now limited to nickel oxide) have been studied by computer simulation. In the diamond structures, besides comparison of the different relative energies of the GBs only the atomic and electronic structures of the interface at 0 K have been studied. In NiO, however, although comparatively less studies have been carried out, more properties have been tentatively computer simulated. These include atomistic structure, formations of intrinsic defects, impurity segregation, space charge analysis and diffusion (Duffy and Tasker, 1985).

In this section we shall quickly look over the different techniques that have been used in semiconductor GB simulations and review their results, the emphasis being on the atomic structure of the grain boundaries. The results dealing with electronic structure are presented more extensively in Sec. 7.4.

7.2.5.1 Computer-Simulated Energy Minimization

Up to now, all semiconductor GB simulations have been static in the sense that they minimize the energy of an assembly of interacting atoms that simulate a perfect GB or a GB with impurities or point defects. No dynamic simulations such as molecular dynamics or Monte Carlo techniques have been employed. Energy minimization has been achieved with a variety methods: the steepest descent method, the conjugate gradient method, metric minimization methods and the molecular dynamics quenching method.

Three problems are encountered in GB simulations. The first is specific to the GB and comprises (i) having to determine the necessary conditions to simulate a GB. The

remaining two difficulties arise from the computer simulation method itself, and these are (ii) the finite size of the box and (iii) the interaction law that exists between atoms. This said, the first two problems really are to define appropriate boundary conditions.

Boundary Conditions

In order that the finite number of atoms considered during the calculation is equivalent to an infinite system, one can apply two different particular boundary conditions to the box. Either the finite volume containing the individual atoms is enclosed in a wider region whose geometry is simply determined (for instance, it is a perfect crystal deformed according to the elastic theory), or periodic boundary conditions are applied.

For time-consuming algorithms, periodic boundary conditions are applied in the three dimensions of the box as this reduces the computation time. Indeed such a supercell approximation introduces undesirable constraints when the system is not periodic. It is however, a well justified approach for periodic high coincidence GBs. The interface plane is then periodic but the artefacts remain in the third direction (x direction in Fig. 7-4) because unrealistic interactions have been introduced.

For simpler calculations, the periodicity in the third direction may be taken away. To do so, the atoms near two opposite planes parallel to the interface are *fixed* in the positions of parts of two correctly disorientated perfect crystals. These *fixed* atoms act as if they belong to two semi-infinite perfect crystals but rigid translation between these two perfect crystals is allowed. These *fixed* atoms at the boundary of the box do not introduce constraints on the interface if the box size perpendicular to the interface is large enough. This can be

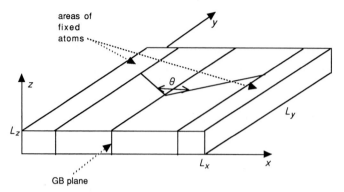

Figure 7-4. Scheme showing the finite box for simulating a tilt GB. The tilt axis lies along z, the misorientation is Θ around z. The normal of the interface is parallel to the x axis. The boundary is assumed to be periodic along y and z. The 2D size (L_y, L_z) of the elementary box is a multiple of the GB periodicity along these two directions. In order to remove artifacts arising on the x axis from the periodization, one generally fixes the atoms close to the planes defined by $x = 0$ and $x = L_x$ in a position corresponding to two perfect crystals in the suitable misorientation.

verified by increasing the box size as long as there are noticeable energy variations.

The translation between two slabs of perfect crystals is an important parameter of relaxation. For simple calculations, this translation and the individual atomic positions relax altogether and the two perfect slabs float rigidly parallel and perpendicular to the boundary while the atoms are relaxing. For time consuming simulations, the energies of different fixed translation state configurations are minimized and compared (Tarnow et al., 1989).

The third problem, that of interactions between the atoms, is a more serious one.

Interaction Laws

Formally, Schrödinger's equation for the complete assembly of atoms (and electrons) must be solved. Ab initio self consistent methods have been developed, but since they require a huge amount of calculation, the number of atoms and thus the box size must be very small (far less than one hundred atoms). This restriction actually limits the application of these accurate techniques to GB structures although, in the future, they should play an increasing role with the development and increased availability of high speed computers. Less demanding techniques like the LCAO techniques (or tight binding methods) have been employed as an alternative (Paxton and Sutton, 1989; Kohyama et al., 1986 a, b and here Schrödinger's equation is still solved, but only by assuming further approximations. Whilst the method is less accurate, the amount of calculations is greatly reduced and as such the box size can be enlarged (up to hundred atoms). These two different approaches have the advantage of solving both the atomic and the electronic structures of the interface and they are comprehensively presented in Chap. 1 of this Volume. Alternatively, if the only interest is in the atomistic structure of the interface, one can consider that the interactions between atoms are represented by a force depending only on the atomic positions. The potential energy U of N atoms included in the box can be expanded in terms of a series of N-body interatomic potentials (Martin, 1975),

$$U(r_1, \ldots, r_n) = \varphi_0 + \sum_{i=1}^{n} \varphi_1(r_i) +$$

$$+ \sum_{i,j} \varphi_2(r_i, r_i) + \ldots + \sum_{i_1, \ldots, i_n} \varphi_n(r_{i_1}, \ldots, r_{i_n})$$

where the N-body interatomic potential φ_n is a function of n atomic positions, and the sums are taken over all combinations of atoms in the structure. In practice, only the first N-body potentials are important. The problem is to determine what kinds of interaction have to be included in the potential and what type of function has to be selected. The solutions of course, depend on the material. We shall not review here all the empirical potentials that have been tested in semiconductors, but rather shall briefly recall the different types of interaction that can be found in semiconductors.

The covalent semiconductors (Si and Ge) are characterized by directional sp^3 hybridized bonds. Both bond-stretching and bond-bending motions must be considered in the energy and as pair potentials cannot properly describe the bond bending that occurs, at least a three-body term must be included. The form of these potentials can be determined using a perturbational method in which the energy is expanded in powers of small atomic displacements around a reference state (which is the perfect crystal). With these so-called *valence-force field potentials* like the famous Keating potential (Keating, 1966), it is impossible to consider rehybridization at a strongly distorted locus and as such they are only useful at sites where the tetrahedral sp^3 bonding is preserved. In order to overcome this limitation, empirical potentials have been developed (Dodson, 1986) whose analytical expressions are derived more or less from *theoretical* considerations. The parameters they introduce are fitted to experimental data such as elastic constants or lattice parameters, or more surprisingly in some cases to computed ab initio energy values (for a short review see Dodson, 1986). The Tersoff potential (1988) is expressed in Morse-like pair potentials but the attractive bond strength is a function of the local environment. Coordination numbers different from four are possible. As Teichler (1989) pointed out, the interatomic potential models must at least reproduce both the elastic properties of the crystal (far from the defect core) and the phonon dispersion (corresponding to the harmonically distorted region closer to the core). Teichler (1989) made comparisons between different potentials and emphasized that the *bond charge model* introduced by Weber (1977) fulfills both conditions and in addition takes the charge screening effect into account.

In zinc-blende semiconductors, the sp^3 bonding is partly ionic. Thus, Coulombic terms have to be added to the valence force field potential (Tewary et al., 1989). An empirical potential has also been introduced (Takai et al., 1990) but to our knowledge, no GB simulation has yet been made in these materials.

In ionic semiconductors, the problem is at first sight simpler, because the Coulombic forces which provide the main energy contribution are known having been determined analytically. Problems however, arise from the facts, that i) the two-body Coulombic energy is a long range potential, ii) an empirical repulsive potential must be added to it and iii) the ion polarization must be considered (Wolf, 1984a, b). In practice, the long range Coulombic energy is not calculated by a direct summation but rather by a plane-wise summation technique. The polarizability of the ions is described by a shell model where the ions are composed of a shell (representing the valence electrons) and a core coupled by an elastic spring (Wolf, 1984 a, b). The shell charge and the spring constants are determined from an empirical fit to the dielectric properties of the crystal. As a consequence of ion polarizabilities, each ion possesses six degrees of freedom in contrast to the three that are available in the covalent semiconductors.

7.2.5.2 Grain-Boundary Computer Simulation Results

Covalent Semiconductors
(Germanium and Silicon)

Tilt Grain Boundaries

High-coincidence $\langle 011 \rangle$ tilt grain boundaries were the first GBs to be simulated in covalent semiconductors, because they were the first covalent semiconductor GBs to be geometrically and experimentally studied. Figure 7-5 shapes the different types of SUs encountered in the description of $\langle 011 \rangle$ or $\langle 001 \rangle$ tilt GBs and these will be referred to in the following. Prior to computer simulations, the energies of some low angle GBs were first evaluated by calculating the elastic field energy (Bourret and Desseaux, 1979 a, b). Then, using the empirical Keating potential, Möller (1982) calculated the energies of several models, mostly those introduced by Hornstra (1959). As this potential can only be used with tetracoordinated models, Möller had to estimate the energies of dangling bonds in order to roughly evaluate the energies of the not entirely tetracoordinated models. Furthermore, he did not take the change in bond length into account. He found that the GB with $(0° < \Theta < 70.53°)$ could be described without any dangling bonds. Moreover, if $\Theta < 26.53°$ $(\Sigma = 19)$, the GB can be considered as a small angle GB with a periodic distribution of edge dislocations

$1/2 [0\bar{1}1]$ characterized by a $5-7$ atom ring SU's (called L; L for Lomer dislocation whose core has been shown to be this $5-7$ atom ring) embedded in the 6 atom ring SU's of the bulk. This can also be described with the help of the SU model using the $\Sigma = 1$ (011) $\langle 011 \rangle$ (perfect crystal) and the $\Sigma = 9$ (122) $\langle 011 \rangle$ GBs as delimiting GBs for a misorientation angle between 0° and 38.94°. When the misorientation angle increases the dislocation cores overlap and the GB can no longer be considered as a subgrain boundary. The SU model is still valid for angles greater than 38.94° but the delimiting GBs are $\Sigma = 9$ (122) $\langle 011 \rangle$ and $\Sigma = 3$ (111) $\langle 011 \rangle$ with the restrictions cited above in Sec. 7.2.3. Furthermore, Vaudin et al. (1983) first mentioned the delimiting $\Sigma = 9$ SU used for describing symmetrical tilt GBs either with Θ up to 38.94° or with Θ between 38.94° and 70.5° is not exactly the same even if it is based on the 5-atom ring and the 7-atom ring. In the first case the $5-7$ atom SU is a symmetrical one (L), whereas in the second case the $5-7$ atom ring is an asymmetrical one (M). This was confirmed by HREM (see Sec. 7.3.1).

Using a band orbital method, Kohyama et al. (1986 a, b) calculated quite a variety of $\langle 011 \rangle$ GBs more thoroughly. In their analysis, they used to the full the structural unit method developed by Sutton and Vitek (1983). In a first paper (Kohyama et al., 1986 a), they studied GBs having a rotation angle in the range $0 \leq \Theta \leq 70.65°$. These GBs can be described with a few structural units and are entirely tetracoordinated. In a second paper (1986 b), using the structural patterns introduced by Papon and Petit (1985) they calculated the energies and proposed several GB models for a rotation range $70.65° \leq \Theta \leq 180°$. These structural patterns permit the construction of any $\langle 011 \rangle$ tilt grain boundaries without any dangling bonds.

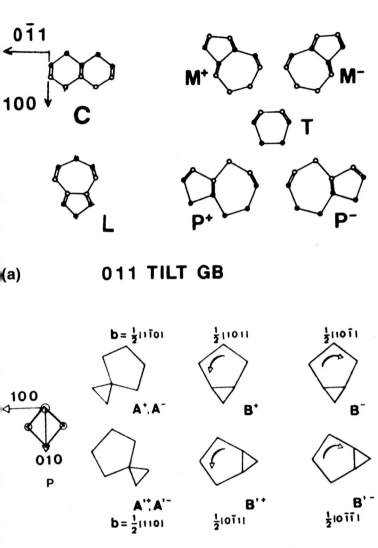

Figure 7-5. Different SUs encountered for describing the (a) ⟨011⟩ and (b) ⟨001⟩ tilt GBs. C is the perfect crystal SU, i.e., (011)⟨011⟩ $\Sigma = 1$, whereas p is the perfect crystal SU linked to (110)⟨001⟩ $\Sigma = 1$. The A, B, and L SUs are linked to the bulk dislocation core: A corresponds to the $1/2$⟨110⟩ edge dislocation, B to the $1/2$⟨101⟩ 45° dislocation aligned along ⟨001⟩ and L to the $1/2$⟨011⟩ Lomer dislocation aligned along ⟨011⟩. T, M, and P SUs are not linked to bulk dislocations but to GBs: T is the SU describing (111)⟨011⟩ $\Sigma = 3$, M describes (122)⟨011⟩ $\Sigma = 9$ and P can describe a highly energetic structure of $\Sigma = 3$.

The same kind of approach can be adopted for ⟨001⟩ GBs (Kohyama, 1987; Bourret and Rouvière, 1989) and with only a few structural units, plenty of tetracoordinated models can be constructed and their energies calculated.

More sophisticated methods were employed in the case of selected and simple GB structures.

The (122) $\Sigma = 9$ symmetrical tilt GB attracted attention because it was the first and at that time the only interface whose structure was experimentally determined. The aims of these studies were firstly to verify that among the two existing models the observed one has the lowest energy and secondly to determine the electronic structure. Thomson and Chadi (1984) carried out the first tight binding study on a GB in silicon. DiVincenzo et al. (1986) used an ab initio local density functional calculation to analyze the electronic properties and they calculated the energies. However the positions of 36 out of 40 atoms of their supercell were determined using a semi-empirical tight binding method.

Later on, the so-called *incoherent* (211) ⟨011⟩ Σ = 3 GB attracted attention because its structure was also experimentally determined. It contains highly distorted bonds and thus was supposed to be electrically active. Starting from the atomic positions determined by a Keating potential, Mauger et al. (1987) analyzed its electronic structure with a tight binding method. Independently Kohyama et al. (1988) and Paxton and Sutton (1989) entirely simulated this grain boundary with equivalent but slightly different tight binding methods. In the same papers they also studied the experimentally determined (310) [001] Σ = 5 tilt GB. Paxton and Sutton calculated the energies for different models of the (211) ⟨011⟩ Σ = 3 interface and found that effectively, the model in agreement with the experimental data was the one

with the lowest energy. Other models however, have nearly equal but actually slightly greater energies.

One interesting aspect of the results of these calculations is that there are deep cusps in the curve representing the energy of the grain boundary versus the rotation angle Θ (Fig. 7-6). Thus the lowest energy model of an interface could be expected to be the most likely model. This has shown to be effectively the case for a number of different interfaces: Σ = 5 (310) (Kohyama, 1987; Bourret and Rouvière, 1989), Σ = 9 (122) (Kohyama et al., 1986 b), Σ = 3 (122) (Paxton and Sutton, 1989). When the difference in energy between the two models is small, this could indicate a multiplicity of structures, a situation that is certainly the case in the (13 3 0) Σ = 65 GB in which two structures coexist experimentally and have nearly the same energy (Rouvière and Bourret, 1990).

However, in view of the fact that these calculations are performed at $T = 0$ K and involve several approximations in solving the Schrödinger equation, their results must be treated with care and mostly require experimental confirmation. One diffi-

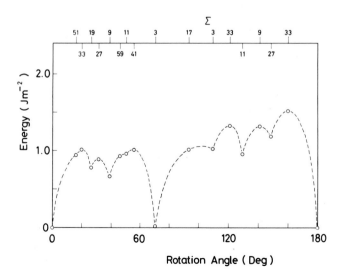

Figure 7-6. Calculated boundary energy versus the rotation angle Θ for different coincidence [011] tilt GBs in Si. The curve is interpolated between the calculated points (Kohyama et al., 1986 c).

culty for instance, is that in certain cases, entropy terms could play an important role (see Sec. 7.2.1.2).

Twist Grain Boundaries

Contrary to the tilt GBs, no atomistic experimental data are available for twist GBs and indeed the data that have been collected for these interfaces were only computer simulated. We provide below some more comments on the results of these simulations.

Kohyama et al. (1986a) relaxed several ⟨111⟩ twist GBs in silicon. They found no cusps in the energy versus Θ relationship. In contrast Phillpot and Wolf (1989) used an empirical potential (Stillinger and Weber, 1985) that enables non-tetracoordinated structures in order to calculate the energies of *point-defect-free twist GBs* on the three densest planes of cubic diamond silicon ({111}, {110} and {001}). The aim of this paper was to determine trends in energy and no attention was paid to the geometry of the GBs which was simply from coincident site twist rotations about the plane normal of two perfect semi-infinite diamond lattices. They found that the energy of the ⟨110⟩ twist GBs are 2 to 3 times higher than the energy of the densest plane {111} twist GBs, and that {100} twist GBs have even higher energies. A strong correlation between low GB energy and larger interplanar spacing was then deduced and considered as a general feature of GBs. It is, however, doubtful that the energies of such point defect free GBs are useful parameters. Only by allowing special reconstruction can the relative boundary energies be obtained.

A more ambitious determination of a grain boundary structure was performed by Tarnow et al. (1989) using an ab initio local density functional calculation. They

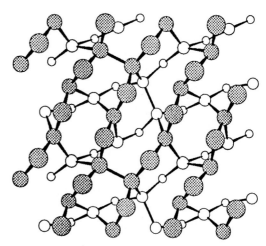

Figure 7-7. Model of the $\Sigma = 5$ twist boundary in Ge after relaxation by an ab initio local density functional calculation (Tarnow et al., 1989). The atomic positions are those for 2 layers above and below the boundary plane normal to the [001] direction. The final structure contains threefold, fivefold-coordinated atoms and [110] dimers.

succeeded in relaxing a twist [001] $\Sigma = 5$ GB in germanium with an elementary supercell box of 60 atoms (40 of which were entirely free to relax). In their preliminary results only the symmetrical translation state between two perfect crystals was explored and their low energy structure was unexpectedly shown to contain 4-atom rings. A more systematic exploration of the translation state did not however confirm this result (Tarnow et al., 1989). Their new ground state has an energy of 0.482 J m^{-2} and contains ring structures of 5 and more atoms though some of the atoms are threefold and some five-fold coordinated (Fig. 7-7).

Ionic Semiconductors (Nickel Oxide)

Studies in these materials are not as developed as they are in the covalent semiconductors for which actual experimental data are available for comparison to the theoretical numerical simulations. Recently some new experimental data (Merkle et al., 1988) became available which helped elucidate the atomic structure of a nickel oxide bicrystal although these new results have not yet lead to new simulations. Thus, results currently available are the fruit of almost blind simulations and as such must be treated with care.

A few $\langle 001 \rangle$ and $\langle 110 \rangle$ symmetric tilt GBs and $\langle 001 \rangle$ twist GBs were simulated by Duffy and Tasker (1984) and Wolf (1984a and 1984b). Thus it has been shown that electrostatic interactions have a dominant influence on the relaxed structure of grain boundaries and that the close proximity of ions of the same charge must be avoided in the construction of a low-energy interface.

As in the case of covalent semiconductors, these GBs have been analyzed using the concept of SU dislocations. For instance, $\langle 001 \rangle$ tilt GBs can be analysed in terms of arrays of [100] dislocations for $\Theta < 53.1°$ or in terms of arrays of [110] dislocations for $\Theta > 53.1°$ (Tasker and Duffy, 1983). The (210) $\Sigma = 5$ GB with $\Theta = 53.1°$ that separates these two domains corresponds to an energy cusp in the $E(\Theta)$ curve.

For the $\langle 011 \rangle$ tilt GBs, the (111) $\Sigma = 3$ GB is a remarkably low energy GB. No other cusps were found in the $E(\Theta)$ curve, contrary to the experimental results determined by Dhalenne et al. (1983). This said, however, it is fair to say that there are certainly too few points in the calculated $E(\Theta)$ curve to find the complete set of minima (Dechamps et al., 1988).

Duffy and Tasker were responsible for one of the most interesting results; they found that by removing Schottky pairs at the interface of (001) twist GBs, the energy of these interfaces can be greatly reduced (Duffy and Tasker, 1985). This is further evidence of the importance of having an awareness of the exact atomic structure when evaluating the energy of the GBs.

7.3 Experimental Results

Many different, powerful techniques such as X-ray diffraction, atom-probe field ion microscopy, scanning tunneling microscopy or electron microscopy have been used to study GB structure on an atomic scale in different materials. Bourret (1990) provided an overview of the varied potentials and limitations of these techniques.

To our knowledge, in contrast to the situation in gold twist GBs (Fitzsimmons and Sass, 1989) where this is the case, in semiconductors no complete study using X-ray techniques has yet been realized. Here, X-ray diffraction has only been used to determine macroscopic rotations between the grains. Only recently have results been obtained with the use of spectroscopic scanning tunneling microscopy on an edge-on GB in boron-doped silicon (Kazmerski, 1989) where composition mapping in the vicinity of an asymmetrical GB ($\Theta = 28°$) has been achieved after fracture in UHV. The interfacial zone exhibits a disordered region about 10 nm thick where the B atoms and H atoms were specifically imaged. The atom probe FIM is now capable of identifying the chemical nature of individual ions from the tip and spectacular results have been obtained looking at impurities segregation at GBs (Karlson and Norden, 1988; Seidman, 1989).

However, the most widely used, and by far the most productive technique has been transmission electron microscopy, employing bright field and dark field techniques, weak beam techniques, α-fringe techniques, diffraction techniques, convergent beam electron microscopy and high resolution electron microscopy (see Vol. 2 A,

Chaps. 1 and 2). This latter variation has given the most spectacular and numerous results. Generally speaking, it could be said that this technique permits the realization of *projections* of atomic structures along low-index axes (for further details on HREM see also Spence, 1988). The first high resolution electron microscopes were only able to provide useful information from images of diamond structure semiconductors taken along the $\langle 011 \rangle$ axis, and within these images only pairs of atomic colums (separated by 0.14 nm in silicon) could be resolved. The new generation of medium voltage microscopes however, has extended this limitation. The resolution of the best microscopes now is about 0.17 nm at Scherzer defocus, allowing new directions of observation and new materials to be examined (Bourret et al., 1988). For example, all the atomic columns of the $\langle 001 \rangle$ projections of nickel oxide (separated by 0.209 nm) or of germanium (separated by 0.2 nm) have now been entirely resolved. Having said this however, as the two crystals of the GB must be simultaneously imaged, only tilt GBs along low index axis can be observed. The other limitations of the technique arise from the fact that (apart from special interface configurations; Ourmazd et al., 1987) no direct chemical information is available using this technology, and that the information is averaged on an atomic column (thus only a *projection* of the structure is obtained).

It is beyond the scope of this chapter to discuss the other electron microscope techniques so we shall merely recall what kind of information each technique can provide. Electron diffraction allows one to determine the periodicity and symmetry of the boundary (Bacmann et al., 1985; Carter et al., 1981). The α-fringe technique allowed Bacmann et al. (1985) to measure precisely the translations between two grains along the same GB plane. CBEM is mainly employed to analyze the symmetry of the GB (Schapink, 1986) though in the case of compound semiconductors it can also determine the polarity of the material (Cho et al., 1988). Weak beam, dark field and bright field techniques are generally used to study the primary dislocations of low angle GBs or the secondary dislocations of high angle GBs.

In this section, we shall review the main experimental results that have been obtained concerning GB structure at an atomic level in the different classes of semiconductors.

7.3.1 Cubic Diamond Semiconductors

The potential for using polycrystalline silicon in solar cells or in electronic circuits has always sustained the studies of silicon grain boundaries. On the other hand, germanium bicrystals have been studied because high purity Ge bicrystals can be easily grown from the melt. The melting temperature of Ge is lower than that of Si and there are fewer contamination problems (carbon and oxygen contamination) in germanium. Numerous grain boundaries have been observed both in Ge and Si and until now it has been found that the structures in these materials are isomorphous.

Two classes of results have been reported for these covalent semiconductors. The first one deals with the exact atomic structure of the GB. These studies have generally used the HREM technique and have mainly concerned high coincidence tilt GBs. The second one is more *macroscopic* and involved studies to determine the set of dislocations called secondary dislocations that make the real deviate from a high coincidence GB. In this chapter a greater emphasis will be put on the exact structure of high coincidence GBs. The GBs are classified in terms of their tilt or twist components, their rotation axis and the rotation angle between the two grains of the bicrystal.

7.3.1.1 Low-Angle Grain Boundaries

Low angle $\langle 001 \rangle$ and $\langle 111 \rangle$ twist GBs in silicon were investigated by Föll and Ast (1979) using conventional electron microscopy. They found simple undissociated lattice dislocation arrays for rotation angles up to $8°$.

$\langle 011 \rangle$ tilt low angle GBs have been comprehensively studied by Bourret and Deseaux (1979 a, b). The situation here is more complicated than the simple Hornstra's models which consist of arrays of $a/2 \langle 011 \rangle$ Lomer edge dislocation, although HREM has confirmed the validity of the O-lattice concept. It was also confirmed that the dislocation-type content of a low angle boundary depends on the GB plane. Moreover, some exotic configurations, although predicted by the O-lattice and b-net theory were found to be periodically distributed along some particular facets. In fact, these dislocations were dissociated into complicated locks with similar configurations stemming from dislocation interactions in the bulk (Thibault-Desseaux et al., 1989). Thus, depending on the GB plane, it has been possible to observe $a/2 \langle 110 \rangle$ edge dislocations, dissociated dislocations with overall Burgers vectors $a/2 \langle 101 \rangle$ ($60°$ dislocations), $a \langle 111 \rangle$ and $a/2 \langle 211 \rangle$ and undissociated dislocations with a $\langle 100 \rangle$ Burger vector.

7.3.1.2 High-Angle Grain Boundaries

A. $\langle 110 \rangle$ Symmetric Tilt GBs

(111) $\Sigma = 3$ (111), $\Theta = 70.53°$

The HREM images of this frequently encountered GB (D'Anterroches and Bour-

ret, 1984) confirm Hornstra's model of boat shaped twin units. The GB plane is a pure mirror. It has been observed that a deviation from the exact coincidence introduces either dissociated or undissociated $a/3 \langle 111 \rangle$ (D'Anterroches and Bourret, 1984) or $a/6 \langle 211 \rangle$ (Föll and Ast, 1979) DSC dislocations. The structure of the period of this GB can be described by the SU sequence (TT). The proposal of another structure by Ichinose et al. (1986) derived from HREM images obtained in polycrystalline Si was unjustified being based on an image artefact that is normally expected, and that arises from the contrast variation due to thickness or defocusing conditions at a (TT) periodical structure (D'Anterroches and Bourret, 1984).

(255) $\Sigma = 27$, $\Theta = 31.59°$

The HREM images of this GB (Vaudin et al., 1983; Bourret and Bacmann, 1987) correlate very well with the atomic model proposed by Vaudin et al. (1983). It can be considered to be formed from structural units of the two limiting GBs $\Sigma = 1$ (C) and (122) $\Sigma = 9$ (L), the period being described by the sequence $L^+ L^- CL^- L^+ C$. (The "$+$" and "$-$" indexes refer here to a mirror symmetry with respect to the GB plane). It could also be described in terms of an equal number of (133) $\Sigma = 19$ and (122) $\Sigma = 9$ structural units i.e. (LC) and (L) respectively.

Small deviations from the (255) plane are possible via the introduction of pure coherent steps or steps associated with dislocations. Bourret and Bacmann (1987) proposed a model for these defects. Thus, if the net boundary plane deviates significantly from (255), the boundary dissociates into two low energy GBs, namely the (221) $\Sigma = 9$ and (111) $\Sigma = 3$ boundaries. The dissociation phenomenon has also been investigated by TEM on polycrystalline Si (Garg and Clark, 1988).

(122) $\Sigma = 9$, $\Theta = 38.94°$

This boundary has been extensively studied (Krivanek et al., 1977; Bacmann et al., 1982b; Papon et al., 1982; D'Anterroches and Bourret, 1984). The structure is clearly that proposed by Hornstra (1959) which has a glide mirror symmetry. As discussed in Section 2.3 the period can be described by either of the two sequences $L^+ L^-$ or $M^+ M^-$.

The defects introduced by deforming this bicrystal have also been studied in detail and will be reviewed in Section 7.5.

(233) $\Sigma = 11$, $\Theta = 50.48°$

Electron diffraction experiments (Papon et al., 1984) and HREM have confirmed (Bourret and Bacmann, 1986, 1987) that the periodicity of the as-grown GB is twice that of the coincidence lattice periodicity in the $\langle 311 \rangle$ direction. The interface has a (233) mirror glide plane.

It has already been stated in Section 7.2.5 that several models based on the structural unit concept have been proposed (Papon et al., 1984), but they are not compatible with the HREM images of as-grown bicrystals. The only compatible model has been that proposed by Bourret and Bacmann (1986), which is characterized by the sequence $(M^+ TM^- P^+ M^- TM^+ P^-)$. A model $(M^+ TM^- T)$ has been found which arises as the result of the deformation of a silicon (122) $\Sigma = 9$ GB in compression at low temperature. At high temperature, the resulting $\Sigma = 11$ is similar to the as-grown bicrystal (Putaux and Thibault, 1990) but this will be discussed in detail in Section 7.5.

Thus it is seen that the same Σ value GB exhibits two different structures. The first one is a mixture of the SUs associated with the two delimiting GBs $\Sigma = 9$ and $\Sigma = 3$ (i.e. M and T, respectively) in equal number. The second one was theoretically unpredictable and is a mixture of M and T SUs plus a special $\Sigma = 3$ SU (called P) corresponding to a high energy structure of $\Sigma = 3$. Nevertheless, both structures could exhibit completed bond reconstruction. Figure 7-8 shows two HREM images of the two structures and their corresponding models.

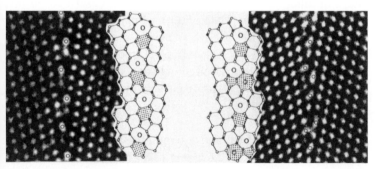

Figure 7-8. HREM of (233) [011] $\Sigma = 11$ tilt GB in Si showing the two different structures detected for the same Σ value. In (a) the GB plane is a glide mirror for the bicrystal; the GB structure contains only two types of SUs, and the period is $(M^+ TM^- T)$. In (b) the GB plane is a pure mirror for the bicrystal, and the period $(M^+ TM^- P^+ M^- TM^+ P^-)$ contains a new SU, known as P. The "+" and "−" indexes indicate mirror-related SUs. The two corresponding models are shown in the frames (Putaux and Thibault, 1990).

(211) $\Sigma = 3$, $\Theta = 70.53°$

The α-fringe method (Vlachavas and Pond, 1981; Fontaine and Smith, 1982) revealed a rigid-body translation along $\langle 111 \rangle$ and a dilatation along $\langle 112 \rangle$. Moreover, high energy electron diffraction (Papon and Petit, 1985) gave clear indication of a period twice as large as the CSL period. HREM confirmed this new model (Bourret et al., 1985) by observing the (211) $\Sigma = 3$ bicrystal along the two directions $\langle 011 \rangle$ and $\langle 231 \rangle$. A large rigid body translation has been found with a component parallel to the GB plane of about $a/11$ $\langle 111 \rangle$ and a component perpendicular to the GB $a/20 \langle 211 \rangle$. The model put forward by Papon and Petit (1985) proposed $a/11 \langle 111 \rangle$ and $a/50 \langle 211 \rangle$ which is comparable to the experimental results. Furthermore, this allows one to conclude that bond reconstruction occurs along the $\langle 011 \rangle$ tilt axis.

Numerous defects have been observed along this GB and their structures determined by HREM (Bourret and Bacmann, 1986). Coherent steps were found, which may play an important role in GB migration because their motion does not require climb, DSC dislocations such as the $1/6 \langle 112 \rangle$ pure edge dislocation which was found to be associated with steps of different heights.

B. $\langle 001 \rangle$ Symmetric Tilt GBs

Some $\langle 001 \rangle$ tilt GBs have been studied in great detail by HREM (Bourret and Rouvière, 1989). The principal results are reported here.

(310) $\Sigma = 5$, $\Theta = 36.87°$

HREM of this GB along two directions $\langle 001 \rangle$ and $\langle 130 \rangle$ allowed confirmation of the model deduced from electron diffrac-

tion and α-fringe observations by Bacmann et al. (1985). This model is based on the structural units proposed by Hornstra (1959) and its period is the CSL period, although two variants exist which arise as a result of symmetry breaking by a rigid body translation parallel to the tilt axis. Figure 7-9 shows the three dimensional analysis of $\Sigma = 5$ in Ge. α-fringes contrast of the $\Sigma = 5$ GB shows two types of defect along the GB; one separating domains with the same RBT and the other separating domains with different RBT. The two HREM images viewed along [001] and [1$\bar{3}$0] clearly reveal the RBT of about $0.13 \, a \langle 001 \rangle$ (i.e. 0.073 nm) $+ 0.006 \, a[1\bar{3}0]$ (i.e. 0.011 nm). This is in good agreement with the corresponding calculated values, namely 0.0783 nm along $\langle 001 \rangle$ plus a dilatation of 0.011 nm along [1$\bar{3}$0].

The SU sequence of one variant is described by $A^+ A'^-$. The "'" refers to a mirror symmetry and the "+" refers to the sign of the rigid translation in contrast to the $\langle 011 \rangle$ tilt GB case. The A unit is associated with a pure edge $a/2 \langle 110 \rangle$ dislocation.

(11 3 0) $\Sigma = 65$, $\Theta = 30.57°$

The periodicity is that of the CSL. The GB structure can be described using the intercalation of perfect crystal SUs (p) in the $\Sigma = 5$ GB and the period is (AA′ApA′AA′p). Numerous defects are encountered along this GB especially coherent steps or facets which make the GB rotate.

(510) and Asymmetric $\Sigma = 13$, $\Theta = 22.62°$ and (320) $\Sigma = 13$, $\Theta = 67.38°$

The principal point is that the (510) GB has several structures which differ notably despite being simultaneously present in the same specimen (Rouvière and Bourret,

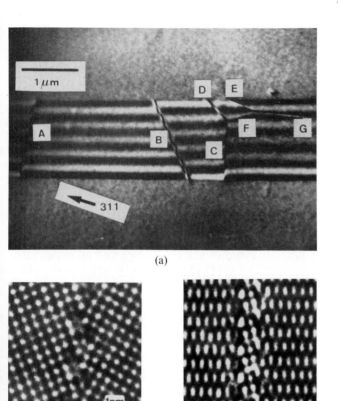

(a)

(b)   (c)

Figure 7-9. 3D structure determination of the (310) [001] $\Sigma = 5$ tilt GB in Ge. (a) α-fringe contrast showing areas of different translation state. A-D-G are dislocations (introducing no RBT) whereas B-C-E-F are defects introducing a RBT (Bacmann et al., 1985). (b) and (c) are HREM pictures showing the projection of the GB along [001] and [1$\bar{3}$0], respectively, and revealing an RBT along the tilt axis of about 0.073 nm (Bourret and Rouvière, 1989).

1989). The first structure is strictly periodic and contains a complex and compact mixture of SUs including A, A', and B, B' where B and B' are SUs linked to the $a/2$ $\langle 011 \rangle$ 45° dislocation core. In fact, this structure is only observed on a small scale and is replaced by a varying structure which destroys the strict GB periodicity. Although rarely observed, it is undubitably the case that not only can the GB structure vary all along the tilt axis but it can also vary from one period to the neighbouring one. A limited number of atoms in the core of the SUs may have two equivalent sites with a low energy barrier in between, so that the SUs have different configurations simultaneously present in the GB plane. The static GB energy calculated from these structures is of the same order of magnitude as that calculated from simpler ones which were never observed. A rough estimation of a configuration entropy term is in favour of the stabilization of a multiple structure.

The (320) GB and the asymmetric $\Sigma = 13$ GB have a simple structure. The period of (320) $\Sigma = 13$ is given by (pppA'$^-$ pppA'$^+$). The asymmetrical GB plane is (810)$_I$ and (740)$_{II}$ and its period can be read easily on the HREM images and involves B SUs.

(710) $\Sigma = 25$, $\Theta = 16.26°$ and
(910) $\Sigma = 41$, $\Theta = 12.68°$

These GBs are at the validity limit for a low angle boundary and can be viewed as discrete dislocation cores. As for the (510) $\Sigma = 13$, the core is extremely variable and segments of different structures can be found. It has the same basic SUs as those used to describe (510) $\Sigma = 13$ but the static energy calculated from different potentials gives generally higher energies than those calculated from simpler models. This was the starting point for the introduction of an entropy term in order to stabilize certain special configurations that have many variants.

7.3.2 Zinc Blende Semiconductors

Only a few experiments have been performed on these materials.

GBs in SiC, which exhibits a strong polymorphism have been observed by HREM (Lancin and Thibault-Desseaux, 1988). No comparisons between experimental images and computer simulations have been attempted, even though some models have been proposed. Hagège et al. (1990) produced HREM pictures of $\Sigma = 3$ and $\Sigma = 9$ GBs in β-SiC although no reliable models were deduced from these images. An attempt was made by Laurent-Pinson et al. (1990) to study the distribution of GBs in hot pressed α-SiC and to characterize the deviation from coincidence positions.

GaAs bicrystals produced by growing GaAs epilayers on substrates cut from Czochralski-grown germanium bicrystals have been observed by Carter et al. (1985) and Cho et al. (1988). In such a material with its sphalerite structure it is possible to change a bicrystal structure without having to rotate or translate one crystal, simply by reversing the polarity of one crystal. Therefore, for a given rotation angle and a given GB plane in diamond structure, two different interfaces exist in GaAs each with different grain polarities (Holt, 1964). The polarity of each grain can be experimentally determined by using the dynamical

coupling effect of Holz's reflections on (200) type convergent beam disks (Tafto, 1979). Using HREM techniques and Holz's reflections of convergent beam patterns, it has been possible to determine the structure of a (111) $\Sigma = 3$ GB. It was found that this coherent twin boundary has no anti-site bonds (Ga–Ga or As–As).

Many microtwins were observed in the sample of the (111) $\Sigma = 3$ GB. Again, the use of HREM experimental images of small areas of these microtwins, lead to the proposal of hypothetical structures for a symmetrical (112) $\Sigma = 3$ and an asymmetrical (111)/(115) $\Sigma = 9$ GB (Cho et al., 1988) although there are currently too few data to allow exact determination of these structures. For example, the proposed (112) $\Sigma = 3$ model which is simpler than the corresponding one observed in the diamond structure, does not fit very well with the relevant experimental images. One large spot in the simulated images corresponds to two separated spots in the experimental image.

7.3.3 Ionic Semiconductors

Several $\langle 001 \rangle$ coincidence tilt GBs in NiO have been observed by electron microscopy and precise information has been obtained (Merkle and Smith, 1987).

Two structures for the (310) $\Sigma = 5$ GB in NiO have been observed (Fig. 7-10a, b) which correspond to two different translations between the grains (Fig. 7-10d, e). These structures can be related to the model proposed by Duffy and Tasker (1985), but they mostly differ from it due to the introduction of an atom row in the open hole of the Duffy and Tasker model (Fig. 7-10c). However, the reduced density in the boundary images could be attributed in large part to the presence of Schottky

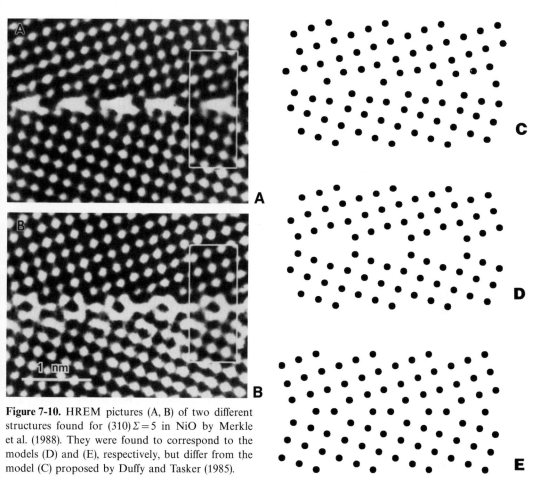

Figure 7-10. HREM pictures (A, B) of two different structures found for $(310)\,\Sigma=5$ in NiO by Merkle et al. (1988). They were found to correspond to the models (D) and (E), respectively, but differ from the model (C) proposed by Duffy and Tasker (1985).

pair vacancies (pairs of anion and cation vacancies) in the atom rows at this core.

The alternative $(210)\,\Sigma=5$ has also been observed (Merkle, 1990). It contains $(410)/(100)$ facets although the images were not compared to the results of atomic simulations. Another asymmetric GB $(\Theta=26.6°)$ has been observed whose plane in fact corresponds to two dense planes $(210)/(100)$ in each adjoining crystal. The author suggested that the asymmetrical configuration is favored owing to the possibility of closed-packed planes wanting to be parallel to each other.

A $(510)\,\Sigma=13$ GB has also been studied (Merkle and Smith, 1988). The core struc-

ture seems to vary from one core to the other. The authors were able to estimate the occupancy of the rows and an approximate concentration of 25% vacancy type defects has been estimated from the contrast change in HREM pictures of the rows.

These vacancy pairs are known from simulation to reduce the GB energy in NiO twist boundaries and it is suggested they might play a similar role in tilt GBs. It seems that in ionic GBs the reduction in energy is provided by a reduced density of the rows in the GB rather than by a lateral expansion.

7.3.4 Reliability of Grain Boundary Structures

In elemental semiconductors, it is known from energetical calculations that the low-energy structures of GBs are those with a complete reconstruction of the bonds. Unfortunately however, HREM is only able to provide a direct experimental answer to the modeling of the average projected structure. So for the case of GB defects the same question remains as yet totally unanswered. The presence of point defects along GBs or along grain boundary dislocations cannot be excluded, although a vacancy content larger than 20% may be detectable as has been shown in NiO GBs.

The atomic simulations discussed in Section 7.2.4 whilst not able to detect the structural multiplicity, still provide a powerful means of confirming the stability of the observed configuration. This can then help in understanding its subsequent evolution under segregation conditions, annealing or mechanical constraints.

7.4 Electrical Properties of Grain Boundaries

7.4.1 Introduction

The change in the electrical properties of a semiconductor induced by a grain boundary (GB) has been for a long time the subject of numerous studies. Since the early work by Taylor et al. (1952) showing that a GB is a barrier for the majority carrier transport, many experiments have shown that GBs are able to induce important electrical effects. It is generally accepted that the disordered region in the GB plane induces new available states inside the gap. In germanium for instance it is easy to produce an inversion layer (a p-layer in an n-type material). This double junction has i) the typical I–V characteristics of a double Schottky barrier, ii) a high sheet conductance and iii) recombination properties for injected minority carriers. In order to interpret these properties the model initially proposed by Taylor et al. (1952) is still widely used. The states existing at the GB plane trap the majority carriers creating a potential barrier at this plane and a depleted zone in the two adjacent regions; thus the barrier is at the origin of the increased resistance and capacitance effect measured perpendicular to the GB plane. The space charge is the source of the low longitudinal resistance and the recombination of electron-hole pairs is due to the presence of deep levels at the GB plane. It is in the framework of this model that all the electrical properties of grain boundaries will be discussed. A great deal of attention over a long period of time has been devoted to precise determination of electrical properties, but in the recent past efforts have been made to better characterize the geometry of the GBs and try to correlate this structure with the observed electrical properties. This is a formidable task which although not yet completed is undoubtedly of prime importance. Of particular interest is the observation that some GBs are active and others not, despite the seeming similarity of their geometries.

The interest in the electrical properties of grain boundaries has been driven by several potential applications:

i) Polycrystalline silicon is interesting as a potential material for low cost solar cells: the presence of GBs tend to decrease the solar cell efficiency. It is important to understand why and how these GBs can be passivated by an appropriate treatment.

ii) Polycrystalline silicon is widely used for interconnections or insulating material in integrated circuit devices. The interaction between dopant and GBs and the possible effect of dopant-induced GB migration is an important topic for study.

iii) Voltage dependent resistors rely heavily on the electrical activity of GBs (ZnO, $SrTiO_3$, TiO_2). The same is true for PTC-resistors based on ferro-electric titanates or chemical sensors and boundary layer capacitors.

The electrical properties of GBs as well as their relation to structure have been the subject of several reviews to which the reader should refer for details: see for example, Grovenor (1985), Mataré (1984) or Harbeke (1985) and Möller et al. (1989). The current chapter will emphasize the relationship between structure and electrical properties and will discuss the very few examples in which this relationship has been clearly defined.

7.4.2 Electrical Properties Induced by Grain Boundaries

Large modifications of bond length as well as bond angle can be present at GBs. In the case of pure tilt GBs with high index common axis the atoms tend to conserve tetracoordination. However, for more general GBs it is conceivable that dangling bonds or more generally non-tetracoordinated atoms could be present in the form of line defects (dislocation-like) or point defects. In addition, impurities could segregate at preferential GB sites. These three types of situation induce local modification of the electronic states which could have several important consequences, each of which is examined in this chapter.

7.4.2.1 Electronic States Associated with a Grain Boundary

For a given GB structure (i.e. knowing all atomic positions) the electronic structure can be calculated using various approximate approaches or by employing ab initio calculation. Therefore, assuming that the structure is correct and includes *all* the defects contained in a GB, the electrical properties could be predicted. A number of general ideas can be summarized from the various calculations which have been performed up to now.

Local Distortions:
Band Tails in the Local Electron Density

Local distortion results in bond length variation as well as in angular variation and produces a variation in the average potential at a length scale of interatomic distances. A similar problem has been solved in amorphous silicon or at Si/SiO_2 interfaces (for a review see Cohen et al., 1988 and Chapter 10 of this Volume). Lattice potential fluctuations result in band tails appearing in the gap at the valence band and the conduction band edges. The density of states decreases exponentially and can extend deeply into the gap. According to Soukoulis et al. (1984), in a three dimensional description, one can say for the width of the exponential decay:

$$\Delta E_0 = \frac{\pi}{2} w^2 \frac{L^2 m_x}{h^2} \qquad (7\text{-}3)$$

for carrier of mass m_x which is localized in Gaussian potential fluctuations with variance w^2 and correlation length L (h is Planck's constant). There is generally an asymmetry in the density of states with a greater degree at the valence band compared to the conduction band. This could be accounted for by the different effective mass values. By analogy with these tails, Werner (1989) has proposed that similar tails are present even on a single GB. This proposition has not yet been supported by calculation, at least not yet in the few examples which have been studied:

i) Di Vicenzo et al. (1986) calculated two different configurations for (122) $\Sigma = 9$ in

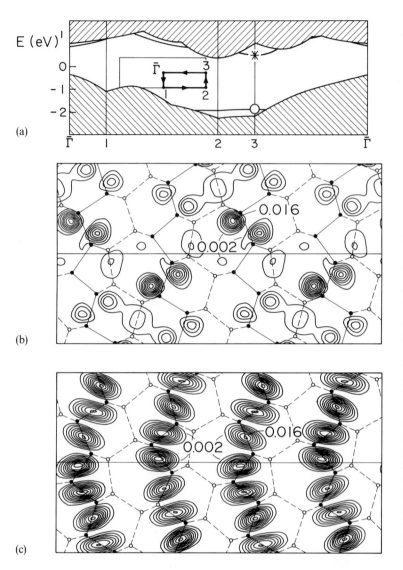

Figure 7-11. Interfacial electronic structure in a cell containing a (112) $\Sigma = 9$ GB. (a) Energy bands as calculated within the local density approximation. The projection of the bulk band is hatched. The inset shows the interfacial Brillouin zone and the wavevectors which have been plotted. Bands appear to emerge from the continuum but never reach the gap at the $\bar{\Gamma}$ point. (b) The charge density of one of the states (*) is represented and shows a good localization at the interface. (c) The charge density of one of the valence-band-edge states (o) indicates that this state is not localized at the GB (from DiVicenzo et al., 1986).

Si, using a first-principles density functional theory, as well as tight-binding approximation. Shallow states do occur close to the conduction and valence band edges. One of these states is clearly localized around atoms having maximum angular variation compared to the bulk (Fig. 7-11).

ii) Mauger et al. (1987) have examined the electronic structure of the $(11\bar{2})\,\Sigma = 3$ in germanium by the recursion method in the framework of the tight-binding approxi-mation. The main perturbation induced by distortions is a 1 eV shift in the peak density of the p states normal to the GB plane. However this shift towards the top of the valence band is too small to push these resonant states into the energy gap. It is interesting to note that the most perturbed atom has an energy 0.118 eV in excess of the perfect lattice atomic energy, and yet no localized states are introduced in the fundamental gap.

iii) Kohyama et al. (1988) calculated the local density of states and the projected band structure of the relaxed $(11\bar{2})\,\Sigma = 3$ and $(1\bar{3}0)\,\Sigma = 5$ GB in silicon. They found localized states at the edges of the conduction and valence bands although they are above the minimum of the conduction band or below the absolute maximum of the valence band respectively. They cannot give rise to band tails in the local density of states.

iv) Paxton and Sutton (1989) studied the local density of states in various configurations of the same $(11\bar{2})\,\Sigma = 3$ and $(1\bar{3}0)\,\Sigma = 5$ GBs in silicon. Unfortunately the resolution of the method was not sufficient to determine whether band tails are associated with the valence or conduction bands or not.

The question arises as to why band tails are not predicted by calculations? Paxton and Sutton (1989) remarked that in a three-dimensional system there is a critical degree of disorder that must be exceeded before localization sets in. This suggests that the calculated results obtained so far have been limited to much too simple GBs. Thus the special twin positions which were used could not be representative of what is a more general GB. The presence of steps, dislocations and facets is inevitable and these would themselves induce other types of defect. It would be highly desirable, in order to clarify this point, to calculate more general configurations either along different axes or to include a large screw component for which the band distortions are large (Tarnow et al., 1989).

Dangling Bonds Introducing Deep Levels

Dangling bonds have been studied for a long time as the possible origin of GB activity. The existence of electron states localized at dangling bond sites can be explained in simple terms using a tight-bind-ing approximation (Lannoo and Bourgoin, 1981). A dangling bond introduces deep levels in the gap and may have different charge states. It also has a spin associated with the lone electron which is detectable using electron spin resonance techniques.

One particular GB containing a dangling bond was calculated by Paxton and Sutton (1989). The GB is a p2'mm' model of the $(11\bar{2})\,\Sigma = 3$ in silicon that has a five-fold coordinated atom (Fig. 7-12) associated with a three-fold coordinated atom. The energy change in the three-fold coordinated atom has a negative value caused by rehybridization with less sp^3 configuration and more $s^2 p^2$ where deviations from the ideal bond angle vary between -40% and $+31\%$. The covalent bond it forms with the five-fold coordinated atom is stronger than in the perfect crystal due to a contraction in the bond length. However, since a large electronic rearrangement of all their neighbours also takes place, it would be invalid to assign to the three fold coordinated atom a dangling bond energy equal to half the energy of a regular bond. The major part of the energy increase is distributed over several other atoms and especially onto the five-fold coordinated atom which sees a 1.9 eV increase. In contrast, a clear state in the band gap can be identified in the local density of states associated to the dangling bond. These states are localized at distances of the order of the lattice parameter. Although the model chosen for the periodic $(11\bar{2})\,\Sigma = 3$ is unlikely to exist as such (it has a much greater energy than all the other models), it serves to illustrate some of the features that a GB would have locally at a defect.

Levels Associated with Segregated Impurities at GBs

The effect of locating an impurity atom at a GB site has not been thoroughly stud-

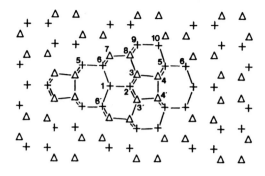

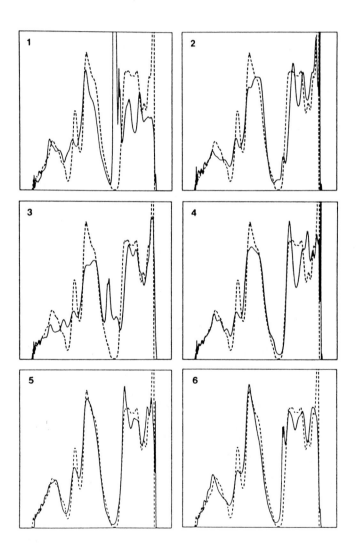

Figure 7-12. Relaxed structure of a $p2'mm'$, $(11\bar{2})$ $\Sigma = 3$ tilt boundary in silicon. Top: projection of the structure showing one dangling bond (or a threefold coordinated atom) at atom 1 and a fivefold coordinated atom at atom 2. The local density of states at points 1 to 6 are represented below. Note the large contribution inside the gap of the Si bulk at atom 1. The broken lines show the local density of states in the perfect diamond cubic crystal for comparison (from Paxton and Sutton, 1989).

ied so far. A first study was made by Masu-da-Jindo (1989) on a (122) $\Sigma = 9$ reconstructed GB in silicon where both doping with shallow levels and transition elements with deep levels were considered. The principal result was that the final configuration of a substitutional impurity has a broken tetrahedral symmetry due to the site distortion inherent at a GB site. As a consequence, the degenerate states are split, the maximum level splitting being of the order of 0.2 eV. This leads also to a substantial change in the defect levels associated with the transition element. For Cr impurity in a perfect silicon crystal, Masuda-Jindo (1989) found (t_2^*, e, t_2) levels at (1.5, 0.23, -0.77) eV with respect to the top of the valence band whilst they are (1.3, 0.2, -0.89) eV for segregated Cr impurity. This behaviour seems to be generally true for other transition elements. Larger effects should even be observed with more distorted bonds since the (112) $\Sigma = 9$ GB is one of the lowest in energy amongst any GBs. Even more drastic effects should occur at non tetra-coordinated sites but as yet no calculations are available to support this idea.

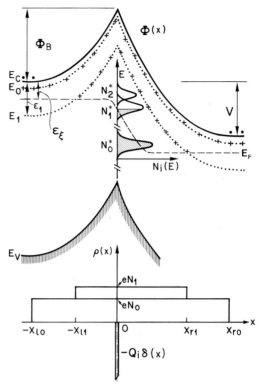

Figure 7-13. Double Schottky barrier at a negatively charged GB in an n-type semiconductor with an applied bias V. The dominating shallow bulk level E_0 is at a density N_0. A second deeper level E_1 is present with density N_1. Electrons are trapped at the GB plane and these charges are screened by a positive space charge distribution in order to establish the overall charge neutrality. N_0^*, N_1^*, N_2^* are the different interface states (from Greuter and Blatter, 1989).

7.4.2.2 Potential Barrier and Transport Properties

The localized states induced by a GB in the band gap or close to it may have a donor or an acceptor-like behavior. They can trap the majority carriers thus creating a potential barrier with a depleted zone in the adjacent grains. In the case of an n-doped material, electrons are trapped giving in extreme cases an n-p-n double junction. The barrier height, Φ_B, and its variation away from the GB plane is given by the solution of Poisson's equation for the charge distribution in the GB vicinity. Blatter and Greuter (1986) studied in detail

the effect of an arbitrary distribution $N_i(E_i)$ of the interface states and a finite number of deep bulk defects beside the dominant shallow bulk level E_0. A typical scheme of a GB barrier with an applied bias voltage V is reproduced in Figure 7-13.

Once the energy barrier and the trap levels are known, the transport properties through or parallel to this barrier can be calculated. One has to be aware however, that the assumed distribution of the interface states and their character will greatly influence the calculated properties. As a consequence, most of the formulae derived below are model dependent and in view of the uncertainties which remain about the origin of these levels one has to be cautious in their use.

Low Field: Orthogonal Conductance

There are several ways for the majority carriers to cross through the potential barrier; by thermoionic emission, by tunnelling or by diffusion. These three approaches yield expressions which are relatively similar.

– By *thermoionic emission* one assumes that only thermally excited majority carriers with potential higher than the barrier will cross. Then the current flow through the GB, I_{GB} in its simplest form is given by:

$$I_{GB} = K e N v_{th} \cdot$$

$$\cdot \left[\exp \left(-\frac{\Phi_B}{kT} \right) - \exp \left(-\frac{\Phi_B + eV}{kT} \right) \right] \quad (7\text{-}4)$$

where the energy barrier height, Φ_B, is measured from the Fermi level, N is the concentration of majority carrier, v_{th} is their thermal velocity. K is a coefficient dependent on the ratio of trapping and emission of the majority carriers at the GB interfacial states. V is the applied voltage, e is the electron charge.

A different but equivalent form is often expressed which includes an effective Richardson constant (Seager and Castner, 1978) such that:

$K e N v_{th}$ is replaced by

$$A^* T^2 \exp \left(\frac{E_F - E_C}{kT} \right) \quad \text{with}$$

$$A^* = \frac{4 \pi m^* e \cdot k^2}{h^3}$$

where m^* is the effective mass of majority carrier and $E_F - E_C$ the energy difference between the Fermi level and the conduction band.

A fit of Eq. (7-4) to an experimental $I(V)$ curve facilitates, in principle, the measurement of the energy barrier Φ_B. This calculation neglects the small currents of electrons being trapped and re-emitted from the interface states. A more precise formula would take this correction into account and an expression for this was proposed by Greuter and Blatter (1989). In addition, the tunneling current must be considered.

The *tunneling current* through the barrier could be important at low temperature and high doping levels at which the space charge region is thinner. A full expression for tunneling and thermoionic emission conductance is given by Seager and Pike (1982).

The *diffusion model* uses a different approach. It considers that the current is due to diffusion induced by a charge gradient in the depleted zones. This model was originally proposed by Taylor et al. (1952) and is physically more appropriate when the carrier mean free path is smaller than the barrier width (low mobility semiconductors). In fact, the final result turns out to be very similar to the thermoionic process except for the pre-exponential factor and there is no clear cut experiment in favor of one model or the other.

Inspection of Eq. (7-4) shows that the I−V characteristics are highly non-linear except with very small voltages. At $V \ll kT/e$ the characteristics are purely ohmic. Then, if Φ_B is constant, the characteristics become sub-ohmic due to the exponential term. For high voltages however, the value of Φ_B could vary as well, due to changes in the filling of the interface states.

High Field: Orthogonal Conductance

When all the interface states are filled up, the charge saturates and the potential barrier drops down with increasing bias. This leads to the region where breakdown occurs. In this region the electric field is so high at the depletion zone that hot electrons are induced. This mechanism has been studied in detail by Blatter and Greuter (1986).

Longitudinal Conductance

The GB sheet resistance (parallel to its plane) is lower than in the bulk. Most of the experimental results show a two-dimensional quasi-metallic conduction with a high carrier density, a result that was interpreted by Landwehr et al. (1985). The GB is a sheet of acceptor states with a strongly localized charge. This charge is neutralized by holes whose density is high enough for complete degeneracy. The conductance is expected to be linear with $1/\log T$.

7.4.2.3 Dynamic Properties and Recombination Properties

All charges at or around the GB plane contribute to the time dependent properties of a GB. For a variable bias, with a frequency f, $V(t) = V_0 + \tilde{V} \exp(i\,2\pi t f)$. The admittance, the capacitance or the conductance as well as the infrared photoconductance are determined by the behavior of several induced currents. These arise due to:

i) the capture and emission of majority carriers,

ii) the trapping and emission of minority carriers,

iii) the electron and hole recombination at the interface.

When the majority carriers are assumed to play the most important role, the two major parameters which govern the dynamic properties are c_n, the characteristic frequency for carrier capture and e_n the

characteristic frequency for emission. Qualitatively, three different frequency ranges may be distinguished:

i) a high frequency range for $f \gg (e_n, c_n)$ for which the boundary charge remains constant during one period of the modulated bias,

ii) a low frequency range for $f \ll (e_n, c_n)$ for which the boundary charge varies in phase with the modulation,

iii) an intermediate range for which the boundary charge is out of phase with the modulation.

The different equations giving the capacitance or the admittance as a function of the frequency (Werner, 1985; Petermann, 1988) allows one to make a link with the density of states.

7.4.3 Experimental Methods for Measuring the Grain Boundary Electrical Activity

7.4.3.1 Methods Based on Transport

Most of the characterization techniques of grain boundary states rely on the analysis of electrical measurements and particularly the I–V characteristics. Pike and Seager (1979) proposed the first deconvolution scheme in order to deduce the energy distribution of traps $N_i(E_i)$ at GBs using the I–V characteristics on a single GB at several temperatures. The starting point was to equate the charge Q (per unit surface) at the GB as deduced from the Poisson's equation, with the charge Q' from the interface density of states, thus:

$$Q = \left(\frac{\Phi_B}{4\gamma}\right)^{1/2} + \left(\frac{\Phi_B + eV}{4\gamma}\right)^{1/2} \quad \text{with}$$

$$\gamma = \frac{1}{8\varepsilon\varepsilon_0 N} \tag{7-5}$$

where N is the doping level and:

$$Q' = e \int_{E_F^{\text{neutral}}}^{E_c} dE_i \, N_i(E_i) \, f_i(E_i) \tag{7-6}$$

where f_i is the Fermi distribution function and $E_F^{neutral}$ the Fermi level in the neutral interface.

By differenting Eq. (7-5) and (7-6) with respect to the applied voltage V, and using expression (7-4) for the current, one has enough information to determine with two I–V characteristics, the $N_i(E_i)$ in a limited energy range above the Fermi level. In order to increase the energy range one has to use different specimens with different bulk Fermi levels. Another, simpler method is accessible if one assumes that the Fermi level is temperature independent. For small voltages, the conductivity can be written as:

$$\sigma = f(T)\exp\left(-\frac{E_c - E_F + \Phi_B}{kT}\right) \qquad (7\text{-}7)$$

thus the measured experimental activation energy is mainly controlled by $E_c - E_F + \Phi_B$. An independent measurement by Hall effect provides $E_c - E_F$ from which Φ_B is deduced. Application of Eq. (7-5) and (7-6) at different doping levels allows one to deduce the energy distribution of traps $N_i(E_i)$.

By exploiting the techniques of *admittance spectroscopy,* this dynamic behavior can also provide information regarding the density of states. The principle here is to measure the admittance of a metal-oxide-silicon capacitor with one GB included in the silicon crystal. The GB charge under a variable bias changes with a finite relaxation time τ leading to characteristic phase shifts. One can deduce the density of states using a series of admittance versus frequency curves with or without bias, and under different doping levels (variable E_F) (see Werner, 1985).

7.4.3.2 Transient Methods

The transient effects associated with carrier capture or emission at the interface give information about the levels associated to the GB states. *Deep level transient spectroscopy* (DLTS) has been the major tool of study for these effects. Firstly, one measures directly the activation energy for carrier emission from deep levels in semiconductors. A p-n junction or a Schottky barrier is then made at the surface and a GB is made accessible in the space charge region of this barrier. The GB traps are filled periodically by means of voltage pulses applied across the junction and the carrier emission produced in the return to equilibrium is recorded by measuring the junction capacitance. In the case of a single GB (bicrystals) the GB itself forms the Schottky junction. The bulk traps have to be recognized and disregarded from this analysis. The distribution $N_i(E_i)$ can be deduced by applying a bias, thus moving the Fermi level across the density of interface states and in addition, the capture cross sections can be measured by combining DLTS with the I–V characteristics (Broniatowski, 1987).

The recombination of electron-hole pairs is directly related to the presence of deep levels. This activity is visible when minority carriers are injected into the material. If the injection is spatially controlled, a local characterization is possible and this forms the basis and the interest in *light* and *electron beam induced current* (respectively LBIC and EBIC). The sample must have a diode at its surface and the light or electron beam is scanned through this diode such that the junction collects the minority carriers emitted by the beam. Any recombination centers at the GB plane would lower the collection efficiency and hence the induced current. The active GB generally appears as a dark region on an LBIC or EBIC image.

The theory of EBIC and LBIC was developed by Zook (1980) and Donolato

(1983) in a way that one can now measure quantitatively the recombination velocity at a given GB. The photo or (electron) conductivity is enhanced when the beam crosses the GB due to neutralization of trapped majority carriers at the GB. This effect can also be exploited to measure the barrier height (photoconductance method).

Finally, the *luminescence* methods which detect radiative recombinations allow one to localize the deep centers. Scanning photoluminescence as well as cathodoluminescence might be used in direct gap semiconductors (GaAs, CdTe ...). The spatial resolution of all these methods is limited by the beam diameter for photons, by diffusion of the incident electrons as well as by the diffusion length of minority carriers. The best resolutions are obtained with a scanning transmission electron microscope (STEBIC technique) and are of the order of ~ 0.1 μm.

7.4.4 Correlation Between Electrical Activity and Structure

The most important works of the last decade have been those devoted to establishing a precise correlation between the electrical activity of a particular GB and its structure. Ideally, the structural determination must be made on the same GB that is electrically characterized. Few attempts have been made to realize these conditions, although it is difficult to exactly determine the defect or impurity content of a specific GB.

7.4.4.1 Transport Experiments in Oxides: Observation of Deep Levels

In view of the essential contribution made by GBs to the desired, strongly nonlinear I–V characteristics, the oxides such as ZnO, $SrTiO_3$, TiO_2 and SnO_2 were mostly studied by transport experiments.

Several studies in ZnO using high resolution electron microscopy have indicated that a thin and homogeneous layer of bismuth ($\lesssim 10$ Å) is segregated at the GB interface (Kanai et al., 1985; Olsson, 1988). The Bi ions are generally believed to be responsible for the electrical activity. However it is also thought that an excess of oxygen at the GB plays an important role. Surface analysis on fractured surfaces shows a surface concentration of approximately one monolayer for both Bi and O at the GB. A loss of oxygen results in a strong reduction of the barrier height. The same observations were reported in $SrTiO_3$ based material in which Na ions play the same role that Bi plays in ZnO (Stucki et al., 1987).

In these materials, all grain boundaries are electrically active, an observation confirmed by a large variety of techniques. Each individual GB shows a varistor-like I–V characteristic and the spread in the electrical properties is remarkably small (Einziger, 1987). A series of typical I–V characteristics obtained by Greuter and Blatter (1989) is represented in Fig. 7-14, with the barrier height. The experimental curves obtained are very well reproduced by the theory, based on Eqs. (7-4) to (7-6). Qualitatively three regimes are observed:

i) In the low-bias regime the behavior is ohmic as predicted from Eq. (7-4).

ii) In an intermediate regime there is a small region of sub-ohmic behavior. This region is, however, very limited because the barrier height Φ_B decreases and this decay is only partially compensated for by the charge increase, Q_i, which results from filling new interface states. This compensation (or pinning of the barrier height) is more effective for higher dopant concentrations.

iii) In the high-bias regime, the barrier height starts to collapse as all states are

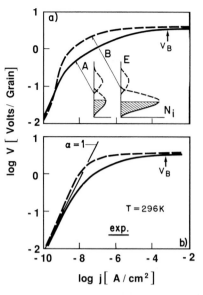
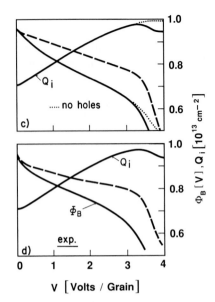

Figure 7-14. I–V characteristics and corresponding barrier height $\Phi_b(V)$ for two ZnO varistors differing by their Mn doping (A = 0.25% and B = 1.5%). Comparison is made between experiment (b, d) and theory (a, c). The bias-dependent interface charge Q_i is shown in (c) for the situation with and without hole creation in the breakdown region (from Greuter and Blatter, 1989).

filled and Q_i can no longer increase. The current drastically rises in a regime known as breakdown. If some holes are created, the collapse is less catastrophic but still visible. Typical values of density and energy of the GB states in ZnO as deduced from modeling are reported in Table 7-1.

7.4.4.2 Transport Experiments in Bicrystals

It has long been determined that GBs in normal semiconductors are electrically active (Siegel et al., 1981; Mataré, 1984; Grovenor, 1985). It has however been rec-

ognized since the early days that this activity could vary with several parameters. Along a given GB some portions were active, others not; in a polycrystalline material, some GBs were active, some were not and it was found that ingot annealing could enhance the electrical activity of most GBs. Thus most of the recent studies have dealt with GBs that have been at least geometrically characterized. For this reason, specially grown bicrystals are particularly suitable. Petermann and Haasen (1989) have shown that deep levels are present in silicon at a $(710)\Sigma = 25$ GB (Fig. 7-15). This bicrystal was grown in

Table 7-1. Characteristics of the two ZnO specimens possessing the I–V curve shown in Fig. 7-14. The density N_i and energy E_i of the GB states are deduced from experiment. Energy gap $E_c - E_r = 3.2$ eV; depletion layer width $X_{ro} \sim 500$–1000 Å; bias before breakdown = 3.3–3.7 volts (from Greuter and Blatter, 1989).

Doping levels	N_0^*	E_0	N_1^*	E_1	N_2^*	E_2
0.25% Mn	$3.9 \cdot 10^{12}$	?	$3.5 \cdot 10^{12}$	$E_c - 1$ eV	$2.5 \cdot 10^{12}$	$E_c - 1$ eV + ε
1.5% Mn	0	?	$1 \cdot 10^{13}$	$E_c - 1$ eV		

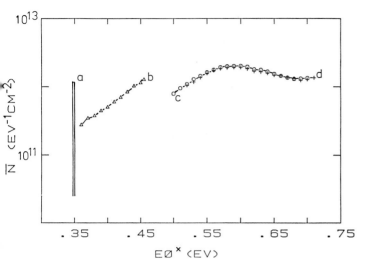

Figure 7-15. Density of states as deduced by deconvolution of current-voltage characteristic and admittance spectroscopy in a FZ silicon bicrystal. The GB is close to a (710) $\Sigma = 25$ orientation. **a** and **b** as observed in as-grown specimen; **c** after annealing 1 h – 730 °C with 1.4×10^{14} cm^{-3} neutron transmutation dopant and **d** same annealing with 1.4×10^{15} cm^{-3} neutron transmutation dopant (from Petermann, 1988).

very good conditions from zone-refined material and was transmutation doped. It has an electrical activity even in as-grown conditions with a discrete level at 0.35 eV above the valence band and a concentration of $\sim 3 \cdot 10^{11}$ cm^{-2}. After annealing, the average state density is 10^{12} eV^{-1} cm^{-2} over at least 0.4 eV extension into the gap. This represents $2.5 \cdot 10^{-4}$ sites on the GB plane and is compatible with the presence of one site every 1700 Å along the *dislocation* lines present in this structure (see above). The authors do not give any interpretations regarding the potential source of these defects. However, according to the structural models proposed recently for (710) $\Sigma = 25$ GB (Bourret and Rouviere, 1989), these levels must be ascribed to a defect in the GB and not to an intrinsic property of the (710) $\Sigma = 25$ GB. Several hypotheses could be envisaged such as a defect in the reconstruction similar to a soliton (Heggie and Jones, 1983) or point defects trapped at the GB or impurities with deep levels segregated at the GB.

On the other hand, the interpretation of the results obtained by Werner and Strunk (1982) and Werner (1985) in the form of band tails in the gap, are more difficult to understand. The potential fluctuations inherent in a specific GB are included in the computer calculations of the electronic structure and yet it does not seem to produce a band tail. These fluctuations must therefore be derived from a different origin. Here again one, is faced with numerous other possibilities amongst which it is difficult to decide on anyone.

Szkielko and Petermann (1985) found in n-doped Ge bicrystal two acceptor levels at 0.2 eV and 0.096 eV above the valence band. Their interpretation was however, later questioned by Labusch and Hess (1989). These authors using Ge n or p bicrystals reported resistivity measurements along or across the GB. They interpreted their result using the model proposed by Landwehr et al. (1985) which differs substantially from the one presented above. The charges trapped at the GB are neutralized by an equal number of holes in two dimensional bands. The density of bound holes is high enough for complete degeneracy and they play the major role in shaping the electrical properties and explain the high conductance parallel to the GB plane. The GB, instead of having acceptor levels, has a 2D-band which contains a large number of free holes at its upper edge so that the Fermi level is within this band. They measured a conductivity have an activation energy of 0.76 eV, which according to Eq. (7-7) facilitates the provision of $E_C - E_F + \Phi_B$. In Ge $E_c = 0.78$ eV and therefore the Fermi level is very close to the valence band in the GB plane. The situation is very different in silicon for which the activation energy of the conductivity depends very much on the doping, a situation that can be explained by a continuum density of localized states.

7.4.4.3 Transient Properties Measured on Bicrystals

The density of states was measured by DLTS on a $(710)\,\Sigma = 25$ silicon bicrystal after different annealing treatments (Fig. 7-16). In as-grown specimens the GB has no detectable DLTS signal. After 10 min at 900 °C the GB is active and has a density of states close to $2 \cdot 10^{12}$ eV^{-1} cm^{-2} in a broad energy range inside the gap (Broniatowski, 1987). With a longer annealing time, a well defined peak is observed at 0.53 eV although this is attributed to copper precipitation at the GB due to a pollution during the annealing process. The copper decorated $\Sigma = 25$ GB has been characterized by TEM (Broniatowski, 1989) and contains microprecipitates, generally in colonies with an average density of 10^{10} cm^{-2}. The interpretation of the spectra proposed by Broniatowski in this case includes a Schottky barrier at each semiconductor-precipitate interface. Thus each precipitate acts as a multiply charged trap and is fundamentally different in nature from the traps considered so far.

In germanium, $(111)\,\Sigma = 3$ and $(1\bar{1}2)\,\Sigma = 9$ GBs are found by DLTS to be electrically inactive (Broniatowski and Bourgoin, 1982) whereas in low angle bicrystal

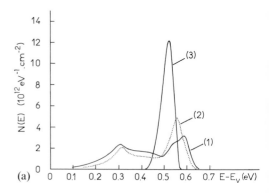

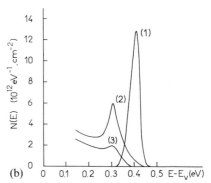

Figure 7-16. Density of states $N(E)$ in a $(710)\,\Sigma = 25$ silicon GB as a function of the energy E measured from the top valence band E_v. This density is deduced from DLTS spectra for: (a) Three different annealing times at 900 °C (1) = 10 min, (2) = 30 min and (3) = 2 hours, followed by rapid cooling. (b) Three different cooling rates (1) 50 °C s^{-1}, (2) 1.5 °C s^{-1}, and (3) 0.15 °C s^{-1} after annealing 24 h at 900 °C (from Aucouturier et al., 1989).

($\Theta = 3.5°$ in germanium) an electrical activity can be detected. The measured density of states is in the range 10^9 to 10^{10} cm^{-3} (see also Werner, 1985), a density that is too low to correspond to periodic sites in the dislocation core structure. This points to an impurity effect with segregated atoms at dislocation cores.

The very complex behavior of electrical properties and the GB annealing temperature indicates that impurities should play a major role. As Broniatowski (1989) showed in the case of Si/Cu, impurities could introduce a new kind of trap which currently is inadequately described by any previous models. One should therefore exercise a degree of care in interpreting both the electrical properties of GBs and the density of states since the character of the traps and their energy are far from being well established.

7.4.4.4 Emission and Capture Properties of Silicon and Germanium Grain Boundaries

Electron or light beam induced current (EBIC or LBIC) techniques have yielded important results on the correlation between structure and electrical activity (Sharko et al., 1982; Buis et al., 1980). The

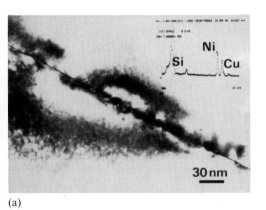

(a)

spatial resolution enables one to distinguish between different GBs or to study the evolution of the signal along the same GB. Unfortunately, however, the spatial resolution of the electrical measurements is several orders of magnitude larger than the transmission electron microscope image and the correlation with a particular defect in the GB is, in most cases, not unique.

On the other hand, Dianteill and Rocher (1982) have shown that dependent upon the $\Sigma = 9$ GB plane, the electrical activity could vary between there being none at the symmetrical (112) twin plane to a high value for a random orientation. On (510) $\Sigma = 3$ and (710) $\Sigma = 25$ in silicon bicrystals, Ihlal and Nouet (1989) measured a complete inactivity in as-grown specimens which becomes a strong uniform or dotted EBIC contrast after annealing between 750 and 950 °C. This has also been observed by Maurice and Colliex (1989) on a (710) $\Sigma = 25$ GB in silicon (Fig. 7-17), as well as on a $\Sigma = 3$ coherent and incoherent twin (Fig. 7-18). Colonies of copper and nickel precipitates are at the source of such electrical activity and the segregation of impurity is itself dependent on the defect content of the GBs. For example, recombination is larger in an asym-

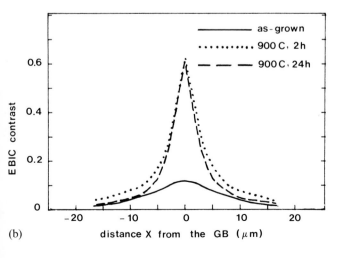

(b)

Figure 7-17. (a) EBIC contrast at a (710) $\Sigma = 25$ in CZ silicon, in as-grown and annealed specimens. The EDX spectrum in the inset indicates the presence of copper and nickel. The contrast increases with the annealing due to the precipitation of metallic impurities as shown in (b) (from Maurice and Colliex, 1989).

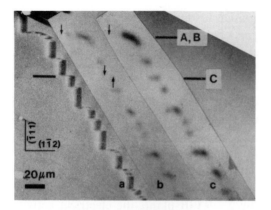

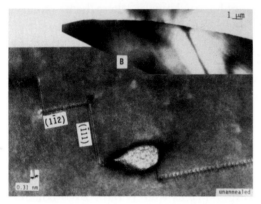

Figure 7-18. Top: scanning electron image of a $\Sigma = 3$ in silicon stepped GB. a: Secondary electrons, b: EBIC in the as-grown sample, c: EBIC same area after annealing at 900 °C. Note that the activity generally appears in the *curved* portion of the GB as well as on the right side of the (112) incoherent parts. Bottom: Transmission electron image of a part of portion B of the above specimen before annealing. Precipitates are visible on the right side of the (112) portion of the GB (from Maurice, 1990).

metric plane or when the misorientation from a twin orientation increases.

Similarly, Martinuzzi (1989) studied polycrystalline material using LBIC. He found that the (111) $\Sigma = 3$ twin is inactive except when it contains additional dislocations. Also, the $\Sigma = 13$ is very active after annealing but exhibits a strong inhomogeneity along the same GB. In more general GBs (polycrystalline materials) the electrical activity is already present in as-

grown specimens, although inhomogeneously, but it is greatly increased by further annealing at temperatures greater than 500 °C for a few hours.

Finally, several authors (Maurice and Colliex, 1989; Broniatowski, 1989) have noted the effect of the cooling rate. At high cooling rate the activity enhancement is greater by far than after a slow cooling. This is further evidence pointing to the importance of the electrical activities of some segregated impurities.

7.4.5 Intrinsic or Extrinsic Origin of Electrical Activity of Grain Boundaries

The origin of the electrical activity of GBs has for a long time been a matter of debate. It could be attributed either i) to the intrinsic nature of the GB structure (distorted bands and/or non-reconstructed bonds), ii) to additional defects such as dislocations or point defects or iii) to segregated impurities. If the first hypothesis were held to be true, the following consequences would necessarily ensue:

i) In perfect twin boundaries with well defined geometry, the structure of which has been well established by HREM, no deep levels should appear. Bond calculations have shown it should have no electrical activity per se.

ii) The thermal history, as well as subsequent thermal annealing, should not change the intrinsic effect of a GB. If there were any change, it would most likely be a decrease in electrical activity due to annealing of unstable situations.

iii) Along a given GB (bicrystal for instance) the electrical activity should be homogeneous.

Experimental results already reported and summarized in Table 7-2 show unambiguously that none of the above conditions are satisfied.

Table 7-2. Electrical activity of grain boundaries for which a direct correlation with the structure has been obtained.

Σ value	Reference	Method	Material type	Electrical activity	Effect of annealing at $T\,°C$ on electrical activity	Additional defects observations
Si (111) = 3	C.S.A. 1982	EBIC	polycrystal	no	none	active dislocations
Si (111) = 3	S.G.T. 1982	EBIC		no	none	
Si (111) = 3	M.C. 1989	EBIC	polycrystal	no	none	
Ge (111) = 3	M. 1989	LBIC	poly cast ingot	no	none	
Ge (111) = 3	B.B. 1982	DLTS	bicrystal	no	none	active secondary dislocations
Si (112) = 3	B.O.S. 1986	EBIC	polycrystal	no	none	
Si (112) = 3	M.C. 1989	EBIC	polycrystal	no		
Si (122) = 9	D.R. 1982	EBIC	bicrystal	no	?	
Ge (122) = 9	B.B. 1982	DLTS	bicrystal	no	?	
Ge random plane = 9	B.B. 1982	EBIC	bicrystal	yes	?	
Si (122) = 9	M. 1989	LBIC	bicrystal	no	poorly active at 750 °C very active at 900 °C	
Si (510) = 13	I.N. 1989	EBIC	bicrystal	low activity	active at 750 °C	precipitates
Si (510) = 13	M. 1989	LBIC	bicrystal	low activity	very active at 900 °C	strong inhomogeneity
Si (013) = 5	M.C. 1989	EBIC	bicrystal	yes, low	active at 900 °C	secondary dislocations
Ge (710) = 25	P.H. 1989	conductance admittance spectrum	bicrystal (zone refined)	yes, 0.35 eV	active at 730 °C (continuous spectrum in the gap)	
Si (710) = 25	I.N. 1989	EBIC	bicrystal	no	active at 450 °C	
Si (710) = 25	B. 1987/89	DLTS	bicrystal	yes	active after quench	copper precipitation
Small angle = 3°5 Ge	B.B. 1982	DLTS	polycrystal	yes	?	

B.:	Broniatowski	C.S.A.:	Cunningham et al.	M.C.:	Maurice and Colliex
B.B.:	Broniatowski and Bourgoin	D.R.:	Dianteil and Rocher	I.N.:	Ihlal and Nouet
B.O.S.:	Buis et al.	M.:	Martinuzzi	P.H.:	Petermann and Haasen

It is true that (111) $\Sigma = 3$ as well as (112) $\Sigma = 3$ or (122) $\Sigma = 9$ have no electrical activity as predicted in i). However, (510) $\Sigma = 13$ or (013) $\Sigma = 5$ as well as (710) $\Sigma = 25$ show activity in as-grown specimens. In addition most of these GBs have their electrical activity enhanced by thermal annealing and moreover, strong inhomogeneity is a prominent feature of very active GBs. Recent results reported on spatially resolved studies of GBs are all in favor of an extrin-

sic origin as the source of the major part of electrical activity.

The question then arises, what type of extrinsic defects? As a first hypothesis, additional dislocations or point defects could be considered since it would appear to be supported by several observations. Additional dislocations in (111) $\Sigma = 3$ serve to activate this twin for instance. However, most of the secondary dislocation cores could have a reconstructed structure (see, for example, the modeling by Bourret and Bacmann, 1985 and 1986) thus if dislocations are at the origin of the electrical activity it should be from defects in the reconstruction or point defects or jogs along these lines. The regularly measured 10^{11} cm^{-2} density of states is compatible with such an explanation.

This said, however, even point defects are inadequate for explaining the annealing behavior. In particular, their effect is rapidly dominated by impurity segregation after thermal annealing. This is especially true in bicrystals in which surfaces and GB planes are often the only available traps for rapidly diffusing impurities. It should be pointed out that a combination of impurities is also possible. Following the scheme responsible for the gettering of metallic impurities by oxygen precipitates (Cerofolini and Maeda, 1989), one could describe the process with two steps in the following way:

– The GB secondary dislocations or extrinsic dislocations attract the interstitial oxygen in their elastic field. The most active dislocations will be those with the largest edge component.

– The oxygen-rich region attracts metallic impurities giving metal-rich particles at the GB surface which are particularly electrically active.

Oxygen gettering is generally optimized between 650 and 900°C which is the temperature range in which GBs are very active. Therefore the most active GBs would be those containing the largest amount of dislocations or steps associated with a dislocation (intersections of $\Sigma = 3$ coherent and incoherent). Even the activity observed in as-grown FZ-germanium bicrystal by Petermann (1988) could be well explained by impurities. 10^{12} states would correspond to the segregation of 0.1 ppb of metallic impurity at the GB plane.

As a conclusion to several years of experimental and theoretical work on GBs it may seem disappointing that the GB is not electrically active per se. However, the clear link with impurity segregation which has recently been established opens a new field of experiment. The different electrical techniques which have been developed for GB studies are now very powerful tools, each able to help in understanding the segregation phenomena with a very high sensitivity. It is clear that this field has a way to go yet and will be largely developed in the future.

7.5 Mechanical Properties of Grain Boundaries in Semiconductors

7.5.1 Introduction

The deformation of polycrystalline materials has been widely and initially studied in metals, alloys and ceramics. The goal was to understand the behavior of materials under different temperature and stress conditions in order to achieve special properties. The role of grain boundaries (GBs) has been emphasized for a long time. The GBs were considered as a barrier to crystal slip and as a consequence they can harden the material; the grain size dependence of the yield stress led to numerous studies. At high temperature, GB behavior was explained by diffusion along the interface and the recrystallization or superplasticity of polycrystalline materials after high temperature deformation was viewed as the consequence of matter transport in both grains and GBs. The role of impurities segregation in GB embrittlement is also a well known phenomenon and led to equally numerous studies either in polycrystals or bicrystals. Theoretical models were elaborated to explain the variable influence of different impurities on GB embrittlement. Although the literature on the role of GBs in the mechanical properties of materials generally refers to metallic or ceramic materials, it is often a good starting point from which to approach the semiconductors case. In 1972, Hirth gave a review paper on the influence of GBs on the mechanical properties of metallic materials. In 1988, the concept of *GB design* was introduced by Watanabe (1988) who emphasized the role of GB character distribution of polycrystalline materials and its link to both the bulk properties and to intrinsic GB properties.

In the case of semiconductor materials, this question of GB mechanical behavior arises during the processing for instance of polysilicon ribbon which is used for low resistivity wires or as isolating material if suitably doped. Performance depends strongly on the stability of the material, and namely on the induced defects, their density, their mobility and the interaction between different more or less welcome defects. Dopant impurities are welcome *defects*, and their spatial distribution has to remain under control during the process i.e. not perturbated by mobile dislocations or migrating GBs. Furthermore, large scale integration leads to specific problems related to the plastic properties of the material. Generally, it induces local stresses which are released by the emission of dislocations whose distribution evolves under further treatments. Furthermore, during recrystallization taking place at high temperature, migrating GBs sweep and absorb dislocations whilst the size of the grains increases. These structural changes may induce unexpected properties and must be understood and if possible controlled.

Several techniques are now available – in situ experiments by synchrotron X-ray topography or by electron microscopy, and the *post-mortem* experiments using conventional and high resolution electron microscopies. As a consequence, the understanding of GB deformation has been clarified and the very early stages of the interaction mechanisms between deformation-induced dislocations and the GB are now better characterized.

7.5.2 Interaction Between Dislocations and Grain Boundaries

Under applied or internal stresses and at suitable temperature, dislocations are created in the bulk and then move within the crystalline material. Consequently they react with all the defects in their path, and in particular with grain boundaries.

The presence of a GB on the dislocation's path generally blocks the slip, and dislocation pile-ups are formed in the vicinity of the GB. When two slip systems are activated, dislocations on one system can be stopped at the GB and then may be intercepted by the second slip system leading to locks. As a consequence, dislocation cross-slipping is highly probable at the GB. As discussed by George et al. (1989), *the neighborhood of a GB is in an advanced state of deformation.* Furthermore, the vicinity of the GB was shown by HREM to be a preferential area for strong interaction between stopped dislocations and point defects created by the deformation (Thibault-Desseaux et al., 1989). Smith (1982) gave a good summary of the different steps of the mechanisms occurring at a GB.

Two major mechanisms for the stress release at the head of the pile-up could be proposed, namely i) the entrance and the dissociation of the dislocations within the GB or ii) their transmission across the GB. As the reaction occurs two conservation rules must be obeyed, firstly, that the Burgers vector is conserved and secondly that the step height associated with the GBD is also conserved.

7.5.2.1 Dislocation Absorption

The dislocations created by deformation are known to be stopped and absorbed by GBs. Thus in the case of a complete absorption the rule is given by

$$(\boldsymbol{b}_0, \boldsymbol{h}_0) = \sum_i (\boldsymbol{b}_i, h_i)$$

where i refers to the ith GBD which the incoming dislocation is decomposed into. The entrance of the incoming dislocation as a whole would have created a global step whose height $\boldsymbol{h}_0$ is the projection on the GB normal of half the Burgers vector $\boldsymbol{b}_0$ (by convention, the reference plane is the mean GB plane, see Sec. 7.2.3.2).

Moreover, one has to keep in mind that the height of the step associated with a given GBD Burgers vector can only take well defined values determined by the DSC lattice geometry. The experimental determination of this was explained in Sec. 7.2.3.2. The appearance of DSC dislocations associated with a step can affect the decomposition of the incoming dislocation. Forwood and Clarebrough (1981) predicted that non-primitive DSC dislocations with no step would be stable in symmetrical tilt GBs despite the unfavorable b^2 criterion.

The application of this rule is simple in relatively low Σ GBs. In fact, this rule is geometrically necessary but it is not so straightforward in polycrystalline materials where the GBs are experimentally difficult to characterize. Experimental evidence for the disappearance of the contrast of dislocations observed by TEM (Pumphrey et al., 1977) suggested to Gleiter (1977) that in the case of general GBs, the cores of external dislocations spread out into dislocations with infinitesimal Burgers vec-

tors. However, Pond and Smith (1977) and Dingley and Pond (1979) pointed out that dislocations can enter both coincidence and non-coincidence GBs where they dissociate into components belonging to the nearest DSC lattice.

The rule of absorption and decomposition in DSC dislocations was experimentally confirmed for the first time using conventional TEM by Bollmann et al. (1972) in a metallic GB and was confirmed using HREM by Elkajbaji and Thibault-Desseaux (1988), and by Skroztky et al. (1987), respectively in Si and Ge bicrystals. The observed GBs were relatively well defined and their Σ equal to or less than 51. Even for such high Σ values, it was possible to detect individual GBDs with localized cores unlike in the Gleiter model.

The absorption rule assumes that dislocations integrate the GB as a whole. However, the deformation induced dislocations $1/2\langle 011 \rangle$ are known to be dissociated in two Shockley partial dislocations $1/6\langle 112 \rangle$ separated by an intrinsic stacking fault. Thus the question arises how to overcome the repulsive interaction between the partials at the entrance event (King and Chen, 1984). This was clarified in the simple case of deformation of the $\Sigma = 9$ Si tilt bicrystal (Thibault et al., 1989). It turns out that the stress at the leading partial is released by decomposition and emission of glissile GBDs into the interface. As a consequence, this occurrence leads to a block of the slip at the GB. On the other hand, with complete integration and decomposition of both the leading and the trailing partials in GBDs, the incoming dislocation generally requires simultaneous glide and climb of the GBDs. Fig. 7-19 shows the integration of a dissociated 60° dislocation $a/2\langle 101 \rangle$ within the (122) $\langle 011 \rangle$ $\Sigma = 9$ GB in silicon. In this case the leading partial decomposed first by glide into the GB. One

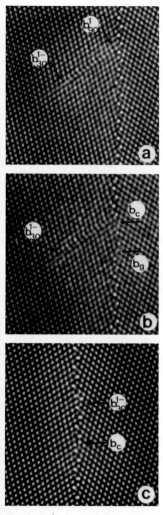

Figure 7-19. Entrance of a 60° dissociated bulk dislocation (a) in a (122) [011] $\Sigma = 9$ tilt GB in Si. Panel (b) shows the decomposition of the leading partial in two GB defects. Three DSC dislocations result from the final decomposition: $b_g = a/18$ [411] is glissile in the GB plane, $b_c = a/9$ [122] is sessile, and b_{30}^1 is both glissile and sessile and carries the screw component. (c) Shows b_c and b_{30}^1. The bond reconstructions along the [011] direction are the same as in the bulk (HREM images from Elkajbaji and Thibault-Desseaux, 1988).

glissile DSC dislocation b_g is ejected in the boundary. After the entrance of the trailing partial, the residual dislocation decomposed by climb (and glide) into two DSC dislocations called b_c and b_{30}^1 on the figure.

This has been observed by experiment to take place at temperatures above 0.5 T_m. Furthermore, at temperatures at which large climb is possible, the dislocations common to both grains can be stopped even at the interface by climb decomposition.

In the case of highly anisotropic GBs, image stresses have to be considered. The image force can make matrix dislocations that are attracted by or repelled from the interface. This helps to pin the dislocations at the GB or promote cross-slip in the vicinity of the boundary (Khalfallah et al., 1990). One has to point out that the core of the DSC dislocations stemming from the entrance of an incoming dislocation is not necessarily well reconstructed. It has been shown (Fig. 7-19) that an in-running 60° dislocation dissociated in 30° and 90° partials decomposes by entrance into a $\Sigma = 9$ $\langle 011 \rangle$ tilt GB in three DSC dislocations, b_{30}^1 bringing the screw component. The structure of the GBD core is not far from that of the bulk dislocation core and can be described with the SU model using 5, 6, and 7 atom rings as in the bulk. In particular, the SU linked to the b_c dislocation is the T SU. The problems linked with bond reconstruction along the GBD line are similar to those occurring in the bulk (Thibault-Desseaux and Putaux, 1989).

7.5.2.2 Dislocation Transmission Across Grain Boundaries

The rule to be applied in the case of the direct transmission of a dislocation from one grain to the adjoining one is expressed by:

$$(\boldsymbol{b}_I, h_I) = (\boldsymbol{b}_{II}, h_{II}) + (\boldsymbol{b}_{gb}, h_{gb})$$

Transmission of dislocations occurs easily if there is no residual GB dislocation left after the transmission i.e. when b_g is zero, otherwise the energetical balance may be unfavorable. In addition, the incoming dislocation line has generally to rotate in order to achieve a suitable slip plane in the second grain. This process involves climb into the boundary and is diffusion limited. Forwood and Clarebrough (1981) provided evidence of direct transmission events across a GB (in stainless steel) containing a line common to the slip planes in both grains. In this case, incursion across the boundary is conservative and is only limited by the residue left in the interface. In-situ 1 MeV and X-ray synchrotron topography showed numerous direct dislocation crossing events common to both grains in well defined germanium tilt bicrystals whose tilt axis was in zone with slip planes in both grains (Jacques et al., 1987; Baillin et al., 1987). This process is likely to occur owing to the absence of GB residue. By the same method, Martinez-Hernandez et al. (1986) showed that transmission of dislocations across the GB is more difficult for $\langle 001 \rangle \Sigma = 25$ than for $\langle 011 \rangle \Sigma = 9$ because in the $\Sigma = 25$ case no slip system is common to both grains in contrast to the $\Sigma = 9$ case. Nevertheless, the passage across the boundary of a dislocation belonging to an uncommon slip system was recently observed during 1 MeV in-situ experiments in Ge bicrystals (Baillin et al., 1990). It was however, difficult to determine whether this was a direct event or an indirect one occurring after the entrance and decomposition of several dislocations. Observations taken of dislocations emitted from an impact point corresponding to a pile-up are in favour of an indirect process because of the numerous dislocations accumulating in the same area.

7.5.2.3 Grain Boundaries as a Dislocation Source

The GBs are a major source of bulk dislocations as well as GBDs during the deformation of polycrystalline materials, although the nature of the dislocation sources is still far from being clear. Nevertheless, since it strongly influences the subsequent evolution of the material under further treatments, it is of fundamental importance to attempt a better understanding of the process.

The direct emission of a dislocation from a GB in a perfect coincidence position seems to be energetically prohibitive. However, in the case of a GB already containing defects this phenomena is more likely. Two kinds of dislocation can be emitted from the GB into the matrix. The first one is a perfect lattice dislocation mainly $1/2 \langle 110 \rangle$ as discussed by Baillin et al. (1990), the second being a partial lattice dislocation $1/6 \langle 112 \rangle$ that remains linked by a staking fault to the boundary (Fig. 7-20). Both conventional TEM and HREM (George et al., 1989) also revealed this second mechanism. It must be noted that conventional TEM shows the formation of extrinsic stacking faults which are in fact the superposition of distinct intrinsic stacking faults emerging from the GB on non-adjacent (111) planes (HREM results). Being close to the GB, this leads to complex configurations which will contribute to locally harden the material and which have a strong influence on the response of the material to chemical or thermal constraints via the presence of a high density of defects. Thus, the emission of dislocations from the GB may be determined by the GB defects already existing within the interface.

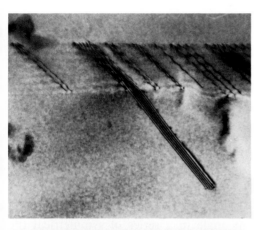

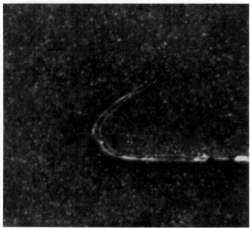

Figure 7-20. Dislocations emission from a GB. TEM shows the emission of a pair of identical Shockley partials from a $\Sigma = 9$ GB in Ge. A pileup of 4 dislocations with $b = a/2$ [110] in grain I on ($\bar{1}11$) has formed against the GB. The weak beam micrograph shows the two partials. The contrast between the fringes is due to an extrinsic stacking fault, each partial having trailed its own intrinsic stacking fault (from George et al., 1989; the marker is 1 μm).

7.5.2.4 Grain Boundary Dislocation Movement

GBDs can move along the boundary as lattice dislocations in the bulk, that is to say by glide and by climb. GBDs with Burgers vector parallel to the GB plane can glide in the interface whereas those with Burgers vector perpendicular to the GB plane move by climb. If the GBD is associated with a step, then the GB migrates laterally as the GBD moves along the interface. If the GBD Burgers vector has a component parallel to the interface, then the GBD motion results in GB sliding.

While moving along the GB plane, a GBD can interact with all the GB defects such as secondary dislocations, steps, or precipitates. All these obstacles are part of the resistance of a GB against the entrance and decomposition of lattice dislocations. This can lead to GBD pile-ups which in turn can leave residual stresses stemming from the lattice and the GB. As Hirth (1972) mentioned, if the resistance is low, complete recovery of the GB can be achieved. GBDs are free to rearrange in order to give a new GB structure without long range stresses. This has been confirmed by HREM observations (Putaux and Thibault, 1990) and details are given in Sec. 7.5.3.2.

7.5.3 Physical Consequences

7.5.3.1 Grain Boundary Migration

It is well known that GB migration controls the grain size and orientation as well as impurities redistribution. In this sense, it plays an important role in the overall properties of the materials and has been extensively studied in metals and ceramics. In the case of semiconductor devices, the occurrence of GBs and their migration might lead to unexpected effects which have to be known even if not well understood.

The phenomenological equation giving the velocity of the GB is at least in the Newtonian regime:

$$v = M f$$

where M is the mobility of the GB and f the driving force. The mobility is controlled by a wide range of intrinsic or extrinsic *obstacles* and it generally is reliant on the temperature. The force can be either an externally applied stress or an internal driving force stemming from the free energy difference of the two grains.

Diffusivity across the GB is one of the thermally activated processes that accounts for the GB mobility. The impurity content influences the GB velocity which can exhibit a non-linear behavior as a function of the driving force; at high velocity, however, the impurities no longer have effect (see the review paper by Bauer, 1982).

Diffusion-induced grain boundary migration (DIGM) is a phenomenon observed in some systems at temperatures where the bulk diffusion is extremely slow. The diffusion of solute atoms into the bulk or from the bulk via the GB result respectively in an alloyed or a de-alloyed region behind the GB path. Two different mechanisms were proposed. Hillert and Purdy (1978) invoked a chemical potential gradient, whereas Balluffi and Cahn (1981) invoked GBD climb. In the first case, a chemical potential gradient would lead to an asymmetrical strain which would be relieved by GB migration, leaving an alloyed region behind the GB. In the second case, the GB would migrate through the movement of GBDs associated with steps; the climbing driving force would come from the supersaturation of point defects arising from a GB Kirkendall effect due to these being no net fluxes of solutes and/or solvent along GB. The first model is based on the anisotropical response of two continuous adjoining grains, whereas in the second model an attempt is made to account for the structure of the GB.

From TEM observation of DIGM in Ni–Cu couple, Liu et al. (1988) argued in favor of the coherency strain model of Hillert and Purdy (1978). A dislocation wall at the original GB position is compensating the misfit between the matrix and the alloyed region, and the movement of the GB in both the forward and backward direction cannot be explained by the dislocation climb model. Vaudin et al. (1988) supported the same idea using the strong evidence of GB migration in MgO by diffusion of NiO along the interfaces.

DIGM was studied as a function of rotation angle in a series of symmetrical tilt GB's by Chen and King (1988) and in a series of asymmetrical tilt GB's by King and Dixit (1990). The response of the GBs to DIGM differs strongly whether they are symmetrical or not. The authors pointed out that apparently contradictory results could be explained if one considers that the coherency strain provides the driving force, whereas the GBD climb is one of the migration mechanisms (occurring exclusively in some cases), with other mechanisms such as atom rearrangement across the interface taking place as well.

During the processing of polycrystalline semiconductors, these considerations must be taken into account even if, compared to the situation in metals, the diffusion and the migration process are slow in covalent materials. Smith and Grovernor (1986) observed the enhancement of boundary mobility in Si and Ge films by doping even at high temperature.

In fact, the GB can migrate under an applied stress or a differential chemical potential. The mechanisms taking place however, could be of two types, namely i) the collective but local motion of the GB primary dislocations or the extension of pure steps perpendicular to the GB plane; this can be viewed as local atomic rearrangements across the boundary in the case of pure steps movement or in the case where the primary dislocations could glide laterally. Or ii) a displacement of individual GBDs (associated with a GB step) over large distances along the interface. Both mechanisms require glide and/or climb depending on the Burgers vector or more precisely on the resultant of the Burgers vector involved in the processes. On the one hand, local and collective glide of GBDs can lead to high energy GB configurations, whilst on the other hand the climb of a GBD over large distances requires high diffusion conditions.

The migration of a pure GB step by local rearrangement of the atoms at the GB was recently observed by HREM in-situ heating experiments (Ichinose and Ishida, 1990). The authors were able to observe both the migration of a $(111)\, \Sigma = 3$ facet in a $(112)\, \Sigma = 3$ GB in silicon and the subsequent structural changes. They showed that through a local rearrangement of the atoms, the facet migrates perpendicular to the (111) plane by steps the size of the CSL unit cell. This mechanism avoids a rigid translation of one grain relative to the other one that would result from the glide of $a/6\langle 112\rangle$ GBD on the (111) plane. However, the portions of the migrating GB are small (111) facets of the $\Sigma = 3$ GB connected by (112) facets, and this provides limiting conditions which might impose on the migration mechanism.

7.5.3.2 Recovery of the Grain Boundary Structure and Cavitation

If the number of extrinsic dislocations increases in a GB, the long-range stresses also increase. As Valiev et al. (1983) and Grabski (1985) recalled, the complete recovery of a GB occurs by motion, by annihilation of GBDs and their decomposition into smaller GBDs, and by rejection or absorption of lattice dislocations in order to obtain a GB in a new equilibrium configuration without long range stresses. The recovery may be accompanied by the formation of subgrain boundaries within the grains.

The recovery of Ge bicrystals by rotation of two adjacent grains was observed macroscopically by Bacmann et al. (1982a). In addition, HREM has allowed the study of the atomic structure evolution undertaken by a $(112)\, \Sigma = 9$ silicon tilt bicrystal during deformation (Putaux and Thibault, 1990). The original rotation angle was found to change from 38.94° to about 56° in compression experiments. The accumulation of deformation induced dislocations and their decomposition within the boundary has led to the observation of different Σ GBs sharing the same (011) mean plane such as $(599)\, \Sigma = 187$ ($\Theta = 42.9°$), $(233)\, \Sigma = 11$ ($\Theta = 50.5°$) and $(344)\, \Sigma = 41$ ($\Theta = 55.9°$) all of whose structures have been clearly identified. Furthermore the formation of subgrain boundaries in the grains has been clearly observed. The junction point between the subgrain boundary and the high-angle boundary delimits two parts of the high-angle GB with two different structures corresponding to two slightly different rotation angles (Thibault-Desseaux et al., 1989). The recovery of GB structure taking place after low strain is accomplished by the accumulation and homogeneous redistribution of one particular structural unit called T in Sec. 7.2.3 linked to the $\Sigma = 9$ primitive DSC dislocation $a/9\, [122]$. However, after high strain the recovery leads to the formation of more complex structures which involve new structural units.

As mentioned in Sec. 7.5.2.1, at high temperatures, slip transfer across the GB becomes unlikely and most lattice dislocations are trapped within the GB where they lower their energy by decomposition into smaller dislocations. These GBDs can then move easily by glide and climb along the boundary. As in the crystal, they can form pile-ups in the interface at triple junctions between GBs or at precipitates and this promotes conditions for cavitation. Lim (1987, 1988) produced a good review on the problems connected with GB cavitation occurring at high temperatures.

7.5.4 Deformation Modelling

As observed in former paragraphs, the deformation mechanisms lead to complex configurations either in the grains or in the GB. Potentially, modelization is a real tool that will help towards a better understanding and more accurate predictions of the evolution of a material under stress. A first attempt to achieve a modelization of dislocation microstructure has been made by Kubin and Canova (1989) who were able to compute the 3-D dislocation distribution and its evolution under a given applied stress. The application of such a modelization to the polycrystal situation would open the way to better predictions of their mechanical properties.

7.6 Conclusion

Grain boundaries in semiconductors are 2-D defects that strongly influence the overall properties of materials through their response to external and internal constraints. However, they are not the only defects present in the materials. Consequently, if external conditions are changing, this results in strong interaction between the different kinds of defects which may lead to large modifications of the material properties. The ultimate purpose of device design would be to keep these interactions under control but this implicitly

first requires a perfect knowledge of the structure and its mechanisms.

As this chapter has shown, the geometrical description is now well established and provides strong support for the corresponding energetical descriptions. However, although these are powerful theoretical tools they are only able to describe the GB atomic and electronic structure of a small cluster of atoms and have failed up to now to predict the structural multiplicity of simple GBs as well as the structure of more general GBs.

Atomic GB structure can be solved experimentally in the favorable cases of GBs with low index tilt axis. Most of the studied GB structures were compatible with completed reconstruction bond structures and furthermore, it was found that the structure of GB defects obeyed the same constraints. A determination of the atomic structure of more general GBs has not up to now been achieved, despite the existence of certain investigation means such as STM or FIM which could provide chemical characterization but which are still not yet used extensively.

We are still, moreover, far from knowing all there is to know about GB interactions with other defects such as point defects, impurities or dislocations. In addition to this, the correlation between the nanoscopic level mechanisms and the macroscopic level properties has never yet been solved clearly either.

Nevertheless, the main point emerges that the electrical activity of GBs does not stem intrinsically from the GB itself but comes essentially from impurities present at the GB which are attracted by the stress field of GBDs. In fact, apart from some low-energy GBs, which are in their lowest energy position only under special limiting conditions (for instance a bicrystal without any stress), most of the GBs encountered either contain linear defects or absorb them. The complex reactions between defects occurring at the GB (in the interface and in the vicinity) may involve more or less long-range stress fields which will have an influence on the properties of the material.

7.7 References

D'Anterroches, C., Bourret, A. (1984), *Phil. Mag. A 49*, 783.
Aucouturier, M., Broniatowski, A., Chari, A., Maurice, J. L. (1989), in: *Polycrystalline Semiconductors: Grain Boundaries and Interfaces:* Möller, H. J., Strunk, H. P., Werner, J. H. (Eds.). Berlin: Springer-Verlag, pp. 64–76.
Bacmann, J. J. (1987), *Jour. de Phys. C6-43*, 93–102.
Bacmann, J. J., Silvestre, G., Petit, M., Bollmann, W. (1981), *Phil. Mag. 43*, 189.
Bacmann, J. J., Gay, M. O., de Tournemine, R. (1982a), *Scripta Met. 16*, 353.
Bacmann, J. J., Papon, A. M., Petit, M. (1982b), *J. de Phys. C1-43*, 15.
Bacmann, J. J., Papon, A. M., Petit, M., Sylvestre, G. (1985), *Phil. Mag. A 51*, 697.
Baillin, X., Pelissier, J., Bacmann, J. J., Jacques, A., George, A. (1987), *Phil. Mag. A 55*, 143.
Baillin, X., Pelissier, J., Jacques, A., George, A. (1990), *Phil. Mag. A 61*, 329.
Balluffi, R. W., Bristowe, P. D. (1984), *Surf. Sci. 144*, 28.
Balluffi, R. W., Cahn, J. W. (1981), *Acta Met. 29*, 493.
Balluffi, R. W., Komen, Y., Schober, T. (1972), *Surf. Sci. 31*, 68.
Bauer, C. H. (1982), *Jour. de Phys. C6-43*, 187.
Bilby, B., Bullough, R., Smith, E. (1955), *Proc. Roy. Soc. London A 231*, 263.
Bishop, G. H., Chalmers, B. (1968), *Scripta Met. 2*, 133.
Blatter, G., Greuter, F. (1986), *Phys. Rev. B 33*, 3952; *Phys. Rev. B 34*, 8555.
Bleris, G., Antaunopoulos, J., Karakostas, Th., Delavignette, P. (1981), *Phys. Stat. Sol. A 67*, 249.
Bleris, G., Nouet, G., Hagège, S., Delavigette, P. (1982), *Acta Cryst. A 38*, 550.
Bollmann, W. (1970), *Crystal Defects and Crystalline Interfaces*, Berlin: Springer.
Bollmann, W. (Ed.) (1982), *Crystal Lattices, Interfaces, Matrices*. Genève.
Bollmann, W., Michaut, B., Sainfort, G. (1972), *Phys. Stat. Sol. a 13*, 637.
Bourret, A. (1990), *Jour. de Phys. C1-51*, 1.
Bourret, A., Bacmann, J. J. (1985), *Inst. Phys. Conf. Ser. 78*, 337.
Bourret, A., Bacmann, J. J. (1986), *Grain Boundary Structure and Related Phenomena – JIMIS-4.* Trans. of the Japan. Inst. of Metals 27, pp. 125–134.
Bourret, A., Bacmann, J. J. (1987), *Rev. Phys. Appl. 22*, 563.
Bourret, A., Desseaux, J. (1979a), *Phil. Mag. 39*, 405.
Bourret, A., Desseaux, J. (1979b), *Phil. Mag. 39*, 419.
Bourret, A., Rouvière, J. L. (1989), in: *Polycrystalline Semiconductors: Grain Boundaries and Interfaces:* Möller, H. J., Strunk, H. P., Werner, J. H. (Eds.). Berlin: Springer Verlag, pp. 8–18.
Bourret, A., Billard, J. L., Petit, M. (1985), *Inst. Conf. Series 76*, 23.
Bourret, A., Rouvière, J. L., Spendeler, J. (1988), *Phys. Stat. Sol. a 107*, 481.
Brandon, D. G. (1966), *Acta Metall. 14*, 1479.
Brandon, D. G., Ralph, B., Ranganathan, S., Wald, M. S. (1964), *Acta Met. 12*, 812.
Broniatowski, A. (1987), *Phys. Rev. B 36*, 5895.
Broniatowski, A. (1989), *Phys. Rev. Lett. 62*, 3074.
Broniatowski, A., Bourgoin, J. C. (1982), *Phys. Rev. Lett. 48*, 424.
Buis, A., Oei, Y. S., Schapink, F. W. (1980), in: *Grain Boundary Structure and Related Phenomena – JIMIS 4.* Trans. of the Japan Inst. of Metals. 27, pp. 221–228.
Carter, C. B., Föll, H. Ast, D. G., Sass, S. L. (1981), *Phil. Mag. A 43*, 441.
Carter, C. B., Cho, N. H., Elgat, Z., Flechter, R., Wagner, D. K. (1985), *Inst. Phys. Conf. 76*, 221.
Cerofolini, G. F., Maeda, L. (1989), in: *Physical Chemistry of, in and on Silicon*. Berlin: Springer Verlag, pp. 81–92.
Chen, F. S., King, A. H. (1988), *Acta Met. 36*, 2827.
Cho, N. H., McKernan, S., Wagner, D. K., Carter, C. B. (1988), *Jour. de Phys. C5-45*, 245.
Christian, J. W. (1965), in: *The Theory of Transformations in Metals and Alloys*. Pergamon Press, p. 322.
Cohen, M. H., Chou, M. Y., Economou, E. N., John, S., Soukoulis, C. M. (1988), *IBM J. Res. Develop. 22*, 82.
Cunningham, B., Strunk, H. P., Ast, D. G. (1982), *Scripta Met. 16*, 349.
Déchamps, M., Dhalenne, G., Barbier, F. (1988), *Phil. Mag. A 57*, 839.
Dhalenne, G., Déchamps, M., Revcolevski, A. (1983), *Adv. Ceram. 6*, 139.
Dianteill, C., Rocher, A. (1982), *J. de Phys. C1-43*, 75.
Dingley, D. J., Pond, R. C. (1979), *Acta Met. 27*, 667.
DiVicenzo, D. P., Alerhand, O. L., Schlüter, M., Wilkins, J. W. (1986), *Phys. Rev. Lett. 56*, 1925.
Dodson, B. W. (1986), *Phys. Rev. B 33*, 7361.
Donolato, C. (1983), *J. Appl. Phys. 54*, 1314.
Duffy, D. M., Tasker, P. W. (1984), *Phil. Mag. A 50*, 143–154.
Duffy, D. M., Tasker, P. W. (1985), *J. de Physique C4-46*, 185.
Einzinger, R. (1987), *Ann. Rev. Mat. Sci. 17*, 299.

Elkajbaji, M., Thibault-Desseaux, J. (1988), *Phil. Mag. A 58*, 325.

Fitzsimmons, M. R., Sass, S. L. (1989), *Acta Metal. 37*, 1009.

Föll, H., Ast, D. (1979), *Phil. Mag. A 40*, 589.

Fontaine, C., Smith, D. A. (1982), *Grain Boundaries in Semiconductors*. MRS Vol. 5, pp. 39.

Forwood, C., Clarebrough, L. (1981), *Phil. Mag. A 44*, 31.

Frank, F. C. (1950), *Report of the Symposium on the Plastic Deformation of Crystalline Solids*. Pittsburg: Carnegie Institute of Technology, pp. 150.

Friedel, C. (1926), *Leçons de Cristallographie*. Paris: Berger Levrault; and (1964) reprint of the 1926 edition, Paris: Blanchard.

Garg, A., Clark, W. A. (1988), *MRS 122*, 75.

George, A., Jacques, A., Baillin, X., Thibault-Desseaux, J., Putaux, J. L. (1989), *Inst. Phys. Conf. 104*, 349.

Gertsman, V. Yu., Nazarov, A., Romanov, A., Valiev, R., Vladimirov, V. (1989), *Phil. Mag. A 59*, 1113.

Gleiter, H. (1977), *Phil. Mag. 36*, 1109.

Goetzberger, A., Shockley, W. (1960), *J. Appl. Phys. 31*, 1821.

Grabski, M. W. (1985), *Jour. de Phys. C 4-46*, 567.

Gratias, D., Portier, R. (1982), *Jour de Phys. C 6-433*, 15.

Greuter, F., Blatter, G. (1989), in: *Polycrystalline Semiconductors: Grain Boundaries and Interfaces:* Möller, H. J., Strunk, H. P., Werner, J. H. (Eds.). Berlin: Springer Verlag, pp. 302–314.

Grimmer, H. (1989), *Acta Cryst. A 45*, 320.

Grimmer, H., Bonnet, R. (1990), Acta Cyst. *A 46*, 510.

Grimmer, H., Bollmann, W., Warrington, D. W. (1974), *Acta Cryst. A 30*, 197.

Grovernor, C. R. M. (1985), *J. Phys. C, Solid State Phys. 18*, 4079.

Hagège, S., Shindo, D., Hiraga, K., Kirabayashi, M. (1990), *J. de Phys. C 1-51*, 167.

Hasson, G., Boos, Y., Herbeuval, I., Biscondi, M., Goux, C. (1972), *Surf. Sci. 31*, 115.

Heggie, M., Jones, R. (1983), *Inst. Phys. Conf. 67*, 45.

Hillert, M., Purdy, G. R. (1978), *Acta Met. 26*, 333.

Hirth, J. P. (1972), *Metallur. Trans. 3*, 3047.

Hirth, J. P., Lothe, J. (1968), *Theory of Dislocations*, McGraw-Hill, p. 671.

Holt, D. B. (1964), *J. Phys. Chem. Solids 25*, 1385.

Hornstra, J. (1959), *Physica 25*, 409.

Hornstra, J. (1960), *Physica 26*, 198.

Ichinose, H., Ishida, Y. (1990), *Jour. de Phys. C 1-51*, 185.

Ichinose, H., Tajima, Y., Ishida, Y. (1986), Trans. Jap. Inst. Metals 27, 253.

Ihlal, A., Nouet, G. (1989), in: *Polycrystalline Semiconductors: Grain Boundaries and Interfaces:* Möller, H. J., Strunk, H. P., Werner, J. H. (Eds.). Berlin: Springer Verlag, pp. 77–82.

Jacques, A., George, A., Baillin, X., Bacmann, J. J. (1987), *Phil. Mag. A 55*, 165.

Kalonji, G., Cahn, J. (1982), *Jour. de Phys. C 6-43*, 25.

Kanai, H., Imai, M., Takahashi, T. (1985), *J. Mat. Sci. 20*, 3957.

Karlson, L., Norden, H. (1988), *Acta Metall. 36*, 13.

Kazmerski, L. L. (1989), in: *Polycrystalline Semiconductors: Grain Boundaries and Interfaces:* Möller, H. J., Strunk, H. P., Werner, J. H. (Eds.). Berlin: Springer Verlag, pp. 96–107.

Keating, P. N. (1966), *Phys. Rev. 145*, 637.

Khalfallah, O., Condat, M., Priester, L. Kirchner, H. O. K. (1990), *Phil. Mag. A 61*, 291.

King, A. H., Dixit, G. (1990), in: *Intergranular and Interphase Boundaries in Materials, Jour. de Phys. C 1-51*, pp. 545–550.

King, A. H., Chen, F. R. (1984), *Mat. Sci. Engi. 66*, 227.

King, A. H., Smith, D. A. (1980), *Acta Cryst. A 36*, 335.

Kohyama, M. (1987), *Phys. Stat. Sol. b 141*, 71.

Kohyama, M., Yamamoto, R., Doyama, M. (1986 a), *Phys. Stat. Sol. b 136*, 31.

Kohyama, M., Yamamoto, R., Doyama, M. (1986 b), *Phys. Stat. Sol. b 137*, 11.

Kohyama, M., Yamamoto, R., Doyama, M. (1986 c), *Phys. Stat. Sol. b 138*, 387.

Kohyama, M., Yamamoto, R., Ebata, Y., Kinoshita, M. (1988), *J. Phys. C: Solid State Physics 21*, 3205; *J. Phys. C: Solid State Physics, L*, 695.

Krivanek, O., Isoda, S., Kobayashi, K. (1977), *Phil. Mag. 36*, 931.

Kröner, E. (1958), in: *Ergebnisse der Angewandten Mathematik, Vol. 5:* Collatz, L., Lösch, F. (Eds.). Berlin: Springer Verlag.

Kröner, E. (1981), in: *Physics of Defects:* Balian, R., Kléman, M., Poirier, J. P. (Eds.). North Holland, pp. 219–315.

Kubin, L., Canova, G. (1989), in: *Electron Microscopy in Plasticity and Fracture Research of Materials:* Messerschmidt, U., Appel, F., Heydenreich, J., Schmidt, V. (Eds.). Berlin: Akademie Verlag, pp. 23–32.

Labusch, R., Hess, J. (1989), in: *Point and Extended Defects in Semiconductors:* Benedek, G., Cavallini, A., Schröter, W. (Eds.). NATO ASI Series, B-202, p. 15.

Lancin, M., Thibault-Desseaux, J. (1988), *J. de Phys. C 5-49*, 305.

Landwehr, G., Bagert, E., Uchida, S. (1985), *Solid State Electronics 28*, 171.

Lannoo, M., Bourgoin, J. (1981), *Point Defects in Semiconductors. Theoretical Aspects*. Berlin: Springer.

Laurent-Pinson, L., Nouet, G., Vicens, J. (1990), *Jour. de Phys. C 1-51*, 221.

Lim, L. C. (1987), *Acta Met. 35*, 163–169.

Lim, L. C. (1988), *Interfacial Structure, Properties and Design, MRS 122*, 317.

Liu, D., Miller, W. A., Aust, K. T. (1988), *J. de Phys. C 5-49*, 635.

Martin, J. W. (1975), *J. Phys. C: Solid State Phys. 8*, 2837; 2858; 2869.

Martinez-Hernandez, M., Jacques, A., George, A. (1986), *Grain Boundary Structure and Related Phenomena, Trans. Jap. Inst. Metals 27*, 813.

Martinuzzi, S. (1989), in: *Polycrystalline Semiconductors: Grain Boundaries and Interfaces:* Möller, H. J., Strunk, H. P., Werner, J. H. (Eds.). Berlin: Springer Verlag, pp. 148–157.

Masuda-Jindo, K. (1989), in: *Polycrystalline Semiconductors: Grain Boundaries and Interfaces:* Möller, H. J., Strunk, H. P., Werner, J. H. (Eds.). Berlin: Springer Verlag, pp. 52–57.

Mataré, H. F. (1984), *J. Appl. Phys. 56*, 2605.

Mauger, A., Bourgoin, J. C., Allan, G., Lannoo, M., Bourret, A., Billard, L. (1987), *Phys. Rev. B 35*, 1267.

Maurice, J. L. (1990), *J. de Physique C1-51*, 581.

Maurice, J. L., Colliex, C. (1989), *Appl. Phys. Lett. 55*, 241.

Merkle, K. L. (1990), *Jour. de Phys. C1-51*, 251.

Merkle, K. L., Smith, D. J. (1987), *Phys. Rev. Lett. 59*, 2887.

Merkle, K. L., Smith, D. J. (1988), *MRS 122*, 15.

Merkle, K. L., Reddy, J. F., Wiley, C. L. (1988), *J. de Phys. C5-49*, 251.

Molinari, E., Bachelet, G., Altarelli, M. (1985), *J. de Phys. C4-46*, 321.

Möller, H. J. (1981), *Phil. Mag. A43*, 1045.

Möller, H. J. (1982), *J. de Phys. C1-43*, 33.

Mykura, H. (1980), in: *Grain Boundary Structure and Kinetics:* ASM, pp. 445–456.

Olsson, E. (1988), *PhD Thesis Göteborg.*

Ourmazd, A., Tsang, W. T., Rentschler, J. A., Taylor, D. W. (1987), *Appl. Phys. Lett. 50*, 1417.

Papon, A. M., Petit, M. (1985), *Scripta Met. 19*, 391.

Papon, A. M., Petit, M., Silvestre, G., Bacmann, J. J. (1982), in: *Grain Boundaries in Semiconductors:* Leamy, H. J., Pike, G. E., Seager, C. H. (Eds.). *MRS 5*, 27–32.

Papon, A. M., Petit, M., Bacmann, J. J. (1984), *Phil. Mag. A49*, 573.

Paxton, A. T., Sutton, A. P. (1989), *Acta Metall. 37*, 1693.

Petermann, G. (1988), *Phys. Stat. Sol. a106*, 535.

Petermann, G., Haasen, P. (1989), in: *Polycrystalline Semiconductors: Grain Boundaries and Interfaces:* Möller, H. J., Strunk, H. P., Werner, J. H. (Eds.). Berlin: Springer Verlag, pp. 332–337.

Phillpot, S. R., Wolf, D. (1989), *Phil. Mag. A60*, 545.

Pike, G. E., Seager, C. H. (1979), *J. Appl. Phys. 50*, 3414.

Pond, R. C. (1981), in: *Dislocation Modeling of Physical Systems:* Ashby, M. F., Bullough, R., Hartey, C. S., Hirth, J. P. (Eds.). Oxford: Pergamon Press, pp. 524–543.

Pond, R. C. (1985), in: *Polycrystalline Semiconductors:* Harbeke, G. (Ed.). *Solid State Sciences 57*, pp. 27–45.

Pond, R. C. (1989), in: *Dislocations in Solids, Vol. 8:* Nabarro, F. (Ed.). North Holland, pp. 1–66.

Pond, R. C., Bollmann, W. (1979), *Phil. Trans. Roy. Soc. London 292*, 449.

Pond, R. C., Smith, D. A. (1976), in: *Proceedings of the 6ᵗʰ European Congress on Electron Microscopy, Israël.* TAL International Publishing Company, pp. 233–238.

Pond, R. C., Smith, D. A. (1977), *Phil. Mag. 36*, 353.

Priester, L. (1989), *Rev. Phys. Appl. 24*, 419.

Pumphrey, P. H., Gleiter, H., Goodhew, P. J. (1977), *Phil. Mag. 36*, 1099.

Putaux, J. L., Thibault, J. (1990), *J. de Phys. C1-51*, 323.

Queisser, H. (1963), *J. Electrochem. Soc. 110*, 52.

Rey, C., Saada, G. (1976), *Phil. Mag. 33*, 825.

Read, W. T. Jr., Shockley, W. (1952), in: *Imperfections in Nearly Perfect Crystals,* New York: John Wiley & Sons, Inc, pp. 352–376.

Rouvière, J. L., Bourret, A. (1989), in: *Polycrystalline Semiconductors: Grain Boundaries and Interfaces:* Möller, H. J., Strunk, H. P., Werner, J. H. (Eds.). Berlin: Springer Verlag, pp. 19–24.

Rouvière, J. L., Bourret, A. (1990), *J. de Phys. C1-51*, 323.

Saada, G. (1979), *Acta Met. 27*, 921.

Schapink, F. W. (1986), *Rev. Phys. Appl. 21*, 747.

Seidman, D. (1989), *MRS 139*, 315.

Seager, C. H., Castner, T. G. (1978), *J. Appl. Phys. 49*, 3879.

Seager, C. H., Pike, G. E. (1982), *Appl. Phys. Lett. 40*, 471.

Sharko, R., Gervais, A., Texier-Hervo, C. (1982), *J. de Phys. C-43*, 129.

Siegel, W., Kuhnel, G., Ziegler, E. (1981), *Phys. Stat. Sol. a64*, 249.

Skrotzki, W., Wendt, H., Carter, C. B., Kohlstedt, D. L. (1987), *Phil. Mag. A57*, 383.

Smith, D. A. (1982), *J. de Phys. C6-12*, 225.

Smith, D. A., Grovernor, C. R. M. (1986), *Trans. Jap. Inst. Metals 27*, 969.

Soukoulis, C. M., Cohen, M. H., Economon, E. N. (1984), *Phys. Rev. Lett. 53*, 616.

Spence, J. (1988), *Experimental High Resolution Microscopy, 2nd. Ed.,* New York: Oxford University Press.

Stillinger, F., Weber, T. (1985), *Phys. Rev. B31*, 5262.

Stucki, F., Brüesch, P., Greuter, F. (1987), *Surface Science 189/190*, 294.

Sumino, K. (1989), in: *Point and Extended Defects in Semiconductors:* Benedek, G., Cavallini, A., Schröter, W. (Eds.). NASO ASI Series B-202, pp. 77–94.

Sutton, A. (1989), *Phil. Mag. Let. 59*, 53–59.

Sutton, A., Balluffi, R. W. (1990), *Phil. Mag. Let. 61*, 91.

Sutton, A., Vitek, V. (1990), *Scripta Met. 14*, 129–132.

Sutton, A., Vitek, V. (1983), *Trans. Roy. Soc. London A309*, (a) 1–36, (b) 37–51, (c) 55–68.

Sutton, A., Balluffi, R. W., Vitek, V. (1981), *Scripta Met. 15*, 989.

Szkielko, W., Petermann, G. (1985), in: *Poly-microcrystalline and Amorphous Semiconductors:* Les Editions de Physique, pp. 379–385.

Tafto, J. (1979), *Proc. 39th Annual Meeting EMSA, San Antonio*, 154.

Takai, T., Choi, D., Thathachari, Y., Halicioglu, T., Tiller, W. A. (1990), *Phys. Stat. Sol. b 157*, K 13.

Tarnow, E., Bristowe, P. D., Joannopoulos, J. D., Payne, M. C. (1989), *J. Phys.: Condens. Matter 1*, 327.

Tasker, P., Duffy, D. M. (1983), *Phil. Mag. A 47*, L45.

Taylor, W. E., Odell, N. Y., Fan, H. Y. (1952), *Phys. Rev. 88*, 867.

Teichler, H. (1989), in: *Polycrystalline Semiconductors: Grain Boundaries and Interfaces:* Möller, H. J., Strunk, H. P., Werner, J. H. (Eds.). Berlin: Springer Verlag, pp. 25–33.

Tersoff, J. (1988), *Phys. Rev. B 37*, 6991.

Tewary, V., Füller Jr., E., Thomson, R. M. (1989), *J. Mat. Res. 4*, 309.

Thibault, J., Putaux, J. L., Bourret, A., Kirchner, H. O. K. (1989), *J. de Phys. 5*, 2525.

Thibault-Desseaux, J., Putaux, J. L. (1989), *Inst. Phys. Conf. 104*, 1.

Thibault-Desseaux, J., Putaux, J. L., Kirchner, H. O. K. (1989a), in: *Point and Extended Defects in Semiconductors:* Benedek, G., Cavallini, A., Schröter, W. (Eds.). Plenum Publ. Corp., pp. 153–164.

Thomson, R. E., Chadi, D. J. (1984), *Phys. Rev. B 29*, 889.

Valiev, R., Gertsman, V., Kaibyshev, O., Khannov, S. (1983), *Phys. Stat. Sol. a 77*, 97.

Vaudin, M., Cunningham, B., Ast, D. (1983), *Scripta Met. 17*, 191.

Vaudin, M., Handwerker, C., Blendell, J. (1988), *J. de Phys. C 5-49*, 687.

Vlachavas, D., Pond, R. C. (1981), *Inst. Phys. Conf. 60*, 159.

Vogel, F. L., Pfann, W., Corey, H., Thomas, E. (1953), *Phys. Rev. 90*, 489.

Watanabe, T. (1988), *J. de Phys. C 5-49*, 507.

Weber, W. (1977), *Phys. Rev. B 15*, 4789.

Werner, J. (1985), in: *Polycrystalline Semiconductors:* Harbeke, G. (Ed.). Berlin: Springer Verlag, pp. 76–87.

Werner, J. (1989), *Inst. Phys. Conf. 104*, 63.

Werner, J., Strunk, H. (1982), *J. de Phys. C 1-43*, 89.

Wolf, D. (1984a), *Acta Metall. 32*, 735.

Wolf, D. (1984b), *Phil. Mag. A 49*, 823.

Zook, J. D. (1980), *Appl. Phys. Lett. 37*, 223.

General Reading

Benedek, G., Cavallini, A., Schröter, W. (Eds.) (1989), *Point and Extended Defects in Semiconductors.* NATO ASI Series B-202.

Grovernor, C. R. M. (1985), *J. Phys. C, Solid State Phys. 18*, 4079.

Harbeke, C. (Ed.) (1985), *Polycrystalline Semiconductors.* Berlin: Springer Verlag.

Leamy, H. J., Pike, G. E., Seager, C. H., (Eds.) (1982), *Grain Boundaries in Semiconductors*, MRS 5.

Möller, H. J., Strunk, H. P., Werner, J. H. (Eds.) (1989), *Polycrystalline Semiconductors: Grain Boundaries and Interfaces.* Berlin: Springer Verlag.

Raj, R., Sass, S. (Eds.) (1988), "Interface Science and Engineering", *J. de Phys. C 5-49.*

Yoo, M. H., Clark, W. A. T., Brian, C. L. (Eds.) (1988), *Interfacial Structure, Properties and Design*, MRS 122.

Proceedings of the 1979 ASM Materials Science Seminar: Grain Boundary Structure and Kinetics (1980). Metals Park, OH: ASM.

"Proceedings of the JIM Int. Symposium on Structure and Properties of Internal Interfaces" (1985), *J. de Phys. C 4-46.*

"Proceedings of the Int. Conf. on Grain Boundary Structure and Related Phenomena" (1986), *Trans. Jap. Inst Metals 27.*

"Proceedings of the Int. Congress on Intergranular and Interphase Boundaries in Materials" (1990), *J. de Phys. C 1-51.*

8 Interfaces

Abbas Ourmazd

AT & T Bell Laboratories, Holmdel, NJ, U.S.A.

Robert Hull and Raymond T. Tung

AT & T Bell Laboratories, Murray Hill, NJ, U.S.A.

List of Symbols and Abbreviations

a	lattice parameter
a_e, a_s	bulk (relaxed) lattice parameters of epilayer and substrate
$\boldsymbol{b}$, b	Burgers vector and modulus of Burgers vector
D	diffusivity constant
E_{loop}	dislocation self energy
E_{strain}	strain energy
E_{total}	total system energy
E_v	glide activation energy
F_{12}	interdislocation force
f	frequency
G	shear modulus
h_c	critical epilayer thickness
k	Boltzmann constant
$k(\theta)$	angular dependent constant
$\bar{l}$	average (straight) dislocation length
N	defect density
N_{he}	heterogeneous source density
R	loop radius
R_c	critical loop radius
$\boldsymbol{R}^t$	template vector
S	geometrical factor
T	temperature
v	Poisson ratio
v	dislocation propagation velocity
ε	strain
σ	standard deviation
σ_{excess}	excess stress
σ_T	restoring stress
σ_ε	biaxial stress
χ_{ev}, χ_{sv}, χ_{es}	epilayer-surface, substrate-surface, epilayer-substrate interface energies
dc	diamond cubic
DT	Dodson-Tsao
EBIC	electron beam induced current
HBT	heterojunction bipolar transistor
HREM	high resolution electron microscopy
HRTEM	high resolution transmission electron microscopy
MB	Matthews, Blakeslee
MBE	molecular beam epitaxy
ML	monolayers
MEIS	medium energy ion scattering
MODFETS	modulation doped field effect transistor
PL	photoluminescence

PLE	photoluminescence excitation
RBS	Rutherford backscattering spectrometry
RE	rare earth
RHEED	reflection high energy electron diffraction
SB	Schottky barrier
SBH	Schottky barrier height
SEXAFS	surface extended X-ray absorption fine structure
SLS	strained layer superlattice
SPE	solid-phase epitaxy
TEM	transmission electron microscopy
UHV	ultra high vacuum
XSW	X-ray standing wave
XRD	X-ray diffraction
zb	zincblende

8.1 Introduction

Any finite system is delimited by interfaces. In this trivial sense interfaces are ubiquitous. However, modern epitaxial techniques seek to modify the properties of materials by stacking dissimilar layers. Band gap engineering, the attempt to tailor the electronic properties of semiconductors by interleaving (many) dissimilar layers is an example of this approach. Many modern materials and devices thus derive their characteristics from the presence of interfaces, sometimes separated by only a few atomic distances.

Interfaces between two solids can be classified into three general categories. 1) Interfaces between lattice matched, isostructural, crystalline systems, differing only in composition (chemical interfaces). The $GaAs/Al_xGa_{1-x}As$ system, with its very small lattice mismatch is the most technologically developed example. 2) Interfaces between isostructural, crystalline systems that differ in composition and lattice parameter. Ge_xSi_{1-x} and $GaAs/In_xGa_{1-x}As$ are representative examples. 3) Interfaces between systems differing in composition and structure. Metal-silicide/semiconductor systems are important representatives of this most general class.

The exploration of the wide variety of possible interfaces and their properties is relatively new. Most extensively studied are the structural properties of interfaces, and much of our discussion will focus on this aspect. Although it has long been appreciated that the structure of an interface can strongly affect the optical and electronic properties of a layer, particularly in thin layers (quantum wells), our understanding of the way this comes about remains qualitative, and perhaps rudimentary. The electronic properties of interfaces have also been the subject of extensive research. An adequate treatment of these requires a separate review. Here, the electronic properties will be considered only in so far as the relationship to the structure is concerned. The reader is invited to consult other sources for a more extensive treatment (Capasso and Margaritondo, 1987). Since the literature concerned with interfaces is extensive and rapidly growing, this chapter does not aim to be an exhaustive review, even of the structural aspects of interfaces. Rather, the overall purpose is to familiarize the reader with some of the key concepts in this dynamic field.

8.2 Experimental Techniques

To probe an interface, information must be extracted from a few monolayers of a sample buried beneath substantial thicknesses of material. This represents a severe experimental challenge.

X-ray diffraction and scattering, and transmission electron microscopy are direct structural probes of buried interfaces. X-rays interact weakly with matter, and are thus capable of deep penetration. For the same reason multiple scattering is essentially absent, resulting in ease of interpretation. However, the interaction of X-rays with a single interface that often consists of only one or two atomic planes is also very weak. X-ray diffraction methods, pioneered by Cook and Hilliard (1969), thus rely on the presence of a periodic multilayer stack to produce sufficiently strong diffraction peaks (satellites), whose intensities can be related to the layer period and the structure of the interfaces present. In this way X-ray diffraction yields highly accurate data about the interfacial configuration, averaged over many interfaces. When the interface itself has a different in-plane periodicity, for example when a periodic array of interfacial dislocations is present, X-ray scattering techniques can be used in conjunction with very bright synchrotron sources to investigate single interfaces, with the information emanating from a large area of the interface. X-rays can also be used to make accurate lattice parameter measurements, and thus explore the accommodation and relief of strain in mismatched systems.

Energetic electrons, on the other hand, interact strongly with matter, but can nevertheless propagate substantial distances, and thus emerge from samples of reasonable thickness (≈ 0.5 µm). This combination makes energetic electrons highly efficient probes of buried interfaces (Suzuki, Okamoto, 1985; Kakibayashi, Nagata, 1986; Ourmazd et al., 1987; Ichinos et al., 1987; Tanaka, Mihama 1988; Ou et al., 1989). The price for this, however, is increased complexity in interpretation of the data, because multiple scattering effects cannot be ignored. The transmission electron microscope (TEM), particularly in the high resolution (lattice imaging) mode, has become a standard tool in the investigation of interfaces (Ourmazd et al., 1987; Kakibayashi, Nagata, 1986). In the case of lattice mismatched systems, the TEM reveals the presence of extended defects, and in the lattice imaging mode can yield information about the atomistic details of strain relaxation. For chemical interfaces, however, the sample structure is of little interest, and chemical information is needed to determine the interface configuration. Recently developed chemical lattice imaging techniques yield quantitative chemical maps of such interfaces at near-atomic resolution and sensitivity (Ourmazd et al., 1989 a, b, 1990).

There is a large variety of techniques that probe the optical or electronic properties of interfaces, and thus indirectly their structure (Weisbach et al., 1981; Tanaka et al., 1986; Tu et al., 1987; Bimberg et al., 1987; Sakaki et al., 1987; Okumura et al., 1987). Most widely used are luminescence techniques (Weisbach et al., 1981; Tanaka et al., 1986; Tu et al., 1987; Bimberg et al., 1987). Due to their inherent simplicity and convenience, they have been extensively applied, and in many instances the results used to optimize growth procedures. More recently, Raman scattering and photoemission spectroscopy have also been used. These techniques are most valuable when the optical or electronic properties of a layer are to be determined. However, the interpretation of such data in terms of the structure is difficult, because no firm understanding of the way the structure influences other properties is yet at hand.

The variety of techniques that have been applied to the study of interfaces precludes a treatment of each individual approach. We will thus describe each method to the extent needed for an adequate discussion of the topic under consideration.

8.3 Interfaces Between Lattice-Matched, Isostructural Systems

8.3.1 Definition

The (chemical) interface between two lattice-matched, isostructural materials can be uniquely defined on all length scales, provided each atom type occupies an ordered set of lattice sites. As an example, consider the GaAs/AlAs system. The interface is simply the plane across which the occupants of the Group III sublattice change from Ga to Al. The interfacial plane thus defined can in principle have a complex waveform, with undulations ranging from atomic to macroscopic length scales (Fig. 8-1a). It is thus convenient to describe an interface in terms of

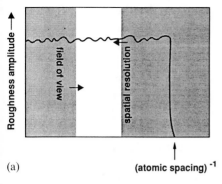

(a)

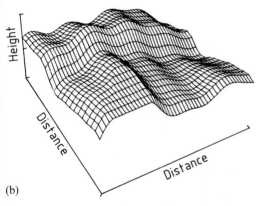

(b)

Figure 8-1. (a) Schematic representation of an interface. (b) Roughness spectrum of a "white noise" interface. This spectrum specifies the amplitude of the roughness vs. the wavelength. The shortest possible wavelength is the atomic spacing. Any experimental technique samples only a limited part of this roughness spectrum. This "window" is bounded by the field of view and the spatial resolution of the technique employed.

its Fourier spectrum, by specifying the amplitude of the undulations as a function of their spatial frequency (Warwick et al., 1990). Figure 8-1 b shows the Fourier spectrum of a "white noise" interface, with a constant roughness amplitude over all possible length scales.

When a given experimental technique is used to investigate an interface, it provides information about the interfacial configu-

ration within a certain frequency window, delimited on the high frequency side by the spatial resolution of the technique, and on the low frequency side by its field of view or spatial coherence. For any experimental technique, this frequency window spans only a small portion of the spatial frequencies needed for a realistic description of the interface. It is thus necessary to collate the information obtained from a large variety of techniques to obtain a complete picture of the interfacial configuration. This is a major challenge, because information from the atomic to the centimeter range, i.e. over eight orders of magnitude is required to provide such a description. However, when only specific properties, such as the optical or electronic properties of an interface are of concern, knowledge of a limited range of frequencies is adequate. In the case of luminescence due to excitonic recombination, for example, roughness over the excition diameter is of primary importance, while for charge transport applications, roughness at the Fermi wavelength is of concern.

The simple definition of an interface in terms of the location of the chemical constituents becomes inadequate when one or both of the parent materials are not chemically ordered, i.e., when some of the atom types are distributed randomly on a set of sites (Warwick et al., 1990; Thomson, Madhukar, 1987; Ogale et al., 1987). In the $GaAs/Al_xGa_{1-x}As$ system, for example, the second material is a random alloy. Thus, the Ga and Al atoms are distributed randomly on the the Group III sublattice, subject to the constraint that the composition, averaged over a sufficiently large region of the $Al_xGa_{1-x}As$ should correspond to the value given by x. In such random alloys, the composition measured in two different regions of the same size will in general not be the same, with the difference being due to statistical departures from the average composition due to the finite sampling volume. Such variations in the local composition grow dramatically as the sampling volume approaches atomic dimensions. In the limit of the sampling volume containing only one Group III sublattice site, the measured composition will of course be 0 or 1, irrespective of the global average composition x.

For this reason, the interface between two materials, at least one of which is a random alloy cannot be satisfactorily defined on all length scales. This is demonstrated in Fig. 8-2, where random alloy $Al_{0.3}Ga_{0.7}As$ has been "deposited" on an atomically flat GaAs surface, and the resulting structure viewed in cross-section. Each panel uses shades of gray to show the composition averaged over a given number of atoms perpendicular to the plane of

the paper. Consider the panel, where only one atomic plane is used, i.e., no averaging has been carried out. A line drawn to contain all the Ga atoms, i.e. to define an interface on an atomic scales would deviate dramatically from the original GaAs surface upon which the random alloy $Al_{0.3}Ga_{0.7}As$ was deposited. This illustrates that in the $GaAs/Al_xGa_{1-x}As$ system no interface can be defined on an atom by atom basis. (Attempts to image interfaces by tunneling microscopy must be viewed in this light). Only as the "thickness" over which the composition is averaged increases, does an isocomposition line approach the initial GaAs surface. For the $GaAs/Al_{0.3}Ga_{0.7}As$ system, the isocomposition line becomes essentially indistinguishable from the original GaAs surface when the composition is averaged over ≈ 30 atoms per column. These considerations apply generally, regardless of

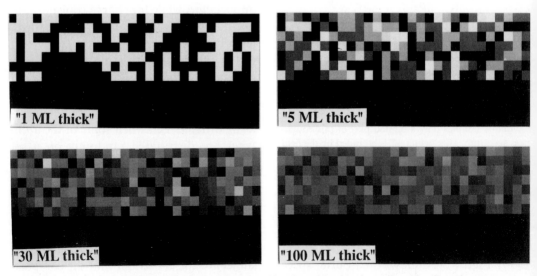

Figure 8-2. Schematic representation of interface formed by depositing random alloy $Al_{0.4}Ga_{0.6}As$ on an atomically smooth GaAs surface (cross-sectional view). Only Group III atoms are shown. White represents pure Al, black pure Ga, other shades of grey intermediate compositions. In each case the composition of the individual atom columns (represented by squares) has been averaged over the "thickness" shown. When the sample is only one monolayer thick, i.e., there has been no averaging, no continuous line can be drawn to contain only Ga (or Al) atoms, illustrating that an interface cannot be defined on an atom-by-atom basis. Only as the "thickness" increases, does the interface become well-defined.

whether the interface is viewed in "cross-section" as in Fig. 8-2, or in "plan-view".

8.3.2 Structure

In this section we attempt to outline how a variety of techniques may be used to gain information about the configuration of a chemical interface over a wide range of spatial frequencies. The discussion is centered on the GaAs/AlGaAs system, because it is technologically advanced and has been extensively investigated. Although the microscopic structure of this interface can now be determined quantitatively, our knowledge of its structure over other length scales remains qualitative. Nevertheless, the discussion illustrates the challenge of describing an interface over a wide frequency range, and the importance of a critical appreciation of the way different techniques provide insight into the properties of an interface.

8.3.2.1 Microscopic Structure

In the absence of catastrophic crystal growth, the *structure* of an interface between two lattice-matched, isostructural crystals is uninteresting. For example, the atoms continue to occupy zinc-blende sites as a perfect semiconductor heterointerface is approached and crossed. In seeking to determine the atomic configuration at such an interface, a *chemical* rather than a structural question is being asked; one is attempting to learn which atom sits where, rather than where the atoms sit.

X-ray diffraction techniques have been applied to systems containing periodic stacks of chemical interfaces. Careful fitting of the satellite peak intensities due to the periodic compositional modulation elucidates the overall features of the interface configuration. Common to many experiments is the finding that the interfacial

region includes a few monolayers whose composition is intermediate between the neighboring materials (Fleming et al., 1980; Vandenberg et al., 1987). In the case of systems such as $In_{0.48}Ga_{0.52}As/InP$, which are lattice-matched only at one composition, the presence of a region of intermediate composition also entails the introduction of strain. In an elegant series of experiments, Vandenberg et al. have hown how the details of the growth procedure can modify the nature of the interfacial layer and the concomitant strain (Vandenberg et al., 1988, 1990).

The TEM in its lattice imaging mode can in principle reveal the local atomic configuration of an interface. Conventional lattice imaging, however, produces a map of the sample structure, and as such is not a useful probe of chemical interfaces. Below, we briefly describe the way the TEM may be used to obtain chemical information at near atomic resolution and sensitivity. The combination of "chemical lattice imaging" and digital pattern recognition quantifies the information content, and hence the composition of individual cells of material $\approx 2.8 \times 2.8 \times 150 \, \text{Å}^3$ in volume.

Chemical Lattice Imaging

In the modern high resolution TEM (HRTEM), a parallel beam of energetic electrons is transmitted through a thin sample to produce a diffraction pattern (Spence, 1980). The phases and amplitudes of the diffracted beams contain all the available information. Part of this information is passed through an aperture and the objective lens, to produce a lattice image. In general, most of the reflections used to form a lattice image come about because of the lattice periodicity, and are relatively insensitive to the exact occupancy of the lattice sites. We name such reflections structural. However, certain reflections, such as the (200) in the zinc-blende system, are due to chemical differences between the occupants of the different lattice sites, and contain significant chemical information (Ourmazd et al., 1986, 1987; Ourmazd, 1989). Such chemical reflections are in general weaker than the "strongly allowed" structural reflections, and the latter usually dominate the information content of lattice images. However, two factors, multiple scattering and lens aberrations, usually considered disadvantages of HRTEM, allow one to select and enhance the relative contribution of the weaker chemical reflections to lattice images.

When the electron beam enters a crystal along a low symmetry direction, a number of reflections are excited, exchanging energy among themselves as they propagate through the sample. To first order, this multiple scattering process may be viewed as the scattering of electrons from the undiffracted beam to each reflection, and their subsequent return. Structural reflections are strongly coupled to the undiffracted beam, and thus exchange energy with it rapidly as they propagate through the sample. This energy exchange is slower for the more weakly coupled chemical reflections. Because of this "pendellösung" effect, at certain sample thicknesses a chemical reflection can actually have a larger amplitude than its structural counterpart (Fig. 8-3). Appropriate choice of sample thickness can thus enhance the chemical information content of the lattice image. Moreover, the severe aberrations of electromagnetic lenses impart the character of a bandpass filter to the objective lens, whose characteristics can be controlled by the lens defocus (Spence, 1980; Ourmazd et al., 1986, 1990). Thus, judicious choice of defocus allows the lens to

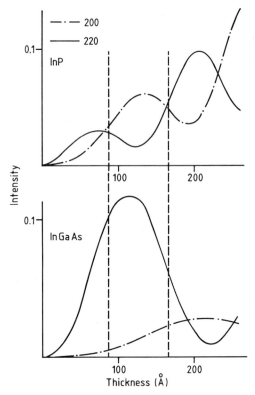

Figure 8-3. Variation of beam intensities with thickness (Pendellösung) for InP and InGaAs. The vertical lines show a suitable thickness window for maximum periodicity change, and hence chemical sensitivity across an InP/InGaAs interface.

select, and thus further enhance the contribution of the chemical reflections to the image.

To obtain chemical lattice images of compound semiconductor heterointerfaces in practice, advantage is taken of the chemical sensitivity of the (200) reflections (Fig. 8-3). The sample is viewed in the $\langle 100 \rangle$ orientation, and the (200) (chemical) and (220) (structural) reflections are used to form an image. The sample thickness and lens defocus are chosen to maximize the change in the frequency content of the lattice image across the interface. Thus the chemical information in the sample is encoded into periodicity information in the lattice image with maximum sensitivity (Ourmazd et al., 1987, 1990).

Atomic Configuration of Chemical Interfaces

Figure 8-4a is a structural lattice image of an InP/InGaAs interface. Such (structural) images are widely used to investigate the nature of chemical interfaces. It is of course true that even these structural images reveal, to some extent, the chemical change across the interface through the change in the background intensity. The question is whether this sensitivity is sufficient to reveal the atomic details of the interfacial configuration. Figure 8-4b is the same as Fig. 8-4a, except that the line marking the position of the interface is removed. The interface position and configuration are now less clear. This emphasizes the limited chemical sensitivity of structural images. Fig. 8-4c is a chemical lattice image of the same atom columns, obtained under optimum conditions for chemical sensitivity (Ourmazd et al., 1987). The InP is represented by the strong (200) periodicity (2.9 Å spacing), while in the InGaAs region the (220) periodicity (2 Å spacing) is dominant. Clearly, the interface is not atomically smooth, the roughness being manifested as the interpenetration of the (200) and (220) fringes.

While this demonstration establishes the inadequacy of normal structural lattice images to reveal the interfacial configuration, it does not necessarily imply that all semiconductor heterointerfaces are rough. In the case of the technologically more mature GaAs/AlGaAs system, the photoluminescence linewidth, and particularly the so-called monolayer splitting of the photoluminescence lines have been interpreted in terms of atomically abrupt and smooth interfaces, with the spacing between interfa-

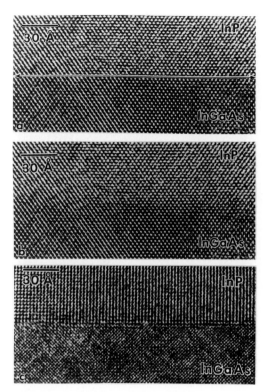

Figure 8-4. (a) $\langle 110 \rangle$ (structural) image of an InP/InGaAs interface. The line draws attention to the interface. (b) Same image without line. (c) Same area of interface imaged along $\langle 100 \rangle$ under chemically sensitive conditions. Note the interpenetration of InP (200) and InGaAs (220) fringes, indicating interfacial roughness.

cial steps estimated at several microns (see below). It is thus important to examine microscopically, GaAs/AlGaAs interfaces purported by luminescence to be of the highest quality.

Figure 8-5 is a chemical lattice image of a high quality $GaAs/Al_{0.37}Ga_{0.63}As$ interface, grown with a two minute interruption at each interface (Tu et al., 1987). The sample thickness and imaging conditions correspond to maximum chemical sensitivity, reflected in the strong change from (220) to (200) periodicity across the GaAs/AlGaAs interface. Visual examination of the image directly reveals the presence of interfacial

AlGaAs   GaAs   5.6Å↓↓

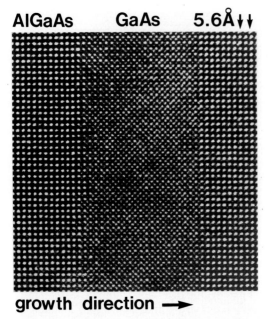

growth direction ⟶

Figure 8-5. Chemical lattice image of GaAs/ $Al_{0.37}Ga_{0.63}As$ quantum well produced after two minutes of growth interruption at each interface. Careful inspection reveals interfacial roughness.

roughness. Thus, even at this qualitative level of inspection, interfaces of the highest optical quality appear microscopically rough. However, the general practice of evaluating lattice images by visual inspection is subjective and unsatisfactory. We describe below a digital pattern recognition approach, which quantifies the local information content of lattice images, leading to their quantitative evaluation (Ourmazd et al., 1989a, b, 1990).

Quantification of Local Information Content of Images

The information content of a lattice image is contained in its spatial frequency spectrum, or alternatively, in the set of patterns that combine in a mosaic to form the image (Ourmazd et al., 1989a, b). In practice, the information content is degraded by the presence of noise. The quantitative

analysis of the information content thus requires three steps: (i) the assessment of the amount and the effect of noise present; (ii) the identification of statistically significant features; and (iii) quantitative comparison with a template. A primary virtue of an image is that it yields spatially resolved information. Thus, whether these tasks are carried out in Fourier- or real-space, the retention of spatial resolution, that is, the *local* analysis of the information content is of paramount importance.

In a chemical lattice image, the local composition of the sample is reflected in the local frequency content of the image, or alternatively, in the local patterns that make up the image (Ourmazd et al., 1987, 1989a, b). Thus local analysis of the image is equivalent to local chemical analysis of the sample. Local analysis of an image can be most conveniently affected by real-space rather than Fourier analysis. Real-space analysis proceeds with the examination of the information content of a unit cell of the image. When the integrated intensity is used to characterize a cell, information regarding the intensity *distribution* within the cell is not exploited. In one dimension this is analogous to attempting to identify a curve from the area under it, which would yield an infinite number of possibilities. Here we describe a simple procedure that exploits the available information more fully.

The task is carried out in several steps. First, perfect models, or templates are adopted from simulation, or developed from the data, which serve to identify the ideal image of each unit cell type. When the template is extracted from experimental images, it is obtained by averaging over many unit cells to reduce the effect of noise. For example, several unit cells of Fig. 8-5 not lying at the interface are averaged to produce the templates for GaAs and $Al_{0.37}Ga_{0.63}As$ shown in Fig. 8-6a. Second, an image unit cell of a particular size is adopted, and divided into an $n \times n$ array of pixels, at each of which the intensity is measured. Typically $n \approx 30$ and thus 900 intensity measurements are made within each unit cell. Third, each unit cell is represented by a multidimensional vector, whose components are the n^2 (usually 900) intensity values obtained from the cell. The ideal image unit cell for each material is now represented by a template, which in turn is represented by a vector R^t. For example, the ideal image unit cells of GaAs and $Al_{0.37}Ga_{0.63}As$ are characterized by the two vectors R^t_{GaAs} and $R^t_{Al_{0.37}Ga_{0.63}As}$, respectively (Fig. 8-6b).

Next, the amount of noise present in the experimental image is deduced from the angular distributions of the real (that is, noisy) unit cell vectors R_{GaAs} and $R_{Al_{0.37}Ga_{0.63}As}$ about their respective templates. The noise in Fig. 8-5 is such that, away from the interface, the R_{GaAs} and $R_{Al_{0.37}Ga_{0.63}As}$ form similar normal distributions around their respective template vectors R^t_{GaAs} and $R^t_{Al_{0.37}Ga_{0.63}As}$. The standard deviation σ of each distribution quantifies the noise present in the images of GaAs and $Al_{0.37}Ga_{0.63}As$ (Fig. 8-6c). A unit cell is different from a given template, with an error probability of less than 3 parts in 10^3, if its vector is separated from the template vector by more than 3σ. With $3.9 \times 3.9 \text{ Å}^2$ image unit cells, the centers of the distributions for the GaAs and $Al_{0.37}Ga_{0.63}As$ unit cells shown in Fig. 8-6 are separated by 12σ, which means that each unit cell of GaAs and $Al_{0.37}Ga_{0.63}As$ can now be correctly identified with total confidence. A representation of the results of the vector pattern recognition analysis of Fig. 8-5 is shown in Fig. 8-7. The image is divided into $2.8 \times 2.8 \text{ Å}^2$ cells, each of which is placed at a height representing the angular position of its vector. The yellow and blue cells lie within 3σ of R^t_{GaAs} and $R^t_{Al_{0.37}Ga_{0.63}As}$ respectively, while the other colors represent 3 to 5, 5 to 7, and 7 to 9σ bands (Ourmazd et al., 1989a, b).

We have now outlined a simple approach capable of quantitatively evaluat-

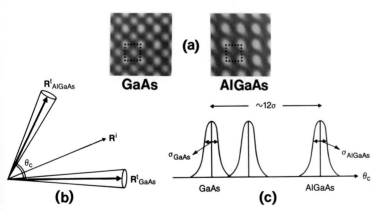

Figure 8-6. (a) Averaged images of GaAs (left) and $Al_{0.37}Ga_{0.63}As$ (right). The unit cells used as templates for pattern recognition are the dotted 2.8 Å squares. (b) Schematic representations of the template vectors R^t_{GaAs} and $R^t_{Al_{0.37}Ga_{0.63}As}$, the distribution of R_{GaAs} and R_{AlGaAs} about them, and an interfacial vector R^i. (c) Schematic representation of the distribution by the GaAs and $Al_{0.37}Ga_{0.63}As$ unit cells about their templates. Note that the angular position of R^i denotes the most likely composition only. The actual composition falls within a normal distribution about this point.

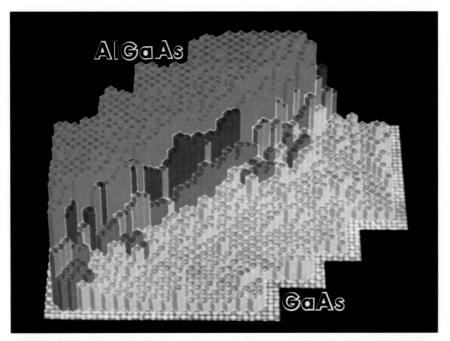

Figure 8-7. Three-dimensional representation of the analyzed lattice image of $Al_{0.37}Ga_{0.63}As$ grown on GaAs after a 2 min. interruption. The unit cells are 2.8 Å squares. The height of each cell represents the angular position of its vector R with respect to the template vectors, which are about $12\,\sigma$ apart. Yellow and blue mark those cells which fall within $3\,\sigma$ of GaAs and $Al_{0.37}Ga_{0.63}As$ templates, respectively. Green, magenta, and red represent 3 to 5, 5 to 7, and 7 to 9 σ bands. Outside the yellow and blue regions, the Al content of each unit cell is intermediate between GaAs and $Al_{0.37}Ga_{0.63}As$, with confidence levels given by normal statistics.

ing the local information content of images made up of mosaics of unit cells. This method exploits all the available information to determine the amount of noise present, is sophisticated in discriminating between noise and signal, identifies statistically significant features, and allows quantitative comparison with templates. Below, we discuss how, in the case of chemical lattice images, the local information content is related to the local composition of the sample.

A lattice image is locally analyzed to gain information about the local atomic potential of the sample. Under general dynamical (multiple) scattering conditions, the electron wavefunction at a point on the exit face of the sample need not reflect the sample projected potential at that point. The emerging electron wave is further convoluted with the aberrations of the lens before forming the image. There is no general relation connecting the local details of a lattice image to the local atomic potential in the sample (Spence, 1980).

In chemical imaging, we are concerned with the way that a compositional inhomogeneity is imaged under conditions appropriate for chemical sensitivity, and how the pattern recognition algorithm extracts information from a chemical lattice image. For samples of reasonable thickness (< 300 Å at 400 kV), as the Al content of homogeneous $Al_xGa_{1-x}As$ is changed from 0 to 0.37, the vector $R^t_{Al_xGa_{1-x}As}$ rotates linearly from R^t_{GaAs} to $R^t_{Al_{0.37}Ga_{0.63}As}$

(Ourmazd et al., 1990). Thus, in homogeneous material, the composition of a unit cell can be directly deduced from the angular position of its vector R with respect to the templates. In general, R deviates from the plane containing the template vectors, and the projection of R on this plane yields the composition. The confidence levels associated with such measurements depend on the amount of noise present, and can be deduced from normal statistics.

In an inhomogeneous sample, this simple procedure requires justification. The problem can be formulated as follows. Given a "chemical impulse" of a specific shape, such as a column of Al atoms imbedded in GaAs (a δ-function), an abrupt interface (a θ-function), or a diffuse interface (say an error function), what is the shape of the impulse on the analyzed chemical image? Or, alternatively, what region of the sample contributes to the information content of an image unit cell? By reciprocity, these two formulations are equivalent. This problem is essentially similar to determining the response function of a system.

The effect of the response function can be determined by analyzing images of samples containing various impulses, simulated under conditions appropriate for chemical imaging (Ourmazd et al., 1989 a, b, 1990). The appropriate conditions are chosen from a bank of simulated images that contain the particular impulse under consideration. For example, the simulated images of an abrupt GaAs/$Al_{0.37}Ga_{0.63}As$ interface (θ-function) show, that in this case, the appropriate conditions correspond to sample thickness and lens defocus values of ≈ 170 Å and ≈ -250 Å, respectively. Such analysis shows that under appropriate chemically sensitive conditions nonlocal effects due to dynamical scattering and lens aberrations are negligible (Ourmazd et al., 1990). This is illustrated in Fig. 8-8, where a the chemical image of a column of Al imbedded in GaAs (a δ-function) is simulated and then analyzed; the input impulse and the analyzed response are identical. The response function is essentially determined by the periodicity of the chemically sensitive reflection, which in the case of the zincblende structure is the (200) periodicity.

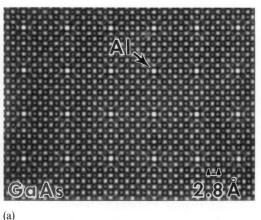

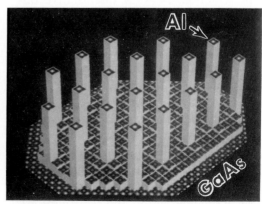

(a)   (b)

Figure 8-8. Simulated image (a) and analyzed image (b) of a series of δ-functions of Al, imbedded in GaAs. Sample thickness: 170 Å, defocus: -250 Å.

This means that in this structure, the composition of a region ¼ of the crystal unit cell in cross-section and ≈ 30 atoms high can be directly determined.

Several other questions regarding the practicality of the approach outlined above, such as the effects of geometrical imperfections in the sample, and photographic non-linearities have been considered elsewhere (Ourmazd et al., 1990). What emerges is the conclusion that the combination of chemical lattice imaging and vector pattern recognition can quantitatively analyze chemical interfaces with near-atomic resolution and sensitivity. Below, we apply these techniques to quantify the composition change across interfaces of the highest quality.

Quantitative Chemical Maps

Figure 8-7 is the analyzed chemical lattice image of the $GaAs/Al_{0.37}Ga_{0.63}As$ interface of Fig. 8-5. The height of a unit cell represents the angular position of its vector $\boldsymbol{R}$ with respect to the template vectors, and the color changes represent statistically significant changes in composition over and above random alloy statistics. This representation allows a quantitative display of the noise and the composition at each Group III atomic column ≈ 30 atoms high. The compositional change from 0 to 0.37 corresponds to changing a column of 30 Ga atoms to one containing ≈ 19 Ga and 11 Al atoms. It turns out that the replacement of one or two Ga atoms with Al can be detected with 60% or 90% confidence, respectively (Ourmazd et al., 1990). This demonstrates that Fig. 8-7 is essentially a spatial map of the composition, at near-atomic resolution and sensitivity. Although luminescence shows this interface to be of the highest quality (Tu et al., 1987), it is clear that its atomic configura-

tion is far from "ideal". The quantitative chemical map of Fig. 8-7, which is typical, shows that the transition from GaAs to $Al_{0.37}Ga_{0.63}As$ takes place over ≈ 2 unit cells, and that the interface contains significant atomic roughness. It is important to note that the region of sample analyzed in Fig. 8-7 is ≈ 30 atoms thick, and thus random alloy roughness is expected to be at a negligible level. Also, because in our analysis the statistical fluctuations in the local composition due to random alloy statistics contribute to the "noise" in the AlGaAs region, only roughness over and above the random alloy component is evaluated as statistically significant. At the level of detail of these composition maps, the assignment of values for interfacial imperfections, such as transition width, roughness, and island size, is a matter of definition. Also, without extensive sampling caution is required in deducing quantitative values for the spacing between interfacial steps, however, they are defined. Nevertheless, it is clear that significant atomic roughness at the ≈ 50 Å lateral scale is present.

8.3.2.2 Mesoscopic and Macroscopic Structure

Due to the limited field of view of direct microscopic techniques, they cannot be used to establish the interfacial configuration over mesoscopic (micron) or macroscopic (millimeter) length scales. To make further progress, it is necessary to use indirect methods to gain insight into the interfacial configuration. Such techniques attempt to determine the interface structure through its influence on other properties of the system, such as its optical or electronic characteristics. Fundamental to this approach is the premise that it is known how the structure affects the particular property being investigated. In practice this is rarely the case. "Indirect" experiments thus face the challenge of simultaneously determining the way a given property is affected by the structure and learning about the structure itself.

Because a direct correlation is thought to exist between the structure of a thin layer and its optical properties, luminescence techniques have been extensively applied to investigate the structure of semiconductor interfaces (Weisbuch et al., 1981; Tu et al., 1987; Bimberg et al., 1987; Thomsen, Madhukar, 1987). In photoluminescence (PL) the carriers optically excited across the band gap form excitions and subsequently recombine, often radiatively. The characteristics of a photon emitted due to the decay of a single free exciton reflect the structural properties of the quantum well, averaged over the region sampled by the recombining excition. In practice, the observed signal stems from a large number of recombining excitions, some of which are bound to defects. The challenge is to extract information about the interfacial configuration from PL measurements, which represent complex weighted averages of the well width and interfacial roughness sampled by a large collection of excitons. The recognition that PL cannot easily discriminate between the recombination of free excitons and those bound at defects has led to the application of photoluminescence excitation spectroscopy (PLE), which is essentially equivalent to an absorption measurement, and thus relatively immune to complications due to defect luminescence.

The photoluminescence spectrum of a typical single quantum well ≈ 50 Å wide, grown under standard conditions, consists of a single line ≈ 4.5 meV wide, at an energy position that reflects the well width and the barrier composition (Tu et al., 1987). This linewidth is significantly larger than that of a free exciton in high quality "bulk" GaAs (≈ 0.2 meV), indicating additional scattering, presumably partly due to interfacial roughness. When the growth of the layer is interrupted at each interface, and the next layer deposited after a period of tens of seconds, the PL spectrum breaks into two or three sharper lines each ≈ 1.5 meV wide. This reduction in the PL linewidth is attributed to a smoothing of the interfaces during the growth interruption. Perhaps more strikingly, however, the two or three lines obtained from a single quantum well are often assigned to excitonic recombination in different regions of the quantum well under the laser spot, within each of which the well is claimed to be an *exact* number of atomic layers thick. Thus the different lines are each thought to arise from recombination within "islands" over which the interfaces are atomically smooth. This model rests essentially on the premise that the several PL and PLE peak energies and their separations correspond to wells exactly an integral number of monolayers (MLs) thick.

This interpretation of the luminescence data thus advocates the existence of atomically perfect (i.e. atomically smooth and abrupt) interfaces. On this basis, a quantum well of nominal thickness n in fact consists of regions (islands), within each of which the thickness is exactly $(n-1)$, n, or $(n+1)$ MLs, between which the interfacial position changes abruptly by 1 ML. These islands have been claimed to be as large as 10 μm in diameter (Bimberg et al., 1987), but are generally thought to lie in the micron range (Miller et al., 1986; Petroff et al., 1987), and in any case to be much larger than the exciton diameter (≈ 15 nm). A consequence of this model is that the PL and PLE peak separations must necessarily correspond to the difference in the energies of excitons that recombine in regions of the well differing in thickness by *exactly* 1 ML. In practice, the splittings rarely correspond exactly to ML changes in well width. Departures from "ML values" are generally ascribed to experimental uncertainties in determining the peak positions, to fluctuations in the composition of the material, to impurities, or to exotic configurations of atomically smooth interfaces (Reynolds et al., 1985).

Quantitative chemical imaging, however, shows these interfaces to be atomically rough. This apparent contradiction is resolved by recalling that each experimental technique probes the interfacial configuration over a limited range of spatial frequencies. Thus, luminescence and chemical imaging results simply reveal different parts of the interfacial roughness spectrum. But the concept that luminescence shows interfaces to be smooth at the *atomic* scale has become so entrenched that it is important to examine the luminescence data carefully to determine whether they can indeed sustain the "atomically smooth model".

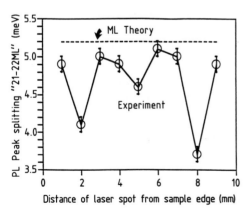

Figure 8-9. Photoluminescence line splitting from a single, nominally 60 Å thick quantum well vs. position on the wafer. The splitting expected from atomically smooth interfaces and a thickness change from 21 to 22 ML is shown by the dashed line. Note the large ($\approx 40\%$) variations in the measured splittings (Warwick et al., 1990).

Warwick et al. (1990) have investigated series of quantum wells grown in different systems and characterized by luminescence to have atomically smooth interfaces (Tu et al., 1987; Bimberg et al. 1987). In their experiments, the laser spot was moved over samples held at 2 K, and a series of PL and PLE spectra obtained from sets of neighboring points on each sample containing single quantum wells. At each point, PL and PLE data were also recorded from the AlGaAs barrier material and the GaAs buffer layer. In this way the energies of the photons emitted from the well, the local mean Al content of the barrier, and the effects of residual stress were directly measured at many points on each sample.

Figure 8-9 shows the separation between the "21 ML" and the "22 ML" PL peaks obtained from a quantum well nominally 60 Å thick, *versus* the position of the laser spot. Note that the observed splitting varies in magnitude by nearly 40%, i.e. over the range 5.1 ± 0.1 to 3.7 ± 0.1 meV. As indicated by the error bars, this variation is an order of magnitude larger than effects expected from the measured compositional fluctuations in the barrier and variations in the residual stress. Similar features are observed in the PLE data (Fig. 8-10).

It is important to note that the observed shifts are substantially larger than those caused by random alloy fluctuations in the barrier. As mentioned in Sec. 8.3.1, when AlGaAs is deposited on an atomically smooth GaAs surface, the resultant interface is atomically rough, because random alloy fluctuations cause the deposited AlGaAs to contain small clusters of GaAs. However, an exciton with a diameter of ≈ 15 nm averages over ≈ 5000 interfacial atoms. Assuming the worst case (Gaussian statistics), random alloy fluctuations can

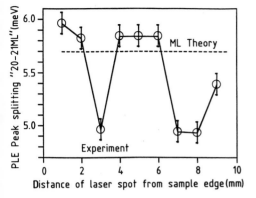

Figure 8-10. Photoluminescence excitation line splitting vs. position on the wafer. The scan line is different from that in Fig. 8-8, because registration between different cryostats was impractical. The splitting expected from atomically smooth interfaces and a thickness change from 20 to 21 ML is shown by the dashed line, although its magnitude is strongly model dependent. Note the large ($\approx 25\%$) variations in the measured splittings (Warwick et al., 1990).

cause the Al content of a ≈ 5000 atom area of the interfacial $Al_{0.37}Ga_{0.63}As$ layer to vary over the range 0.370 ± 0.006. This gives rise to part of the observed PL linewidth (≈ 1.0 meV). The peak position can be determined to within $\pm 2\%$ of the width, and thus the splitting to within ± 0.03 meV, which is much smaller than the observed fluctuations.

The variations in the PL *and* PLE peak separation are clearly incompatible with the model of an atomically smooth and abrupt interface. This illustrates the inadequacy of attempting to describe the complex waveform of an interface in terms of a single island size, i.e. a δ-function in the frequency spectrum. However, the compilation of information over the necessary frequency range, extending from the nm to the cm range is a formidable task. Warwick et al. have thus attempted a qualitative compilation of data from a variety of techniques as follows (Warwick et al., 1990).

From the roughness revealed by chemical lattice imaging (Ourmazd et al., 1989 a), they surmise the presence of "significant" roughness at the atomic scale. On the other hand, the relatively sharp luminescence lines indicate "little" roughness at wavelengths comparable with the exciton diameter. The occurrence of several sharp PL lines when a ≈ 100 μm diameter laser spot is used to excite luminescence, the observation of "islands" in cathodoluminescence (Bimberg et al., 1987; Miller et al. 1986; Petroff et al., 1987), and the observed variations in the PL peak splittings indicate "substantial" low frequency roughness. These observations are schematically summarized in Fig. 8-11. It is argued that such an interfacial configuration can in principle give rise to multiple peaks with a splitting which *may* be approximately one ML.

If such qualitative remarks can be taken seriously, the interfacial roughness spec-

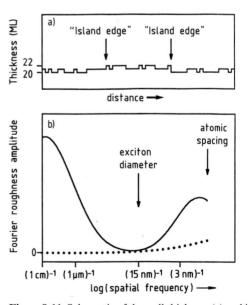

Figure 8-11. Schematic of the well thickness (a) and its Fourier transform (b). Roughness due to random alloy fluctuations is shown dotted in (b) but is too fine to be represented in (a).

trum is at least bimodal, with a minimum in the vicinity of the exciton diameter. This may be no accident; growth procedures have been so optimized as to give sharp PL lines, thus pushing roughness away from length scales comparable with the exciton diameter. While this may indeed appropriate for materials intended for optical applications, it may not be ideal for transport experiments and electronic devices, where roughness at the Fermi wavelength must be minimized (Sakaki et al., 1987).

It is interesting to note that the bimodal roughness spectrum, if indeed present, does not arise by a simple relaxation of the surface by diffusion. In a simple Monte Carlo simulation, where a "white noise" interface is allowed to relax by diffusion to produce (immobile) islands, a "$1/f$" roughness spectrum is produced (Warwick et al., 1990). It appears that the use of luminescence to optimize crystal growth leads to the selection of growth conditions that produce a highly novel roughness spectrum.

8.3.3 Relaxation of Chemical Interfaces

Because semiconductor multilayers are becoming increasingly familiar, it is easy to overlook the fact that they are highly inhomogeneous systems far from equilibrium. On crossing a modern GaAs/AlGaAs interface, the Al concentration changes by several orders of magnitude in a few lattice spacings. As pointed out by Cahn (1961), such systems relax by interdiffusion, sometimes in unusual ways. It is thus scientifically interesting and technologically important to investigate the stability of chemical interfaces against interdifusion. In semiconductors, the modest diffusivities of point defects limit substantial relaxation at room temperature. However, an interface can relax during thermal annealing,

in-diffusion of dopants, or ion-implantation. The extensive literature concerned with such phenomena will not be summarized here, because excellent reviews already exist (Deppe, Holonyak, 1988). Rather, we describe the new understanding that emerges when the chemical relaxation of interfaces is studied at the atomic level.

Using the quantitative chemical mapping techniques described above, it is straightforward to make sensitive measurements of interdiffusion at single interfaces. The composition profile across a given interface is measured in two pieces of the same sample, one of which has been annealed in bulk form (Fig. 8-12). Starting with the initial profile and using the diffusion coefficient D as free parameter, the diffusion equation is solved to fit the final (annealed) profile, thus deducing D as a function of temperature and interface depth (Kim et al., 1989, 1990a, b). In the case of ion-implanted samples, the composition profiles are characterized by fitting an erfc-profile to the data, with the profile width L as the free parameter. Intermixing

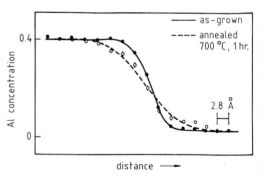

Figure 8-12. Composition profiles of a C: GaAs/Al-GaAs interface at a depth of ≈ 300 Å, as-grown (solid line), and after 700 °C, one hour anneal (dotted line). One standard deviation error bars are shown. Each measurements refers to a single atomic plane, and is obtained by averaging the composition over a $\approx 30 \times 2.8$ Å^2 segment of an individual atomic plane.

due to implantation is thus quantified in terms of changes in the interfacial width L (Bode et al., 1990).

8.3.3.1 Interdiffusion Due to Thermal Annealing

Figure 8-13 is an Arrhenius plot of D vs. $1/kT$ for C-doped $GaAs/Al_{0.37}Ga_{0.63}As$ interfaces at three different depths beneath the surface. Each measurement is made in a region $\approx 10^{-19}$ cm^3 in volume. Remarkably, the magnitude of the interdiffusion coefficient, and the activation energy for intermixing change strongly with depth. Since this behavior is observed both in the GaAs/AlGaAs and the HgCdTe/CdTe systems (Kim et al., 1989, 1990 a, b), it is likely that the depth-dependence of the interdiffusion coefficient is a general effect. This is more clearly displayed in Fig. 8-14, where $\ln D$ is plotted as a function of the interface depth. At the lower temperatures (700 °C and particularly at 650 °C), $\ln D$ initially decreases linearly with increasing distance from the surface, but appears to drop exponentially beyond a certain critical depth.

This effect has been shown to be related to the injection of point defects from the sample surface during the anneal. In particular, interdiffusion in these systems is assisted by the presence of native point defects (interstitials or vacancies), whose concentrations are often negligible in as-grown samples. For interdiffusion to occur, such native defects must be injected from the sample surface during the anneal. The interdiffusion coefficient is a sensitive function of the concentration of these defects at the particular interface studied, and thus can be used to investigate the microscopics of native point defect diffusion in multilayered systems. Indeed, it should be possible to measure the forma-

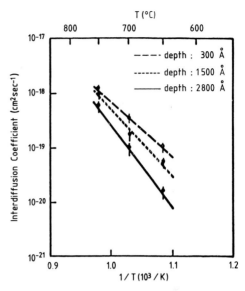

Figure 8-13. Arrhenius plot of the interdiffusion coefficient D at C:GaAs/AlGaAs interfaces at three different depths.

tion energy and migration energy of a given native defect (interstitial or vacancy) as a function of its charge state.

Returning to interdiffusion, two important points emerge. First, the interdiffusion coefficient varies strongly with depth. Thus a measurement of this parameter is meaningful only if it refers to a single inter-

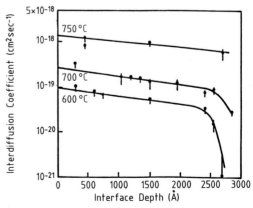

Figure 8-14. Plot of the $\ln D$ vs. interface depth (z) at three different temperatures, for C:GaAs/AlGaAs interfaces.

face at a known depth. Second, it follows that the interface stability is also depth-dependent. Thus the layer depth must be regarded as an important design parameter in the fabrication of modern devices. This effect assumes additional importance when interdiffusion is also concentration dependent, leading to strong intermixing at very low temperatures (Kim et al., 1990a). These phenomena highlight the importance of a microscopic understanding of the relaxation of systems far from equilibrium.

8.3.3.2 Intermixing Due to Ion Implantation

The passage of energetic particles through an inhomogeneous solid deposits sufficient energy in the solid to cause intermixing, even at very low temperatures. Using chemical mapping techniques, it is possible to detect the intermixing due to the passage of a single energetic ion.

Consider a GaAs/AlAs multilayer, held at 77 K and implanted with 320 keV Ga$^+$

ions to a dose of 5×10^{12} cm^{-2}, i.e. about one ion per 2000 Å^2 area of each interface. Figure 8-15 shows a chemical lattice image of an unimplanted, 50 Å thick GaAs layer between its two adjacent AlAs layers, together with the composition profiles for each interface. The GaAs layer is situated 1400 Å beneath the surface, and is thus close to the depth where the maximum damage during subsequent implantation is expected to occur. The growth direction is from bottom to top, the (subsequent) implantation direction from top to bottom. Each point on the profiles of Fig. 8-15 represents the average composition of a 1 μm segment of a given atomic plane before implantation. Both top and bottom interfaces (A and B) display excellent lateral uniformity, and can be characterized by similar characteristic widths L ($L_A = 2.4 \pm 0.1$ Å, $L_B = 2.7 \pm 0.1$ Å).

After implantation to a dose of 5×10^{12} cm^{-2}, chemical analysis of individual interfaces located at depths between 1000 and 1700 Å beneath the surface reveals significant intermixing across the top

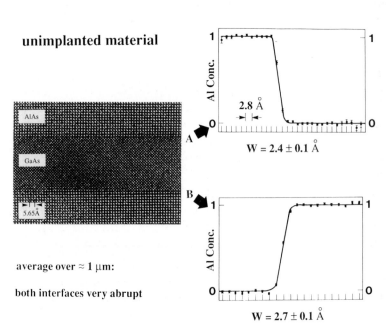

unimplanted material

Al Conc.

2.8 Å

A

W = 2.4 ± 0.1 Å

AlAs

GaAs

5.65Å

B

Al Conc.

W = 2.7 ± 0.1 Å

average over ≈ 1 μm:

both interfaces very abrupt

Figure 8-15. Chemical lattice image of GaAs layer between two AlAs layers, with composition profiles across the interfaces. Growth direction is from bottom to top. One standard deviation error bars are shown.

interfaces, although (on average) only one Ga^+ ion has passed through each 2000 $Å^2$ area of the interface. The intermixing is not unfiorm along the top interface, but shows large fluctuations. In Fig. 8-16 three concentration profiles for adjacent 50 Å segments of an interface are shown. Profiles a and c display similar degrees of intermixing, characterized by a width L of $\approx 6.2 \pm 0.2$ Å, while profile b is characterized by $L = 4.6 \pm 0.2$ Å.

Kinematic implantation simulations using the TRIM program (Biersack, 1987) show, that under the conditions used in these experiments, a single implanted Ga^+ ion creates a damage track ≈ 50 Å wide. This agrees qualitatively with the width of the observed fluctuations in the degree of intermixing along the interface. After implantation at a higher dose $(1 \times 10^{13}$ $cm^{-2})$, the intermixing along the interface is uniform. It is thus likely that the intermixing caused by the passage of single energetic ions is being directly imaged.

In these experiments, a series of chemical interfaces is used as a stack of photographic emulsion layers, to record the passage of energetic ions, or native point defects, implanted at or injected from the surface. Thus chemical interfaces can be used to reveal the microscopics of defect processes at the atomic level.

8.3.4 Summary

At present, interfaces of the highest perfection, and thus widest application are those between lattice-matched, pseudomorphic, crystalline solids, differing only in composition. In Sec. 8.3, we attempted to outline the concepts needed to define such "chemical" interfaces. Two concepts emerge as fundamentally important. First, the definition of an interface is most conveniently affected in terms of its roughness spectrum, where the amplitudes of the interfacial undulations are specified as a function of their spatial frequency. Second, when one of two materials forming

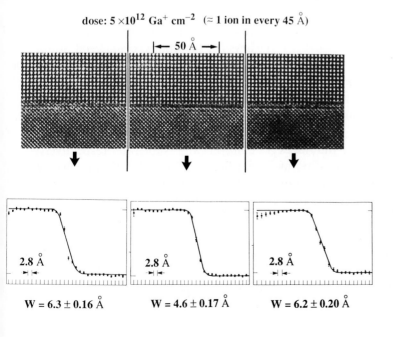

dose: 5×10^{12} Ga^+ cm^{-2} (≈ 1 ion in every 45 Å)

|← 50 Å →|

2.8 Å   2.8 Å   2.8 Å

W = 6.3 ± 0.16 Å   W = 4.6 ± 0.17 Å   W = 6.2 ± 0.20 Å

Figure 8-16. Chemical lattice image of GaAs/AlAs sample implanted with 320 keV Ga^+ ions to a dose of 5×10^{12} cm^{-2}. This dose corresponds to the implantation of $\approx$ one ion per 45 Å length of the interface. The composition profiles refer to adjacent ≈ 50 Å segments of the top interface. Note the large local variations in intermixing on the 50 Å lateral scale.

the interface is a random alloy, the interfacial configuration cannot be defined at the atomic level. Only when the length scale for the definition of the interface is so large that random alloy fluctuations are at a sufficiently low level, can an interface be adequately defined.

Experimentally, it is essential to realize that any technique probes only a small part of the roughness spectrum. This "window" is delimited on the high frequency side by the spatial resolution of the technique, and on the low frequency side by the field of view. Moreover, a technique may possess an intrinsic length scale, the exciton diameter for luminescence, the Fermi wavelength for transport, which plays a crucial role in determining the wavelength of the interfacial roughness that can be most sensitively probed.

To gain a realistic impression of the interfacial configuration, information over a wide range of frequencies is needed. It is thus necessary to collate the data produced by a variety of techniques. The description of an interface in terms of an "island size" is essentially an attempt to replace the real roughness spectrum by a single frequency component. This is too naive to be realistic.

Chemical interfaces and multilayers are systems far from equilibrium, able to relax through interaction with point defects. This allows them to be modified by suitable processing for device applications. Equally importantly, chemical interfaces can be used to track the passage of point defects, providing a microscopic view of the processes that govern the elementary structural excitations of solids.

8.4 Interfaces Between Mismatched, Isostructural Systems

Although it is now possible to grow highly perfect semiconductor heterostructures when the constituent materials have identical (or very similar) lattice parameters, strained layer epitaxy – the art of synthesizing single crystal structures from materials with different lattice parameters – is still evolving. In this section, we will summarize progress to date in understanding and eliminating those problems peculiar to lattice-mismatched growth. It should be emphasized at the outset, however, that this is a field very much in flux, with theories and perceptions that change rapidly. Nevertheless, an attempt will be made to emphasize those concepts or models which are widely accepted and appear durable. More controversial or recent experiments or theories will be identified as such.

8.4.1 Fundamental Principles – Critical Layer Thickness

An epitaxial layer can be grown upon a substrate with a different lattice parameter, such that below a "critical" epilayer thickness h_c, the epitaxial material adopts the lattice parameter of the substrate parallel to the interfacial plane. This results in a biaxial strain within the interfacial plane of magnitude $\varepsilon \approx (a_e - a_s)/a_s$, where a_e and a_s are the bulk (relaxed) lattice parameters of epilayer and substrate respectively. According to classical elasticity theory, this results in a (tetragonal) distortion normal to the interfacial plane of magnitude $\varepsilon(1+v)/(1-v)$, where v is the Poisson ratio of the epilayer material. For a typical value of $\varepsilon = 10^{-2}$ and a typical (bulk) value of $v = 1/3$ this gives a 2% tetragonal distortion of the epilayer unit cell. Such distortions of the inter-atomic bonds induce, particularly in the strongly directional co-valent bonding of many semiconductors, a large elastic strain energy in the epitaxial layer. (We assume here and in subsequent discussion that the substrate is of infinite thickness, with the lattice mismatch distortion entirely accommodated by the epilayer.) For epilayer thicknesses below h_c, the interfacial atomic sites are relatively close to their bulk positions (as for the isostructural lattice-matched interfaces discussed in the last section), but small displacements of atoms in the epilayer produce high configurational energies due to stretching and shearing of inter-atomic bonds.

The biaxial interfacial strain will also induce a biaxial stress, of magnitude $\sigma_\varepsilon = 2\varepsilon G(1+v)/(1-v)$, where G is the epilayer shear modulus. For a typical system with $\varepsilon = 0.01$, $v = 1/3$, $G = 5 \times 10^{10}$ Pa, we obtain a stress of 2 GPa. With the exception of the diamond anvil cell, this stress is well in excess of values attainable by conventional loading techniques on bulk samples.

The elastic strain energy stored in the epitaxial layer provides a strong driving force for strain relief. This is effected by the introduction of "misfit" dislocations into the interfacial plane, which allow the epilayer to relax towards its bulk (unstrained) lattice parameter by reducing the average bond distortion. If the dislocation Burgers vectors have a finite component parallel to the interfacial plane and perpendicular to their line directions of magnitude b_p, they will produce a net strain relief of $\Delta\varepsilon = b_p/p$ where p is the average distance between misfit dislocations. These dislocations will generally lie at or near the interface between the two materials.

Most semiconductors of practical interest are diamond cubic (dc) or zincblende (zb) in structure. In these structures the dominant glide plane system is the $\{111\}$

set. Considering the prevalent (100) interfacial plane orientation, these glide planes intersect the interface along orthogonal [011] and [0$\bar{1}$1] directions. The misfit dislocations thus form a square mesh in the interfacial plane, as shown in Fig. 8-17. The Burgers vectors of perfect dislocations in zb and dc structures are almost invariably of the $a/2 \langle 011 \rangle$ type, this being the minimum length lattice translation vector in these structures (Fig. 8-18). These dislocations may be either glissile, with their Burgers vectors lying within a (111) slip plane, or they may be sessile with their Burgers vectors lying in the interfacial plane. Dislocations are characterized by the angle between their Burgers vector and their line direction, such that for a (100) interface the glissile dislocations are said to be of 60°-type and the sessile dislocations of 90°-type. The 60°-dislocations are of mixed edge and screw character, and move by the relatively rapid glide process. Their projection onto the interfacial plane is of magnitude $a/(2\sqrt{2})$. The 90°-dislocation direction is of edge type, and its projection onto the interfacial plane is $a/\sqrt{2}$. Although it relieves strain twice as effectively as the 60°-type, the latter predominate as they may move by glide, whilst the edge dislocations have to be moved by (typically) far slower climb processes, involving mass transport of point defects.

With typical values of $\varepsilon = 0.01$, $a = 5.6$ Å and $b_p = 2.0$ Å for 60°-dislocations, we would expect a mean dislocation spacing of $p = 200$ Å for complete relaxation of lattice strain. This would massively disturb the interface structure, which would now have superimposed upon it a quasi-periodic perturbation of period p and amplitude ≈ 0.2 times the inter-atomic bond length (Hirth, Lothe, 1968). Considering the two orthogonal orientations of interfacial dislocation line ([011] and [01$\bar{1}$] for a (100)

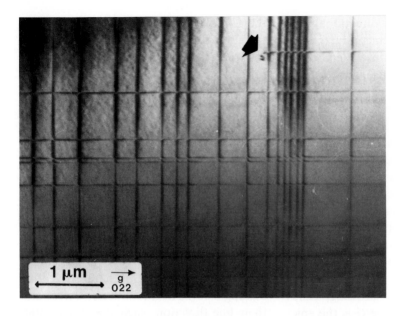

Figure 8-17. Plan view TEM image of the orthogonal misfit dislocation array at a $Ge_{0.15}Si_{0.85}/Si(100)$ interface. A threading dislocation is arrowed.

interface), the required value of p would necessitate a dislocation line length of 10^4 m in each cm^2 area of the interface, or about 10^6 m over a 10 cm diameter wafer!

Despite the extremely high stresses and elastic strain energies in pseudomorphic lattice-mismatched systems, these enormous dislocation lengths are not energetically favored until the critical epilayer thickness h_c is exceeded. This is because a dislocation has a self-energy arising from both the extremely high distortions at its core (where classical elasticity theory breaks down as inter-atomic bonds are disturbed beyond Hooke's limit), and from the radially decaying strain field in the surrounding material outside the core region (where classical elasticity theory may be reasonably applied). Thus, the critical thickness could be defined as that thickness at which the energy of a dislocation array is equal to the elastic strain energy stored in the strained epilayer. This concept is the essence of the approach of Frank, van der Merwe and coworkers (Frank, van der Merwe, 1949 a, b; van der Merwe, Ball, 1975) in calculating the critical thickness. In this approach the energy of a misfit dislocation array at an interface is minimized with respect to the mean dislocation spacing. This is not straightforward, because at higher dislocation densities interaction energies due to overlapping dislocation strain fields have to be considered. However, in principle it can predict dislocation densities as a function of epilayer thickness. In general, numerical methods must be used, but analytical solutions are obtained in the limits of very thin and very thick epilayers.

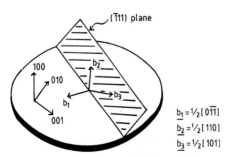

Figure 8-18. Possible orientations of misfit dislocation Burgers vectors at a (100) interface for one {111} slip plane in diamond cubic or zincblende material.

$b_1 = \frac{1}{2}[0\bar{1}1]$
$b_2 = \frac{1}{2}[110]$
$b_3 = \frac{1}{2}[101]$

The strained-dislocated transition is perhaps easier to understand conceptually in terms of stresses in the epitaxial layer. Increasing the length of an individual interfacial dislocation line induces a back stress due to the increase in dislocation self energy. This restoring stress acts on the dislocation segment which threads through the epitaxial layer from the interface to the epilayer surface and is loosely termed the dislocation line tension, σ_T. (Note that the presence of this "threading" arm is an essential geometrical property of a dislocation – the interfacial segment cannot simply terminate in the bulk of the crystal, but must terminate at a free surface or at a node with another defect.) Equality of the applied and restoring stresses, $\sigma_\varepsilon = \sigma_T$, is the essence of the Matthews and Blakeslee (MB) model of critical thickness (Matthews, 1975; Matthews, Blakeslee, 1974, 1975, 1976) (Fig. 8-19). Standard expressions for the applied and restoring stresses (Hirth, Lothe, 1968; Matthews, 1975; Matthews, Blakeslee, 1974, 1976) then yield:

$$2S\varepsilon \frac{(1+v)}{(1-v)} G$$
$$= \frac{Gb\cos\varphi(1-v\cos^2\theta)}{4\pi(1-v)h_c} \ln\left(\frac{\alpha h}{b}\right) \quad (8\text{-}1)$$

The term on the left hand side of the equation is essentially σ_ε with S an geometrical factor resolving the applied stress onto the glide plane and onto the dislocation Burgers vector. The term on the right of

the equation is the dislocation line tension, with θ the angle between the interfacial normal to the dislocation line and its Burgers vector, φ is the angle between the glide plane and the interfacial normal, h_c the critical epilayer thickness, b the magnitude of the Burgers vector, and α a factor related to the magnitude of the dislocation core energy. The values of S, θ and the exact form of the logarithmic term in the expression for σ_T are configuration dependent, and exact values of α are not known (typical estimates from atomistic calculations vary from 1 in metals, to 4 for screw dislocations and 6 for edge dislocations in covalent semiconductors (Hirth, Lothe, 1968)). Nevertheless, the MB model represents a conceptually correct determination of that epilayer thickness at which it becomes favorable for pre-existing dislocations to move, so as to generate extra interfacial misfit dislocation line length.

For III–V compound semiconductors, experimental determinations of critical thickness have shown reasonable agreement with the MB model (Fritz et al., 1985; Gourley et al., 1988; Temkin et al., 1989). An example for the $In_xGa_{1-x}As/GaAs$ (100) system is shown in Fig. 8-20. However, many sets of measurements of the critical thickness in the Ge_xSi_{1-x}/Si (100) system (Bean et al., 1984; Kohama et al., 1988; Tsao et al., 1987) have shown experimental values of h_c substantially above the predictions of the MB model, (Fig. 8-21). This suggests the existence of a large metastable regime in the covalently bonded Ge_xSi_{1-x}/Si system, which is absent in the mixed ionic and covalent bonding of III–V compounds.

The existence of a metastable region may be understood in terms of the limiting kinetics of strain relaxation due to finite misfit dislocation nucleation, propagation and interaction rates. Although the MB

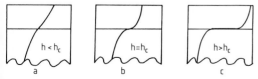

Figure 8-19. Illustration of the Matthews-Blakeslee mechanical equilibrium model for critical thickness.

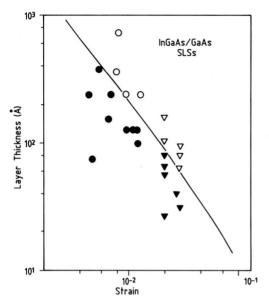

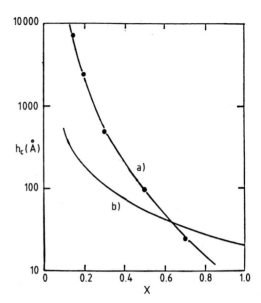

Figure 8-20. Critical thickness data for the In$_x$Ga$_{1-x}$/ GaAs(100) system. Solid points correspond to dislocated structures, open points to undislocated structures. Solid line shows the prediction of the Matthews-Blakeslee model for this system. From Fritz et al. (1985).

Figure 8-21. Experimental data for critical thickness in the Ge$_x$Si$_{1-x}$/Si(100) system at a growth temperature of 550 °C, from Bean et al. (1984). Also shown is the prediction of the Matthews-Blakeslee equilibrium model. (a) Dotted curve represents experimental values at 550 °C. (b) Solid curve represents theoretical values for $a/2 \langle 110 \rangle$ glide dislocations.

model should provide an accurate estimate of the epilayer thickness at which interfacial dislocation motion first takes place, it says nothing about the time-scale on which the transition from strained to relaxed epilayers should occur. In a seminal paper, Dodson and Tsao (DT) (1987) first attempted to model accurately kinetic relaxation rates in strained layers. They assumed that dislocations propagate with a velocity proportional to the epilayer "excess stress," $\sigma_{excess} = \sigma_\varepsilon - \sigma_T$, which is limited by thermal activation over the Peierl's barrier:

$$v = v_0 (\sigma_\varepsilon - \sigma_T) \, e^{-E_v/kT} \qquad (8\text{-}2)$$

where v_0 is a constant and E_v is the glide activation energy. This equation was justified by analogy with detailed work on dislocation motion in bulk semiconductors

(Alexander, Haasen, 1968; Imai, Sumino, 1983; George, Rabbier, 1987). It should be pointed out that work on bulk materials has often revealed a more complex stress-velocity relationship than summarized in Eq. (8-2). A more general relationship would be:

$$v = v_0 (T) [\sigma_\varepsilon - \sigma_T]^{m(T)} \, e^{-E_v(\sigma_\varepsilon - \sigma_T)/kT} \qquad (8\text{-}3)$$

for the dislocation velocity, v, as a function of temperature T. The exponent $m(T)$ is generally of the order 1.0 to 1.6 (George, Rabier, 1987; Alexander, Haasen, 1968) for stresses of the order of tens to hundreds of MPa. There is also experimental evidence that as the applied stress in bulk materials approaches a substantial fraction of a GPa, the activation energy, E_v, becomes stress dependent (George, Rabier, 1987; Kuesters, Alexander, 1983;

Dodson, 1988). Dodson has also predicted this effect theoretically as the excess stress approaches the Peierls' stress of the material.

The somewhat simpler DT approximation, Eq. (8-2), might be expected to be semiquantitatively valid, at least in the regime of moderate lattice mismatch and growth temperature. The main characteristics of elastic strain relaxation in the DT model are then as follows: The original source of misfit dislocations is assumed to be pre-existing defects in the substrate or epilayer, as originally suggested by Matthews and Blakeslee (1975). As the critical layer thickness is exceeded during growth, it becomes energetically favorable for dislocations to extend in the interfacial plane, increasing the misfit dislocation line length. For epilayer thicknesses only slightly greater than h_c, the excess stress and dislocation velocities will be low. The initial stages of strain relaxation will thus be sluggish, and the interfacial dislocation density remains low. As the epilayer thickness increases substantially beyond its critical value, the excess stress and misfit dislocation velocities will dramatically increase. In the original DT formulation, it was also argued that the active dislocation source density will increase due to operation of dislocation multiplication mechanisms. The rate of strain relief thus increases until the introduction of sufficient interfacial dislocation line length has substantially lowered the excess stress, and the residual lattice mismatch asymptomatically approaches zero with increasing epilayer thickness.

The large "metastable regime" of Fig. 8-21 for $x < 0.6$ (above this composition other dislocation microstructures, e.g. edge dislocations, become operative invalidating the theoretical 60° MB curve) thus corresponds to an "incubation period" in which dislocation velocities and densities are relatively low. As first pointed out by Fritz (1987) and experimentally verified by Gourley et al. (1988), the experimental measurement of critical thickness depends crucially upon the detection limit of the technique applied. Typical characterization tools such as Rutherford Backscattering Spectroscopy (RBS), X-ray Diffraction (XRD) and TEM can detect strain relaxation of ~ 1 part in 10^3–10^4. Since the experimental curves of Figs. 8-20 and 8-21 were established using these techniques, they simply delineate a given, measurable amount of strain relaxation. This strain state is achieved far more slowly in Si-based materials, because the glide activation energy is a factor of 2 higher than in e.g. GaAs (George, Rabier, 1987) (≈ 2 eV vs. 1 eV). Also, initial substrate defect densities are a factor of $\approx 10^2$ lower. The much more rapid strain relief in III–V systems causes closer agreement between MB predictions and experimental measurements of h_c.

By incorporating into Eq. (8-2) estimates of the original "source" dislocation density and dislocation multiplication mechanisms, Dodson and Tsao (1987) used their kinetic model to predict critical thickness in the $Ge_x Si_{1-x}/Si$ (100) system. The activation energy for dislocation glide was extrapolated from bulk measurements of 2.2 eV in Si and 1.6 eV in Ge, and the velocity prefactor in Eq. (8-2) and a dislocation multiplication rate were extracted by fitting experimental data. The results of their model, assuming the experimental techniques used to measure critical thickness to be sensitive to 1 part in 10^3 strain relaxation are shown in Fig. 8-22. Impressive agreement between experiment and theory is obtained. The concept of kinetically limited strained layer relaxation thus appears to be firmly established.

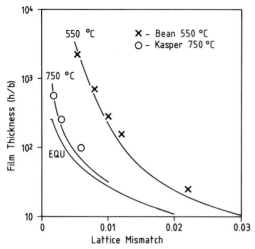

Figure 8-22. Comparison of the Dodson-Tsao model with experimental data for critical thickness in the Ge$_x$Si$_{1-x}$/Si(100) system, from Dodson and Tsao (1987). Experimental data is shown from the work of Bean et al. (1984) (MBE growth temperature of 550 °C) and Kasper et al. (1989) (MBE growth temperature of 750 °C). Predictions of the Dodson-Tsao model are shown for these two growth temperatures. Also shown is a curve labelled EQU which describes the equilibrium critical thickness.

8.4.2 Experimental Results

8.4.2.1 Experimental Techniques

Many experimental techniques have been used to probe the evolution of interfacial structure during strain relaxation in lattice-mismatched epitaxy. Experimental understanding of details of misfit dislocation nucleation, propagation and interaction mechanisms is starting to emerge.

He$^+$ ion channeling in the MeV range is a powerful tool for detecting the degree of strain relaxation, as it is sensitive to the slightly different channeling axes in epilayer and substrate due to the tetragonal distortion of the epitaxial layer. (This distortion causes a given set of lattice planes not parallel to the interfacial plane to be slightly out of coincidence in epilayer and sub-

strate, Fiory et al. (1984).) X-ray diffraction can measure directly the lattice parameter difference between the two materials, as well as the tetragonal distortion of the epilayer. Reflection High Energy Electron Diffraction (RHEED) can be used to probe lattice parameter variations in-situ during MBE growth, although with less sensitivity than the previous two techniques (Whalen, Cohen, 1990). Optical and charge-sensitive imaging techniques such as photoluminescence microscopy (Gourley et al., 1988) and Electron Beam Induced Current (EBIC) (Kohama et al., 1988) can reveal very low densities of electrically or optically active defects, perhaps as few as 1 cm^{-2}. They generally cannot resolve defect separations less than ≈ 1 micron, however, and thus complement higher resolution imaging techniques such as TEM.

Transmission electron microscopy has probably proved to be the most useful technique for studying dislocation microstructures and the more detailed processes of strain relaxation. This is because TEM is an imaging technique, able to resolve individual dislocations with separations as small as tens of Å. The image contrast of each individual dislocation is sensitive to the magnitude and the symmetry of the strain field around it, and by standard techniques it is possible to deduce the Burgers vector of each dislocation (Hirsch et al., 1977). The main disadvantage of this technique is its inability to detect the initial stages of strain relaxation: standard specimen thinning processes used to ensure electron transparency (typical incident electron energies are of the order of hundreds of kilovolts), provide sufficiently thin viewing areas of perhaps tens of square microns. This limits the minimum detectable dislocation density to 10^5 to 10^6 cm^{-2}.

In order to understand the overall process of strain relaxation by the increase of interfacial dislocation line length, it is necessary to understand in detail dislocation nucleation, propagation and interaction processes. The manner in which these individual processes combine to produce the overall relaxation kinetics will be highly dependent on the strain dimensions and geometry of the heterostructure under consideration. Conventional ex-situ characterization techniques such as RBS, XRD and TEM allow the magnitude of strain relaxation as a function of epilayer thickness to be determined. This is achieved by growing a number of samples with varying epilayer thicknesses, in effect producing a series of "snapshots" during the growth process and characterizing each sample individually. This reveals the combined effects of dislocation nucleation, propagation and interaction. TEM imaging also allows the detailed dislocation microstructure in each snapshot to be analyzed, which may in principle allow some information about separate dislocation processes to be inferred. TEM observations by many groups have revealed that the details of the strain relaxation process are very complex. For example, in a wide range of systems, defect microstructures vary from the expected orthogonal array of 60° dislocations at lower lattice mismatches ($\approx 2\%$) to a more disordered array of 90° dislocations at higher mismatch ($> 2\%$) (Chang et al., 1988; Kvam et al., 1988). This transition is generally accompanied by a shorter average dislocation length and denser and more irregular dislocation distributions. Such detailed microstructural changes would not be obvious from diffraction or lower spatial resolution imaging techniques.

8.4.2.2 Strain Relaxation After Growth

An important question in lattice-mismatched systems is their thermal stability. As discussed earlier, finite dislocation nucleation and propagation rates allow lattice-mismatched films to be grown substantially beyond the MB critical thickness, especially in the Ge_xSi_{1-x}/Si system, with only partial (often negligible) relaxation of the lattice strain. Post-growth annealing of these structures may then induce further strain relaxation by generation of new defects and/or extension of existing dislocations. This is illustrated in Fig. 8-23 for annealing of a 350 Å $Ge_{0.25}Si_{0.75}$/Si(100) structure grown at 550 °C.

Post-growth thermal relaxation may substantially limit practical applications of mismatched systems. Layer thicknesses required for devices can exceed the critical thickness, resulting in the formation of substantial dislocation densities during growth. Even if it is possible to reduce the dislocation density to acceptable levels by employing a relatively low growth temperature (cf. the 550 °C temperature of the data in Fig. 8-21), subsequent processing may cause unacceptable increases in defect density. For these reasons it is important to understand relaxation mechanisms during post-growth annealing as well as during epilayer growth. Post-growth relaxation may also provide an experimental opportunity for understanding misfit dislocation processes.

8.4.2.3 In-Situ Transmission Electron Microscopy Experiments

By annealing "metastable" films inside a TEM, it has been possible to directly observe dynamic misfit dislocation processes (Hull, Bean, 1989 a, b, c; Hull et al., 1989 a, b). The basic experimental geometry is shown in Fig. 8-24. By reference to Fig. 8-

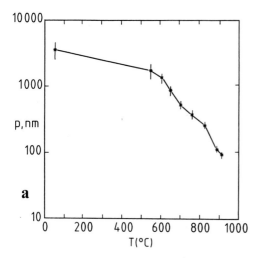

Figure 8-23. (a) Graphical illustration of in-situ TEM relaxation of a 350 Å $Ge_{0.25}Si_{0.75}/Si(100)$ structures as function of annealing temperature (~ 4 min. anneals at each temperature). The quantity p is the average distance between misfit dislocations. (b)–(d) Plan view TEM images of the relaxing structure for annealing temperatures of 550 °C, 750 °C and 900 °C.

21, we employ $Ge_xSi_{1-x}/Si(100)$ structures in the composition range $x = 0.15$ to 0.35, grown by MBE at 550 °C, such that they lie between the experimental and MB curves for h_c. These structures would thus be expected to relax by the generation of misfit dislocations, but have been kinetically prevented from doing so during the growth process. In-situ annealing in the TEM allows the generation of interfacial dislocations to be observed in real-time.

As indicated in Fig. 8-24, these experiments are done in the plan-view geometry (electron beam approximately perpendicular to the interfacial planes), because in the cross-sectional geometry (electron beam parallel to interfacial planes), surface diffusion across exposed interfaces might play a substantial role. In plan-view samples care has to be taken to avoid seriously violating the rigid substrate approximation. This requires that the thickness of Si substrate remaining in the area of observation be substantially larger than the Ge_xSi_{1-x} thickness (Hull et al., 1988). If we assume a maximum sample thickness of 1 to 2 μm for the 200 kV electrons employed here, and aim to keep at least one order of magnitude greater thickness of Si than Ge_xSi_{1-x}, this limits epilayer thicknesses to the order 1000 Å for quantitative work.

To understand in detail the mechanisms by which the interface transforms from coherently strained to dislocated, we need to consider separately how dislocations nucleate, propagate and interact. Each of these processes will now be considered.

8.4.2.4 Dislocation Nucleation

An important question in strained systems is the nature of the sources for the very high defect densities present in relaxed films. Unless a sufficient density of

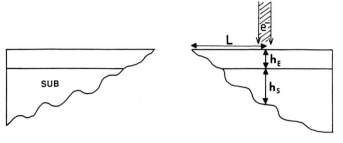

Figure 8-24. Experimental geometry for in-situ TEM relaxation experiments.

nucleation sources is present, the interface will never be able to relax to its equilibrium state; systems do exist in which relaxation is *nucleation limited* in this fashion. As discussed earlier, to relieve a lattice mismatch of 1% across a 10 cm wafer requires $\approx 10^6$ m of dislocation line length. Matthews and Blakeslee originally assumed that these defects originated in the substrate – a more realistic assumption for III–V wafers in the early 70's than for Si (or even GaAs) substrates today. For a substrate defect density of N cm^{-2}, the above analysis implies an average (straight) dislocation length, $\bar{l}$, of $(10^4/N)$ cm. On a 10 cm diameter wafer, the maximum value of $\bar{l}$ is clearly 10 cm, giving $N \approx 10^5$ cm^{-2}. Contemporary Si and GaAs substrates may have $N \approx 10^1$ to 10^2 and 10^3 to 10^4 cm^{-2} substrate threading dislocations respectively. Additional sources of misfit dislocations are thus clearly required.

Three general classes of misfit dislocation sources may operate in strained layer systems: (i) multiplication mechanisms, (ii) heterogeneous sources arising from growth or substrate non-uniformities and (iii) homogeneous sources. Multiplication mechanisms are particularly attractive to invoke as they could drastically reduce the required density of dislocation sources, and are generally invoked in plastic deformation in bulk semiconductors (Alexander, Haasen, 1968; George, Rabier, 1987).

An example of such a multiplication mechanism has been proposed by Hagen and Strunk (1978) based upon intersection of dislocations with equal Burger vectors. Essentially, the dislocation intersections glide from the interface to the surface under the influence of image forces, and on intersecting the free surface form new dislocation segments. This process can act repeatedly to produce bunches of parallel dislocations with equal Burgers vectors. This mechanism was originally reported in Ge/GaAs (Hagen, Strunk, 1978) and also since in InGaAs/GaAs (Chang et al., 1988), and GeSi/Si, (Rajan, Denhoff, 1987) although other groups (including ours) have consistently failed to observe this mechanism in the GeSi/Si system. Indeed, detailed examination of the Hagen-Strunk mechanism suggests that it could only operate within a narrow range of epilayer thickness and strain (Eaglesham et al., 1989) and it is difficult to imagine it as a ubiquitous and efficient source for misfit dislocations in mismatched systems.

Careful etching experiments by Tuppen et al. (1990) for relatively thick layers of dilute $(x < \approx 10\%)$ Ge$_x$Si$_{1-x}$/Si layers have demonstrated the existence of very efficient multiplication mechanisms in the early stages of strain relaxation. Interestingly, the multiplication event was not associated with dislocation intersections, but rather was ascribed to a variant of the Frank-Read mechanism (see Hirth, Lothe, 1968).

By homogeneous nucleation, we mean nucleation at points not associated with any specific site or fault in the lattice. If such sites exist, they would thus exist in very high densities. The energetics of nucleation of complete dislocation loops within a layer or of half loops at the free surface, have been discussed by a number of authors (Hull, Bean, 1989 a; Eaglesham et al., 1989; Matthews et al., 1976; Fitzgerald, 1989; Fitzgerald et al., 1989). Surface nucleation is generally expected to dominate due to the lower line length (and hence self energy) of a half loop at the surface versus a full loop within the epilayer. The total system energy is calculated by balancing the dislocation self energy with the strain energy relieved and the energy of surface steps created or destroyed, as a function of loop radius, R:

$$E_{\text{total}} = E_{\text{loop}} - E_{\text{strain}} \pm E_{\text{step}} =$$
$$= AR \ln(R) - BR^2 \pm CR \quad (8\text{-}4)$$

The dislocation self-energy, E_{loop}, varies as $R \ln R$ and dominates at low R, while the strain energy relieved, E_{strain}, varies as R^2 and thus dominates at high R. As illustrated in Fig. 8-25, the total system energy, E_{total}, generally passes through a maximum δE, at a critical loop radius R_c and then decreases. The quantity δE represents an activation barrier to loop nucleation, and depending upon the elastic constants of a particular system is typically very high (tens to hundreds of eV) for strains $< \approx 1\%$, but is plausibly surmountable at typical crystal growth temperatures for strains $\approx 2 \to 4\%$. Thus, homogeneous surface half loop nucleation can act as a very efficient source in systems with high lattice mismatch, but will not operate at lower mismatches.

In the low mismatch regime, strained systems are thus limited to heterogeneous

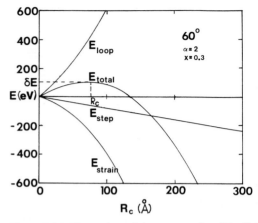

Figure 8-25. Illustration of the energy of a glide dislocation half loop nucleating at a free surface in $Ge_{0.3}Si_{0.7}/Si(100)$ vs. loop radius R. The total loop energy passes through a maximum, δE at a critical loop radius, R_c.

dislocation sources arising at growth non-uniformities (particulates, residual substrate surface contamination, precipitates, stacking faults, etc.) which can locally generate the stress required to nucleate dislocations. The density of such sources is presumably relatively low (say in the range 10^2 to 10^4 cm^{-2}), although each source could produce a number of dislocations (but not an arbitrarily large number, as this would be limited by the local dislocation density that reduces the local stress to zero). The paucity of available nucleation sites at low strain is consistent with the large metastable region of layer thicknesses often observed at low lattice mismatches (e.g., in the $Ge_xSi_{1-x}/Si(100)$ system, Fig. 8-21).

8.4.2.5 Dislocation Propagation

Individual dislocation velocities in relaxing strained systems may be measured from video images recorded during in-situ TEM relaxation experiments (see Sec. 8.4.2). Although dislocation velocities are known to be affected by electron irradiation during TEM examination (particularly of III–V compounds), e.g. Kuester et al. (1985), we do not observe any systematic difference in dislocation velocities in these experiments with and without electron irradiation – presumably because the major driving force causing defect motion is the enormous excess stress, and because we are typically observing at electron irradiation intensities below those at which enhanced glide velocities are observed in Si (Maeda et al., 1991). For the structure imaged in Fig. 8-23, dislocation velocities of the order 1 μm sec^{-1} are measured at 550 °C, in reasonable agreement with extrapolation from lower stress regimes in bulk semiconductors (Alexander, Haasen, 1968; Imai, Sumino, 1983; George, Rabier 1987) assuming a linear dependence of the dislocation velocity upon excess stress. Dislocation velocities at higher annealing temperatures in this structure increase more slowly than predicted by activation energies for dislocation glide in bulk Si and Ge (Hull et al., 1989 a, b), yielding an apparent activation energy of 1.1 ± 0.2 eV in the temperature range 550 to 800 °C. Recent experiments indicate that much of this activation energy lowering is due to dislocation interactions reducing velocities at higher temperatures (see below), where the film is more relaxed and defect densities and hence interaction probabilities are higher. In-situ TEM measurements of dislocation velocities in comparable structures at lower temperatures (≈ 450 to 550 °C) yield higher activation energies in the range 1.6 to 1.8 eV, still significantly lower than linear interpolation of bulk Ge and Si values (≈ 2.0 eV at Ge$_{0.25}$Si$_{0.75}$). It has been suggested (Hull et al., 1989a, b; Hull, Bean, 1989c) that this apparent reduction in the activation energy in very thin uncapped GeSi layers is due either to (i) the nucleation of kinks (by which dislo-

cation glide occurs (Hirth, Lothe, 1968)) at the free surface instead of the bulk, or (ii) a stress dependent lowering of the glide activation energy (Dodson, 1988). Kuesters and Alexander (1983), have also measured an activation energy of only 1.8 eV in bulk Si at 300 MPa.

Dislocation velocities in strained Ge$_x$Si$_{1-x}$/Si(100) heterostructures have also been measured by Tupen and Gibbings (1989, 1990) by post-growth annealing of structures in which dislocation nucleation sites had been induced by scratching. In relatively thick ($> \sim 1$ μm), dilute ($x \leq 0.2$) uncapped Ge$_x$Si$_{1-x}$ layers, dislocation velocites were found to agree well with extrapolations of bulk measurements with a constant prefactor and a linear composition dependence of the activation energy. These trends were repeated in buried Ge$_x$Si$_{1-x}$ layers, except that as the layer thickness became $< \sim 0.5$ μm the dislocation velocity was observed to decrease with the layer thickness, as predicted by the Hirth-Lothe (Hirth and Lothe, 1968) theory of kink nucleation and propagation.

Houghton (1991 a, b) has measured dislocation velocities inferred from maximum dislocation lengths after annealing, deducing an activation energy of 2.25 eV in Ge$_x$Si$_{1-x}$/Si epilayers, quantum wells and superlattices ($0 < x < 0.25$), and a square power law dependence of velocity upon excess stress ($m=2$ in Eq. 8-3). Hull et al. (1991) have reported extensive in-situ TEM measurements of dislocation velocities in capped and uncapped (Si)/Ge$_x$Si$_{1-x}$/Si epilayers, observing an exponential dependence upon composition and stress, and markedly different velocities in equivalent capped vs. uncapped structures.

In compound semiconductor systems such as In$_x$Ga$_{1-x}$As/GaAs, dislocation velocities are expected to be far higher than in Si based systems due to the lower Peierls barrier. This has been borne out by preliminary in-situ TEM experiments (Bonar et al., 1990).

This recent ability to directly measure misfit dislocation velocities is of considerable significance, as it refines experimental determination of interfacial relaxation rates and allows comparison with the Dodson-Tsao kinetic model.

8.4.2.6 Dislocation Interactions

The final process which must be considered is that of dislocation interactions. In general, these will act to limit strain relaxation. The elastic strain field around a dislocation core exerts a stress on another dislocation producing a general interdislocation force per unit length of:

$$F_{12} = k(\theta)\frac{\boldsymbol{b}_1 \cdot \boldsymbol{b}_2}{R} \tag{8-5}$$

where $k(\theta)$ is an angular-dependent constant and $\boldsymbol{b}_1$ and $\boldsymbol{b}_2$ are the individual dislocations' Burgers vectors. For the case of infinitely long parallel screw dislocations, for example, $k(\theta) = 1/2\pi$. The quantity R is the spatial separation of the two interacting segments.

The inter-dislocation force is maximally repulsive for parallel Burgers vectors, maximally attractive for anti-parallel Burgers vectors and zero for orthogonal Burgers vectors. For the case of parallel Burgers vectors the inter-dislocation force may be of sufficient magnitude to cancel the Dodson-Tsao "excess stress", $\sigma_\varepsilon - \sigma_T$, driving dislocation motion. Thus, dislocations may pin each other, as illustrated for intersection of orthogonal dislocations in Fig. 8-26. We note that as F_{12} is inversely proportional to R, dislocation pinning events are more likely in thinner than thicker films, because the magnitude of F_{12} is more likely to counter-balance the

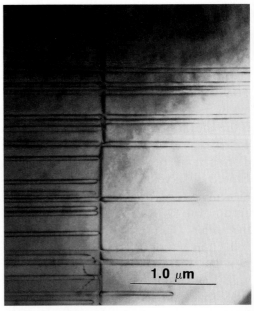

1.0 μm

Figure 8-26. Illustration of the pinning of misfit dislocation (running horizontally in the figure) by a pre-existing orthogonal defect (running vertically in the figure) in a Si/Ge$_x$Si$_{1-x}$/Si(100) heterostructure.

excess stress along the entire threading defect arm. For fuller discussions of these concepts see (Hull and Bean, 1989 b; Freund, 1990).

A consequence of these dislocation pinning mechanisms is that much higher threading defect densities can be expected in higher mismatch systems. This is because for a given materials system, the higher the lattice mismatch the lower the epilayer thickness at which strain relaxation begins. Dislocation pinning events are therefore more likely, and the average length to which a misfit dislocation can grow is substantially reduced. To achieve a given amount of strain relaxation (i.e. to attain a given amount of interfacial dislocation line length for given wafer dimensions) requires a higher number of individual dislocation segments, and threading defect densities increase accordingly (Hull, Bean, 1989 b).

8.4.2.7 Overall Strain Relaxation Rates

We can now summarize our understanding of the strained layer relaxation process for single epilayers by reference to the $Ge_xSi_{1-x}/Si(100)$ system in Fig. 8-21. The figure may be divided into two composition regimes: a low strain regime at $\varepsilon < \sim 0.01$ ($x < \sim 0.25$) and a high strain regime at $\varepsilon > \sim 0.01$ ($x > \sim 0.25$). In the low strain regime, misfit dislocations nucleate at a relatively low density of heterogeneous sites, and the strain relaxation rate at thicknesses greater than the equilibrium critical thickness is relatively slow. A large range of metastable layer thicknesses thus exists, particularly at lower growth temperatures. In the high strain regime, relaxation is not so severely nucleation limited, as a larger density of homogenoeus sites becomes active. Experimental measurements and equilibrium predictions of critical thickness thus agree relatively well in this regime.

Throughout both strain regimes, relaxation rates are limited by finite dislocation velocities, but again in the particular instance of the Ge_xSi_{1-x}/Si system this effect is less marked in the high strain regime due to the inherently lower activation energy for dislocation glide in pure Ge as opposed to pure Si. (In the Ge_xSi_{1-x}/Ge system, glide activation energies should become higher with increasing strain, as the alloy becomes more Si-rich.)

The effect of dislocation interactions, however, is greater in the high strain regime. This is because of the lower epilayer thicknesses at which strain relaxation takes place; as indicated above, interactions are more important in thinner films (of the order hundreds of Å) than thicker ones. This makes thin, highly mismatched films more thermally stable, because pinning events prevent the growth of individual dislocation segments, producing slower strain relaxation rates and higher threading dislocation densities.

All of the above trends apply to compound semiconductor (III–V and II–VI) systems, but the difference between the two strain regimes is less marked. This is because Si and Ge are mechanically harder and have higher bonding energies than most other semiconductors. The activation barrier for surface half-loop nucleation is directly proportional to the epilayer shear modulus and thus is lower in softer materials. The Peierls barrier to dislocation glide is also substantially lower, allowing dislocations to move more rapidly in compound semiconductors. The regime of metastable growth is thus correspondingly smaller.

8.4.2.8 Double Interface Systems

Understanding strain relaxation mechanisms at a single interface is important, because it represents the simplest system in which to study the fundamental misfit dislocation processes. However, strained layer geometries of practical importance for electronic devices generally involve more than one interface. For example the simplest high speed heterojunction transistors, such as modulation doped field effect transistors (MODFETS) or heterojunction bipolar transistors (HBTs) generally utilize the interface sequence A/B/A (with appropriate doping transitions), with B the heteroepitaxial strained layer. Thus misfit dislocation processes in double and multiple interface systems are of great interest.

In the simplest case of an A/B/A structure, it might be expected that growing dislocation loops should simultaneously relax the top B/A and bottom A/B interfaces. If only the bottom interface were relaxed, as in the single interface B/A case, then al-

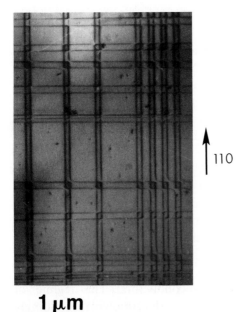

1 µm

Figure 8-27. Dislocation microstructure in a partially relaxed $Si/Ge_{1-x}Si_x/Si(100)$ heterostructure. Pairs of closely spaced dislocation lines correspond to segments of the same dislocation loop simultaneously relaxing top and bottom Ge_xSi_{1-x}/Si interfaces.

though it would be possible to relax the B layer towards its own natural lattice parameter, the top A layer would be forced to adopt this same value and would itself become strained. Thus, as illustrated in Fig. 8-27, both interfaces generally relax simultaneously. This situation may be reasonably approximated by a simple modification to the Matthews-Blakeslee model:

$$\sigma_{excess} \simeq \sigma_\varepsilon - 2\sigma_T \qquad (8\text{-}6)$$

where the factor of two before σ_T arises from the need to generate two misfit dislocation line lengths at top and bottom interfaces. Only in the limiting case where the capping layer becomes very thin (say substantially thinner than the buried B layer) is the situation reached where relaxation of the bottom interface only becomes favorable (Twigg, 1990). This will occur if the strain energy involved in distorting the A

capping layer is less than the self energy of the top dislocation line.

An equivalent strained B layer grown in the buried (A/B/A) configuration as opposed to the free surface (B/A) configuration will generally be more resistant to strain relaxation because:

(i) The Matthews-Blakeslee excess stress is smaller.

(ii) Dislocation nucleation is inhibited, as any generation within the strained layer requires full loop nucleation, as opposed to the half loop nucleation possible at a free surface. Thus the dislocation loop self energy term, E_{loop}, increases by a factor ≈ 2, and the loop nucleation activation barrier increases.

(iii) It has been suggested (Hull et al., 1989 b; Louchet, 1981), that dislocation propagation in very thin uncapped epilayers could be aided by dislocation kink nucleation at the free surface, as previously discussed. Such processes would not be possible in a buried layer, and dislocation velocities might be correspondingly slower.

The extra stability of buried layer structures is a great boon in device processing, where the strained structure may have to be exposed to temperatures substantially above the original growth temperature. Substantial increases in interfacial stability have been reported (Hull, Bean, 1989 b; Noble et al., 1989; Scott et al. 1989).

8.4.2.9 Multiple Interface Systems (Strained Layer Superlattices)

It can easily be shown (Hull et al., 1986; Freund et al., 1989) that the equilibrium limit of a strained layer in a superlattice A/B/A/B/A ... is equivalent to a single layer of the average superlattice composition, weighted over the elastic constants of the individual layers of constant thicknesses d_A and d_B. For growth on an A substrate, therefore, the overall "superlattice critical thickness", H_c may be substantially greater than the single layer critical thickness h_c, particularly if $d_A \gg d_B$, provided each B layer is thinner than the appropriate h_c for B/A growth. If the superlattice thickness exceeds H_c under these conditions, the relaxation of the superlattice occurs primarily at the substrate/superlattice interface, i.e. between the substrate and the first superlattice strained layer (Fig. 8-28). If individual layers exceed the appropriate h_c, then relaxation occurs at individual interfaces within the superlattice.

If the substrate is a different material C with a lattice parameter corresponding to the average (weighted over layer thicknesses and elastic constants) of the superlattice layers A and B, then providing the individual layers of A and B do not exceed the A/C or B/C single layer critical thicknesses, arbitrarily thick superlattices may be grown. This has been demonstrated for example for ultra-thin (≈ 10 Å) GaAs-InAs superlattices on InP substrates (Tamargo et al., 1985). Although the lattice mis- match of GaAs and InAs to InP are – and +3.6% respectively, the average lattice parameter of GaAs and InAs is very close to that of InP and many thin bi-layers (with equal GaAs and InAs layer thicknesses) may be grown without interfacial dislocations appearing.

It should also be stressed that relaxation in strained layer superlattices may be very sluggish. As pointed out by several authors (e.g. Miles et al., 1988), misfit dislocations have to traverse many interfaces in these structures and their net velocities may be very low (see Sec. 8.4.6).

8.4.2.10 Interfaces in Clustered Epitaxy

We have so far implicitly assumed that heteroepitaxial growth is two-dimensional,

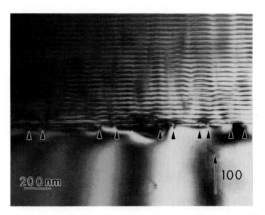

Figure 8-28. Relaxation of the interface between a Si substrate and first layer of a $Ge_x Si_{1-x}/Si$ superlattice. Misfit dislocations are arrowed.

or layer-by-layer, whereas in the general case it is three-dimensional, causing the formation of clusters. Heteroepitaxial systems generally prefer to minimize the interfacial area, as may be simply visualized by analogy to the equilibrium model of the contact angle for a liquid drop on a planar surface:

$$\chi_{sv} = \chi_{es} + \chi_{ev} \cos \theta \qquad (8\text{-}7)$$

where χ_{ev}, χ_{sv} and χ_{es} are the epilayer-surface, substrate-surface and epilayer-substrate interface energies respectively and θ is the contact angle of the epitaxial island. In the general heteroepitaxial case, the deposited overlayer forms clusters as epilayer surface is created in preference to epilayer-substrate interface. Note that elastic strain is also predicted to encourage clustered over layer-by-layer growth (Grabow, Gilmer, 1987; Bruinsma, Zangwill, 1987).

In practical applications, one is clearly attempting to minimize any tendency towards clustering. This may be achieved by:

(i) Reducing chemical dissimilarity across the interface, e.g. for an alloy $A_x B_{1-x}$ grown on B, by reducing x. This reduces the interfacial energy χ_{es}.

(ii) Reducing the growth temperature: this has the effect of reducing the surface mobilities of deposited atoms preventing them from achieving their equilibrium state. Clustered growth is therefore effectively "frozen out".

(iii) Reducing lattice mismatch: clustering may in itself be regarded as a strain-relieving mechanism, as the greater number of atoms in the vicinity of a free surface are more able to relax towards their bulk lattice parameter (Grabow, Gilmer, 1987; Bruinsma, Zangwill, 1987). Surface reconstruction energies should also be considered.

The above trends are illustrated experimentally in Fig. 8-29, which shows experimental regimes of clustered vs. layer-by-layer growth in the $Ge_xSi_{1-x}/Si(100)$ system (from Bean et al., 1984).

A further complication in compound semiconductor systems is possible significant differences in energy between {111} faces terminated on the different atomic species (e.g. for GaAs, the {111}-Ga and {111}-As faces). This can lead, for example, to the formation of anisotropic island dimensions (e.g. Pirouz et al., 1988).

Lattice mismatch relief mechanisms can be significantly different in clustered growth from planar growth. Only a brief description of possible differences will be given here, as practical heteroepitaxy generally suppresses clustering. However, some potentially important heterointerfaces do involve clustering at early stages of the growth process, e.g. the GaAs/Si system. In homoepitaxy, GaAs is generally grown in the temperature range 500 to 600 °C, but growth of GaAs on a Si substrate at this temperature can lead to clustered growth which persists up to very thick (microns) layers. Clustering is thus generally suppressed in the initial stages of heteroepitaxy by using an unusually low GaAs growth temperature, say 350 to 400 °C. This reduces surface diffusion, producing a higher density of smaller nuclei than would be expected at higher temperatures. This is illustrated by the work of Biegelsen et al., 1987, 1988 (Fig. 8-30). By the time a layer thickness of say 1000 Å is reached, the GaAs layer, although not planar, is at least continuous with no bare Si substrate remaining (Harris et al., 1987). Subsequent higher temperature (in the normal homoepitaxial deposition regime) GaAs growth causes the layer to planarize as homoepitaxial growth conditions have now effectively been established.

The GaAs cluster/Si substrate interface is initially coherently strained (Hull, Fischer-Colbrie, 1987). As the cluster dimensions become larger, dislocations appear; this occurs at a critical transition which depends upon all the island dimensions (Luryi, Suhir, 1986). In particular, for island widths not much greater than the equilibrium inter-dislocation spacing required to fully relax strain in an equivalent planar structure, the interface remains coherent for island heights substantially greater than the equivalent planar critical

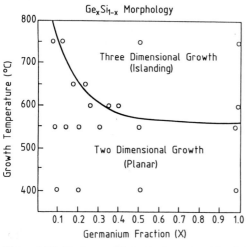

Figure 8-29. Regimes of layer-by-layer vs. islanded growth in $Ge_xSi_{1-x}/Si(100)$, from Bean et al. (1984).

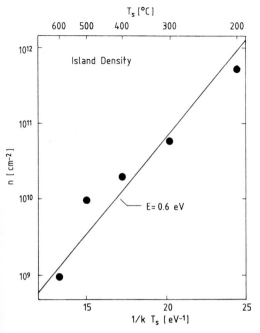

Figure 8-30. Nucleation density vs. inverse substrate temperature for GaAs/Si(100), from Biegelsen et al. (1988).

thickness (Hull, Fischer-Colbrie, 1987; Luryi, Suhir, 1986). If the island width is actually below a critical dimension approximately equal to the equilibrium inter-dislocation spacing in the equivalent relaxed planar structure, the interface can in principle remain coherent for columns of infinite thickness (Luryi, Sukir, 1986). Since for a lattice mismatch of a few percent, this critical width is only of the order of a hundred Å, this concept may be difficult to realize in practice. Attempts at utilizing this approach, however, have been made by growing GaAs on porous Si substrates (e.g. Lin et al., 1987).

When islands exceed the critical size for misfit dislocation introduction, extra degrees of freedom exist for nucleation paths. For example on (100) surfaces, it has been suggested that it is easier to introduce interfacial edge dislocations into islands than into planar structures (Eaglesham

et al., 1988). Another route to accommodating lattice mismatch is by generation of stacking faults (Pirouz et al., 1988). These are equivalent to partial dislocations with Burgers vectors $1/6 \langle 211 \rangle$ or $1/3 \langle 111 \rangle$ for diamond cubic or zincblende structures. Such faults can arise due to incorrect atom placement during growth (Pirouz et al., 1988), and as they have a component of their Burgers vectors parallel to the interface can be effective in relieving strain. Note that such partial dislocations can also arise from dissociation of total or perfect dislocations. Finally, a novel strain relaxation mode for islanded epitaxy of Ge/Si(100) has been demonstrated by Eaglesham and Cerullo (1991) in which the substrate is itself distorted in the vicinity of the interface. This again effectively increases the critical island dimensions for dislocation introduction.

In summary of this section, due to relatively high interfacial energies and the ability to relax strain, three-dimensional cluster growth may be regarded as the general heteroepitaxial growth mode, although techniques for encouraging layer-by-layer growth exist. The presence of clusters modifies the energetic arguments and limits for interfacial stability against dislocation introduction, in general allowing dislocation free islands to be grown to greater "thicknesses" than equivalent planar structures. Kinetic effects are liable to be less dominant in clustered growth, however, as additional mechanisms exist for introduction of interfacial dislocations.

8.4.3 Techniques for Reducing Interfacial and Threading Dislocation Densities

Although some heterojunction devices (e.g. HBTs in GeSi/Si or InGaAs/GaAs) can be fabricated with layer dimensions below the appropriate critical thicknesses, it is often necessary to exceed the critical thickness. The question then becomes not how to avoid misfit and threading dislocations, but how to minimize their impact.

The primary technique for reducing interfacial dislocation densities at given interfaces within a structure is to grow a "sacrificial" template upon the substrate. For example, in the work of Kasper et al. (1989) superlattices consisting of ultra-thin pure Ge and Si layers were grown onto buffer layers of the average superlattice composition, the buffer layers in turn grown onto a Si(100) substrate. The intent was to grow the superlattice onto a *relaxed* buffer layer with the same lattice parameter as the average superlattice lattice parameter. If all misfit dislocations can be confined to the substrate/buffer layer interface, the superlattice can in principle be defect free. (In practice, high densities of threading defects are still observed in the superlattice.) Note that this is essentially the approach adopted in GaAs/Si growth, where the original low temperature "buffer layer" may be regarded as a low structural quality, sacrificial epilayer.

This buffer layer approach can be adapted to most practical device structures, so the question then becomes how many threading dislocations remain, how many can be tolerated, and how their density can be reduced. As a general rule of thumb, electronic devices relying upon majority carrier transport can tolerate perhaps as many as 10^7 to 10^8 dislocations cm^{-2}, whereas for minority carrier and optoelectronic devices dislocation densities below 10^3 to 10^4 cm^{-2} are necessary. At present, threading defect densities in the highest structural quality GaAs/Si(100), a system which has been exhaustively studied for a decade, are of the order 10^6 to 10^7 cm^{-2} for layer thicknesses of the order microns.

The techniques adopted for reducing theading dislocation densities are:

(i) Increasing epilayer thicknesses. Once a structure has reached its equilibrium strain state at the growth temperature, threading dislocations actually increase the system energy due their own self-energy. In the lowest energy state, there would be no threading defects. Thus, as the layer grows thicker, threading dislocations can interact and annihilate each other (Fig. 8-31). The major problem here, as with thermal annealing and strained layer superlattice filtering to be discussed below, is that these annihilation processes are effective only at high defect densities; as the dislocation density decreases, the average distance between threading dislocations increases and the probability of dislocations meeting and annihilating eventually becomes vanishingly small.

(ii) Thermal annealing. Threading defect mobility, and hence the probability of their meeting and annihilating, can be increased by thermal annealing during or after growth. These techniques lead to significant defect reduction in the GaAs/Si system (e.g. Chand et al., 1986), but the same caveat regarding lower defect interaction probabilities at lower densities applies.

(iii) Strained Layer Superlattice Filtering. Threading defect interaction probabil-

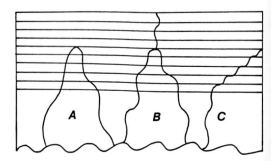

Figure 8-31. Schematic illustration of dislocation interaction/annihilation mechanisms during growth of a thick epilayer or superlattice structure.

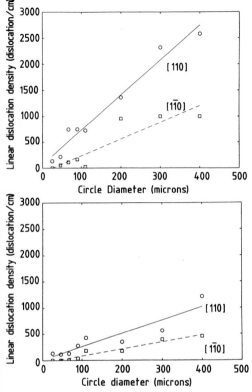

Figure 8-32. Graphs of measured average linear misfit dislocation density vs. mesa size in the $In_xGa_{1-x}/$ GaAs(100) ($x = 0.05$) system, with (a) 1.5×10^5 cm^{-2} and (b) 10^4 cm^{-2} pre-existing dislocations in the substrate. From Fitzgerald et al. (1989).

ities may be increased by providing a specific vector for defect motion, and by increasing the length of each individual defect. This has been achieved, as first suggested by Matthews and Blakeslee (1974, 1976), Matthews (1975), by incorporating strained layer superlattices (SLSs) into the epitaxial system. The principle is that at the SLS interfaces, threading dislocations become interfacial misfit dislocations and propagate along individual interfaces for relatively large distances. As they do so, they may meet other defects and annihilate, or even propagate all the way to the edge of the wafer. The SLS has to be incorporated with sufficient strain to deflect

existing threading dislocations, but not so much strain as to generate substantial densities of new defects. This can in principle be achieved if the threading defects can act as the necessary misfit dislocation "sources" and several groups (Olsen et al., 1975; Lilliental-Weber et al., 1987; Fischer et al., 1986; Dupuis et al., 1986) have claimed substantial success at reducing threading defect densities by this technique. Nevertheless, the same caveat of reduced interaction probabilities at reduced densities still applies, and as analyzed by Hull et al. (1989c) for the $Ge_xSi_{1-x}/$ Si(100) system, final attainable defect densities depend upon layer growth times and temperatures, and misfit dislocation velocities. For $Ge_xSi_{1-x}/Si(100)$ SLS growth at 550 °C, final defect densities of the order 10^7 cm^{-2} were predicted. In compound semiconductor systems where dislocation velocities are higher, the minimum attainable density should be somewhat lower.

All three of the above techniques can be, and often are, used in conjunction. Threading defect densities can thereby be impressively reduced, but a "floor" of $\approx 10^6$ cm^{-2} is generally encountered.

A technique which can reduce densities of both interfacial and threading dislocation densities is "finite area" or "patterned" epitaxy, in which growth is on substrate mesas, or where the epilayer is patterned after growth. The small mesa approach has been explored in some detail by Fitzgerald et al. (1988, 1989) and Fitzgerald (1989) in both the InGaAs/GaAs and GeSi/Si systems. As illustrated in Fig. 8-32, patterned growth at relatively low strains in the $In_xGa_{1-x}As/GaAs(100)$ system causes significant reductions in both threading and misfit dislocation densities with decreasing mesa size. The interfacial misfit defect density is reduced due to the finite density of heterogeneous dislocation sources in the "low strain" regime. For a heterogeneous source density N_{he} cm^{-2}, mesa areas $< \sim 1/N_{he}$ cm^{-2} will on average contain no dislocation sources, therefore no misfit dislocations are expected. Threading dislocation densities may be reduced, or even eliminated, for mesa areas substantially above this, because the dislocations which do form may have sufficient velocities and time at temperature during growth to reach the mesa edge. This approach, however, is primarily beneficial in the "low" strain regime; in the "high strain" regime, homogeneous dislocation nucleation at the mesa surfaces (or edges) is likely to become significant.

Several groups (e.g. Lee et al., 1988; Matyi et al., 1988) have also reported patterned epitaxy in GaAs/Si by defining small mesas either before or after growth. Post growth thermal annealing, especially if SLSs are incorporated into the structure, may allow existing threading dislocations to propagate to the mesa edge and annihilate. The possibility of growth onto mesas that are so small that it is not energetically favorable for any dislocations to form (the Luryi-Suhir model (1986)) was discussed earlier.

8.4.4 Summary

We have attempted to summarize those structural properties which are peculiar to lattice-mismatched interfaces. Satisfactory equilibrium models for the critical thickness of a strained epilayer at which it becomes favourable to form interfacial misfit dislocations were first developed by Matthews-Blakeslee and van der Merwe over a decade ago. Experimental studies have supported these equilibrium models in the limit where experimental techniques are able to detect sufficiently small dislocation densities. Kinetic effects greatly limit interfacial relaxation rates, particularly at lower strains where only few dislocation sources exist, and in materials such as Si and Ge which have high activation barriers for dislocation glide. The Dodson-Tsao model provides a powerful framework for understanding and modelling these kinetically-limited processes. This model assumes dislocation velocities, and hence interfacial relaxation rates, to be proportional to the Matthews-Blakeslee excess stress and to be controlled by thermal activation over the Peierl's barrier. In-situ and ex-situ experimental studies have begun to probe these kinetic regimes and extract the parameters necessary for accurate application of kinetic modelling. Mechanisms for strain relief in clustered epitaxy have also been studied and modelled. Several techniques for controlling or even removing deleterious threading and interfacial misfit dislocations for strained layer thicknesses above the critical thickness have been developed, but threading dislocation densities below the 10^6 cm^{-2} range have not generally been achieved for unpatterned epitaxy of highly mismatched systems. Patterned epitaxy appears to offer the best prospect for eliminating or reducing defects densities in mesas with dimensions in the microns to tens of microns range.

8.5 Interfaces Between Systems Differing in Composition and Structure

8.5.1 Introduction

Up to this point, we have limited our discussion to epitaxial semiconductor systems involving materials with identical structures. For certain applications, it is useful to fabricate epitaxial structures involving metals and insulators on a semiconductor substrate. This invariably involves the growth of a material with a crystal structure different from that of the substrate. Such growth faces lattice mismatch and interface roughness problems as for the systems discussed previously, but is considerably more difficult because of the following additional issues. (1) Differences in the structures of the substrate and overlayer. (2) Significant interfacial free energy. This may have a strong influence on the morphology of the epitaxial layer. (3) Heterogeneous nucleation, leading to multiple orientations within the epitaxial layer. (4) The change of crystal symmetry across the heterointerface, requiring the presence of specific interfacial defects. In this section, we will address some of these issues, using epitaxial metal-semiconductor (M-S) structures as an example.

M-S structures are an essential part of virtually all electronic and optoelectronic devices. Recently, much progress has been made in our understanding of the chemistry and metallurgy of these interfaces. However, detailed physical and electronic properties at these interfaces are still poorly understood. For instance, the formation mechanism of the Schottky barrier (SB) at a M-S interface, despite much investigation and debate, is still not well understood. This is because the atomic structure at a usual M-S interface is too complicated, often due to polycrystallinity, to be determined experimentally or used in theoretical calculation of the M-S interface electronic properties. SB theories are therefore based on bulk physical properties, or are essentially phenomenological in nature. Neither approach has been particularly successful. The recent fabrication of *epitaxial* M-S interfaces represents an unprecedented opportunity to understand, experimentally and theoretically, the formation of SB from first principles. Much attention has been placed on a few single

crystal, epitaxial M-S interfaces, whose interfacial atomic structure has been determined experimentally. Sophisticated calculations of the electronic properties based on the atomic structures of these interfaces have been carried out and have shown good agreement with experimental Schottky barrier height (SBH) results. Epitaxial M-S structures are quickly becoming the arena for the development of a correct SB theory. In addition to advantages from a fundamental point of view, epitaxial M-S structures have numerous advantages over non-epitaxial structures in terms of performance in microelectronic devices, such as the possibility of monolithic vertical integration, higher stability, and the possibility of hot-electron high-speed devices. In this section, we examine our understanding of general M-S interfaces resulting from the investigation of epitaxial interfaces. The focus will be placed on the interfaces. Other interesting findings, notably the electronic, optical, and magnetic properties of thin single-crystal metal layers will not be treated.

The majority of epitaxial M-S structures are fabricated by the growth of a metallic thin film on a semiconductor substrate. As in all heteroepitaxial growth, close matching in the crystal structures and lattice parameters between the metal and the semiconductor is crucial to high quality epitaxial growth. Interfacial stability is also an important consideration in choosing suitable systems for epitaxial growth. Table 8-1 provides some background information on the most common epitaxial M-S systems. Because of the favorable conditions for epitaxy within these M-S systems, it is relatively easy to fabricate epitaxial metallic films. However, the quality of the resultant interfaces depends critically on the precise way these films are prepared. By quality we mean the unifor-

Table 8-1. Crystal structures and lattice constants.

Material	Crystal structure	Lattice constant Å
Al	f.c.c.	4.05
Ag	f.c.c.	4.09
α-Fe	b.c.c.	5.74
CoGa	CsCl	2.88
NiGa	CsCl	2.89
CoAl	CsCl	2.86
NiAl	CsCl	2.89
LuAs	NaCl	5.68
ScAs	NaCl	5.46
ErAs	NaCl	5.74
$NiSi_2$	CaF_2	5.41
$CoSi_2$	CaF_2	5.37
Si	diamond	5.43
GaAs	zincblende	5.65
InP	zincblende	5.89

mity, abruptness, structural order, and chemical stability of the interface. Obviously, the densities of structural defects and chemical impurities should be as low as possible. In order to study the properties of a perfect M-S system, the first task is the creation of high quality structures.

8.5.2 Fabrication of Epitaxial Metal-Semiconductor Interfaces

The highest quality epitaxial M-S interfaces have thus far been fabricated under very clean conditions, by careful preparation of atomically clean semiconductor surfaces and the use of molecular beam epitaxy (MBE) in an ultrahigh vacuum (UHV) environment. With few exceptions, the semiconductor surface becomes, or is in the close proximity of, the eventual M-S interface. As there is usually a change of symmetry across the M-S interface, partial dislocations are sometimes topologically required at steps at the interface. A well-known example of such symmetry-prescribed defects are the phase domain boundaries in GaAs layers grown on Si(100) as a result of Si surface steps. Other imperfections on the semiconductor surface may also lead to interfacial defects. Therefore, extreme care should be taken in the preparation of clean surfaces for epitaxial growth. This includes the use of precisely oriented crystals, or intentionally offcut crystals, the removal of surface damage and contamination, careful handling, the avoidance of particulates, and common UHV practices such as degassing, stripping of the native oxides and the growth of a buffer layer.

Unlike heteroepitaxy involving materials with the same crystal structure, growth of a metal layer on a differently structured semiconductor requires the nucleation of the metallic crystals and the creation of heterointerfaces with high interface energies. Precise control of the nucleation conditions at the initial stages of metal deposition is crucial in determining epitaxial orientation and layer morphology. The lattice mismatch between the semiconductor and the overgrown metal leads to additional energy due to misfit strain and eventual relaxation through the generation of misfit dislocations. These problems with heterogeneous nucleation and misfit strain often lead to layer nonuniformity and a high density of defects. To overcome these problems, two-step growth schemes are usually preferred over the growth of an entire metallic layer under a single set of growth condition. The idea is to deposit the initial metal, usually < 50 Å thick, under conditions particularly suited for uniform nucleation of the metallic layer. This first step, which usually occurs at a lower temperature than the next growth step, results in an abrupt interface and a uniform epitaxial layer. This layer is then used as a template for subsequent growth under optimum conditions for high quality metallic homoepitaxy. Examples of such two-step growths include epitaxial silicides, GaAs on Si and SiC on Si. The properties of the thin templates have in themselves been the subject of many interface studies.

8.5.2.1 Epitaxial Silicides

Silicides are metal-silicon compounds with specific compositions and crystal structures. Most of these are metallic, but a few display semiconducting or semi-metallic behavior. Preferred growth along certain orientations has been observed for many silicide-Si systems and many more are being discovered. Since the Si substrate is an infinite source of Si for the formation of silicides, stoichiometric silicides can be grown by providing only metal, or both metal and silicon. Deposition of an elemental metal on Si and annealing at suitable temperatures, known as solid-phase epitaxy (SPE), (Tu, Mayer, 1978; Ishiwara, 1980) can lead to the growth of a desired epitaxial silicide phase (Saitoh et al., 1981; Chiu et al., 1981; Tung et al., 1982a). For thick layers (>100 Å) of deposited metal, this reaction usually starts with the growth of metal-rich silicides and progressively, different silicides may be grown, ending with silicon-rich silicides (Tung et al., 1982b). The reader is referred to existing reviews for a discussion of SPE grown epitaxial silicide structures (Tung et al., 1982b; Tung, 1988). Co-deposition of metal and silicon, in a stoichiometric ratio on heated silicon substrate, generally referred to as MBE, may also be employed to grow epitaxial silicides (Bean, Poate, 1980). Thick (>500 Å) silicide layers grown by SPE or MBE alone are frequently non-uniform and contain multiple orientations. As mentioned above, most of these problems may be solved by the proper application of the template technique (Tung et al.,

1983a, b). The problem of epitaxial silicide growth is therefore often reduced to the question of growing a high quality template layer. Depending on the particular circumstances, the growth of an optimum, thin, silicide template layer may require depositions of only metal, layered metal and silicon, co-deposited metal silicide of a particular stoichiometry (which may be different from the stoichiometry of the intended silicide), or a combination of the above. With thinner layers of deposited metal, the silicide reaction may not follow the sequence observed in thick films.

The two silicides $NiSi_2$ and $CoSi_2$, with their fluorite lattice structure and good lattice matching with silicon, are the most-studied among silicides. As a result of these favorable conditions for epitaxy, thin films of these two silicides have the highest degree of crystalline perfection amongst epitaxial silicides. On Si(111), two epitaxial orientations are possible: The type A silicide has the same orientation as the silicon substrate; The type B silicide shares the surface normal $\langle 111 \rangle$ axis with Si, but is rotated 180° about this axis with respect to the Si (Tung et al., 1982b). High quality single crystals of $NiSi_2$ may be grown on Si(111), with either type A or type B orientation, by a proper choice of template growth condition (Tung et al., 1983a; van Loenen et al., 1985a; Hunt et al., 1986; Känel et al., 1987). As shown in Fig. 8-33, the epitaxial orientation of thin $NiSi_2$ layers depends on the amount of deposited nickel. When $\approx 16–20$ Å nickel is deposited at room temperature, subsequent annealing leads to the growth of type A $NiSi$ (Tung et al., 1983a). If a small amount (<5 Å) of nickel is deposited on Si(111) at room temperature, annealing leads to the growth of $NiSi_2$ with majority type B orientation. In reality, type B $NiSi_2$ layers grown by this technique are not very uni-

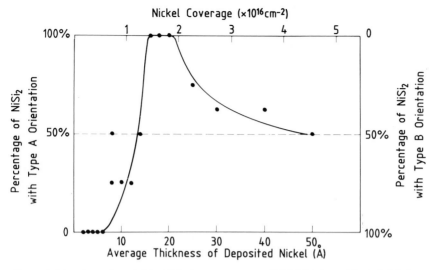

Figure 8-33. Orientation of thin NiSi$_2$ layers grown on Si(111) by deposition of nickel at room temperature and annealing to $\approx 450\,^\circ$C.

form in thickness and often contain a small fraction of type A grains. So in practice, growth techniques involving deposition of Si or co-deposition of NiSi$_2$ are more often employed for the growth of uniform single crystal type B NiSi$_2$, with very reliable outcome (Tung, 1987; Tung and Schrey, 1989a). Moreover, it is now clear that nucleation and growth of type B NiSi$_2$ can occur at room temperature. For instance, Fig. 8-34d shows a plan view transmission electron microscope (TEM) image of a single crystal NiSi$_2$ layer grown at room temperature on Si(111) by deposition of ≈ 2 Å of Ni and the co-deposition of NiSi$_2$ (Tung and Schrey, 1989a). One major difference between the epitaxial structures of a type B NiSi$_2$ and a type A NiSi$_2$ interface is the characteristics of interfacial defects. At a type B interface, symmetry requires the existence of a partial dislocation (or an antiphase domain boundary, which is rarely seen) at a step of single (3.13 Å) or double atomic height. For instance, the dislocations seen in Fig. 8-34 are due to steps on the original Si(111) surface. These dis-

locations may have a Burgers vector of $1/6\langle 11\bar{2}\rangle$ or $1/3\langle 111\rangle$. So the dislocation density at a type B NiSi$_2$/Si interface may be strongly influenced by the (accidental) misorientation of the original Si(111) wafer. For this reason, even ultrathin (<20 Å thick, supposedly below the critical thickness for misfit dislocation generation) type B NiSi$_2$ layers contain a significant density of dislocations. Of course, when thick layers of type B NiSi$_2$ are grown, dislocations are also generated to relieve misfit stress. In this latter case, partial $1/6\langle 11\bar{2}\rangle$ type dislocations are most commonly observed. At a type A NiSi$_2$ interface, no partial dislocations are allowed and no dislocations are required at steps of any height. So the generation of dislocations at a type A NiSi$_2$ interface is driven by stress due to lattice mismatch. Below a certain critical thickness, it is possible to grow type A NiSi$_2$ layers essentially free of any dislocations (this is not possible for type B NiSi$_2$). Misfit dislocations at a type A NiSi$_2$ interface are most frequently $1/2\langle 110\rangle$ in nature.

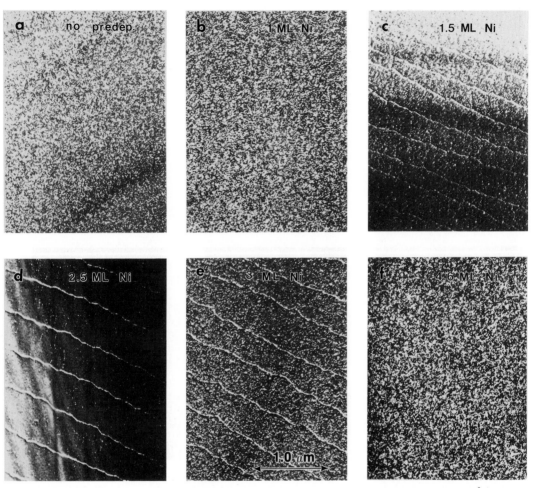

Figure 8-34. Plan-view dark-field TEM images of six NiSi$_2$ layers grown by co-deposition of ≈ 72 Å NiSi$_2$ at room temperature. Prior to co-deposition, 1, 1.5, 2.5, 3, and 4.6 ML Ni was deposited onto clean Si(111) for (b), (c), (d), (e) and (f), respectively. No pre-deposition was used in the growth of layer (a).

Type B is the dominant epitaxial orientation observed for CoSi$_2$ layers grown on Si(111) (Tung et al., 1982a; Kao et al., 1985a; Arnaud D'Avitaya et al., 1985; Sheng et al., 1987; Phillips et al., 1987). Epitaxial type B CoSi$_2$ may also nucleate on Si(111) at as low a temperature as $\approx 20\,°$C, by deposition of <4 ML Co (Pirri et al., 1984; Comin et al., 1983). If more than 4 ML Co is deposited, more cobalt-rich compounds are formed on the surface at room temperature and annealing even-

tually leads to the growth of CoSi$_2$ by a lateral progression of the reaction front (Gibson et al., 1987). There are two common problems with the CoSi$_2$ layers grown by this thin film SPE process. One is the non-uniformity of the silicide layers, most notably the presence of pinholes (Tung et al., 1986a). The other is a high density of dislocations. By depositing Si at room temperature, following the deposition of cobalt, or by co-deposition of CoSi$_x$ onto Si(111), it has been shown that much more

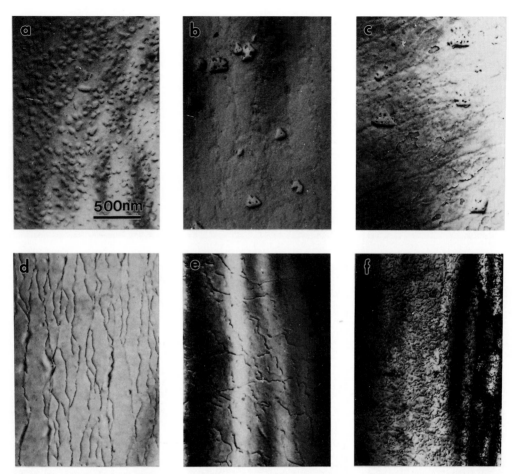

Figure 8-35. Plan-view bright-field TEM images, taken near the (111) zone axis, of assorted thin $CoSi_2$ layers grown by deposition of cobalt (and silicon) at room temperature and annealing to 500 °C. All layers have an average thickness of 14 Å. The depositions used were (a) 4 Å Co only, (b) 4 Å Co and 4 Å Si, (c) 2 Å Co and a coevaporated $CoSi_{1.8}$ layer ($t_{Co} = 2$ Å, is the equivalent thickness of cobalt contained in a coevaporated layer of $CoSi_x$). (d) a coevaporated $CoSi_{1.2}$ layer ($t_{Co} = 4$ Å), (e) a coevaporated $CoSi_{1.2}$ layer ($t_{Co} = 2$ Å) and a coevaporated $CoSi_{1.8}$ layer ($t_{Co} = 2$ Å), (f) a coevaporated $CoSi_2$ layer ($t_{Co} = 4$ Å). The layer shown in (a) has a high density of pinholes. That shown in (f) has almost no pinholes, but contains a high density of dislocations.

uniform $CoSi_2$ layers may be grown after annealing (Tung et al., 1986a; Hunt et al., 1987; Henz et al., 1987; Lin et al., 1988; Fischer et al., 1988a). However, the dislocation problem is not solved by these techniques which deposit Si for the $CoSi_2$ reaction. In fact, when an excess of Si is supplied, the dislocation density is considerably increased (Tung, Schrey, 1988) (Fig. 8-35). An important driving force for pin-hole formation in epitaxial $CoSi_2$ films was identified as a change in the energetics associated with different surface structures of $CoSi_2(111)$ (Tung, Batstone, 1988a; Tung, Schrey, 1989b). As a result of this discovery, pinholes may be totally suppressed in a fashion completely independent of the deposition schedule for the thickness of the entire $CoSi_2$ film. This allows the growth of pinhole-free films with-

out an increase in the density of dislocations. Recently, it was demonstrated that
thick layers of single crystal CoSi$_2$ can be
grown at room temperature by precisely
controlled deposition and co-deposition
schedules (Tung, Schrey, 1989 b). Dislocations found at these silicide interfaces are
only those required by symmetry, due to
the presence of steps on the original
Si(111) surface. By using the one-to-one
correspondence of dislocations seen at
CoSi$_2$ (or NiSi$_2$) layers grown at room
temperature with surface steps, surface
topographical changes due to various
processes may be studied (Tung, Schrey,
1989 c). In Fig. 8-36, a change of Si(111)
surface topography due to Si homoepitaxial growth is illustrated. The improvement
in the quality of epitaxial type B ÇoSi
layers, gauged by the control of pinholes
and dislocations, leads to an increased
level of strain that thin layers can accommodate without nucleation of dislocations,
in addition to those required by symmetry.
Novel features were discovered in the
TEM images of these layers, as shown in

Fig. 8-37, indicating lateral distortion of
some parts of the CoSi$_2$ crystals (Tung,
Schrey, 1988). These features are shown to
be associated with a novel, stress-driven,
structural phase transformation at the interface, involving a few ML's of ÇoSi
(Eaglesham et al., 1990). By a slight change
in the procedures of silicide growth, it is
possible to prepare CoSi$_2$ layers which undergo, or do not undergo such a phase
transition upon cooling to room temperature. This interfacial phase transformation
can be removed by ion beam bombardment. The exact nature of this transformation is still not fully understood.

There is no evidence for the formation
of epitaxial NiSi$_2$ at room temperature on
clean Si(100). In order to grow NiSi$_2$,
some form of annealing has to take place.
The growth issues concerning NiSi$_2$ on
Si(100) are illustrated by the example
shown in Fig. 8-38. First, single orientation NiSi$_2$(100)[011]/Si(100)[011] can be
easily produced. There may be problems
with layer non-uniformity in the following
two forms: (1) incomplete coverage, rang-

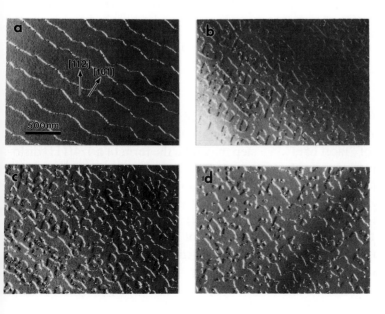

Figure 8-36. Dark-field
TEM images of CoSi$_2$ layers grown on Si surfaces
after Si MBE growth.
(a) substrate surface, and
surfaces after the growth
of (b) 1 ML, (c) 2 ML, and
(d) 3 ML, respectively, at
650 °C, at a rate of 0.4 Å/s.

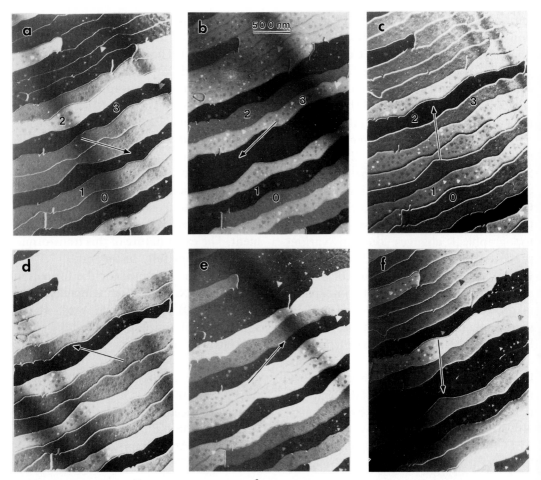

Figure 8-37. Dark-field $\langle 2\bar{2}0 \rangle$ TEM images of a 25 Å thick CoSi$_2$ layer on Si(111) which had been annealed to 600 °C. This layer was originally grown at room temperature by cobalt pre-deposition and co-deposition of CoSi$_2$. The approximate direction of the g vector is indicated on each micrograph. All pictures were taken in the weak-beam, with $\langle 6\bar{6}0 \rangle$ close to Bragg condition. The reason for these domain-like contrasts is an interfacial phase transformation.

ing from the existence of small pinholes to the more severe case of isolated NiSi$_2$ islands, and (2) the existence of "facet bars". (A facet bar is a NiSi$_2$ protrusion at the NiSi$_2$/Si interface, bounded by two inclined $\langle 111 \rangle$ facets, and long in the direction, either [011] or [01$\bar{1}$], parallel to the facets.) In addition, there are misfit dislocations and anti-phase domains. The latter is a result of interface steps of odd number of atomic planes, similar to those seen in

GaAs on Si(100). In principle, for NiSi$_2$ on Si(100), the interface defect at a single step may take the form of a 1/4[111] dislocation, which is frequently observed, an antiphase domain boundary which has not been observed experimentally, or a "coreless" defect (Batstone et al., 1988). The existence of "coreless defects", which is largely based on plan-view TEM observations, has not been confirmed in studies of cross-sectional samples. Various deposi-

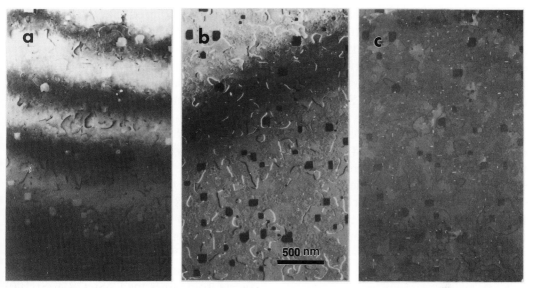

Figure 8-38. Plan-view TEM images of a 50 Å thick NiSi$_2$ layers on Si(100). (a) $\langle 022 \rangle$ bright field, (b) $\langle 022 \rangle$ weak-beam, and (c) $\langle 020 \rangle$ dark field. Rectangular shaped areas, dark in (b) and (c), are pinholes. "Facet bars" are most clearly revealed in (c), as short straight streaks which are lighter than background.

tion schedules have been tried to determine the optimum conditions for the formation of the most perfect NiSi$_2$ template layers on Si(100), i.e. ones with the smallest densities of facet bars and phase domain boundaries. One may deposit pure Ni, layered Ni and Si, co-deposited NiSi$_x$, or a combination of these layers. A proper choice of the thickness of deposited Ni may considerably reduce the faceting problem in NiSi$_2$ (Tung et al., 1983a). However, for a complete elimination of the facet bars, deposition of Si is necessary (Tung, Schrey, 1991). Shown in Fig. 8-39 are TEM images of a uniform NiSi$_2$ layer grown on Si(100) by the use of a co-de-

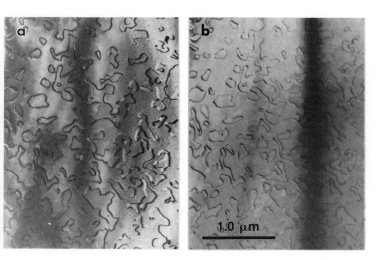

Figure 8-39. A uniform, 110 Å thick, NiSi$_2$ layer on Si(100). (a) $\langle 022 \rangle$ weak-beam and (b) $\langle 020 \rangle$ dark field.

posited template layer. The main issue for the growth of NiSi$_2$ on Si(110) is also interface faceting (Tung et al., 1985). Essentially flat NiSi$_2$/Si(110) interface may be grown by low temperature ($< 500\,°C$) annealing of deposited Ni and Si. However, upon annealing to high temperature ($> 550\,°C$), the interface breaks up into inclined $\langle 111 \rangle$ facets. Indeed, interfacial faceting of this system is so complete that, in most well-annealed samples, the entire NiSi$_2$/Si is made up of inclined $\langle 111 \rangle$ facets. (Fig. 8-40) $1/4\langle 111 \rangle$ type dislocations are required at the boundaries between two facets (Tung et al., 1985). It

should be recognized that the reaction kinetics often play a role in determining silicide layer morphology. But the fact that not even a small portion of flat [110] NiSi$_2$/Si interface has been observed gives a clear indication that this interface may well be unstable.

Unlike NiSi$_2$ epitaxy on Si(100), the issue for epitaxial CoSi$_2$ on Si(100) is not faceting, but multicrystallinity. There are two major competing orientations for CoSi$_2$ epitaxy on Si(100): CoSi$_2$(100)//Si(100) and CoSi$_2$(110)//Si(100) (Tung et al., 1988; van Ommen et al., 1988a; Yalisove et al., 1989a). Probably due to nucleation of type B CoSi$_2$ on inclined $\langle 111 \rangle$ facets, sometimes the CoSi$_2$(22$\overline{1}$)//Si(100) relationship is also observed for some grains (Jimenez et al., 1990). There are two variants of the [110] epitaxy, related by a rotation of 90°. TEM image of a mixed orientation CoSi$_2$ layer is shown in Fig. 8-41. Most of the dislocations seen in (100)-oriented CoSi$_2$ areas are $1/4\langle 111 \rangle$ in character. The previous discussion regarding steps of odd atomic height at a NiSi$_2$/Si(100) interface is also applicable here. Epitaxial CoSi$_2$ films on Si(100) may be grown with very large phase domains which seem to mimic the terrace structure of the original Si surface (Tung, Schrey, 1991). This observation is suggestive of some form of "lateral" CoSi$_2$ reaction, originating from isolated nucleation sites and sweeping across the surface. Various deposition schedules have been studied in order to reduce the fraction of [110] grains in epitaxial CoSi$_2$ films. It is discovered that wafer roughness and surface cleanliness also have an effect on the observed areal fraction occupied by [110] grains (Yalisove et al., 1989a). The reason for the nucleation of [110] oriented grains is not clear at time of this writing, but it is very likely that the heterogeneous nature of the

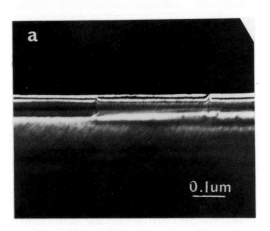

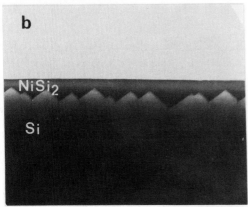

Figure 8-40. TEM image of a single crystal NiSi$_2$ layer on Si(110) grown by the use of a template, (a) viewed in the [001] direction and (b) a [1$\overline{1}$0] cross-sectional view.

0.5 μm

Figure 8-41. Plan-view, ⟨022⟩ weak-beam, TEM images of two CoSi₂ layers on Si(100). The majority of these films has [100] orientation. Small dark areas have [110] orientation.

Co/Si reaction is responsible. It is possible to eliminate completely [110] grains using deposited cobalt and silicon, but reproducibility is poor from run to run. Best results are obtained from precisely oriented Si(100) wafers which have been carefully prepared by Si buffer layer growth.

CoSi₂ grows with the regular epitaxial orientation on Si(110) (Yalisove et al., 1989b). Again, in sharp contrast to NiSi₂ epitaxy on this surface, faceting is not a problem for CoSi₂. Uniform layers of CoSi₂ may be grown with cobalt deposition and CoSi₂ co-deposition at room temperature, followed by annealing. A TEM image of a thin CoSi₂ layer grown on Si(110) is shown in Fig. 8-42. There is no phase difference across a step at an otherwise planar (110) silicide/Si interface. However, most defects seen at the CoSi₂/Si(110) interfaces, e.g. Fig. 8-42, are phase domain boundaries (Eaglesham et al., 1991). The different phases arise from inequivalent lateral shifts at the silicide/Si interface, rather than from interfacial steps (Eaglesham et al., 1991). The dechanneling peak from thin CoSi₂ layers in ion scattering experiments, arising only from Si atoms at the silicide interface, is clear evidence for these lateral relaxations. Typical ion channeling and random spectra are shown in Fig. 8-43b, which are to be compared with those shown in Fig. 8-43a from a CoSi₂/Si(100) layer which does not contain any rigid lateral shifts between the two lattices. As the CoSi₂ film thickness increases, additional dislocations are generated at the interface to relieve misfit stress. The kinetics of strain relief along the two orthogonal directions [1̄10] and [001] are markedly different (Yalisove et al., 1989b). For a fixed growth and annealing sched-

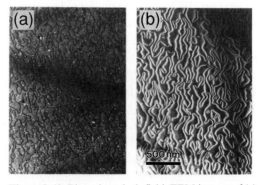

(a)   (b)

500 nm

Figure 8-42. Plan-view, dark-field, TEM images of (a) a 18 Å thick annealed CoSi₂ template layer on Si(110), and (b) a 110 Å thick CoSi₂ layer grown at room temperature on (a).

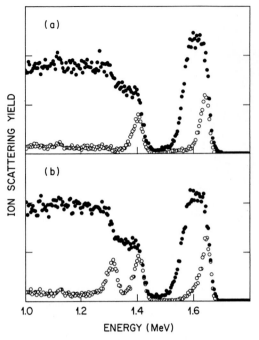

Figure 8-43. Channeling (open circles) and random (closed circles) ion scattering spectra of CoSi$_2$ layers grown at room temperature. (a) a 100 Å thick layer on Si(100) and (b) a 110 Å thick layer on Si(110).

ule, the generation of dislocations occurs as if different critical thicknesses were operative for these two directions; strain along [001] is relieved at a smaller film thickness than that along [1$\bar{1}$0] (Yalisove et al., 1989b). Epitaxial CoSi$_2$ has been grown on the less common Si(311) surface where two epitaxial orientations are observed. These are the regular CoSi$_2$(311)//Si(311) and a CoSi$_2$(177)//Si(311) which is due to type B relationship with respect to the least-inclined [111] plane (Yu et al., 1988).

It is interesting to note that the nucleation of CoSi$_2$ and NiSi$_2$ phases on Si(100) and (110) requires temperatures in excess of 300 °C, in contrast to the nucleation behavior on Si(111). The (homo-) epitaxial growth of CoSi$_2$ and NiSi$_2$ occur at room temperature along all directions (Tung

et al., 1989a). It is also important to note that although CoSi$_2$ and NiSi$_2$ have identical crystal structures and similar lattice constants, the problems facing the epitaxial growth of these two silicides are very different. Regarding the formation of ultrathin (<20 Å) silicide layers, a slight difference in lattice mismatch plays only a minor role. Therefore, the observed difference of the silicide reactions is possibly due to two factors. One is the fundamental difference between the natures of the Ni-Si and the Co-Si bonds (van den Hoek et al., 1988), leading to differences in energetics at the silicide/silicon interface. A second factor may be the dramatically different point defect densities of the two crystals, as implied by their electrical transport properties (Hensel et al., 1984), leading to very different diffusivities at typical growth temperatures (Tung, Schrey, 1988).

Other silicides grown epitaxially on Si include Pd$_2$Si (Saitoh, 1981; Freeouf et al., 1979), YSi$_2$ (Knapp, Picraux, 1986; Gurvitch et al., 1987; Siegal et al., 1989), ErSi$_2$ (Arnaud d'Avitaya et al., 1989; Kaatz et al., 1990a), TbSi$_2$ (Kaatz et al., 1990b), ReSi$_2$ (Mahan et al., 1991), and PtSi (Ishiwara et al., 1979). Partially epitaxial or textured CrSi$_2$ (Shiau et al., 1984), FeSi$_2$ (Cheng et al., 1984), and other platinum group (Chang et al., 1986) and refractory silicides (Chen et al., 1986) have also been grown. Si(111) is the most common substrate for the epitaxial growth of these silicides. Readers are referred to review articles on the epitaxial orientations of these semi-epitaxial films (Chen et al., 1986).

The growth of high quality epitaxial Si on CoSi$_2$ and NiSi$_2$ has been demonstrated. Best results were obtained on the (111). Thick double heterostructures (>500 Å) were demonstrated in some early works by both MBE and SPE (Saitoh et al., 1980;

Tung et al., 1982b; Ishizaka, Shiraki, 1984). In more recent studies of double heterostructures, focus was placed on thinner (<100 Å) silicide layers which have potential application in high speed devices (Tung et al., 1986a; Rosencher et al., 1984). The use of a Si template is critical in ensuring uniform Si overgrowth on silicides (Tung et al., 1986a; Hunt et al., 1987; Henz et al., 1987; Tung et al., 1986b; Henz et al., 1989). Without it, the growth of Si occurs three dimensionally and the crystalline quality is poor. Epitaxial Si may be grown in either type A or type B orientation on $CoSi_2(111)$, by using specially designed Si templates based on the knowledge of the two surface structures of $CoSi_2$ (Tung, Batstone, 1988c). If no special attention is paid to the surface structure of $CoSi_2$, the overgrown Si contains both orientations (d'Anterroches, Arnaud d'Avitaya, 1986). Superlattices consisting of layers of epitaxial $CoSi_2$ and epitaxial Si have also been fabricated, although the structural quality degrades rapidly with superlattice thickness (Hunt et al., 1987; Henz et al., 1989). On $NiSi_2(111)$, epitaxial Si always occupies the type B orientation (Ishizaka, Shiraki, 1984; Tung et al., 1986b; Rizzi et al., 1989). Double heteroepitaxial structures fabricated on (111) are of rather high structural quality. Transistor action was observed in planar Si/$CoSi_2$/Si structures. The measured electronic transmission through the thin ÇoSi base layer was once thought to be due to ballistic transport through the epitaxial silicide layer (Rosencher et al., 1984, 1986; Delage et al., 1986). More careful studies of the morphology of the buried silicide base layers revealed the true mechanism of the observed transmission – through pinholes in the silicide layers (Tung et al., 1986a; Hensel et al., 1985; Hunt, 1987). The absence of ballistic transport through

thin $CoSi_2(111)$ layers is due to a lack of electronic states at the crystal momenta of the silicon conduction band minima, as explained by recent theoretical calculations (Mattheiss, Hamann, 1988; Stiles, Hamann, 1989). Because of this fundamental limitation, device fabrication effort based on epitaxial silicides has since focused on the use of pre-patterned structures (Ishizaka, Shiraki, 1984; Ishibashi, Furukawa, 1984; Glastre et al., 1988; Miyao et al., 1990). The epitaxial growth of Si on other orientations of $CoSi_2$ and $NiSi_2$, e.g. (100) and (110), is much more difficult than the (111). A properly chosen Si template seems to improve the quality of epitaxial Si on $NiSi_2(100)$ (Tung et al., 1986b; Itho et al., 1989), but the upper interface is unstable against faceting. As for Si epitaxial growth on $CoSi_2(100)$, finely textured films are often observed (Tung, Schrey, 1991). A likely explanation for the difficulty of Si epitaxial growth is the large free energies associated with the (110) and (100) silicide interfaces.

Recently, an entirely different approach to silicide growth, "mesotaxy", has emerged and shown great potential, especially in the fabrication of double M-S heterostructures (White et al., 1987, 1989). High-energy high-dose implantations of metal into silicon single crystal and high temperature annealing has led to the formation of buried silicide layers with excellent structural and electrical properties. The nucleation of silicide (precipitates) inside the Si crystal is apparently more favorable than the heterogeneous nucleation processes near a free surface. Remarkable success was immediately achieved by this technique, as high quality Si/$CoSi_2$/Si(100) and (111) structures were fabricated using hot implantation and post-annealing (White et al., 1987). Very abrupt and flat interfaces were observed in these structures. However, a critical implantation dose, of $\approx 1 \times 10^{17}$ cm^{-2}, is required for the formation of a continuous buried layer of CoSi$_2$. This sets a lower limit for the buried silicide thickness of ≈ 400 Å. In contrast to MBE, buried thick CoSi$_2$ layers usually have pure type A orientation in Si(111) (White et al., 1987; Barbour et al., 1988; van Ommen, 1988 b). The nucleation and coalescence of CoSi$_2$ crystallites has been studied in detail (Bulle-Lieuwma et al., 1989 b). Other silicides, including CrSi$_2$, CaSi$_2$, YSi$_2$, and NiSi$_2$, have also been mesotaxially fabricated (White et al., 1989; Lindner, Kaat, 1988). Because of the coalescence of silicide crystallites during post-annealing, not only is there a sharpening of the vertical metal concentration profile, but the lateral dimension of the silicided regions after the annealing is much smaller than the original pattern size of the implantation. This has lead to the fabrication of fine wires of single crystal CoSi$_2$, buried in Si (Berger et al., 1989). High energy implantation is thus far the only technique capable of producing the technologically important structure Si/CoSi$_2$/Si(100). However, the drawbacks of this approach are its inability to grow uniform buried metal layers with thicknesses less than ≈ 400 Å and residual damage due to implantation.

8.5.2.2 Epitaxial Elemental Metals

The aluminum/Si interface has many important applications in integrated circuits. The growth of epitaxial Al on Si has largely been motivated by its superior thermal stability (Missous et al., 1986; Yapsir et al., 1988 a). Al has a very large lattice mismatch with Si, $\approx -25\%$. If one considers the matching condition of four Al lattice planes to three Si lattice planes then a small, $\sim 0.56\%$, effective mismatch is found. Good epitaxy has been demonstrated for Al(111) on Si(111). Al is usually deposited at room temperature. As-deposited films contain both (111) and (100) oriented grains, which convert to pure (111) upon annealing to 400 °C (Le Goues, et al., 1986). The Al/Si interface is usually rough, probably as a result of poor vacuum conditions and the incomplete removal of oxide layer from the Si surface (Le Goues et al., 1986). However, by using partially ionized Al beams, flat interfaces between single crystal Al/Si have been fabricated (Yapsir et al., 1988 b; Lu et al., 1989). There is evidence that under these conditions, the epitaxy may be incommensurate in nature (Lu et al., 1989). Silver has a similar four-to-three type lattice matching condition with Si. Recently, thick layers of Ag(111) have been grown on Si(111) under clean MBE conditions (Park et al., 1988; Park et al., 1990).

Elemental metals have also been grown on III–V compound semiconductors. Most notable of these are Al, Ag, and Fe. A large nominal lattice mismatch exists between GaAs and either Al or Ag. With a 45° azimuthal rotation, good lattice matching conditions may be established for these two metals on GaAs(100). Such is indeed the most common epitaxial orientation for Al(100) grown on GaAs(100) (Cho, Dernier, 1978; Ludeke et al., 1980). However, (110) oriented Al has also been observed (Ludeke et al., 1973). It has been recently pointed out that the reconstruction of the initial GaAs surface structure may influence the epitaxial orientation of the Al films (Donner et al., 1989). Silver grows with (100) orientation on GaAs(100) at elevated temperatures, and with pure (110) orientation at room temperature (Ludeke et al., 1982; Massies et al., 1982). Surprisingly, there is no azimuthal rotation for epitaxial Ag(100) on

GaAs(100). The azimuthal orientation of the (110)-oriented Ag and Al appears to be related to the prevalent dangling bond direction on GaAs(100) surface. As-stabilized and Ga-stabilized surfaces are known to have dangling bonds point at directions 90° apart, leading to Ag(110) which are also rotated by this angle (Massies et al., 1982). (110)-oriented Al has also been grown on GaAs(110), with the expected 90° azimuthal rotation for better matching of the in-plane lattice parameters (Prinz et al., 1982). α-Fe (b.c.c.) has a lattice constant about one half of that of GaAs and ZnSe. This effective lattice matching condition has lead to the growth of single crystal Fe on these two semiconductors (Prinz, Krebs, 1981, 1982; Krebs et al., 1987; Jonker et al., 1987; Farrow et al., 1988). Fe grows with relative ease on GaAs(110) with regular epitaxial orientation (Prinz, Krebs, 1981, 1982), but the growth has proved to be more difficult on GaAs(100) (Krebs et al., 1987). ZnSe(100) appears to be a better substrate for epitaxial Fe films (Jonker et al., 1987). High quality epitaxial Fe(100) films were also grown on lattice-matched (two-to-one) InGaAs surface (Farrow et al., 1988). Interesting magnetic properties have been observed from epitaxial Fe films. Other metals and superlattices have been grown on GaAs, e.g. metastable b.c.c. Co on GaAs(110) (Prinz, 1985) and metal superlattices on GaAs(100) (Baibich et al., 1988; Lee et al., 1989).

8.5.2.3 Epitaxial Metallic Compounds on III–V Semiconductors

Except for refractory metals, most elemental metal/III–V compound semiconductor structures are thermodynamically unstable. Annealing, even at moderate temperatures, often leads to interdiffusion and formation of other compound phases.

Two groups of intermetallic compounds are found to have stable interfaces with III–V compound semiconductors with good lattice matching conditions (Sands, 1988). Hence, these are good candidates for the formation of stable, single crystal, M-S interfaces with GaAs. NiAl (Sands, 1988; Harbison et al., 1988), CoGa (Palmstrom et al., 1989a), NiGa (Guivarc'h et al., 1987), CoAl, and related compounds form one group. They have the cubic CsCl crystal structure and lattice parameters ≈ 1 to 2% larger than one half the lattice parameter of GaAs (Sands, 1988). The exact lattice parameter of the gallides and aluminides may vary slightly with the stoichiometry (Wunsch, Wachtel, 1982). The orientation of epitaxial films on GaAs or AlAs(100) is usually (100). However, (110) oriented growth has also been observed on (100) GaAs (Harbison et al., 1988; Palmstrom et al., 1989a; Guivarc'h et al., 1988). Careful control of the initial reaction on AlAs(100) has led to the successful growth of single crystal (100) CoAl and NiAl (Harbison et al., 1988). The non-zero lattice mismatch results in a network of closely spaced misfit dislocations at the interface (Sands, 1988; Zhu et al., 1989). Overgrowth of GaAs or AlAs on these metals requires exposing the gallides or aluminides to As_4, which can lead to instability and formation of transition-metal arsenides. This problem can be overcome by careful adjustment of the surface composition prior to the growth of GaAs or AlAs (Zhu et al., 1989; Sands et al., 1990). Buried, thin (down to 15 Å) NiAl layers have been grown between AlAs cladding layers inside GaAs. However, the change of symmetry across the $AlAs_{over}/NiAl$ interface leads to a phase difference at every step of the interface which is an odd number of atomic planes in its height. (This situation is analogous to the anti-phase domain boundaries in GaAs films grown on Si(100), due to surface steps. Also note that, at least in principle, Si films may be grown without antiphase domain boundaries on a stepped GaAs surface.) This causes a high density of stacking faults to be present in the overgrown AlAs/GaAs.

Rare-earth (RE) monopnictides, mostly with the NaCl structure, form the other group of metallic compounds suitable for epitaxial growth on III–V semiconductors. Here, an abundance of systems exists which are closely lattice-matched to GaAs (Palmstrom et al., 1990). ErAs (Palmstrom et al., 1988), YbAs (Richter et al., 1988), LuAs (Palmstrom et al., 1989 b), as well as alloyed (ternary) compounds, $ErP_{1-x}As_x$ (Le Corre et al., 1989), $Sc_{1-x}Er_xAs$ (Palmstrom et al., 1990), have been grown on GaAs(100). The usual (100) epitaxial orientation is found for these epitaxial systems. The alloyed compounds are investigated for the obvious reason of obtaining exact lattice-matching. Growth on InP(100) has also been studied (Guivarc'h et al., 1989). RE metals are highly reactive which leads to uniform growth of their arsenides. Similar to the epitaxial growth of GaAs, REAs may be grown in an As over pressure and the exact As to RE ratio is not critical. RHEED intensity oscillations have been observed in the growth of $Sc_{0.32}Er_{0.68}As$, indicating uniform layer-by-layer growth. Stability and lattice-match conditions seem also to be satisfied for a few RE-chalcogenides/II–VI compound semiconductor systems (Palmstrom et al., 1990). These have yet to be tested. The overgrowth of III–V compound semiconductors on RE monopnictide layers has been much more difficult (Palmstrom et al., 1988, 1990) than overgrowth on the aluminides (Zhu et al., 1989. GaAs grows with very rough surface morphology on REAs (Palmstrom et al., 1988).

8.6 Structure, Energetics, and Electronic Properties of Metal-Semiconductor Interfaces

8.6.1 Epitaxial Silicide-Si Interfaces

Unquestionably, epitaxial $CoSi_2$-Si and $NiSi_2$-Si interfaces are structurally the most perfect M-S interfaces. Not surprisingly, they are also the best characterized. High resolution electron microscopy (HREM) has been frequently used to study the atomic structure of epitaxial silicide interfaces. Both type A and type B $NiSi_2/Si(111)$ interfaces were found to have the 7-fold model (Cherns et al., 1982; Föll, 1982; Gibson et al., 1983), while $NiSi_2/Si(100)$ interface was found to be 6-fold structured (Gibson et al., 1983; Cherns et al., 1984; d'Anterroches, Arnaud d'Avitaya, 1986). (The terminology for the structure of a silicide interface, e.g. 7-fold, is based on the number of nearest Si neighbor atoms to a metal atom at the interface. In a bulk disilicide lattice, each metal atom has a coordination number of 8.) These results are in agreement with those from X-ray standing wave (XSW) and medium energy ion scattering (MEIS) studies (van Loenen et al., 1985 b; Vlieg et al., 1986; Zegenhagen et al., 1987 a; Robinson et al., 1988). Models of these interfaces are shown in Fig. 8-44. For the type B $CoSi_2/Si(111)$ interface, early studies by HREM (Gibson et al., 1982), XSW (Fischer et al., 1987; Zegenhagen et al., 1987 b), and MEIS (Fischer et al., 1988 b) suggested the 5-fold model. At the time of these experiments, comparisons were only made with the two (111) models originally proposed by Cherns et al. (1982) the 5-fold and the 7-fold structures, and other possibilities were not considered. Since the surface structures of $CoSi_2$ became known, the tendency for Co atoms to maintain full

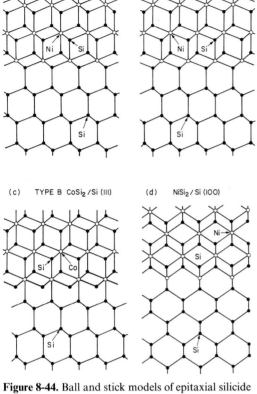

Figure 8-44. Ball and stick models of epitaxial silicide interfaces.

8-fold coordination, even at a surface/interface, was clearly revealed, and thus the notion that Co maintains a 5-fold coordination at the type B interface was challenged. Two theoretical papers (Fischer et al., 1987; Hamann, 1988a) pointed to the high interfacial free energy for the 5-fold model in comparison to the 8-fold model. A SEXAFS study claimed to provide direct evidence for 8-fold coordination at this interface (Rossi et al., 1989), but these results are most probably fortuitous because of the very poor preparation conditions chosen for the growth of silicide layers (Zegenhagen et al., 1990). Detailed investigations by HREM showed that 8-fold model is more likely the structure

experimentally observed (Bulle-Lieuwma et al., 1989b). However, it is also known that the interfacial structure of type B $CoSi_2/Si(111)$ may vary according to preparation – evidence for 7-fold coordinated structure has been observed. Type A $CoSi_2/Si(111)$ is thought to be 7-fold coordinated (Bulle-Lieuwma et al., 1989b). A 2×1 reconstruction (two domains) is often observed at the annealed interfaces of $CoSi_2/Si(100)$ and the likely atomic structure responsible for this interface dimerization has been discussed (Loretto et al., 1989). One notes that, with this reconstruction, full 8-fold coordination is achieved at the $CoSi_2/Si(100)$ interface. The occurrence of this reconstruction is dependent on the preparation of the silicide layer, being most prominently seen in samples which have been annealed at higher temperatures. A similar reconstruction is thought to occur at the $NiSi_2/Si(100)$ interface, although this has not been observed consistently. The interface between epitaxial $CoSi_2$ and $Si(110)$ always shows a dechanneling peak in ion scattering experiments and domain contrasts in TEM (Eaglesham et al., 1991). Experimental results are in agreement with the existence of a rigid shift between the $CoSi_2$ and the Si, with an in-plane component along the $[1\bar{1}0]$ direction (Eaglesham et al., 1991). It should be noted that, with the proposed rigid shift, interfacial Co atoms are situated on bridge-sites over the Si lattice, again achieving an 8-fold coordination. Other silicide interfaces, notably the $Pd_2Si/Si(111)$, have also been studied by HREM (Krakow, 1982). However, the interface is rough and the structure nonuniform.

An intriguing dependence of the SBH on the epitaxial orientation has been observed at epitaxial silicide interfaces. Type A and type B $NiSi_2$ have distinctively different

SBH's (Tung, 1984a, b) on Si(111), and the epitaxial NiSi$_2$/Si(100) interface has yet a different SBH (Levi et al., 1987). The original (111) results that type B NiSi$_2$ has a SBH about 0.14 eV higher than type A NiSi$_2$ on n-type Si(111), were initially challenged (Liehr et al., 1985). The reason for this disagreement is now understood (Tung et al., 1986c) to be a surface boron contamination problem associated with certain experimental conditions. The effect of epitaxial orientation on the SBH has now been fully confirmed (Hauenstein et al., 1985; Ospelt et al., 1988; Vrijmoeth et al., 1990). A low SBH, ≈ 0.65 eV, for type A NiSi$_2$ on n-type Si(111) is consistently measured by every laboratory. As for SBH of type B NiSi$_2$, there is some fluctuation of data which depends on preparation, diode processing, and the method of SBH measurements (Hauenstein et al., 1985; Ospelt et al., 1988; Vrijmoeth et al., 1990). The reported observation of a dependence of type B NiSi$_2$ SBH on film thickness (Kikuchi, 1988, 1989) has not been observed by other laboratories. As we noted earlier, the dislocation density at a type B NiSi$_2$ interface is not a simple function of the film thickness, but also depends on the substrates employed. Some of the inconsistencies on type B SBH are likely related to the existence of small type A regions in the samples. Leakage current from perimeters of the diodes, especially when a high SBH is concerned, is another problem. Our experience is that, with proper diode processing, well-annealed, single crystal layers with low density of dislocations show a high SBH, ≈ 0.79 eV. The SBH of NiSi$_2$ on Si(100) shows considerable scatter with preparation. One recalls that this is the interface with a faceting problem under most growth conditions. The most uniform NiSi$_2$ layers show a rather low SBH on n-type Si(100),

< 0.5 eV (Levi et al., 1987; Tung, Schrey, 1991). With an increased degree of faceting, the SBH increases (Tung, Schrey, 1991). The observation of a higher SBH (n-type) is not an indication of a higher quality interface, as previously claimed (Kikuchi et al., 1988).

The dependence of SBH on interface orientation, three different SBH's for three differently structured interfaces between the same two materials, has not been observed for non-epitaxial M-S systems and appears to be in disagreement with many existing SBH theories. In order to understand the difference in SBH's of type A and type B NiSi$_2$, there have been a number of attempts to measure the density of interface electronic states experimentally. Forward bias capacitance techniques have been used to derive a density of interface states for these Schottky diodes (Ho et al., 1986). However, such measurements ignore effects arising from minority carriers, which may completely dominate the admittance at a forward biased Schottky diode (Werner et al., 1988). Interface state measurements by capacitance methods do not have a strong basis. Infrared absorption measurements on thin layers of type A and type B NiSi$_2$ have shown no evidence of interface electronic states (Flohr et al., 1987), in sharp contrast to results from polycrystalline silicide films. These measurements set an upper limit of $\approx 10^{14}$ cm^{-2} eV^{-1} for the electronic state density at epitaxial silicide interfaces. Initial theoretical calculations were rather unsophisticated and generally yield identical SBH for the unrelaxed A and B structures (Yongnian et al., 1986; Rees, Matthai, 1991; Bisi, Ossicini, 1987). If a difference in the relaxation of the interface bond length is assumed, then a difference in the SBH is brought about somewhat artificially. Recent calculations have shed some light on the origin of the observed difference of A and B type SBH. Very large supercell sizes were found necessary to observe the difference in interface electronic properties due to the subtle difference in the interface atomic structures (Fujitani, Asano, 1988, 1989; Das et al., 1989; Hamann, 1988b). The experimentally observed difference in A and B type SBH is qualitatively reproduced in these first-principle calculations. Quantitative agreement seems also reasonable with the most sophisticated calculations. At this stage, it seems appropriate to attribute the mechanism for SBH formation, at least for these near perfect M-S interfaces, to intrinsic properties associated with the particular interfacial atomic structure.

The SBH at type B $CoSi_2$/Si(111) interface is also expected to depend on the interface atomic structure (Rees, Matthai, 1988). Films grown by room temperature deposition of ≈ 10 to $50\,\text{Å}$ thick cobalt and annealing to $> 600\,^\circ\text{C}$ usually show a SBH in the range 0.65 to 0.70 eV on n-type Si (Tung, 1984b; Rosencher et al., 1985; Kao et al., 1985b). These interfaces have the 8-fold structure. However, recent experiments on the growth of $CoSi_2$ layers at lower temperatures and with lower dislocation densities have produced interfaces which show considerable variation in SBH (Tung et al., 1989b). A low SBH, <0.5 eV, of some type B $CoSi_2$ layers on n-type Si(111) has been observed which have been tentatively explained as originating from the 7-fold structured portions of the interface. With the expected variation of atomic structure and the existence of a phase transformation at this M-S interface, both discussed above, the situation is expected to be quite complicated. This is a particular system where theoretical calculations may be especially illuminating. The SBH's of single crystal $CoSi_2$ layers on

Table 8-2. Schottky barrier heights of epitaxial silicides.

Silicide	Orientation	Substrate	Interface structure	SBH (eV)	
				n-type	p-type
$NiSi_2$	Type A	Si(111)	7-fold	0.65	0.47
$NiSi_2$	Type B	Si(111)	7-fold	0.79	0.33
$NiSi_2$	(100)	Si(100)	6-fold[b]	0.40	0.72
$CoSi_2$	Type B	Si(111)	8-fold	0.67	0.44
$CoSi_2$	(100)	Si(100)	8-fold[a]	0.71	0.41
$CoSi_2$	(110)	Si(110)	8-fold[b]	0.70	0.42

[a] 1×2 reconstructed.
[b] Tentative.

Si(110) and Si(100) are similar, both at ≈ 0.7 eV on n-type substrate. A summary of the SBH's from epitaxial silicide interfaces is given in Table 8-2.

8.6.2 Epitaxial Elemental Metals

The epitaxial interfaces Al/Si(111) and Ag/Si(111) have been studied by HREM (Le Goues et al., 1986, 1987). The high density of interface steps and defects has so far prevented a conclusive understanding of the interface structures. Al/Si(111) interfaces formed by partially ionized beam appear to be smoother, but incommensurate (Lu et al., 1989).

Depending on the growth conditions, an Al-Ga exchange reaction takes place at the Al/GaAs interface (Landgren, Ludeke, 1981). There is also a noted reaction at the interface of Fe/GaAs (Krebs et al., 1987). The interfaces between epitaxial Al and GaAs(100) have been studied by a number of techniques (Marra et al., 1979; Kiely et al., 1988). MBE prepared GaAs(100) surface may have a variety of reconstructions which are associated with different surface stoichiometries. The periodicities of some of these superstructures are found to be preserved at the Al/GaAs interface (Mizuki et al., 1988). It has been claimed that the Schottky barrier height between epitaxial Al or Ag layers and GaAs is a function of the original GaAs surface reconstruction/stoichiometry (Cho, Dernier, 1978; Ludeke et al., 1982; Wang, 1983). However, different conclusions have been drawn in other studies (Barret, Massies, 1983; Svensson et al., 1983; Missous et al., 1983).

8.6.3 Intermetallic Compounds on III–V Semiconductors

The interfaces between GaAs(AlAs) and a number of epitaxial intermetallic compounds have been examined by HREM (Palmstrom et al., 1988, 1989 b; Zhu et al., 1989; Sands et al., 1990; Tabatabaie et al., 1988). However, the atomic structures of these interfaces have not been properly modelled. Evidence for a reconstruction has been observed at the interface between NiAl and overgrown GaAs (Sands et al., 1990). Negative differential resistance has been observed at a discrete bias in electrical transport perpendicular to the interfaces of a GaAs/AlAs/NiAl/AlAs/GaAs structure (Tabatabaie et al., 1988). This is thought to indicate quantization of states in the thin NiAl layer (Tabatabaie et al., 1988).

8.7 Summary

We have briefly reviewed important developments in the field of epitaxial metal-semiconductor structures. Obviously, much has been accomplished in the fabrication and characterization of these structures. But the need still exists for better structures and better electrical properties. Further studies should prove beneficial both to our understanding of the SBH mechanisms and, perhaps, to next generation devices.

8.8 Future Directions

In this chapter, we have attempted to outline the major structural characteristics of three representative interface systems.

While the structure of lattice matched interfaces is now known over a substantial range of spatial frequencies, our understanding of the correlation between the interface structure and its electronic and optical properties remains qualitative, and in some cases controversial. Establishing a firm link between interfacial structure and properties would have important scientific and technological implications, and merits serious effort. Future challenges in the area of lattice matched systems consist of atomic level control of their growth and relaxation modes, so as to optimize them for specific applications. A more difficult task is that of forming high quality *lateral* interfaces, to fabricate low dimensional structures such as quantum wires and quantum dots.

For lattice mismatched systems, we have learnt to grow high quality pseudomorphic, strained systems, which in many cases exhibit exciting properties. However, we must also learn to achieve strain *relaxation* whilst controlling threading dislocation densities. This would allow the exploitation of epitaxy in a more general sense, whereby highly perfect layers with different lattice parameters could be stacked indefinitely. In many instances, it appears essential to grow layers under specific amounts of tensile or compressive strain, which cannot be produced by growth on common substrate materials. The growth of relaxed buffer layers, whose lattice parameter can be controlled through their composition may provide the degrees of freedom needed for general control of electronic properties through strain.

For systems consisting of dissimilar materials, we must learn to overcome problems associated with wetting, and find means of manipulating interfacial energies. It is only in this way that we can free ourselves from present epitaxial constraints, and grow "anything on anything".

Our understanding of the non-structural properties of interfaces is still relatively primitive. We must learn to determine and control the electronic properties of interfaces themselves, and not of the layers that are grown. In this way we may hope to achieve and exploit the true two-dimensional limit.

Much remains to be done in interface physics. The bewildering variety of possible interfaces can be systematically controlled only if we seek to distill from the data guiding principles of general applicability. This is our ultimate challenge.

8.9 Acknowledgements

Many expert colleagues have contributed to the work described here. We particularly wish to thank F.H. Baumann, D. Batrick, J.L. Batstone, M. Bode, J.C. Bean, J.M. Bonar, J. Cunningham, D.J. Eaglesham, J.M. Gibson, Y. Kim, A.F.J. Levi, J.A. Rentschler, F. Schrey, C.W. Tu, C.A. Warwick, and S.M. Yalisove, and acknowledge valuable discussions with A. Bourret, L.C. Feldman, J.P. Harbison, T.M. Lu, J.R. Kwo, C.J. Palmstrom, and W. Schröter.

8.10 References

Alexander, H., Haasen, P. (1968), in: *Solid State Physics*, Vol. 22, p. 27.

Arnaud D'Avitaya, F., Delage, S., Rosencher, E., Derrien, J. (1985), *J. Vac. Sci. Technol,* B3, 770.

Arnaud d'Avitaya, F., Perio, A., Oberlin, J.-C., Campidelli, Y., Chroboczek, J.A. (1989), *Appl. Phys. Lett. 54*, 2198.

Baibich, M.N., Broto, J.M., Fert, A., Nguyen Van Dau, F., Petroff, F., Eitenne, P., Creuzet, G., Friederich, A., Chazelas, J. (1988), *Phys. Rev. Lett. 61*, 2472

Barbour, J.C., Picraux, S.T., Doyle, B.L. (1988), *Mat. Res. Soc. Symp. Proc. 107*, 269.

Barret, C., Massies, J. (1983), *J. Vac. Sci. Technol. B 1*, 819.

Batstone, J.L., Gibson, J.M., Tung, R.T., Levi A.F.J. (1988), *Appl. Phys. Lett. 52*, 828.

Bean, J.C., Poate, J.M. (1980), *Appl. Phys. Lett. 37*, 643.

Bean, J.C., Feldman, L.C., Fiory, A.T., Nakahara, S., Robinson, I.K. (1984), *J. Vac. Sci. Technol. A2*, 436.

Berger, S.D., Huggins, H.A., White, A.E., Short, K.T., Loretto, D. (1989), *Nanostructure Physics and Fabrication*, London: Academic Press, p. 401.

Biegelsen, D.K., Ponce, F.A., Smith, A.J., Tramontana, J.C. (1987), *J. Appl. Phys. 61*, 1856.

Biegelsen, D.K., Ponce, F.A., Krusor, B.S., Tramontana, J.C., Yingling, R.D., Bringans, R.D., Fenner, D.B. (1988), *Proc. Mat. Soc. 116*: Choi, H.K., Hull, R., Ishiwara, H., Memanich, R.J. (Eds.). Pittsburg, PA, p. 33.

Biersack, J.P. (1987), *Nucl. Instr. Methods B 19*, 32.

Bimberg, D., Christen, J., Fukunaga, T., Nakashima, H., Mars, D.E., Miller, J.N. (1987), *J. Vac. Sci. Technol. B 5*, 1191.

Bisi, O., Ossicini, S. (1987), *Surface Sci. 189/190*, 285.

Bode, M., Ourmazd, A., Rentschler, J.A., Hong, M., Feldman, L.C., Mannaerts, J.P. (1990), *Proc. Mat. Res. Soc. 157*: Knapp, J.A., Børgesen, P., Zuhn, R. (Eds.). Pittsburgh, PA, p. 197.

Bonar, J.M., Hull, R., Malik, R.J., Ryan, R.W., Walker, J.F. (1990), *Proc. Mat. Res. Soc. 160*: Dodson, B.W., Schowalter, L.J., Cunningham, J.E., Pollak, F.H. (Eds.). Pittsburgh, PA, p. 117.

Bruinsma, R., Zangwill, A. (1987), *Europhys. Lett. 4*, 729.

Bulle-Lieuwma, C.W.T., van Ommen, A.H., van Ijzendoorn, L.J. (1989a), *Appl. Phys. Lett. 54*, 244.

Bulle-Lieuwma, C.W.T., de Jong, A.F., van Ommen, A.H., van der Veen, J.F., Vrijmoeth, J. (1989b), *Appl. Phys. Lett. 55*, 648.

Cahn, J.W. (1961), *Acta Met. 9*, 525.

Capasso, F., Margaritondo, G. (Ed.) (1987), *Heterojunction Band Discontinuities*, Amsterdam: North Holland.

Chand, N., People, R., Baiocchi, F.A., Wecht, K.W., Cho, A.Y. (1986), *Appl. Phys. Lett. 49*, 815.

Chang, Y.S., Chu, J.J., Chen, L.J. (1986), *Mat. Res. Soc. Symp. Proc. 54*, 57.

Chang, K.H., Berger, P.R., Gibala, R., Bhattacharya, P.K., Singh, J., Mansfield, J.F., Clarke, R. (1988), *Proc. of TMS/AIME Symposium on "Defects and Interfaces in Semiconductors"*. Rajaj,

K.K., Narayan, J., Ast, D. (Eds.). Warrendale: The Metallurgical Society, p. 157.

Chen, L.J., Cheng, H.C., Lin, W.T. (1986), *Mat. Res. Soc. Symp. Proc. 54*, 245.

Cheng, H.C., Chen, L.J., Your, T.R. (1984), *Mat. Res. Soc. Symp. Proc. 25*, 441.

Cherns, D., Anstis, G.R., Hutchison, J.L., Spence, J.C.H. (1982), *Philos. Mag. A 46*, 849.

Cherns, D., Hetherington, C.J.D., Humphreys, C.J. (1984), *Philos. Mag. A 49*, 165.

Chiu, K.C.R., Poat, J.M., Rowe, J.E., Sheng, T.T., Cullis, A.G. (1981), *Appl. Phys. Lett. 38*, 988.

Cho, A.Y., Dernier, P.D. (1978), *J. Appl. Phys. 49*, 3328.

Comin, F., Rowe, J.E., Citrin, P.H. (1983), *Phys. Rev. Lett. 51*, 2402.

Cook, H.E., Hilliard, J.E. (1969), *J. Appl. Phys. 40*, 2191.

d'Anterroches, C., Arnaud d'Avitaya, F. (1986), *Thin Solid Films 137*, 351.

Das, G.P., Blöchl, P., Andersen, O.K., Christensen, N.E., Gunnarsson, O. (1989), *Phys. Rev. Lett. 63*, 1168.

Delage, S., Badoz, P.A., Rosencher, E., Arnaud d'Avitaya, F. (1986), *Electron. Lett. 22*, 207.

Deppe, D.G., Holonyak, N., Jr. (1988), *J. Appl. Phys. 64*, R93.

Dodson, B.W. (1988), *Phys. Rev. B38*, 12383.

Dodson, B.W., Tsao, J.Y. (1987), *Appl. Phys. Lett. 51*, 1325.

Donner, S.K., Caffey, K.P., Winograd, N. (1989), *J. Vac. Sci. Technol. B7*, 742.

Dupuis, R.D., Bean, J.C., Brown, J.M., Macrander, A.T., Miller, R.C., Hopkins, L.C. (1986), *J. Elec. Mat. 16*, 69.

Eaglesham, D.J., Cerullo, M. (1991), to be published in: *Proc. Mat. Res. Soc. 198.*

Eaglesham, D.J., Aindow, M., Pond, R.C. (1988), in: *Mat. Res. Proc. Soc. 116*: Choi, H.K., Hull, R., Ishiwara, H., Nemanich, R.J. (Eds.). Pittsburgh: MRS, p. 267.

Eaglesham, D.J., Kvam, E.P., Maher, D.M., Humphreys, C.J., Bean, J.C. (1989), *Phil. Mag. A 59*, 1059.

Eaglesham, D.J., Tung, R.T., Headrick, R.L., Robinson, I.K., Schrey, F. (1990), *Mat. Res. Soc. Symp. Proc. 159,* 141.

Eaglesham, D.J., Tung, R.T., Yalisove, S.M. (1991), to be published.

Farrow, R.F.C., Parkin, S.S.P., Speriosu, V.S., Wilts, C.H., Beyers, R.B., Pitner, P., Woodall, J.M., Wright, S.L., Kirchner, P.D., Pettit, G.D. (1988), *Mat. Res. Soc. Symp. Proc. 102*, 483.

Fiory, A.T., Bean, J.C., Feldman, L.C., Robinson, I.K. (1984), *J. Appl. Phys. 56*, 1227.

Fischer, R., Morkoc, H., Neumann, D.A., Zabel, H., Choi, C., Otsuka, N., Longerbone, M., Erickson, L.P. (1986), *J. Appl. Phys. 60*, 1640

Fischer, A.E.M.J., Vlieg, E., van der Veen, J.F., Clausnitzer, M., Materlik, G. (1987), *Phys. Rev. B36*, 4769.

Fischer, A.E.M.J., Slijkerman, W.F.J., Nakagawa, K., Smith, R.J., van der Veen, J.F., Bulle-Lieuwma, C.W.T. (1988a) *J. Appl. Phys. 64*, 3005

Fischer, A.E.M.J., Gustafsson, T., van der Veen, J.F. (1988b), *Phys. Rev. B37*, 6305.

Fitzgerald, E.A. (1989), *J. Vac. Sci. Tech. B7*, 782.

Fitzgerald, E.A., Kirchner, P.D., Proano, R., Petit, G.D., Woodall, J.M., Ast, D.G. (1988), *Appl. Phys. Lett. 52*, 1496.

Fitzgerald, E.A., Watson, G.P., Proano, R.E., Ast, D.G., Kirchner, P.D., Pettit, G.D., Woodall, J.M. (1989), *J. Appl. Phys. 65*, 2688.

Fleming, R.M., McWhan, D.B., Gossard, A.C., Wiegmann, W., Logan, R.A. (1980), *J. Appl. Phys. 51*, 357.

Flohr, Th., Schulz, M., Tung, R.T. (1987), *Appl. Phys. Lett. 51*, 1343.

Föll, H. (1982), *Phys. Stat. Sol. (a) 69*, 779.

Frank, F.C., van der Merwe, J.H. (1949a), *Proc. Roy. Soc. A198*, 205, 216, A200 (1949b), 125.

Freeouf, J.L., Rubloff, G.W., Ho, P.S., Kuan, T.S. (1979), *Phys. Rev. Lett. 43*, 1836.

Freund, L.B. (1990), *J. Appl. Phys. 68,* 2073.

Fritz, I.J. (1987), *Appl. Phys. Lett. 51*, 1080.

Fritz, I.J., Picreaux, S.T., Dawson, L.R., Drummond, T.J., Laidig, W.D., Anderson, N.G. (1985), *Appl. Phys. Lett. 46*, 967.

Fujitani, H., Asano, S. (1988), *J. Phys. Soc. Jpn. 57*, 2253.

Fujitani, H., Asano, S. (1989), *Appl. Surface Sci. 41/42*, 164.

George, A., Rabier, J. (1987), *Rev. Phys. Appl. 22*, 1941.

Gibson, J.M., Bean, J.C., Poate, J.M., Tung, R.T. (1982), *Appl. Phys. Lett. 41*, 818.

Gibson, J.M., Tung, R.T., Poate, J.M. (1983), *MRS Symp. Proc. 14*, 395.

Gibson, J.M., Batstone, J.L., Tung, R.T. (1987), *Appl. Phys. Lett. 51*, 45.

Glastre, G., Rosencher, E., Arnaud d'Avitaya, F., Puissant, C., Pons, M., Vincent, G., Pfister, J.C. (1988), *Appl. Phys. Lett. 52*, 898.

Gourley, P.L., Fritz, I.J., Dawson, L.R. (1988), *Appl. Phys. Lett. 52*, 377.

Grabow, M., Gilmer, G. (1987), in: *Proc. Mat. Res. Soc. 94*: Hull, R., Gibson, J.M., Smith, D.A. (Eds.). Pittsburgh: MRS, p. 13.

Guido, L.J., Holonyak, N., Jr., Hsieh, K.C., Baker, J.E. (1989), *Appl. Phys. Lett. 54*, 262.

Guivarc'h, A., Guerin, R., Secoue, M. (1987), *Electron. Lett. 23*, 1004.

Guivarc'h, A., Secoue, M., Guenais, B. (1988), *Appl. Phys. Lett. 52*, 948.

Guivarc'h, A., Caulet, J., LeCorre, A. (1989), *Electron. Lett. 25*, 1050.

Gurvitch, M., Levi, A.F.J., Tung, R.T., Nakahara, S. (1987), *Appl. Phys. Lett. 51*, 311.

Hagen, W., Strunk, H. (1978), *Appl. Phys. 17*, 85.

Hamann, D.R. (1988a), *Phys. Rev. Lett. 60*, 313.

Hamann, D.R. (1988b), in: *Metallization and Metal-*

Semiconductor Interfaces: Batra, J.P. (Ed.). New York: Plenum.

Harbison, J.P., Sands, T., Tabatabaie, N., Chan, W.K., Florez, L.T., Keramidas, V.G. (1988), *Appl. Phys. Lett. 53*, 1717.

Harris, J.S., Koch, S.M., Rosner, S.J. (1987), in: *Heteroepitaxy on Si II, Proc. Mat. Res. Soc. 91*: Fan, J.C.C., Phillips, J.M., Tsaur, B.-Y. (Eds.). Pittsburgh: MRS, p. 3.

Hauenstein, R.J., Schlesinger, T.E., McGill, T.C., Hunt, B.D., Schowalter, L.J. (1985), *Appl. Phys. Lett. 47*, 853.

Hellman, F., Tung, R.T. (1988), *Phys. Rev. B37*, 10786.

Hensel, J.C., Tung, R.T., Poate, J.M., Unterwald, F.C. (1984), *Appl. Phys. Lett. 44*, 913.

Hensel, J.C., Levi, A.F.J., Tung, R.T., Gibson, J.M. (1985), *Appl. Phys. Lett. 47*, 151.

Henz, J., Ospelt, M., von Känel, H. (1987), *Solid State Commun. 63*, 445.

Henz, J., Ospelt, M., von Känel, H. (1989), *Surface Sci. 211/212*, 716.

Hirsch, P.B., Howie, A., Nicholson, R.B., Pashley, D.W., Whelan, M.J. (1977), *Electron Microscopy of Thin Crystals*, 2nd ed. Malabar, Florida: Robert E. Krieger.

Hirth, J.P., Lothe, J. (1968), *Theory of Dislocations*. New York: McGraw-Hill.

Ho, P.S., Yang, E.S., Evans, H.L., Wu, X. (1986), *Phys. Rev. Lett. 56*, 177.

Houghton, D.C. (1991a), *Appl. Phys. Lett. 57*, 1434.

Houghton, D.C. (1991b), *Appl. Phys. Lett. 57*, 2124.

Hull, R., Bean, J.C. (1989a), *J. Vac. Sci. Techn. A7*, 2580.

Hull, R., Bean, J.C. (1989b), *Appl. Phys. Lett. 54*, 925.

Hull, R., Bean, J.C. (1989c), *Appl. Phys. Lett. 55*, 1900.

Hull, R., Fischer-Colbrie, A. (1987), *Appl. Phys. Lett. 50*, 851.

Hull, R., Bean, J.C., Cerdeira, F., Fiory, A.T., Gibson, J.M. (1986), *Appl. Phys. Lett. 48*, 56.

Hull, R., Carey, K.W., Reid, G.A. (1987), *Mat. Res. Soc. Proc. 77*, 455, Warwick, C.A., private communication.

Hull, R., Bean, J.C., Werder, D.J., Leibenguth, R.E. (1988), *Appl. Phys. Lett. 52*, 1605.

Hull, R., Bean, J.C., Werder, D.J., Leibenguth, R.E. (1989a), *Phys. Rev. B40*, 1681.

Hull, R., Bean, J.C., Eaglesham, D.J., Bonar, J.M., Buescher, C. (1989b), *Thin Solid Films, 183*, 117.

Hull, R., Bean, J.C., Leibenguth, R.E., Werder, D.J. (1989c), *J. Appl. Phys. 65*, 4723.

Hull, R., Bean, J.C., Bahnck, D., Petticolas, L., Short, K.T., Unterwald, F.C. (1991), *Appl. Phys.*, in press.

Hunt, B.D. (1987), *Air Force report*, unpublished.

Hunt, B.D., Schowalter, L.J., Lewis, N., Hall, E.L., Hauenstein, R.J., Schlesinger, T.E., McGill, T.C., Okamoto, M., Hashimoto, S. (1986), *Mat. Res. Soc. Symp. Proc. 54*, 479.

Hunt, B.D., Lewis, N., Schowalter, L.J., Hall, E.L., Turner, L.G. (1987), *Mat. Res. Soc. Symp. Proc. 77*, 351.

Ichinose, H., Ishida, Y., Furuta, T., Sakaki, H. (1987), *J. Electron Microsc. 36*, 82.

Imai, M., Sumino, K. (1983), *Phil. Mag. A47*, 599.

Ishibashi, K., Furukawa, S. (1984). *Extended Abstracts 16th Conf Solid State Devices and Materials*, Kobe, p. 35.

Ishiwara, H., Hikosaka, K., Furukawa, S. (1979), *J. Appl. Phys. 50*, 5302.

Ishiwara, H. (1980), in: *The Electrochem. Soc. Symp. Proc. 80-2*, 159.

Ishizaka, A., Shiraki, Y. (1984), *Japan. J. Appl. Phys. 23*, L499.

Itoh, M., Kinoshita, M. Ajioka, T., Itoh, M., Inada, T. (1989), *Appl. Surface Sci. 41/42*, 262.

Jimenez, J.R., Hsiung, L.M., Thompson, R.D., Hashimoto, S., Ramanathan, K.V., Arndt, R., Rajan, K., Iyer, S.S., Schowalter, L.J. (1990), *MRS Symp. Proc. 160*, 237.

Jonker, B.T., Krebs, J.J., Prinz, G.A., Quadri, S.B. (1987), *J. Crst. Growth, 81*, 524.

Kaatz, F.H., Siegal, M.P., Graham, W.R., Van der Spiegel, J., Santiago, J.J. (1990a), *Thin Solid Films*, in press.

Kaatz, F.H., Graham, W.R., van der Spiegel, J. (1990b), *Appl. Phys. Lett.*, in press.

Kakibayashi, H., Nagata, F. (1986), *Jap. J. Appl. Phys. 25*, 1644.

Kao, Y.C., Tejwani, M., Xie, Y.H., Lin, T.L., Wang, K.K. (1985a), *J. Vac. Sci. Technol. B3*, 596.

Kao, Y.C., Wu, Y.Y., Wang, K.L. (1985b), *Proc. 1st Int'l Symp. Si MBE*: Bean, J.C. (Ed.). The Electrochem. Soc., p. 261.

Kasper, E., Kibbel, H., Presting, H. (1989), in: *Silicon Molecular Beam Epitaxy*: Kasper, E., Parker, E.H.C. (Eds.). Amsterdam: North Holland, p. 87.

Kiely, C.J., Cherns, D., Eaglesham, D.J. (1988), *Philos. Mag.*

Kikuchi, A., Ohshima, T., Shiraki, Y. (1988), *J. Appl. Phys. 64*, 4614.

Kikuchi, A. (1989), *Phys. Rev. B40*, 8024.

Kim, Y., Ourmazd, A., Bode, M., Feldman, R.D. (1989), *Phys. Rev. Lett. 63*, 636.

Kim, Y., Ourmazd, A., Feldman, R.D. (1990a), *J. Vac. Sci. Technol., A8*, 1116.

Kim, Y., Ourmazd, A., Malik, R.J., Rentschler, J.A. (1990b), *Proc. Mat. Res. Soc. 159*: Bringans, R.D., Feenstra, R.M., Gibson, J.M. (Eds.). Pittsburgh, PA, p. 351.

Knapp, J.A., Picraux, S.T. (1986), *Appl. Phys. Lett. 48*, 466.

Kohama, Y., Fukuda, Y., Seki, M. (1988), *Appl. Phys. Lett. 52*, 380.

Krakow, W. (1982), *Thin Solid Films, 93*, 109.

Krebs, J.J., Jonker, B.T., Prinz, G.A. (1987), *J. Appl. Phys. 61*, 2596.

Kuesters, K.H., Alexander, A. (1983), *Physica 116B*, 594.

Kuesters, K.H., De Cooman, B.C., Carter, C.B. (1985), *Proc. 13th Int. Conf. on Defects in Semiconductors*, 1984: Kimmerling, L.C., Parsey, J.M., Jr. (Eds.). Warrendale: AIME, p. 351.

Kvam, E.P., Eaglesham, D.J., Maher, D.M., Humphreys, C.J., Bean, J.C., Green, G.S., Tanner, B.K. (1988), *Proc. Mat. Res. Soc. 104*: Tung, R.T., Dawson, L.R., Gunshor, R.L. (Eds.). Pittsburgh: MRS, p. 623.

Landgren, G., Ludeke, R. (1981), *Solid State Commun. 37*, 127.

Le Corre, A., Caulet, J., Guivarc'h, A. (1989), *Appl. Phys. Lett. 55*, 2298.

Lee, C.H., He, H., Lamelas, F., Vavra, W., Uher, C., Clarke, R. (1989), *Phys. Rev. Lett. 62*, 653.

Le Goues, F.K., Krakow, W., Ho, P.S. (1986). *Philos. Mag. A53*, 833.

Le Goues, F.K., Liehr, M., Renier, M. (1987), *MRS Symp. Proc. 94*, 121.

Lee, H.P., Huang, Y.-H., Liu, X., Lin, H., Smith, J.S., Weber, E.R., Yu, P., Wang, S., Lilliental-Weber, Z. (1988), in: *Proc. Mat. Res. Soc. 116*: Choi, H.K., Hull, R., Ishiwara, H., Nemanich, R.J. (Eds.). Pittsburgh: MRS, p. 219.

Levi, A.F.J., Tung, R.T., Batstone, J.L., Gibson, J.M., Anzlowar, M., Chantre, A. (1987), *MRS Symp. Proc. 77*, 271.

Liehr, M., Schmidt, P.E., LeGoues, F.K., Ho, P.S. (1985). *Phys. Rev. Lett. 54*, 2139.

Lilliental-Weber, Z., Weber, E.R., Washburn, J., Liu, T.Y., Kroemer, H. (1987), *Proc. Mat. Res. Soc. 91*: Fan, J.C.C., Phillips, J.M., Tsaur, B.Y. (Eds.). Pittsburgh; MRS, p. 91.

Lin, T.L., Sadwick, K., Wang, K.L., Kao, Y.C., Hull, R., Nieh, C.W., Jamieson, D.N., Liu, J.K. (1987), *Appl. Phys. Lett., 51*, 814.

Lin, T.L., Fathauer, R.W., Grunthaner, P.J. (1988), *Appl. Phys. Lett. 52*, 804.

Lindner, J.K.N., te Kaat, E.H. (1988), *J. Mater. Res. 3*, 1238.

Loretto, D., Gibson, J.M., Yalisove, S.M. (1989), *Phys. Rev. Lett. 63*, 298.

Louchet, F. (1981), *Inst. Phys. Ser. Conf. 60*. Bristol, G.B. Institute of Physics.

Lu, T.-M., Bai, P., Yapsir, A.S., Chang, P.-H., Shaffner, T.J. (1989), *Phys. Rev. B39*, 9584.

Ludeke, R., Chang, L.L., Esaki, L. (1973), *Appl. Phys. Lett. 23*, 201.

Ludeke, R., Landgren, G., Chang, L.L. (1980), *Vide. Couches Minces, Suppl., 201*, 579.

Ludeke, R., Chiang, T.-C., Eastman, D.E. (1982), *J. Vac. Sci. Technol. 21*, 599.

Luryi, S., Suhir, E. (1986), *Appl. Phys. Lett. 49*, 140.

Maeda, K., Yamashita, Y., Maeda, N., Takeuchi, S. (1991), to be published in: *Proc. Mat. Res. Soc. 184*.

Mahan, J.E., Geib, K.M., Robinson, G.Y., Long, R.G., Yan, X.H., Bai, G., Nicolet, M.-A., Nathan, M. (1991), to be published.

Marra, W.C., Eisenberger, P., Cho, A.Y. (1979), *J. Appl. Phys. 50*, 6927.

Massies, J., Delescluse, P., Etienne, P., Linh, N.T. (1982), *Thin Solid Films 90*, 113.

Mattheiss, L.F., Hamann, D.R. (1988), *Phys. Rev. B37*, 10623.

Matthews, J.W. (1975), *J. Vac. Sci. Technol. 12*, 126 (and references contained therein).

Matthews, J.W., Blakeslee, A.E. (1974), *J. Cryst. Growth 27*, 118.

Matthews, J.W., Blakeslee, A.E. (1976), *J. Cryst. Growth 32*, 265.

Matthews, J.W., Blakeslee, A.E., Mader, S. (1976), *Thin Solid Films 33*, 253.

Matyi, R.J., Shichijo, H., Tsai, H.L. (1988), *J. Vac. Sci. Technol. B6*, 699.

Miles, R.H., McGill, T.C., Chow, P.P., Johnson, D.C., Hauenstein, R.J., Nieh, C.W., Strathman, M.D. (1988), *Appl. Phys. Lett. 52*, 916.

Miller, R.C., Tu, C.W., Sputz, S.K., Kopf, R.F. (1986), *Appl. Phys. Lett. B49*, 1245.

Missous, M., Rhoderick, E.H., Singer, K.E. (1983), *J. Appl. Phys. 54*, 4474.

Missous, M., Rhoderick, E.H., Singer, K.E. (1986), *J. Appl. Phys. 59*, 3189.

Miyao, M., Ohshima, T., Nakamura, N., Nakagawa, K. (1990), *MRS Symp. Proc. 160, 275*.

Mizuki, J., Akimoto K, Hirosawa, I., Hirose, K., Mizutani, T., Matsui, J. (1988), *J. Vac. Sci. Technol. B6*, 31.

Noble, D.B., Hoyt, J.L., Gibbons, J.F., Scott, M.P., Ladermann, S.S., Rosner, S.J., Kamins, T.I. (1989), *Appl. Phys. Lett. 55*, 1978.

Ogale, S.B., Madhukar, A., Voillot, F., Thomsen, M., Tang, W.C., Lee, T.C., Kim, J.Y., Chen, P. (1987), *Phys. Rev. B36*, 1662.

Olsen, G.H., Abrahams, M.S., Buiocchi, G.J., Zamerowski, T.J. (1975), *J. Appl. Phys. 46*, 1643.

Okumura, H., Yoshida, I., Misawa, S., Yoshida, S. (1987), *J. Vac. Sci. Technol. B5*, 1622.

Ospelt, M., Henz, J., Flepp, L., von Känel, H. (1988), *Appl. Phys. Lett. 52*, 227.

Ou, H.-J., Tsen, S.-C.Y., Tsen, K.T., Cowley, J.M., Chyi, J.I., Salvador, A., Morkoç, H. (1989), *Appl. Phys. Lett. 54*, 1454.

Ourmazd, A. (1989), *J. Crys. Growth 98*, 72.

Ourmazd, A., Spence, J.C.H. (1987), *Nature 329*, 425.

Ourmazd, A., Rentschler, J.A., Taylor, D.W. (1986), *Phys. Rev. Lett. 57*, 3073.

Ourmazd, A., Tsang, W.T., Rentschler, J.A., Taylor, D.W. (1987), *Appl. Phys. Lett. 50*, 1417.

Ourmazd, A., Taylor, D.W., Cunningham, J., Tu, C.W. (1989a), *Phys. Rev. Lett. 62*, 933.

Ourmazd, A., Taylor, D.W., Bode, M., Kim, Y. (1989b), *Science 246*, 1571.

Ourmazd, A., Baumann, F.H., Bode, M., Kim, Y. (1990), *Ultramicroscopy 34, 237*.

Palmstrom, C.J., Tabatabaie, N., Allen, S.J., Jr. (1988), *Appl. Phys. Lett. 53*, 2608.

Palmstrom, C.J., Fimland, B.-O., Garrison, K.C., Bartynski, R.A. (1989a), *J. Appl. Phys. 65*, 4753.

Palmstrom, C.J., Garrison, K.C., Mounier, S., Sands, T., Schwartz, C.L., Tabatabaie, N., Allen, S.J., Jr., Gilchrist, H.L., Miceli, P.F. (1989b), *J. Vac. Sci. Technol. B 7*, 747.

Palmstrom, C.J., Mounier, S., Finstad, T.G., Miceli, P.F. (1990), *Appl. Phys. Lett. Jan. 22.*

Park, K.H., Jin, H.-S., Luo, L., Gibson, W.M., Wang, G.-C., Lu, T.-M. (1988), *Mat. Res. Soc. Symp. Proc. 102*, 271.

Park, K.-H., Smith, G.A., Rajan, K., Wang, G.-C. (1990), *Met. Trans. A*, in press.

Petroff, P.M., Cho, A.Y., Reinhart, F.K., Gossard, A.C., Wiegmann, W. (1982), *Phys. Rev. Lett. 48*, 170.

Petroff, P.M., Cibert, J., Gossard, A.C., Dolan, G.J., Tu, C.W. (1987), *J. Vac. Sci. and Technol. B 5*, 1204.

Phillips, J.M., Batstone, J.L., Hensel, J.C., Cerullo, M. (1987), *Appl. Phys. Lett. 51*, 23.

Pirouz, P., Ernst, F., Cheng, T.T. (1988), in: *Proc. Mat. Res. Soc. 116*: Choi, H.K., Hull, R., Ishiwara, H., Nemanich, R.J. (Eds.). Pittsburgh: MRS.

Pirri, C., Peruchetti, J.C., Gewinner, G., Darrien, J. (1984), *Phys. Rev. B 29*, 3391.

Prinz, G.A. (1985), *Phys. Rev. Lett. 54*, 1051.

Prinz, G.A., Krebs, J.J. (1981), *Appl. Phys. Lett. 39*, 397.

Prinz, G.A., Krebs, J.J. (1982), *J. Appl. Phys. 53*, 2087.

Prinz, G.A., Ferrari, J.M., Goldenberg, M. (1982), *Appl. Phys. Lett. 40*, 155.

Rajan, K., Denhoff, M. (1987), *J. Appl. Phys. 62*, 1710.

Rees, N.V., Matthai, C.C. (1988), *J. Phys. C 21*, L 981.

Rees, N.V., Matthai, C.C. (1991), to be published.

Reynolds, D.C., Bajaj, K.K., Litton, C.W., Yu, P.W., Singh, J., Masselink, W.T., Fisher, R., Morkoç, H. (1985), *Appl. Phys. Lett. 46*, 51.

Richter, H.J., Smith, R.S., Herres, N., Seelmann-Eggebert, M., Weenekers, P. (1988), *Appl. Phys. Lett. 53*, 99.

Rizzi, A., Förster, A., Lüth, H., Slijkerman, W. (1989), *Surface Sci. 211/212*, 620.

Robinson, I.K., Tung, R.T, Feidenhans'l, R. (1988), *Phys. Rev. B 38*, 3632.

Rosencher, E., Delage, S., Campidelli, Y., Arnaud d'Avitaya, F. (1984), *Electron. Lett. 20*, 762.

Rosencher, E., Delage, S., Arnaud D'Avitaya, F. (1985), *J. Vac. Sci. Technol. B 3*, 762.

Rosencher, E., Glastre, G., Vincent, G., Vareille, A., Arnaud d'Avitaya, F. (1986), *Electron. Lett. 22*, 699.

Rossi, G., Jin, X., Santaniello, A., De Padova, P., Chandesris, D. (1989), *Phys. Rev. Lett. 62*, 191.

Saitoh, S., Ishiwara, H., Furukawa, S. (1980), *Appl. Phys. Lett. 37*, 203.

Saitoh, S., Ishiwara, H., Asano, T., Furukawa, S. (1981), *Jpn. J. Appl. Phys. 20*, 1649.

Sakaki, H., Noda, T., Hirakawa, K., Tanaka, M., Matsusue, T. (1987), *Appl. Phys. Lett. 51*, 1934.

Sands, T. (1988), *Appl. Phys. Lett. 52*, 197.

Sands, T., Harbison, J.P., Tabatabaie, N., Chan, W.K., Gilchrist, H.L., Cheeks, T.L., Florez, L.T., Keramidas, V.G. (1990), *Surface Sci.*, in press.

Scott, M.P., Laderman, S., Kamins, T., Reid, G., Rosner, S.J., Gronet, C., King, C., Hoyt, J., Noble, D., Gibbons, J.F. (1989), *Proc. Mat. Res. Soc. 130*: Bravman, J.C., Nix, W.D., Barnett, D.M., Smith, D.A. (Eds.). Pittsburgh: MRS.

Sheng, H.C., Wu, I.C., Chen, L.J. (1987), *Appl. Phys. Lett. 50*, 174.

Shiau, F.Y., Cheng, H.C., Chen, L.J. (1984), *Appl. Phys. Lett. 45*, 524.

Siegal, M.P., Kaatz, F.H., Graham, W.R., Santiago, J.J., van der Spiegel, J. (1989), *J. Appl. Phys. 66*, 2999.

Spence, J.C.H. (1980), *Experimental High Resolution Electron Microscopy*, New York: Oxford Univ. Press.

Stiles, M.D., Hamann, D.R. (1989), *Phys. Rev. B 40*, 1349.

Suzuki, Y., Okamoto, H. (1985), *J. Appl. Phys. 58*, 3456.

Svensson, S.P., Landgren, G., Andersson, T.G. (1983). *J. Appl. Phys. 54*, 4474.

Tabatabaie, N., Sands, T., Harbison, J.P., Gilchrist, H.L., Keramidas, V.G. (1988), *Appl. Phys. Lett. 53*, 2528.

Tamargo, M.C., Hull, R., Greene, L.H., Hayes, J.R., Cho, A.Y. (1985), *Appl. Phys. Lett. 46*, 569.

Tanaka, N., Mihama, K. (1988), *Ultrramicroscopy 26*, 37.

Tanaka, M., Sakaki, H., Yoshino, J. (1986), *Jap. J. Appl. Phys. 25*, L 155.

Temkin, H., Gershoni, D.G., Chu, S.N.G., Vandenberg, J.M., Hamm, R.A., Panish, M.B. (1989), *Appl. Phys. Lett. 55*, 1668.

Thomsen, M., Madhukar, A. (1987), *J. Crys. Growth, 84*, 98.

Tsao, J.Y., Dodson, B.W., Picreaux, S.T., Cornelison, D.M. (1987), *Phys. Rev. Lett. 59*, 2455.

Tu, K.N., Mayer, J.W. (1978), in: *Thin Films Interdiffusion and Reactions*: Poate, J.M., Tu, K.N., Mayer, J.W. (Eds.). New York: Wiley.

Tu, C.W., Miller, R.C., Wilson, B.A., Petroff, P.M., Harris, T.D., Kopf, R.F., Sputz, S.K., Lamont, M.G. (1987), *J. Crys. Growth 81*, 159.

Tung, R.T. (1984a), *Phys. Rev. Lett. 52*, 461.

Tung, R.T. (1984b), *J. Vac. Sci. Technol. 2*, 465.

Tung, R.T. (1987), *J. Vac. Sci. Technol. A 5*, 1840.

Tung, R.T. (1988), in: *Silicon Molecular Beam Epitaxy*: Bean, J.C., Kasper, E. (Eds.). Boca Raton, FL, CRC Press.

Tung, R.T. (1989), *J. Vac. Sci. Technol. A 7*, 599.

Tung, R.T., Batstone, J.L. (1988a), *Appl. Phys. Lett. 52*, 648.

Tung, R.T., Batstone, J.L. (1988c), *Appl. Phys. Lett. 52*, 1611.

Tung, R. T., Schrey, F. (1988), *Mat. Res. Soc. Symp. Proc. 122*, 559.

Tung, R. T., Schrey, F. (1989a), *Appl. Phys. Lett. 55*, 256.

Tung, R. T., Schrey, F. (1989b), *Appl. Phys. Lett. 54*, 852.

Tung, R. T., Schrey, F. (1989c), *Phys. Rev. Lett. 63*, 1277.

Tung, R. T., Schrey, F. (1991), to be published.

Tung, R. T., Gibson, J. M., Bean, J. C., Poate, J. M., Jacobson, D. C. (1982a), *Appl. Phys. Lett. 40*, 684.

Tung, R. T., Poate, J. M., Bean, J. C., Gibson, J. M., Jacobson, D. C. (1982b), *Thin Solid Films 93*, 77.

Tung, R. T., Gibson, J. M., Poate, J. M. (1983a), *Phys. Rev. Lett. 50*, 429.

Tung, R. T., Gibson, J. M., Poate, J. M. (1983b), *Appl. Phys. Lett. 42*, 888.

Tung, R. T., Nakahara, S., Boone, T. (1985), *Appl. Phys. Lett. 46*, 895.

Tung, R. T., Levi, A. F. J., Gibson, J. M. (1986a), *Appl. Phys. Lett. 48*, 635.

Tung, R. T., Gibson, J. M., Levi, A. F. J. (1986b), *Appl. Phys. Lett. 48*, 1264.

Tung, R. T., Ng, K. K., Gibson, J. M., Levi, A. F. J. (1986c), *Phys. Rev. B33*, 7077.

Tung, R. T., Batstone, J. L., Yalisove, S. M. (1988), *Mat. Res. Soc. Symp. Proc. 102*, 265.

Tung, R. T., Schrey, F., Yalisove, S. M. (1989a), *Appl. Phys. Lett. 55*, 2005.

Tung, R. T., Levi, A. F. J., Schrey, F., Anzlowar, M. (1989b), *NATO ASI Series B: Physics, 203*, 167.

Tuppen, C. G., Gibbings, C. J. (1989), in: *Silicon Molecular Beam Epitaxy.* Kasper, E., Parker, E. H. C. (Eds.). Amsterdam: North Holland.

Tuppen, C. G., Gibbings, C. J., Hockly, M., Roberts, S. G. (1990), *Appl. Phys. Lett. 56*, 54.

Tuppen, C. G., Gibbings, C. J. (1990), *J. Appl. Phys. 68*, 1526.

Twigg, M. E. (1990), *J. Appl. Phys. 68*, 5109.

Vandenberg, J. M., Hamm, R. A., Parish, M. B., Temkin, H. (1987), *J. Appl. Phys. 62*, 1278.

Vandenberg, J. M., Parish, M. B., Temkin, H., Hamm, R. A. (1988), *Appl. Phys. Lett. 53*, 1920.

Vandenberg, J. M., Parish, M. B., Hamm, R. A., Temkin, H. (1990), *Appl. Phys. Lett. 56*, 910.

van den Hoek, P. J., Ravenek, W., Baerends, E. J. (1988), *Phys. Rev. Lett. 60*, 1743.

van der Merwe, J. H., Ball, C. A. B. (1975), in: *Epitaxial Growth, Part b*: Matthews, J. W. (Ed.). New York: Academic, pp. 493–528.

van Loenen, E. J., Fischer, A. E. M., van der Veen, J. F., LeGoues, F. (1985a), *Surface Sci. 154*, 52.

van Loenen, E. J., Frenken, J. W. M., van der Veen, J. F., Valeri, S. (1985b), *Phys. Rev. Lett. 54*, 827.

van Ommen, A. H., Bulle-Lieuwma, C. W. T., Langereis, D. (1988a), *J. Appl. Phys. 64*, 2706.

van Ommen, A. H., Ottenheim, J. J. M., Theunissen, A. M. L., Mouwen, A. G. (1988b), *Appl. Phys. Lett. 53*, 669.

Vlieg, E., Fischer, A. E. M. J., van der Veen, J. F., Dev, B. N., Materlik, G. (1986), *Surface Sci. 178*, 36.

von Känel, H., Graf, T., Henz, J., Ospelt, M., Wachter, P. (1987), *J. Cryst. Growth 81*, 470.

Vrijmoeth, J., van der Veen, J. F., Heslinga, D. R., Klapwijk, T. M. (1990), *Phys. Rev. B42*, 9598.

Wang, W. I. (1983), *J. Vac. Sci. Technol. B1*, 574.

Warwick, C. A., Jan, W. Y., Ourmazd, A., Harris, T. D. (1990), *Appl. Phys. Lett., 56*, 2666.

Weisbuch, C., Dingle, R., Gossard, A. C., Wiegmann, W. (1981), *Solid State Comm. 38*, 709.

Werner, J., Levi, A. F. J., Tung, R. T., Anzlowar, M., Pinto, M. (1988), *Phys. Rev. Lett. 60*, 53.

Whaley, G. J., Cohen, P. I. (1990), *Appl. Phys. Lett. 57*, 144.

White, A. E., Short, K. T., Dynes, R. C., Garno, J. P., Gibson, J. M. (1987), *Appl. Phys. Lett. 50*, 95.

White, A. E., Short, K. T., Dynes, R. C., Hull, R., Vandenberg, J. M. (1989), *Nuc. Instr. Meth. Phys. Res. B39*, 253.

Wunsch, K. M., Wachtel, E. (1982), *Z. Metallkde. 73*, 311.

Yalisove, S. M., Tung, R. T., Loretto, D. (1989a), *J. Vac. Sci. Technol. A7*, 599.

Yalisove, S. M., Eaglesham, D. J., Tung, R. T. (1989b), *Appl. Phys. Lett. 55*, 2075.

Yapsir, A. S., Bai, P., Lu, T.-M. (1988a), *Appl. Phys. Lett. 53*, 905.

Yapsir, A. S., Choi, C.-H., Yang, S.-N., Lu, T.-M., Madden, M., Tracy, B. (1988b), *MRS Symp. Proc. 116*, 465.

Yongnian, Xu, Kaiming, Z., Xide, X. (1986), *Phys. Rev. B33*, 8602.

Yu, I., Phillips, J. M., Batstone, J. L., Hensel, J. C., Cerullo, M. (1988), *MRS EA-18*, 11.

Zegenhagen, J., Kayed, M. A., Huang, K.-G., Gibson, W. M., Phillips, J. C., Schowalter, L. J., Hunt, B. D. (1987a), *Appl. Phys. A44*, 365.

Zegenhagen, J., Huang, K.-G., Hunt, B. D., Schowalter, L. J. (1987b), *Appl. Phys. Lett. 51*, 1176.

Zegenhagen, J., Tung, R. T., Patel, J. R., Freeland, P. E. (1990), *Phys. Rev. Lett. 64*, 980.

Zhu, J. G., Carter, C. B., Palmstrom, C. J., Garrison, K. C. (1989), *Appl. Phys. Lett. 55*, 39.

9 The Hall Effect in Quantum Wires

Albert M. Chang

AT & T Bell Laboratories, Holmdel and Murray Hill, NJ, U.S.A.

List of Symbols and Abbreviations

A	vector potential
B	magnetic field
c	speed of light
E_F	Fermi energy
$E_{n,k}$	energy of the state n, k
E_r	resonant energy
e	fundamental charge
H	Hamilton operator
h	Planck constant
I_i	current of lead i
k	quantum number
k	Boltzmann constant
l_0	magnetic length
l_{in}	inelastic length
l_Φ	coherence length
m^*	effective mass
N_H	numerator of the Hall resistance formula
n	electron density
n	quantum number
p	linear momentum
q^*	quasi-particle charge
R	reflection probability
R_c	cyclotron radius
R_H	Hall resistance
s, s_0	slope of the Hall resistance at $B=0$, conventional slope
T	temperature
T	transmission probability
V	voltage
V_g	backgate voltage
V_H	Hall voltage
Γ_i	coupling of the i-th state to the resonant state
μ	local chemical potential
ν, ν_b	Landau level filling factor, filling factor in the barrier region
τ_{tun}	tunneling time
ϕ_n	Hermite polynomial
$\Psi_{n,k}$	eigenstate n, k
ω_c	cyclotron frequency
ω_0	natural frequency of the confining parabolic potential
MOSFET	metal oxide semiconductor field effect transistor
PMMA	polymethylmethacrylate
QHE	quantum Hall effect
RF	radio frequency

9.1 Introduction

Quantum transport in semiconductor microstructures has enjoyed a period of tremendous development in the past few years. By exploiting the forefront of technology in microfabrication and crystal growth, it has been possible to explore an entirely new regime of transport. In this new regime, transport properties are determined by the scattering of electron waves from well-controlled features, such as Hall junctions of different geometries, narrow constrictions, single impurity quasi-bound states, or spatial regions exhibiting the integer or fractional quantum Hall effects, rather than from a large number of random impurities. Therefore transport is qualitatively different from diffusive systems such as quasi-one-dimensional metallic wires or silicon MOSFET's, for which the elastic length for scattering from impurities is still smaller than any geometric size and determines to a large extent the transport properties. Many new phenomena have been observed. Outstanding examples include the quantized resistance of ballistic point contacts at zero magnetic field (van Wees et al., 1988; Wharam et al., 1988) electron focusing of ballistic electrons by a weak magnetic field (van Houten et al., 1988) quenching of the Hall resistance in ballistic junctions at low magnetic fields (Roukes et al., 1987; Ford et al., 1988a; Chang et al., 1989; Ford et al., 1989) non-local bend resistances (Timp et al., 1988; Takagaki et al., 1988) deviation of the quantum Hall effect from exact quantization in narrow wires (Chang et al., 1988a) scattering experiments involving barriers exhibiting the integral (Haug et al., 1988; Washburn et al., 1988) and fractional quantum Hall effects (Chang and Cunningham, 1989; Haug and von Klitzing, 1989a) and selective population of edge channels in the integral (van Wees et al., 1989; Komiyama et al., 1989; Haug and von Klitzing, 1989b; Alphenaar et al., 1990; McEuen et al., 1990) and fractional quantum Hall regimes (Kouwenhoven et al., 1990; Chang, 1990a). The experimental development of this field has progressed hand in hand with theory. At present all experiments are analyzed in terms of the Büttiker-Landauer formulas for quantum transport which are based on a scattering approach to the transport process (Büttiker, 1986; Landauer, 1970; Büttiker, 1988a; Baranger and Stone, 1989a). These formulas have proven extremely successful for non-interacting electron systems. Attempts to generalize these formulas to the fractional quantum Hall regime where electron-electron interaction plays a dominant role have been made (Beenakker, 1990; MacDonald, 1990; Chang, 1990b). However, a generally accepted picture has yet to emerge.

In this review article we focus on three experiments on the Hall effect in quantum wires: 1) quenching of the Hall resistance in a novel geometry (Chang et al., 1989), 2) deviation of the quantum Hall effect from exact quantization in narrow wires (Chang et al., 1988a), and 3) scattering experiment involving the transmission and reflection probabilities of an electron or fractionally-charged quasi-particle incident upon a barrier exhibiting a fractional quantum Hall effect (Chang and Cunningham, 1989; Khurana, 1990). In the first experiment on quenching of the Hall resistance, we experimentally demonstrate the necessary and sufficient conditions required of a ballistic junction geometry in order to observe low magnetic field Hall resistance anomalies such as the suppression of the resistance to nearly zero. In the second experiment, we investigate the quantum Hall effect (von Klitzing et al.,

1980) in narrow wires to determine the conditions under which exact quantization no longer holds. We find that in wires of width $\approx 2000\,\text{Å}$, deviations of a few percent is observable. Other unusual behaviors are observed as well, such as a negative longitudinal resistance and reduced activation energies. We identify a resonant tunneling mechanism for these effects and provide experimental evidence for its existence. These two experiments at low and intermediate magnetic fields are readily understood within the framework of the Büttiker-Landauer resistance formulas for non-interacting electrons. The third experiment probes a new regime of quantum transport occurring at high fields in which electron-electron interaction plays a dominant role. Here electrons or quasi-particles are scattered from a barrier region exhibiting a fractional quantum Hall effect (Tsui et al., 1982), such as the $v = 1/3$ or $v = 2/3$ effects. In one interpretation, the deduced transmission and reflection probabilities provide a direct measure of the quasi-particle fractional charge (MacDonald, 1990; Chang, 1990b; Khurana, 1990). The experimental system we use, as in all experiments to date, is the quasi-one-dimensional electron gas system in lithographically patterned $GaAs-Al_xGa_{1-x}As$ heterostructure devices.

This chapter will be organized as follows: Sec. 9.2 contains a discussion of the Büttiker-Landauer formalism which forms the basis of the analysis of our data. A subsection is devoted to the Hall resistance within this theoretical framework. Sec. 9.3 deals with the fabrication technique for submicron devices in $GaAs-Al_xGa_{1-x}As$ heterostructures in order to achieve a quasi-one-dimensional electron gas system (Sec. 9.2.4). The three experiments mentioned above are described in Secs. 9.4, 9.5, and 9.6, respectively. For reviews on other

experiments such as ballistic point contacts or Aharonov-Bohm quantum interference effects, please see van Houten et al. (1990), Timp (1990), and Chang et al. (1988b).

9.2 Theoretical Framework – The Büttiker-Landauer Formulas

Conventionally, electron transport properties in metals or semiconductors are understood in terms of the semiclassical Drude type conductivity which is proportional to a scattering length equal to a scattering time multiplied by the Fermi velocity. Although this scattering length is calculated quantum-mechanically, the scattering process is otherwise viewed as classical and no attempt is made to account for the phase of the electron wave and possible interference effects between successive scattering events. Further, it is assumed reasonable to define a local conductivity or resistivity based on a local determination of the voltages for a given current, where at each lead the voltage is related to a local chemical potential via the relationship $\mu = -eV$. However, if the size of a conductor is sufficiently reduced so that the conducting system is comparable or smaller than the quantum-mechanical phase coherence length, l_Φ, as illustrated in Fig. 9-1, transport is instead more appropriately described as a wave scattering process involving electron waves, much like the scattering of surface waves in a pond by rocks, where the measured resistances are related to the transmission and reflection probabilities, T and R. In addition, due to the wave nature of the electrons and the *long energy* and *phase relaxation lengths*, it is not longer possible to define a local chemical potential and local conductivity (resistivity) within a small conductor during transport. Rather, it is necessary to

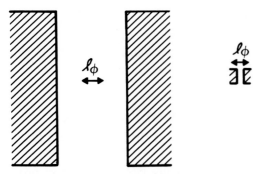

Figure 9-1. A large device of size much larger than the phase coherence length, l_ϕ, for which it is possible to define a local conductivity, left. A small device of size smaller than l_ϕ for which local conductivity is not well defined, right. Shaded regions represent electrical contacts.

consider the total conductance or resistance of the system which includes the current and voltage leads as well. The important idea emphasizing the wave scattering nature of transport was put forth long ago by Landauer (1970), and more recently by Enquist and Anderson (1981) who showed independently that the resistance can be expressed as $(h/e^2)(R/T)$ under suitable conditions. The proper treatment of current and voltage leads on an equal footing was recently achieved by Büttiker (1986), following a series of earlier attempts by Büttiker, Imry, Landauer, and others (Büttiker et al., 1985; Büttiker and Imry, 1985). This treatment has provided the framework to make direct contact with experiments. The resulting Büttiker-Landauer formulas for direct current transport has been successful in explaining a variety of novel effects observed in semiconductor and metallic quantum wires, including the lack of symmetry of the four terminal resistance under magnetic field reversal discovered in seminal experimental work in metallic wires (Umbach et al., 1984; Benoit et al., 1986). More recently, Baranger and Stone (1989 a) have rederived these formu-

las from linear response theory using the Kubo formulas.

9.2.1 Basic Ideas

To obtain an intuitive understanding of the Büttiker-Landauer formulas, we bring forth the basic ideas by considering a conductor with two leads (Fig. 9-2). For simplicity, we envision a quantum coherent region of interest connected to two ideal leads each modeled by a perfect region free of all scattering, leading into a chemical potential reservoir in which phase and energy relaxation processes can take place (Büttiker, 1986; Stone and Szafer, 1988;

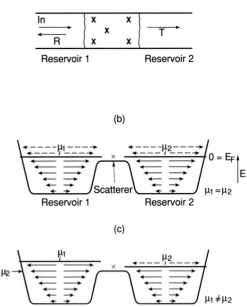

Figure 9-2. (a) A two lead conductor with a coherent scattering region between the reservoirs in the leads. For electrons sourced from a reservoir incident upon the scattering region, there is a probability T for transmission and a probability R for a reflection. (b) In equilibrium, the chemical potentials μ_1 and μ_2 of the reservoirs are both equal to E_F. There is no current flow. (c) By raising the chemical potential μ_1 within reservoir 1 relative to μ_2 by applying a voltage, a net current flows from leads 1 to lead 2.

Baranger and Stone, 1989 a). In the coherent region, *there are no phase or energy relaxations.* It is therefore of size $\approx l_\phi$ which we take to be approximately equal to l_{in}, the inelastic length for energy relaxation, for the time being. The voltage measured with an ideal voltmeter is equal to the chemical potential divided by the electron charge, $V = \mu/(-e)$, which in general is different from the electrostatic voltage due to the non-local nature of transport (Landauer, 1970; Büttiker, 1988 b). Note that an arbitrary shift of zero in the voltage is inconsequential. We specialize to the case where only one quantum channel (mode) (Sec. 9.2.4) is occupied below the Fermi energy, the strictly one-dimensional limit. In an equilibrium situation, the chemical potential of the two leads is in balance and is equal to μ_0, or equivalently the Fermi energy, E_F. For each occupied energy level, the corresponding electron wave injected from lead 1 can scatter from the coherent region, part of it is transmitted to lead 2, part of it reflected back into lead 1. The transmission and reflection probabilities at each energy satisfy the current conservation condition $T(E) + R(E) = 1$. The current carried by the transmitted wave is however exactly balanced by a current injected from lead 2 to lead 1. No net current re-

sults. If we now raise the chemical potential in lead 1 relative to lead 2 by a small amount $\Delta\mu = \mu_1 - \mu_0 \ll \mu_0$ while maintaining $\mu_1, \mu_2 \approx \mu_0$, there is a net imbalance of current flowing from lead 1 to lead 2 given by the following expression:

$$
\begin{aligned}
I = nev &= \frac{dn}{dE}\, \Delta\mu\, e\, \frac{1}{\hbar}\, \frac{dE}{dk}\, T = \\
&= \frac{-e}{h}\, [(1-R)\,\mu_1 - T\mu_2] = \\
&= \frac{e^2}{h}\, [(1-R)\,V_1 - TV_2] \qquad (9\text{-}1)
\end{aligned}
$$

where the transmission and reflection probabilities are evaluated at the Fermi energy, E_F, and we have made use of *the exact cancellation of the density of states for forward propagating waves and the group velocity, up to a constant of h^{-1}, valid for every quantum channel.* This relationship between the current and the measured voltages, expressed in terms of the fundamental unit of conductance, e^2/h, and the transmission and reflection probabilities, completely determines the conductance (or resistance) of the conducting system including leads. The generalization of this relationship to a situation where an arbitrary number of leads are connected to the conductor is straight forward. Fig. 9-3 illustrates a model system for a four lead

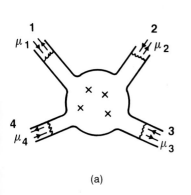

(a)

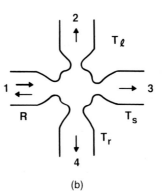

(b)

Figure 9-3. Model four-lead conductors: (a) general case, (b) a four-fold symmetric Hall junction with the probabilities: T_l for transmission to the left, T_r for transmission to the right, T_s for transmission straight through, and R for reflection.

conductor. The generalized relation for the outgoing current in the i-th lead is then given by:

$$I_i = -\frac{e}{h}\left[(1-R_{ii})\mu_i - \sum_{j\neq i} T_{ji}\mu_j\right] \qquad (9\text{-}2)$$

where the unit probability of 1 indicates the number of quantum channels which equals 1 injected, R_{ii} refers to the reflection probability of carriers injected in lead i to be reflected back into its reservoir, and T_{ji} to the probability for carriers injected in lead j to be transmitted into lead i therefore contributing to the current with a negative sign. Note that again we have the sum rules associated with current conservation –

$$R_{ii} + \sum_{j\neq i} T_{ji} = 1 \quad \text{and} \quad R_{ii} + \sum_{j\neq i} T_{ij} = 1$$

We have sketched the main ideas for the Büttiker-Landauer formulation based on a simplified model in which we artificially divide the conductor into spatial regions with or without phase and energy relaxing processes. In a real system these processes occur statistically throughout the conductor with exponentially small probabilities ($\sim e^{-L/l_\Phi}$, $e^{-L/l_{in}}$, respectively, where L is the size of the scattering region). We thus expect the formulation to contain the essential physics, and to become more accurate at low temperatures where the relaxation lengths become long. The assumption of approximately equal l_Φ and l_{in} was made for convenience. Of course, strictly speaking, the idea of a chemical potential reservoirs acting as a source and sink of carriers applies even in the classical limit where electron wave phase coherence can be neglected. Therefore, it is useful to distinguish between two regimes of transport: $l_\Phi \simeq l_{in} \simeq L$ and $l_\Phi \ll l_{in} \simeq L$. The third possibility of $l_\Phi > l_{in}$ does not arise since an inelastic event necessarily randomizes the phase. The first possibility describes fully quantum coherent transport where quan-

tum interference plays a major role. The second describes situations where either interference effects are not relevant or are averaged out. Nevertheless, the concept of transmission and reflection of carriers still has validity.

9.2.2 Special Symmetry Property under Magnetic Field Reversal

In part, the Büttiker-Landauer formulas were developed to account for the observed *lack of symmetry* of the four-terminal longitudinal resistance under magnetic field reversal (Umbach et al., 1984; Benoit et al., 1986), unaccounted for by earlier formulations. The generalized resistances obtained from the Büttiker-Landauer formulas are not expressible in terms of the diagonal matrix elements of the scattering matrix alone but involve the off-diagonal elements as well. Therefore, they are symmetric only under total time reversal, corresponding to interchange of voltage and current leads as well as magnetic field reversal. This fundamental prediction has now been verified experimentally in semiconductor systems as well (van Houten et al., 1990).

To see this symmetry explicitly, it is instructive to consider the special case of a four-lead conductor as depicted in Fig. 9-3. The resistance with current passed between lead i and j, and voltage measured between k and l, $R_{ij,kl}$, is given by (Büttiker, 1986):

$$R_{ij,kl} = \frac{h}{e^2}\frac{T_{ik}T_{jl}-T_{il}T_{jk}}{D} \qquad (9\text{-}3)$$

$$\begin{aligned}D = &\; T_{12}[T_{23}T_{34}+T_{24}(1-R_{33})] + \\ &+ T_{13}[T_{24}T_{32}+T_{34}(1-R_{22})] + \\ &+ T_{14}[T_{24}(1-R_{33})+T_{21}(1-R_{33}) + \\ &+ T_{23}T_{34}+T_{23}T_{31}]\end{aligned} \qquad (9\text{-}4\,\text{a})$$

$$= \tfrac{1}{4} [(1 - R_{11})(1 - R_{22})(1 - R_{33}) +$$
$$+ (1 - R_{11})(1 - R_{22})(1 - R_{44}) +$$
$$+ (1 - R_{11})(1 - R_{33})(1 - R_{44}) +$$
$$+ (1 - R_{22})(1 - R_{33})(1 - R_{44})] -$$
$$- \tfrac{1}{4} [T_{12} T_{21} (1 - R_{33}) +$$
$$+ T_{13} T_{31} (1 - R_{22}) + T_{23} T_{32} (1 - R_{11}) +$$
$$+ T_{12} T_{21} (1 - R_{44}) + T_{14} T_{41} (1 - R_{22}) +$$
$$+ T_{24} T_{42} (1 - R_{11}) + T_{13} T_{31} (1 - R_{44}) +$$
$$+ T_{14} T_{41} (1 - R_{33}) + T_{34} T_{43} (1 - R_{11}) +$$
$$+ T_{23} T_{32} (1 - R_{44}) + T_{24} T_{42} (1 - R_{33}) +$$
$$+ T_{34} T_{43} (1 - R_{22})] -$$
$$- \tfrac{1}{4} [(T_{12} T_{23} T_{31} + T_{13} T_{21} T_{32}) +$$
$$+ (T_{12} T_{24} T_{41} + T_{14} T_{21} T_{42}) +$$
$$+ (T_{13} T_{34} T_{41} + T_{14} T_{31} T_{43}) +$$
$$+ (T_{23} T_{34} T_{42} + T_{24} T_{32} T_{43})] \qquad (9\text{-}4\,\text{b})$$

The transmission and reflection probabilities obey total time reversal symmetry: $T_{ij}(B) = T_{ji}(-B)$, and $R_{ii}(B) = R_{ii}(-B)$. Therefore, the denominator D is manifestly symmetric under magnetic field reversal and any permutation of the indices i, j, k and l. On the other hand, the numerator $(T_{ik} T_{jl} - T_{il} T_{jk})$ is symmetric under B reversal only under the simultaneous interchange of the current and voltage leads. This fundamental symmetry can be shown to hold for a conductor with an arbitrary number of leads (Büttiker, 1988 b).

9.2.3 Generalization to an Arbitrary Magnetic Field

The development of the Büttiker-Landauer formulas was motivated initially by experiments which were performed at low magnetic fields where the effect of the field is simply to contribute an Aharonov-Bohm phase factor (Aharonov and Bohm, 1959) to the zero field wave function. The full power of the formulas was not brought out until their validity was demonstrated for an arbitrary magnetic field. Theoretically, this important result was arrived at by Büttiker (1988 a) with a proper treatment of the interface between a metallic contact region, in which conduction occurs via a very large number of channels, and the semiconductor region, where only a few Landau channels contribute to conduction. It has also been derived from linear response theory via the Kubo formula by Baranger and Stone (1989 a).

9.2.4 Quantum Channels or Modes

In this subsection we introduce the concept of channels or modes in order to generalized the above discussion for the strictly one-dimensional limit to so-called quasi-one-dimensional situation where a few channels are occupied below the Fermi energy. In an n-doped GaAs–Al$_x$Ga$_{1-x}$As heterostructure before lithographic patterning is performed, the electron system is two-dimensional in nature. The band gap discontinuity and band offset at the interface are such that the electrons reside on the GaAs side. For the degree of freedom perpendicular to the interface, defined as the z-direction, the electrons occupy the lowest available quantum level with a binding energy of ≈ 150 meV and an excitation of ≈ 15 meV to the first excited level. In the remaining degrees of freedom parallel to the interface, the $x - y$ directions, the electrons are freely moving (see Fig. 9-4).

As a result of this energy structure, for accessible Fermi energies of ≈ 10 meV above the lowest level, the z-degree of freedom is frozen out, especially at low temperatures $(T < 77\,\text{K})$ where quantum transport experiments are carried out, and the system is two-dimensional in nature.

By patterning the heterostructure surface together with chemical etching or electrostatic side-gating to introduce lateral

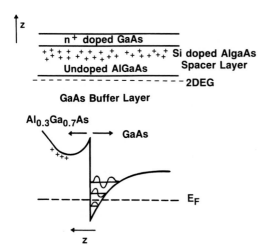

Figure 9-4. Top: Layer structure of the modulation-doped n-type GaAs–$Al_xGa_{1-x}As$ heterostructure. z represents the growth direction and the 2D electron gas resides on the GaAs side. Bottom: The conduction band diagram. The band discontinuity at the GaAs–$Al_xGa_{1-x}As$ interface gives rise to a confining well on the GaAs side. In this well along the z-direction, discrete eigen levels are formed. At typical Fermi energies of ≈ 10 meV, only the lowest level is occupied. In the $x-y$ plane parallel to the interface, the electrons are freely moving. Therefore the electron system is two dimensional in nature.

channels are occupied below the Fermi energy. In the simplest approximation to the y-confinement potential, a square well potential is assumed. In this case the situation is entirely analogous to what occurs in a rectangular microwave waveguide where only allowed modes can propagate.

In the presence of a perpendicular magnetic field, particularly a strong one, the electron eigenstates and wavefunctions are substantially modified from the zero field case described above. The magnetic field introduces a new length scale, the magnetic length, $l_0 = [\hbar c/e B]^{1/2} = [\hbar/m^* \omega_c]^{1/2}$, which is proportional to the cyclotron radius of the electron, R_c. When R_c is much larger than the width, W, introduced by the y-confinement potential, the zero field wavefunctions and energy levels represent good approximations. At intermediate fields where $R_c \simeq W$, the wavefunctions exhibit a mixed character reflecting both the electrostatic and magnetic confinements. At strong fields where $R_c \ll W$, the eigenstates acquire Landau-level like behavior

confinement within the $x-y$ plane, the electrons are further confined to a width of ≤ 3000 Å in the direction defined as the y-direction. The eigenstates associated with this direction become discretized with typical energy spacings of order 1 meV. These eigenstates are readily resolved at temperatures $T \leq 4$ K where $kT \leq 0.3$ meV. The third degree of freedom remains along the x-direction for which there is a continuous spectrum of available forward and backward propagating plane-wave states with energy which depend on momentum quadratically. Thus for each eigenstate in the y-direction, there is a continuous spectrum of k-states in the x-direction as depicted in Fig. 9-5. Together, they are referred to as a channel or mode. In the illustrations of Figs. 9-5a and 9-5b, three

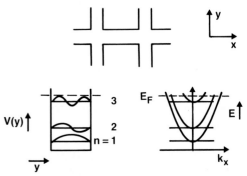

Figure 9-5. Channels or modes. In the k direction across the conductor, there is a confinement potential well in which the eigen levels are discrete. In narrow wires these levels are readily resolved at low temperatures. For each level, there is a set of continuous k-states in the x-direction with energy which depend quadratically on momentum, $\hbar k$. Each y-eigenstate, together with the continuous k-states along x, constitutes a channel or mode. In the figure, three channels are occupied below E_F.

where the wavefunctions are still plane waves in the x-direction but are Gaussian localized at different spatial points along y with a localization length l_0. Because the electrostatic potential energy is higher at the boundaries of the y-confinement, the energy of the states situated at the boundaries or edges are higher than those situated in the bulk region. Note that at all magnetic fields, k along the x-direction remains a good quantum number.

To illustrate these changes in the character of the eigenstates and wavefunctions, we consider the case of a parabolic confinement potential $V(y) = \frac{1}{2} m^* \omega_0^2 y^2$ for which analytic solutions can be obtained at all magnetic fields (Kaplan and Warren, 1986; Berggren et al., 1986). We choose the Landau gauge, $A = -B y x$, where x is the unit vector in the x-direction, to preserve the x-translation invariance of the Hamiltonian. We have:

$$H = \frac{1}{2m^*} \left(p + \frac{e}{c} A \right)^2 + \frac{1}{2} m^* \omega_0^2 y^2$$

$$A = -B y x$$

$$\Psi_{n,k} = e^{ikx} \phi_n(y - y_k) e^{\frac{-(y-y_k)^2}{2l^2}} \quad (9\text{-}5)$$

$$E_{n,k} = (n + \tfrac{1}{2}) \hbar \omega + \tfrac{1}{2} m^* \omega_0^2 y_k^2 + \\ + (1/2m^*) \hbar^2 k^2 (\omega_0/\omega)^4$$

where m^* is the effecting mass of the electron, $k = 2\pi n/L_x (L_x \to \infty)$, $\omega_c = eB/m^* c$, $\omega^2 = \omega_0^2 + \omega_c^2$, $l = [\hbar/m^* \omega]^{1/2}$, $y_k = (\omega_c/\omega) \cdot k l^2$, and ϕ_n is the Hermite polynomial associated with the harmonic oscillator wave functions of the order n. The expectation value of the velocity operator, $v = (1/m^*)[p + (e/c) A]$, is:

$$\left\langle \Psi_{n,k} \left| \frac{1}{m^*} \left(p + \frac{e}{c} A \right) \right| \Psi_{n,k} \right\rangle = $$

$$= \frac{1}{m^*} \hbar k \left(1 - \frac{\omega_c^2}{\omega^2} \right) x \quad (9\text{-}6)$$

At low magnetic fields where $\omega_c \ll \omega_0$, the eigenstate wavefunctions are just the harmonic oscillator wavefunctions in the parabolic well, $V(y) = \frac{1}{2} m^* \omega_0^2 y^2$, for the y-degree of freedom. For each harmonic oscillator level of given index n, there is a continuous spectrum of plane wave states in the x-direction. The wave function is spread out over the entire potential well in the y-direction, and has an expected velocity of $\hbar k/m^*$ along x. As B is increased, the vector potential A contributes a confinement term which is quadratic in y to the Hamiltonian. For each k, the total parabolic confinement is no longer centered at zero but displaced to $y_k = (\omega_c/\omega) k l^2$. At strong B fields where $\omega_c \gg \omega_0$, the energy spectrum is determined by the cyclotron energy, $\hbar\omega_c$, associated with the Landau level spacing, plus a kinetic energy term due to the drift motion produced by the cross electric and magnetic fields, with a drift velocity of $v_d = (\nabla V \times B)/B^2$. Within a channel of index n, the states closer to the boundaries have higher energy. The wavefunction is Gaussian localized on the order of l_0 about y_k, and drifts with opposite velocities on opposite boundaries.

9.2.5 Multi-Channel Transport Formulas

The generalization of the Büttiker-Landauer formulas to the multi-channel case is straight forward. One only needs to replace $(1 - R_{ii})$ by $(N - \Sigma_{m,n} R_{ii,mn})$ and T_{ji} by $\Sigma_{m,n} T_{ji,mn}$, where N refers to the total number of channels, m and n indices to the m-th or n-th channel within the i-th of j-th lead, respectively.

9.2.6 The Hall Resistance

We are now in a position to discuss the Hall resistance within the framework of the new transport formulas. Classically, the Hall voltage arises from a charge separation process across the current flow induced by the Lorentz force, $F_L = -(e/c) \, v \times B$, as depicted in Fig. 9-6. With the flow of current, the deflection of electrons to one edge continues until the force due to the electric field produced by the accumulated charges exactly balances the Lorentz force, resulting in a Hall voltage across the current flow. The Hall resistance, R_H, is thus defined to be the Hall voltage, V_H, divided by the current, I. Similarly one defines the familiar longitudinal resistance, R_{xx}, as the voltage drop along the current flow divided by I.

To obtain an intuitive understanding of the Hall resistance within the Büttiker-Landauer formalism, consider a geometry which readily brings out the physics, that

of a four-fold symmetric Hall junction shown in Fig. 9-3b. Because of the four-fold symmetry, we have $R_{11} = R_{22} = R_{33} = R_{44} \equiv R$, $T_{12} = T_{23} = T_{34} = T_{41} \equiv T_L$, $T_{21} = T_{32} = T_{43} = T_{14} \equiv T_R$, and $T_{13} = T_{31} = T_{24} = T_{42} \equiv T_S$, where T_L, T_R, and T_S refer to the transmission probabilities to the left, to the right, and straight through, respectively. Inserting these probabilities into Eq. 9-3, we find for $R_H = R_{13,24}$:

$$R_H = \frac{h}{e^2} \frac{(T_L - T_R)}{[(N-R)^2 + T_S^2 - 2 T_L T_R]} \qquad (9\text{-}7)$$

This formula indicates that the Hall resistances arises from the difference between the probability for an electron to turn a corner to the left versus to the right. In a magnetic field, this difference arises from the Lorentz force which bends the electron motion to one side. This formula is intuitively sensible. In subsequent sections, we will make use of it to analyze our results in several of the experiments. In the general case where no special symmetry exists at the Hall junction, R_H is still expressible in terms of the differences between transmission probabilities to the left versus to the right.

9.3 Device Fabrication

The method for the fabrication of submicron $GaAs-Al_xGa_{1-x}As$ heterostructure devices has developed independently in several research laboratories around the world. To obtain submicron resolution, electron-beam or X-ray lithography is necessary. Electron beam lithography is versatile and relatively easy to use and is the technique of choice for most research groups. To define larger features such as lead extensions from the submicron region to larger areas for electrical contact, photolithography is needed. The method

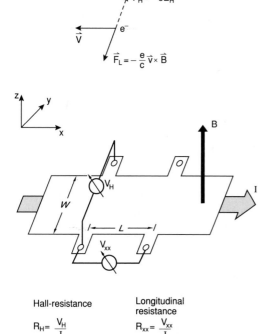

Hall-resistance

$$R_H = \frac{V_H}{I}$$

Longitudinal resistance

$$R_{xx} = \frac{V_{xx}}{I}$$

Figure 9-6. The Hall resistance, R_H, and longitudinal resistance, R_{xx}. Classically, the Hall voltage, V_H, across current flow arises from the Hall electric field, E_H, which is determined by the condition that the Lorentz force, F_L, is balanced by the force due to E_H.

varies for lateral confinement of the electron gas into the lithographically defined patterns. It can be classified into three types: 1) shallow chemical etching (van Houten et al., 1986; Behringer et al., 1987), 2) electrostatic side-gating (Thornton et al., 1986; Zheng et al., 1986), and 3) shallow ion damage (Scherer et al., 1987). The confinement mechanism for the first two techniques is electrostatic in origin, while it is thought to be due to dramatically reduced electron mobility in the ion-damaged region in the third technique. The devices used in the experiments to be discussed are all fabricated using the first technique, which we will describe below.

The fabrication proceeds in five steps: 1) indium contact alloying for ohmic contact to the electron gas, 2) electron beam lithography to define the narrow pattern, 3) metalization and lift-off to produce an etch mask for the narrow pattern, 4) photolithography to define the lead frame extending the narrow pattern to the indium contacts, and 5) chemical etching to laterally confine the electron gas underneath the etch mask. These steps are schematically illustrated in Fig. 9-7.

The detailed procedure is as follows (Owusu-Sekyere et al., 1988): First, for each contact a thin layer of indium is put down on the surface of an approximately 3 mm × 3 mm piece of $GaAs-Al_xGa_{1-x}As$ heterostructure crystal by soldering iron. Alloying is performed at 450 °C in N_2 atmosphere for five minutes. A 1000 Å thick layer of electron beam resist, polymethylmethacrylate (PMMA), is spun onto the surface and baked for two hours at 160 °C. The desired pattern containing the narrow submicron features is written onto the resist by exposure to 30 keV electrons at a dosage of $\approx 100\ \mu C\ cm^{-2}$, followed by development of the resist in a mixture of 3 parts ethylene glycol monoethyl ether (cel-

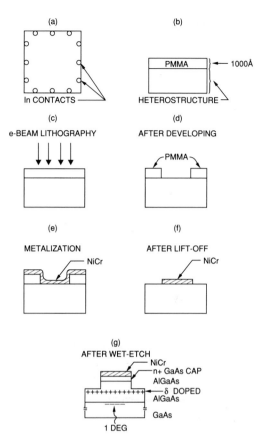

Figure 9-7. Schematic illustration of the fabrication procedure for submicron devices (Owusu-Sekyere et al., 1988).

lusolve) and 7 parts methanol in volume ratio for 15 seconds, resulting in the removal of the exposed portions. A metal, typically nichrome (250 Å) or gold (150 Å), is thermally evaporated onto the sample. Lift-off is carried out in acetone under ultrasound agitation, leaving the metal etch mask in the shape of the written pattern while elsewhere on the surface it is free of resist or metal. Next, photolithography is performed to extend the electron beam pattern out to the indium contacts to provide the leads necessary for electrical measurement. Chemical etching follows. This is the critical step, since an etch which is too shallow will not isolate the different features in the pattern, while too deep an etch will completely deplete the narrow regions rendering them non-conducting. The etching typically proceeds in two steps: the first to isolate the lead extensions, and the second, more delicate etch to isolate the narrow region. Before etching the semiconductor, however, an oxygen plasma etch is used to remove residual PMMA or photoresist polymers left on the surface in the unmasked regions; a thin layer of polymer can greatly reduce the etch rate.

The first etch is performed with the electron beam written region protected under photoresist. A rather deep etch, typically $\geq 400\,\text{Å}$ deep and well below the position of the silicon dopants in the $Al_xGa_{1-x}As$, is carried out. The etchant can either be a non-selective wet etchant such as H_3PO_4: $H_2O_2 : H_2O$ (1:1:38) with a nominal rate of $1200\,\text{Å}\,\text{min}^{-1}$, or an RF (radio frequency) plasma etch in a CCl_2F_2 and inert gas (e.g. He) mixture at typical rates of $100\,\text{Å}\,\text{min}^{-1}$ for AlGaAs and $200\,\text{Å}\,\text{min}^{-1}$ for GaAs under conditions of 5 mTorr pressure, 20 sccm (standard cubic centimeter per minute) flow rate, and 15 W RF excitation. The second etch, performed after a second photolithography step which uncovers the electron beam pattern while protecting the leads, is a shallow etch carried out to the top of the silicon dopant layer, typically $\approx 250\,\text{Å}$ deep. The difference in depth of etch between a lack of isolation (too shallow) and total depletion of submicron features (too deep) is about 15 to 40 Å. After a successful etch, the depletion width from the etched wall is about $1000-1500\,\text{Å}$. Therefore a wire which is lithographically $5000\,\text{Å}$ (0.5 μm) side would exhibit an electrical width of $\approx 2000-3000\,\text{Å}$.

9.4 Quenching of the Hall Resistance in a Novel Geometry

Conventionally, the Hall resistance at low magnetic field is well known to be proportional to B, with a coefficient equal to $1/[n(-e)c]$ in cgs units for a single carrier system, where n is the carrier density, and $-e$ the electron charge. Consequently, the measurements of the Hall resistance has developed into a standard technique for determining the carrier density of metals or semiconductors. However, recently, Roukes et al. (1987), Chang et al. (1988 b), and Timp et al. (1987 b) discovered that in ballistic $GaAs-Al_xGa_{1-x}As$ heterostructure Hall junctions of width $\approx 2000\,\text{Å}$, certain anomalies occur in the Hall resistance at low B. The most striking features are the suppression of the Hall resistance to a value near zero for $B \leq 2\,\text{kG}$ for both positive and negative fields (Roukes et al., 1987), the so-called "quenching" of the Hall resistance, and a "last-plateau" feature reminiscent of a quantized Hall plateau at fields just above the quenching region (Timp et al., 1987 b; Chang et al., 1988 b; Roukes et al., 1987). Initially, it was thought that a sharp four lead junction of uniformly narrow width is necessary for the occurrence of the quenching anomaly. This notion led to intense theoretical and numerical calculations based on the scattering of electrons by such a sharp junction (Baranger and Stone, 1989 b; Kirczenow, 1989; Ravenhall et al., 1989; Schult et al., 1990). The results were puzzling at best, since in this model junction quenching was produced only at very special values of electron density or Fermi energy, in direct contrast to experiment (Ford et al., 1988 a). The understanding of the quenching phenomenon came only after parallel break throughs in experiment (Chang et al., 1989; Ford et al., 1989) and theory (Baranger and Stone, 1989 b; Beenakker and van Houten, 1989 a) in which flaring of the junction region was introduced in novel junction geometries. In this section, we review our contribution to understanding the Hall resistance quenching and last plateau anomalies. In particular, we experimentally determine some sort of necessary and sufficient condition for observing the quenching behavior by studying a series of novel ballistic junctions consisting of a wide junction region with incoming leads out of which four, three, two or one of the leads can contain a narrow constriction. We demonstrate that only the geometry of four constrictions leading into a "quantum dot" gives rise to a full quenching of the Hall resistance over a substantial range of electron density, while junctions containing fewer constrictions exhibit little or no quenching behavior. Our results provide strong evidence that the anomalies arise from the scattering of electrons by the junction geometry, and not other mechanisms such as electron-electron interaction.

9.4.1 The Experiment

9.4.1.1 Device Characteristics

In Fig. 9-8 we show our device containing the Hall junctions. The solid lines indicate the pattern defined by lithography and the dotted lines approximate the conducting region. The lithographic dimensions are 1 µm in the wide region and 0.5 µm in the constrictions while the conducting widths are approximately 0.8 µm and 0.2 µm, respectively. The starting material for the device has a 4.2 K mobility of $320\,000\ \mathrm{cm^2\,V^{-1}\,s^{-1}}$ and a density of $3.3 \times 10^{11}\ \mathrm{cm^{-2}}$. The device itself has a mobility of $180\,000\ \mathrm{cm^2\,V^{-1}\,s^{-1}}$ and a reduced density of $2.75 \times 10^{11}\ \mathrm{cm^{-2}}$, with a corresponding elastic scattering length of

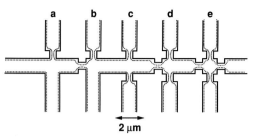

Figure 9-8. Schematic of the device containing Hall junctions of novel geometries. Solid lines indicate device as defined by lithography. Dotted lines depict approximately the conducting structure (Chang et al., 1989).

1.6 µm, which is significantly larger than the junction region. At the temperature of 4.2 K where the measurements were carried out, the inelastic scattering length, l_{in}, is expected to exceed 5 µm, while the phase coherence length, l_{Φ}, is of the order of 1.2 µm. Therefore, we are in a regime where the Büttiker-Landauer formulas directly apply, but quantum interference effects due to multiple reflections from the junction are washed out. Therefore, transport across our junctions is ballistic.

9.4.1.2 Data Presentation – Quenching in the Four-Constriction Junction and Absence of Quenching in Other Junctions

We carry out our measurements with a lock-in amplifier at 23 Hz and 5 nA excitation current where the current is passed through the main horizontal channel. Fig. 9-9 shows the Hall resistance, $R_{\mathrm{Hall}} = V_{\mathrm{Hall}}/I$, for the four-constriction junction between $\pm 3\,\mathrm{kG}$ magnetic field, at different values of the backgate voltage, V_g. A more positive gate voltage corresponds to a higher electron density. At $-200\,\mathrm{V}$, the Hall resistance is not quenched in the region about $B = 0$. A full quenching is observed commencing at $-125\,\mathrm{V}$ and extending to $+25\,\mathrm{V}$, beyond which the quenching behavior again disappears. The

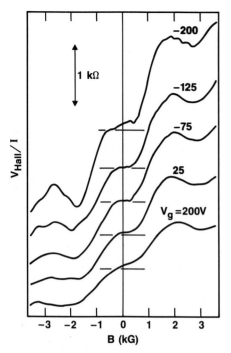

Figure 9-9. The Hall resistance, $R_{Hall} = V_{Hall}/I$, of the four-constriction junction, versus magnetic field between ± 3 kG at $T = 4.2$ K. The traces correspond to different backgate voltages (Chang et al., 1989).

appearance and disappearance of the quenching behavior has also been observed by other workers (Ford et al., 1988a and 1989; Roukes et al., 1990). The behavior versus V_g or electron density, n, can be summarized by plotting the ratio of the slope of the Hall resistance at $B = 0$, s, to the conventional slope deduced from the 3 to 6 kG region, s_0, as is shown in Fig. 9-10. It is clear that between the densities 1.6×10^{11} cm^{-2} and 2.2×10^{11} cm^{-2} quenching is observed, while at higher or lower densities, the Hall slope approaches the conventional value. A rough estimate of the number of quantum channels in the narrow constrictions yields five for the lower density limit of the quenching region and nine for the higher limit. These estimates are arrived at from the two terminal resistance of an individual constriction, which is ex-

pected to be roughly given by $h/(2ie^2)$, i being the number of channels (van Wees et al., 1988; Wharam et al., 1988). The quenching result has been duplicated in a four-constriction junction of a separate device. These results demonstrate conclusively that quenching of the Hall resistance can occur in Hall junctions of this novel geometry.

To contrast the four-constriction junction to junctions with fewer constrictions, we plot in Fig. 9-11 s/s_0 for these fewer-constriction junctions versus electron density over a similar density range. With the exception of the three-constriction case, all other junctions show no quenching at all. Even in the three-constriction case, the quenching is minimal, occurring only in a very small region about 2.25×10^{11} cm^{-2} and a small $|B|$ region (< 30 Gauss) as well. In the two opposite constriction junction, interestingly, s is nearly equal to s_0, while it is considerably reduced in the two adjacent constriction junction. Moreover, the single-constriction junction shows substantial enhancement of s over s_0. In the

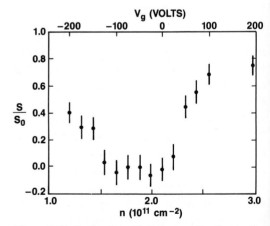

Figure 9-10. Ratio of the Hall slope at $B = 0$, s, to the conventional slope, s_0, versus electron density and backgate voltage, for the four-constriction junction (Chang et al., 1989).

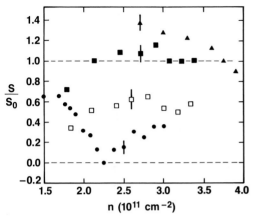

Figure 9-11. s/s_0 versus electron density for the Hall junctions containing: three constrictions, circles; two opposite constrictions, solid squares; two adjacent constrictions, open squares; and one constriction, triangles (Chang et al., 1989).

discussion below, we will show that all these behavior can be understood in terms of the scattering of electrons by the geometries of the junctions.

9.4.1.3 The Last Plateau

The second anomalous feature we would like to point out is the last plateau feature. In Fig. 9-9, a relatively flat portion reminiscent of a plateau is present in R_{Hall} above ≈ 1.5 kG. In wires of nominally uniform and narrow widths, Roukes et al. (1987) and Chang et al. (1988 b) independently attempted to relate this feature to the number of quantum channels occupied. This feature is apparently observable in our novel four-constriction geometry as well. More surprisingly, it is present even in the three- and two-constriction geometries, as can be seen in Fig. 9-12. In the case of two-constrictions, the last plateau is observed in the adjacent-constriction geometry in one B field direction only, while the opposite constriction geometry shows no plateau feature at all!

9.4.2 Discussions

9.4.2.1 Mechanisms for Quenching – Collimation and Reflection off 45° Walls

The key to understanding the behavior of the quenching for the different geometries rests on two observations. Firstly, because l_{in} is large, local equilibrium cannot be established in the junction region during transport. It is necessary to make use of the Büttiker-Landauer multi-lead formulas. Secondly, since transport is ballistic, the geometric features determine the scattering of electrons. Quenching of the Hall resistance is then a consequence of the special scattering property of particular junctions. Consider the four-constriction case. According to Eq. (9-7) for a four-fold symmetric junction, the Hall resistance, R_H, is proportional to $(T_L - T_R)$, the asymmetry between the probability for transmission to the left versus to the right. To suppress R_H to nearly zero it is necessary to either sig-

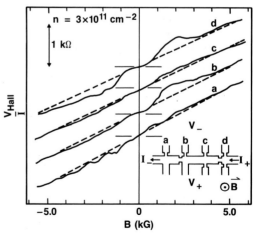

Figure 9-12. The Hall resistance, $R_{Hall} = V_{Hall}/I$, as a function of magnetic field for the junctions containing three, two, and one constriction. Note that the three-constriction and two-adjacent constriction junctions show a last plateau feature between 1 and 3 kG in $|B|$, the latter in positive B direction only. In contrast, the junctions containing two opposite and one constriction show no last plateau feature (Chang et al., 1989).

nificantly reduce the magnitude of both T_L and T_R, to equalize them, or a combination of these two. The denominator for R_H, $[(N-R)^2 + T_S^2 - 2T_L T_R]$, is insensitive to T_L or T_R provided R does not become large and approach N, the number of channels, which is the case for our structure. The four-constriction geometry achieves both magnitude reduction and equalization through the respective processes of 1) collimation of the injected electron beam to within a cone of forward directed angles (Baranger and Stone, 1989b; Beenakker and van Houten, 1989b), thereby reducing the probabilities to turn a corner, and 2) reflection of the electrons off of the approximately 45° angle walls in the wide junction region, causing some electrons to enter the opposite, or "wrong", lead, thereby equalizing T_L and T_R (Ford et al., 1989). These processes are illustrated in Fig. 9-13.

Collimation occurs under the condition where the change in width of the device is sufficiently gradual upon entering and exiting a constriction region so as to give rise to adiabatic transport (Baranger and Stone, 1989b; Beenakker and van Houten, 1989b; Glazman et al., 1988). From a quantum-mechanical view point, the basic

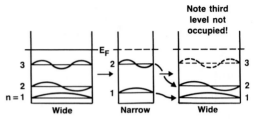

Non-Equilibrium Mode Occupation

Figure 9-14. Collimation from a quantum-mechanical view point, illustrated for a situation where three channels are occupied in the wide lead region. When the change in width is adiabatic, the transverse channel index, n, is preserved for each channel, leading to a non-equilibrium population in the wide junction region where only the lowest two channels are occupied. For these low lying channels most of the energy and momentum are associated with the forward direction, resulting in collimation.

idea for the collimation of electron waves involves an adiabatic preservation of the transverse channel index, *resulting in a selective population of the lower-lying transverse channels in the wide junction, while the higher-lying modes are no longer occupied.* Because the scattering process is elastic and conserve the energy which equals E_F, the electrons end up with most the energy and momentum in the degree of freedom associated with the forward direction, and are consequently collimated in this forward direction. Fig. 9-14 illustrates the process for a case where three channels are occupied below the Fermi energy in the wide lead region. Here, when electrons are injected from the wide region on the left into the constriction, the constriction acts as a filter and completely reflects the highest lying third channel, which has an energy above E_F within the constriction, while transmitting the two lower transverse channels. As electrons exit the constriction and enters the wide junction, the channel indices are adiabatically preserved, resulting in a population which is

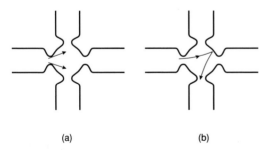

(a)　　　　　　　　　(b)

Figure 9-13. The two mechanisms that lead to quenching: (a) collimation of the electron beam by the constriction-horn structure, leading to an enhanced T_S and reduced T_L and T_R, and (b) reflection off of the 45° wall causing electrons to enter the opposite lead, leading to an equalization of T_L and T_R.

selectively populated in the two lowest modes, while the third mode remains unoccupied. Since forward-directed electrons do not turn corners as readily, even in the presence of a weak magnetic field, T_L and T_R are both reduced. Numerical calculations indicate the reduction to be about 60% (Baranger and Stone, 1989 b). A similar argument can be advanced based on semiclassical considerations, relying on an adiabatic invariant quantity (van Houten et al., 1990). In addition to acting as a collimator, the horn structure consisting of the constriction and the 45° walls is also an efficient collector of incoming electrons when the process is reserved. The end result is that T_S is enhanced at the expense of the turning probabilities. The second process of scattering off the walls is intuitively plausible. However, the quantitative demonstration that it actually equalizes T_L and T_R can only be achieved numerically (Baranger and Stone, 1989 b; Beenakker and van Houten, 1989 a). In fact, it is possible to design geometries where an "over equalization" occurs, giving rise to negative Hall slopes instead of the usual positive slope (Ford et al., 1989; Beenakker and van Houten, 1989 a; Baranger and Stone, 1989 b).

We are able to understand the results of the other geometries based on junction scattering as well. Let us examine the numerator, N_H, of the Hall resistance formula given by Eq. (9-3). For the three-constriction, two opposite constrictions, two adjacent constrictions, and one constriction junctions depicted in Fig. 9-8, N_H may be written respectively as:

$$N_H = \tfrac{1}{2}[(T_{3L} + T_{3R})(T_{1L} - T_{1R}) +$$
$$+ (T_{1L} + T_{1R})(T_{3L} - T_{3R})] \quad (9\text{-}8)$$

$$N_H = (T_{2L} + T_{2R})(T_{2L} - T_{2R}) \quad (9\text{-}9)$$

$$N_H = T_{12}\,T_{34} - T_{14}^2 = T_{1L}\,T_{3L} - T_{1R}^2 \quad (9\text{-}10)$$

$$N_H = \tfrac{1}{2}[(T_{2L} + T_{2R})(T_{4L} - T_{4R}) +$$
$$+ (T_{4L} + T_{4R})(T_{2L} - T_{2R})] \quad (9\text{-}11)$$

arrived at by making use of the relationships $T_{ij}(B) = T_{ji}(-B)$, $R_{ii}(B) = R_{ii}(-B)$, and exploiting the symmetries of each junction. We first note the behavior of R_H under B field reversal. With the exception of the two adjacent constriction junction, N_H and hence R_H are manifestly antisymmetric under field reversal, since the denominator D is always B symmetric. These predictions are in agreement with experimental data presented in Fig. 9-12. This antisymmetry property is a consequence of the presence of a mirror symmetry through either the current or voltage leads. On the other hand, the junction of two adjacent constrictions is not expected to be antisymmetric, precisely as observed (see Figs. 9-12 and 9-15). Particularly striking is the presence of a last plateau feature in one field direction and not the other and the different Hall slopes near $|B| = 0$ in the two directions.

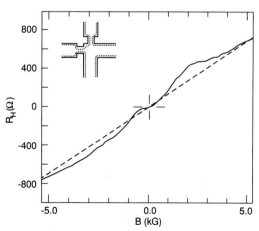

Figure 9-15. The Hall resistance for the two adjacent constriction junction at an electron density of 1.85×10^{11} cm^{-2}. Note the asymmetry of the Hall slope for $|B| > 0$, and the appearance of a last plateau in the $+B$ direction (pointing out of the page).

Next we discuss the quenching behavior. The three constriction junction comes closest to exhibiting a quenching. However, in N_H (Eq. 9-8), the term $(T_{1L} - T_{1R})$ benefits from collimation and is reduced, but there is no $45°$ wall to equalize T_{1L} and T_{1R}, while the term $(T_{3L} - T_{3R})$ does not benefit from collimation since lead 3 is without a constriction-horn structure but does have $45°$ walls for equalization. Quenching is therefore incomplete. For the two opposite constriction junction in which $T_{2L} = T_{4L}$ and $T_{2R} = T_{4R}$ (Eq. (9-9)), the horn structure is only partially in place in leads 2 and 4 and is absent in leads 1 and 3. The term $(T_{2L} - T_{2R})$ only enjoys partial reduction from incomplete collimation/collection while there is a negligible $45°$ wall section for equalization. Quenching is not expected. Furthermore, it turns out the small reduction is compensated by a comparable reduction in D. Therefore, R_H shows no reduction from the conventional result. Similar reasoning applied to the one-constriction junction (Eq. (9-11)) indicates a lack of quenching as well. However, the constriction in lead 2 reduces the probability T_{42} while enhancing T_{4L} and T_{4R}. This partially enhancement is compensated for by a reduction in T_{2L} and T_{2R}. The net result is in favor of an enhanced Hall slope. The case of the two adjacent constrictions (Eq. (9-10)) is more complicated. In a positive B field pointing out of the page, T_{1L} is reduced due to partial collimation whereas there is no collimation effect for T_{3L} or T_{1R}, thereby reducing N_H. In the opposite direction, T_{1L} is still reduced from collimation, but T_{3L} is enhanced from the expected value due to reflection from the $45°$ wall between leads 1 and 2. Because the effect of the wall tends to be larger than that of partial collimation, the product $T_{1L} T_{3L}$ is enhanced, while T_{1R} is not affected, again resulting in a reduced N_H and R_H.

9.4.2.2 Numerical Results on Junctions of Similar Geometries

The experimental result on the necessity of four constrictions to give rise to the quenching of the Hall resistance is supported by numerical calculations of Baranger and Stone (1990). Figs. 9-16 and 9-17 show the Hall resistance versus magnetic field for the four-constriction geometry, and the normalized Hall slope, s/s_0, versus Fermi energy and channel number for junctions with a different number of constrictions for different geometries, respectively. The inset in Fig. 9-16 shows the potential contours for electrostatic confinement in the four-constriction junction. It is clear that only the four-constriction geometry shows quenching over a range of Fermi energies. The trend for the other geometries are also similar to experiment.

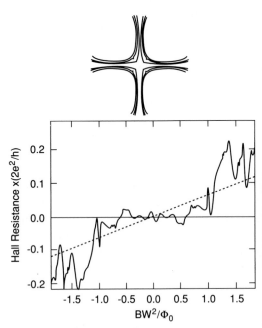

Figure 9-16. Numerical results showing quenching of the Hall resistance in a four-constriction geometry. Dashed lines show the conventional result. Inset shows the confinement potential contours of the junction used in the calculation (Baranger and Stone, 1990).

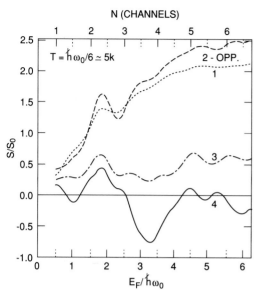

N (CHANNELS)

Figure 9-17. Numerical results for the Hall slope at $B = 0$, s, normalized to the conventional slope, s_0, versus Fermi energy on the bottom axis and channel number within the constrictions at the top. Solid line – four constrictions, dashed dotted line – three constrictions, dashed line – two opposite constrictions, and dotted line-one constriction. Note that only the four-constriction geometry shows quenching behavior, in agreement with experiment (Baranger and Stone, 1990).

9.4.2.3 The Last Plateau

The occurrence of the last plateau can now be understood. Its presence appears correlated with the constriction-horn structure which functions as collimator and collector. The two opposite constriction and one constriction junctions have no horn structure and hence do not exhibit the last plateau feature. For the other junctions, the last plateau occurs at a magnetic field for which the collimated beam from an injecting lead is bent by the Lorentz force so that most of it can be collected by a side lead which contains a horn structure (see Fig. 9-18). The magnetic field range in which this effect should occur is determined approximately by the requirement

that the radius of curvature, or cyclotron radius, be equal to the size of the junction (0.5 μm). This gives a rough characteristic field of 1.7 kG which is consistent with experiment. For the two adjacent constriction junction, this effect is only present in T_{1L} in the positive B direction pointing out of the page, and not in T_{3L} nor when B points into the page. Therefore the plateau is less developed than for the four or three constriction junctions, and is present only in one B direction.

The above discussion demonstrates that the low magnetic field Hall resistance anomalies observed in our experiment can be related to the electron scattering properties of the junction geometry. These anomalies do not involve quantum interference of electron waves, but only ballistic transport of electrons across the Hall junction. At low temperatures (≈ 100 mK), l_Φ becomes long and quantum interference effects become observable (Ford et al., 1988 a). Here, fluctuations are observed in the Hall resistance on top of the anomalies. These fluctuations arise from interference between complex trajectories involving multiple-scattering off of the junction side

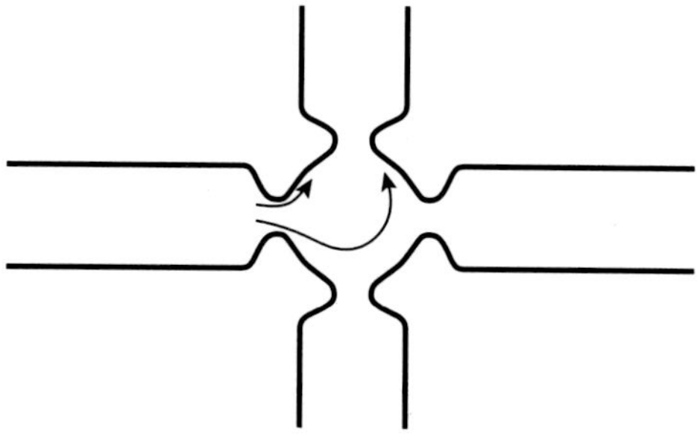

$$|\vec{F}| = m^* \frac{v^2}{R} = \frac{evB}{c} \qquad R \approx 0.5\mu m \implies B \approx 1.7 \text{ kG}$$

Figure 9-18. Mechanism for the last plateau – efficient collection of the collimated beam by the top voltage lead. The magnetic field at which this occurs is determined by the requirement that the cyclotron radius R is approximately equal to half the junction width.

walls, and are related to classical chaotic behavior of the trajectories as a function of incident angle of the electrons (Jalabert et al., 1990).

9.5 Deviation of the Quantum Hall Effect from Exact Quantization in Narrow GaAs−Al$_x$Ga$_{1-x}$As Heterostructure Wires

The most outstanding characteristic of the quantum Hall effect discovered by von Klitzing, Dorda and Pepper (1980), is the exact quantization of the Hall resistance to units of $h/e^2 \approx 25\,812.80\,\Omega$, the fundamental unit of resistance, divided by an integer i. The quantization is accurate to better than 2 parts in 10^7 when compared to known standards which are now believed to drift, and to the experimental resolution limit of 1 part in 10^8 when comparing different samples exhibiting the effect (Cage, 1987). In this section, we describe experiments on submicron quantum wires to determine conditions under which the exact quantization of the Hall resistance no longer holds, and our identification of a viable mechanism for the breakdown of exact quantization, under the condition of low current excitation in the transport process.

The quantum Hall effect is found in effectively two-dimensional electron gas systems in semiconductor devices such as the Si MOSFET or GaAs−Al$_x$Ga$_{1-x}$As heterostructure when a strong magnetic field is applied perpendicular to the 2d layer. The integer i in the quantized Hall resistance indexes the number of Landau levels occupied and can be varied by tuning either the magnetic field or electron density.

The quantization of the Hall resistance occurs with a concomitant dissipation-free transport along the current flow, giving rise to the zero resistance states in the longitudinal resistance, R_{xx}. It was recognized early on that the existence of a zero resistance state signifies an absence of back scattering of the electrons (Prange, 1987), whereas the presence of back scattering would give rise to a non-zero R_{xx}, and likely a deviation of R_H from exact quantization as well. Since back-scattering in the bulk of a 2D sample is highly suppressed, it is necessary to bring the two edges which lie parallel to current flow close together by narrowing the sample, so that back scattering between counter propagating states located at the edges can be enhanced (Niu and Thouless, 1987; Shapiro, 1986; MacDonald and Streda, 1984; Levine et al., 1984; Khmel'nitskii, 1983). The situation is illustrated in Fig. 9-19 for an impurity-free conductor in the case where the lowest Landau level is occupied. Here the eigenfunctions in the Landau gauge, $\Psi_{n,k_x}(x,y)$ are indexed by n, the Landau level (channel) index, and k_x, the wavenumber in the x-direction, see Sec. 9.2.4 and Eq. (9-5). $\Psi_{n,k_x}(x,y)$ is a plane wave along x and is Gaussian localized about $y_k = k_x l_0^2$ on the length scale of $l_0 = (\hbar c/e B)^{1/2}$ along y. The energy of the states is higher at the edges compared to states in the bulk, and the edge states on opposite edges propagate in opposite directions with a drift velocity given by $\boldsymbol{v}_d = \nabla V(x,y) \times \boldsymbol{B}/B^2$ for smooth confinement potentials $V(x,y)$. When the channel is wide ($W \gg l_0$), the opposite edge states at the Fermi energy are spatially well separated. Under this condition, a quantized Hall resistance is observed when a net current is made to flow along x by displacing the chemical potentials at the edges (Halperin, 1982). As the width is reduced until W is of the order of l_0, tunneling between opposite edge states can now take place, leading to back-scattering and a possible non-quantized value

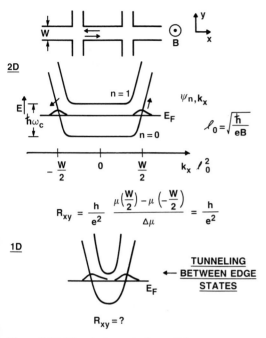

2D

$$R_{xy} = \frac{h}{e^2} \frac{\mu\left(\frac{W}{2}\right) - \mu\left(-\frac{W}{2}\right)}{\Delta\mu} = \frac{h}{e^2}$$

1D

TUNNELING
← BETWEEN EDGE
STATES

$R_{xy} = ?$

Figure 9-19. The electron single-particle energy spectrum in the presence of a strong perpendicular magnetic field, and confinement in the y direction. The width of the wires is W. For each Landau channel, n, there is a continuous set of k_x states which are plane waves along x and Gaussian localized along y within l_0, centered about $y_k = k_x l_0^2$. At the Fermi energy, two edge states propagate in opposite directions. If the channel is wide, $W \gg l_0$, the edge states do not overlap and a quantized Hall resistance is observed (Halperin, 1982). However, if W approaches l_0, the two edge states overlap and tunneling can occur across the conductor, leading to a *non-quantized* Hall resistance.

9.5.1 The Experiment

9.5.1.1 Device Characteristics

The devices we study are depicted in Fig. 9-20. The lithographic line widths are all nominally 0.5 μm, with a corresponding conducting width of 2000–3000 Å (0.2–0.3 μm). These patterns were designed to study quantum transport properties at low and high magnetic fields. At low fields, quantum interference effects dominate. Both the periodic Aharonov-Bohm effect associated with magnetic flux penetrating the area enclosed by the ring structure (Chang et al., 1988c and 1988d; Timp et al., 1987a; Ford et al., 1988b), and the aperiodic fluctuations associated with magnetic flux penetrating the area within a narrow wire (Timp et al., 1987b; Chang et al., 1988b and 1988e) are observed, the latter being related to the universal conductance fluctuations in metal and Si MOSFET wires (Umbach et al., 1984; Skocpol et al., 1987; Lee and Stone, 1985; Al'tshuler, 1985). Here we concentrate on the high magnetic field quantum Hall region. We study samples made from two different crystals, one of low mobility and the other of higher mobility. Samples of different mobilities are important since back scattering should occur more readily

of R_H. These ideas formed the basis for our experimental work. At the time the experiments were performed, the Büttiker-Landauer formulas for high magnetic field transport were as yet undeveloped. Therefore the role of the voltage leads and the need for equilibration among different Landau channels in order to observe exact quantization (Büttiker, 1988a) were not appreciated. In our analysis here, however, we will make full use of the multi-lead transport formulas.

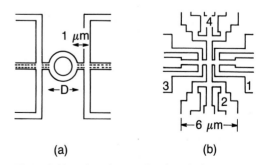

(a)   (b)

Figure 9-20. The ring and wire device patterns. Dashed line indicates conducting width (Chang et al., 1988a).

with greater amount of impurities and imperfections. The lower mobility devices were made from a crystal with mobility of $40\,000\ cm^2\ V^{-1}\ s^{-1}$ and electron density of $4.5 \times 10^{11}\ cm^{-2}$, and the higher mobility devices from a crystal of mobility $300\,000$ $cm^2\ V^{-1}\ s^{-1}$ and density $4 \times 10^{11}\ cm^{-2}$.

9.5.1.2 Data Presentation – Deviation of the Hall Resistance from Quantization, Negative Longitudinal Resistance, and Reduced Activation Energy

In Fig. 9-21, panel (a), we show the resistances R_{xx} and $R_H(R_{xy})$ as a function of magnetic field for two narrow devices in the Fig. 9-20 a pattern, and in panel (b) for a 1 mm wide device made from the same low mobility crystal in panel (b). The measurements were made at a temperature of 50 mK and 350 mK, respectively. In the narrow devices, noise-like fluctuations are observed on top of the familiar Schubnikov-de Haas oscillations in R_{xx} and Hall plateaus in R_H, in direct contrast to the smooth behavior for the wide device. These fluctuations are not temporal noises, but are exceedingly reproducible under repeated cycling of the magnetic field. They represent the magneto-finger print of a device reflecting the specific distribution of random impurities. At low magnetic fields, they are related to the universal conductance fluctuations. At high fields, they arise from a resonant tunneling mechanism discussed below (Jain and Kivelson, 1988; Büttiker, 1988 c and 1989; Sivan and Imry, 1988). These fluctuations persist up to 4.2 K in temperature and are characterized by large spacings in magnetic field between successive peaks (≈ 1 kG). Therefore they are unrelated to the Aharonov-Bohm oscillations associated with the ring structure which are expected to be in the 10 to 50 G range.

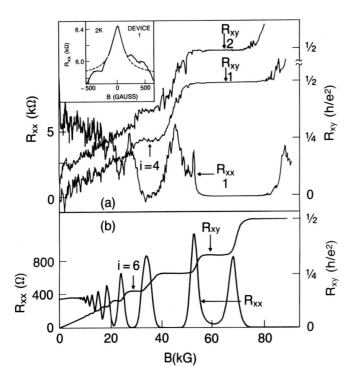

Figure 9-21. (a) R_{xx} and $R_H (R_{xy})$ versus magnetic field for the narrow devices 1 and 2, at 50 mK. Inset shows negative magneto-resistance peak in device 1 at low B from which the width is deduced. (b) Results for a wide device made from the same crystal, at 350 mK (Chang et al., 1988 a).

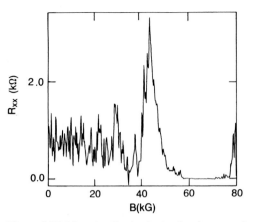

Figure 9-22. Negative R_{xx} in the $i = 4$ resistance minimum for a low mobility device of the geometry shown in Fig. 9-20(b).

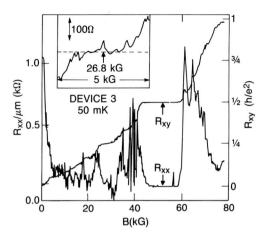

Figure 9-23. R_{xx} and R_H (R_{xy}) versus B for the higher mobility device of the pattern in Fig. 9-20 (b). Inset shows the $i = 4$ plateau region (Chang et al., 1988 a).

We focus on the $i = 4$ and $i = 2$ Hall plateaus in the narrow devices. The $i = 4$ quantum Hall plateau is no longer flat. Instead, the Hall resistance fluctuates about the quantized value of $6453.2\ \Omega$ with a peak to peak amplitude of about $250\ \Omega$. The fluctuations occur in spite of the presence of a deep minimum in R_{xx}. Similar fluctuations are also present in R_{xx}. In fact, these R_{xx} fluctuations fall below zero resistance, giving rise to a negative R_{xx}. In contrast to the behavior for $i = 4$, the $i = 2$ Hall plateau is quantized to one part in 10^3, with a concomitant zero resistance state in R_{xx}. These unusual behaviors of a negative R_{xx} and of fluctuations in R_H are observed in other devices as well as shown in Figs. 9-22 and 9-23. These devices are fabricated in the pattern of Fig. 9-20b, where the device of Fig. 9-22 is made from the low mobility crystal, while the device of Fig. 9-23 is made from the higher mobility crystal. For the latter, as a result of the higher mobility, the size of the fluctuations in the $i = 4$ plateau and in R_{xx} is smaller, of the order of $70\ \Omega$ peak to peak.

To further characterize the behavior of the resistances, we study their temperature

dependences. In Fig. 9-24 a we show the temperature evolution of the $i = 4$ Hall plateau in device 1 of Fig. 9-21 a, and in (b) the evolution of R_{xx} for the $i = 4$ and $i = 2$ minima. We observe that the aperiodic fluctuations in R_H (R_{xy}) grows in size with decreasing temperature, and contains

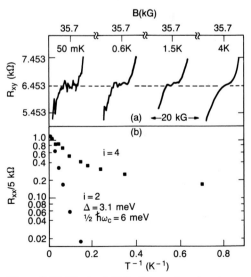

Figure 9-24. The $i = 4$ Hall plateau at various temperatures, (a), and semilog plot of R_{xx} versus $1/T$ at the $i = 2$ and $i = 4$ minima, (b), for device 1 in Fig. 9-21 (Chang et al., 1988 a).

sharper features at lower temperatures. The development of the R_{xx} resistance minima is substantially weaker in this narrow device than what it would be in this wide device. In particular the activation energy for the $i = 2$ minimum is ≈ 3.1 meV compared to the expected value of $\frac{1}{2}\hbar\omega_c = = 6$ meV (Tausenfreund and von Klitzing, 1984), and the $i = 4$ minimum approaches zero even more slowly compared to the expected activated dependence with a corresponding $\frac{1}{2}\hbar\omega_c$ of 3 meV. In the wide device, on the other hand, we have verified that the activation energy is close to $\frac{1}{2}\hbar\omega_c$ ($\approx 80\%$). Although it is not possible to directly compare the narrow and wide samples due to the reduction in electron density in the narrow device and the shift in each corresponding resistance minimum to a lower magnetic field, it is nevertheless clear that the weaker dependence is characteristic of the behavior of the narrow device. This conclusion is reached based on the observation that the $i = 2$ minimum which occurs at 72 kG shows a reduction in the activation energy, while the $i = 4$ minimum in the wide device occurring at a substantially low field of 45 kG still behaves as expected and shows no reduction in the activation energy. This unusual temperature dependence can also be understood as resulting from the tunneling of electrons between edge states on opposite sides of the wire. One additional and related feature of the data needs to be pointed out. The $i = 3$ Hall plateau is completely absent in the narrow device even though it is fully developed in the wide device.

9.5.1.3 Evidence for Resonant Tunneling via a Single Quasi-Bound State

In the original work, the cause of all these unusual effects, including 1) deviation of the $i = 4$ Hall plateau from exact quantization, 2) negative R_{xx} in the $i = 4$ minimum, and 3) reduced activation energy, were attributed to a scattering mechanism between opposite edge states by Chang et al., 1988 a, nevertheless it was puzzling why the scattering probability would be so large in view of the fact that the device width is of the order of 2300 Å compared to an l_0 of order 130 Å. A naive estimate given by the overlap of two Gaussian wavefunctions centered on opposite edges yields a probability $\approx \exp[-2(W/2l_0)^2] \approx 10^{-68}$ which is prohibitively small. The resolution of this difficulty was provided by Jain and Kivelson (1988), Büttiker (1988 c and 1989), and Sivan and Imry (1988), who proposed a resonant tunneling mechanism for back scattering from one edge to the other. The idea is that if a quasi-bound state is situated within the bulk which is resonant in energy with the edge states at the Fermi energy, back scattering can occur through such a state by first tunneling from one edge to it, and then to the other edge. In the event the size of this state is comparable to the width of the conductor, the probability for tunneling across can be dramatically enhanced. This idea is illustrated in Fig. 9-25. Here, the term quasi-bound state signifies a bound state from which an electron has a small but finite probability for tunneling out. The conductor in Fig. 9-25 a has a large quasi-bound state placed in the main conducting channel. At the position of this state the potential energy across the channel shows a repulsive hump (a hill) between the edges. The hump is assumed to be roughly circular in shape. (The potential as drawn neglects the presence of the two voltage probes.) In a strong magnetic field, the eigenstates are running waves along equipotential contours satisfying the Bohr-Sommerfeld quantization condition of the enclosure of an integral number of flux

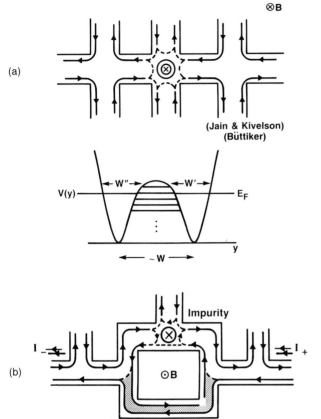

(a)

(Jain & Kivelson)
(Büttiker)

(b)

Figure 9-25. Resonant tunneling via a quasi-bound state giving rise to back scattering. (a) A wire geometry device and the potential or single-particle energy spectrum across the wire at the position of the quasi-bound state. The potential shows a hump between the two edges. Note that in drawing the potential, the voltage leads have been omitted. (b) A ring geometry device with a quasi-bound state.

quanta, $n(hc/e)$, and are Gaussian localized in the direction perpendicular to the contours (Tsukada, 1976; Kazarinov and Luryi, 1982; Iordansky, 1982; Trugman, 1983). The energy of a particular state is approximately given by the Landau level energy plus the potential energy of the contour. Within the potential hump, quasi-bound states circular in shape therefore exist, even though the potential is repulsive. Let us follow the evolution of a given state as the magnetic field is increased. To preserve the enclosed flux, the area of the state must decrease. This is accomplished by shifting the orbit toward higher potential thereby raising the energy of the state. In this manner, successive states will come in and out of resonance with the edge

states at E_F. There are two consequences of this scenario: 1) When a quasi-bound state comes in resonance with the edge states, the tunneling probability is given by $\exp[-2(W'/2l_0)^2]\exp[-2(W''/2l_0)^2]$ instead of $\exp[-2(W/2l_0)^2]$, where W' and W'' denote the distances of closest approach between the quasi-bound state equipotential contour and the two edge states. If the diameter of the contour approaches the width W, W' and W'' become small and approach l_0, the tunneling probability can then approach unity. 2) For a given potential hump (hill) resonance occurs approximately periodically in magnetic field, with a period determined by the condition that the area enclosed by the contour at E_F multiplied by the magnetic field equals nhc/e, where n is an integer which increases in units of 1, provided E_F is reasonably field independent.

At the peak of a resonance, the probability for back scattering is at a maximum. This gives rise to a maximum in R_{xx} and an extremum in R_H. The probability falls off toward zero to each side of the peak. At low temperatures, the resonance structures are expected to be sharp spikes, provided the line width is smaller than the spacing between successive resonances. The energy line width of a resonance is determined by $\hbar/\tau_{tun}$, where τ_{tun} is the tunneling time (Jain and Kivelson, 1988). The energy width can be translated into a magnetic field width by scaling the change in energy of the quasi-bound state with B. At higher temperatures, the resonances are thermally broadened. There are direct evidences that such a tunneling mechanism is present in narrow wires. In Figs. 9-26 and 9-27, we show sharp resonance structures in R_{xx} versus magnetic field for two devices made from different materials. The resonances occur on the high magnetic field side of the $i = 2$ zero resistance state, and are approximately periodic in B with a period of 0.8 kG and 2.2 kG, respectively. In Fig. 9-28, we show the Fourier power spectrum for $R_{1,11}$ (upper curve) and $R_{11,9}$ (lower curve) for the data in the inset of Fig. 9-27. Both spectra show a pronounced peak at 0.45 kG^{-1}. Since the resonances are observable in *adjacent segments* at the same magnetic field positions, the quasi-bound state most likely is situated at the junction of the middle voltage lead, as illustrated in Fig. 9-25a and b respectively for the samples of Figs. 9-26 and 9-27. From the magnetic field periods, we estimate the equipotential contour of the quasi-bound state to have a diameter of 3000 Å and 1600 Å, respectively, where we assume a roughly circular orbit and neglecting the variation of

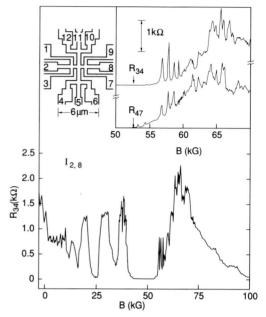

Figure 9-26. Evidence for resonant tunneling observable on the high B field side of the $i = 2$ R_{xx} zero resistance state. Inset shows the spikes in R_{xx} in an expanded view. Note the four spikes between 55 to 60 kG are nearly periodic in B.

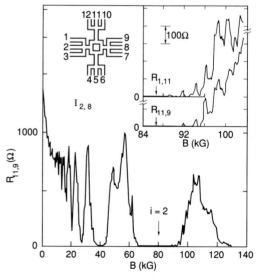

Figure 9-27. Evidence for resonant tunneling in a second device. In this device, seven spikes nearly periodic in B are visible in R_{xx}, on the high B side of the $i = 2$ zero resistance state.

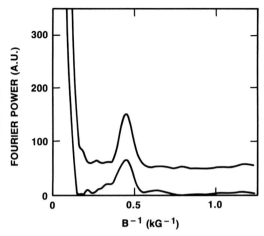

Figure 9-28. Fourier power spectra of the data in the inset of Fig. 9-27. Top curve corresponds to $R_{1,11}$, and bottom curve to $R_{11,9}$. Note the pronounced Fourier peak at $0.45\,\text{kG}^{-1}$.

E_F with B. Similar resonances have also been observed by Simmons et al. (1989). The inferred diameters are in the range of typical widths of the devices. At first, it may appear surprising to find quasi-bound states of such large sizes. However, numerical calculations of potential fluctuations due to remote ionized impurities in the $Al_xGa_{1-x}As$ dopant region indicate that hills (humps) and valleys of sizes in the 1000–2000 Å are typical, with energy fluctuations of order 10–30 meV (Nixon and Davies, 1990). Quasi-bound states of many sizes other than the width W are likely present as well. However, if the size is smaller, the tunneling probability decreases rapidly and the resonances become unobservably small. On the other hand, if many states of size $\approx W$ are present, the resonances overlap and a continuum background is observed. Individual resonances can be resolved only when one or a few states of size $\approx W$ are available to mediate tuneling.

In our discussion thus far, we have only discussed states within the same Landau

level as the edge state. It is also possible to have resonant states in the higher Landau levels bound by an attractive potential (Kircznow, 1990; Büttiker, 1990). However, these states tend to be further away from the edges.

9.5.2 Analysis and Discussions

9.5.2.1 Deviation of the Hall Resistance from Quantization

We are now in a position to explain all the unusual features of our data using this resonant tunneling mechanism. We analyze the quantum Hall resistance using the Büttiker-Landauer formulas for a four lead geometry. First consider the ideal case where no back scattering occurs, in the simplest case where only the lowest Landau channel is occupied, as shown in Fig. 9-29 a. With the magnetic field pointing into the page, the transmission and reflection probabilities are given by: $T_{14} = = T_{21} = T_{32} = T_{43} = 1$ and all others zero. The current in the leads in terms of the voltages $V_i = \mu_i/(-e)$:

$$I_1 = \frac{e^2}{h}[V_1 - V_2] = I$$

$$I_3 = \frac{e^2}{h}[V_3 - V_4] = -I$$

$$I_2 = \frac{e^2}{h}[V_2 - V_3] = 0 \tag{9-12}$$

$$I_4 = \frac{e^2}{h}[V_4 - V_1] = 0$$

yielding an exactly quantized Hall resistance of $R_H = [V_4 - V_2]/I = h/e^2$, and $V_1 = V_4$, $V_2 = V_3$. Following Büttiker (1989) the simplest scattering which gives rise to a deviation of the Hall resistance is illustrated in Fig. 9-29 b. Tunneling occurs via an elongated quasi-bound state between the edge states which are diagonally oppo-

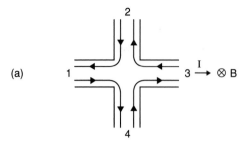

$$T_{14} = T_{21} = T_{32} = T_{43} = 1,\text{ all others are 0!}$$

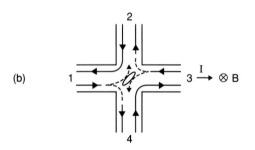

$$T_{43} = T_{21} = 1,\ T_{14} = T_{32} = 1 - \epsilon,\ T_{12} = T_{34} = \epsilon,$$
all others are 0!

Figure 9-29. The transmission and reflection probabilities in the single Landau channel limit for (a) an ideal conductor with no scattering between edges yielding a quantized Hall resistance, and (b) the simplest scattering process leading to a *non-quantized* Hall resistance (Büttiker, 1989).

site to each other originating from leads 1 and 3. We have the following transmission and reflection probabilities: $T_{43} = T_{21} = 1$, $T_{14} = T_{32} = 1 - \varepsilon$, $T_{12} = T_{34} = \varepsilon$, and all others zero. The currents are given by:

$$I_1 = \frac{e^2}{h}[V_1 - V_2] = I$$

$$I_3 = \frac{e^2}{h}[V_3 - V_4] = -I$$

$$I_2 = \frac{e^2}{h}[V_2 - \varepsilon V_1 - (1 - \varepsilon)V_3] = 0 \qquad (9\text{-}13)$$

$$I_4 = \frac{e^2}{h}[V_4 - (1 - \varepsilon)V_1 - \varepsilon V_3] = 0$$

yielding a Hall resistance of $R_{\mathrm{H}} = (h/e^2)$ $[(1 - 2\varepsilon)/(1 - \varepsilon)]$ which is no longer quantized for $\varepsilon > 0$! Let us next examine a situation where two Landau levels are occupied, corresponding to the $i = 4$ quantum Hall effect with electron-spin degeneracy which is of direct relevance to our experiment. In the simplest case, the edge states associated with the lowest Landau channel do not scatter and the edge states of the first excited channel located more into the bulk scatter in accordance to the model just discussed. We have $T_{21} = T_{43} = 1 + 1 = = 2$, $T_{14} = T_{32} = 1 + (1 - \varepsilon) = 2 - \varepsilon$, and $T_{12} = T_{34} = \varepsilon$, yielding $R_{\mathrm{H}} = (h/2e^2)[(2 - 2\varepsilon)/(2 - \varepsilon)]$, which again is not quantized to the expected value of $h/2e^2$. In fact this type of coupling can only yield a reduction in the Hall resistance from the quantized value for a magnetic field into the page. It is also possible to achieve an enhancement with a proper choice of scattering. For example, the following probabilities:

$$T_{14} = T_{21} = T_{32} = T_{43} = T_{\mathrm{R}} = 2 - \varepsilon,$$
$$T_{12} = T_{23} = T_{34} = T_{41} = T_{\mathrm{L}} = 0,$$
$$T_{13} = T_{24} = T_{31} = T_{42} = T_{\mathrm{S}} = 0, \quad \text{and}$$
$$R_{11} = R_{22} = R_{33} = R_{44} = R = \varepsilon$$

give $R_{\mathrm{H}} = (h/2e^2)[2/(2 - \varepsilon)]$ which exceeds $(h/2e^2)$. In general, it is found that exact quantization occurs only when there is no scattering in either the two current leads or the two voltage leads, or both (Büttiker, 1988 a). In the experimental data, the deviation of the $i = 4$ Hall plateau from quantization manifests itself as aperiodic fluctuations rather than periodic oscillations. This likely arises due to the presence of more than one quasi-bound state available to mediate tunneling, each with a difference enclosed area and oscillation period. The superposition of oscillations of different periods brings about the appearance of aperiodicity. In any event, the identification of a resonant tunneling mechanism for the deviation of the Hall resistance from quantization is strongly bolstered by the typical spacing of the peaks of the aperiodic fluctuations. This spacing is of the order of 1 kG, which translates into a diameter of order 2000 Å, which is nearly the width of the devices as required.

9.5.2.2 Negative Longitudinal Resistance

Using this type of analysis, it is possible to obtain a negative longitudinal resistance, R_{xx}, as well. Büttiker has shown that by assuming a Breit-Wigner form for the resonance tunneling probabilities, a negative R_{xx} can be obtained for the $i = 4$ minimum (Büttiker, 1988 c). Fig. 9-30 illustrates the required scattering, where in panel (a) we show the geometry in the actual device, and in panels (b) and (c), we illustrate successively simplified geometries which are equivalent to (a) when measuring $R_{xx} = R_{14, 23} = (V_2 - V_3)/I$, due to the fact that $V_5 = V_6 = V_1$. In the Breit-Wigner model, the transmission and reflection probabilities are given by (Büttiker, 1988 b):

$$T_{ji} = S_{j+3, i} \tag{9-14a}$$

$$R_{ii} = S_{i+3, i} \tag{9-14b}$$

where the indices are taken to be modulo 4, and

$$S_{ij} = \frac{\Gamma_i \Gamma_j}{\Delta} \quad i \neq j \tag{9-15a}$$

$$S_{ii} = N - \frac{\Gamma_i(\Gamma - \Gamma_i)}{\Delta} \tag{9-15b}$$

Here Γ_i characterizes the coupling of the i-th lead edge state to the resonant state, $\Delta = (E - E_r)^2 + \Gamma^2/4$, $\Gamma = \Sigma \Gamma_i$, N is the number of Landau channels occupied, E_r is the resonant energy, and i refers to the leads in the geometry of Fig. 9-30c. As an example, consider the $i = 4$ resistance

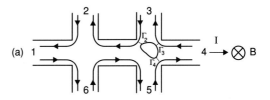

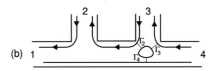

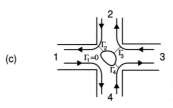

Figure 9-30. The mechanism for a negative longitudinal resistance, $R_{xx} = R_{14, 23}$. (a) Actual device configuration, (b) and (c) successively simplified equivalent configurations.

minimum for which $N = 2$. Choosing $\Gamma_1 = 0$, $\Gamma_2 = \Gamma_4 = 0.025 E_0$, $\Gamma_3 = 0.15 E_0$, we find the following transmission and reflection probabilities at resonance: $T_{21} = 2$, $T_{43} = 1.25$, $T_{12} = T_{34} = 0.0625$, $T_{14} = T_{32} = 1.5625$, $T_{13} = T_{42} = 0.375$, $T_{23} = T_{24} = T_{31} = T_{41} = 0$, $R_{11} = R_{22} = 0$, and $R_{33} = R_{44} = 0.375$, yielding a negative resistance of 230 Ω. Note that the presence of a negative resistance in a four terminal measurement does not violate thermodynamic laws. Two terminal resistances are always positive as requires.

9.5.2.3 Reduction of the Activation Energy

The last feature of our data we would like to explain is the reduced activation energy of the quantum Hall R_{xx} minima. In a wide two dimensional sample, it is generally accepted that essentially all states in the impurity-broadened Landau band are localized within the sample bulk, with the exception of a very small number of states at the band center. This is illustrated in Fig. 9-31 a where the shaded portions represent localized states. The localization length of the localized states are expected to depend on energy. It should be smallest at midgap between two Landau bands, increases toward the band center on either side, and diverges at the mobility edge corresponding to the boundary between localized and extended states. In this two dimensional situation, the activation energy for transport is approximately given by $\frac{1}{2}\hbar\omega_c$ when the Fermi energy resides at midgap, neglecting the small Zeeman energy associated with the electron spin. However, as the device width is made small, some of the states which appear localized on the scale of a wide 2D device would appear extended instead. Such states would be able to mediate tunneling between opposite edges with high probability when E_F resides in their energy range, giving rise to back scattering and an increase in R_{xx}. Therefore the region of extended states near the Landau band center is enlarged, as shown in Fig. 9-31 b. In this situation, when E_F resides at midgap the activation energy required to access the extended region is reduced, precisely as observed in the experiment. One note of caution is in order. Strictly speaking, the idea

of a smoothly increasing localization length with energy toward band center is correct only in the limit of a very large collection of states and impurities, or in effect very long wire. However, due to a finite l_Φ, it is likely only a relatively small number of impurities and states are involved in experimental situations. Nevertheless, the tendency of the localization length, or more appropriately, tunneling distance, to increase toward band center is still expected to hold. We can intuitively understand this point by observing that, for instance in the potential energy hump of Fig. 9-25 a, the size of the quasi-bound states is maximum at the bottom, corresponding to the band center (cf. Fig. 9-19), and decreases with increasing energy away from band center.

The reduced widths of the narrow devices causing states localized on the scale of 2D samples to appear extended is also responsible for the disappearance of the $i = 3$ quantum Hall plateau and *the difficulty in observing the fractional quantum Hall effect.* For these effects, the energy gap, i.e. the Zeeman energy and a fraction of the Coulomb energy are much smaller in comparison to the cyclotron energy, $\hbar\omega_c$; a small enlargement of the region of extended states will reduce the activation energy to nearly zero, rendering the temperature development exceedingly weak.

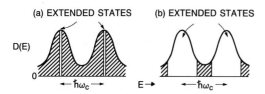

Figure 9-31. Density of bulk states in the impurity-broadened Landau bands. (a) The 2D limit where only a small number of extended states exist at the band centers, and (b) the situation for a narrow device where some states localized on the scale of a wide 2D device now appear extended.

9.6 Experiments on the Scattering of Electrons and Quasi-Particles from Regions Exhibiting the Fractional Quantum Hall Effect: Evidence for Quasi-Particle Fractional Charge

In the two experiments described in Secs. 9.4 and 9.5, the electron system is treated as non-interacting. Specifically, we demonstrated that novel effects observed in the Hall resistance at low magnetic fields in small ballistic junctions, and in the quantum Hall regime in narrow wires, can be understood in terms of the Büttiker-Landauer multi-lead transport formulas for *independent electrons*. The scattering of single-particle electron waves at the Fermi energy coupled with properties of the electron chemical potential reservoirs are responsible for the unusual effects observed, while electron-electron interaction plays an inconsequential role. Armed with the success of experiment and theory in the non-interacting regime we pursue a qualitatively different regime of quantum transport in which electron-electron interaction plays a dominant role, the regime of the fractional quantum Hall effect (Tsui et al., 1982). The simplest experiment to perform is a scattering experiment in which electrons are scattered from a spatial region exhibiting a fractional quantum Hall effect. Since the fractional quantum Hall effect arises from the formation of a novel correlated electron fluid (Laughlin, 1983), in this experiment, the electrons actually scatter off of a novel fluid. Conceptually, we may envision the fractional QHE region as a black box, and from the transmission and reflection probabilities of electron waves impinging on this black box, we deduce information about what lies within. In this sense, this experiment is akin to scattering experiments in particle and nuclear physics in which known particles are scattered from unknown particles to deduce information about their structures and properties. In terms of an actual experiment, it is conceptually most clear-cut and technically most feasible as well to scatter electrons in the single channel limit, in edge states of the lowest Landau level within the $i = 1$ integral quantum Hall effect regime. The idea is illustrated in Fig. 9-32, where a

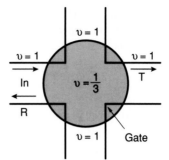

Figure 9-32. Scattering experiment of electrons in edge states of the $i = 1$, lowest spin-polarized Landau channel, off a $v = 1/3$ fractional quantum Hall region at the junction. The transmission and reflection probabilities T and R deduced should yield information about the fractional Hall state.

$v = 1/3$ fractional effect region resides in the Hall junction region while elsewhere in the leads, the electron system exhibits the $i = 1$ integral effect. Within these $i = 1$ regions in the leads, transport occurs via edge states of the lowest spin-polarized Landau channel. Edge state electrons injected in a lead toward the junction are scattered by the fractional effect region, giving rise to transmission and reflection probabilities which are related to the properties of the fractional quantum Hall state.

To achieve adjacent spatial regions of different quantum Hall effects, a gating technique is employed. Starting with a narrow device of a uniform electron density, the magnetic field is tuned to a value at which the $i = 1$ integral effect is obtained throughout. At this fixed magnetic field, a negative gate voltage is applied to the Hall junction region to reduce the density until a desired fractional quantum Hall effect, e.g. $v = 1/3$ or $v = 2/3$ effect, is obtained underneath the gate. The application of a negative gate voltage increases the electrostatic potential energy thereby introducing a potential barrier to this junction region relative to the ungated regions. Consequently, this experiment can be viewed as a barrier scattering experiment. This barrier scattering technique was pioneered in the work of Haug et al. (1988) and Washburn et al. (1988) in which scattering of Landau channel edge states in an integral quantum Hall effect by a barrier region containing fewer occupied Landau channels was achieved. In these experiments, it was demonstrated that the *longitudinal resistance* across the entire barrier region can be quantized as well as the Hall resistance, in accordance with the Büttiker-Landauer formulas. In the devices of our experiments, voltage leads are placed both inside and outside the barrier region, as shown in the schematic diagram in Fig. 9-33 a. Therefore we are able to measure the boundary resistances across a single interface between the $i = 1$ region and the barrier region at a fractional quantum Hall effect, rather than across the entire barrier region, and the Hall resistance within the barrier as well. From these resistances, it is possible to deduce the transmission and reflection probabilities defined in Fig. 9-33 b by use of the Büttiker-Landauer formulas for the $i = 1$ integral quantum Hall effect. In this Section, we will describe scattering experiments between the $i = 1$ integral quantum Hall effect and the $v_b = 1/3$ effect

in the barrier region, and between the $i = 1$ effect and the $2/3$ effect. In addition, we describe a scattering experiment between two fractional effects, $2/3$ and $1/3$, in which edge state quasi-particles in the $2/3$ effect are scattered off of the barrier region at $v = 1/3$. The following results were obtained: For transmission between the $i = 1$ and $v_b = 1/3$ effects, the total transmission probability is $1/3 \times (1.018 \pm 0.013)$ and the reflection probability is $1/3 \times (0.993 \pm 0.008)$. For transmission between the $i = 1$ and $v_b = 2/3$ effects, they are $2/3 \times (0.999 \pm 0.002)$ and $1/3 \times (0.995 \pm 0.005)$. For transmission between the $v = 2/3$ and $v_b = 1/3$ effects, they are 1.003 ± 0.005 and 0.989 ± 0.010, where these probabilities refer to $-e/3$ fractionally charged, quasi-particle edge channels. To interpret these results on the scattering of electrons and to deduce the probabilities for the third experiment, it is necessary to generalize the Büttiker-Landauer formulas to the fractional quantum Hall regime (Beenakker, 1990; MacDonald, 1990; Chang, 1990 b). According to MacDonald, the results of the experiment between the $i = 1$ and $v_b = 1/3$ effects can be interpreted as a direct measure of the charge of the quasi-particles to be $-e/3 \times (1.000 \pm 0.010)$ (Khurana, 1990). Following this idea, Chang (1990 b) has proposed that the results of the other experiments may be similarly interpreted. This interpretation is based on a picture according to which there are two $-e/3$ quasi-particle edge channels available for conduction in the $2/3$ effect. Alternatively, MacDonald (1990) has proposed one electron channel plus one $+e/3$ channel for the $2/3$ effect. The two pictures may in the end be equivalent. In our analysis we will assume the existence of edge channels. The experimental demonstration that fractional edge channels actually exist has been achieved for the $v = 2/3$ effect within a

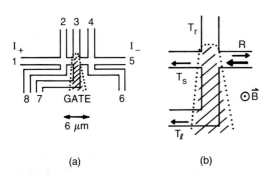

(a)   (b)

Figure 9-33. (a) Schematic of the device geometry. Shaded region indicates the top gate/barrier region. (b) Transmission and reflection probabilities for the barrier experiment (Chang and Cunningham, 1989).

bulk $i = 1$ region by Kouwenhoven et al. (1990), and for the $v = 1/3$ effect within a bulk $v = 2/3$ effect by Chang (1990a).

9.6.1 The Experiments – Device Characteristics and Data Presentation

Our experiments were carried out in two devices. Device 1 was fabricated from a crystal of mobility 300000 cm²/Vs and density 1.94×10^{11} cm⁻², and device 2 from a crystal of mobility 2 000 000 cm²/Vs and density 2.6×10^{11} cm⁻². In Fig. 9-34 we show magnetic field traces of the Hall resistance, $R_{2,8}$ and $R_{3,7}$ and the longitudinal resistance $R_{2,4}$ for device 1, at a temperature of 70 mK. The $i = 1$ integral quantum Hall effect is observed between 65 and 89 kG and the $v = 2/3$ fractional effect between 120 and 125 kG. By fixing the magnetic field at 87 kG where $i = 1$ is observed throughout the sample and varying the gate voltage at the Hall junction of voltage leads 3 and 7, we obtain the traces shown in Fig. 9-35 for the resistance pairs $R_{i,j}$ indicated on the left. Note that $R_{3,7}$ measures the Hall resistance of the gated region. $R_{2,4}$ and $R_{8,6}$ measure the longitudinal resistances across the entire gated barrier region on the top and bottom edges, respectively. $R_{2,4}$ equals the sum of the two boundary resistances $R_{2,3}$ and $R_{3,4}$ each across a single interface of the gated and ungated regions, and similarly $R_{8,6} = R_{8,7} + R_{7,6}$. $R_{2,8}$ measures the Hall resistance outside the gated region. At zero voltage, we have the $i = 1$ integral effect throughout. All Hall resistances are quantized to h/e^2 and longitudinal resistances equal to zero. As we reduce the gate voltage, $R_{3,7}$ shows a plateau at $h/(2/3) e^2$ between -100 and -120 mV, with a deep minimum in $R_{2,3}$ and $R_{7,6}$ at -107 mV. The longitudinal resistances $R_{2,4}$, $R_{8,6}$, $R_{3,4}$, and $R_{8,7}$ also show a plateau with a

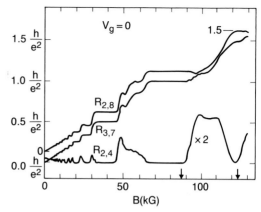

Figure 9-34. The resistances $R_{3,7}$, $R_{2,8}$, and $R_{2,4}$ as a function of magnetic field, at zero barrier gate voltage, for device 1. The temperature is 70 mK. Note the zero for $R_{2,8}$ is displaced upward (Chang and Cunningham, 1989).

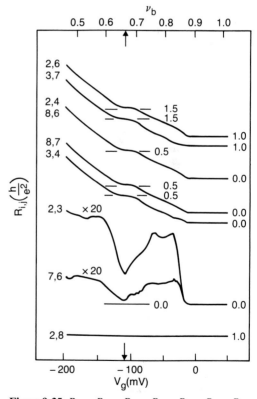

Figure 9-35. $R_{2,6}$, $R_{3,7}$, $R_{2,4}$, $R_{8,6}$, $R_{8,7}$, $R_{3,4}$, $R_{2,3}$, $R_{7,6}$, and $R_{2,8}$, versus V_g, the barrier gate voltage at the bottom axis, and the barrier filling factor, v_b, at the top axis, for the $i = 1$ to $v_b = 2/3$ scattering experiment, in device 1. The magnetic field is 87 kG (Chang and Cunningham, 1989).

nearly quantized value around $(1/2) \times h/e^2$ $= h/(2/3) e^2 - h/e^2$. Outside the barrier region, $R_{2,8}$ remains quantized to h/e^2. At 107 mV, all resistances are quantized to units of h/e^2, i.e. $h/(2/3) e^2$, $0 \times h/e^2$, or $(1/2) \times h/e^2$, to 2%. Fig. 9-36 shows similar results for device 2 obtained at 114 kG magnetic field. In addition to the structures in the resistances between 80 and 105 mV corresponding to scattering off of a barrier at $v_b = 2/3$, plateaus and minima are observed between 210 and 230 mV corresponding to scattering off of a barrier at the $v_b = 1/3$ fractional effect. At the voltages of the $R_{2,3}$ ($R_{7,6}$) minimum, the resistance quantizations are to better than 0.5% for $v_b = 2/3$ at 90 mV, and to 1% for $v_b = 1/3$ at 220 mV. In Fig. 9-37, we present results for device 1 on the $v = 2/3$ to $v_b = 1/3$ scattering experiment at the magnetic field of 123 kG. Again plateaus are observed in $R_{3,7}$, $R_{2,4}$, $R_{8,6}$, $R_{3,4}$, and

$R_{8,7}$, and minima in $R_{2,3}$ and $R_{7,6}$. The plateau in $R_{3,7}$ corresponds to the $v_b = 1/3$ fractional effect with a value near $h/(1/3) e^2$ and for the plateaus in other resistance pairs, near $(3/2) \times h/e^2 = h/(1/3) e^2 - h/(2/3) e^2$. Similar results are obtained for device 2.

9.6.2 Analysis and Discussions

We first analyze the scattering experiments for electrons in the $i = 1$ integral effect. For these experiments we readily apply the Büttiker-Landauer formulas to the spatial regions outside the barrier "black box" region. Assuming a four-fold symmetric gated junction with transmission and reflection probabilities indicated in Fig. 9-33 b, a reasonable assumption provided all regions are quantized, we find the following expressions for the resistances

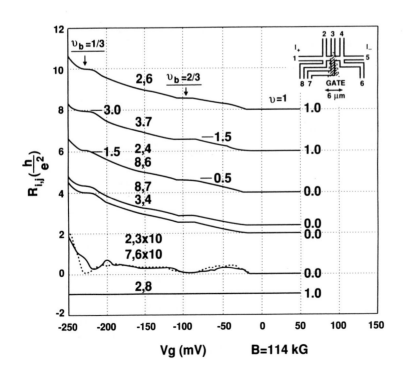

Figure 9-36. $R_{2,6}$, $R_{3,7}$, $R_{2,4}$, $R_{8,6}$, $R_{8,7}$, $R_{3,4}$, $R_{2,3}$, $R_{7,6}$, and $R_{2,8}$, versus V_g for the $i = 1$ to $v_b = 2/3$ and $v_b = 1/3$ scattering experiments in device 2. The magnetic field is 114 kG and the temperature is 80 mK.

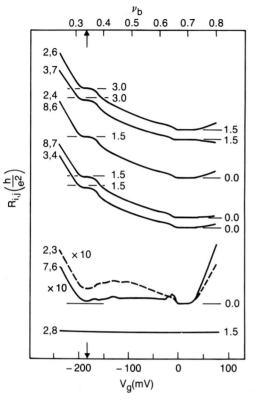

Figure 9-37. $R_{2,6}$, $R_{3,7}$, $R_{2,4}$, $R_{8,6}$, $R_{8,7}$, $R_{3,4}$, $R_{2,3}$, $R_{7,6}$, and $R_{2,8}$, versus V_g at the bottom axis, and v_b at the top, the $v = 2/3$ to $v_b = 1/3$ scattering experiment in device 1. The magnetic field is 123 kG (Chang and Cunningham, 1989).

(Chang and Cunningham, 1989):

$$R_{3,7} = \frac{h}{e^2} \frac{2d}{D^2 + d^2}$$

$$R_{2,6} = \frac{h}{e^2} \frac{2D}{D^2 + d^2}$$

$$R_{2,4} = \frac{h}{e^2} \left[\frac{2D}{D^2 + d^2} - 1 \right] =$$

$$= R_{2,6} - \frac{h}{e^2} = R_{8,6}$$

$$R_{2,3} = \frac{h}{e^2} \frac{D-d}{D^2 + d^2} = R_{7,6}$$

$$R_{3,4} = R_{2,4} - R_{2,3} = R_{8,7}$$

(9-16)

where $d = T_L - T_R$ and $D = 1 - R + T_S$. From these expressions and the measured resistances, we are able to deduce the probabilities, T_L, T_R, T_S, and R. We find for the $v_b = 2/3$ scattering experiment, $T_L = 2/3 \times (0.999 \pm 0.002)$, $R = 1/3 \times (0.995 \pm 0.005)$, and $(T_R, T_S) \leq 1/3 \times (0.005 \pm 0.005)$, at the $R_{2,3}$ minimum. For the $v_b = 1/3$ experiment, we find $T_L = 1/3 \times (1.013 \pm 0.010)$, $R = 2/3 \times (0.993 \pm 0.008)$, and $(T_R, T_S) \leq 1/3 \times (0.010 \pm 0.010)$. These results characterize the scattering properties of the $v_b = 2/3$ and $v_b = 1/3$ barriers, and are summarized in Table 9-1.

To interpret these results, it is necessary to generalize the quantum transport formulas into the regime of the fractional quantum Hall effect. The simplest approach is to follow the proposal first put forth by Chang and Cunningham (1989), for conduction in a pure fractional quantum Hall phase, that of replacement of the electron charge by q^*, the quasi-particle charge. The Büttiker-Landauer formulas become:

$$I_i = \frac{-e\, q^*}{h} \left[(N - R_{ii}) V_i - \sum_{j \neq i} T_{ji} V_j \right] \quad (9\text{-}17)$$

where the remaining charge $(-e)$ is associated with the definition of the chemical potential in the normal metal electrical contact for which $\mu_i = (-e) V_i$. A justification of these formulas will be presented below. According to these formulas, in the $v = 1/3$ fractional quantum Hall effect, transport occurs via one $q^* = -e/3$ fractionally charged quasi-particle edge channel, while in the 2/3 effect, it occurs via two $-e/3$ quasi-particle edge channels. The experimentally deduced transmission and reflection probabilities can now be interpreted as follows. In the case of scattering off of the 1/3 effect, as the electron in the $i = 1$ effect encounters the interface with the $v_b = 1/3$ barrier region, it breaks up into

Table 9-1. Summary of the transmission and reflection probabilities for scattering between: 1) the $i=1$ integral quantum Hall effect and $v_b=1/3$ fractional effect, 2) the $i=1$ and $v_b=2/3$ effects, and 3) the $v=2/3$ and $v_b=1/3$ effects.

$i=1$, $v_b=\frac{1}{3}$	$i=1$, $v_b=\frac{2}{3}$	$v=\frac{2}{3}$, $v_b=\frac{1}{3}$
$T_L = \frac{1}{3}(1.013 \pm 0.010)$	$T_L = \frac{2}{3}(0.999 \pm 0.002)$	$T_L = 1.003 \pm 0.005$
$R = \frac{2}{3}(0.993 \pm 0.008)$	$R = \frac{1}{3}(0.995 \pm 0.005)$	$R = 0.989 \pm 0.010$
$(T_R, T_S) \leq \frac{1}{3}(0.013 \pm 0.010)$	$(T_R, T_S) \leq \frac{1}{3}(0.005 \pm 0.005)$	$(T_R, T_S) \leq 0.010 \pm 0.010$
$q^* = \dfrac{-e}{3}(1.000 \pm 0.010)$	$q^* = \dfrac{-e}{3}(1.000 \pm 0.005)$	

three $-e/3$ charged quasi-particles. Since only one quasi-particle channel is available for conduction in the $1/3$ region, one of the three quasi-particles is transmitted while the other two are reflected. This gives rise to a transmission probability of exactly $1/3$ and a reflection probability of $2/3$ (Mac-Donald, 1990; Chang, 1990 b). See Fig. 9-38. In this sense, our experimental result represents a direct measure of the quasi-particle charge q^* (Khurana, 1990). In a similar way, we interpret the results for the $v_b=2/3$ experiment, although such an interpretation is somewhat controversial due

to the possibility of viewing the $2/3$ effect as having one electron channel plus an $e/3$ quasi-hole channel instead (MacDonald, 1990). It is the belief of the author that in the end, the two pictures are equivalent. Taking the view of two $-e/3$ quasi-particle channels for the $2/3$ effect, we expect for scattering between $i=1$ and v_b that two quasi-particle channels are transmitted and one reflected, giving rise to a transmission probability of $2/3$ and a reflection probability of $1/3$, in precise agreement with experiment. We are also able to analyze the results for scattering between the $v=2/3$ and $v_b=1/3$ effects. Since the $v=2/3$ fractional effect has two $-e/3$ quasi-particle channels while the $v=1/3$ effect has one $-e/3$ quasi-particle channel, situation is completely analogous to scattering between the $i=2$ and $i=1$ integral effects. We expect one quasi-particle channel to be transmitted and one reflected, yielding exactly the measured values for the resistances, when the generalized transport formulas (Eq. 9-17) are applied.

9.6.3 Generalized Büttiker-Landauer Formulas for the Fractional Quantum Hall Regime

To generalize the Büttiker-Landauer formulas to the fractional quantum Hall regime, we rely on two key ingredients: 1) a microscopic wavefunction for a pure

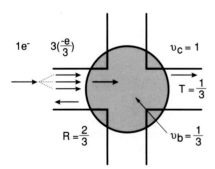

Figure 9-38. Interpretation of the results of the $i=1$ to $v_b=1/3$ scattering experiment in terms of $-e/3$ fractionally-charged quasi-particles. As an electron approaches the $v_b=1/3$ interface, it breaks up into three $-e/3$ charged quasi-particles. Since in the $1/3$ effect, only one quasi-particle channel is available for conduction, one quasi-particle is transmitted while the two others are reflected, leading to a total transmission probability of $1/3$ and a reflection probability of $2/3$.

phase of an incompressible fractional quantum Hall state, in the presence of a confinement potential and hence edges or boundaries (Chang, 1990b), and 2) phase separation of the electron system into different fractional quantum Hall incompressible fluids and possible compressible fluids as well for soft confinement (Beenakker, 1990; MacDonald, 1990; Chang, 1990b) here we follow the treatment of Chang (1990b).

To construct a microscopic wavefunction for a pure phase in the presence of a confinement potential with or without impurity fluctuation potential, we propose to simply write down a Jastrow type wavefunction involving the single particle eigenstates in manner analogous to what Laughlin (1983) has done in the 2D case without confinement or fluctuation potentials. However, in our case we only include states with energy below the equilibrium chemical potential, μ_0, which equals the Fermi energy in the metallic contacts. Here, the single particle eigenstates are states in a self-consistent potential including the background bare potential and the potential due to the other electrons. As a result of the presence of confinement and fluctuations, the energy is no longer identical for all states within a Landau level; the single particle states situated at the edges are on the average more energetic than those in the bulk. This prescription for constructing a pure phase state is valid provided μ_0 is less than the gap for creation of quasi-particles or holes. As an example, consider the case of the $v = 1/m$ fractional quantum Hall state in the presence of a parabolic confinement potential, $V(y) = 1/2\, m^* \omega_0^2 y^2$, without fluctuations. The conductor has a length of L_x in the x-direction with $L_x \to \infty$. This limit is taken to obtain a system which is open and therefore able to conduct electricity. Making use

of the single particle wavefunctions in the Landau gauge given by Eq. (9-5), we write the following Jastrow wavefunction (Chang, 1990b):

$$\Psi_{\frac{1}{m}} = N^{-1} \prod_{i>j} \sin^m \left[\frac{k_0}{2} (z_i - z_j) \right] e^{-\sum_i \frac{y_i^2}{2 l^2}}$$

$$z_j = x_j - i\alpha y_j \quad \alpha = \frac{\omega_c}{\omega} \quad k_0 = \frac{2\pi}{L_x} \quad (9\text{-}18)$$

where the product only involves states of the lowest Landau channel with energy below μ_0, N is a normalization constant, and α parametrizes the ellipticity of the wavefunctions introduced by confinement. A symbolic representation of this wavefunction for the 1/3 state is depicted in Fig. 9-39a where each circle represents a single particle state corresponding to a given k centered at $y_k = (\omega_c/\omega) k l^2$, and the notation of every third state occupied indicates an average density of $1/3 (e B/h)$ but *does*

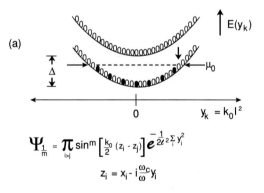

$$\Psi_{\frac{1}{m}} = \prod_{i>j} \sin^m \left[\frac{k_0}{2} (z_i - z_j) \right] e^{-\frac{1}{2 l^2} \sum y_i^2}$$

$$z_i = x_i - i\frac{\omega_c}{\omega} y_i$$

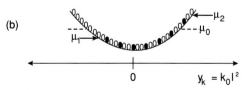

Figure 9-39. (a) Schematic representation of the Jastrow wavefunction proposed for the 1/3 fractional quantum Hall state in the presence of parabolic confinement. (b) An excited state wavefunction which carries current obtained from the ground state by a displacement in the y direction.

*not signify that only these states are in-
cluded in the product.* This wavefunction is
incompressible and approaches that written down by Thouless for the 2D situation
for $\alpha \to 1$ (Thouless, 1984). This ground
state wavefunction has zero expectation for
the x-velocity, $v_x = (1/m^*)[p_x + (e/c)A_x]$,
and therefore carries no current. To obtain
an excited state which carries current that
is not a gapped excitation, we displace the
ground state wavefunction in Eq. (9-18)
by the operator $\prod_j \exp \{i[x_j/\alpha - (\alpha l_0^2/\hbar)p_{y_j}$
$- i(\alpha^2 l_0^2/\hbar)p_{x_j}]\Delta y/l^2\}$. The wavefunction
which results corresponds to a situation
with unbalanced chemical potentials on
the top and bottom edges with $\Delta \mu =$
$= [\partial E/\partial y|_{y_{top}} - \partial E/\partial y|_{y_{bottom}}]\Delta y$, and has
a conductance of $v\,e^2/h$, as desired. See Fig.
9-39b. In a similar manner, a wavefunction
can be written down for other fractional
quantum Hall states once the 2D wavefunction is known. For example, the
$v = 1 - 1/m$ state wavefunction can be obtained in the second quantized representation by substituting the creation operators,
a_i^+, operating on the vacuum state, with
the destruction operators, $b_i^+ = a_i$, operating on the filled lowest Landau level, the
"hole vacuum state". The operator a_i^+ acting on the vacuum state creates a state in
which the i-th single particle state is occupied. Similarly, b_i^+ acting on the "hole vacuum state" creates a hole in the i-th state in
an otherwise filled Landau level. The fractional quantum Hall wavefunction is a
weighted sum of single-particle product
wavefunctions. Therefore to make a
$v = 1 - 1/m$ state in the presence of confinement and/or fluctuations, we first fill all
the single particle states with energy below
μ_0, and then operate on it with the
weighted sum of products of b_i^+'s corresponding to the weighted sum of products
of a_i^+'s for the $v = 1/m$ state.

Now, imagine a simple scattering situation involving a pure phase of a $v = 1/m$
fractional effect where there is a single infinitely massive impurity within the conducting channel which causes back scattering in an otherwise impurity-free device.
Just as in the case of the integral quantum
Hall effect for single particle states, momentum is no longer conserved. Naively
one may expect that back scattering would
couple coherent fractional quantum Hall
states at the same energy but different total
x-momentum each obtained from the
ground state (Eq. (9-18)) by operating on it
with the appropriated displacement operator. These two states may be labeled $+K$
and $-K$. It appears reasonable to define
transmission and reflection probabilities
for these two states. However, a finite reflection probability would lead to a density
of $(1 + R)\,v\,(e\,B/h)$ everywhere except at the
edges, and would not be admissible. *Instead, we construct a coherent Jastrow
wavefunction from the single particle scattering states.* In this situation, the net current carried still arises from an imbalance
in the chemical potentials of the reservoirs
injecting right versus left going scattering
states, and the transmission and reflection
probabilities refer to those of the single-
particle states near μ_0. Note however these
single-particle states enter into the wavefunction with definite amplitude and phase
relations. In this manner, the generalized
transport formulas of Eq. (9-17) are obtained.

For the case of fractional effects other
than the $v = 1/m$ effects the situation is
more complicated. Here we discuss the
$v = 1 - 1/m$ effects and in particular the 2/3
effect only. There are two possibilities –
that of a pure 2/3 phase for abrupt
boundaries, and a phase separated situation with 2/3 into the bulk and 1/3 states
close to the edges for soft (slowly rising

potential) boundaries. In either case, since the potential increases towards the edges, *the channels for conduction must be electron-like*! It is therefore sensible to view the conduction as occurring via *quasi-particle channels*. In fact, if we write down the free energy density of the system under appropriate assumptions, the 2/3 state appears as a state of two filled channels each accommodating a density equal to 1/3 the density of a filled Landau level. However, in the case of a pure phase, the two channels are collapsed on each other at the boundary and cannot be separated. In the case of soft boundary, separation is possible and is experimentally observed (Chang, 1990a). In a scattering situation, Jastrow wavefunctions again can be written in terms of the scattering states, yielding a generalized Büttiker-Landauer formulas of the form of Eq. (9-17).

With the success of these generalized formulas in explaining the experiments (Chang and Cunningham, 1989; Kouwenhoven et al., 1990; Chang, 1990a), we are beginning to have a good understanding of quantum transport in this new regime where electron-electron interaction plays a dominant role.

9.7 Acknowledgements

This work is the outgrowth of my collaboration with many colleagues. I would like to thank Harold Baranger and Markus Büttiker for deepening my understanding of the subject, Tao Chang and Jack Cunningham for the excellent crystals they have provided me, Kofi Owusu-Sekyere, R. Behringer, G. Timp, and P. Mankiewich for fabrication and measurements and R. Howard for discussions at the early stages of some experiments.

9.8 References

Aharonov, Y., Bohm, D. (1959), *Phys. Rev. 115*, 485.

Alphenaar, B. W., McEuen, P. L., Wheeler, R. G., Sacks, R. N. (1990), *Phys. Rev. Lett. 64*, 677.

Al'tshuler, B. L. (1985), *Pis'ma Zh. Eksp. Teor. Fiz. 41*, 530 (JETP Lett. 41, 648).

Baranger, H. U., Stone, A. D. (1989a), *Phys. Rev. B 40*, 8169.

Baranger, H. U., Stone, A. D. (1989b), *Phys. Rev. Lett. 63*, 414.

Baranger, H. U., Stone, A. D. (1990), *Surf. Science 229*, 212.

Beenakker, C. W. J. (1990), *Phys. Rev. Lett. 64*, 216.

Beenakker, C. W. J., van Houten, H. (1989a), *Phys. Rev. B 39*, 10445.

Beenakker, C. W. J., van Houten, H. (1989b), *Phys. Rev. Lett. 63*, 1857.

Behringer, R. E., Mankiewich, P. M., Howard, R. E. (1987), *J. Vac. Sci. Technol. B 5*, 326.

Benoit, A., Washburn, S., Umbach, C. P., Laibowitz, R. B., Webb, R. A. (1986). *Phys. Rev. Lett. 57*, 1765.

Berggren, K. F., Thornton, T. J., Newson, D. J., Pepper, M. (1986), *Phys. Rev. Lett. 57*, 1769.

Büttiker, M. (1986), *Phys. Rev. Lett. 57*, 1761.

Büttiker, M. (1988a), *Phys. Rev. B 38*, 9375.

Büttiker, M. (1988b), *IBM J. Res. Dev. 32*, 317.

Büttiker, M. (1988c), *Phys. Rev. B 38*, 12724.

Büttiker, M. (1989), *Phys. Rev. Lett. 62*, 229.

Büttiker, M. (1990), unpublished.

Büttiker, M., Imry, Y. (1985), *J. Phys. C 18*, L467.

Büttiker, M., Imry, Y., Landauer, R., Pinhas, S. (1985), *Phys. Rev. B 31*, 6207.

Cage, M. (1987), in: *The Quantum Hall Effect*: Prange, R. E, Girvin, S. M. (Eds.), Berlin, New York: Springer-Verlag.

Chang, A. M. (1990a), unpublished.

Chang, A. M. (1990b), *Solid State Comm. 74*, 871.

Chang, A. M., Cunningham, J. E. (1989), *Solid State Comm. 72*, 651.

Chang, A. M., Timp, G., Chang, T. Y., Cunningham, J. E., Mankiewich, P. M., Behringer, R. E., Howard, R. E. (1988a), *Solid State Comm. 67*, 769.

Chang, A. M., Timp, G., Howard, R. E., Behringer, R. E., Mankiewich, P. M., Cunningham, J. E., Chang, T. Y., Chelluri, B. (1988b), *Superlattices and Microstructures 4*, 515.

Chang, A. M., Timp, G., Chang, T. Y., Cunningham, J. E., Chelluri, B., Mankiewich, P. M., Behringer, R. E., Howard, R. E. (1988c), *Surf. Science 196*, 46.

Chang, A. M., Owusu-Sekyere, K., Chang, T. Y. (1988d), *Solid State Comm. 67*, 1027.

Chang, A. M., Timp, G., Cunningham, J. E., Mankiewich, P. M., Behringer, R. E., Howard, R. E., Baranger, H. U. (1988e), *Phys. Rev. B 37*, 2745.

Chang, A. M., Chang, T. Y., Baranger, H. U. (1989), *Phys. Rev. Lett. 63*, 996.

Enquist, H.-L., Anderson, P. W. (1981), *Phys. Rev. B 24*, 1151.

Ford, C. J. B., Thornton, T. J., Newbury, R., Pepper, M., Ahmed, H., Peacock, D. C., Ritchie, D. A., Frost, J. E. F., Jones, G. A. C. (1988 a), *Phys. Rev. B 38*, 8518.

Ford, C. J. B., Thornton, T. J., Newbury, R., Pepper, M., Ahmed, H., Foxon, C. T., Harris, J. J., Roberts, C. (1988 b), *J. Phys. C 21*, L325.

Ford, C. J. B., Washburn, S., Büttiker, M., Knoedler, C. M., Hong, J. M. (1989), *Phys. Rev. Lett. 62*, 2724.

Glazman, L. I., Lesovick, G. B., Khmel'nitskii, D. E., Shekhter, R. I. (1988), *Pis'ma Zh. Eksp. Teor. Fiz. 48*, 218 (JETP Lett. 48, 238).

Halperin, B. I. (1982), *Phys. Rev. B 25*, 2185.

Haug, R., von Klitzing, K. (1989 a), *Europhys. Lett. 10*, 489.

Haug, R., von Klitzing, K. (1989 b), unpublished.

Haug, R., MacDonald, A. H., Streda, P., von Klitzing, K. (1988), *Phys. Rev. Lett. 61*, 2797.

Iordansky, S. V. (1982), *Solid State Comm. 43*, 1.

Jain, J. K., Kivelson, S. A. (1988), *Phys. Rev. Lett. 60*, 1542.

Jalabert, R. A., Baranger, H. U., Stone, A. D. (1990), *Phys. Rev. Lett. 65*, 2442.

Kaplan, S. B., Warren, A. C. (1986), *Phys. Rev. B 34*, 1346.

Kazarinov, R. F., Luryi, S. (1982), *Phys. Rev. B 25*, 7626.

Khmel'nitskii, D. E. (1983), *Pis'ma Zh. Eksp. Teor. Fiz. 38*, 669 (JETP Lett. 38, 553).

Khurana, A. (1990), *Physics Today, Search and Discovery Section, 43*, 19.

Kirczenow, G. (1989), *Phys. Rev. Lett. 62*, 2993.

Kirczenow, G. (1990), unpublished.

Komiyama, S., Hirai, H., Sasa, S., Hiyamizu, S. (1989), *Phys. Rev. B 40*, 12 586.

Kouwenhoven, L. P., van Wees, B. J., van der Vaart, N. C., Harmans, C. J. P. M., Timmering, C. E., Foxon, C. T. (1990), *Phys. Rev. Lett. 64*, 685.

Landauer, R. (1970), *Philos. Mag. 21*, 863.

Laughlin, R. B. (1983), *Phys. Rev. Lett. 50*, 1395.

Lee, P. A., Stone, A. D. (1985), *Phys. Rev. Lett. 55*, 1622.

Levine, H., Libby, S. B., Pruisken, A. M. M. (1984), *Nucl. Phys. B 240*, 30.

MacDonald, A. H. (1990), *Phys. Rev. Lett. 64*, 220.

MacDonald, A. H., Streda, P. (1984), *Phys. Rev. B 29*, 1616.

McEuen, P. L., Szafer, A., Richter, C. A., Alphenaar, B. W., Jain, J. K., Stone, A. D., Wheeler, R. G., Sacks, R. N. (1990), *Phys. Rev. Lett. 64*, 2062.

Niu, Q., Thouless, D. J. (1987), *Phys. Rev. B 35*, 2188.

Nixon, J. A., Davies, J. H. (1990), *Phys. Rev. B 41*, 7929.

Owusu-Sekyere, K., Chang, A. M., Chang, T. Y. (1988), *Appl. Phys. Lett. 52*, 1246.

Prange, R. E. (1987), in: *The Quantum Hall Effect.*

Prange, R. E., Girvin, S. M. (Eds.). Berlin, New York: Springer-Verlag.

Ravenhall, D. G., Wyld, H. W., Schult, R. L. (1989), *Phys. Rev. Lett. 62*, 1780.

Roukes, M. L., Scherer, A., Allen, S. J., Craighead, H. G., Ruthen, R. M., Beebe, E. D., Harbison, J. P. (1987), *Phys. Rev. Lett. 59*, 3011.

Roukes, M. L., Scherer, A., Van der Gaag, B. P. (1990), *Phys. Rev. Lett. 64*, 1154.

Scherer, A., Roukes, M. L., Craighead, H. G., Ruthen, R. M., Beebe, E. D., Harbison, J. P. (1987), *Appl. Phys. Lett. 51*, 2133.

Schult, R. L., Wyld, H. W., Ravenhall, D. G. (1990), *Phys. Rev. B 41*, 12760.

Shapiro, B. (1986), *J. Phys. C 19*, 4709.

Simmons, J. A., Wei, H. P., Engel, L. W., Tsui, D. C., Shayegan, M. (1989), *Phys. Rev. Lett. 63*, 1731.

Sivan, U., Imry, Y. (1988), *Phys. Rev. Lett. 61*, 1001.

Skocpol, W. J., Mankiewich, P. M., Howard, R. E., Jackel, L. D., Tennant, D. M., Stone, A. D. (1987), *Phys. Rev. Lett. 56*, 2865.

Stone, A. D., Szafer, A. (1988), *IBM J. of Res. and Dev. 32*, 384.

Takagaki, Y., Gamo, K., Namba, S., Ishida, S., Takaoka, S., Murase, K., Ishibashi, K., Aoyagi, Y. (1988), *Solid State Comm. 68*, 1051.

Tausenfreund, B., von Klitzing, K. (1984), *Surf. Science 142*, 220.

Thornton, T. J., Pepper, M., Ahmed, H., Andrews, D., Davies, G. J. (1986), *Phys. Rev. Lett. 56*, 1198.

Thouless, D. J. (1984), *Surf. Science 142*, 147.

Timp, G., Chang, A. M., Cunningham, J. E., Chang, T. Y., Mankiewich, P., Behringer, R., Howard, R. E. (1987a), *Phys. Rev. Lett. 58*, 2814.

Timp, G., Chang, A. M., Mankiewich, P., Behringer, R., Cunningham, J. E., Chang, T. Y., Howard, R. E. (1987b), *Phys. Rev. Lett. 59*, 732.

Timp, G., Baranger, H. U., de Vegvar, P., Cunningham, J. E., Howard, R. E., Behringer, R. E., Mankiewich, P. M. (1988), *Phys. Rev. Lett. 60*, 2081.

Timp, G. (1990), to appear in: *Semiconductors and Semimetals*. Reed, M. (Ed.). New York: Academic Press.

Trugman, S. A. (1983), *Phys. Rev. B 27*, 7539.

Tsui, D. C., Stormer, H. L., Gossard, A. C. (1982), *Phys. Rev. Lett. 48*, 1559.

Tsukada, M. (1976), *J. Phys. Soc. Japan 41*, 1466.

Umbach, C. P., Washburn, S., Laibowitz, R. B., Webb, R. A. (1984), *Phys. Rev. B 30*, 4048.

van Houten, H., van Wees, B. J., Heijman, M. G. J., Andre, J. P. (1986), *Appl. Phys. Lett. 49*, 1781.

van Houten, H., van Wees, B. J., Mooji, J. E., Beenakker, C. W. J., Williamson, J. G., Foxon, C. T. (1988), *Europhys. Lett. 5*, 721.

van Houten, H., Beenakker, C. W. J., van Wees, B. J. (1990), to appear in: *Semiconductors and Semimetals*. Reed, M. (Ed.). New York: Academic Press.

van Wees, B. J., van Houten, H., Beenakker, C. W. J., Williamson, J. G., Kouwenhoven, L. P., van der Marel, D., Foxon, C. T. (1988), *Phys. Rev. Lett. 60*, 848.

van Wees, B. J., Willems, E. M. M., Harmans, C. J. P. M., Beenakker, C. W. J., van Houten, H., Williamson, J. G., Foxon, C. T., Harris, J. J. (1989), *Phys. Rev. Lett. 62*, 1181.

von Klitzing, K., Dorda, G., Pepper, M. (1980), *Phys. Rev. Lett. 45*, 494.

Washburn, S., Fowler, A. B., Schmid, H., Kern, D. (1988), *Phys. Rev. Lett. 61*, 2801.

Wharam, D. A., Thornton, T. J., Newbury, R., Pepper, M., Ahmed, H., Frost, J. E. F., Hasko, D. G., Peacock, D. C., Ritchie, D. A., Jones, G. A. C. (1988), *J. Phys. C 21*, L209.

Zheng, H. Z., Wei, H. P., Tsui, D. C., Weimann, G. (1986), *Phys. Rev. B 34*, 5635.

General Reading

Beaumont, S. P., Sotomayor-Torres, C. M. (Eds.) (1989), *Science and Engineering of 1- and 0-Dimensional Semiconductors*. London: Plenum.

Buttiker, M. (1988), *IBM J. Res. Dev. 32*, 317.

Chakraborty, T., Pietilainen, P. (Eds.) (1988), *The Fractional Quantum Hall Effect*. Berlin: Springer.

Chamberlain, Eaves, L., Portal, J. C. (Eds.). *Electronic Properties of Multilayers and Low-Dimensional Semiconductor Structures*. London: Plenum, to be published.

Chang, A. M., Timp, G., Howard, R. E., Behringer, R. E., Mankiewich, P. M., Cunningham, J. E., Chang, T. Y., Chelluri, B. (1988), *Superlattices and Microstructures 4*, 515.

Heinrich, H., Bauer, G., Kuchar, F. (Eds.) (1988), *Physics and Technology of Submicron Structures*. Berlin: Springer.

Landauer, R. (1985), in: *Localization, Interaction, and Transport Phenomena*, Kramer, B., Bergmann, G., Bruynseraede, Y. (Eds.). Berlin: Springer.

Prange, R. E., Girvin, S. M. (Eds.) (1987), *The Quantum Hall Effect*. Berlin: Springer.

Reed, M., Kirk, W. P. (Eds.) (1989), *Nanostructure Physics and Fabrication*. New York: Academic Press.

Timp, G., Chang, A. M., de Vegvar, P., Howard, R. E., Behringer, R., Cunningham, J. E., Mankiewich, P. (1988), *Surf. Science 196*, 68.

10 Material Properties of Hydrogenated Amorphous Silicon

R. A. Street and K. Winer

Xerox Palo Alto Research Center, Palo Alto, CA, U.S.A.

List of Symbols and Abbreviations

A	average recombination constant
a_D	distance between sites
c_d	creation probability of defects
d	sample thickness
d_0	characteristic distribution coefficient
D^-	negatively charged defect states
D_0	prefactor of diffusion coefficient
D_H	hydrogen diffusion coefficient
$d(P)$	phosphorus distribution coefficient
e	electron charge
E	energy from the mobility edge
E	energy
E_a	activation energy of conductivity
E_B	activation energy barrier of a thermal defect creation process
E_C	mobility edge energy of conduction band
E_{0C}	slope of conduction band tail
E_{0V}	slope of valence band tail
E_D	gap state energy associated with U_{D0}
E_D	demarcation energy
E_F	Fermi energy
E_{HD}	energy to release hydrogen from a Si–H bond
E_P	gap state energy associated with donor
E_T	trap depth below conduction band
E_V	mobility edge energy of valence band
E_{WB}	gap state energy of the valence band tail state
E_X	energy provided by recombinations
E_σ	conductivity activation energy
ΔE	shift of defect band
ΔE_F	shift in Fermi energy
ΔE_{ion}	change of energy of the ion core interaction with and without a defect
$f(E,T)$	Fermi function
F	electric field
g	Landé-g-factor
G	illumination intensity
ΔH	hyperfine splitting
I	nuclear spin
k	crystal momentum
k	Boltzmann constant
$k(t)$	time dependent rate constant
K	reaction constant
n	charge carrier density
n_0	carrier density prefactor
N	number of valence electrons

$N(E)$	density of states distribution function
$N(E_C)$	distribution of conduction band energies
$N(E_T)$	distribution of trap energies
N_0	concentration of the 4-fold silicon sites
$N_0(U)$	distribution function of the formation energy
n_{BT}	density of excess electrons in band edge states
N_C	effective conduction band density of states
N_D	defect concentration
$N_D(t)$	time dependent defect density
Δn_{BT}	decrease of excess electron density
N_{D0}	neutral defect density
N_{D^-}	negatively charged defect density
N_{DB}	equilibrium defect density
N_{DB0}	neutral defect density for unshifted Fermi energy
N_{don}	dopant concentration
N_H	hydrogen concentration
N_T	trap concentration
N_{V0}	band tail density-of-states prefactor
p	hole concentration
R_1	Si–Si bond length
R_C	cooling rate
t	time
T	temperature
T_0	effective temperature of trap distribution
T_C	slope of the exponential conduction band tail
T_E	freezing temperature
T_G	glass transition temperature
T_m	parameter describing the Meyer-Neldel rule
T_V	slope of the exponential valence band tail
U_C	defect correlation energy
U_D	formation energy of the defect
U_{D0}	formation energy of the neutral defects
U_P	formation energy of the donor
U_{P0}	formation energy of the neutral dopant
V	volume
V_0	disorder potential
V_0	glass volume
W	r.f. plasma power
X	gas phase mole fraction
X_I	mole fraction of impurity I
X_P	mole fraction of P
Z	coordination number, number of neighbors in amorphous material

α	dispersion parameter, T/T_0
α	deposition rate per unit r.f. plasma power
β	pyrolytic rate constant
β	stretched exponential parameter, T/T_E
γ	temperature coefficient of Fermi energy
θ	bond angle between neighboring Si-atoms
η	doping efficiency, $[P_4]/[P]$
φ	dihedral angle
$\mu(E)$	charge carrier mobility
μ_0	free mobility of charge carriers
μ_D	effective drift mobility of charge carriers
μ_D^0	mobility prefactor
σ	width of gaussian distribution of defect states
$\sigma(E)(T)$	conductivity
σ_0	conductivity prefactor
σ_{min}	minimum metallic conductivity
σ_{00}	conductivity constant
τ	decay time
τ_{free}	free carrier lifetime
τ_{trap}	lifetime of carrier in traps
τ_R	relaxation time
ω_0	rate prefactor, attempt-to-escape frequency
$\xi(T, T_V, N_H)$	entropy factor
a-Ge:H	hydrogenated amorphous germanium
a-Si	amorphous silicon
a-Si:H	hydrogenated amorphous silicon
a-SiC:H	amorphous silicon carbide
a-SiGe:H	amorphous silicon-germanium alloy
a-SiO$_x$:H	amorphous silicon oxide
a-SiN$_x$:H	amorphous silicon nitride
c-Si	crystalline silicon
d.c.	direct current
ESR	electron spin resonance
MBE	molecular beam epitaxy
PECVD	plasma-enhanced chemical vapor deposition
RDF	radial distribution function
r.f.	radio frequency
r.m.s.	root mean square
TFT	thin film transistors
u.v.	ultra violet
vppm	parts per million of volume

10.1 Introduction

Hydrogenated amorphous silicon (a-Si:H) has been actively studied for about 20 years. The unhydrogenated material (a-Si) has such a large defect density that it is unusable for electronic devices, although there is continuing interest in its atomic structure. The beneficial effects of hydrogen were discovered at least in part by accident, when the material was deposited from silane (SiH_4) gas in a plasma discharge. This method of growth results in a reduction of the defect density by about four orders of magnitude compared to unhydrogenated a-Si, giving material which is of device quality. It is now recognized that hydrogen removes defects by bonding to unterminated silicon atoms. The first demonstration of substitutional doping, made possible by the low defect density, was reported in 1975 and opened the way to many device applications (Spear and LeComber, 1975). Since that time the research effort has greatly expanded so that this material now dominates studies of amorphous semiconductors, and many technology applications have developed.

The first photovoltaic solar cells made from a-Si:H were described by Carlson and Wronski (1976), and the conversion efficiency has steadily improved to its present value of 13–14%. In 1979 the plasma deposition of silicon nitride was used in conjunction with a-Si:H to produce field effect transistors (LeComber et al., 1979). Large area arrays of these thin film transistors (TFT) are now important in liquid crystal displays (Miki et al., 1987) and monolithic circuits for printing and input scanning applications (Thompson and Tuan, 1986). The combination of doping and transistor action means that essentially all of the circuit elements used in crystalline silicon electronics can be repro-

duced in a-Si:H, giving it a broad versatility in electronic circuit design. The principal advantage of using a-Si:H is that it can be deposited over large areas on low cost substrates such as glass.

10.1.1 Plasma-Enhanced Chemical Vapor Deposition Growth of Hydrogenated Amorphous Silicon

The usual method of depositing a-Si:H is by plasma decomposition of silane gas, SiH_4, with other gases, such as PH_3, B_2H_6, GeH_4 etc., added for doping and alloying. Silane decomposes thermally above about 450 °C and amorphous films can be grown in this way at temperatures less than about 550 °C. However, these films are of limited utility because the temperature is too high to retain the hydrogen. The deposition of hydrogenated films at lower temperatures requires a source of energy to dissociate the SiH_4, which is the role of the plasma. The first plasma deposition system for amorphous silicon was a radio frequency (r.f.) inductive system developed by Chittick et al. (1969). Most subsequent reactors are in a diode configuration in which the plasma is confined between two parallel electrodes. This type of reactor is illustrated in Fig. 10-1, and

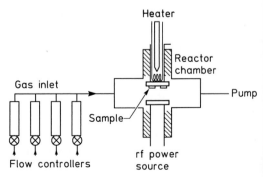

Figure 10-1. Schematic diagram of the main components of a typical r.f. diode plasma reactor for depositing a-Si:H and its alloys.

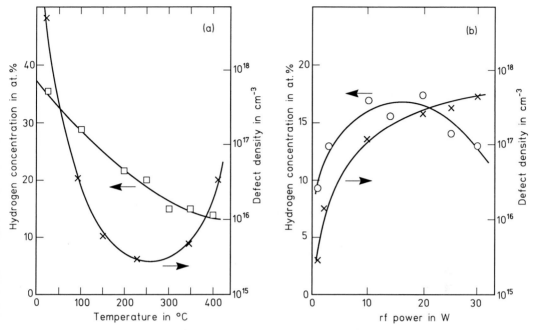

Figure 10-2. Illustration of the dependence of material properties on PECVD deposition conditions, showing variations of the hydrogen concentration and defect density on (a) substrate temperature, and (b) r.f. power (Knights and Lucovsky, 1980).

consists of a gas inlet arrangement, the deposition chamber which holds the heated substrate, a pumping system, and the source of power for the discharge. The deposition process is usually referred to as *plasma-enhanced chemical vapour deposition* (PECVD). There are numerous variations on this basic deposition process, but the resulting material is largely independent of the choice of technique.

The structural and electronic properties of the film depend on the conditions of growth, particularly substrate temperature, r.f. power and gas composition. Figure 10-2 summarizes some typical measurements of PECVD a-Si:H films (Knights and Lucovsky, 1980). The hydrogen content varies between 8 and 40 at.% and decreases slowly as the substrate temperature is raised. In addition, the hydrogen content depends on the r.f. power in

the plasma and on the composition of the gas; Fig. 10-2b shows the variation when SiH_4 is diluted to 5% concentration in argon. The defect density also depends on the substrate temperature and power with variations of more than a factor 1000. The minimum defect densities of $10^{15}-10^{16}$ cm^{-3} usually occur between 200 and 300 °C and at low r.f. power densities (< 100 mW/cm^2), giving material with the most useful electronic properties. Typical growth rates range between 1 and 10 Å/s and are approximately proportional to the r.f. power, provided that the SiH_4 is undiluted. Heavy dilution with argon and high r.f. power give depletion effects in which the SiH_4 is completely consumed in the plasma and the deposition rate saturates (i.e. does not increase with power). Good quality material is grown far away from this condition.

The physical morphology of the films also depends on the deposition conditions. Low r.f. powers and undiluted SiH_4 result in smooth conformal film growth and material with a low defect density. In contrast, high r.f. power and SiH_4 diluted in argon give films with a strongly columnar microstructure oriented in the direction of growth and a high defect density (Knights and Lujan, 1979). The columns have dimensions ranging from about $100\,\text{Å}$ to $1\,\mu m$. The smooth growth habit is important for a-Si:H-based devices and multilayer structures whose layer thicknesses are typically only a few hundred Ångstrom.

Film growth occurs because SiH_4 is dissociated by the plasma and the fragments condense out of the gas onto the substrate and other reactor surfaces. Although the principle of deposition is quite simple, the physical and chemical processes which take place during a-Si:H film growth are complex. A plasma is sustained by the acceleration of electrons due to the alternating electric field. The electrons collide with the molecules of the gas, causing ionization (amongst other processes) and releasing more electrons. There is negligible acceleration of the ions in the plasma by the electric field because of their large mass, so that the energy of the plasma is acquired by the electrons. In addition to ionization, the collisions of the energetic electrons with the gas molecules cause dissociation of the gas, creating either neutral radicals or ions which are the precursors to deposition. Examples of silane dissociation reactions which require low energies are

$$SiH_4 \rightarrow SiH_2 + H_2 \quad (+2.2\,\text{eV}) \qquad (10\text{-}1)$$

$$SiH_4 \rightarrow SiH_3 + H \quad (+4.0\,\text{eV}) \qquad (10\text{-}2)$$

In addition, many other higher energy silane dissociation and reaction processes have been identified.

At the normal deposition pressures of $0.1-1$ Torr, the mean free path of the gas molecules is $10^{-3}-10^{-2}$ cm, which is much smaller than the dimensions of the reactor. Many intermolecular collisions take place, therefore, in the process of diffusion to the substrate. An understanding of the growth is complicated by these secondary reactions because they greatly alter the mix of radicals within the plasma. Those radicals with a high reaction rate have a low concentration and short diffusion length and so are less likely to reach the growing surface. The least reactive species tend to survive the collisions longest, and have the highest concentrations, irrespective of the initial formation rates. SiH_x radicals with $x \leq 2$ readily react with SiH_4 to form Si_2H_{4+x}, whereas SiH_3 does not react with SiH_4, because Si_2H_7 is not a stable structure. Thus, it is believed that SiH_3 is an important precursor to growth, particularly in low power plasmas (Gallagher, 1986).

The growth of a-Si:H is completed when molecular fragments are adsorbed onto the growing surface, with the concomitant release of atoms or molecules from the surface. Figure 10-3 illustrates some of the processes which have been proposed to occur at the growing surface of an a-Si:H film (Gallagher, 1986). Of all the gas species near the surface, atomic hydrogen can penetrate furthest into the material. At the normal deposition temperature of $200-300\,°C$, interstitial hydrogen can move quite rapidly into the bulk where it readily bonds to silicon dangling bonds. Hydrogen, therefore, has the fortunate property of being able to remove most of the subsurface defects left by the deposition process.

The microstructure of the film reflects the growth process. Columnar growth is caused by a shadowing effect and occurs

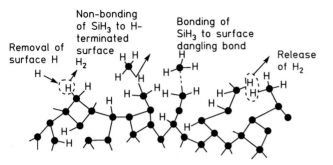

Figure 10-3. Illustration of some possible processes taking place at the a-Si:H surface during growth (Gallagher, 1986).

when the impinging molecular fragments have a very low surface mobility. Any depression in the surface is shadowed from the plasma by the surrounding material, which suppresses the growth rate and accentuates the depression. The result is a columnar structure containing many voids and defects. The shadowing is prevented when the growth species have a high surface mobility. SiH$_3$ is believed to have this property because it is less reactive than the other radicals.

10.1.2 Molecular Structure

Bonding is primarily covalent in a-Si and the tetrahedral arrangement of bonds characteristic of crystalline Si (c-Si) is essentially preserved in the amorphous phase. The bulk electronic and vibrational densities of states are, to a good approximation, broadened versions of their crystalline counterparts. Disorder in a-Si is best characterized, therefore, in terms of deviations from the perfect diamond structure of c-Si. Diffraction measurements can provide only average measures of these deviations, which has led to the development of sophisticated model building techniques to uncover the microscopic structure of a-Si (Polk, 1971). A computer-generated picture of the result of such a model building exercise is shown in Fig. 10-4. This hand-built, 105-atom ball-and-stick model

of a-Si:H with periodic boundaries permits the calculation of a-Si:H properties and provides a wealth of microscopic structural information unavailable by other means (Winer and Wooten, 1984). It is clear from Fig. 10-4 that a-Si:H is both structurally and compositionally disordered. The hydrogen in a-Si:H reduces the network strain which allows the material to be doped both n- and p-type.

Experimental information about the local order of silicon atoms comes from the radial distribution function (RDF) which is the average atomic density at a distance r from any atom. The RDF of a-Si:H shown in Fig. 10-5 (Schulke, 1981), has sharp structure at small interatomic distances,

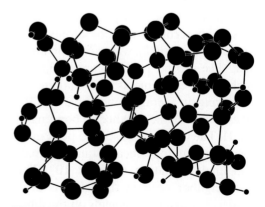

Figure 10-4. Computer generated picture of a hand-built, 105 atom model of a-Si:H with periodic boundaries.

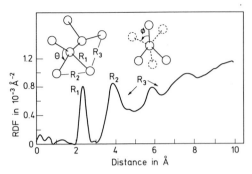

Figure 10-5. Example of the radial distribution function (RDF) of a-Si:H obtained from X-ray scattering. The atomic spacings which correspond of to the RDF peaks are indicated (Schulke, 1981). The arrows indicate the range of values of the third nearest neighbor as the dihedral angle ϕ is varied.

progressively less well-defined peaks at larger distances, and is featureless beyond about 10 Å. This reflects the common property of all covalent amorphous semiconductors, that the short-range order of the crystal is preserved but the long-range order is lost.

The Si–Si bond length, R_1, given by the first peak in the RDF is the same as in crystalline silicon, 2.35 Å, and the intensity of the first peak confirms the expected 4-fold coordination of the atoms. The bond length distribution is evidently small and is estimated to be about 1% (Etherington et al., 1984). The second peak in the RDF arises from second neighbor atoms at a distance $2 R_1 \sin(\theta/2)$, where θ is the bond angle. The second peak occurs at 3.85 Å, which is also the same as in crystalline silicon, giving an average bond angle of 109°, and establishing the tetrahedral bonding of a-Si:H. The width of the second peak indicates that the root mean square (r.m.s.) bond angle deviation is about 10°. The third neighbor peak in the RDF is even broader. The distance depends on the dihedral angle, φ (see Fig. 10-5) and a very broad distribution of angles about the

crystalline silicon values is deduced from the position and width of the peak. The third neighbor peak and the more distant shells overlap, so that no detailed information can be deduced.

An idealized, though useful, way to view a-Si:H is as a homogeneous collection of Si–Si and Si–H bonds rather than as a collection of individual atoms. The non-equilibrium nature of the typical a-Si:H growth process prevents the structure from attaining its thermodynamic ground state (i.e. the crystal) resulting in a large (≈ 0.2 eV/atom) excess enthalpy of disorder. Under typical growth conditions, large-scale heterogeneities can also form such as voids and cracks. Nevertheless, the structure of a-Si:H is remarkably close to the diamond structure with most bond lengths within 1% and most bond angles within 10% of their c-Si values. These relatively small strains are responsible for the localization of gap states that are the distinguishing feature of the electronic structure of a-Si:H.

10.1.3 Chemical Bonding

A molecular orbital representation of the conduction and valence bands of a-Si:H is illustrated in Fig. 10-6. These s and p states combine to form the sp^3 hybrid orbitals characteristic of ideal tetrahedral bonding (see Chap. 1). These orbitals are split by the bonding interactions to form the valence and conduction bands. Any non-bonding silicon orbitals, such as dangling bonds, are not split by the bonding interaction and give states in the band gap. The model in Fig. 10-6 is sufficient to predict the general features of the density-of-states distribution, but much more detailed calculations are needed to obtain an accurate distribution. Many calculations have been performed (Allen and Joanno-

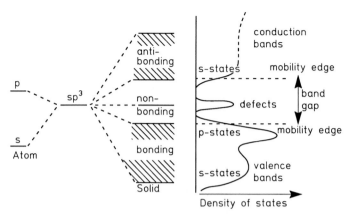

Figure 10-6. Schematic molecular orbital model of the electronic structure of amorphous silicon and the corresponding density of states distribution.

poulos, 1984), but present theories are not yet as accurate as the corresponding results for the crystalline band structure. The Si–H bonds give further contributions to the electronic structure, with bonding states deep in the valence band and anti-bonding states near the conduction band edge (Ching et al., 1980).

The atomic structure of a covalent amorphous material is described by a continuous random network as illustrated in Fig. 10-4, in which each atom is bonded to a well-defined number of neighbors – the coordination number, Z. In the absence of the topological constraints of the crystalline lattice, the local bonding of each atom is determined by the optimum chemical configuration. A covalently-bonded atom has minimum energy when the maximum number of electrons are in bonding states. The most favorable coordination of atoms from group IV to VII of the periodic table is

$$Z = 8 - N \qquad (10\text{-}3)$$

where N is the number of valence electrons ($N = 4$ for Si etc.). This "$8 - N$" rule is a useful description of the bonding chemistry of amorphous networks containing multiple elements (Mott, 1969).

The effects of the bonding chemistry in amorphous and crystalline materials are different, due to the absence of a periodic structure. This is illustrated by considering impurity atoms. Since every atomic site in a crystal is defined by the periodic lattice, the impurity either substitutes for the host, adapting itself to the chemistry of the host, or occupies a position which is not a lattice site, forming a defect. A substitutional impurity such as phosphorus is 4-fold coordinated in crystalline silicon and acts as a donor by releasing one of its electrons into the conduction band. An amorphous material has no rigidly defined array of lattice sites, so that an impurity can adapt the local environment to optimize its own bonding configuration, while also remaining a part of the host atomic network. The $8 - N$ rule predicts that phosphorus in amorphous silicon should be 3-fold coordinated and, therefore, inactive as an electronic dopant. Indeed, it seems to follow from the $8 - N$ rule that substitutional doping must be impossible in an amorphous semiconductor. Actually, the chemical bonding does not forbid, but does severely constrain doping in a-Si:H, as discussed in Sec. 10.4.2.

The possibility of electronically-induced structural reactions is also suggested by the $8 - N$ rule, which predicts that the excitation of an electron out of a state can desta-bilize the chemical bond and, thereby, induce a change in coordination. Such reactions are usually prevented in crystalline semiconductors by the long-range order of the lattice and the extended electronic wavefunctions. Structural changes are promoted in amorphous materials by the adaptibility of the continuous random network and by the localization of electronic carriers. The various metastable and equilibrium processes described in Sec. 10-4 follow from this property.

10.1.4 Localization of Electronic States

The periodic potential in a crystal leads to the familiar Bloch wavefunctions which are extended states defined by the crystal momentum, k. The solutions to Schrödinger's equation do not apply to an amorphous semiconductor because the potential is not periodic. The structural disorder of an amorphous semiconductor causes such frequent electron scattering that the wavefunction loses phase coherence over a distance of a few atomic spacings. The uncertainty in the momentum arising from the scattering is of order k, so that momentum is not a good quantum number and is not conserved in electronic transitions. Consequently the energy bands are no longer described by the energy-momentum dispersion relations, but instead by a density-of-states distribution, $N(E)$, illustrated in Fig. 10-6, and the symmetry selection rules do not apply to optical transitions. The disorder also reduces the carrier mobility and causes localization of the electron wavefunctions.

Anderson (1958) showed that all the electronic states are localized when there is a large enough site-to-site variation in the atomic potentials, V_0. The condition for complete localization is that V_0 exceeds about three times the electronic band width. The disorder in the amorphous

semiconductors is not sufficient to localize all the electron states, primarily because the short range order limits the magnitude of the disorder potential. Instead, the center of the band comprises extended electron states at which there is strong scattering, while states at the extreme edges of the bands are localized (Mott and Davis, 1979; Chap. 1). The density-of-states distribution in the region of the band edges is shown in Fig. 10-7. The extended and localized states are separated by a mobility edge, at energy E_C and E_V, which derives its name because at zero temperature, only electrons above E_C are mobile and contribute to the conduction. The energy of the mobility edge within the band depends on the degree of disorder and is typically $0.1-0.5$ eV from the band edge in all the amorphous semiconductors. The detailed properties of states near the mobility edge are much more complicated than in this simple model of an abrupt mobility edge (Abra-

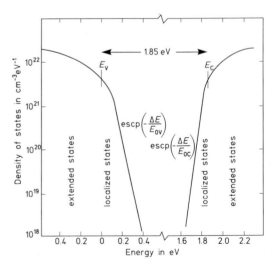

Figure 10-7. Density of states distribution near the band edges, showing localized band tail states extending into the band gap of a-Si:H. E_{0V} and E_{0C} are the slopes of the exponential band tails; E_C and E_V are the energies of the valence and conduction band mobility edges.

hams et al., 1979; Mott and Kaveh, 1985). However, the simple model is sufficient to interpret most experimental measurements. The reason that electronic conduction in a-Si:H is not metallic is therefore that electronic states in the gap are localized and conduction of charged carriers through these states occurs at elevated temperatures via thermal excitation to the mobility edges or by hopping from site to site. As the conducting states become less localized (toward the band edges) and the site separation decreases, the carrier mobility increases until extended state conduction is achieved. The region between E_V and E_C in Fig. 10-7 is referred to as the "mobility gap", and is of magnitude $1.8 - 1.9$ eV (Wronski et al., 1989).

10.2 Electronic Structure and Localized States

The distribution of electronic states in a-Si:H is similar to that of crystalline Si except that critical point structure is absent in the amorphous phase due to the lack of translational symmetry and a well-defined band structure. Similarly, transitions between electronic states in a-Si:H are not subject to symmetry selection rules or crystal momentum conservation requirements, which leads to the direct nature of band gap transitions in a-Si:H and higher absorption coefficients than c-Si in the visible. The most important distinction between the electronic structure of a-Si:H and that of c-Si, however, is the localization of a small but important fraction ($\approx 0.02\%$) of the electronic states in a-Si:H. The distributions of bond lengths and bond angles which characterize a-Si:H disorder lead to the localization of electronic states near the valence and conduction band edges as well as states associated with defects and impurities. The densities and distributions of localized states within the fundamental band gap of a-Si:H are of primary importance in determining the optical and electronic limitations of any a-Si:H-based technology.

10.2.1 Band Tail States

The distribution of band tail states is most easily, although indirectly, observed in the optical absorption spectrum, which in the case of a-Si:H and a host of other glassy and crystalline materials exhibits a characteristic exponential absorption or Urbach edge as shown in Fig. 10-8. Several explanations of the exponential nature of the Urbach edge have been proposed, many relying on strong electron-phonon or excitonic mechanisms of band edge broadening (Mott and Davis, 1979; Chap. 10-6). While such explanations apply to the case of ionic crystalline materials such as the alkali halides, it is now well established that the densities of electronic states themselves tail exponentially into the band gap in a-Si:H as indicated in Fig. 10-7. The conduction band tail decreases exponentially into the gap with a characteristic inverse slope E_{0C} between 20 and 30 meV, or equivalently, with a characteristic temperature T_C ($E_{0C} = k T_C$) near 300 K. The valence band slope is generally broader with a characteristic inverse slope E_{0V} between 40 and 50 meV, in the best material, corresponding to T_V near 500 K. Lower quality material has E_{0C} up to 150 meV.

For a long time it was not clear why the disordered nature of a-Si should lead to a gap in the density of states. Weaire (1971) used a simple two-parameter tight-binding Hamiltonian to model a-Si and showed that for reasonable levels of off-diagonal disorder forbidden energies always appear in the density-of-states. In real a-Si:H the valence and conduction band tails inter-

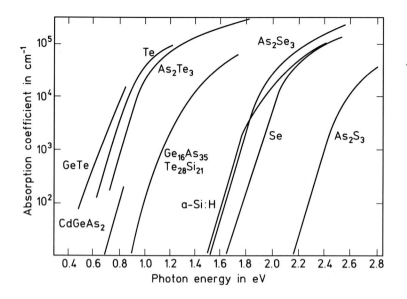

Figure 10-8. The optical absorption edges of various amorphous semiconductors including a-Si:H (Mott and Davies, 1979).

sect at some very low density-of-states and the Fermi energy lies deep in the gap.

The band tails are the most prominent feature in the mobility gap and have a dominant effect on important processes such as optical absorption, electron-hole recombination, and electronic transport in a-Si:H. Their exponential distribution arises from the distribution of weak (strained) Si–Si bonds whose bonding and anti-bonding states make up the valence and conduction band tails, respectively. If a Si–Si bond becomes too strained, it can break to reduce the system free energy and, thereby, form a network defect whose state is localized deep in the mobility gap. The electronic states of substitutional impurities are also localized similar to the shallow donor and acceptor states in c-Si. In fact, the behavior of dopants and defects in a-Si:H is qualitatively the same as that in c-Si, but the uniform disorder of a-Si:H leads to several interesting differences that have only recently become understood, and which are discussed in Sec. 10-4.

10.2.2 Doping and Dopant States

A-Si:H, like c-Si, can be doped either n- or p-type by the incorporation of phosphorus or boron, respectively, during growth or by ion-implantation after growth. The solubility of impurities in a-Si:H is higher than in c-Si and P or B concentrations up to 10^{21} cm^{-3} are easily achieved. In c-Si, every P or B atom is incorporated at a substitutional lattice site which constrains all impurities to be 4-fold coordinated and, therefore, active donors or acceptors. In a-Si:H, the network disorder provides a certain structural flexibility which allows a large fraction of the impurities to achieve their preferred valency; i.e. 3-fold coordination in the cases of P or B.

For a long time this flexibility was thought to preclude the significant doping of amorphous semiconductors. However, the strain relief provided by hydrogenation permits doping efficiencies as high as 1% in a-Si:H. The best evidence of doping is the increase of the d.c. conductivity of a-Si:H observed when phosphine or diborane is added to the silane plasma during film growth (Spear and LeComber, 1975). The room temperature dark d.c. conductivity increases rapidly from $\approx 10^{-9}\,\Omega^{-1}\,\mathrm{cm}^{-1}$ in intrinsic a-Si:H up to $\approx 10^{-2}\,\Omega^{-1}\,\mathrm{cm}^{-1}$ in highly P-doped a-Si:H as shown in Fig. 10-9. The exponential dependence of the conductivity on inverse temperature is characteristic of the thermally-activated charge transport and demonstrates that the Fermi energy lies within the mobility gap.

The "kinks" in the a-Si:H conductivity curves and thermopower near 150 °C are similar to those observed in the conductivity of crystalline silicon or even gold (Hannay, 1967). In these crystals, the conductivity is limited by carrier scattering from point defects, and "kinks" occur in the conductivity curves when the temperature drops below the point where the defect formation reactions can no longer maintain equilibrium and non-equilibrium point defect concentrations are frozen. In the case of a-Si:H, the "kinks" occur at temperatures below which bulk dopant activation reactions, such as

$$P_3^0 \leftrightarrows P_4^+ + e^- \qquad (10\text{-}4)$$

can no longer reach equilibrium and non-equilibrium carrier concentrations are frozen. The equilibrium concentration of charge carriers and the equilibration kinetics of such reactions are discussed in detail in Sec. 10-4.

10.2.2.1 The Doping Efficiency

The fraction of bulk impurities which are active dopants, termed the doping efficiency, is shown for arsenic-doped a-Si:H in Fig. 10-10. These results are derived from measurements of the concentration of excess electrons introduced by the doping. We might expect that only a limited number of sites are available at which impurities in a-Si:H can be constrained to 4-fold coordination, so that the doping efficiency should increase with decreasing impurity concentration. This is exactly what is observed under most plasma-enhanced deposition conditions, where the doping efficiency decreases as the square-root of the gas-phase mole fraction X, of impurity, I ($\equiv X_I$) (Stutzmann et al., 1987). However, at very low a-Si:H growth rates (low r.f. power), another, pyrolytic process of inactive impurity incorporation acts to reduce the doping efficiency at low X_I.

This behavior can be understood with the following analysis. The doping effi-

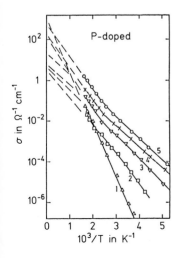
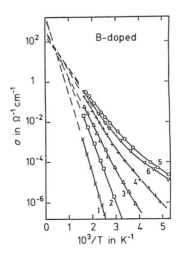

Figure 10-9. Temperature dependence of the conductivity of phosphorus- and boron-doped a-Si:H. The doping levels range from 10^{-6} to 2×10^{-2} (Beyer and Overhof, 1984).

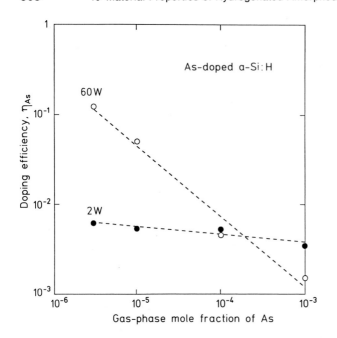

Figure 10-10. Doping efficiency of arsenic in a-Si:H as a function of the mole fraction of As in the gas during deposition (Winer and Street, 1989), for different r.f. powers.

ciency of P, for example, in a-Si:H is defined by $\eta \equiv [P_4]/[P]$, where the square brackets denote the solid phase atomic density. In order to interpret the data in Fig. 10-10, we want to rewrite η in terms of the gas-phase P mole fraction X_P. The P distribution coefficient $d(P) \equiv [P]/[Si]X_P$ relates the solid-phase P density $[P]$ to X_P. A simple two-parameter analytical expression for $d(P)$ has been fitted to a large set of data to obtain (Winer and Street, 1989)

$$d(P) = d_0 + \frac{\beta \cdot X_P^{-1/2}}{\alpha \cdot W} \qquad (10\text{-}5)$$

where d_0 is a constant which is characteristic of the relative plasma-enhanced dissociation rates of phosphine and silane, β is the pyrolytic rate constant, and α is the deposition rate per unit r.f. plasma power W. This relation describes the competition between impurity incorporation via pyrolysis ($\propto \beta X_P^{-1/2}$) and via plasma-enhanced decomposition ($\propto \alpha W$). When the latter dominates (at high r.f. plasma power and large X_P), the distribution coefficient is es-

sentially constant between 5 and 10. When thermal decomposition of the impurity gas (i.e., B_2H_6, AsH_3, PH_3, etc.) dominates impurity incorporation (low r.f. plasma power and small X), the distribution coefficient can attain values near 30 for P (β small) and near 300 for As (β large).

Finally, active dopant incorporation can be described by the chemical reaction

$$P_{gas} \leftrightarrows P_4^+ + e^- \qquad (10\text{-}6)$$

so that by the law of mass action at equilibrium and $[P_4^+] = [e^-]$ one obtains $[P_4^+] \propto X_P^{1/2}$ as observed experimentally (Street, 1982). We can now express η completely in terms of X_P to obtain

$$\eta \propto \frac{X_P^{1/2}}{d(P)X_P} = \frac{X_P^{-1/2}}{d_0 + \dfrac{\beta X_P^{-1/2}}{\alpha W}} \qquad (10\text{-}7)$$

The same competition between plasma-enhanced decomposition and pyrolysis determines the doping efficiency in a-Si:H. Under conditions where plasma decomposition dominates such as high growth rates or small β's (i.e. B or P doping) the doping efficiency is simply proportional to $X^{-1/2}$. Under conditions where thermal decomposition is significant such as low growth rates and large β's (i.e. As doping), η is constant and very small, as observed in Fig. 10-10.

10.2.2.2 Dopant States

Four-fold coordinated dopants such as B_4 and P_4 form effective mass donors and acceptors with localized dopant states near the band edges, as in c-Si. The variations in disorder potential broaden the dopant levels, similar to the band tailing of the silicon states. Donor states have been observed through their electron spin resonance (ESR) hyperfine interaction with the spin of donor electrons (Stutzmann and Street, 1985). Donor electrons in a-Si:H are more localized than in c-Si and the hyperfine interaction is correspondingly stronger. The donor band is deeper in the gap and broader than the shallow, well-defined donor levels in c-Si. Because the valence band is much broader than the conduction band tail in a-Si:H, hyperfine measurements have not been able to detect the corresponding B_4 donor states in B-doped a-Si:H. The probable distributions of P_4, As_4 and B_4 dopant states in highly-doped a-Si:H are shown in Fig. 10-11. The arsenic donor is deeper than the phosphorus donor because of differences in size, electronegativity and bond strength.

The P_4^0 ESR hyperfine signal in a-Si:H exhibits two lines split by $\Delta H = 24.5$ mT with a peak-to-peak line width of about 6 mT for each line (Stutzmann and Street, 1985). The two hyperfine lines are due to the interaction of the spins of electrons localized at P_4 sites with the spin $I = 1/2$ of the ^{31}P nucleus. The ^{31}P hyperfine signal increases with the gas-phase P mole frac-

tion X_P, but the concentration of neutral donors is much smaller than that of ionized donors, because of compensation by defects.

The distribution of electrons which occupy the conduction band tail states can be measured by photoemission yield spectroscopy (Winer et al., 1988). In undoped or slightly P-doped a-Si:H, the expected Boltzmann fall-off in the density of occupied states above the Fermi energy is observed. However, as the gas-phase P mole fraction increases beyond 10^{-4}, the Fermi tail broadens considerably. Dividing the measured occupied density of states by the Fermi-Dirac occupation function reveals an exponentially increasing total density of states above the Fermi energy consistent with the leading edge of the P_4 donor band of states. The density and location of donor states observed by this method agrees with those inferred from the ESR hyperfine measurements.

The carriers near the band edge include those occupying dopant states as well as the intrinsic band tail states, and in n-type a-Si:H there are roughly equal concentrations of each, because the band tail over-

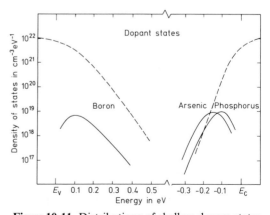

Figure 10-11. Distributions of shallow donors states in doped a-Si:H. Such states are broadened relative to their crystalline silicon counterparts by the network disorder (Street, 1987).

laps the donor band. The density of shallow carriers is quite low because of the low doping efficiency and the compensation by defects. The doping effect in a-Si:H is consequently rather weak. For example, at a gas phase doping level of 1%, less than 1% of the phosphorus is in the form of 4-fold donor states, and of these donors, about 90% are compensated by deep defects. Of the remaining 10% of the donor electrons at the band edge, most occupy band tail states, so that the free electron concentration is only 10^{-4} of the impurity concentration. The low mobility of free carriers in a-Si:H compared to c-Si, results in the conductivity of n-type a-Si:H being more than five orders of magnitude below that of c-Si.

10.2.3 Native Defects and Defect States

Unsatisfied or "dangling" bonds are the dominant native defect in a-Si:H; the amphoteric nature of this defect gives rise to either unoccupied (positively-charged), singly-occupied (neutral), or doubly-occupied (negatively-charged) electronic states. These localized states lie near the middle of the mobility gap and greatly affect the movement of the Fermi energy and transport and recombination processes. An understanding of the character and distribution of defect states and their dependence on growth conditions and doping level is essential for the successful application of a-Si:H and its alloys. Not surprisingly, a multitude of experimental methods have been developed and intensive effort has been expended in the pursuit of this understanding.

10.2.3.1 Microscopic Character of Defects

Defect states are generally the most localized of all electronic states in a-Si:H, lying deep in the mobility gap at low densi-

ties. Most of our knowledge concerning the microscopic character of defects is derived from electron spin resonance (ESR) measurements. ESR measures both the concentration and the local environment of the neutral paramagnetic defects in a-Si:H. The concentration of charged defects can also be determined by ESR by illuminating the material with band gap light during the measurement to depopulate charged defects and make them spin active (Street and Biegelsen, 1982). The dominant defect in undoped a-Si:H has a nearly symmetric spin signal at a g-value of 2.0055 and a peak-to-peak width of ≈ 0.75 mT. These characteristics are similar to those of the ESR signal from unterminated bonds at the c-Si/SiO$_2$ interface, which has led to the identification of the $g = 2.0055$ signal in a-Si:H as due to undercoordinated or "dangling" bonds.

It has also been proposed that native defects are better described by overcoordinated or "floating" bonds (Pantelides, 1986). Such floating bonds have many of the same properties as dangling bond except that they are predicted to be very mobile whereas dangling bonds should be essentially immobile. Recent calculations and defect diffusion measurements have all but ruled out the floating bond hypothesis (Fedders and Carlssen, 1989; Jackson et al., 1990); the dangling bond remains the only defect model that can consistently account for the properties of native a-Si:H defects.

10.2.3.2 Dependence of the Defect Concentration on Doping

The incorporation of substitutional impurities during growth leads to an increased charge carrier concentration in both c-Si and a-Si:H. In a-Si:H, the majority of extrinsic charge carriers are taken up to form compensating charged defects. This process can be described schematically in the case of P doping by the following chemical reactions

$$P_3 \leftrightarrows P_4^+ + e^- \qquad (10\text{-}8)$$

$$Si_4 + e^- \leftrightarrows Si_3^- \qquad (10\text{-}9)$$

where the subscripts refers to the coordination. The reactions are independent but they are linked by their dependence on the carrier concentration. The required coordination changes are accomplished via hydrogen bond diffusion, and the kinetics of this and similar reactions is determined by the kinetics of dispersive hydrogen diffusion. Applying the law of mass action to reactions (10-8) and (10-9) leads to the square-root dependence of the defect $[Si_3^-]$ and active dopant $[P_4^+]$ concentrations on the solid-phase P concentration $[P]$, which is proportional to the gas-phase phosphorus mole fraction under most deposition conditions. This square-root dependence of $[Si_3^-]$ on X_1 is observed for all substitutional dopants, as shown in Fig. 10-12. However, the high intrinsic defect density in a-Ge:H masks the doping dependence in this material.

10.2.3.3 Distribution of Defect States

The defect concentration in undoped a-Si:H determined by ESR or optical absorption typically lies between 10^{15} and 10^{16} cm^{-3} for optimized deposition conditions. In heavily-doped a-Si:H, the defect concentrations can increase to 10^{18} cm^{-3}. The detection of such small defect concentrations by ESR and optical absorption measurements (Jackson and Amer, 1982) is relatively simple and has become so standardized that the defect concentration dependences on growth conditions and doping determined by different laboratories usually agree. However, the number of dif-

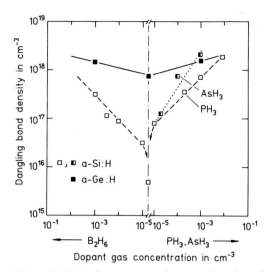

Figure 10-12. Defect concentrations as a function of gas phase mole fraction of boron, phosphorus, and arsenic (Stutzmann et al., 1987).

ferent methods for measuring defect state distributions is nearly as large as the number of different laboratories interested in this problem! The defect distributions extracted or inferred from these many methods do not always agree and at least two distinct schools have developed whose disagreement in the placement of defect state energies in the gap amounts to several tenths of an electron volt (Lang et al., 1982, Okushi et al., 1982). The results of three typical experimental methods which provide a more or less consistent picture of the defect state distribution in a-Si:H are shown in Fig. 10-13.

The luminescence transition observed in P-doped a-Si:H at 0.8–0.9 eV (Fig. 10-13) has been interpreted as a band-tail-to-defect transition (Street, 1982). The 0.9 eV energy, therefore, is a measure of the separation between the charged defect band peak and, most probably, the conduction band tail. The shape and the 0.3 eV width of the luminescence band should be characteristic of the shape of the defect band. The defect distribution inferred from deep-level

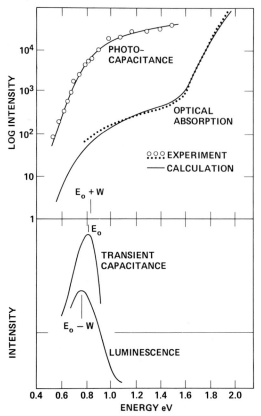

Figure 10-13. Typical data from which a consistent distribution of gap states has been inferred; obtained from luminescence, deep level transient spectroscopy, and optical absorption experiments (Street et al., 1985). E_0 is the defect level measured from E_C and W is the contribution to the optical transition energies from lattice relaxation. The calculated absorption is a convolution of the conduction band and the defect level distribution.

transient spectroscopy (Lang et al., 1982) and optical absorption measurements is consistent with the luminescence measurements; negatively-charged D^- defect states lie in a broad band $0.8-1$ eV below the conduction band mobility edge in P-doped a-Si:H. The optical absorption data indicate a defect level at about the same energy. Similar measurements on p-type material find a defect level which is well separated from the valence band edge (Kocka et al.,

1987). In summary, the majority of doping-induced states and the Fermi energy lie opposite each other in the gap; the Fermi energy of doped a-Si:H lies in a relative minimum in the gap state distribution. This picture has been confirmed by several other types of capacitance and absorption measurements, as well as by photoemission spectroscopy measurements (Winer et al., 1988). A schematic description of the distribution of gap states in p-type, n-type, and intrinsic a-Si:H is shown in Fig. 10-14.

The $0.1-0.3$ eV difference between the negatively-charged (doubly-occupied) and neutral (singly-occupied) defect state energies determined from optical absorption data is a measure of the defect correlation energy, U_C, defined as the energy cost to place a second electron on a singly-occupied, localized defect level (Jackson, 1982). Electron spin resonance and optical absorption measurements are also consistent with a small, positive value of U_C (Street and Biegelsen, 1982; Dersch et al., 1980; Jackson, 1982). The doubly occupied defect gap state energy level lies above the singly occupied level by an energy U_C, as depicted in Fig. 10-14. This view of the distribution of defect states in intrinsic a-Si:H is generally accepted. However, if these deep defect states were fixed in energy, movement of the Fermi energy via doping would not result in the reversal of the defect energy ordering, which suggests instead a defect correlation energy which is negative. The apparently conflicting requirements of a small, positive correlation energy in intrinsic a-Si:H and the reversal of the defect energy ordering in a doped a-Si:H can be resolved by invoking a broadly-distributed pool of virtual defect states within a chemical equilibrium framework of defect formation from which the system can choose to form defects in order to minimize the system free energy (Winer,

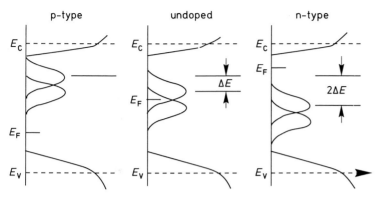

Figure 10-14. Schematic distribution of defect states in the gap for n-type, intrinsic and p-type a-Si:H. The two levels of the dangling bond are separated by the correlation energy U_C. The shift, ΔE, of the defect levels with doping is given by Eq. (10-48) and is discussed in the text.

1989). This defect formation framework is discussed in detail in Sec. 10-4.

10.2.3.4 Dependence of the Defect Concentration on Growth Conditions

Thin films of a-Si:H are typically deposited onto heated (250 °C) glass substrates by the r.f. plasma decomposition of silane at gas pressures near 100 mTorr. Under these conditions, growth rates of about 1 Å per second are typically achieved. Ion bombardment, u.v. exposure, and strong electric fields accompany such plasma-enhanced chemical vapor deposition, which can enhance defect formation while complicating the analysis of growth processes. The dominant growth parameter under normal deposition conditions, however, is the substrate temperature, whose effect on the defect concentration is shown schematically in Fig. 10-2 a.

The minimum defect concentration near 3×10^{15} cm^{-3} is obtained for substrate temperatures between about 200 and 300 °C. Growth in this temperature range results in reasonably homogeneous, highly photoconductive, high resistivity a-Si:H films suitable for photovoltaic and photoelectronic applications. On either side of this optimal growth temperature range, the material properties degrade for reasons that depend on the details of the growth processes, which are not yet completely understood. In simple terms, the defect concentration increases as the growth temperature is reduced because strain-relieving chemical processes that depend on surface diffusion are kinetically limited. As the growth temperature increases beyond the optimal range, increased desorption of strain-relieving species like hydrogen again lead to an increased defect concentration. More detailed models which attempt to account for the behavior of Fig. 10-2a are discussed elsewhere (Tanaka and Matsuda, 1987; Winer, 1990).

10.2.4 Surfaces and Interfaces

The electronic states specific to semiconductor surfaces and interfaces can affect device performance when the active layers are very thin. Although adsorption of water, oxygen, and other gases has been shown to lead to large changes in the measured conductivity of a-Si:H under certain conditions (Tanielian, 1982), the effects of surface or interface states on a-Si:H device performance are usually negligible. This is due, in part, to two factors. Firstly, there is little distinction between localized bulk and localized surface or interface states in terms of their distribution in the gap which might otherwise affect electronic transport and recombination processes. Secondly, the concentration of intrinsic surface states in a-Si:H is quite low: typically 5×10^{11} cm^{-2} compared to 5×10^{14} cm^{-2} on c-Si. This is due to the effective passivation of the surface by hydrogen at growth termination which makes the a-Si:H surface relatively stable against oxidation in air compared to c-Si (Ley et al., 1981).

10.2.4.1 Surface States

The concentration of intrinsic surface states has been determined by a variety of methods, usually by measuring the defect volume concentration as a function of film thickness and extrapolating to zero thickness (Jackson et al., 1983). The distribution of surface states has been measured by photoemission spectroscopy (Winer and Ley, 1987). Most methods find an intrinsic surface defect concentration between 1 and 10×10^{11} cm^{-2} in optimally-grown intrinsic a-Si:H which increase upon n-type doping and initial oxidation. Though small by c-Si standards, this surface defect concentration is equivalent to quite a high volume defect concentration when the layer of a-Si:H is thin. This excess density of gap states at the surface relative to the bulk can be removed by slight boron doping. In fact, the lowest conductivity a-Si:H films are usually obtained by 10 vppm gas-phase boron doping which might be related to the removal of surface or interface states.

The reduction in occupied surface gap states allows the valence band tail to be directly observed over a wide energy range by surface-sensitive photoemission spectroscopy. The exponential nature of the tail is maintained over several orders of magnitude in the density of states, and the slope of the tail agrees with values inferred from bulk transport and optical absorption measurements. In general, the distribu-

tions of localized surface and bulk states in the band gap are essentially the same. The surface properties of a-SiGe:H alloys are similar to those of a-Si:H (Aljishi et al., 1990), while little is known concerning the surface properties of a-SiC:H or other quaternary alloys systems.

10.2.4.2 Oxidation

In the initial stages of oxidation (0–1 monolayers), surface defect states are induced by activated oxygen adsorption which have been attributed to chemisorbed 3-fold coordinated surface oxygen atoms (Winer and Ley, 1987). The distribution of these oxygen-induced states is similar to that of defects induced by n-type doping by overcoordinated phosphorus atoms incorporated from the plasma into the a-Si:H film during growth. This suggests that the overcoordinated chemisorbed oxygen is likewise a surface donor, contrary to the normally strongly electronegative character of adsorbed oxygen. Upon further oxidation (for example, air exposure for long times) a true oxide layer forms with a change to slightly upward band bending and a correspondingly wider band gap than a-Si:H (Berner et al., 1987; Ley et al., 1981).

In clean undoped a-Si:H, the free surface is under electron accumulation corresponding to an 0.5 eV downward band bending. Upon either n- or p-type doping, this downward band bending is removed. Activated oxygen adsorption pins the surface Fermi level just above midgap which results in an 0.25 eV downward band bending. On fully oxidized surfaces, however, the bands appear to be bent in the upward direction, as expected for the strongly electronegative surface oxide (Street et al., 1985 b; Berner et al., 1987).

10.2.4.3 Interfaces

Interfaces are an integral part of all a-Si:H-based devices and their properties can greatly affect device performance. For example, the gate field that extends from the silicon nitride layer into the undoped a-Si:H active layer in an a-Si:H thin film transistor not only enhances the channel conductivity but can lead to defect creation near the a-SiN$_x$:H/a-Si:H interface, which changes the transistor off voltage. Interface properties become even more important in multilayer structures.

A multitude of phenomena initially observed in crystals have been observed in a-Si:H multilayer systems as well and generally result from the same physical origins. Examples are sub-band optical absorption (Hattori et al., 1988), resonant tunneling (Miyazaki et al., 1987), acoustic phonon zone-folding (Santos et al., 1986), and persistent photoconductivity (Kakalios and Fritzsche, 1984). Multilayers of alternating n- and p-type a-Si:H layers (so-called "nipi" doping superlattices) and alternating alloy heterostructures (i.e. a-Si:H/a-Ge:H...) have been grown and studied for many years. Such structures display a wide variety of phenomena similar to the crystalline analogs. After some initial uncertainties, there is now good evidence for carrier quantization in ultrathin a-Si:H heterostructures (Hattori et al., 1988; Miyazaki et al., 1987). Interface abruptness is a key problem for substantiating claims of quantized behavior in and exploitation of a-Si:H heterostructure multilayers. The low growth rates typical for a-Si:H (0.5–1.0 Å/s, comparable to MBE growth rates) allow ultrathin multilayer growth with near atomic resolution (Yang et al., 1987).

10.2.5 Alloys

The largest application of a-Si:H is for large-area solar cell arrays. In this application, the ability to lower the band gap of a-Si:H by alloying with Ge in order to more effectively match the solar spectrum has led to more efficient multijunction devices. On the other hand, alloying a-Si:H with carbon has led to efficient electroluminescent pin diodes with emission in the red, yellow, and green (Kruangam et al., 1987). In addition, amorphous silicon nitride (a-SiN$_x$:H) or oxide (a-SiO$_x$:H) is used as insulating layers for thin film transistor devices. A-Si:H alloys are usually grown in the plasma-enhanced mode as pure a-Si:H but with the addition of methane, germane, ammonia, etc. as appropriate. The study of alloy properties has proceeded in parallel with that of a-Si:H. Our discussion is limited to weak alloys with the Group IV elements C and Ge, since the properties of these are the most extensively studied. As might be expected, these alloy properties deviate little from those of a-Si:H at low alloy levels with a few interesting exceptions.

Dilution of silane with methane for C incorporation generally slows the growth rate with a particularly steep drop occurring near 90% gas-phase methane mole fraction. This is due to the small incorporation probability of C for the PECVD growth mode at 250°C; the distribution coefficient for C under these conditions is typically less than 0.1. There is a large incorporation probability of Ge into a-SiGe:H films. However, growth of good quality a-SiGe:H is usually made at reduced growth rates, but this is due to the high hydrogen dilution which has been found to be necessary in order to reduce the otherwise large defect concentrations in undiluted a-SiGe:H films. Significant changes in the band gap of a-Si:H occur for C or Ge concentrations above about 10%.

Infrared absorption spectra of such alloys show a Si–H stretching mode near 2100 cm^{-1} in addition to the stretching mode at 2000 cm^{-1} normally observed in pure, optimally-grown a-Si:H. The 2100 cm^{-1} mode increases in strength with increased C or Ge incorporation, similar to the increase observed with decreasing growth temperature in unalloyed a-Si:H. This behavior is believed to be due to increased hydrogen incorporation, perhaps in the form of clusters, which results from growth far from optimal conditions.

The neutral defect (spin) concentration and Urbach edge (valence band tail) slope in undoped a-Si:H alloys also increase with increasing C or Ge incorporation. Post-deposition annealing near the deposition temperature can reduce the defect concentration by up to an order of magnitude with a concomitant decrease in the Urbach edge slope. However, the infrared absorption spectra of a-SiGe:H and a-SiC:H alloys are unaffected by such thermal annealing. Alloy defect concentrations are always higher than that of unalloyed a-Si:H, usually by between 10 and 100 times. This is probably due to the inadequate knowledge of optimal growth conditions of a-Si:H alloys. Electron spin resonance data show that the dominant spin-active ($g = 2.0055$) defect in a-Si:H and its alloys are the same for moderate alloy compositions (Stutzmann et al., 1989). The spin centers change to $g = 2.0037$ for C concentrations above about 40%, which probably is due to the emergence of C–C bonds as the dominant structural unit near this concentration. In a-Si:Ge:H alloys the Ge dangling bond dominates when the Ge concentration is above 20–30 at%.

10.3 Electronic Transport

10.3.1 Electrical Conductivity, Thermopower and Hall Effect

The conductivity, σ, is the product of the carrier density, n, and the carrier mobility, μ, and the electron charge, e.

$$\sigma = n e \mu \qquad (10\text{-}10)$$

The total conductivity is an integral over the density-of-states, $N(E)$

$$\sigma = \int N(E) e \mu(E) f(E, T) \, \mathrm{d}E \qquad (10\text{-}11)$$

where $f(E, T)$ is the Fermi function. The integral contains contributions from electron transport above the Fermi energy E_F and hole transport below E_F. When conductivity takes place far from E_F by a single type of carrier, non-degenerate statistics can be applied and so

$$(10\text{-}12)$$

$$\sigma = \int N(E) e \mu(E) \exp\left[-\frac{(E - E_F)}{kT}\right] \mathrm{d}E$$

This equation is usually written as

$$\sigma = \frac{1}{kT} \int \sigma(E) \exp\left[-\frac{(E - E_F)}{kT}\right] \mathrm{d}E$$

where

$$(10\text{-}13)$$

$$\sigma(E) = N(E) e \mu(E) kT \qquad (10\text{-}14)$$

$\sigma(E)$ is the conductivity that would be observed when $E_F = E$.

The conductivity is determined by the density-of-states, the carrier mobility and the Boltzmann factor. When there is a sufficiently high defect density at the Fermi energy, as in unhydrogenated amorphous silicon, conduction takes place by variable range hopping, with a temperature dependence $\exp[-A/T^{1/4}]$ (Mott, 1968). The low defect density in a-Si:H prevents this mechanism from contributing significantly, and instead conduction takes place by electrons or holes at the band edges where both the density of states and the mobility increase rapidly with energy. For the particular case in which $\sigma(E)$ increases abruptly from zero to a finite value σ_{min} at the mobility edge energy, E_C, evaluation of Eq. (10-12) gives

$$\sigma(T) = \sigma_{min} \exp[-(E_C - E_F)/kT] \quad (10\text{-}15)$$

σ_{min} is referred to as the minimum metallic conductivity and is given by

$$\sigma_{min} = N(E_C)\, e\, \mu_0\, kT \quad (10\text{-}16)$$

where μ_0 is the free carrier mobility at E_C. There is considerable doubt about the sharpness or even the existence of a mobility edge (Abrahams et al., 1979; Mott and Kaveh, 1985). Nevertheless, virtually all the conductivity experiments are analyzed in terms of Eq. (10-15) which is a reasonable approximation even if $\mu(E)$ does not change abruptly, provided that it increases rapidly over a limited energy range.

The conductivity of a-Si:H is usually thermally activated, at least over a limited temperature range, and is described by,

$$\sigma(T) = \sigma_0 \exp[-E_\sigma/kT] \quad (10\text{-}17)$$

Comparison of Eqs. (10-15) and (10-17) suggests that a measurement of $\sigma(T)$ immediately gives the location of the mobility edge ($E_C - E_F$), and that the prefactor gives the conductivity at the mobility edge. In actual fact there is a huge variation in the values of σ_0 which is shown by the results in Fig. 10-15 for undoped and doped a-Si:H. The correlation between σ_0 and the activation energy, E_σ, is referred to as the Meyer-Neldel rule (after its first observation in poly-crystalline materials by Meyer and Neldel (1937)), and is described by

$$\ln \sigma_0 = \ln \sigma_{00} + E_\sigma/kT_m \quad (10\text{-}18)$$

where σ_{00} is a constant with a value of about $0.1\,\Omega^{-1}\,cm^{-1}$ and kT_m is ~ 50 meV.

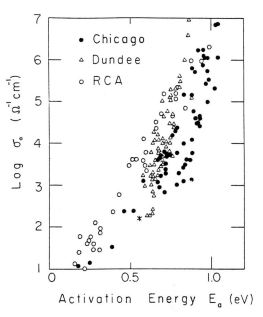

Figure 10-15. Measured values of the conductivity prefactor σ_0 versus the conductivity activation energy, showing the Meyer-Neldel rule (Tanielian, 1982).

A substantial part of the variation in the conductivity prefactor is due to the temperature dependence of the Fermi energy (Beyer and Overhof, 1984)

$$(E_C - E_F) = (E_C - E_F)_0 - \gamma T \quad (10\text{-}19)$$

from which it follows that

$$\sigma(T) = \sigma_0\, e^{\gamma/k} \exp[-(E_C - E_F)_0/kT] \quad (10\text{-}20)$$

Various experiments show that σ_0 is $100-200\,\Omega^{-1}\,cm^{-1}$, and that γ can vary in the range $\pm 5 \times 10^{-4}$ eV/K, due to the shape of the density of states distribution and the temperature dependence of the band gap energy. The free mobility according to Eq. (10-16) is

$$\mu_0 = \sigma_0/N(E_C)\, e\, kT = 10-15\ cm^2/Vs \quad (10\text{-}21)$$

The thermopower measures the average energy of the transport with respect to the Fermi energy. The sign of the thermopower determines whether there is electron

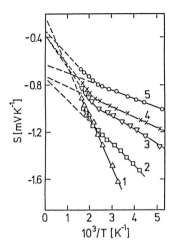

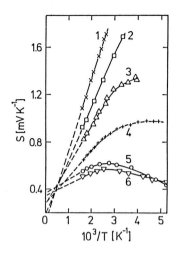

Figure 10-16. Temperature dependence of the thermopower of n-type and p-type a-Si:H. The doping levels range from 10^{-6} to 2×10^{-2} (Beyer and Overhof, 1984).

or hole transport, and confirms that n-type and p-type doping occurs in a-Si:H. This observation is important because the Hall effect, which is the more common measure of the doping type in crystalline semiconductors, has an anomalous sign in amorphous semiconductors as described shortly. The thermopower is expected to exhibit the same activation energy as the conductivity when conduction occurs above a well defined mobility edge (Cutler and Mott, 1969). Figure 10-16 shows examples of the temperature dependence of the thermopower in both n-type and p-type a-Si:H. The thermopower energy does decrease with doping, as expected from the shift of the Fermi energy by doping, but is always smaller than the conductivity energy by about 0.1 eV. This shows that the simple conduction model of an abrupt mobility edge is not exact, although the reason for the difference in energy of conductivity and thermopower is not resolved. Possible mechanisms include long range potential fluctuations of the band edges due to charged localized states, a gradual rather than abrupt increase in the mobility near E_C, or polaron conduction. The charged defects and dopants in doped material make the potential fluctuations

the most probable mechanism, but this model may not apply to undoped a-Si:H which has many fewer charged defects.

A curious aspect of carrier transport in a-Si:H is the anomalous sign reversal in the Hall mobility coefficient. In doped a-Si:H the magnitude of the Hall mobility usually lies between 0.01 and 0.1 cm²/Vs (LeComber et al., 1977) much lower than typical values ($\approx 100-1000$ cm²/Vs) observed in c-Si, and also lower than the drift mobility of a-Si:H. However, the sign of the Hall coefficient is opposite to the sign of the thermopower (i.e. positive in n-type and negative in p-type a-Si:H). Upon crystallization of a-Si:H, the Hall coefficients revert to their proper sign. The origin of this anomalous sign reversal is not completely understood, but is presumed to be related to the very short scattering lengths (Friedman, 1971; Emin, 1977).

10.3.2 The Drift Mobility

Conduction of electron and holes occurs by frequent trapping in the tail states followed by excitation to the higher energy conducting states. The drift mobility, μ_D, is the free carrier mobility reduced by the fraction of time that the carrier spends in the traps

$$\mu_D = \mu_0 \, \tau_{free}/\tau_{trap} \tag{10-22}$$

When there is a single trapping level, with density N_T, at energy E_T below E_C,

$$\mu_D = \mu_0 \, N_C/[N_T \exp(E_T/kT) + N_C] \approx$$
$$\approx \mu_0(N_C/N_T)\exp(-E_T/kT) \tag{10-23}$$

where N_C is the effective conduction band density of states, and the approximate expression applies when $\mu_D \ll \mu_0$. The drift mobility is thermally activated with the energy of the traps.

A distribution of trap energies, $N(E_T)$, arising from the band tail of localized states, gives a drift mobility which reflects the average release time of the carriers. When the band tail distribution is sufficiently broad, the drift mobility becomes dispersive and is time dependent, following the relation (Scher and Montroll, 1975),

$$\mu_D = \mu_D^0 \, t^{\alpha-1} \tag{10-24}$$

where

$$\alpha = T/T_0, \tag{10-25}$$

is the dispersion parameter. The unusual time-dependence occurs because the probability that a carrier is trapped in a very deep trap increases with time.

Of the many theoretical studies of dispersive transport by multiple trapping, the analysis of Tiedje and Rose (1980) and similarly Orenstein and Kastner (1981), is particularly instructive because the physical mechanism is easy to understand. The approach in this model is to consider an exponential band tail of traps with density proportional to $\exp(-E/kT_0)$, where E is the energy from the mobility edge. A demarcation energy, E_D, which varies with the time, t, after the start of the experiment is defined by

$$E_D = kT \ln(\omega_0 t) \tag{10-26}$$

E_D is the energy at which the average release time of the carrier from the trap is just equal to the time t. Provided the temperature is less than T_0, electrons in traps which are shallower than E_D will have been excited to the mobility edge and trapped many times, but electrons in states deeper than E_D have a very low probability of release within the time t. Thus the states deeper than E_D are occupied in proportion to the density of states, but the states above E_D have had time to equilibrate and follow a Boltzmann distribution for which the electron density decreases at smaller trap energies. The electron distribution therefore has a peak at E_D, and from the definition of E_D, this peak moves to larger trapping energies as time progresses. The approximation is made that all the electrons reside at E_D, so that the problem can be treated as trapping at a single level, with the added property that the trap energy is time dependent. Substituting E_D from Eq. (10-26) into Eq. (10-23), with the assumed exponential density of states gives

$$\mu_D(t) = \mu_0 \, \alpha (1-\alpha)(\omega_0 t)^{\alpha-1} \tag{10-27}$$

which has a power law time dependence in agreement with the measurements.

In the time-of-flight experiment, the drift mobility is deduced from the time taken for a carrier to cross the sample. It is easily shown that the measured drift mobility is

$$\mu_D^{exp} = \omega_0 \left[\frac{\omega_0}{2\mu_0(1-\alpha)} \right]^{-\frac{1}{\alpha}} \left[\frac{F}{d} \right]^{\frac{1}{\alpha}-1} \tag{10-28}$$

where d is the sample thickness and F is the applied electric field. Thus the dispersive mobility is time, thickness and field dependent, and its magnitude is given in terms of the slope of the exponential band tail.

Figure 10-17 shows the field dependence of the electron and hole drift mobilities at

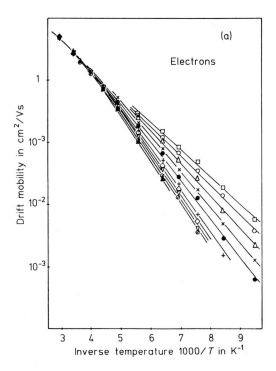

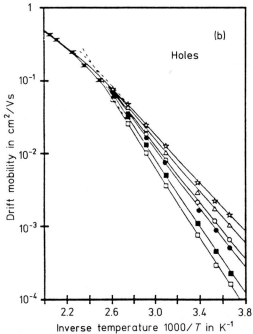

Figure 10-17. Temperature dependence of the (a) electron and (b) hole drift mobility at different applied fields ranging from 5×10^2 V/cm to 5×10^4 V/cm. The field dependence of μ_D is caused by the dispersion (Marshall et al., 1986; Nebel et al., 1989).

different temperatures (Marshall et al., 1986, Nebel et al., 1989). The electron mobility is thermally activated, as expected for a trap limited process and at room temperature there is no field dependence and no dispersion because $T_0 \lesssim 300$ K. However, in the low temperature dispersive regime, there is a large field dependence of μ_D and a time-dependent dispersion parameter α described by T/T_0. The data for holes is qualitatively similar; the only difference is that the temperature scale is changed, so that the hole transport is dispersive up to 400–500 K. The difference occurs because the valence band tail is wider than the conduction band tail. The mobility measurements find a dispersion parameter $T_0 = T_C$ of 250–300 K for electrons and $T_0 = T_V$ of 400–450 K for holes (Tiedje et al., 1981).

10.4 Defect Equilibrium and Metastability

Chemical bonding rearrangements are an important influence on the electronic properties of a-Si:H. Defect and dopant states are created and annihilated either thermally or by external excitations such as illumination, leading to metastable structures. The thermal changes are described by thermodynamic equilibria with defect and dopant concentrations determined by minimization of the free energy. It is perhaps surprising to be able to apply equilibrium concepts to a-Si:H, because the amorphous phase of a solid is not the lowest free energy phase. However, subsets of network constituents may be in equilibrium with each other even if the network structure as a whole is not in its lowest energy state. The collective motion of many atoms is required to achieve long-range order, and there are strong topological constraints which usually prevent such ordering. However, chemical bonding transformations of defects or dopants require the cooperation of only a small number of atoms. Therefore, the small concentrations of defects or impurities in a-Si:H may be expected to participate in local thermodynamic equilibrium which takes place within the more or less rigid Si random network. Evidence that such equilibration is mediated by hydrogen motion is presented and discussed in Sec. 10.4.5.2.

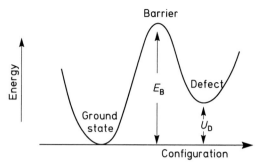

Figure 10-18. Configurational coordinate diagram of the equilibration between two states separated by a potential energy barrier.

10.4.1 The Hydrogen Glass Model

The properties needed to describe the chemical bonding changes in a-Si:H are the equilibrium state and the kinetics of the process. Equilibrium is calculated from the formation energies of the various species, by minimizing the free energy, or equivalently, by applying the law of mass action to the chemical reactions describing the changes. For example, the reaction may correspond to a change in atomic coordination which causes the creation and annihilation of defect or dopant states

$$Si_4 \leftrightarrows Si_3 \quad \text{or} \quad P_3 \leftrightarrows P_4 \qquad (10\text{-}29)$$

The kinetics of the reaction are described by a relaxation time, τ_R, required for the structure to overcome the bonding constraints which inhibit the reaction. τ_R is associated with an energy barrier, E_B, which arises from the bonding energies and is illustrated in Fig. 10-18 by a configurational coordinate diagram. The energy difference between the two potential minima is the defect formation energy, U_D, and determines the equilibrium concentrations of the two species. The equilibration time is related to the barrier height by

$$\tau_R = \omega_0^{-1} \exp(-E_B/kT) \qquad (10\text{-}30)$$

where ω_0 is a rate prefactor of order 10^{13} sec^{-1}. A larger energy barrier obviously

requires a higher temperature to achieve equilibrium in a fixed time. The formation energy U_D and the barrier energy E_B are often of very different magnitudes.

There is a close similarity between the defect or dopant equilibration of a-Si:H and the behavior of glasses near the glass transition, which is useful to keep in mind in the analysis of the a-Si:H results. Configurations with an energy barrier of the type illustrated in Fig. 10-18 exhibit a high temperature equilibrium and a low temperature frozen state when the thermal energy is insufficient to overcome the barrier. The temperature, T_E, at which freezing occurs is calculated from Eq. (10-30) by equating the cooling rate, R_C, with $dT/d\tau_R$,

$$T_E \ln(\omega_0 k T_E^2/R_C E_B) = E_B/k \qquad (10\text{-}31)$$

The approximate solution to Eq. (10-31) for a freezing temperature in the vicinity of 500 K is

$$k T_E = E_B/(30 - \ln R_C) \qquad (10\text{-}32)$$

from which it is readily found that an energy barrier of 1–1.5 eV is needed for $T_E = 500$ K for a normal cooling rate of 10–100 K/s. An order of magnitude increase in cooling rate raises the freezing temperature by about 40 °C. Below T_E, the equilibration time is observed as a slow

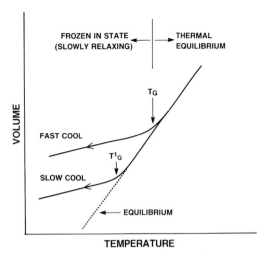

Figure 10-19. Illustration of the properties of a normal glass near the glass transition. The low temperature frozen state is kinetically determined and depends on the cooling rate.

relaxation of the structure towards the equilibrium state.

Figure 10-19 illustrates the properties of a normal glass by showing the temperature dependence of the volume, V_0. There is a change in the slope of $V_0(T)$ as the glass cools from the liquid state, which denotes the glass transition temperature T_G. The glass is in a liquid-like equilibrium above T_G, but the structural equilibration time increases rapidly as it is cooled. The glass transition occurs when the equilibration time becomes longer than the measurement time, so that the equilibrium can no longer be maintained so that the structure is frozen. The transition temperature is higher when the glass is cooled faster, and the properties of the frozen state depend on the thermal history. Slow structural relaxation is observed at temperatures just below T_G.

The glass-like characteristics are exhibited by the electronic properties of a-Si:H. However, a-Si:H is not a normal glass; it cannot be quenched from the melt. In a glass, all network constituents that contribute to the electronic structure participate in the structural equilibration. In a-Si:H, the disordered Si network is more or less rigid and the majority of Si atoms are fixed in a non-equilibrium configuration which persists up to the crystallization temperature (600 °C). As is discussed in later sections, the kinetics of the equilibration of defects and dopants are governed by the motion of hydrogen, which mediates the coordination changes necessary in the approach to dopant or defect equilibrium. Virtually all hydrogen incorporated into the a-Si:H network participates in defect formation and dopant activation reactions, which are in turn governed by a hydrogen chemical potential. It is the kinetics of hydrogen motion which determines the kinetics of defect formation and dopant activation reactions in a-Si:H. The analogy between dopant activation kinetics in a-Si:H and the kinetics of structural relaxation in glasses can be interpreted in terms of the glassy behavior of the hydrogen subnetwork in a-Si:H, which has been termed the hydrogen glass model (Street et al., 1987a).

10.4.2 Thermal Equilibration of Electronic States

Figure 10-20 shows the temperature dependence of the d.c. conductivity of n-type a-Si:H for different thermal treatments, and the features of the data are obviously similar to those of glasses in Fig. 10-19. There is a change of slope at about $T_E = 130$ °C which distinguishes the high and low temperature regimes. Fast quenching from high temperature results in a higher conductivity than slow cooling, at a given temperature below 100 °C. Above T_E, the conductivity has a different activation energy and is independent of the thermal history. This is the equilibrium regime in which the defect and dopant densities are temperature dependent according to free energy minimization. The structure is frozen at lower temperature and has the metastable structure characteristic of a glass below the glass transition temperature. If the temperature is not too low, then there is slow relaxation of the structure which is described in Sec. 10.4.5.

Metastable defects are also created thermally in undoped a-Si:H. Figure 10-21 shows the temperature dependence of the defect density between 200 °C and 400 °C for material deposited under different plasma conditions. The defect density increases with temperature with an activation energy of about 0.2 eV. Although the defect density is reversible, a high metastable density is maintained by rapid quenching from the anneal temperature. Prolonged annealing at a lower tempera-

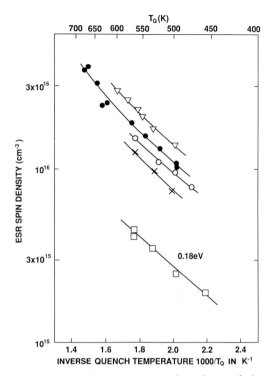

Figure 10-21. The temperature dependence of the equilibrium neutral defect density in undoped a-Si:H deposited with different deposition conditions (Street and Winer, 1989).

ture reduces the defect density back to its original value.

Thus both doped and undoped a-Si:H have the glass-like property of a high temperature equilibrium and a low temperature frozen state. The equilibration temperature of undoped a-Si:H is higher than that of n-type material, indicating a slower relaxation process, arising from a higher barrier energy. The relaxation of p-type material is faster, yielding a lower equilibration temperature. Metastable defect creation is a related phenomenon. Here defects are created by an external stress such as illumination or bias. The defects are metastable provided that the temperature is well below the equilibration temperature, but are removed by annealing (see Sec. 10.4.6.).

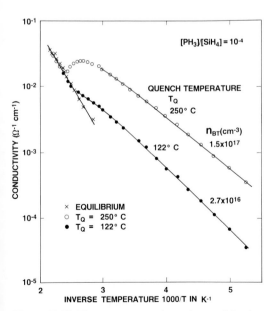

Figure 10-20. The temperature dependence of the d.c. conductivity of n-type a-Si:H, after annealing and cooling from different temperatures, and in a steady state equilibrium. The measurements are made during warming (Street et al., 1988 b).

10.4.3 The Defect Compensation Model of Doping

The reversible changes in the conductivity of doped a-Si:H arise because the equilibration of defects and dopants alters the electrical conductivity. The charged defects act as compensating centers for the dopants, so that the density, n_{BT}, of excess electrons occupying band edge states is

$$n_{BT} = N_{don} - N_D \qquad (10\text{-}33)$$

where N_{don} and N_D are the dopant and defect concentrations. The conductivity is

$$\sigma = n_{BT}\, e\, \mu_D \qquad (10\text{-}34)$$

where μ_D is the effective drift mobility. Thus, changes in the density of donor and defect states are reflected in the conductivity.

Equilibration occurs between the different bonding states of the silicon and dopant atoms, which can both have atomic coordination 3 or 4. The lowest energy states are Si_4^0 and P_3^0, as indicated by the $8-N$ rule (see Sec. 10.1.3). The formation energies of P_4 and Si_3 are large enough that neither would normally be expected to have a large concentration. However, when both states are formed, the electron liberated from the donor is trapped by the dangling bond, liberating a substantial energy and promoting their formation. The compensation of phosphorus donors by defects is described by the chemical reaction

$$P_3^0 + Si_4^0 \leftrightarrows P_4^+ + Si_3^- \qquad (10\text{-}35)$$

The equilibrium state of doped a-Si:H is calculated by applying the law of mass action

$$N_{P_4} N_{Si_3} = K\, N_0 N_{P_3}; $$
$$K = \exp[-(U_P + U_D)/kT] \qquad (10\text{-}36)$$

where the different N's denote the concentrations of the different species, N_0 is the concentration of 4-fold silicon sites, and K is the reaction constant. U_P and U_D are the formation energies of donor and defect. Single values of the formation energies are assumed for simplicity; the next section includes a distribution of formation energies, which is more appropriate for a disordered material.

When the doping efficiency is sufficiently low that $N_{P_4} \ll N_{P_3}$, and N_{P_4} is equated to N_{Si_3} as required by charge neutrality, then Eq. (10-36) becomes

$$N_{P_4} = N_{Si_3} = \qquad (10\text{-}37)$$
$$= (N_0 N_P)^{1/2} \exp[-(U_P + U_D)/2kT]$$

This equation predicts the square root law for defect creation which is observed in the data of Fig. 10-12. The thermodynamics also predicts that the doping efficiency is temperature dependent and explains the high metastable d.c. conductivity which is frozen in by quenching (Street et al., 1988a).

Defect equilibration also occurs in undoped a-Si:H. In the absence of dopants, the model predicts that the temperature dependence of the defect density is

$$N_{D0} = N_0 \exp(-U_{D0}/kT) \qquad (10\text{-}38)$$

The defect density does increase with temperature as is seen in the data of Fig. 10-21. However, the temperature dependence has an activation energy of only about 0.2 eV and a small N_0, whereas the model for doping just described indicates a considerably larger value of the formation energy (Street et al., 1988a). The difference originates from the distribution of formation energies, which must be included to get the correct defect density. This is discussed in the next section.

This type of defect reaction also provides a general explanation of all the other metastable phenomena described in Sec. 10.4.6. The formation energies of charged defects and dopants depend on the position of the Fermi energy, E_F

defect: $U_D = U_{D0} - (E_F - E_D)$ (10-39)

dopant: $U_P = U_{P0} - (E_P - E_F)$ (10-40)

U_{D0} and U_{P0} are the formation energies of the neutral states and E_D and E_P are the associated gap state energies. The second terms in Eqs. (10-39) and (10-40) are the contributions to the formation energy from the transfer of an electron from the Fermi energy to the defect or from the donor to E_F. The negatively charged defect density is given by a Boltzmann expression (Shockley and Moll, 1960)

$$N_{D-} = N_0 \exp\left[-U_{D0}/kT\right] \cdot$$
$$\cdot \exp\left[(E_F - E_D)/kT\right] =$$
$$= N_{D0} \exp\left[(E_F - E_D)/kT\right]   (10-41)$$

and there are similar expressions for positive defects and donors. Equation (10-41) assumes that the defects have the same formation energy and gap state levels in doped and undoped a-Si:H. The defect density is, therefore, a function of the position of the Fermi energy, and Eq. (10-41) expresses the interaction between the electronic properties and the bonding structure. The equilibrium defect density in Eq. (10-41) increases exponentially as E_F moves from the dangling bond gap state energy.

Thus, doping increases the defect density, as does any other process which moves the Fermi energy from mid-gap, such as illumination or voltage bias. The doping efficiency is suppressed by doping, but enhanced by compensation. All of these effects are observed in a-Si:H.

10.4.4 The Weak Bond Model

The random network of an amorphous material such as a-Si:H implies that the formation energy varies from site to site. A full evaluation of the equilibrium must include this distribution and also the disorder broadening of the defect energy levels. The generalized form of Eq. (10-38) is

$$N_D = \int \frac{N_0(U_D) \exp\left(-U_D/kT\right)}{1 + \exp\left(-U_D/kT\right)} \, dU_0   (10-42)$$

where $N_0(U_D)$ is the distribution function of the formation energy.

Calculations of the distribution of formation energies have been addressed by the weak bond model (Stutzmann, 1987; Smith and Wagner, 1987). Figure 10-22a shows a schematic model of a weak Si–Si bond, and a pair of dangling bonds. When the weak bond is converted into a neutral dangling bond, the electron energy increases from that of a bonding state in the valence band to that of a non-bonding state in the gap. The formation energy of a neutral defect is

$$U_{D0} = E_D - E_{WB} +$$
$$+ \left[\sum_{V \neq WB} (E_V' - E_V) + \Delta E_{ion} \right]   (10-43)$$

E_D and E_{WB} are the gap energies of the defect electron and of the valence band tail state associated with the weak bond. E_V and E_V' are energies of the valence band electrons, before and after the bond is broken and the sum represents the change in energy of all the valence band states other than from the broken weak bond. ΔE_{ion} is the change in energy of the ion core interaction for the structure with and without the defect. The weak bond model assumes that the terms in the square bracket in Eq. (10-43) are small, so that

$$U_{D0} \simeq E_D - E_{WB}   (10-44)$$

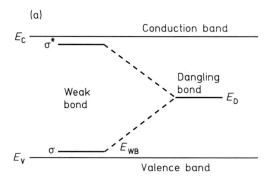

(a)

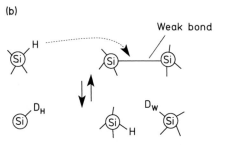

(b)

Figure 10-22. (a) Energy level diagram showing the conversion of a weak Si–Si bond into a dangling bond: (b) Illustration of the hydrogen-mediated weak bond model in which a hydrogen atom moves from a Si–H bond and breaks a weak bond, leaving two defects (D_H and D_W) (Street and Winer, 1989).

The distribution of defect formation energies is therefore described by the density of valence band tail states. Most states have a high formation energy, but an exponentially decreasing number have lower formation energies. In equilibrium, virtually all the band tail states which are further from the valence band than E_D convert into defects, while only a small temperature dependent fraction of the states between E_D and the mobility edge convert. The weak bond model is useful because the distribution of formation energies can be evaluated from the known valence band and defect density of states distributions.

A calculation of the defect density requires a specific physical model for defect

creation. Dangling bond defects form by the breaking of silicon bonds, and several specific models have been proposed (Smith and Wagner, 1987; Street and Winer, 1989; Zafar and Schiff, 1989). We analyze a model in which the bonds are broken by the motion of hydrogen. Figure 10-22 b shows hydrogen released from an Si–H bond, breaking a Si–Si bond to give two separate defects. Experimental evidence for the involvement of hydrogen in the equilibration is described in a later section. The hydrogen-mediated weak bond model of Fig. 10-22 b is described by the defect reaction

$$Si-H + (\text{weak bond}) \leftrightarrows D_H + D_W \quad (10\text{-}45)$$

The two defects may be electrically identical but make different contributions to the entropy. The law of mass action solution for the defect density, including the distribution of formation energies, is (Street and Winer, 1989)

$$N_D = 2 N_H N_{V0} \cdot \quad (10\text{-}46)$$
$$\cdot \exp(-E_D/k T_V) \int \frac{\exp(U_{D0}/k T_V)\, dU_{D0}}{N_H + N_D \exp(2 U_{D0}/k T)}$$

where T_V is the slope of the exponential valence band tail. N_{V0} is the band tail density of states at the assumed zero of the energy scale for E_D. Numerical integration of Eq. (10-46) gives an excellent fit to the data of Fig. 10-21, for band tail and defect parameters which are consistent with the known electrical properties. The weak temperature dependence of N_D follows directly from the distribution of formation energies (Street and Winer, 1989).

The weak bond model also explains the variation of the defect density with growth conditions in the plasma reactor. In material with more disorder, the valence band tail is broader (i.e. larger T_V) and N_D increases according to Eq. (10-46). The defect density is conveniently expressed as

$$N_D = \xi(T, T_V, N_H) N_{V0} k T_V \cdot$$
$$\cdot \exp[-E_D/k T_V] \qquad (10\text{-}47)$$

where $\xi(T, T_V, N_H)$ represents the entropy factor which differs for each specific defect creation model, but which is a slowly varying function. The equilibrium defect density is primarily sensitive to T_V and E_D, through the exponential factor. For example, raising T_V from 500 K to 1000 K increases the defect density by a factor of about 100.

The sensitivity of the defect density to the band tail slope accounts for the large change in defect density with deposition conditions and annealing. Figure 10-23 shows the correlation between the valence band tail slope and the defect density for undoped a-Si:H deposited by different methods and under different deposition conditions. The data show that a high defect density is correlated with a wide band tail slope, and is explained by the equilibrium model. The band tails are much broader at low deposition temperatures, so that Eq. (10-47) predicts the higher defect density which is observed. The defect density is reduced when the low deposition temperature material is annealed, and the band tail slope is correspondingly reduced. Similarly, the slope of the Urbach tail and the defect density both increase at deposition temperatures well above 300 °C, and are associated with a lower hydrogen concentration in the film. Both E_D and T_V may change in alloys of a-Si:H and this is perhaps the origin of the different defect densities in these materials. There is a larger defect density in low band gap a-SiGe:H alloys, which is predicted from the reduced value of E_D accompanying the shrinking of the band gap. In the larger gap alloys, such as a-SiC:H, the predicted reduction in N_D

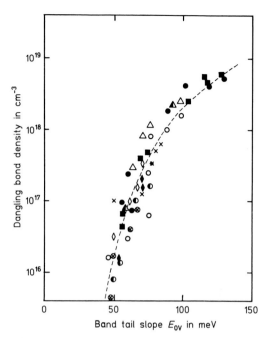

Figure 10-23. Dependence of the defect density on the slope of the Urbach absorption edge for undoped a-Si:H deposited under a variety of conditions (Stutzmann, 1989).

due to the larger E_D seems to be more than offset by a larger T_V, so that the defect density is again greater than in a-Si:H. There is, however, no complete explanation of why a-Si:H has the lowest defect density of all the alloys which have been studied.

10.4.4.1 The Distribution of Gap States

The theory described above only considers defects with a single energy in the gap. Neutral defects in undoped a-Si:H are known to be distributed in an approximately Gaussian band ~ 0.1 eV wide. Bar-Yam and Joannopoulos et al. (1986) first pointed out that the minimization of the free energy of the broadened defect band causes a shift of the defect gap state energy level. The reason is that the gap state energy is contained within Eq. (10-44) for the defect formation energy. Thus, states that are at a lower energy in the band gap will have a lower formation energy and, therefore, a higher equilibrium density. This is the basis of the defect pool concept, in which there is a distribution of available states where defects can be formed, which are selected on the basis of energy minimization.

The interesting feature of this dependence on gap state energy is that it leads to different defect state distributions depending on the charge state of the defect. The formation energy of positively charged defects is not influenced by the energy of the gap state, because the defect is unoccupied. On the other hand, negative defects contain two electrons and the gap state energy enters twice. For a Gaussian distribution of possible defect state energies, there is a shift of the defect band to low energy by (Street and Winer, 1989)

$$\Delta E = \frac{\sigma^2}{k\,T_V} \tag{10-48}$$

for each electron in the defect, where σ is the width of the Gaussian distribution. Measured values of the defect band width are imprecise but lie in the range 0.2–0.3 eV, corresponding to $\sigma \approx 0.1$ eV. The predicted shift of the peak is therefore about 0.2 eV, when $k\,T_V \approx 45$ meV. This shift of the defect energy with doping ex-

plains some of the differences in the measured defect energies. The equilibration process has the effect of removing gap states from near the Fermi energy. Either unoccupied state above E_F, or occupied states below E_F are energetically preferred to partially occupied states at E_F. The shift of the defect levels with doping is illustrated in Fig. 10-14.

10.4.5 Defect Reaction Kinetics

There is a temperature dependent equilibration time associated with the chemical bonding changes. The very long time constant at room temperature is responsible for the metastability phenomena, because the structure is frozen. An example of the slow relaxation towards equilibrium in n-type a-Si:H is shown in Fig. 10-24 by the time dependence of the electrons occupying shallow states, n_{BT}, following a rapid quench from 210 °C. n_{BT} decays slowly to a steady state equilibrium, with the decay taking more than a year at room temperature, but only a few minutes at 125 °C. The temperature dependence of the relaxation time, τ, is plotted in Fig. 10-25, and has an activation energy of about 1 eV, which measures the energy barrier for bonding rearrangement. The relaxation is faster in p-type than in n-type a-Si:H and has a

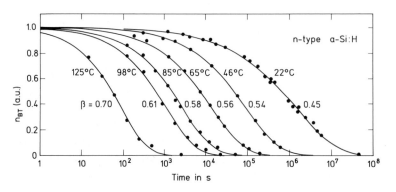

Figure 10-24. The time-dependent relaxation of the band tail carrier density plotted in normalized form. The solid lines are fits to the stretched exponential with parameter β as indicated (Kakalios et al., 1987).

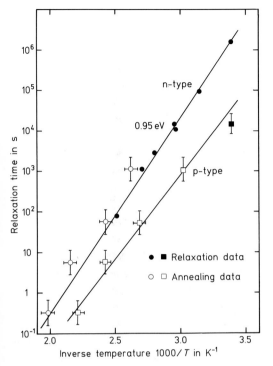

Figure 10-25. Temperature dependence of the relaxation time in n-type and p-type a-Si:H (Street et al., 1988a).

slightly lower activation energy. A similar relaxation occurs in undoped a-Si:H, with a larger activation energy and a longer relaxation time (Street and Winer, 1989). Equations (10-30) and (10-31) relate the relaxation time to the equilibration temperature.

10.4.5.1 Stretched Exponential Decay

The relaxation kinetics follow a stretched exponential relation

$$\Delta n_{BT} = n_0 \exp[-(t/\tau)^\beta] \qquad (10\text{-}49)$$

with $0 < \beta < 1$ (Kakalios et al., 1987). This type of decay also describes the equilibration of a wide class of disordered materials, particularly the structural relaxation of a normal glass. The origin of the stretched

exponential is the disorder which leads to a distribution of activation barriers for structural change. However, the mathematical description of the relaxation is complicated and is not completely resolved. The relaxation may be understood from a simple rate equation with a time dependent rate constant, $k(t)$.

$$dn/dt = -k(t)n \qquad (10\text{-}50)$$

The time dependence of $k(t)$ reflects a structural relaxation occurring within a system which is itself time dependent. If k has a time dependence of the form

$$k(t) = k_0 t^{\beta-1} \qquad (10\text{-}51)$$

then the integration of Eq. (10-50) gives the stretched exponential

$$n(t) = n_0 \exp[-(t/\tau)^\beta] \qquad (10\text{-}52)$$

The temperature dependence of the parameter β, measured from the relaxation data of Fig. 10-24, follows the relation

$$\beta = T/T_E \qquad T_E \simeq 600 \text{ K} \qquad (10\text{-}53)$$

The equilibration rate is determined by the motion of either silicon, hydrogen or the dopants, which cause the structural relaxation. Equation (10-51) implies that the atomic motion has a power law time dependence. Hydrogen is known to diffuse fairly easily in a-Si:H, as is obvious from the fact that it is evolved from a-Si:H at a fairly low temperature (400–500°C). On the other hand, the silicon network is rigid and stable, and both silicon and the dopants have a much lower diffusion rate. Hydrogen motion is therefore an obvious candidate for the mechanism by which the thermal equilibration takes place. The next section shows that the diffusion properties of hydrogen within a-Si:H quantitatively account for the relaxation rates and the stretched exponential decay.

10.4.5.2 Hydrogen Diffusion

The hydrogen diffusion coefficients shown in Fig. 10-26 are thermally activated

$$D_H = D_0 \exp(-E_{HD}/kT) \qquad (10\text{-}54)$$

where the energy E_{HD} is about 1.5 eV, and the prefactor D_0 is about 10^{-2} cm^2 s^{-1}. The diffusion is also quite strongly doping dependent, as seen in the data for p-type, n-type and compensated material. The greatest change is with the p-type material for which D_H is larger by a factor 10^3 at 250 °C. The increase is less in n-type material, while the diffusion rates in compensated and undoped a-Si:H are similar. The low diffusion in compensated material suggests that the doping effect is electronic rather than structural in origin. The diffusion activation energy is similar to the energy for structural relaxation, and the trends with doping mirror the doping dependence of the equilibration rates (compare Figs. 10-25 and 10-26).

Thermally-activated diffusion is explained by a trapping mechanism. There is no significant diffusion of silicon at these temperatures, so that hydrogen diffusion occurs by breaking a Si–H bond and reforming the bond at a new site. Hydrogen is trapped by bonding to the silicon dangling bonds and the mobile hydrogen moves between higher energy interstitial sites. Highly distorted Si–Si bonds also act as traps for hydrogen which can break the weak bond by the reaction shown in Fig. 10-22. The interstitial is the Si–Si bond center site in crystalline silicon and is most probably the same in a-Si:H (Johnson et al., 1986). For this model the diffusion is given by

$$D_H = a_D^2 \omega_0 \exp(-E_{HD}/kT) \qquad (10\text{-}55)$$

where $\omega_0 \sim 10^{13}$ s^{-1} is the attempt-to-escape frequency for the excitation out of the trapping site, a_D is the distance between sites, and E_{HD} is the energy to release hydrogen from a Si–H bond. A diffusion prefactor of 10^{-2} cm^2 s^{-1} results from a hopping distance of about 3 Å. However, the prefactor is difficult to interpret because there are other hydrogen traps in addition to the dangling bonds and the measurement is influenced by the time-dependence of the diffusion.

The hydrogen diffusion coefficient is not constant, but decreases with time (Street et al., 1987b). The data in Fig. 10-27 show a power law decrease in p-type a-Si:H of the form $t^{\beta-1}$, with $\beta \simeq 0.8$ at the measurement temperature of 200 °C. The time dependence is associated with a distribution of traps originating from the disorder. The effect is similar to the dispersive transport of electrons and holes in an exponential

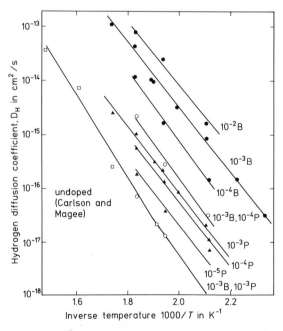

Figure 10-26. The temperature dependence of the hydrogen diffusion coefficient at different doping levels as indicated (Street et al., 1987 b), including data from Carlson and Magee (1978).

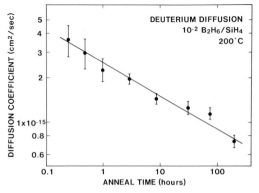

Figure 10-27. The power law decrease in the time dependence of the hydrogen diffusion coefficient of p-type a-Si:H (Street et al., 1987b).

band tail. The time dependence of the hydrogen diffusion is of the form of Eq. (10-51), and accounts for the stretched exponential decay, since the temperature dependence of the diffusion dispersion parameter agrees with the value from the relaxation measurements. These results indicate that the motion of hydrogen is the rate limiting process in the bonding equilibration. Hydrogen can change the coordination of a silicon or impurity atom by breaking and forming bonds, causing the creation and annihilation of defects and dopants.

10.4.6 Metastability

A serious limitation to some applications of a-Si:H is the phenomenon of reversible metastable changes which are observed in the electronic properties when stress is applied. Such changes are induced by illumination, voltage bias or an electric current, and are reversed by annealing to 150–200 °C. The most widely studied metastability is the creation of defects by prolonged illumination, known as the Staebler-Wronski effect (Staebler and Wronski, 1977). In addition, space-charge-limited current flow in p-i-p structures induces defects (Kruhler et al., 1984), and studies of the current-voltage characteristics of thin film transistors find a threshold shift which is due to defects in the a-Si:H layer, induced by the electron accumulation at the interface (Jackson and Kakalios, 1989). In common with all the metastable changes, the annealing process is thermally activated with an energy about 1 eV, so that annealing of undoped a-Si:H takes a few minutes at 200 °C, several hours at 150 °C, and an indefinitely long time at room temperature.

The external excitation may change both the reaction rate and the equilibrium state of a defect reaction. Figure 10-28 illustrates the expected temperature dependence of the reaction rate in different situations. The activation energy of a purely thermal defect creation process is E_B, which is about 1 eV in a-Si:H and associated with hydrogen diffusion. An external

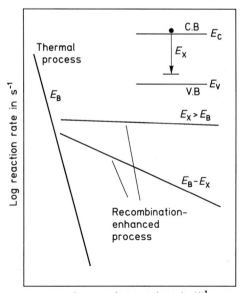

Figure 10-28. The expected temperature dependence of the reaction rate for a recombination-enhanced defect creation process (Kimerling, 1978).

excitation can provide a non-thermal source of energy to overcome the potential barrier – the non-radiative recombination of excess carriers is an example. The recombination-enhanced reaction rate has a lower activation energy of $E_B - E_X$, where E_X is the energy provided by recombination (Kimerling, 1978). The reaction rate is athermal when E_X is equal to or larger than E_B, but is thermally activated when E_X is smaller.

Defects or dopants are created because the external excitation drives the defect reaction away from the initial equilibrium. The formation energy of charged defects, and, therefore, their equilibrium density, depends on the position of the Fermi energy according to Eq. (10-39). Thus, the equilibrium defect density is

$$N_{DB} = N_{DB_0} \exp(\Delta E_F / k T) \qquad (10\text{-}56)$$

where ΔE_F is the shift of the Fermi energy, and N_{DB_0} is the neutral defect density corresponding to an unshifted Fermi energy. Any external stress which causes the Fermi energy to move disturbs the equilibrium of the states and tends to change the defect density. The examples of metastability discussed below cover most of the possible ways in which the Fermi energy (or quasi-Fermi energy) can be moved by an external influence. Others include bulk doping (Sec. 10.4.3) and oxidation (Sec. 10.2.4.2).

10.4.6.1 Defect Creation by Illumination

The most widely studied example of metastability in a-Si:H is the creation of defects by prolonged exposure to light. The defects are dangling silicon bonds and are detected by ESR and other defect spectroscopy techniques. Figure 10-29 shows the time dependent increase in the defect density, which follows a $t^{1/3}$ law (Stutzmann et al., 1985). The defect density increases

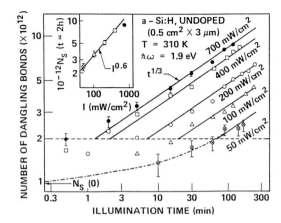

Figure 10-29. The time dependent increase of the defect density by illumination of undoped a-Si:H for different illumination intensities. The lines show the fit to a (time)$^{1/3}$ dependence. The intensity dependence at constant illumination time is shown in the insert (Stutzmann et al., 1985).

from 10^{16} cm^{-3} to about 10^{17} cm^{-3} after an hour of sunlight illumination. Although Fig. 10-29 shows no sign of saturation after 2–3 hours of strong illumination at room temperature, a steady state is obtained at higher light intensities or longer illumination times (Park et al., 1990). The defect creation results from the non-radiative recombination of electrons and holes, which releases enough energy to cause bonding rearrangement. The sub-linear creation kinetics occurs because the additional defects act as recombination centres and suppress the density of excited carriers. The excess defects are metastable and are removed by annealing above about 150°C, so that the defect creation process is reversible.

Defects are created by the recombination of photoexcited carriers, rather than by the optical absorption. The experimental evidence is that defect creation also results from charge injection without illumination, and that defect creation by illumination is suppressed by a reverse bias across the sample which removes the excess carriers (Swartz, 1984). The defect creation rate is almost independent of temperature, indicating that all the energy needed to overcome the barrier is provided by the recombination. The kinetics of defect creation are explained by a recombination model which assumes that the defect creation is initiated by the non-radiative band-to-band recombination of an electron and hole. The recombination releases about 1.5 eV of energy which breaks a weak bond and generates a defect. In terms of the configurational coordinate model of Fig. 10-18, the recombination energy completely overcomes the barrier E_B. Stutzmann et al. (1985) propose the following model to explain the defect creation. The creation rate is proportional to the band-to-band recombination rate

$$dN_D/dt = c_d n p \qquad (10\text{-}57)$$

where c_d is a constant describing the creation probability, and n and p are the electron and hole concentrations. The defect creation process represents only a small fraction of the recombination, most of which is by trapping at the dangling bond defects. The recombination is quite complicated, but provided the illumination intensity G is high enough, the carrier densities

n and p at the conduction and valence band edges are given to a fair approximation by

$$n = p = G/A N_D \qquad (10\text{-}58)$$

where A is an average recombination constant. Photoconductivity experiments, which measure the carrier concentration n, confirm the predicted dependence on G and N_D, although at low defect density there are deviations from a monomolecular decay.

The time dependence of the defect density is obtained by combining Eqs. (10-57) and (10-58) and integrating, to give

$$N_D^3(t) - N_D^3(0) = 3 c_d G^2 t/A^2 \qquad (10\text{-}59)$$

where $N_D(0)$ is the initial equilibrium defect density. At sufficiently long illumination times such that $N_D(t) > 2 N_D(0)$, Eq. (10-59) approximates to

$$N_D(t) = (3 c_d/A^2)^{1/3} G^{2/3} t^{1/3} \qquad (10\text{-}60)$$

which agrees well with the intensity and time dependence of the experimental results.

The annealing kinetics of the light induced defects are shown in Fig. 10-30. Several hours at 130 °C are needed to anneal the defects completely, but only a few min-

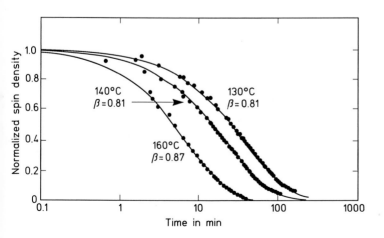

Figure 10-30. The decay of the normalized light induced defect density at different anneal temperatures, showing the stretched exponential behavior (Jackson and Kakalios, 1989).

utes at 200 °C. The relaxation is non-exponential, and in the initial measurements of the decay the results were analysed in terms of a distribution of time constants (Stutzmann et al., 1986), which is centered close to 1 eV with a width of about 0.2 eV. Subsequently it was found that the decay fits a stretched exponential. The parameters of the decay – the dispersion, β, and the temperature dependence of the decay time, τ, – are similar to those found for the thermal relaxation data, and so are consistent with the same mechanism of hydrogen diffusion mediating the chemical reactions. The annealing is therefore the process of relaxation to the equilibrium state with a low defect density.

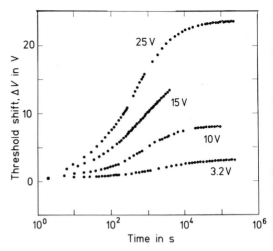

Figure 10-31. Time dependence of the defect density created near a dielectric interface by voltage bias. The measurement is of the threshold shift, ΔV, which is proportional to the defect density (Jackson and Kakalios, 1989).

10.4.6.2 Defect Creation by Bias and Current

Similar defect creation occurs when excess space charge is introduced in undoped a-Si:H. This is achieved by biasing into electron accumulation the interface of a-Si:H with a dielectric. Figure 10-31 shows the time dependent change in the threshold voltage of the junction capacitance, which is proportional to the induced defect density. In this case, as many as 10^{18} cm^{-3} defects are created over an extended time. The defect creation rate is thermally activated, and the rate increases by a factor 100 between 360 K and 420 K. The defect creation kinetics follow a stretched exponential time dependence with parameters which are similar to those found for the equilibration kinetics discussed in Sec. 10.4.5.1. The defect creation rate is low at room temperature but is thermally activated with an energy of about 1 eV, and has a time constant of a few minutes at 150 °C.

A space charge limited current flow in a p$^+$-p-p$^+$ structure also results in a meta-

stable increase in the defect density (Kruhler et al., 1984). The presence of extra defects is inferred from the reduction in the current seen in the current-voltage characteristics. As with the other metastable changes, the effect is reversible by annealing to about 150 °C. Both of these experiments show that electron-hole recombination is not necessary to create defects, since there are virtually no minority carriers in either case. The energy needed to form the defects must instead come from one carrier.

10.5 Summary

Much has been learned about hydrogenated amorphous silicon in the past two decades. Most of its interesting properties follow from the disorder of the atomic structure, which influences the material in several different ways. For example, the disorder inherent in the bond length and bond angle variations causes band tailing, strong carrier scattering and localization. The characteristic transport properties of a-Si:H follow, including the thermally activated electrical conduction and the dispersive carrier mobility. The random network structure also allows bonding configurations which are not allowed in the corresponding crystalline material. Coordination defects, in which a single atom has a different coordination from normal, are the elementary point defects in an amorphous semiconductor, of which the dangling bond is the primary example. The elementary defects in a crystalline material are the vacancy and interstitial, and the isolated dangling bond cannot occur, primarily because of the constraints of the long range order.

Defect reactions and metastability are a general phenomenon in a-Si:H. Electronic excitations of all types induce structural changes, which are metastable at room temperature and anneal out above 100–150 °C. The phenomena are broadly explained by the chemical bonding of the random network which allows a change of atomic coordination when an extra electron occupies a localized state. Consequently the equilibrium defect and dopant concentrations depend on the electron distribution and Fermi energy position. The hydrogen plays an important role in the metastability through its property of diffusing at moderate temperatures. Slow relaxation near the equilibrium temperature follows a stretched exponential time dependence, common to many disordered systems.

10.6 References

Abrahams, E., Anderson, P. W., Licciardello, D. C., Ramakrishman, T. V. (1979), *Phys. Rev. Lett. 42*, 673.

Allen, D. C., Joannopoulos, J. D. (1984), in: *The Physics of Hydrogenated Amorphous Silicon II:* Joannopoulos, J. D., Lucovsky, G. (Eds.). Heidelberg: Springer Verlag.

Aljishi, S., Shu Jin, Ley, L., Wagner, S. (1990), *Phys. Rev. Lett. 65*, 629.

Anderson, P. W. (1958), *Phys. Rev. 109*, 1492.

Bar-Yam, Y., Joannopoulos, J. D. (1986), *Phys. Rev. Lett. 56*, 2203.

Berner, H., Munz, P., Bucher, E., Kessler, F., Paasche, S. M. (1987), *J. Non-Cryst. Solids 97 & 98*, 847.

Beyer, W., Overhof, H. (1984), *Semiconductors and Semimetals, 21 C.* Orlando: Academic Press, Chap. 8.

Carlson, D. E., Magee, C. W. (1978), *Appl. Phys. Lett. 33*, 81.

Carlson, D. E., Wronski, C. R. (1976), *Appl. Phys. Lett. 28*, 671.

Ching, W. Y., Lam, D. J., Lin, C. C. (1980), *Phys. Rev. B 21*, 2378.

Chittick, R. C., Alexander, J. H., Sterling, H. F. (1969), *J. Electrochemical Soc. 116*, 77.

Cutler, M., Mott, N. F. (1969), *Phys. Rev. 181*, 1336.

Dersch, H., Stuke, J., Beichler, J. (1980), *Appl. Phys. Lett. 38*, 456.

Emin, D. (1977), *Proc. 7th Int. Conf. on Amorphous and Liquid Semiconductors:* Spear, W. E. (Ed.). Edinburgh: CICL, 249.

Etherington, G., Wright, A. C., Wenzel, J. T., Dove, J. C., Clarke, J. H., Sinclair, R. N. (1984), *J. Non-Cryst. Solids 48*, 265.

Fedders, P. A., Carlssen, A. E. (1989), *Phys. Rev. B,* in press.

Friedman, L. (1971), *J. Non-Cryst. Solids 6*, 329.

Gallagher, A. (1986), *Mat. Res. Soc. Symp. Proc. 70*, 3.

Hannay, N. B. (1967), *Solid State Chemistry.* New Jersey: Prentice Hall.

Hattori, K., Mori, T., Okamoto, H., Hamakawa, Y. (1988), *Phys. Rev. Lett. 60*, 825.

Jackson, W. B. (1982), *Solid State Commun. 44*, 477.

Jackson, W. B., Amer, N. M. (1982), *Phys. Rev. B 25*, 5559.

Jackson, W. B., Kakalios, J. (1989), in: *Amorphous Silicon and Related Materials* Vol. 1: Fritzsche H. (Ed.). Singapore: World Scientific, 247.

Jackson, W. B., Biegelsen, D. K., Nemanich, R. J., Knights, J. C. (1983), *Appl. Phys. Lett. 42*, 105.

Jackson, W. B., Tsai, C. C., Thompson, R. (1990), *Phys. Rev. Lett.,* in press.

Johnson, N. M., Herring, C., Chadi, D. J. (1986), *Phys. Rev. Lett. 56*, 769.

Kakalios, J., Fritzsche, H. (1984), *Phys. Rev. Lett. 53*, 1602.

Kakalios, J., Street, R. A., Jackson, W. B. (1987), *Phys. Rev. Lett. 59*, 1037.

Kimerling, L. (1978), *Solid State Electronics 21*, 1391.

Knights, J. C., Lucovsky, G. (1980), *CRC Critical Reviews in Solid State and Materials Sciences, 21*, 211.

Knights, J. C., Lujan, R. A. (1979), *Appl. Phys. Lett. 35*, 244.

Kocka, J., Vanacek, M., Schauer, F. (1987), *J. Non-Cryst. Solids 97 & 98*, 715.

Kruangam, D., Deguchi, M., Hattori, Y., Toyama, T., Okamoto, H., Hamakawa, Y. (1987), *Proc. MRS Symp. 95*, 609.

Kruhler, W., Pfleiderer, H., Plattner, R., Stetter, W. (1984), *AIP Conf. Proc. 120*, 311.

Lang, D. V., Cohen, J. D., Harbison, J. P. (1982), *Phys. Rev. B 25*, 5285.

LeComber, P. G., Jones, D. I., Spear, W. E. (1977), *Philos. Mag. 35*, 1173.

LeComber, P. G., Spear, W. E., Ghaith, A. (1979), *Electronics Letters 15*, 179.

Ley, L., Richter, H., Karcher, R., Johnson, R. L., Reichardt, J. (1981), *J. Phys. (Paris) C 4*, 753.

Marshall, J. M., Street, R. A., Thompson, M. J. (1986), *Philos. Mag. 54*, 51.

Meyer, W. von, Neldel, H. (1937), *Z. Tech. Phys. 12*, 588.

Miki, H., Kawamoto, S., Horikawa, T., Maejima, H., Sakamoto, H., Hayama, M., Onishi, Y. (1987), *MRS Symp. Proc. 95*. Boston: Materials Research Society, 431.

Miyazaki, S., Ihara, Y., Hirose, M. (1987), *Phys. Rev. Lett. 59*, 125.

Mott, N. F. (1968), *J. Non-Cryst. Solids 1*, 1.

Mott, N. F. (1969), *Philos. Mag. 19*, 835.

Mott, N. F., Davis, E. A. (1979), *Electronic Processes in Non-Crystalline Materials*. Oxford: Oxford University Press.

Mott, N. F., Kaveh, M. (1985), *Adv. Phys. 34*, 329.

Nebel, C. E., Bauer, G. H., Gorn, M., Lechner, P. (1989), *Proc. European Photovoltaic Conf.*, to be published.

Okushi, H., Tokumaru, Y., Yamasaki, S., Oheda, H., Tanaka, K. (1982), *Phys. Rev. B 25*, 4313.

Orenstein, J., Kastner, M. (1981), *Phys. Rev. Lett. 46*, 1421.

Pantelides, S. T. (1986), *Phys. Rev. Lett. 57*, 2979.

Park, H. R., Liu, J. Z., Wagner, S. (1990), *Appl. Phys. Lett.*, in press.

Polk, D. E. (1971), *J. Non-Cryst. Solids 5*, 365.

Santos, P., Hundhausen, M., Ley, L. (1986), *Phys. Rev. B 33*, 1516.

Scher, H., Montroll, E. W. (1975), *Phys. Rev. B 12*, 2455.

Schulke, W. (1981), *Philos. Mag. B 43*, 451.

Shockley, W., Moll, J. L. (1960), *Phys. Rev. 119*, 1480.

Smith, Z. E., Wagner, S. (1987), *Phys. Rev. Lett. 59*, 688.

Spear, W. E., LeComber, P. G. (1975), *Solid State Commun. 17*, 1193.

Staebler, D. L., Wronski, C. R. (1977), *Appl. Phys. Lett. 31*, 292.

Street, R. A. (1982), *Phys. Rev. Lett. 49*, 1187.

Street, R. A. (1987), *Proc. SPIE Symp. 763*, 10.

Street, R. A., Biegelsen, D. K. (1982), *Solid State Comm. 44*, 501.

Street, R. A., Winer, K. (1989), *Phys. Rev. B 40*, 6263.

Street, R. A., Biegelsen, D. K., Jackson, W. B., Johnson, N. M., Stutzmann, M. (1985 a), *Philos. Mag. B 52*, 235.

Street, R. A., Thompson, M. J., Johnson, N. M. (1985 b), *Philos. Mag. B 51*, 1.

Street, R. A., Kakalios, J., Tsai, C. C., Hayes, T. M. (1987 a), *Phys. Rev. B 35*, 1316.

Street, R. A., Tsai, C. C., Kakalios, J., Jackson, W. B. (1987 b), *Philos. Mag. B 56*, 305.

Street, R. A., Hack, M., Jackson, W. B. (1988 a), *Phys. Rev. B 37*, 4209.

Street, R. A., Kakalios, J., Hack, M. (1988 b), *Phys. Rev. B 38*, 5603.

Stutzmann, M. (1987), *Philos. Mag. B 56*, 63.

Stutzmann, M. (1989), *Philos. Mag. B 60*, 531.

Stutzmann, M., Street, R. A. (1985), *Phys. Rev. Lett. 54*, 1836.

Stutzmann, M., Jackson, W. B., Tsai, C. C. (1985), *Phys. Rev. B 32*, 23.

Stutzmann, M., Jackson, W. B., Tsai, C. C. (1986), *Phys. Rev. B 34*, 63.

Stutzmann, M., Biegelsen, D. K., Street, R. A. (1987), *Phys. Rev. B 35*, 5666.

Stutzmann, M., Street, R. A., Tsai, C. C., Boyce, J. B., Ready, S. E. (1989), *J. Appl. Phys. 66*, 569.

Swartz, G. A. (1984), *Appl. Phys. Lett. 44*, 697.

Tanaka, K., Matsuda, A. (1987), *Mater. Sci. Reports 3*, 142.

Tanielian, M. (1982), *Philos. Mag. B 45*, 435.

Thompson, M. J., Tuan, H. C. (1986), *IEDM Tech. Digest*. Los Angeles: IEDM, 192.

Tiedje, T., Rose, A. (1980), *Solid State Commun. 37*, 49.

Tiedje, T., Cebulka, J. M., Morel, D. L., Abeles, B. (1981), *Phys. Rev. Lett. 46*, 1425.

Weaire, D. (1971), *Phys. Rev. Lett. 26*, 1541.

Winer, K. (1989), *Phys. Rev. Lett. 63*, 1487.

Winer, K. (1990), *Phys. Rev. B 41*, 7952.

Winer, K., Ley, L. (1987), *Phys. Rev. B 36*, 6072.

Winer, K., Street, R. A. (1989), *Phys. Rev. Lett. 63*, 880.

Winer, K., Wooten, F. (1984), *Phys. stat. sol. (b) 124*, 473.

Winer, K., Hirabayashi, I., Ley, L. (1988), *Phys. Rev. B 38*, 7680.

Wronski, C. R., Lee, S., Hicks, M., Kumar, S. (1989), *Phys. Rev. Lett. 63*, 1420.

Yang, L., Abeles, B., Eberhardt, W., Stasiewski, H., Sondericker, D. (1987), *Phys. Rev. B 35*, 9395.

Zafar, S., Schiff, E. A. (1989), *Phys. Rev. B 40*, in press.

General Reading

Elliott, S. R. (1990), *Physics of Amorphous Materials*. New York: Longman.

Joannopoulos, J. D., Lucovsky, D. (Eds.) (1984), *The Physics of Hydrogenated Amorphous Silicon I and II*. Berlin: Springer-Verlag.

Mott, N. F., Davis, E. A. (1979), *Electronic Processes in Non-Crystalline Materials*. Oxford: Oxford University Press.

Pankove, J. (Ed.) (1984), *Semiconductors and Semimetals* Vol. 21: *Hydrogenated Amorphous Silica*. Orlando: Academic Press.

Street, R. A. (1991), *Hydrogenated Amorphous Silicon*. Cambridge: Cambridge University Press.

Zallen, R. (1983), *The Physics of Amorphous Solids*. New York: Wiley.

11 High-Temperature Properties of 3d Transition Elements in Silicon

Wolfgang Schröter, Michael Seibt, Dieter Gilles

IV. Physikalisches Institut der Georg-August-Universität, Göttingen, Federal Republic of Germany

List of Symbols and Abbreviations

$a_M([M])$	activity of metal atoms M in silicon at the concentration [M]
$A_S^{(-)}$	negatively charged shallow acceptor
d	specimen thickness
d_0	distance between two neighboring lattice sites in silicon
D_M	diffusion coefficient of metal atoms M in silicon
$D_s^{(\sigma)}$	diffusion coefficient of substitutional metal atom in charge state σ
D_0	pre-exponential factor of diffusion coefficient
D_{eff}	effective diffusion coefficient
$D_i^{(+)}, D_i^{(0)}$	diffusion coefficient of positively charged (+), neutral (0) interstitial metal atom in silicon
e	electron charge
E_C	conduction band edge
E_F	Fermi energy referred to the conduction band edge
E_V	valence band edge
f	Debye frequency
f_{cubic}	fraction of cobalt atoms on cubic lattice sites
F^{Si}, F^{ic}	free energy of silicon and intermetallic compound
Δf_{chem}	gain of chemical free energy per metal atom due to precipitation
$\Delta F(n)$	change of free energy due to the formation of a precipitate containing n metal atoms
$\Delta F_S(n)$	contributions reducing the gain in free energy due to the formation of a precipitate containing n metal atoms; $\Delta F_S(n)$ contains, e.g., interfacial energy and strain energy
$G^{(0/+)}, G^{(-/0)}$	defect donor, acceptor level
$\Delta G^{ic/Si}$	excess partial free enthalpy of mixing
$\Delta H_{MSi_2}^f$	partial enthalpy of formation of metal silicide MSi_2 per metal atom
$\Delta H_M^{(mi)}$	migration enthalpy of metal atoms M in silicon
$\Delta H_M^{(Si)}$	partial solution enthalpy of metal M in silicon
$\Delta H_M^{ic/Si}$	partial heat of mixing of metal atoms M in silicon with the intermetallic compound as reference state
ic	intermetallic compound
K	elastic modulus
k_B	Boltzmann constant
L_D	diffusion length of charge carriers
M	metal atom
$M_{i,s}^{(\sigma)}$	metal atoms on interstitial (i) or substitutional (s) sites in the charge state $\sigma = ..., 2-, -, 0, +, 2+, ...$
$[M_{i,s}^{(\sigma)}]$	concentration of $M_{i,s}^{(\sigma)}$
$[M]_{eq}(T)$	equilibrium concentration of M in silicon as a function of T
$[M]_{eq}(E_F)$	equilibrium concentration of M in silicon as a function of E_F
$[M^{(mob)}]_{eq}$	solubility of mobile metal atoms M in silicon
$[M_{i,s}^{(\sigma)}]_{eq}$	equilibrium concentration of $M_{i,s}^{(\sigma)}$ in silicon
$[M^{(im)}]_{eq}$	solubility of immobile metal atoms M in silicon

$[M]_{sink}$	concentration of metal atoms at a sink of silicon self interstitials
n	precipitate density $(=N_V)$
n_i	concentration of intrinsic charge carriers
n^*	number of metal atoms in a precipitate nucleus of critical size
N_i	volume density of tetrahedral interstitial sites in silicon $(=5 \times 10^{22} \, \text{cm}^{-3})$
N_V	volume density of precipitates $(=n)$
p	pressure
Q	total amount of predeposited phosphorus
Q_M	slope of Arrhenius plot of solubility of metal M in silicon
Q_{el}	amount of electrically active phosphorus
r	concentration ratio of non-cubic and cubic cobalt species
r_M	hard sphere radius of metal atom M
r_{Si}	atomic radius of Si-atom
r_0	radius of a precipitate
$R^{(T, H)}$	hard sphere radius of tetrahedral (T) or hexagonal (H) interstitial site
Si_I	silicon self-interstitial
$\Delta S_M^{(mi)}$	migration entropy of metal atom M in silicon
$\Delta S_M^{(ic/Si)}$	partial entropy of solution of transition element M referred to intermetallic compound
T	temperature
t_a	annealing time
T_d	symmetry group of tetrahedral sites in silicon
T_{eut}	eutectic temperature
$T^{(f)}$	melting point of silicon
ΔU_{el}	elastic contribution to the migration enthalpy
$\Delta U_{el}^{(T, H)}$	elastic energy of metal atom on tetrahedral (T) or hexagonal (H) interstitial site
V	vacancy
$\langle v_{PSG} \rangle$	growth velocity of PSG-layer
$x_{M, eq}$	solubility of metal M in silicon expressed in atomic fractions
z	coordination number
$z^{(T, H)}$	number of nearest neighbors on tetrahedral (T) or hexagonal (H) interstitial site
α	central force constant
β	normalized derivate of elastic modulus K with respect to temperature T
ε	static dielectric constant
ε_0	permittivity of free space
μ_M^0	concentration-independent part of $\mu_M^{(Si)}$
$\mu_M^{(ic)}$	chemical potential of metal atoms M in an intermetallic compound
$\mu_M^{(Si)}$	chemical potential of metal atoms M in silicon
σ	charge state of metal atom
τ	time constant of precipitation
τ_P	relaxation time constant of pairing reaction
b.c.c.	body-centered cubic

Cz	Czochralski
DLTS	deep-level-transient spectroscopy
EBIC	electron beam induced current
EPR	electron paramagnetic resonance
f.c.c.	face-centered cubic
FTIR	Fourier transform infrared
FZ	floating zone
HRTEM	high resolution transmission electron microscopy
H-site	hexagonal interstitial site
MOS	metal-oxide semiconductor
PDG	phosphorus-diffusion gettering
PSG	phosphorsilicate glass
SEM	scanning electron microscopy
SF	stacking fault
TEM	transmission electron microscopy
T-site	tetrahedral interstitial site

11.1 Introduction

The 3d elements (the row from Sc to Cu) form a class of impurities in silicon, which combine unusual physical properties with detrimental effects in device technology. Both aspects have initiated intensive research activities. Initial concepts have been developed to model electrical properties (at low and high temperatures), thermodynamics, transport behavior, precipitation and gettering of those impurities.

Due to extensive experimental and theoretical investigations, the origin and background of multiple deep levels introduced into the bandgap of silicon by 3d elements are now well understood. Basic concepts which lead to this electronic structure are described in Chapter 4 of this volume.

In this chapter an outline will be given of what is known about solubility and diffusion (Sec. 11.2), electronic structure at high temperature (Sec. 11.3), precipitation (Sec. 11.4) and gettering (Sec. 11.5) of 3d elements. Experimental data and models describing these properties are more recent and less advanced than the low-temperature electronic structure. However, they already give a consistent overall picture of this class of impurities and specify the ingredients for basic theoretical treatments. They also elucidate the role which the 3d elements play in device fabrication as described by Graft (1991). The implications from a technological point of view of 3d transition element precipitates located in device-active regions are briefly discussed in Sec. 11.4.3. Gettering techniques, dealt with in Sec. 11.5, have been developed to remove 3d elements from the device active area. Our main concern in that section is to model gettering mechanisms and to specify the experimental conditions under which these mechanisms operate.

Extremely small quantities of 3d elements are sufficient to produce malfunctions of devices. It is reported in the literature, that a concentration of 10^{11} Ti atoms/cm^3 cannot be tolerated in a solar cell (Chen et al., 1979). The ubiquitous presence of 3d transition elements such as Fe, Ni, and Cu along with high diffusivities makes them highly undesirable in device processes. This is one reason why clean room facilities are imperative in device manufacturing. Up to now, effective removal of metallic impurities during or after any device step to regions outside of the device-active area, termed gettering, has been implemented to ensure sufficient yield of devices.

The diamond lattice is a rather open structure. Its large interstices provide the same space for atoms on substitutional and on tetrahedral interstitial sites (T sites). Both sites also have the same symmetry (T_d).

Metallic atoms usually prefer one of those sites since they have to adjust their electronic configuration to the bond structure of the host. 3d elements in intrinsic silicon prefer the T site. However, partial solution enthalpies $\Delta H_M^{(Si)}$ are unusually large (1.5–2.1 eV, see Sec. 11.2) and maximum solubilities are extremely small (about 0.2 ppm for Cr and about 20 ppm for Ni). The major part of this enthalpy must be of electronic origin, since strain energy can only account for a small fraction of $\Delta H_M^{(Si)}$. On the other hand, the migration enthalpy of the 3d elements, decreasing from about 1.8 eV for Ti to about 0.4 eV for Co, is mainly of elastic origin and can be accounted for using a simple hard-sphere model (Sec. 11.2).

A slight shift of the Fermi level from its position in intrinsic silicon toward the conduction band creates a new situation: the solubility of the substitutional species becomes comparable to that of the interstitial species. It strongly increases further with the Fermi level approaching the conduction band, the effective diffusivity of the metallic atoms decreasing to the same extent (see Sec. 11.3). The solubility enhancement is caused by acceptor levels of the substitutional species, whose occupation contributes to the partial enthalpy of solution. A position of the Fermi energy slightly above midgap marks the transition from the interstitial to the substitutional species determining the behavior of Mn, Fe, Co and Cu.

Solubility, together with diffusion data measured in extrinsic silicon, can be used to study the electronic structure of metallic atoms at high temperatures. It was found that the low-temperature defect configuration of interstitial Mn, Fe and Co becomes unstable above 1000 K, which possibly is analogous to a point defect-droplet transition of the silicon self interstitial, as conjectured by Seeger and Chik (1968) (Sec. 11.3).

Precipitation of supersaturated solid solutions of 3d elements in silicon proceed with a large gain of free energy of the order of 1 eV per metal atom, an unknown situation in metallic systems. Descriptions of precipitation phenomena in terms of classical nucleation theory seem no longer to be appropriate. Instead, particle compositions or morphologies are selected which realize maximum rates of free energy degradation. Initial results are discussed in Sec. 11.4.

To remove 3d elements from device-active parts of silicon wafers, annealing procedures have been developed by which segregation or precipitation is induced in predetermined parts (gettering, see Sec. 11.5). Three basically different gettering mechanisms are discussed. Gettering may occur by precipitation at extended defects (oxide particles, dislocations, stacking faults, etc.) generated by a preceding treatment from a supersaturated solution (relaxation-induced gettering). It may also result from diffusion of metallic atoms into parts of the wafer, whose thermodynamic properties have been modified (segregation-induced gettering). Finally, it may come about by a flux of self interstitials towards a sink, which induces a flux of solute atoms and their precipitation at the sink when the solute concentration at the sink exceeds the solubility (injection-induced gettering).

11.2 3d Transition Elements in Intrinsic Silicon

11.2.1 Solubility

In 1983, solubility data of 3d transition elements in Si were published by Weber (1983), covering transition elements from Cr to Ni within the periodic table. His results on Mn, Fe and Co have been confirmed by other authors (Gilles et al., 1986; Isobe et al., 1989; Utzig and Gilles, 1989; Zhu et al., 1989). In addition, new data on the solubility of Ti have become available (Hocine and Mathiot, 1988). For Sc and V, there have not been any systematic studies yet, but the solubility appears to be comparable to Ti (Lemke, 1981).

As can be seen in Fig. 11-1, the solubilities of 3d elements investigated so far show a strong decrease with decreasing temperature. Compared to group III (B, Al, Ga) or group V (P, As, Sb) impurities, the maximum solubility is quite small. For Cu and Ni, up to about 10^{18} atoms/cm^3 are soluble, but for Cr, Mn, Fe and Co it is less than 4×10^{16} atoms/cm^3. These values can be compared with a maximum solubility of about 10^{21} atoms/cm^3 for phosphorus and about 5×10^{20} atoms/cm^3 for boron in silicon.

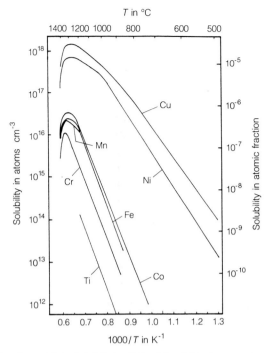

T in °C

Solubility in atoms cm⁻³

Solubility in atomic fraction

1000/T in K⁻¹

Figure 11-1. Solubilities $[M]_{eq}(T)$ of 3d-elements in silicon (Ti: after Hocine and Mathiot, 1988; Cr to Cu: data collected by Weber, 1983) given in atoms/cm³ and in atomic fractions on the left and right ordinate, respectively. For details and for the analysis of these data, see text and Table 11-1.

On the silicon-rich side of the phase diagram some systems Si−M have a eutectic temperature well below the melting point of Si. They show a retrograde solubility, i.e. that the temperature of maximum solubility is higher than the eutectic temperature T_{eut} (Mn, Fe, Co, Ni, Cu).

The solubility of M in Si depends on enthalpies and entropies. Therefore it has to be measured with the external phase that is given by the phase diagram as being in equilibrium with the solid solution of M in Si (Dorward and Kirkaldy, 1968). When starting with a pure metal on top of silicon, one must make certain that the correct phase is formed and that it is equilibrated with the solid solution at the annealing temperature.

With the exception of copper, solubilities for the 3d elements are defined by the equilibrium between silicon and solid MSi_2 (for Cu: Cu_3Si), if $T < T_{eut}$, and between silicon and the liquid solution, if $T > T_{eut}$.

For all 3d elements the solubility $[M]_{eq}(T)$ below T_{eut} shows an Arrhenius type behavior with a slope Q_M/k_B (k_B: Boltzmann constant) and with Q_M around 2.9 eV for Ti, Cr, Mn, Fe and Co and around 1.5 eV for Ni and Cu (see Table 11-1).

As will be shown in Sec. 11.3, for Mn, Fe and Co, and presumably also for Ti and Cr, the dominant species at high temperatures is $M_i^{(0)}$. The measured values of Q_M and also the prefactor can then be given a clear thermodynamic meaning, which follows from equal chemical potentials of the two phases with respect to M:

$$\mu_M^{(ic)}(T,p) = \mu_M^{(Si)}(T,p,x_{M,eq}) \qquad (11\text{-}1)$$

The index ic denotes the intermetallic phase (MSi_2 for the elements from Ti to Ni and Cu_3Si for Cu), and $\mu_M^{(ic)}$ and $\mu_M^{(Si)}$ are the chemical potentials (partial free enthalpies per atom) of M in the intermetallic phase and $M_i^{(0)}$ in Si, respectively. In the formulas concentration is expressed in atomic fractions $x_{M,eq} = [M]_{eq}/N_i = [M_i^{(0)}]_{eq}/N_i$ with $N_i = 5 \times 10^{22}$ cm⁻³ being the density of tetrahedral interstitial sites in silicon.

Describing the deviations from the ideal solution of $M_i^{(0)}$ in Si by an excess partial free enthalpy of mixing gives

$$\Delta G_M^{(ic/Si)} = \Delta H_M^{(ic/Si)} - T\Delta S_M^{(ic/Si)} \qquad (11\text{-}2)$$

which leads to

$$\mu_M^{(Si)}(T,p,x_{M,eq}) = \qquad (11\text{-}3)$$
$$= \mu_M^0(T,p) + k_B T \ln(x_{M,eq}) + \Delta G_M^{(ic/Si)}$$

where μ_M^0 is the reference state of metal in solid solution and is chosen here to be

MSi_2; $\Delta G_M^{(ic/Si)}$ is the free enthalpy required to transfer one metal atom from the disilicide to an interstitial site in silicon as a neutral species $M_i^{(0)}$; $\Delta H_M^{(ic/Si)}$ is the partial heat of mixing, which is zero for the ideal solution; and $\Delta S_M^{(ic/Si)}$ is the excess partial entropy of mixing. It is in excess to the partial entropy of mixing. $\Delta S_M^{(ic/Si)}$ results from a variation of the phonon frequencies in mixing M into Si.

Another approach derives those mixing quantities from the data measured for $T > T_{eut}$, i.e., from the equilibrium between solid silicon and the liquid solution. The procedure is slightly more complicated, since the liquid solution is not ideal, and therefore a numerical fitting of the liquidus line must be carried out. However, the advantage of this procedure is its reliance on experimental data; the liquid boundary phase should have the composition corresponding to the liquidus line at the annealing temperature, since diffusion in liquids is usually fast.

The formation of the solid silicide by reaction of the metallic film on top of silicon at annealing temperatures below the eutectic temperature has not been con-

firmed by an independent method. An exception is the experiment of Dorward and Kirkaldy (1968), who equilibrated silicon with solid Cu_3Si via the vapor phase (see Table 11-1).

In all other cases we can only perform a consistency check between the results obtained for $T > T_{eut}$ with the second analysis and those obtained for $T < T_{eut}$ with the first analysis. The partial solution enthalpies and entropies should be the same if the external solid phase is the correct one.

Weber (1983), analyzed data for $T > T_{eut}$ using the second analysis and assumed that the solution of the metal in the melt can be treated as ideal (which had been justified for Cu by Dorward and Kirkaldy, 1968). For the partial enthalpy of solution of Ni and Cu, he obtained

$$\Delta H_M^{(Si)} = 1.5 \pm 0.1 \text{ eV}$$
$$\Delta S_M^{(Si)} = (2.0 \pm 0.5) \, k_B \tag{11-4}$$

For these quantities the reference states are pure metal and pure silicon. The measured quantities Q_M, which should refer to the phase in equilibrium with the solid solu-

Table 11-1. Solubility data of 3d-transition elements in silicon.

	$\Delta H_M^{(ic/Si)}$ in eV	$\Delta S_M^{(ic/Si)}$ in k_B	Temp. range in °C	Reference	$\Delta H_{MSi_2}^f$ in eV	$\Delta H_M^{(Si)}$ in eV
Ti	3.05	4.2	950 – 1330	Hocine and Mathiot (1988)	1.39	1.66
Cr	2.79	4.8	900 – 1335	Weber (1983)	1.28	1.51
Mn	2.81	7.3	900 – 1142	Weber (1983)	≥ 1.0	≤ 1.81
Fe	2.94	8.2	900 – 1206	Weber (1983)	0.85	2.09
Co	2.83	7.6	700 – 1260	Weber (1983)	1.07	1.76
Ni	1.68	3.2	500 – 993	Weber (1983)	0.91	0.76
Cu	1.49	2.4	500 – 802	Weber (1983)		
	1.75 ± 0.02	4.9 ± 0.3	650 – 802	Dorward and Kirkaldy (1968)		

$\Delta H_M^{(ic/Si)}$: Partial enthalpy of solution of transition element M refered to intermetallic compound; $\Delta S_M^{(ic/Si)}$: Partial entropy of solution of transition element M refered to intermetallic compound; $\Delta H_{MSi_2}^f$: Partial enthalpy of formation of metal silicide MSi_2 per metal atom (data taken from Murarka, 1983); $\Delta H_M^{(Si)}$: Partial enthalpy of solution of transition element in silicon.

tion ($NiSi_2$ and Cu_3Si) and therefore should be identical with $\Delta H_M^{(ic/Si)}$ are

$$Q_{Ni} = 1.7\,eV \quad and \quad Q_{Cu} = 1.5\,eV \quad (11\text{-}5)$$

Dorward and Kirkaldy (1968) obtained $Q_{Cu} = 1.75\,eV$.

Since the heat of formation of $NiSi_2$ is $0.9\,eV/Ni$-atom (Murarka, 1983), the first analysis leads to $\Delta H_{Ni}^{(Si)} = 0.8\,eV$, which is not in agreement with the results given by Eq. (11-4).

To resolve this discrepancy, Weber (1983) proposed that the solubility for $T < T_{eut}$ had actually been measured in contact with pure nickel. Inspection of the experimental results (see Fig. 11-1) shows that this proposal cannot be correct, since there is no discontinuity in the measured curve at $T = T_{eut}$ (as has been observed, e.g., for Fe in Si by Struthers, 1956).

Studies of the formation kinetics of metal silicides (see d'Heurle and Gas, 1986) have shown a sudden formation of $NiSi_2$ around 800 °C, which is well below the eutectic temperature (993 °C). Since the solubility of Ni in Si has been measured down to 500 °C, there have probably been several solid Ni_xSi_y boundary phases. Thus Q_{Ni} has no thermodynamic meaning. Another possible reason for the inconsistencies between the results of the first and second analysis could be a change in the atomic configuration of Ni and Cu around 900 °C similar to that observed for Mn, Fe and Co (see Sec. 11.3, Fig. 11-12).

For the elements Ti, Cr, Mn, Fe and Co the activation energies Q_M obtained for $T < T_{eut}$ and also – after subtraction of the partial formation enthalpies of the disilicides – the partial solution enthalpies $\Delta H_M^{(Si)}$ are summarized in Table 11-1. At present, studies of formation kinetics of silicides indicate that $TiSi_2$ (Bentini et al., 1985) and $CoSi_2$ (Lien et al., 1984) may form in the entire temperature range of sol-

ubility measurements, so that in these cases the values are true partial solution enthalpies. Three data points for $T > T_{eut}$ are also available for Mn in Si. Weber (1983) applied the second analysis to them and obtained

$$\Delta H_{Mn}^{(Si)} = 2.1 \pm 0.1\,eV \quad and$$
$$\Delta S_{Mn}^{(Si)} = (2.0 \pm 0.5)\,k_B \quad (11\text{-}6)$$

In conclusion, the available data, although carefully measured, are either incomplete (Ti, Cr, Fe, Co) or inconsistent, if compared for $T < T_{eut}$ and $T > T_{eut}$. Considering those cases which have either been measured in equilibrium with the liquid solution (Mn, Ni, Cu) or – according to independent investigations – in equilibrium with the solid disilicide (Ti, Co), the partial solution enthalpies lie between 1.5 and 2.1 eV. In comparison to usual solubilities in metals, the solution enthalpies found for the 3 d elements in silicon are very large. As a consequence the solid solution of M in Si shows a retrograde solubility and, when cooled or quenched, decomposes by precipitation under unusually large driving forces.

11.2.2 Diffusion

Compilations of diffusion data (Weber, 1983; Graff, 1986; Weber and Gilles, 1990) show that 3 d transition elements are among the fastest diffusing impurities in intrinsic Si with diffusion coefficients as high as $10^{-4}\,cm^2/s$ (Cu at 1100 °C, see Table 11-3). This explains why these impurities are harmful during device processing. For instance, for a wafer with a thickness of typically 500 μm penetration might take less than 10 s for metal impurities. High diffusion coefficients along with low migration enthalpies are characteristics of simple interstitial diffusion, in which case no diffusion vehicle is needed and the formation of covalent bonds with Si neighbors does not take place.

For a simple interstitial diffusion mechanism, an Arrhenius-type behavior of the diffusion coefficient is expected:

$$D = D_0 \exp\left(-\frac{\Delta H_M^{(mi)}}{k_B T}\right) \tag{11-7}$$

where D_0 is the preexponential factor, and $\Delta H_M^{(mi)}$ the migration barrier.

Electron paramagnetic resonance measurements (EPR) have confirmed that Cr, Mn and Fe, which can be quenched to room temperature, predominantly occupy the tetrahedral interstitial site (Ludwig and Woodbury, 1962). In addition, the electronic structure of the dominant defect, as measured by deep-level-transient spectroscopy (DLTS) (for review see Graff, 1986), is in agreement with total energy calculations of the interstitial metal atom (Beeler et al., 1985; Zunger, 1986; cf Chapter 4 of this volume). There have been reports about a substitutional component of Ni (Kitagawa and Nakashima, 1987, 1989), but it should be emphasized that its concentration is several orders of magnitude smaller than the total Ni solubility. For Mn and Cr, a substitutional component has been found after co-diffusion with Cu (Ludwig and Woodbury, 1962). The mechanism of M_s formation upon Cu precipitation is not yet understood.

As can be seen from Fig. 11-2, there is a clear distinction between the diffusivities of impurities which occupy substitutional sites (e.g., shallow dopants) and those on interstitial sites. However, there is also a considerable variation in the diffusivities within the 3d row. From Ti to Ni the diffusivity at 1000 °C increases by about six orders of magnitude and the migration enthalpy decreases by about a factor of four.

Surprisingly, a rather simple hardsphere model, which takes into account the

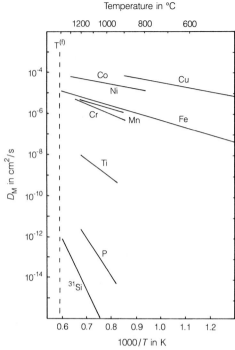

Figure 11-2. High-temperature diffusion coefficients of the 3d-elements in silicon and for comparison of P and ^{31}Si. There is a clear distinction between the diffusivities of impurities, which occupy substitutional sites and the 3d-elements, which occupy tetrahedral interstitial sites. There is also a considerable variation of D within the 3d-row, which has been explained in a simple hardsphere model (see text); for references, see Table 11-3.

variation of atomic size within the 3d-row, satisfactorily explains these trends, which have been discussed as "diffusion puzzle" for a long time (Zunger, 1986). In this model Utzig (1989) calculates the elastic energy required to move M_i from one tetrahedral interstitial (T site) site to the next via the hexagonal interstitial site (H site) as the saddle point (see Fig. 11-3). The pre-exponential factor is calculated as the migration entropy $\Delta S_M^{(mi)}$ using an approach by Zener (1952).

The T site is in the center of the tetrahedron formed by four Si atoms ($z^{(T)} = 4$). The distance between the center of the tetrahedron and its corners is the same as that

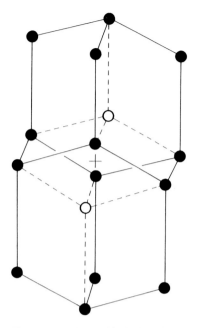

Figure 11-3. The tetrahedral (T-site: o) and the hexagonal interstitial site (H-site: +) in the diamond lattice. 3d-elements occupy the T-site and diffuse presumably via the H-site as the saddle point.

between two neighboring silicon lattice sites, d_0. The hard sphere radius of the T site is then $R^{(T)} = d_0 - r_{si}$, $r_{si} = d_0/2$ being the atomic radius of silicon, so that $R^{(T)} = 0.5 \cdot d_0 = 1.17 \, \text{Å}$. The H site is in the center of a hexagonal ring formed by six Si atoms $(z^{(H)} = 6)$ and has a hard-sphere radius $R^{(H)} = 0.95 \cdot d_0 - r_{si} = 0.45 \cdot d_0 = 1.05 \, \text{Å}$.

Inserting an atom of radius r_M into the interstitial site requires an elastic energy given by

$$\Delta U_{el}^{(T, H)} \qquad (11\text{-}8)$$

$$= \begin{cases} \alpha (r_M - R^{(T, H)})^2 \, z^{(T, H)} & \text{if } r_M > R^{(T, H)} \\ 0 & \text{else} \end{cases}$$

where α is the central force constant per next Si-neighbor ($z^{(T)}$ for the T site and $z^{(H)}$ for the H site): $\alpha = 3.02 \, \text{eV/Å}$ (Keating, 1966). The difference $\Delta U_{el} = \Delta U_{el}^{(H)}(r_M) - \Delta U_{el}^{(T)}(r_M)$ is a contribution to $\Delta H_M^{(mi)}$.

The atomic radii of the 3d elements, taken as the distance of closest approach in

the metal (see Table 11-2; Hall, 1967), have been corrected following an empirical rule detected by Goldschmidt (1928). Depending on the coordination number z, the interatomic distance was found to be about 3% less, if z is 8 instead of 12, 4% less if it is 6 (H site) and 12% less if it is 4 (T site).

In Fig. 11-4 calculated values of ΔU_{el} are compared with measured values of $\Delta H_M^{(mi)}$ for 3d elements. The agreement is quite satisfactory. The sharp drop of $\Delta H_M^{(mi)}$ between Ti and Cr and the subsequent slow decrease are quite well reproduced by this simple model. Note that the estimation of ΔU_{el} for copper is for $Cu_i^{(0)}$, while the experimental data are for $Cu_i^{(+)}$.

Under condition that the diffusion barrier is primarily due to elastic strain, which appears to be fulfilled for the 3d elements, Zener (1952) has given an approximate formula to calculate the migration entropy:

$$\text{(a)} \quad \Delta S_M^{(mi)} = \beta \, \frac{\Delta H_M^{(mi)}}{T^{(f)}},$$

$$\text{(b)} \quad \beta = \frac{d\,(K/K_0)}{d\,(T/T^{(f)})} \qquad (11\text{-}9)$$

Table 11-2. Comparison of experimental migration barriers $\Delta H_M^{(mi)}$ with calculated elastic contributions ΔU_{el} (after Utzig, 1989); r_M is the atomic radius of the respective atom.

	r_M in nm	ΔU_{el} in eV	$\Delta H_M^{(mi)}$ in eV	Reference, see Table
Ti	0.1467	2.15	1.79	11-3
Cr	0.1357	1.15	0.81	11-5
	0.1267	0.50		
Mn	0.1306	0.75	0.70	11-5
	0.1261	0.46		
Fe	0.1260	0.46	0.68	11-5
Co	0.1252	0.42	0.37	11-3
Ni	0.1244	0.38	0.47	11-3
Cu	0.1276	0.55	0.43	11-3

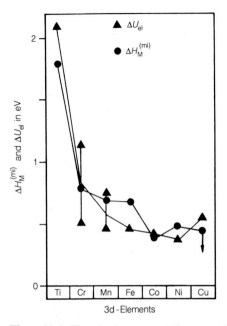

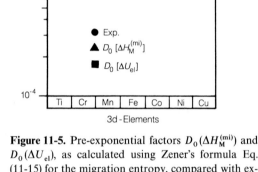

Figure 11-4. The elastic energy difference ΔU_{el} between the H-site and the T-site, as calculated using Eq. (11-14), and experimental values of the migration enthalpy $\Delta H_M^{(mi)}$ for the 3 d-elements.

Figure 11-5. Pre-exponential factors $D_0 (\Delta H_M^{(mi)})$ and $D_0 (\Delta U_{el})$, as calculated using Zener's formula Eq. (11-15) for the migration entropy, compared with experimental D_0-values for the 3 d-elements.

where K is the elastic modulus, and $T^{(f)}$ the melting point of silicon.

Taking $D_0 = 1/6 \cdot d_0^2 \cdot f \cdot \exp (\Delta S_M^{(mi)}/k_B)$, where f is the Debye frequency, measured values of the pre-exponential factor of diffusion D_0 can also be compared with the prediction of the simple hardsphere model (see Fig. 11-5), again showing remarkable agreement.

Diffusion coefficients D_M estimated from the time necessary to saturate Si samples with a 3d element M (Weber, 1983), showed that care must be taken if D_M is evaluated from a concentration profile. Under the usual experimental conditions, a specimen with a radioactive tracer of M on its surface is annealed at the diffusion temperature for a certain period of time t_a. The dilemma of this experiment is that for a small solubility a low detection limit of metal concentrations can easily be accom-

plished only with an infinitesimally thin layer at the boundary (finite source experiment). However, since the Si surface tends to act as a sink for metallic impurities (see Sec. 11-4), only an unspecified fraction of metal atoms is able to diffuse into the crystal, leading to unknown boundary conditions.

The alternative boundary condition is a constant source realized by the deposition of several layers of the transition element M on the specimen surface and inducing the formation of the equilibrium silicide. In this case only a small fraction of the evaporated amount of M is radioactive. With the advent of germanium detectors with high spectral resolution and efficiency for radiochemical analysis, and of DLTS, techniques have become available which allow to determine those diffusion profiles. Weber's solubility data obtained for

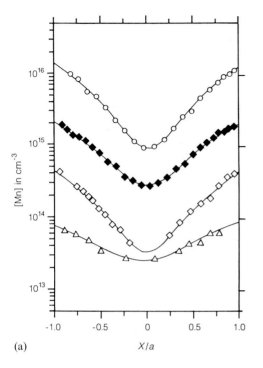

(a)

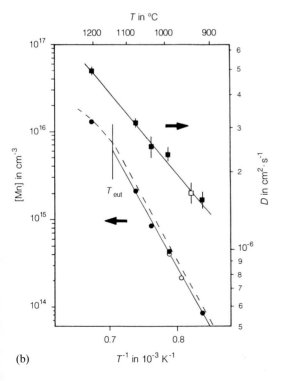

(b)

$(D_M \cdot t_a)^{1/2} \gg d$ (d: specimen thickness) have become a critical check of well-defined boundary conditions. The extrapolated surface concentrations of diffusion profiles must coincide with his solubility data at the diffusion temperature.

Let us consider as an example the diffuson of Mn in Si. In the temperature range between 900 °C and 1200 °C symmetric diffusion profiles were obtained after in-diffusion from opposite surfaces of Si samples (Fig. 11-6a; Gilles et al., 1986). Surface concentrations were found in good agreement with solubility data (Fig. 11-6b), which means that boundary conditions independent of time have been achieved. Such checks are indispensible to accurately determine migration enthalpies below 1 eV. Note, that this agreement does not imply that the equilibrium phase has formed, as discussed in Sec. 11.2.1.

An independent method for measuring D_M below 100 °C is the result of the study of the pairing reaction $M_i^{(+)} + A_s^{(-)} = (M_i A_s)$, where A_s is a shallow acceptor. The kinetics of pair formation is determined by the diffusion of $M_i^{(+)}$. According to Reiss et al. (1956) the diffusing species $M_i^{(+)}$ is captured by $A_s^{(-)}$ via electrostatic attraction as soon as the electrostatic energy exceeds the thermal energy, $k_B T$, of the ion. The identity of those two energy terms defines the capture radius, R, such that from

Figure 11-6. Solubility and diffusion coefficient of Mn in silicon (after Gilles et al., 1986). (a) Concentration profiles determined by the tracer method after diffusion from both surfaces into specimens of thickness $2a$ for different temperatures: 1200 °C (o), 1080 °C (■), 990 °C (◇), 920 °C (△). (b) Temperature dependence of diffusion coefficients, D, calculated from diffusion profiles of the total Mn-concentration (tracer method, solid square, cp. (a)) or those of interstitial Mn (DLTS, open squares). The surface concentration of those profiles (tracer method, solid circles) agree well with the saturation concentrations.

the relaxation time constant, τ_P, the diffusivity, $D_i^{(+)}$, of $M_i^{(+)}$ can be calculated:

$$\tau_P = \frac{1}{4\pi[A_s^{(-)}]R D_i^{(+)}} = \\ = \frac{\varepsilon\varepsilon_0 k_B T}{e^2[A_s^{(-)}]D_i^{(+)}} \qquad (11\text{-}10)$$

where $\varepsilon = 11.7$ is the static dielectric constant of Si, ε_0 the permittivity of free space and e the unity of electric charge.

In order to determine the relaxation time, isolated M_i has to be quenched to a temperature below which the ratio $[(M_i^{(+)}A_s^{(-)})]/[M_i^{(+)}] > 1$, and at which the relaxation time is experimentally accessible. While for Cr, Fe and Mn both conditions can be fulfilled below 100°C, for Co, relaxation is already completed during or immediately after quenching (Bergholz,

1983). In Tables 11-3 and 11-4 a comparison of high- and low-temperature diffusion data is given. Note that for such low migration enthalpies, variations in diffusivity in the experimentally accessible high-temperature range are less than a factor of five, while in the low-temperature region it differs by up to three orders of magnitude. Hence, there is a higher accuracy of low-temperature migration enthalpies.

Differences in the migration barrier between high and low temperatures are less than 0.2 eV. Extrapolation of high-temperature data to low temperatures and vice versa shows satisfactory agreement for Mn and Fe, whereas the variations are larger for Cr. However, the error bar to $\Delta H_M^{(mi)}$ given by Bendik et al. (1970) is quite large $(+/-0.3 \text{ eV})$. A simultaneous fit to high- and low-temperature diffusion data, as

Table 11-3. High-temperature diffusion data of interstitial 3d-transition elements in silicon; $\Delta H_M^{(mi)}$ is the migration barrier and D_0 the pre-exponential factor (compare Eq. (11-13)).

	$\Delta H_M^{(mi)}$ in eV	D_0 in cm²/s	Temp. range in °C	$D(T=1100°C)$ in cm²/s	Reference	$D(20°C)$ [a] in cm²/s
Ti	1.79	1.5×10^{-2}	950–1200	4×10^{-9}	Hocine and Mathiot (1988)	2×10^{-33}
Cr	1.0	1×10^{-2}	900–1250	2.1×10^{-6}	Bendik et al. (1970)	6×10^{-20}
Mn	0.63	7×10^{-4}	900–1200	3.4×10^{-6}	Gilles et al. (1986)	1×10^{-14}
Fe	0.65	9.5×10^{-4}	800–1070	3.9×10^{-6}	Isobe et al. (1989)	6×10^{-15}
Co	0.37	9×10^{-4}	700–1100	3.9×10^{-5}	Utzig and Gilles (1988)	4×10^{-10}
Ni	0.47	2×10^{-3}	800–1300	3.8×10^{-5}	Bakhadyrkhanov et al. (1980)	2×10^{-11}
Cu [b]	0.43	5×10^{-3}	400–700	1.2×10^{-4} [a]	Hall and Racette (1964)	2×10^{-10}

[a] extrapolated values; [b] measured in 5×10^{20} cm⁻³ B-doped silicon.

Table 11-4. Low-temperature diffusion data of 3d-transition elements in silicon (for explanation of symbols, see Table 11-3).

	$\Delta H_M^{(mi)}$ in eV	D_0 in cm²/s	Temp. range in °C	$D(T=20°C)$ in cm²/s	Reference	$D(1100°C)$ [a] in cm²/s
Cr	0.9	6×10^{-2}	24–96	2×10^{-17}	Zhu et al. (1989)	3×10^{-5}
Mn	0.72	2.4×10^{-3}	14–90	1.1×10^{-15}	Nakashima and Hashimoto (1991)	5×10^{-6}
Fe	0.77	2.3×10^{-2}	30–85	1.3×10^{-15}	Shepard and Turner (1962)	3×10^{-5}
Cu			20	$>1 \times 10^{-7}$	Zundel et al. (1988)	

[a] extrapolated values.

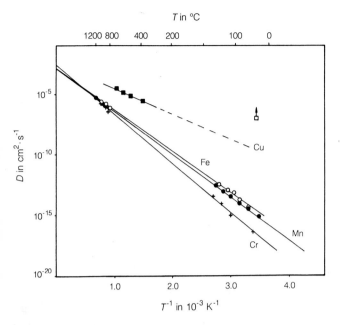

Figure 11-7. High- and low-temperature diffusion data of Cr, Mn, Fe, and Cu in silicon (reference see Table 11-5). At high temperatures data were measured for intrinsic silicon, with the exception of Cu (5×10^{20} B-atoms/cm^3). At low temperatures, data were derived from pairing kinetics of $M_i^{(+)}$ with boron. The low-temperature value of Cu is a lower limit estimated from the time necessary to penetrate a silicon wafer after chemomechanical polishing at room temperature. It is speculated that the discrepancy between high- and low-temperature data of Cu is related to Cu$_i$B-pairing at high temperatures.

proposed by Weber (1983) for Fe in Si and plotted in Fig. 11-7 for Cr and Mn as well, strongly suggests a single Arrhenius-type relationship for the diffusivity of Cr, Fe and Mn. Least square fits to the data are shown in Table 11-5. This is not to be expected in all cases, since at high temperatures different charge states and even configurations are involved (see Sec. 11.3). On the other hand, Shepard and Turner (1962) proposed from precipitation kinetics of neutral Fe$_i$, that the impact of charge states on diffusion is rather small below 200 °C. In the following section, an analysis is presented that allows one to derive charge states and energy levels within the bandgap at high temperatures.

It has recently been discovered that by chemomechanical polishing at room temperture, Cu is able to penetrate a Si wafer within a couple of hours (Schnegg et al., 1988; Prescha et al., 1989; Keller et al., 1990). Although the mechanism is not yet fully understood, the formation of CuB complexes, which is evident from the increase in the resistivity, plays a major role

in the apparent "solubility enhancement". Disregarding electronic contributions to the solubility which might be quite strong near room temperature, the room-temperature solubility of Cu is found to be less than 1 atom/cm^3 by extrapolation from solubility data above 400 °C. This is a puzzle that has yet to be solved. The Cu diffusivity appears to be unexplained as well. Extrapolation of Cu diffusion data of Hall and Racette (1964) to room temperature yields a diffusion coefficient of about 10^{-10} cm^2/s is expected. With such a diffusion coefficient it would take more than a month to saturate a wafer with Cu. We

Table 11-5. Simultaneous fit of high- and low-temperature diffusion data (20 – 1200 °C) (for explanation of symbols, see Table 11-3).

	$\Delta H_M^{(mi)}$ in eV	D_0 in cm^2/s
Cr	0.81	2.5×10^{-3}
Mn	0.70	1.3×10^{-3}
Fe [a]	0.68	1.3×10^{-3}

[a] data after Weber (1983).

think that this discrepancy of room temperature diffusion studies and the results of extrapolation from high-temperature diffusion data is attributable to the fact that Hall and Racette (1964) performed their studies on Si doped with 5×10^{20} B-atoms/cm^3. As was shown for Mn and Co in Si (Gilles et al., 1990a) (see Sec. 11.3), such high doping might result in pairing thereby affecting the effective diffusion coefficient of Cu by pairing. At lower temperatures the effect of pairing with B is more pronounced. If CuB pairing plays a role in the diffusion of Cu in Si with such high B doping, the activation enthalpy of 0.43 eV is higher than the actual migration enthalpy of Cu$_i$. Lemke (1981, 1983) reported binding energies of 3d elements with boron ranging beween 0.4 and 0.6 eV. Due to such variations it is difficult to calculate the fraction [Cu$_i$B]/[Cu$_i$] and the diffusion coefficient of Cu$_i$, as attempted by Keller et al. (1990).

In summary, diffusion data of 3d transition elements in intrinsic Si have been improved to such an extent that even these small migration enthalpies are quite reliable. More studies on the diffusion of Cu, and probably Ni, are necessary. It is interesting to note that the pre-exponential factor of the impurities Cr, Mn, Fe (a fit to high- and low-temperature data; Table 11-5) and Co (Table 11-3) differ only up to a factor of two, whereas diffusion coefficients differ by more than one order of magnitude. Thus it is mainly the migration barrier which determines differences in diffusion coefficients of 3d transition elements. Unlike the solubility data, which show an abrupt increase going from Co to Ni, such an abrupt increase in the diffusion coefficients occurs between Fe and Co. It helps to understand why Co, Ni and Cu cannot be quenched on interstitial sites: the lower migration barrier of those impurities

favors precipitation during quenching, since their diffusivities near room temperature are several orders of magnitude higher than those for Fe, Mn and Cr.

11.3 3d Transition Elements in Extrinsic Silicon

11.3.1 Introduction

A semiconductor is termed extrinsic if the dopant concentration (B, P) exceeds the thermal equilibrium concentration of the intrinsic carriers, n_i. In Chapter 5 of this volume the temperature dependence of n_i is shown for Si, Ge and GaAs. At room temperature Si is normally extrinsic due to doping or residual impurities ($n_i = 10^{10}$ cm^{-3}), whereas at temperatures above 700 °C a minimum dopant concentration of 10^{18} atoms/cm^3 is required to affect the position of the Fermi level.

As will be shown below, by shifting the Fermi level, the concentration of charged impurity species can be enhanced to dominate the solubility, such as $M_i^{(+)}$ by doping with B, or $M_s^{(-)}$ by doping with P. Hence information about charge states at high temperature can be obtained from solubility data, but no detailed information can be obtained about the defect species.

We are dealing now with a ternary system and the solubility of 3d elements is composed of additional terms, e.g., pairs of substitutional or interstitial impurities with dopant atoms. In this situation a partial classification of impurity species involved in the solubility enhancement can be derived from an analysis of diffusion profiles. Unlike gold diffusion in intrinsic silicon (see Chapter 5 of this volume), transport of 3d metal atoms in extrinsic Si was found not to be limited by diffusion of intrinsic defects (vacancies or Si self inter-stitials), but by diffusion of the interstitial metal impurity. The effective diffusion co-efficient of the impurity, D_{eff}, is determined by the ratio of the average times that an atom stays on interstitial and substitu-tional sites or in complexes. The solubility enhancement of charged impurities by Fermi level shift and the simultaneous variation of D_{eff} are experimentally acces-sible quantities and can be combined to separate electronic properties of interstitial species and of immobile, i.e., substitutional species and complexes.

A more refined classification of immo-bile species is provided by Mössbauer spectroscopy of Co and by Rutherford Backscattering combined with channeling of Cu. Mössbauer spectroscopy discrimi-nates between different Co species accord-ing to their s electron densities at the nucleus site (isomer shift) and to the elec-tric field gradient produced by distortions of cubic environment (quadrupole split-ting).

In the following sections an analysis of solubility and diffusion data is presented. It gives clear evidence that above 600 °C in-terstitial metal impurities are deep donors with a level related to that calculated from first principles at zero temperature (Beeler et al., 1985). It further indicates a rather abrupt change of the electronic configura-tion of these impurities around 900 °C.

11.3.2 Solubility

The dependence of the solubility on the position of the Fermi level results from an electronic contribution to the partial en-thalpy of solution $\Delta H_M^{(Si)}$ for the charged species (see Sec. 11.2.1). Shockley and Moll (1960) argued that this contribution is the electron transfer energy from the defect do-nor level $G^{(0/+)}$ to the Fermi level E_F, or from E_F to the defect acceptor level $G^{(-/0)}$.

Since the solubility of the neutral species is independent of E_F, the concentration ratio of two species which differ in their charge states by $+1$ or -1 is given by

$$\frac{[M^{(\sigma)}]}{[M^{(\sigma-1)}]} = \exp\left(\frac{G^{(\sigma/(\sigma-1))} - E_F}{k_B T}\right) \quad (11\text{-}11)$$

Consider, e.g., an impurity ionization level, $G^{(0/+)}$ in the band gap of Si above the Fermi level.

$E_C - G^{(0/+)}$ (E_C being the conduction band edge) is the change in the standard chemical potential for the reaction to form a positively charged donor and a free elec-tron in the conduction band (Van Vechten and Thurmond, 1976). The variation of $[M^{(\sigma)}]$ as predicted by Eq. (11-11) is de-picted schematically in Fig. 11-8. The solu-bility is the sum over all impurity species and their charge states, i.e.,

$$[M]_{eq} = \sum_\sigma [M_i^{(\sigma)}]_{eq} + \sum_\sigma [M_s^{(\sigma)}]_{eq} \quad (11\text{-}12)$$

It is obvious from Fig. 11-8 that charge states of impurity species can be derived directly from the slope of $\ln[M]_{eq}$ with Fermi energy if one species and one charge state are dominant. In order to vary the Fermi energy in the temperature regime of interest (700–1200 °C), the dopant concen-tration must exceed intrinsic carrier den-sity, n_i (cf. Chapter 5 of this volume). At 700 °C a minimum dopant concentration of 10^{18} atoms/cm³ is necessary to affect the position of the Fermi level, at 1200 °C it is 2×10^{19} atoms/cm³.

Figure 11-9 shows an example of how the solubility of Mn in Si is affected by doping with B or P for three different tem-peratures. The Fermi energy was calcu-lated using Ehrenberg's approximation (Ehrenberg, 1950). All dopants were as-

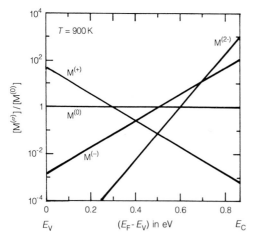

Figure 11-8. Solubility at 900 K of the charged species $M^{(+)}$, $M^{(-)}$, and $M^{(2-)}$ relative to that of the uncharged species $M^{(0)}$ for a point defect M as a function of the Fermi energy.

The electronic contributions to the partial enthalpy of solution $\Delta H_M^{(Si)}$ are quite similar for Cu (data from Hall and Racette, 1964), Co, Fe and Mn. As illustrated in Fig. 11-10 these 3 d elements show a solubility enhancement in B-doped Si which is caused mainly by a single positively charged species. In P-doped Si multiple negative charge states are involved.

It is interesting to note that the Cu solubility first drops on doping with P before the solubility enhancement sets in. This clearly shows that Cu is positively charged in intrinsic Si at 700 °C and 600 °C (Hall and Racette, 1964). Therefore, part of the enthalpy Q_{Cu} (see Sec. 11.2.1) of the Cu solubility is indeed of electronic origin.

sumed to be ionized at the diffusion temperature. As can be seen in Fig. 11-9, at 1040 °C there is hardly any variation in the solubility with Fermi energy. As the interstitial species of manganese, Mn_i, is the dominant one in intrinsic Si (see Sec. 11.2), it must be neutral at that temperature ($[Mn]_{eq} \approx [Mn_i^{(0)}]_{eq}$), independent of the position of the Fermi level.

At lower diffusion temperatures, solubility variations with Fermi energy become more pronounced. At 850 °C and by doping with B up to 10^{20} atoms/cm^3, the solubility of Mn is increased tenfold compared to its value in intrinsic silicon, while at 700 °C the solubility is enhanced by two orders of magnitude. By doping with P up to 10^{20} atoms/cm^3, there is an even stronger solubility enhancement of up to four orders of magnitude. Similar effects were also observed for Co and Fe (Gilles et al., 1990a). These interstitial impurities are neutral in intrinsic Si above 800 °C and show a pronounced solubility enhancement in B- and P-doped Si at 700 °C.

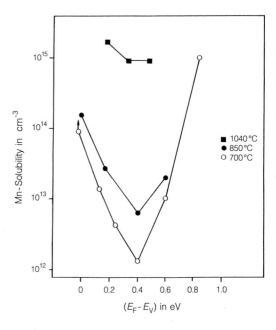

Figure 11-9. Solubility of manganese $[Mn]_{eq}$ versus Fermi energy for three different temperatures: 1040 °C (■), 850 °C (●), and 700 °C (○). Data are from Gilles et al. (1990a). The Fermi energy was calculated using Ehrenberg's approximation (Ehrenberg, 1950).

11.3.3 Diffusion

Although the solubility data of Fig. 11-10 clearly show that charged species are the dominant defects in extrinsic silicon at 700 °C they do not allow the determination of the lattice site occupied by the respective impurity. A decision between $M_i^{(+)}$ and $M_s^{(+)}$ or $M_i^{(2-)}$, $M_s^{(2-)}$ and $M_s P$ is not possible.

However, it is generally accepted that interstitial species of 3 d elements are much more mobile than substitutional species and pairs with shallow dopants. Transport of transition metal impurities is determined by diffusion via interstitial sites. Hence measurements on the influence of

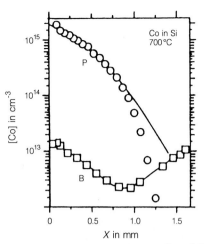

Figure 11-11. Concentration profiles of Co in silicon showing the influence of dopants on the solubility (taken as the surface concentration) and diffusion of Co at 700 °C: 8×10^{19} B-atoms/cm^3, $t_{diff} = 9$ min ($\square$), 1×10^{20} P-atoms/cm^3. $t_{diff} = 7$ d ($\circ$). Note that the solubility of Co in intrinsic silicon is 2×10^{11} Co-atoms/cm^3 at 700 °C (after Gilles et al., 1990 a).

doping on the diffusion coefficient of transition elements can supply valuable information about the identity of the lattice site predominantly occupied by the impurity. Hall and Racette (1964) reported fast diffusion of Cu in intrinsic and B-doped Si, but a strong retardation in P- and As-doped Si.

A more detailed analysis was possible for Co in Si. In Fig. 11-11 the influence of doping on solubility (taken as the surface concentration) and diffusion of Co at 700 °C are shown. Note, that samples of intrinsic Si would have been saturated with Co at a concentration of 2×10^{11} atoms/cm^3 for either diffusion time. In intrinsic and B-doped Si an annealing time of 9 min was sufficient to nearly saturate the Si specimens (diffusion from both sides of the specimen), but seven days were necessary to obtain a comparable depth in the P-doped sample (diffusion from one side of the specimen only).

Diffusion coefficients for Co and other 3 d elements are summarized in Table 11-6.

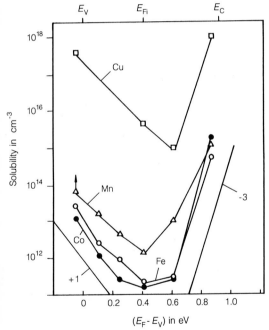

Figure 11-10. Solubility of Cu, Mn, Fe, and Co versus Fermi energy at 700 °C, demonstrating the electronic contribution to the partial enthalpy of solution $\Delta H_M^{(Si)}$. The slopes expected for a single positively charged species (1+) and a threefold negatively charged species (3−) are indicated (data for Cu: after Hall and Racetter, 1964; and for Mn, Fe, Co: after Gilles et al., 1990a).

Table 11-6. Comparison of effective diffusion coefficient D_{eff} of 3 d-transition elements in extrinsic and intrinsic silicon at 700 °C.

Doping concentration in cm^{-3}	D_{eff} in cm^2/s			
	Mn	Fe	Co	Cu
P: 1×10^{20}	$< 4 \times 10^{-11}$	4×10^{-11}	2×10^{-9}	$< 10^{-8}$
P: 8×10^{14}	7×10^{-8}	4×10^{-7}	1.1×10^{-5}	
B: 8×10^{19}	$> 2 \times 10^{-8}$	2.4×10^{-7}	2.3×10^{-6}	3×10^{-5}

For all 3 d elements investigated so far (Cu, Co, Mn and Fe) the trends are the same:

(1) solubility enhancement in P-doped Si coupled with a strong retardation of the effective diffusion coefficient, and

(2) solubility enhancement in B-doped Si which does not affect the diffusion coefficient significantly.

(3) For Co, Mn and Fe those electronic contributions to the solubility and diffusivity disappear towards higher temperatures.

11.3.4 High-Temperature Electronic Structure

In order to identify charge states and impurity species, one has to go further into the classification of immobile species by applying appropriate experimental techniques. In their early work Hall and Racette (1964) interpreted the solubility and diffusion data of Cu in extrinsic silicon as being due to positively charged interstitial Cu in B-doped Si and triply negatively charged substitutional Cu in As- and P-doped Si. Meek and Seidel (1975) pointed out that pairs of substitutional Cu and P (As) might be energetically favorable and dominate the solubility for high P (As) dopings.

For Co in Si Gilles et al. (1990a) presented experimental evidence that such a pairing reaction does in fact take place. Using Mössbauer spectroscopy the authors identified two immobile species involved in the solubility enhancement in P-doped Si, one having cubic symmetry and the other having an electric field gradient at the site of the metal impurity. Both species transform into each other reversibly on annealing. The concentration ratio, r, of the noncubic and the cubic species in silicon doped with 10^{20} P-atoms/cm^3 shows an Arrhenius-type behavior:

$$r = 4.2 \cdot 10^{-7} \exp \left(\frac{1.5 \, eV}{k_B T} \right) \qquad (11\text{-}13)$$

In a consistent interpretation two immobile species showing cubic and non-cubic symmetry were identified as substitutional Co (Co$_s$) and its pair with P (Co$_s$P), respectively. From Eq. (11-13) it can be calculated that at 700 °C the concentration of those pairs amounts to 96% of the solubility, at 600 °C it is above 99%. Taking pairs with P into account, substitutional Co is at least doubly negatively charged in highly P-doped Si at 700 °C.

In heavily B-doped silicon, room-temperature Mössbauer spectroscopy of Co showed a spectrum characteristic of Co$_i$B pairs (Bergholz, 1983). However, no conclusion about the dominant defect in this material at 700 °C can be drawn, since Co$_i$ may form those pairs during quenching because of its high mobility. In this situation, immobile species such as M$_s$P, M$_s$ or M$_i$B prevent the charge state of interstitial impurities to be directly derived from the Fermi energy dependence of the solubility,

$[M]_{eq}(E_F)$. However, one can separate electronic contributions of interstitial impurities using the Fermi energy dependence of the product of solubility and diffusivity, $[M]_{eq}(E_F) \cdot D_{eff}$. The solubility can be separated into a mobile and an immobile component. The former is the sum of the concentrations of differently charged interstitial species

$$[M^{(mob)}]_{eq} = \sum_{\sigma} [M_i^{(\sigma)}]_{eq} \qquad (11\text{-}14)$$

The latter is the sum of concentrations of substitutional impurities and of complexes like $M_i B$- or $M_s P$-pairs and their various charge states:

$$[M^{(im)}]_{eq} = \begin{cases} \sum_{\sigma} [M_s^{(\sigma)}]_{eq} + \sum_{\sigma} [(M_i B)^{(\sigma)}]_{eq}, & \text{in B-doped Si} \\[2mm] \sum_{\sigma} [M_s^{(\sigma)}]_{eq} + \sum_{\sigma} [(M_s P)^{(\sigma)}]_{eq}, & \text{in P-doped Si} \end{cases} \qquad (11\text{-}15)$$

It is experimentally confirmed that the so-called immobile species do contribute to the solubility, but not to $[M]_{eq} \cdot D_{eff}$, i.e.,

$$\sum_{\sigma} [M_i^{(\sigma)}] D_i^{(\sigma)} \gg \sum_{\sigma} [M_s^{(\sigma)}] D_s^{(\sigma)} +$$

$$+ \text{(terms of M-complexes)} \qquad (11\text{-}16)$$

Therefore the product of diffusivity and solubility can be expressed as

$$([M^{(mob)}]_{eq} + [M^{(im)}]_{eq}) D_{eff} =$$

$$= [M]_{eq} D_{eff} = \sum_{\sigma} [M_i^{(\sigma)}]_{eq} D_i^{(\sigma)} \qquad (11\text{-}17)$$

In the simple case of two charge states 0 and $+1$ Eq. (11-17) simplifies to

$$[M]_{eq} D_{eff} = [M_i^{(0)}]_{eq} D_i^{(0)} + [M_i^{(+)}]_{eq} D_i^{(+)} =$$

$$= [M_i^{(0)}]_{eq} D_i^{(0)} \left\{ 1 + \exp\left(\frac{G^* - E_F}{k_B T}\right) \right\} \qquad (11\text{-}18)$$

The product stays constant with a shift of the Fermi energy if the interstitial species is neutral. It varies exponentially with the position of the Fermi energy if the interstitial species is positively charged. $\tilde{G} = G + k_B T \ln D_i^{(+)}/D_i^{(0)}$ is the donor level G of the interstitial species modified by the ratio of the diffusivities of positively charged and neutral species which were found to be close to 1, i.e., $\tilde{G} = G$.

Table 11-7 shows the product $[M]_{eq} \cdot D_{eff}$ of solubility and diffusivity for Co at 700 °C and 800 °C. The product is enhanced tenfold by doping with B, and it is independent of doping with P, which gives clear evidence of a deep donor level of interstitial Co. The analysis was applied to Mn and Fe and led to the same conclusion.

Figure 11-12 shows the analysis of the position of deep donor levels with respect

Table 11-7. Total solubility $[Co]_{eq}^{tot}$ and product with effective diffusion coefficient D_{eff} of Co in Si for $T = 700\,°C$ and $T = 800\,°C$.

Doping	$T = 700\,°C$		$T = 800\,°C$	
in cm^{-3}	$[Co]_{eq}^{tot}$ in cm^{-3}	$[Co]_{eq}^{tot} D_{eff}$ in 1/cm s	$[Co]_{eq}^{tot}$ in cm^{-3}	$[Co]_{eq}^{tot} D_{eff}$ in 1/cm s
$[P] = 1 \times 10^{20}$	1.8×10^{15}	3.6×10^6	4.5×10^{14}	5.4×10^7
$[P] = 8 \times 10^{14}$	4×10^{11}	4.4×10^6	5.1×10^{12}	8.4×10^7
$[B] = 8 \times 10^{19}$	1.5×10^{13}	3.4×10^7		

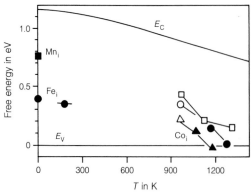

Figure 11-12. Donor levels G (solid symbols) and values $\tilde{G}$ modified by the ratio of the diffusivities of the positively charged and neutral species (open symbols) of Mn_i, Fe_i, and Co_i in the bandgap of silicon as a function of temperature. At low temperatures, data were obtained by DLTS, at high temperatures they were obtained from an analysis of solubility versus Fermi energy, using Eq. (11-16) (after Gilles et al., 1990 a).

to the valence band edge. Above 1000 K the donor level of each impurity species shifts towards the valence band much stronger than the conduction band edge E_c. The large entropy of ionization at high temperatures was considered evidence for a new point defect configuration, as compared to low temperatures. Hence there is experimental evidence that a point-defect configuration becomes unstable at high temperatures, which is possibly analogous to an instability of the Si self-interstitial conjectured by Seeger and Chik (1968). Obviously, this finding prevents an extrapolation of low-temperature energy levels to diffusion temperatures.

11.4 Precipitation of 3d Transition Elements in Silicon

11.4.1 General Considerations

Metal silicide precipitates in silicon are one of the most serious problems of mod-

ern device technology. This is partly due to the fact that most metal silicides exhibit metallic conductivity (Murarka, 1983). Hence, metal silicide precipitates located in active device regions (e.g., a pn-junction) may cause short circuits leading to destruction of the device (e.g., Augustus et al., 1980). In addition, it has been inferred that transition metal particles located at Si/SiO$_2$ interfaces catalyse the decomposition reaction $Si + SiO_2 \rightleftharpoons 2\,SiO\uparrow$ of SiO$_2$ leading to local thinning of the oxide and hence poor gate oxide quality (Liehr et al., 1988 a, b). Furthermore, the precipitates are believed to serve as preferred sites for the formation of extended defects e.g. stacking faults or dislocations (for review see Kolbesen et al., 1989). These aspects will be discussed in Sec. 11.4.3 of this chapter.

From a more basic point of view, 3d elements in silicon may serve as model systems to investigate how relaxation occurs under conditions of high supersaturation. Undercooling of a solution of a 3d element in silicon results in an extremely large supersaturation. As a consequence their precipitation proceeds with a large gain of chemical free energy (see below).

As has been shown in Sect. 11.2.1 and 11.2.2 some 3d transition elements in silicon (namely Co, Ni and Cu) exhibit a small solid solubility with partial enthalpies of solution between 1.5 and 3 eV and high (interstitial) diffusivity with migration enthalpies as low as 0.4 eV. It is the combination of these properties which favours the formation of precipitates even during rapid quenching from high temperatures. In fact, by applying high cooling rates the 3d elements from Ti to Fe can be kept on interstitial sites, whereas Co, Ni and Cu precipitate during or immediately after quenching.

For a detailed description of precipitation phenomena the reader is referred to

Chapter 4 (Wagner and Kampmann, 1991) of Volume 5 of this series. Like these authors we approximate the Gibbs free energy G of the system by its free energy F. To illustrate the effect of the above mentioned properties of 3d elements in silicon consider the change $\Delta F(n)$ in free energy due to the formation of a precipitate containing n metal atoms:

$$\Delta F(n) = -n\,\Delta f_{\text{chem}} + \Delta F_{\text{S}}(n) \qquad (11\text{-}19)$$

where $n\,\Delta f_{\text{chem}}$ denotes the gain of chemical free energy. Δf_{chem} is frequently termed chemical driving force for precipitaton (Wagner and Kampmann, 1991). $\Delta F_{\text{S}}(n)$ contains energy contributions reducing the gain of free energy (e.g., interfacial energy, coherency strains). The latter depends on the shape and size of the precipitate as well as the number n of atoms contained in it. Equation (11-19) is usually used to calculate the number n^* of atoms contained in the critical nucleus. This is given by the maximum positive value of $\Delta F(n) = \Delta F(n^*)$, which is called the nucleation barrier.

For an evaluation of Eq. (11-19) Δf_{chem} has to be estimated. For a given temperature, pressure and composition of a binary alloy, equilibrium between two phases 1 and 2 implies equality of the chemical potentials of the two components in either phase. Inspection of binary phase diagrams of the type M:Si shows that below the eutectic temperatures intermetallic compounds (ic) of the type MSi_2 are likely to form upon precipitation, with the exception of Cu where Cu_3Si is expected (Hansen and Anderko, 1958).

Figure 11-13 shows free energy curves of the two phases 1 (Si) and 2 (ic). The equilibrium concentration $[\text{M}]_{\text{eq}}$ of the metal M in Si is determined by the common tangent to the free energy curves F^{Si} and F^{ic}. For an initial concentration $[\text{M}]_0$ equilibrium is

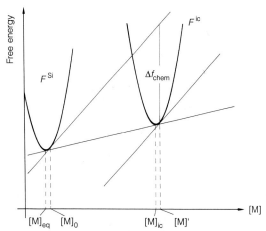

Figure 11-13. Free energy curves of silicon (F^{Si}) and of a metal silicide (F^{ic}) for the calculation of the gain of chemical free energy Δf_{chem} due to precipitation formation note that the diagram only schematically reproduces the true situation since $[\text{M}]_0$ and $[\text{M}]_{\text{eq}}$ are less than 100 ppm and $[\text{M}]_{\text{ic}}$ is of the order of 30%.

maintained with a precipitate whose composition $[\text{M}]'$ is illustrated by parallel tangents to F^{Si} and F^{ic} at $[\text{M}]_0$ and $[\text{M}]'$, respectively. Considering the change in free energy due to decomposition of the supersaturated solution shows that the gain of chemical free energy per metal atom Δf_{chem} can be visualized by the vertical distance of the tangent to F^{Si} from F^{ic} at the composition $[\text{M}] = [\text{M}]'$ (e.g. Haasen, 1978). To estimate Δf_{chem} it is assumed that the composition of the precipitate is that of the silicide (rather than $[\text{M}]'$). Bearing in mind that concentrations of 3d elements in silicon are well below 100 ppm a good approximation of Δf_{chem} is given by

$$\Delta f_{\text{chem}} = k_{\text{B}}\,T \ln\!\left(\frac{a_{\text{M}}([\text{M}]_0)}{a_{\text{M}}([\text{M}]_{\text{eq}})} \right) \qquad (11\text{-}20)$$

in which $a_{\text{M}}([\text{M}])$ is the activity of metal atoms in silicon for a given concentration $[\text{M}]$, and $[\text{M}]_{\text{eq}}$ is the solubility of metal atoms in *equilibrium with the silicide*. Within this model for a regular solution, the activities can be replaced by concentrations giving

$$\Delta f_{\text{chem}} = k_{\text{B}} T \ln \left(\frac{[\text{M}]_0}{[\text{M}]_{\text{eq}}} \right) \qquad (11\text{-}21)$$

Suppose a silicon crystal saturated with a 3d transition element (e.g., Co) at a given temperature T_s (say 1100 °C) is cooled to $T < T_s$. This leads to a chemical driving energy Δf_{chem} for precipitation.

In Fig. 11-14 Δf_{chem} is plotted as a function of the undercooling $(T_s - T)$ for a solution of Co in Si saturated at $T_s = 1100$ °C for the case of $CoSi_2$ precipitates. To illustrate the basic difference between 3d elements in silicon and usual metal–metal binary alloys Δf_{chem} for $[\text{M}]_0/[\text{M}]_{\text{eq}} = 100$ is indicated, which corresponds to the maximum supersaturation attainable in metallic systems. It can be seen that this value of Δf_{chem} is reached on undercooling the solution by only *215 K*. Further reduction of the temperature leads to driving forces of a few tens of $k_{\text{B}} T$. According to Eq. (11-19), this implies that even larger values of $\Delta F_{\text{S}}(n)$ can be counterbalanced by cooling the solution to sufficiently low temperatures. Hence, a variety of structures, morphologies and possibly compositions not representing equilibrium states are likely to form during precipitation of 3d transition elements in silicon, as these metal atoms are mobile even near room temperature.

In classical nucleation theories *energetics* and *kinetics* of the nucleation process are considered separately, i.e., relations like Eq. (11-19) are first used to calculate the nucleation barrier, and then kinetic properties (diffusivity) of the solute atoms are used to estimate nucleation rates and growth of the precipitates. For systems with such large driving forces as Si:M this treatment appears to be insufficient. In a recent paper, Bené (1987) proposed that during the reaction of metal films on silicon substrates silicides are formed, which *maximize* the *energy degradation rate* instead of those related to a *minimum free energy*. As we shall see in the following section some of the phenomena related to 3d transition element precipitation in silicon are indeed in favour of Bené's conjecture, i.e., it is the *combination* of kinetic and energetic arguments, which determines the structure (Seibt, 1990), morphology (Utzig, 1988; Seibt and Schröter, 1989) or composition (Baumann and Schröter, 1991) of the precipitates. This may easily lead to the selection of relaxation paths which result in the formation of so-called "kinetically determined" non-equilibrium states (Seibt and Schröter, 1991).

11.4.2 Experimental Observations

It was recognized in the 1960s that after high temperature annealing, cooling to room temperature and subsequent preferential etching silicon wafers (or parts of them) frequently appear bright when illuminated with a spotlight and viewed along nonreflecting directions (e.g., Pomerantz, 1967). This phenomenon is called "haze"

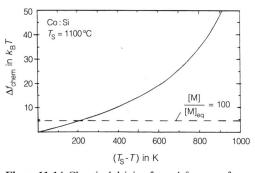

Figure 11-14. Chemical driving force Δf_{chem} as a function of undercooling $(T_s - T)$ for a solution of Co in Si saturated at $T_s = 1000$ °C; the horizontal line indicates Δf_{chem} for the maximum supersaturation attainable in metallic systems.

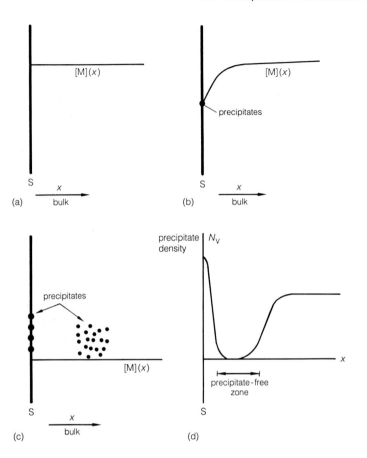

Figure 11-15. Basic process of haze formation of Si wafer surfaces (S); in (a)–(c) concentration profiles ($[M](x)$) of dissolved metal impurities are indicated; (d) shows the resulting depth profile of the precipitate density N_V; for details, see text.

(or "fog") and is due to scattering of the incident light by an accumulation of small etch pits (sometimes also called "s-pits"). Haze is also observed after intentional contamination of silicon wafers with transition metals or handling of samples with stainless steel tweezers (e.g., Hill, 1981). Graff (1983) showed that the appearance of haze is strongly dependent on the cooling rate after the high-temperature treatment. He also found that haze-forming metals are iron, cobalt, nickel, copper, palladium and rhodium.

High cooling rates could not completely prevent haze formation, with the exception of iron, for which no haze appeared after quenching.

Investigations by Seibt and Graff (1988 a, b) using transmission electron microscopy (TEM) revealed that in all cases metal silicides are the haze-forming precipitates.

The basic process responsible for haze formation is schematically shown in Fig. 11-15: Suppose a silicon wafer at a temperature $T = T_0$, with a homogeneous concentration of a fast diffusing 3 d element (Fig. 11-15 a), is cooled to $T = T_1 < T_0$, triggering precipitation at the wafer surfaces, but not in the bulk (Fig. 11-15 b). This imposes a concentration gradient of dissolved metal atoms directed towards the surfaces and results in a precipitate-free zone below the surface in which the supersaturation is not sufficient for homogeneous nucleation of precipitates during further cooling. At a still lower temperature, $T = T_2 < T_1$, precipitate nucleation also starts in the bulk of the wafer (Fig. 11-15c) leading to depletion of dissolved impurities within the whole sample. The resulting depth dependence of the precipitate density N_V is given in Fig. 11-15d: Below the subsurface region containing a high density of particles a *precipitate free zone* (not to be confused with a *denuded zone* as discussed for internal gettering in Sec. 11.5) is observed, whereas precipitates are still present in the bulk of the wafer. The width of the precipitate free zone may vary from a few microns after quenching (cooling rate 1000 K/s) to hundreds of microns after slow cooling (cooling rate < 5 K/s). Distributions like that of Fig. 11-15d can be obtained using preferential etching of wafer cross sections, or by x-ray section topography (Graff, 1983).

The reason why precipitate nucleation occurs first at wafer surfaces (Fig. 11-15a) has not been clarified yet. It has been proposed that the temperature gradient produced during cooling leads to higher supersaturations at the surfaces (Graff, 1983). However, wafer surfaces (being usually Si/SiO_2 interfaces) can also provide preferential nucleation sites, which seems to be the main reason for haze formation (Seibt, 1990).

We shall return to haze formation in Sec. 11.5 since this phenomenon usually competes with gettering procedures based on relaxation induced gettering. Hence the intensity of haze can be used as a qualitative test of gettering efficiency (Graff et al., 1985; Falster and Bergholz, 1990).

We now briefly discuss how experiments suitable for the investigation of 3d transition element precipitation have to be designed. TEM is the only experimental technique available for structural analysis.

Precipitate densities are extremely small. While in metal–metal decomposition studies densities are at least $10^{17} - 10^{18}\,\mathrm{cm}^{-3}$, those obtained for 3d elements in silicon are smaller by at least five orders of magnitude. From our preceding description of haze formation it is clear that utmost care has to be taken from which *part* of the silicon samples the TEM foils are prepared. Hence the use of cross-section geometry is highly favoured. A second parameter which has to be controlled is the initial concentration of the dissolved metal atoms. This can be done by either directly measuring the concentration using techniques like neutron-activation analysis or by applying experimental conditions for which the solubility is known from literature. The latter method uses in-diffusion from films containing several layers of metal atoms deposited on the wafer surfaces by evaporation or sputtering (Weber, 1983). Since the macroscopic spatial distribution as well as precipitate size, morphologies and composition are critically dependent on the *cooling rate* after in-diffusion of the 3d element under investigation this parameter is of utmost importance and should be specified in any study of precipitation of fast diffusing impurities.

However, due to the technological importance most studies have been performed under conditions suitable for device production. These usually involve slow cooling (< 10 K/s) of the samples to room temperature so that the precipitates observed are located at wafer surfaces and the initial concentration of metal impurities is usually unknown. Hence these studies are of limited use from a basic point of view.

In what follows we shall describe phenomena related to the precipitation of nickel and cobalt on the one hand and copper on the other. It will be seen that nickel and cobalt precipitate almost without a volume change and form metastable particle morphologies upon quenching. For copper a considerable volume expansion is associated with precipitation inducing the formation of metastable stacking-fault-like defects, when rapidly cooled from high temperatures. Both these observations are discussed in terms of Bené's hypothesis.

11.4.2.1 Precipitation Without Volume Change: Nickel and Cobalt in Silicon

For cobalt and nickel the silicides in equilibrium with silicon are $CoSi_2$ and $NiSi_2$, respectively. They have the face centered cubic CaF_2 structure with lattice parameters differing from that of silicon by less than 1.2% (for a detailed description of these silicides see Chapter 8 of this volume). The specific volume of silicon atoms is nearly equal for both silicon and the silicides. Hence the volume change associated with $NiSi_2$ and $CoSi_2$ precipitate formation is small and coherency strains are only important in the case of large precipitates. Furthermore, both disilicides grow epitaxially on $Si\{111\}$ planes indicating small interfacial energies. All these properties along with the high mobility of interstitially dissolved nickel and cobalt atoms imply that precipitate nucleation and growth can occur at small supersaturations and lead to low-energy configurations. However, as we shall see below, both 3d impurities form metastable precipitates upon quenching from high temperatures.

Using high resolution TEM (HRTEM) Seibt and Schröter (1989) investigated the early stages of nickel precipitation in silicon. In their experiments a nickel film was evaporated on one surface of the samples before annealing at temperatures between

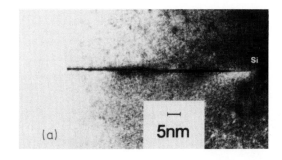

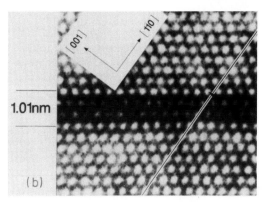

Figure 11-16. (a) High resolution electron micrograph of a $NiSi_2$ platelet obtained after rapid quenching from high temperatures; (b) magnified detail showing a relative shift of the Si lattice above and below the platelet (Seibt and Schröter, 1989).

850 and 1050 °C. The in-diffusion was terminated by quenching the samples in ethylene glycol leading to cooling rates of about $500-1000$ K/s. Fig. 11-16a is a direct lattice image of a typical plate-shaped precipitate obtained in the bulk of the samples, i.e., at least 100 μm away from any surface. These platelets are parallel to $Si\{111\}$ planes and have diameters between 0.4 and 0.9 μm.

Analysis of lattice images revealed that the precipitates consist of only two $\{111\}$ planes $NiSi_2$ (Fig. 11-16b). This implies that each nickel atom in these platelets belongs to the precipitate/matrix interface. The atomic structure of this interface could be determined by means of micrographs

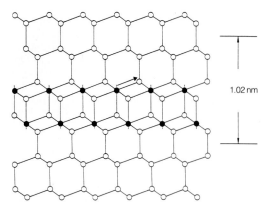

1.02 nm

Figure 11-17. Ball and stick model of NiSi$_2$ platelets observed after quenching; the precipitate/matrix interfaces are built up by Si–Si bonds; the arrow indicates the rigid shift of the Si lattice above and below the platelet (after Seibt and Schröter, 1989).

like Fig. 11-16b, leading to the balls and sticks model shown in Fig. 11-17. Both {111} interfaces are built up by Si–Si bonds leaving nickel atoms in sevenfold coordination (compared to eight silicon nearest neighbors as in bulk NiSi$_2$). Hence, since no nickel atom in these precipitates has cubic surroundings, as would be the case for particles with a low energy morphology, these structures are metastable. The interfacial structure described above agrees with that obtained from NiSi$_2$ films epitaxially grown on Si{111} substrates (Cherns et al., 1982; see also Chapter 8 of this volume). For geometrical reasons platelets with the observed interfacial atomic arrangement introduce a displacement in the silicon lattice as is indicated by the arrow in Fig. 11-17. The displacement has to be compensated by a dislocation bounding the precipitate. These dislocations, which have Burgers vectors of type $a/4\langle 111\rangle$ inclined with respect to the platelet normal, have been predicted on the basis of crystallographic considerations (Pond, 1985) and have been observed using conventional TEM (Seibt and Schröter, 1989).

Why does this particle morphology form? It has been proposed that it can grow much faster compared to more compact particles, because the bounding dislocation is the most efficient channel for the incorporation of nickel atoms into the precipitates (Seibt and Schröter, 1989). For example consider the transformation from the diamond structure (silicon) to the fluoride structure (NiSi$_2$). Basically this process can be viewed as the replacement of a silicon atom by a nickel atom pushing the silicon atom onto the antibonding site (nearest tetrahedral interstitial site) as shown in Fig. 11-18a. This process is essentially different from that of interstitial diffusion, where the atoms are jumping from one interstitial site to another. Hence, one might speculate that the process depicted in Fig. 11-18a involves a high barrier, which controls precipitate growth and leads to a small energy degradation rate. In Fig. 11-18b a possible core structure of the bounding $a/4\langle 111\rangle$ dislocation is shown, indicating that the dislocation provides distorted sites where replacement of a silicon atom by a nickel atom can occur more easily so leading to diffusion controlled precipitation. Accepting that the bounding dislocations establish a high incorporation rate of nickel atoms into the platelets, re-

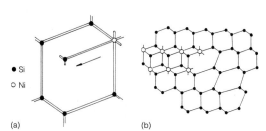

● Si
○ Ni

(a)   (b)

Figure 11-18. (a) Basic process of NiSi$_2$ formation from Si: replacement of a Si atom by a nickel atom pushing the silicon atom to the tetrahedral interstitial site; (b) possible core structure of the bounding $a/4\langle 111\rangle$ dislocation providing sites for easy incorporation of Ni atoms.

laxation of the supersaturation takes place faster for large platelet diameters. Thus the observed precipitates consisting of only two {111} layers of $NiSi_2$ can be understood on the basis of Bené's hypothesis, i.e., the formation and growth of platelets realize a maximum energy degradation rate.

In the case of cobalt precipitation in silicon a similar behavior can be expected since $CoSi_2$ and $NiSi_2$ are isomorphous and the diffusivities are comparable (see Fig. 11-2). Mössbauer spectroscopy (Utzig, 1988) was used to separate different cobalt species in silicon. In low-doped p- and n-type samples all the cobalt atoms were found on non-cubic sites directly after quenching. In complete analogy to the case of nickel in silicon, this species is considered as cobalt atoms in the precipitate/matrix interface of the platelets, implying that thin platelets are also formed in this case.

On additional annealing at medium temperatures ($200 - 600\,°C$) a transformation of cobalt atoms from non-cubic to cubic sites is observed due to an increase of the platelet thickness. Furthermore, it could be shown that the fraction f_{cubic} of cobalt atoms on cubic sites only depends on the annealing temperature (Fig. 11-19a) and a constant value of f_{cubic} is reached quite rapidly (e.g., 4 min at $400\,°C$; Fig. 11-19b). Both these observations led to the conclusion that the transformation process occurs without long range diffusion, i.e., it is due to rearrangement of a single precipitate in order to establish a minimum energy configuration. This configuration is determined by a compromise in the minimum surface energy (realized by spherical precipitates) and the elastic energy due to coherency strain. The latter favours plate-shaped precipitates (Nabarro, 1940). The temperature dependence of the lattice parameters of silicon, $NiSi_2$ and $CoSi_2$ (data taken from Murarka, 1983), shown in

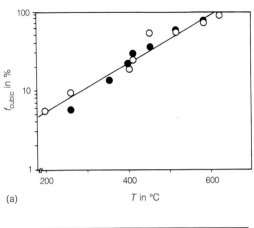

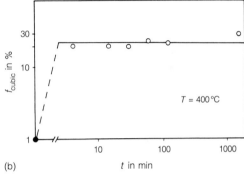

Figure 11-19. (a) Fraction f_{cubic} of cobalt atoms on cubic lattice sites as a function of annealing temperature; (b) f_{cubic} as a function of annealing time for $T = 400\,°C$ (data after Utzig, 1988); open and closed circles refer to Cz and FZ silicon, respectively.

Fig. 11-20, demonstrates that the misfit of $CoSi_2$ with respect to silicon decrease with increasing temperature. Thus thicker platelets are favoured at higher temperatures in agreement with the behavior of $CoSi_2$ platelets (Fig. 11-19a). Although a consistent description of Mössbauer data was given, the above model has to be verified by TEM investigations.

The platelets observed after quenching from high temperatures exhibit a large surface energy in addition to the energy of the bounding dislocation. Hence there is a considerable driving force for particle ripening as observed for cobalt by means of Mössbauer spectroscopy.

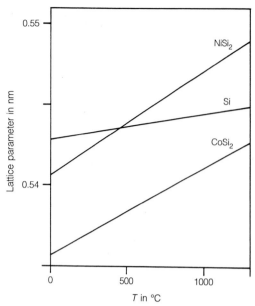

Figure 11-20. Temperature dependence of the lattice parameters of $NiSi_2$, $CoSi_2$ and silicon (data taken from Murarka, 1983).

For the case of $NiSi_2$ platelets it has been shown (Seibt and Schröter, 1989) that during additional annealing in the temperature range of 500–900 °C the precipitate density is reduced by a factor of 10–100 (depending on the annealing temperature). The platelet thickness increases by the same factor so that the precipitate diameter remains almost unaffected. This leads to a decrease in the interfacial energy per precipitated nickel atom by at least one order of magnitude. This kind of particle ripening is accompanied by a process, which is shown in Fig. 11-21 in a series of electron micrographs. The platelets now contain regions of different orientation relationship of $NiSi_2$ and silicon. In both types $NiSi_2$ shares a $\langle 111 \rangle$ direction with silicon and is either aligned with silicon (type A) or rotated about this $\langle 111 \rangle$ axis by 180°, i.e., twinned.

Type A and type B orientations are frequently observed in epitaxial $NiSi_2$ films

grown on $\{111\}$ silicon substrates (Cherns et al., 1982; Tung et al., 1983; see also Chapter 8 of this volume). Fig. 11-21 a–c shows type A regions at the border of a $NiSi_2$ platelet and a central type B grain. A change in the platelet thickness is often observed at A–B boundaries (Fig. 11-21 d). In contrast to epitaxially grown films, where small initial nickel coverages lead to type B formation (Tung et al., 1983), no correlation is observed between the platelet thickness and the type of orientation. Instead, the fraction of type B grains was found to depend on the annealing temperature and was interpreted in terms of an A to B transformation occurring at temperatures below 800 °C (Seibt and Schröter,

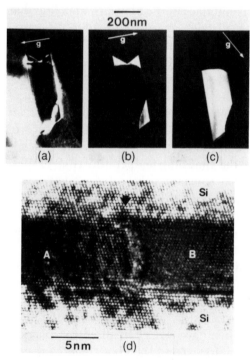

Figure 11-21. (a) Bright-field TEM micrograph showing an inclined $NiSi_2$ platelet after quenching and annealing at 500 °C; (b) and (c) show type B and type A regions of the platelet, respectively; (d) lattice image of an A–B boundary associated with a thickness change (Seibt and Schröter, 1989).

1989). However, the reason for this behavior has not yet been clarified.

Hitherto, we have not considered the influence of elastic strains due to lattice mismatch between the matrix and the precipitate. This is only justified for small precipitates. Under conditions of slow cooling (cooling rates < 50 K/s), however, platelets with thicknesses of up to 100 nm and diameters of up to 100 μm have been observed in near surface regions of silicon wafers in the case of nickel (Stacy et al., 1981; Augustus, 1983a, b; Picker and Dobson, 1972; Seibt and Graff, 1988a, b; Cerva and Wendt, 1989a; Kola et al., 1989). Another precipitate morphology is also observed, i.e., modified pyramidal shapes (octahedral, tetrahedral) which always exhibit type A orientation. Fig. 11-22 shows an example of such a $NiSi_2$ precipitate, which is heavily deformed, as indicated by the dislocations within the particle. Part of the misfit strain has also relaxed by producing dislocations in the silicon matrix, which often form closed loops (see arrows in Fig. 11-22). These loops have been shown to be extrinsic in nature (Augustus, 1983b), leading to the conclusion, that the precipitates form above 400 °C, which is where the misfit changes sign (compare Fig. 11-20). However, it should be noted that this conclusion is only valid for cases of slow cooling from high temperatures.

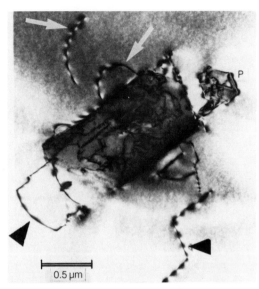

Figure 11-22. Heavily deformed polyhedral $NiSi_2$ precipitate of type A orientation, and type B orientated platelet (P) after slow cooling from high temperatures; part of the lattice mismatch has relaxed by the production of dislocations in the Si matrix (arrows).

11.4.2.2 Precipitation With Volume Expansion: Copper in Silicon

The investigation of the precipitation behavior of copper in silicon started as early as 1956, when Dash used copper to decorate dislocations in silicon crystals. Since then numerous investigations have been performed mostly stimulated by the technological importance of copper as a common contamination in silicon device production. From TEM studies it is well-known that copper precipitates in the form of star-shaped particle colonies if medium or slow cooling rates are applied (an extensive list of references is given in Seibt and Graff, 1988a). The colonies consist of small copper silicide particles forming planar arrangements parallel to Si{110} or Si{001} planes (Nes and Washburn, 1971), which are bounded by edge-type dislocation loops. Fig. 11-23a is a low magnification electron micrograph of such a star-shaped agglomeration. A magnified detail showing the bounding dislocations is given in Fig. 11-23b. A prominent feature of the almost spherical precipitates inside the dislocation loops is the occurrence of Moiré fringes with a spacing of about 4 nm when Si{220} reflections are used. This "fingerprint" may be used to identify copper related precipitates with high confidence in cases

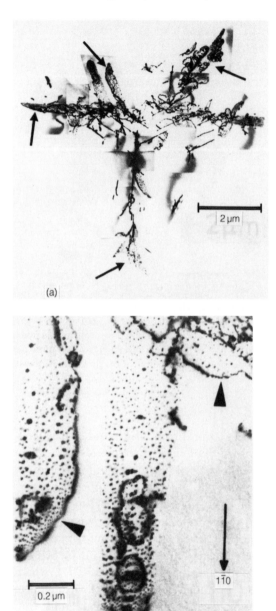

(a)

(b)

Figure 11-23. (a) Low-magnification TEM micrograph of a Cu particle colony; (b) magnified detail showing bounding edge-type extrinsic dislocation loops (arrows).

where the type of contaminant is not known.

In the late 1970's a model was developed by Nes and co-workers (Nes and Washburn, 1973; Nes, 1974; Solberg and Nes, 1978 a) which describes the colony growth. The model is based on earlier observations of NbC precipitation at stacking faults in niobium containing austenitic stainless steels (Silcock and Tunstall, 1964).

Before we briefly describe the basic features of this model, we have to consider the composition of the small precipitates. Early suggestions that the particles consist of b.c.c. CuSi (Nes and Lunde, 1972) or f.c.c. CuSi (Das, 1972) have been disproven on the basis of electron diffraction studies (Solberg, 1978; Seibt and Graff, 1988 a). Extensive investigations using TEM led to the proposal that the precipitates have the Cu_3Si structure (Solberg, 1978), which is the phase in equilibrium with silicon below the eutectic temperature (see Sec. 11.2.1). Although not proven beyond doubt, we shall, however, adopt this proposal by Solberg. The formation of Cu_3Si precipitates is associated with a volume expansion of about 150%, which can be relaxed by the emission of one silicon self-interstitial (Si_I) per two precipitated copper atoms, or by absorption of one vacancy (V). One should note that the following description remains valid for other precipitate compositions as long as their formation leads to comparable Si_I emission or V absorption. We shall discuss the processes involved in terms of Si_I emission only, and neglect contributions from V.

Within the model of Nes and co-workers the first stage of precipitation is the nucleation of particles at a dislocation (Fig. 11-24 a). Subsequent growth of the Cu_3Si precipitates leads to the emission of Si_I, which force the dislocation to climb (Fig. 11-24 b) by the absorption of Si_I. During this process the precipitates are dragged by the climbing dislocation, which is indicated in Fig. 11-24b by the dashed line showing the original position of the dislocation. Particle dragging has been observed by means of in-situ TEM for Cu_3Si particles (Solberg and Nes, 1978b). As the Cu_3Si particles grow their mobility decreases and dislocation segments between them bow out until dislocation unpinning occurs (Fig. 11-24c). Now the situation of the first stage of precipitation is restored and new particles can nucleate at the dislocation. This mode of precipitation has been termed repeated nucleation and growth on climbing dislocations, which is an autocatalytic process.

The model describes the formation of star-shaped colonies once dislocations are present. However, float zone silicon crystals are virtually dislocation free and, for them at least, the model does not account for the very early stages of precipitation. In a quite recent paper the question of how the particle colonies nucleate has been answered by means of HRTEM (Seibt, 1990).

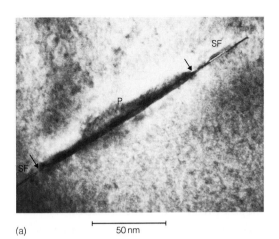

(a)   50 nm

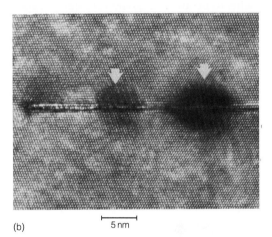

(b)   5 nm

Figure 11-25. (a) Cu silicide platelet (P) associated with a stacking-fault-like defect (SF) after quenching from high temperatures; (b) lattice image of a stacking-fault-like defect and of spherical Cu silicide precipitates (arrows) (after Seibt, 1990).

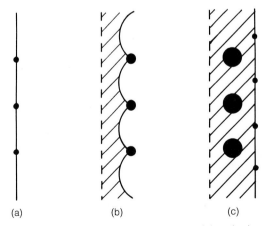

(a)   (b)   (c)

Figure 11-24. Growth model for Cu particle colonies (after Nes, 1974): (a) nucleation of precipitates at a dislocation; (b) growth of the particles leading to dislocation climb accompanied by particle dragging; (c) dislocation unpinning restoring the original situation (for details, see text).

The investigation of bulk regions of copper diffused samples quenched from high temperatures (cooling rate 1000 K/s) revealed plate shaped defects parallel to Si{111} planes. Fig. 11-25a is a HRTEM micrograph showing a central platelet (P) with a thickness of 3 nm, which is surrounded by a stacking-fault-like defect. This term has been used because lattice images of these defects (Fig. 11-25b) reveal contrasts considerably different from that of Frank-type stacking faults. In addition, Fig. 11-25b shows small Cu_3Si particles, which have nucleated at the SF-like defects (arrows). The processes leading to colony nucleation have been modelled on these observations. These early stages are drawn schematically in Fig. 11-26a–d. Figure 11-26a shows the homogeneous nucleation of a precipitate. For small particle sizes the volume expansion is expected to be compensated by elastic strains leading to plate-shaped precipitates (Fig. 11-26b).

The formation of plate-shaped particles seems to occur during the very early stages of copper precipitation, since platelets with diameters between 10–40 nm and thicknesses below 1 nm have been observed recently in silicon devices unintentionally contaminated with copper impurities (Cerva, 1991).

Further growth of the platelets initiates strain relaxation via Si_I emission, which finally leads to the nucleation of SF-like defects (Fig. 11-26c). These in turn can act as nucleation sites for further Cu_3Si particles

resulting in configurations which can be viewed as particle colonies in the embryonic stage (Fig. 11-26d). During subsequent growth the SF-like defects are transformed into Frank-type SFs (Seibt, 1990). The latter are rather stable and are sometimes observed in large colonies (Fig. 11-27).

The process described above is very similar to the precipitation of oxygen in silicon (see Gösele and Tan, Chapter V of this volume). However, one basic difference is due to the fact that oxygen in silicon has a much lower mobility than Si_I, whereas copper diffusion is at least two orders of magnitude faster than Si_I diffusion. Unless sinks are present for Si_I, this implies that copper precipitation kinetics are limited by the mobility of Si_I (Marioton and Gösele, 1988). Hence, the formation of Si_I sinks via SF-like defect nucleation (Fig. 11-26c) is a *prerequisite* for fast relaxation of the copper supersaturation. Obviously, as was discussed for nickel and cobalt precipitation, a metastable structure is formed in order to establish fast relaxation. Thus the SF-like defects can be understood on the basis of Bené's hypothesis.

At the end we briefly want to mention that copper silicide precipitates of tetrahedral shape have been observed at $Si\{001\}$ surfaces after in-diffusion of copper using rapid thermal annealing (Kola et al., 1989). These authors argue that the free surface acts as a sink for Si_I so that dislocations or stacking faults are not required to realize fast precipitation.

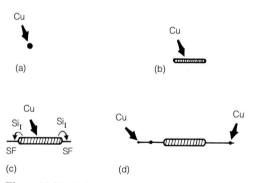

Figure 11-26. Early stages of colony formation (after Seibt, 1990): (a) homogeneous nucleation of a Cu silicide precipitate; (b) growth leads to plate-shaped particles in order to minimize the strain energy; (c) further growth initiates the emission of Si_I, which condense into stacking-fault-like defects; (d) nucleation of spherical Cu silicide precipitates at the stacking-fault-like defect; this configuration may be viewed as a particle colony in the embryonic stage.

11.4.3 Technological Implications

As we have seen in the preceding sections some 3d transition elements (viz. cobalt, nickel and copper) in silicon readily precipitate upon cooling from high temperatures. Since device processing usually

Figure 11-27. Copper particle colony containing an extrinsic Frank-type stacking fault.

involves numerous high temperature annealings followed by slow coolings to lower temperatures, the formation of the precipitates is likely to occur in near surface regions of silicon wafers (see Sec. 11.4.2), where active regions of devices are produced. In addition, transition elements have a strong tendency to decorate extended crystal defects like dislocations (Dash, 1956; Tice and Tan, 1976) stacking faults, grain boundaries (Maurice and Colliex, 1989; Broniatowski and Haut, 1990), oxygen (Ourmazd and Schröter, 1984, 1985) or even other transition element precipitates (e.g., Seibt, 1990). On the one hand, this property of fast diffusing metal impurities to decorate crystal defects during cooling from high temperatures is exploited in gettering techniques based on relaxation gettering (see Sec. 11.5). On the other hand, crystal defects are often unintentionally introduced during processing in active device regions. If decorated these defects are electrically active (see below) and have detrimental effects on the device performance. Up to now there are only a few studies that have systematically investigated the interaction of extended crystal defects with a certain impurity. Hence the question concerning what properties make a crystal defect a preferred nucleation site for a given metal silicide precipitate cannot be answered at present. However, from the discussion of the precipitation behavior of copper in silicon in Sec. 11.4.2.2 it may be easily understood that stacking faults or dislocations are preferred nucleation sites because of their ability to act as sinks for Si_I. The latter property is unlikely to be responsible for the decoration of stacking faults with nickel or cobalt, since both these elements precipitate without considerable volume change (see Sec. 11.4.2.1). Keeping in mind that the 3d transition elements treated in Sect. 11.4.2.1 and 11.4.2.2 precipitate homogeneously via the formation of metastable structures there is some evidence that alternative relaxation paths involving equilibrium defect configurations are accessible to the systems in the presence of crystal defects. This might be viewed as the underlying principle of defect decoration.

As has been mentioned in Sec. 11.4.1 most 3d metal silicides exhibit metallic conductivity (Murarka, 1983). In particular, the properties of $NiSi_2$ and $CoSi_2$ Schottky barriers on silicon have been extensively studied by means of electrical and structural techniques (see Chapter 8 of this volume). There is some evidence that metal silicide precipitates in silicon are strong recombination centers and that the electrical properties of some crystal defects are governed by the metallic precipitates decorating them (Broniatowski, 1989; Broniatowski and Haut, 1990; Maurice and Colliex, 1989; Lee et al., 1990). Grain boundaries in silicon with a large number of coincidence sites are highly reconstructed and are thus electrically inactive (see also Chapter 7 of this volume). After decoration with transition metal precipitates (nickel or copper related) these grain boundaries act as a barrier for charge carriers and exhibit a large recombination activity (Maurice and Colliex, 1989). Using capacitance transient spectroscopy (DLTS) Broniatowski (1989) investigated the electrical properties of a twin boundary decorated with copper silicide particle colonies. He assumed that the origin of the observed barrier effect was due to the small spherical particles acting as a Schottky barrier and described the properties of the precipitates in terms of thermionic emission. The barrier heights in n- and p-type material determined in the framework of this model add up to the band gap of silicon, which confirms the validity of this approach. However, the barrier heights depend on the morphology of the copper silicide precipitates (Broniatowski and Haut, 1990). This cannot be understood in terms of Schottky-Mott theory of metal semiconductor contacts unless different metallic phases are involved or there is a structural change of the precipitate/matrix interfaces.

The electrical activity of crystal defects has often been ascribed to silicide particle decoration. Bearing in mind that sec-ondary defects like dislocations or stacking faults are likely to form during precipitation a separation of electrical effects due to both types of defects is rather difficult. In a recent paper experimental conditions suitable for nickel precipitation without secondary defect formation have been applied to study the recombination properties of $NiSi_2$ platelets (Kittler et al., 1991) by means of the EBIC-mode electron beam induced current of SEM. It turned out that the diffusion length L_D of minority carriers as measured by EBIC was governed by the precipitates and depended on the precipitate density N_V according to $L_D = 0.7 N_V^{-1/3}$. Furthermore, extremely large EBIC contrasts of up to 40% were obtained indicating that $NiSi_2$ precipitates are efficient recombination centers. This observation demonstrates that even slight decoration of extended defects with metal silicide precipitates might appreciably change their electrical properties.

Owing to the fact that the dielectric properties of SiO_2 films are of utmost importance in silicon device technology, the tendency of 3d transition elements to form precipitates at wafer surfaces (i.e., Si/SiO_2 interfaces) gives rise to serious problems. For example, iron silicide precipitates are known to degrade reverse junction breakdown characteristics of bipolar transistor devices (Cullis and Katz, 1974; Augustus et al., 1980). In what follows we shall focus on the influence of silicide precipitates on the performance of silicon oxides.

It has been shown that one of the main problems concerning the dielectric quality of SiO_2 films is due to local decomposition into Si and volatile SiO (Rubloff et al., 1987). This decomposition is catalyzed by metallic impurities or decorated stacking faults at the Si/SiO_2 interface (Liehr et al., 1988a, b). In a series of TEM studies Honda and co-workers demonstrated that nickel, copper, and especially iron silicide precipitates located at the Si/SiO_2 interface lead to low breakdown voltages with an increasing effect from nickel relative to iron (Honda et al., 1984, 1985 and 1987). Quite recently, cross-sectional TEM revealed that copper silicide precipitates located at the Si/SiO_2 interface protrude into and hence thin the oxide (Wendt et al., 1989; Cerva and Wendt, 1989 b), which leads to a poor quality of gate oxides in MOS devices.

Conversely $NiSi_2$ platelets at Si/SiO_2 interfaces do not grow into the oxide, which is correlated with high breakdown voltages (Cerva and Wendt, 1989 a). Thus it has been demonstrated that growth behaviour of silicide precipitates and electrical effects in device technology may be closely related.

11.5 Gettering of 3d Transition Elements in Silicon

11.5.1 Introduction to Gettering Mechanisms

Gettering means the removal of impurities from the device-active area by transport to pre-designed regions of the wafer or by evaporation. The latter is termed chemical gettering, which makes use of atoms (e.g., Cl) attacking the silicon surface via the gas phase to form volatile complexes with the transition elements (for a recent review, see Shimura, 1988). The definition of gettering excludes passivation as a technique because in the case of passivation it is not the impurity that diffuses, but an additional impurity to form an electrically inactive complex (see e.g., reviews on hydrogen in silicon: Pearton et al., 1987, 1989).

Basically, two types of gettering techniques are conceivable. For illustration, we

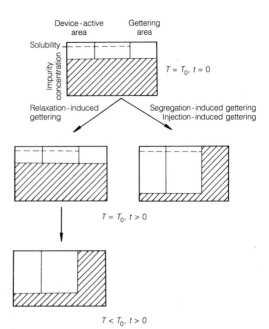

Figure 11-28. Schematic illustration of gettering mechanisms showing cross-sections of a silicon wafer divided into a device-active area and a gettering area. The latter is located in the bulk of the wafer (internal gettering) or at the backside of the wafer (external gettering). The hatched areas indicate the impurity concentrations in the different regions. During annealing at temperatures T_0 metal impurities are distributed homogeneously in the wafer up to the solubility limit (dashed line, top part of the figure). For segregation- and injection-induced gettering (right branch of the figure) metal impurity redistribution takes place during high-temperature annealing. Relaxation-induced gettering (left branch of the figure) is not effective as long as the impurity concentration is below its solubility. During cooling ($T_1 < T_0$) metal impurities predominantly precipitate in the gettering layer, inducing diffusion from the bulk to the gettering layer.

divide a silicon wafer into a device-active area (frontside of the wafer) and a gettering area, which may be the wafer backside (external gettering) or the bulk of the wafer (internal gettering) (see top of Fig. 11-28). One gettering mechanism (shown on the left in Fig. 11-28) requires supersaturation of metal impurities, along with a sufficient diffusivity and higher density of heterogeneous nucleation sites, for metal impurity precipitation in the gettering layer rather than the device-active area. During high-temperature treatment there is no gettering as long as the metal impurity concentration is below the solubility, defined by the actual reference phase (which does not need to be the equilibrium silicide). Supersaturation of metal impurities is achieved during cooling. A high density of heterogeneous nucleation sites in the gettering layer provides metal impurity precipitation at smaller supersaturations in the gettering layer than in other parts of the wafer, resulting in a concentration gradient of the diffusing metal impurity species. It is this concentration gradient that determines the efficiency of this type of gettering. We might call it *relaxation-induced gettering* because the nucleation sites allow supersaturated solutions to relax, which takes longer or might be inhibited, if no heterogeneous nucleation sites were available.

The other gettering mechanism (shown on the right in Fig. 11-28) does not require a supersaturation of metal impurities in the bulk. In this case the gettering layer collects metal impurities during the high-temperature treatment to reduce metal impurity concentrations well below the solubility. Such gettering action may come about by an enhanced solubility of metal impurities or by stabilization of a new metallic compound in the gettering layer (Seidel et al., 1975; Lescronier et al., 1981; Cerofolini and Ferla, 1981). It is called *segregation-induced gettering* in the following. An alternative process operating during high-temperature treatment as well, is based on the coupling of metal impurities to other diffusion currents, e.g., Si self-interstitials (Si_I). Such processes have to be described by concepts of irreversible thermodynamics, and are conceivable for coupling of substitutionally dissolved metal impurities

to Si_I. This mechanism will be termed *injection-induced gettering*. Such concepts have been introduced to explain phosphorus-diffusion gettering (Schröter and Kühnapfel, 1990).

Quite often, a classification of gettering techniques as to whether relaxation-induced gettering, segregation-induced gettering or injection-induced gettering is operative has not been done or has not been possible on the basis of the experiments.

With respect to gettering techniques in silicon, one distinguishes between internal and external gettering (sometimes referred to as intrinsic or extrinsic). External gettering means gettering of impurities by extrinsic or intrinsic defects after an external treatment, e.g., P-diffusion (Meek and Seidel, 1975; Lescronier et al., 1981; Cerofolini and Ferla, 1981; Shaikh et al., 1985; Falster, 1985), ion implantation (Meek and Seidel, 1975), mechanical abrasion (Mets, 1965), poly-Si deposition (Chen and Silvestri, 1982; Falster, 1989) to name some of them. In the case of internal gettering, extrinsic or intrinsic defects are already present in as-grown material or they are produced and activated by an annealing treatment.

In the following sections, we shall discuss the mechanism of internal gettering and then describe the present understanding of the most effective external gettering technique, namely P-diffusion gettering.

11.5.2 Internal Gettering

11.5.2.1 Oxygen Precipitation and Denuded Zone Formation

The concept of internal gettering was first introduced by Tan et al. (1977) to improve the so-called leakage-limited yield of bipolar transistors in Cz–Si. The authors argued that metal impurities precipitate at dislocations punched out by silicon oxide precipitates (Tice and Tan, 1976). Due to the technological importance of Cz-silicon, the use of oxygen precipitation-induced defects as efficient gettering centers has stimulated a great deal of investigations, and many details about oxygen in silicon have been uncovered (for a recent review, see Corbett et al., 1989).

Oxygen is incorporated into single crystalline silicon grown by the Czochralski method by reaction of the Si melt with the quartz crucible leading to typical oxygen concentrations of 7×10^{17} to 1×10^{18} atoms/cm^3, which can be adjusted with an accuracy of $+/-10\%$. For typical processing temperatures of $1000\,°C$, the oxygen solubility is about 10^{17} atoms/cm^3 (Hrostowski and Kaiser, 1959; Mikkelsen, 1982 a, b).

Internal gettering makes use of defects resulting from the formation of silicon oxide precipitates in silicon. This process is associated with a large volume expansion which is relaxed by the emission of silicon self-interstitials (leading to stacking faults growth) or by plastic deformation of the silicon matrix (prismatic punching) (for reviews, see Hu, 1981; Bourret, 1986). For internal gettering special annealing cycles have been designed, which allow the production of oxygen precipitation-induced defects in the bulk of silicon wafers but not in the device active area. For this purpose, wafers are first treated at a temperature (usually above $1050\,°C$), which is sufficiently high to prevent the formation of stable nuclei and allow out-diffusion of oxygen from a subsurface layer with a thickness of several microns. In contrast to the bulk of the wafer, the oxygen supersaturation in this area is not sufficient to initiate oxygen precipitate nucleation during a subsequent low-temperature treatment (usually about $700\,°C$). Finally, a second

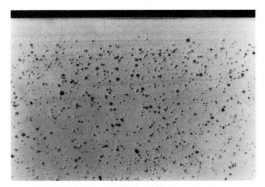

Figure 11-29. Cross-section of a wafer that was defect-etched to reveal extended defects after the internal gettering process (magnification: $50 \times$). Denuded zone and bulk microdefects are clearly visible (with kind permission of Wacker Chemitronic GmbH, Burghausen).

high-temperature annealing (about $1050\,°C$) is used to initiate oxide precipitate growth and formation of secondary defects. Figure 11-29 shows the resulting defect distribution as obtained by preferential etching of a wafer cross-section: Below a several microns thick denuded zone bulk microdefects are clearly visible.

11.5.2.2 Gettering Mechanism of Interstitial Impurities

Supersaturation of fast-diffusing metal impurities is relaxed by homogeneous or heterogeneous nucleation of metal precipitates. Heterogeneous nucleation takes place at crystal imperfections in the bulk and at the wafer surface.

In FZ–Si with an oxygen content too low to form oxygen-related extended defects, preferred nucleation occurs at the wafer surface, and can be observed as so-called haze after a defect delineating etch (compare Sec. 11.4.2).

In the past the internal gettering mechanism was mainly derived from correlation studies of haze formation and distribution of microdefects inside the wafer as obtained from preferential etching (Tan et al., 1977; Shimura et al., 1981; Graff et al., 1985, 1988). In most cases, the type of gettering mechanism the internal gettering belonged to was not specified. Colas et al. (1986) proposed that segregation-induced gettering is the basic mechanism. A silicate formed by precipitation with oxygen would result in a lower solubility of metal impurities than with respect to the equilibrium silicide. However, Graff et al. (1988) were able to show that the amount of electrically active Fe after internal gettering depends on the cooling rate of the wafer, providing clear evidence of internal gettering being a relaxation-induced gettering phenomenon.

Hence from the discussion about oxygen-precipitation-induced defects, oxygen precipitates, dislocations, and stacking faults may act as nucleation sites. Which type of defect is the most important one depends on the density of those defects and the critical supersaturation of metal impurities required for each type of defect to nucleate metal precipitates.

Graff et al. (1988) have found that with the disappearance of stacking faults during high temperature annealing the efficiency of internal gettering approaches zero. From this finding they concluded that stacking faults are the dominant gettering sites. However, at high temperatures the density and strain of oxygen precipitates are reduced, as well. As those extended defects are interrelated, a conclusive identification of dominant gettering sites requires a quantitative analysis of the density of metal precipitate nucleation sites and the density of oxygen precipitation-induced defects.

Gilles et al. (1990 b, c) applied Ham's theory (Ham, 1958) of diffusion-limited precipitation from a supersaturated solution to the precipitation kinetics of oxygen and iron and determined the product $n \cdot r_0$ of both (n and r_0 are the density and radius of the precipitates, respectively). The time constant of the exponential precipitation kinetics can be expressed as

$$\tau = \frac{1}{4\pi D n r_0} \tag{11-22}$$

where D denotes the diffusion coefficient of the diffusing species.

From an application of Eq. (11-22) to the precipitation kinetics of oxygen, size and density of oxygen precipitates can be determined separately. Approximating the precipitates by spherical particles and their radius by the maximum value after precipitation, the loss of interstitial oxygen with annealing time, as measured by infrared absorption spectroscopy, determine the product $n \cdot r_0^3$. Livingston et al. (1984) reported good agreement of oxygen precipitate densities calculated in such a way (Binns et al., 1983, 1984) with those derived from etch pit densities after application of a defect-delineating etch or by small-angle neutron scattering.

Figure 11-30 shows a plot of the iron precipitation kinetics at 235 °C in Cz–Si pre-annealed at 700 °C to nucleate oxygen precipitates and then annealed at 1050 °C for 2 h to saturate the specimen with Fe. For longer pre-annealing times a higher density of oxygen precipitates is expected after the high-temperature step. Fast quenching was chosen to keep Fe on interstitial sites. The concentration of interstitial Fe was determined by EPR.

By variation of the temperature it is found that the precipitation kinetics are thermally activated by an activation enthalpy of 0.7 eV, which is in good agreement with the migration barrier of interstitial Fe. Hence the application of Ham's

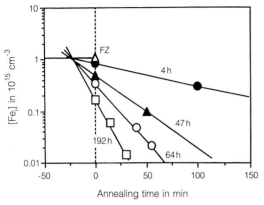

Figure 11-30. Precipitation kinetics of Fe in silicon at 235 °C after thermal processing (700 °C, 4 h (47 h, 64 h, 192 h) and 1050 °C, 2 h) and quenching. FZ denotes floating zone silicon with an oxygen concentration below the detection limit of FTIR. The other data were obtained from Cz–Si ([P] = 1.5 × 10^{16} atoms/cm^3, [O_i] = 1.1 × 10^{18} atoms/cm^3 (IOC 88-standard) and [C] = 1.5 × 10^{16} atoms/cm^3). Preannealing time at 700 °C is indicated in the figure.

theory modelling diffusion-limited precipitation is justified.

An analysis of the precipitation time constant of oxygen and Fe is shown in Table 11-8, in which known values of diffusivities and measured values of the time constant were inserted using Eq. (11-22) to determine $n \cdot r_0$ for oxygen as well as iron.

Table 11-8. Comparison of the products of density n and size r_0 for oxygen precipitates (calculated from the loss of interstitial oxygen, Eq. 11-22) and nucleation sites for Fe precipitation (calculated from the time constant of Fe precipitation at 235 °C) after thermal processing (700 °C, 4 h (32 h, 128 h) and 950 °C, 4 h (1050 °C, 2 h)).

Pre-anneal time in h	Diffusion temperature			
	950 °C		1050 °C	
	$n r_0$(O) in 10^4 cm^{-2}	$n r_0$(Fe)	$n r_0$(O) in 10^4 cm^{-2}	$n r_0$(Fe)
4	8.6	5.9	<2	0.76
32	25	18	4.1	4.8
128	75	41	29	28

For different annealing conditions overall agreement between $n \cdot r_0$ for iron and oxygen precipitates is found. Another remarkable feature is the exponential precipitation kinetics from the very beginning in Cz-silicon, which is not found in FZ-silicon. This means that $n \cdot r_0$ for the growing Fe-precipitates is independent of annealing time in Cz-silicon. From these observations Gilles et al. (1990b) concluded that

(1) oxygen precipitates are the nucleation sites for iron precipitation, and

(2) the iron precipitation kinetics are determined by the SiO$_2$/Si interface of the oxygen particles.

These conclusions are valid within the temperature range from 200 °C to 300 °C. However, for effective internal gettering the above process has to take place at or above 500 °C, since a sufficiently high mobility is needed for Fe$_i$ to diffuse out of the device active area. From a detailed analysis of high- and low-temperature iron precipitation kinetics, Gilles et al. (1990 a, c) confirmed that the conclusions remain valid under the conditions of internal gettering (higher temperatures corresponding to lower driving forces for Fe-precipitation) as well. As can be seen in Fig. 11-30, the Fe concentration when quenched to room temperature at a rate of 1000 K/s, shows a correlation with low-temperature precipitation kinetics: The faster the Fe precipitation kinetics, the lower is the Fe concentration as quenched. From the intersection of the extrapolation of those data at a concentration identical to the solubility of Fe in Si, it is concluded that there is no variation of the solubility in differently preannealed Cz–Si, but that different fractions of Fe have precipitated during quenching.

The role of oxygen precipitates in internal gettering has recently been addressed by Falster and Bergholz (1990), who found effective internal gettering for very low amounts of precipitated oxygen. For such small precipitates, stacking faults and dislocations are not expected.

An additional piece of information about gettering by oxygen precipitates came from a comparison of the Fe precipitation kinetics in high- and low-carbon-doped Cz–Si. It is reported in the literature that the addition of carbon results in a high density of *strain-free* oxygen precipitates after two-step annealing and a low density of secondary defects (stacking faults, dislocations) (Kanamori and Tsuya, 1985; Hahn et al., 1988). It has been speculated that this phenomenon can be understood if one takes into account that C is a small atom on a substitutional site. Co-precipitation of C and O is believed to reduce the strain of oxygen precipitates and, as a consequence of lower strain, stacking fault and dislocation densities are reduced as well.

Figure 11-31 shows the results of Fe precipitation kinetics in Cz–Si doped with 1.5×10^{17} carbon atoms/cm^3 during annealing at 235 °C. Precipitation rates are

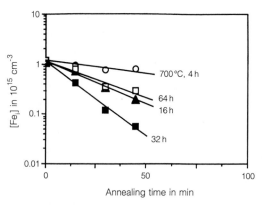

Figure 11-31. Precipitation kinetics of Fe in silicon at 235 °C after thermal processing (700 °C, 4 h (16 h, 32 h, 64 h) and 1050 °C, 2 h) and quenching (Cz–Si, [P] = 1×10^{15} atoms/cm^3, [O$_i$] = 1×10^{18} atoms/cm^3, [C] = 1.5×10^{17} atoms/cm^3). Preannealing time at 700 °C is indicated in the figure. Retardation of Fe precipitation kinetics for long annealing time can be explained by similar effects of oxygen precipitation in high-carbon Cz–Si (Ogino, 1982).

strongly enhanced compared to oxygen-free FZ–Si. Unlike low-carbon Cz–Si, strong enhancement is observed for short pre-annealing times at 700 °C, consistent with fast precipitation of oxygen in that material. The results are further corroberation of the previous conclusion about oxygen precipitates as nucleation sites for metal precipitates. However, there is one decisive difference in the precipitation kinetics: In high-C Cz–Si all of the Fe can be quenched on interstitial sites independent of its low-temperature precipitation kinetics. In low-C Cz–Si it is found that for high oxygen precipitate densities precipitation takes place even during cooling. Thus, in high-C material a high supersaturation of metal impurities is required to nucleate metal precipitates. It is proposed that in low-C Cz–Si it is the strain field of oxygen precipitates that reduces the critical supersaturation of metal impurities and makes precipitation possible at small undercooling. This allows precipitation to occur at high temperatures when metal impurities are sufficiently mobile to diffuse out of the denuded zone. Therefore the strain field of oxygen precipitates is expected to be necessary for effective internal gettering.

11.5.3 Phosphorus-Diffusion Gettering and Other External Gettering Techniques

11.5.3.1 Relaxation-Induced Gettering

External gettering is employed in the device manufacturing process as a protection against metal contamination, in addition to internal gettering. It can be effective from the very beginning of the device process, while internal gettering requires a low temperature step to nucleate oxygen precipitates.

External gettering sites are created predominantly in the subsurface layer of the back surface of the wafer. In addition to various kinds of damage (mechanical abrasion, laser irradiation, ion implantation), growth of poly-Si films or diffusion of a high concentration of phosphorus have been implemented.

Initial results have recently been reported on external gettering close to the front surface of a wafer after C-implantation (Wong et al., 1988) or growth of GeSi-alloy layers inbetween the Si substrate and an epitaxially grown layer of silicon (Salih et al., 1985). Whereas in the former process gettering sites in the buried layer have not been identified yet, in the latter process a network of misfit dislocations is created to relax the strain produced by lattice mismatch between the GeSi-alloy and Si.

From the discussion of the mechanism of internal gettering, it is inferred that for interstitial impurities most external gettering techniques operate by the relaxation mechanism as well. In the case of mechanical damage, a high density of stacking faults and dislocations are created which may nucleate metallic precipitates after supersaturation of metallic impurities. It is difficult to apply results about gettering efficiency of dislocations and stacking faults (Sec. 11.5.2) to external gettering, as in addition to those extended defects, a strain field is created by the damage, which is believed to assist metallic precipitate nucleation.

In the case of poly Si deposition on the backside of a wafer recrystallization during subsequent annealing processes leads to the formation of grain boundaries and dislocations, the densities of which depend on the details of the thermal treatment. According to Falster (1989), gettering of interstitially dissolved impurities by a poly Si-film and by stacking faults after mechanical damage is similar: gettering efficiency could not be improved by longer diffusion annealings, in support of relaxation-in-

duced gettering. In the case of Ni gettering by a misfit dislocation network of a GeSi alloy, Lee et al. (1990) showed that the size and density of Ni precipitates depended on annealing temperature and not on annealing time, again a confirmation of relaxation-induced gettering by dislocations.

11.5.3.2 Phosphorus-Diffusion Gettering, Segregation- and Injection-Induced Gettering

The final section about external gettering is devoted to phosphorus-diffusion gettering, PDG. Not only does it show the highest gettering efficiency of all the external gettering techniques, but it is unique with respect to the gettering mechanism of interstitial metal impurities, as it is the only known technique that operates by the segregation-induced gettering and injection-induced gettering mechanisms.

In-diffusion of P at a temperature above 900 °C results in a layer of less than 1 μm, doped with P at a concentration greater than 10^{20} P atoms/cm^3, which is called the *gettering layer*. It is achieved by depositing a layer of phosphosilicate glass (PSG) onto the back surface of a wafer. P activity depends on temperature, gas flow, and the volume fractions of the gas constituents, namely nitrogen, oxygen and POCl$_3$ (Negrini et al., 1975).

High Resolution Electron Microscopy shows that in addition to P on substitutional sites, SiP particles may extend into the silicon. Bourret and Schröter (1984) concluded that because of the lower density of Si in SiP large currents of Si self-interstitials come about, which in part are directed towards the interface Si/PSG and induce there epitaxial growth of silicon. Therefore, three different mechanisms of PDG are conceivable:

1) relaxation-induced gettering;

2) segregation-induced gettering by enhanced solubility in P-doped Si (cf. Sec. 11.3), or by stabilization of a new metal-containing phase;

3) injection-induced gettering involving currents of Si self-interstitials.

The phosphorus glass layer was found to contain a negligible amount of metal impurities after gettering (Kühnapfel, 1990). From out-diffusion profiles convincing evidence is reported that PDG is not relaxation gettering (Gilles, 1987). Thus gettering takes place during high-temperature treatment, such that the longer the annealing time the lower is the residual bulk contamination. Reemission of impurities upon annealing, observed for internal gettering and expected for all relaxation gettering techniques, occurs – if at all – very slowly.

In Sec. 11.3.2 it was shown that all 3d elements investigated so far show a solubility enhancement in P-doped silicon. This was attributed partly to the Fermi level shift, which gives rise to a solubility increase of negatively charged point defects. Another solubility enhancement results from pair formation between the two so-

lutes M_s and P. Since in the gettering layer we are dealing with a ternary Si–M–P system, also the stabilisation of a $M_x P_y$-phase (Meek et al., 1975) or a ternary phase could lead to a solubility reduction in the wafer and thereby contribute to PDG. However, such a phase has never been identified within the gettering layer.

For Cu Meek et al. (1975) deduced from Rutherford backscattering and channeling experiments, that in highly P-doped silicon a major fraction of Cu stays on substitional sites, and concluded that PDG comes about by solubility enhancement of negatively charged substitutional Cu and its pairs with P.

Applying this model to 3d elements with lower solubility in intrinsic silicon, e.g., Co, it can be calculated from the results presented in Sec. 11.3 that a solubility enhancement from 1×10^{14} Co-atoms/cm^3 at 920°C in the wafer to about 10^{16} Co-atoms/cm^3 within the gettering layer is expected by the Fermi level shift and the presence of P as a second solute. However, Kühnapfel et al. (1986) reported a Co concentration exceeding this estimation a hun-

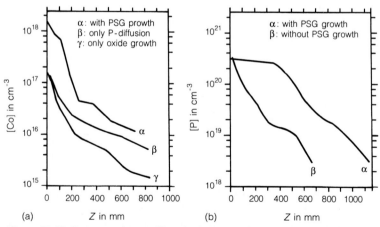

(a) Z in mm   (b) Z in mm

Figure 11-32. Concentration profiles of cobalt (a) and of phosphorus (b) for the boundary conditions as described in the inserts, all other parameters kept constant, $T = 920°C$, $t = 108$ min. Injection-induced gettering (curve α) was always found to be associated with a plateau in the phosphorus concentration.

dredfold giving evidence that an additional gettering mechanism is operative.

The accumulation of Co within the gettering layer is demonstrated in Fig. 11-32. It is interesting to note that a plateau in the electrically active phosphorus concentration develops under the conditions, which induce the set-in of this additional gettering (Schröter and Kühnapfel, 1989, 1990). Another interesting feature about the mechanism responsible for the additional gettering is presented in Fig. 11-33. The gettering efficiency, measured as the cobalt concentration within the first 35 nm of the gettering layer, is shown as a function of the PSG-layer growth rate $\langle v_{PSG} \rangle$ (Kühnapfel and Schröter, 1990). As can be seen, the gettering efficiency sharply rises as soon as $\langle v_{PSG} \rangle$ reaches a critical value. This has been taken as additional evidence that the injection rate of self interstitials might be responsible for this additional gettering.

The model proposed to explain injection-induced gettering is based on theoretical concepts worked out in the seventies to describe radiation-induced phase instability, solute redistribution and precipitation (Wiedersich et al., 1977; Martin, 1978). Under irradiation vacancies and interstitials are generated in equal numbers. They may react with each other or with solutes to form complexes or to transform one solute species into another one. They may also migrate to point defect sinks, like dislocations, stacking faults, the surface, or the precipitate/matrix interface, and vanish there.

In a simplified model Martin (1978) has considered the influence of point defect fluxes on solute atoms. The basic ideas of his model are summarized in two assumptions:

(1) The flux of point defects (towards the point defect sinks) induces a flux of solute atoms.

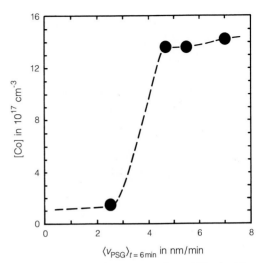

Figure 11-33. Cobalt concentration within the 35 nm thick layer below the PSG/Si interface, after PDG at 920 °C for 6 min, as a function of the growth velocity of the PSG-layer $\langle v_{PSG} \rangle$ averaged over 6 min. The ratio of the total amount of predeposited phosphorus Q to that of electrically active phosphorus Q_{el} was chosen $Q/Q_{el} = 2.3$ (after Negrini et al., 1975).

(2) When the solute concentration at the sink exceeds the solubility $[M]_{eq}$, of solute in the solvent in absence of irradiation, a new phase appears at the sink.

The first assumption is justified by basic principles of thermodynamics of irreversible processes. The second assumption is less evident, since a point defect supersaturation might affect the equilibrium between two phases. Indeed, an incompatibility with this assumption arises, when two point defect fluxes arrive at the interface between the two phases, which both establish a solute concentration gradient there. Therefore it should be kept in mind that the quantitative formulation of this model should carefully check assumption (2) for the case studied.

During PDG point defect fluxes are not driven by radiation, but by injection of silicon interstitials Si_i. The flux arises from the PSG/Si interface and, if the P-activity

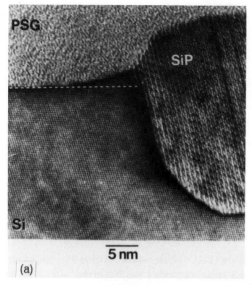

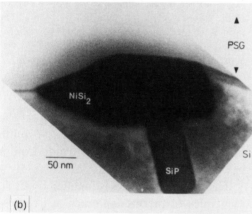

Figure 11-34. (a) Lattice image of PSG/Si interface in the neighborhood of a SiP particle growing into the silicon. Note the protrusion of the SiP particle and of the PSG/Si interface near the particle. The undisturbed interface is indicated by the dotted line. (b) Bright-field TEM image of the PSG/Si interface showing a SiP particle and a precipitate identified as NiSi$_2$ (according to Ourmazd and Schröter, 1985).

is sufficiently large to induce SiP-growth, from the SiP/Si interface (Bourret and Schröter, 1984). Sinks for self interstitials are stacking faults and – according to TEM-images (see Fig. 11-34) – so are those parts of the PSG/Si interface surrounding SiP-particles (Bourret and Schröter, 1984; Ourmazd and Schröter, 1984).

The self interstitial flux couples to 3d elements on substitutional sites via the kick-out reaction: $Si_i + M_s \rightleftharpoons M_i$, but not directly to metal atoms on interstitial sites. Therefore, the high P-doping within the gettering layer, leading to a strong increase of $[M_s]_{eq}$, is a prerequisite for this coupling to be effective.

The flux of M consists of two terms: one is a diffusion flux proportional to the gradient of [M], the second is a drift flux proportional to [M] and to the self interstitial flux (Schröter and Kühnapfel, 1989). The solute drift flux is directed towards the sink, since M migrates as an interstitial species, and the diffusion flux is directed away from the sink. The balance between the two establishes a concentration of $[M]_{sink}$ at the sink. According to assumption (2) of Martin's model the metal atoms precipitate as soon as $[M]_{sink} > [M]_{eq}$.

One important issue of the proposed model is the prediction that the additional gettering should result in silicide formation. Indeed, after PDG of nickel and iron doped wafers NiSi$_2$- and FeSi$_2$-particles have been observed in the gettering layer by TEM, respectively (Ourmazd and Schröter, 1984, 1985). In cobalt doped silicon local Mössbauer spectroscopy revealed a quadrupole doublet in the gettering layer (Shaikh et al., 1985), which has been tentatively associated with very thin CoSi$_2$-platelets (Utzig, 1988; Schröter and Kühnapfel, 1989; see also Sec. 11.4.2.1). This identification relies on the comparison of line parameters of spectra observed in cobalt doped wafers after quenching from high temperature.

A second important issue of the proposed model is the prediction that the onset of silicide formation, i.e., of the additional gettering, should occur, when a critical self interstitial flux, i.e., a critical injection rate, has been reached, which has indeed been observed (see Fig. 11-33) (Kühnapfel and Schröter, 1990). In spite of these strong indications, direct evidence for the injection-induced gettering mechanism is still lacking.

11.6 Summary and Outlook

In this chapter we have outlined the high-temperature characteristics of 3 d elements in silicon. The solubilitiy and diffusivity of these elements have been investigated in intrinsic and extrinsic silicon. In intrinsic silicon all 3 d atoms dissolve predominantly on tetrahedral interstitial sites, which means that $[M_i]_{eq} > [M_s]_{eq}$. For Mn_i, Fe_i, and Co_i also the charge state of the dominant species has been determined to be $M_i^{(0)}$. This implies that for them $[M_i^{(0)}]_{eq} > [M_s^{(0)}]_{eq}$. Solubility data are available for almost all 3 d elements. However, their thermodynamic interpretation is still uncertain in some cases (e.g. Ni) due to unknown boundary conditions.

For temperatures below 1100 K the interstitial species of Mn, Fe, Co, and Cu have been shown to be donors, the substitutional species to be multiple acceptors. Consequently in extrinsic n-type silicon one has $[M_s]_{eq} > [M_i]_{eq}$ in this temperature range. For Mn_i, Fe_i, and Co_i a strong shift of the donor level towards the valence band above 1100 K indicates a transition from a low-temperature to a high-temperature atomic configuration.

Compared to usual solubilities in metallic systems, partial solution enthalpies found for Ti_i, Mn_i, Co_i, Ni_i, and Cu_i are very large (1.5–2.1 eV). Co_i, Ni_i, and Cu_i are among the fastest diffusing impurities in silicon with migration enthalpies below 0.5 eV. For the lighter 3 d elements diffusivities decrease and $H_M^{(mi)}$ increases to about 1.8 eV for Ti_i.

The systematics and for the heavier 3 d elements also the absolute values of these interstitial diffusivities have been explained by a simple hard sphere model. The difference of the elastic energy between the tetrahedral and hexagonal interstitial site is found to be a major contribution to the migration enthalpy of diffusion. Compared with this the elastic energy of 3 d atoms on interstitial sites is only a small fraction of the solution enthalpy. At present it is an open question how these various aspects of 3 d atoms in silicon can be combined in a single concept.

Concerning the precipitation behavior of the fast diffusing 3 d elements cobalt, nickel and copper, there is now some detailed knowledge which precipitate structure, morphology and composition forms for different experimental conditions. We have seen that precipitation of these impurities is closely related to the more general question of how systems with large driving forces relax toward thermal equilibrium. The precipitate morphology in the case of cobalt and nickel and the extrinsic stacking-fault-like defect in the case of copper have been related to kinetic arguments, i.e. these configurations merely serve as a tool to establish fast degradation of free energy. The concept of *maximum energy degradation rates* – originally introduced by Bené (1987) to describe silicide nucleation in metal-silicon diffusion couples – may be viewed as a first attempt to deal with relaxation of highly supersaturated solutions. This rather qualitative description should serve as a basis for the development of more predictive approaches.

There is evidence that metal silicide precipitates may strongly affect the electrical behavior of silicon. Modelling the *electronic* properties of such many-particle systems and determining their relationship to *structural* properties of the precipitates is a challenge for future investigations.

The fundamental knowledge of thermodynamic and transport properties of 3d transition elements was applied to the problem of gettering, i.e. the question of how those impurities can be located away from the device-active area to improve device properties. Gettering techniques were classified into relaxation-induced, segregation-induced and injection-induced gettering, according to the different mechanisms by which they are governed. For interstitially dissolved 3d transition elements relaxation-induced gettering is dominant for internal as well as for various types of external gettering techniques. Other types of gettering mechanisms have been identified for phosphorus-diffusion gettering. We consider such phenomenological classification as a prerequisite for microscopic models. Development of such models will allow one to evaluate gettering efficiency especially with respect to processing temperature and time and to determine interactions of internal and external gettering.

11.7 Acknowledgements

The authors would like to thank Prof. P. Haasen, Dipl. Phys. H. Hedemann, Dr. K. Graff, and Dipl. Phys. A. Koch for their critical comments on this chapter, Profs. G. Borchardt and H. Feichtinger for helpful remarks concerning Sec. 11.2 and K. Heisig for preparing the drawings.

11.8 References

Augustus, P. D. (1983 a), *Inst. Phys. Conf. Ser. No. 67*, 229.
Augustus, P. D. (1983 b), in: *Proc. Electrochem. Soc. 'Defects in Silicon'*. Pennington: The Electrochemical Society, p. 414.
Augustus, P. D., Knights, J., Kennedy, L. W. (1980), *J. of Microscopy 118*, 315.
Bakhadyrkhanov, M. K., Zainabidov, S., Khamidov, A. (1980), *Sov. Phys. Semicon. 14*, 243.
Baumann, F. H., Schröter, W. (1991), *Phys. Rev. B43*, 6510.
Beeler, F., Anderson, O. K., Scheffler, M. (1985), *Phys. Rev. Lett. 55*, 1498.
Bendik, N. T., Garnyk, V. S., Milevskii, L. S. (1970), *Sov. Phys. Solid State 12*, 150.
Bené, R. W. (1987), *J. Appl. Phys. 61*, 1826.
Bentini, G. G., Nipoti, R., Armagiati, A., Berti, M., Drigo, A. V., Cohen, C. (1985), *J. Appl. Phys. 57*, 270.
Bergholz, W. (1983), *Physica 116 B*, 312.
Binns, M. J., Brown, W. P., Wilkes, J. G., Newman, R. C., Livingston, F. M., Messoloras, S., Stewart, R. J. (1983), *Appl. Phys. Lett. 42*, 525.
Binns, M. J., Brown, W. P., Wilkes, J. G. (1984), *J. Phys. C 17*, 6253.
Bourret, A. (1986), in: *Oxygen, Carbon, Hydrogen and Nitrogen in Crystalline Silicon:* Mikkelsen, Jr., J. L., Pearton, S. J., Corbett, J. W., Pennycook, S. J. (Eds.). Pittsburg: Materials Research Society, pp. 223–240.
Bourret, A., Schröter, W. (1984), *Ultramicroscopy 14*, 97.
Broniatowski, A. (1989), *Phys. Rev. Lett. 62*, 3074.
Broniatowski, A., Haut, C. (1990), *Phil. Mag. Lett. 62*, 407.
Cerofolini, F., Ferla, G. (1981), in: *Semiconductor Silicon:* Huff, H. R., Kriegler, J., Takeishi, Y. (Eds.). Pennington: The Electrochemical Society, p. 724.
Cerva, H. (1991), *Proc. 7th Internat. Conf. of Def. in Semic.*, Oxford.
Cerva, H., Wendt, H. (1989 a), *Inst. Phys. Conf. Ser. No. 100*, 587.
Cerva, H., Wendt, H. (1989 b), *Mat. Res. Soc. Symp. Proc. 138*, 533.
Chen, J. W., Milnes, A. G., Rohatgi, A. (1979), *Solid State Electron. 22*, 801.
Chen, M. C., Silvestrii, V. J. (1982), *J. Electrochem. Soc. 129*, 1294.
Cherns, D., Anstis, G. R., Hutchinson, J. L., Spence, J. C. H. (1982), *Phil. Mag. A 46*, 849.
Colas, E., Weber, E. R. (1986), *Appl. Phys. Lett. 48*, 1371.
Corbett, J. W., Deak, P., Lindström, J. L., Roth, L. M., Snyder, L. C. (1989), *Mater. Sci. Forum 38–41*, 579.
Cullis, A. G., Katz, L. E. (1974), *Phil. Mag. 30*, 1419.
Das, G. (1972), *J. Appl. Phys. 44*, 4459.
Dash, W. C. (1956), *J. Appl. Phys. 27*, 1193.
d'Heurle, F. M., Gas, P. (1986), *J. Mat. Res. 1*, 205.
Dorward, R. C., Krikaldy, J. S. (1968), *Trans. AIME 242*, 2055.
Ehrenberg, W. (1950), *Proc. Phys. Soc. London 63 A*, 75.
Falster, R. (1985), *Appl. Phys. Lett. 46*, 737.

Falster, R. (1989), in: *Gettering and Defect Engineering in the Semiconductor Technology:* Kittler, M. (Ed.). Vaduz, Liechtenstein: Sci. Tech. Publ., p. 13.

Falster, R., Bergholz, W. (1990), *J. Electrochem. Soc. 137*, 1548.

Gilles, D. (1987), Thesis, Göttingen.

Gilles, D., Bergholz, W., Schröter, W. (1986), *J. Appl. Phys. 59*, 3590.

Gilles, D., Schröter, W., Bergholz, W. (1990a), *Phys. Rev. B41*, 5770.

Gilles, D., Weber, E. R., Hahn, S. K. (1990b), *Phys. Rev. Lett. 64*, 196.

Gilles, D., Weber, E. R., Hahn, S. K., Monteiro, O. R., Cho, K. (1990c), in: *Semiconductor Silicon 1990:* Huff, H. R., Barraclough, K. G., Chikawa, J.-I. (Eds.). Pennington: Electrochemical Society, p. 697.

Goldschmidt, B. M. (1928), *Z. Phys. Chem. 133*, 397.

Graff, K. (1983), in: *Aggregation of Point Defects in Silicon:* Sirtl, E., Goorissen, J. (Eds.). Pennington: The Electrochemical Society, p. 121.

Graff, K. (1986), in: *Semiconductor Silicon:* Huff, H. R., Abe, T., Kolbesen, B. (Eds.). Pennington: The Electrochemical Society, p. 751.

Graff, K. (1991), *Metal Impurities in Silicon Device Fabrication.* Heidelberg: Springer Verlag, in press.

Graff, K., Hefner, H. A., Pieper, H. (1985), *Mat. Res. Symp. Proc. Vol. 36*, 19.

Graff, K., Hefner, H.-A., Hennerici, W. (1988), *J. Electrochem. Soc. 135*, 952.

Haasen, P. (1978), *Physical Metallurgy.* Cambridge: Cambridge University Press.

Hahn, S., Arst, M., Ritz, K. N., Shatas, S., Stein, H. J., Rek, Z. U., Tiller, W. A. (1988), *J. Appl. Phys. 64*, 849.

Hall, J. J. (1967), *Phys. Rev. 161*, 756.

Hall, R. N., Racette, H. (1964), *J. Appl. Phys. 35*, 379.

Ham, F. S. (1958), *J. Phys. Chem. Solids 6*, 335.

Hansen, M., Anderko, K. (1958), *Constitution of Binary Alloys.* New York: McGraw-Hill.

Hill, D. E. (1981), in: *Semiconductor Silicon 1981:* Huff, H. R., Kriegler, R. J., Takeishi, Y. (Eds.). Pennington: The Electrochemical Society, p. 354.

Hocine, S., Mathiot, D. (1988), *Appl. Phys. Lett. 53*, 1269, 1989, and *Mat. Science Forum (Trans. Tech.) 38–41*, 725.

Honda, K., Ohsawa, A., Toyokura, N. (1984), *Appl. Phys. Lett. 45*, 270.

Honda, K., Ohsawa, A., Toyokura, N. (1985), *Appl. Phys. Lett. 46*, 582.

Honda, K., Nakanishi, T., Ohsawa, A., Toyokura, N. (1987), *Inst. Phys. Conf. Ser. No. 87*, 463.

Hrostowski, H. J., Kaiser, R. H. (1959), *J. Phys. Chem. Solids 9*, 214.

Hu, S. M. (1981), *J. Appl. Phys. 52*, 3974.

Isobe, T., Nakashima, H., Hashimoto, K. (1989), *Jpn. J. Appl. Phys. 28*, 1282.

Kanamori, M., Tsuya, H. (1985), *J. Appl. Phys. 24*, 557.

Keating, P. N. (1966), *Phys. Rev. 145*, 637.

Keller, R., Deicher, M., Pfeiffer, W., Skudlik, H., Steiner, D., Wichert, Th. (1990), *Phys. Rev. Lett. 65*, 2023.

Kitagawa, H., Hashimoto, K. (1977), *Jpn. J. Appl. Phys. 16*, 173.

Kitagawa, H., Nakashima, H. (1987), *Phys. Stat. Sol. (a) 102*, K25.

Kitagawa, H., Nakashima, H. (1989), *Jpn. J. Appl. Phys. 28*, 305.

Kittler, M., Lärz, J., Seifert, W., Seibt, M., Schröter, W. (1991), *Appl. Phys. Lett. 58*, 911.

Kola, R. R., Rozgonyi, G. A., Li, J., Rogers, W. B., Tan, T. Y., Bean, K. E., Lindberg, K. (1989), *Appl. Phys. Lett. 55*, 2108.

Kolbesen, B. O., Bergholz, W., Wendt, H. (1989), in: *Materials Science Forum:* Ferenczi, G. (Ed.). Aedermannsdorf: TransTech. Publ. 38–41, p. 1.

Kühnapfel, R. (1990), Thesis, Göttingen.

Kühnapfel, R., Schröter, W. (1990), in: *Semiconductor Silicon 1990:* Huff, H. R., Barraclough, K. G., Chikawa, Y.-I. (Eds.). Pennington: The Electrochemical Society, p. 651.

Kühnapfel, R., Schröter, W., Gilles, D. (1986), in: *Materials Science Forum:* von Barderleben, H. J. (Ed.). Aedermanndorf: TransTech. Publ. Vol. 10–12, p. 151.

Lee, D. M., Maher, D. M., Shimura, F., Rozgonyi, G. A. (1990), in: *Semiconductor Silicon 1990:* Huff, H. R., Barraclough, K. G., Chikawa, J.-I. (Eds.). Pennington: The Electrochemical Society, p. 639.

Lemke, H. (1981), *Phys. Stat. Sol. (a) 64*, 549.

Lemke, H. (1983), *Phys. Stat. Sol. (a) 76*, 223.

Lescronier, D., Paugam, J., Pelous, G., Richou, F., Salvi, M. (1981), *J. Appl. Phys. 52*, 5090.

Liehr, M., Bronner, G. B., Lewis, J. E. (1988a), *Appl. Phys. Lett. 52*, 1892.

Liehr, M., Dallaporta, H., Lewis, J. E. (1988b), *Appl. Phys. Lett. 53*, 589.

Lien, C. D., Nicolet, M.-A., Lau, S. S. (1984), *Appl. Phys. A34*, 249.

Livingston, F. M., Messoloras, S., Newman, R. C., Pike, B. C., Stewart, R. J., Binns, M. J., Brown, W. P., Wilken, J. G. (1984), *J. Phys. C17*, 6263.

Ludwig, G. W., Woodbury, H. H. (1962), *Solid State Physics 13*, 223.

Marioton, B. P. R., Gösele, U. (1988), *J. Appl. Phys. 63*, 4661.

Martin, G. (1978), *Phil. Mag. A38*, 131.

Maurice, J. L., Colliex, C. (1989), *Appl. Phys. Lett. 55*, 241.

Meek, R. L., Seidel, T. E. (1975), *J. Phys. Chem. Solids 36*, 731.

Meek, R. L., Seidel, T. E., Cullis, A. G. (1975), *J. Electrochem. Soc. 112*, 786.

Mets, E. J. (1965), *J. Electrochem. Soc. 112*, 420.

Mikkelsen, J. C. (1982a), *Appl. Phys. Lett. 41*, 671.

Mikkelsen, J. C. (1982b), *Appl. Phys. Lett. 40*, 336.

Murarka, S. P. (1983), in: *Silicides for VLSI Application.* Orlando: Academic Press.

Nabarro, F. R. N. (1940), *Proc. Roy. Soc. A175*, 519.

Nahashima, H., Hashimoto, K. (1991), *J. Appl. Phys. 69*, 1440.

Negrini, P., Nobili, D., Solmi, S. (1975), *J. Electrochem. Soc. 122*, 1254.

Nes, E. (1974), *Acta Metall. 22*, 81.

Nes, E., Washburn, J. (1971), *J. Appl. Phys. 42*, 3562.

Nes, E., Lunde, G. (1972), *J. Appl. Phys. 43*, 1835.

Nes, E., Washburn, J. (1973), *J. Appl. Phys. 44*, 3682.

Ogino, M. (1982), *Appl. Phys. Lett. 41*, 847.

Ourmazd, A., Schröter, W. (1984), *Appl. Phys. Lett. 45*, 781.

Ourmazd, A., Schröter, W. (1985), *Mat. Res. Soc. Symp. Proc. Vol. 36*, 25.

Pearton, J., Corbett, J. W., Sliv, T. S. (1987), *Appl. Phys. A43*, 153.

Pearton, J., Stavola, M., Corbett, J. W. (1989), *Mater. Sci. Forum:* Ferenczi, G. (Ed.). Aedermannsdorf: TransTech. Publ. Vol. 38–41, p. 25.

Picker, C., Dobson, P. S. (1972), *Crystal Lattice Defects 3*, 219.

Pomerantz, D. (1967), *J. Appl. Phys. 38*, 5020.

Pond, R. C. (1984), in: *Polycrystalline Semiconductors:* Harbeke, G. (Ed.). Berlin: Springer Verlag (1985), p. 27.

Prescha, Th., Zundel, T., Weber, J., Prigge, H., Gerlach, P. (1989), *Mater. Sci. Eng. B4*, 79.

Reiss, H., Fuller, C. S., Morin, F. J. (1956), *Bell Syst. Techn. J. 35*, 535.

Rubloff, G. W., Hofmann, K., Liehr, M., Young, D. R. (1987), *Phys. Rev. Lett. 58*, 2379.

Salih, A. S., Kim, H. J., Davies, R. F., Rozgonyi, G. A. (1985), *Appl. Phys. Lett. 46*, 419.

Schnegg, A., Prigge, H., Grundner, M., Hahn, P. D., Jacob, H. (1988), *Mater. Res. Soc. Symp. Proc. 104*, 291.

Schröter, W., Kühnapfel, R. (1989), in: *Point and Extended Defects in Semiconductors:* Benedek, G., Cavallini, A., Schröter, W. (Eds.). NATO ASI Series B 202. New York: Plenum Press, p. 95.

Schröter, W., Kühnapfel, R. (1990), *Appl. Phys. Lett. 56*, 2207.

Seeger, A., Chik, K. P. (1968), *Phys. Stat. Sol. 29*, 455.

Seibt, M. (1990), in: *Semiconductor Silicon 1990:* Huff, H. R., Barraclough, K. G., Chikawa, Y. I. (Eds.). Pennington: The Electrochemical Society, p. 663.

Seibt, M. (1991), in: *Proc. of the 4th Intern. Autumn Meeting on Gettering and Defect Engineering in Semiconductor Technology – Solid State Phenomena:* Kittler, M. (Ed.). Vaduz: Sci-Tech. Publ., in press.

Seibt, M., Graff, K. (1988a), *J. Appl. Phys. 63*, 4444.

Seibt, M., Graff, K. (1988b), *Mat. Res. Soc. Symp. Proc. Vol. 104*, 215.

Seibt, M., Schröter, W. (1989), *Phil. Mag. A59*, 337.

Seibt, M., Schröter, W. (1991), in: *Proc. of the 4th Intern. Autumn Meeting on Gettering and Defect Engineering in Semiconductor Technology – Solid*

State Phenomena: Kittler, M. (Ed.). Vaduz: Sci-Tech. Publ., in press.

Seidel, T. E., Meek, R. L., Cullis, A. G. (1975), *J. Appl. Phys. 46*, 600.

Shaikh, A. A., Schröter, W., Bergholz, W. (1985), *J. Appl. Phys. 58*, 2519.

Shepard, W. H., Turner, J. A. (1962), *J. Phys. Chem. Solids 23*, 1697.

Shimura, F. (1988), *Semiconductor Silicon Crystal Technology* (Academic Press, San Diego), p. 344.

Shimura, F., Tsuya, H., Kawamura, T. (1980), *J. Appl. Phys. 51*, 269.

Shimura, F., Tsuya, H., Kawamura, T. (1981), *J. Electrochem. Soc. 128*, 1579.

Shockley, W., Moll, J. L. (1960), *Phys. Rev. 119*, 1480.

Silcock, J. M., Tunstall, W. J. (1964), *Phil. Mag. 10*, 361.

Solberg, J. K. (1978), *Acta Cryst. A34*, 684.

Solberg, J. K., Nes, E. (1978a), *J. Mat. Sci. 13*, 2233.

Solberg, J. K., Nes, E. (1978b), *Phil. Mag. A37*, 465.

Stacy, W. T., Allison, D. F., Wu, T. C. (1981), in: *Semiconductor Silicon 1981:* Huff, H. R., Kriegler, R. J., Takeishi, Y. (Eds.). Pennington: The Electrochem. Soc., p. 354.

Struthers, J. D. (1956), *J. Appl. Phys. 27*, 1560.

Tan, T. Y., Gardner, E. E., Tice, W. K. (1977), *Appl. Phys. Lett. 30*, 175.

Tice, W. K., Tan, T. Y. (1976), *J. Appl. Phys. 28*, 564.

Tung, R. T., Gibson, J. M., Poate, J. M. (1983), *Phys. Rev. Lett. 50*, 429.

Utzig, J. (1988), *J. Appl. Phys. 64*, 3629.

Utzig, J. (1989), *J. Appl. Phys. 65*, 3868.

Utzig, J., Gilles, D. (1989), *Materials Science Forum:* Ferenczi, G. (Ed.). Aedermannsdorf: TransTech. Publ. Vol. 39–41, p. 729.

Van Vechten, J. A., Thurmond, C. D. (1976), *Phys. Rev. B14*, 3539.

Wagner, R., Kampmann, R. (1991), in: *Materials Science and Technology, Vol. 5: Phase Transformation in Materials:* Cahn, R. W., Haasen, P., Kramer, E. J. (Eds.). Weinheim: VCH Verlagsgesellschaft, p. 212.

Watkins, G. D., Corbett, J. W., McDonald, R. S. (1982), *J. Appl. Phys. 53*, 7097.

Weber, E. R. (1983), *Appl. Phys. A30*, 1.

Weber, E. R., Gilles, D. (1990), in: *Semiconductor Silicon 1990:* Huff, H. R., Barraclough, K. G., Chikawa, Y.-I. (Eds.). Pennington: The Electrochemical Society, p. 387.

Wendt, H., Cerva, H., Lehmann, V., Pamler, W. (1989), *J. Appl. Phys. 65*, 2402.

Wiedersich, H., Okamoto, P. R., Lam, N. Q. (1977), in: *Radiation Effects in Breeder Reactor Structural Materials:* Bleiberg, M. L. (Ed.). New York: AIME, p. 801.

Wong, H., Cheung, N. W., Chu, P. K. (1988), *Appl. Phys. Lett. 52*, 889.

Zener, C. (1952), in: *Imperfections of Nearly Perfect Crystals:* Shockley, W. (Ed.). New York: Wiley, p. 289.

Zhu, J., Diz, J., Barbier, D., Langner, A. (1989), *Materials Science and Engineering B4*, 185.

Zundel, T., Weber, J., Benson, B., Hahn, P. O., Schnegg, A., Prigge, H. (1988), *Appl. Phys. Lett. 53*, 1426.

Zunger, A. (1986), *Solid State Physics 39*, 275.

General Reading

Benedek, G., Cavallini, A., Schröter, W. (Eds.) (1989), *Point and Extended Defects in Semiconductors*, NATO ASI Series B 202. New York: Plenum Press.

Frank, W., Gösele, U., Mehrer, H., Seeger, A. (1984), "Diffusion in Silicon and Germanium", in: *Diffusion in Crystalline Solids:* Murch, G. E., Nowick, A. S. (Eds.). Orlando: Academic Press, p. 64–142.

Graff, K. (1991), *Metal Impurities in Silicon Device Fabrication*. Heidelberg: Springer Verlag, in press.

Shimura, F. (1988), *Semiconductor Silicon Crystal Technology*. San Diego: Academic Press.

Weber, E. R. (1983), "Transition Metals in Silicon", *Appl. Phys. A 30*, 1.

Index

© VCH Verlagsgesellschaft mbH, D-6940 Weinheim (Federal Republic of Germany), 1991

Distribution:

VCH, P.O. Box 101161, D-6940 Weinheim (Federal Republic of Germany)

Switzerland: VCH, P.O. Box, CH-4020 Basel (Switzerland)

United Kingdom and Ireland: VCH (UK) Ltd., 8 Wellington Court, Cambridge CB1 1HZ (England)

USA and Canada: VCH, Suite 909, 220 East 23rd Street, New York, NY 10010-4606 (USA)

ISBN 3-527-26817-0 (VCH, Weinheim)
Set ISBN 3-527-26813-8 (VCH, Weinheim)

ISBN 0-89573-692-6 (VCH, New York)
Set ISBN 1-56081-190-0 (VCH, New York)